→ Indef[illegible] 316
→ Definite " 338
Notation 138, 155f
Substitution 327, 328f
Limit does not exists 174
Segment partition graph 336

Applied Calculus

SECOND EDITION

SECOND EDITION

Applied Calculus

Stefan Waner
Steven R. Costenoble
Hofstra University

Australia • Canada • Mexico • Singapore • Spain • United Kingdom • United States

Sponsoring Editor: *Curt Hinrichs*
Marketing Manager: *Karin Sandberg*
Marketing Team: *Samantha Cabaluna and Beth Kroenke*
Editorial Assistant: *Emily Davidson*
Production Editor: *Mary Vezilich*
Production Service: *Susan Graham*
Manuscript Editor: *Carol Reitz*
Permissions Editor: *Sue Ewing*
Interior Design: *Susan Schmidler*
Cover Design: *Roy R.Neuhaus*
Cover Photo: *Walter Bibikow/FPG International*
Interior Illustration: *Network Graphics*
Print Buyer: *Vena Dyer*
Typesetting: *TSI Graphics*
Printing and Binding: *Transcontinental Printing, Inc.*

For more information about this or any other Brooks/Cole products, contact:
BROOKS/COLE
511 Forest Lodge Road
Pacific Grove, CA 93950 USA
www.brookscole.com
1-800-423-0563 (Thomson Learning Academic Resource Center)

Printed in Canada.

10 9 8 7 6 5 4 3 2

Library of Congress Cataloging-in-Publication Data

Waner, Stefan, [date–]
Applied calculus/Stefan Waner, Steven R. Costenoble.—2nd ed.
p. cm.
Rev. ed. of: Finite mathematics & calculus applied to the real world. 1st ed. c1996.
ISBN 0-534-36631-7 (alk. paper)
1. Mathematics. 2. Calculus. I. Costenoble, Steven R., [date–] II. Title.

QA37.2.W34 2000b
515—dc21 00-046758

Contents

chapter 1 **Functions and Linear Models 2**

1.1 Functions from the Numerical and Algebraic Viewpoints 3
1.2 Functions from the Graphical Viewpoint 15
1.3 Linear Functions 27
1.4 Linear Models 38
1.5 Linear Regression (optional) 54
You're the Expert Modeling Spending on Internet Advertising 63
chapter 1 review test 67
Optional Internet Topic
New Functions from Old: Scaled and Shifted Functions

chapter 2 **Nonlinear Models 70**

2.1 Quadratic Functions and Models 71
2.2 Exponential Functions and Models 79
2.3 Logarithmic Functions and Models 94
2.4 Trigonometric Functions and Models (optional) 103
You're the Expert Epidemics 116
chapter 2 review test 119
Optional Internet Topics
Linear and Exponential Regression
Using and Deriving Algebraic Properties of Logarithms

chapter 3 **Introduction to the Derivative 122**

3.1 Average Rate of Change 124
3.2 The Derivative As Rate of Change: A Numerical Approach 133
3.3 The Derivative As Slope: A Geometric Approach 143
3.4 The Derivative As a Function: An Algebraic Approach 152
3.5 A First Application: Marginal Analysis 163
3.6 Limits and Continuity: Numerical and Graphical Approaches (optional) 172
3.7 Limits and Continuity: Algebraic Approach (optional) 183

You're the Expert Reducing Sulfur Emissions 191

chapter 3 review test 194

Optional Internet Topics

Sketching the Graph of the Derivative

Proof of the Power Rule

Continuity and Differentiability

chapter 4 **Techniques of Differentiation 198**

4.1 The Product and Quotient Rules 199

4.2 The Chain Rule 209

4.3 Derivatives of Logarithmic and Exponential Functions 220

4.4 Derivatives of Trigonometric Functions (optional) 232

4.5 Implicit Differentiation 239

You're the Expert Projecting Market Growth 247

chapter 4 review test 254

Optional Internet Topic

Linear Approximation and Error Estimation

chapter 5 **Applications of the Derivative 256**

5.1 Maxima and Minima 257

5.2 Applications of Maxima and Minima 269

5.3 The Second Derivative and Analyzing Graphs 283

5.4 Related Rates 294

5.5 Elasticity of Demand 302

You're the Expert Production Lot Size Management 308

chapter 5 review test 312

chapter 6 **The Integral 314**

6.1 The Indefinite Integral 315

6.2 Substitution 326

6.3 The Definite Integral As a Sum: A Numerical Approach 335

6.4 The Definite Integral As Area: A Geometric Approach 345

6.5 The Definite Integral: An Algebraic Approach and the Fundamental Theorem of Calculus 354

You're the Expert Wage Inflation 367

chapter 6 review test 370

Optional Internet Topic

Numerical Integration

chapter 7 **Further Integration Techniques and Applications of the Integral 372**

7.1 Integration by Parts 373

7.2 Area Between Two Curves and Applications 380

7.3 Averages and Moving Averages 393

7.4 Continuous Income Streams 402
7.5 Improper Integrals and Applications 407
7.6 Differential Equations and Applications 415
You're the Expert Estimating Tax Revenues 423
chapter 7 review test 428

chapter 8 **Functions of Several Variables 430**
8.1 Functions of Several Variables from the Numerical and Algebraic Viewpoints 431
8.2 Three-Dimensional Space and the Graph of a Function of Two Variables 444
8.3 Partial Derivatives 455
8.4 Maxima and Minima 463
8.5 Constrained Maxima and Minima and Applications 473
8.6 Double Integrals 484
You're the Expert Modeling Household Income 493
chapter 8 review test 500

chapter S **Calculus Applied to Probability and Statistics (Optional Internet Topic)**
S.1 Continuous Random Variables and Histograms
S.2 Probability Density Functions: Uniform, Exponential, Normal, and Beta
S.3 Mean, Median, Variance, and Standard Deviation
You're the Expert Creating a Family Trust

APPENDIX **Algebra Review A 1**
A.1 Real Numbers A 1
A.2 Exponents and Radicals A 7
A.3 Multiplying and Factoring Algebraic Equations A 14
A.4 Rational Expressions A 19
A.5 Solving Polynomial Equations A 21
A.6 Solving Miscellaneous Equations A 27

Answers to Selected Exercises A 31

Index I 1

Preface

Applied Calculus, Second Edition, is a substantial revision of the First Edition. The book is intended for a one- or two-term course for students majoring in business, the social sciences, or the liberal arts. Like the First Edition, the Second Edition is designed to address the considerable challenge of generating enthusiasm and developing mathematical sophistication in an audience that is often underprepared for and disaffected by traditional mathematics courses. Unlike the First Edition—and indeed unlike any other text on the market to date—the book is supported by, and linked with, an extensive and highly developed web site containing interactive tutorials, on-line technology, optional supplementary material, and much more. Instructors using graphing calculators, spreadsheets, or on-line computer technology will be pleased to find plenty of support in the text and at **www.AppliedCalc.com**. Like the First Edition, this text is usable by itself, independent of the Web or any other technology.

Our Approach to Pedagogy

Real World Orientation We are particularly proud of the diversity, breadth, and abundance of examples and exercises we have been able to include in this edition. A large number of these are based on real, referenced data from business, economics, the life sciences, and the social sciences. For a list of the companies used as sources for these examples and exercises, turn to the inside cover.

Adapting real data for pedagogical use can be tricky; available data can sometimes be numerically complex, intimidating for students, or incomplete. In this edition we have modified and streamlined many of the real world applications, rendering them as tractable as any "made-up" application. At the same time, we have been careful to strike a pedagogically sound balance between applications based on real data and more traditional "generic" applications. Thus the density and selection of real data–based applications has been tailored to the pedagogical goals and appropriate difficulty level for each section.

Readability We would like students to read this book. We would like students to *enjoy* reading this book. Thus we have written the book in a conversational and student-oriented style, and have made frequent use of a question-and-answer dialogue format in order to encourage the development of the student's mathematical curiosity and intuition. We hope that this text will give the student insight into how a mathematician develops and thinks about mathematical ideas and their applications.

Five Elements of Mathematical Pedagogy The "Rule of Three" is a common pedagogical theme in reform-oriented texts. Accordingly, we discuss many of the central concepts numerically, graphically, and algebraically, and go to some lengths to clearly delineate these distinctions. As a fourth element, we emphasize verbal communication of mathematical concepts through translation of English sentences into mathematical statements, and with our communication and reasoning exercises at the end of each section.

The fifth element, interactivity, is new to the Second Edition and, we believe, unique to the market. This active learning element is implemented within the printed text through expanded use of question-and-answer dialogues, but most dramatically through the web site. At the web site, students can interact with the material in several ways: through true–false quizzes on every topic; through interactive review exercises, and questions and answers in the tutorials (including multiple choice items with detailed feedback and Javascript-based interactive elements); and through on-line utilities that automate a variety of tasks, from graphing to regression and matrix algebra.

Exercise Sets Our substantial collection of exercises is one of the best features of the text. The exercises provide a wealth of material that can be used to challenge students at almost every level of preparation, and include everything from straightforward drills to interesting and rather challenging applications. These exercise sets have been carefully graded to move from the straightforward to the challenging. We have also included, in virtually every section of every chapter, applications based on real data, communication and reasoning exercises useful for writing assignments, exercises ideal for the use of technology, and amusing exercises.

Many of the scenarios used in application examples and exercises are revisited several times throughout the book. Thus, for instance, students will find themselves using a variety of techniques, from graphing through the use of derivatives to elasticity of demand, to analyze the same application. Reuse of scenarios and important functions provides unifying threads and shows students the complex texture of real-life problems.

Additional Features

Extended Applications and Projects Each chapter begins with the statement of an interesting problem that is returned to at the end of that chapter in a section entitled "You're the Expert." This extended application uses and illustrates the central ideas of the chapter. The themes of these applications are varied and they are designed to be as nonintimidating as possible. We avoid pulling complicated formulas out of thin air, but focus instead on the development of mathematical models appropriate to the topics. These applications are ideal for assignment as projects, and to this end we have included groups of exercises at the end of each.

Question-and-Answer Dialogue We frequently use informal question-and-answer dialogues to guide the student through the development of new concepts. These dialogues often anticipate students' questions.

Before We Go On Most examples are followed by supplementary interpretive discussions under the heading "Before we go on." These discussions may include a check on the answer, a discussion of the feasibility and significance of a solution, or an in-depth look at what the solution means.

Quick Examples Most definition boxes include one or more straightforward examples that a student can use to solidify each new concept as soon as it is encountered.

Communication and Reasoning Exercises for Writing and Discussion These are exercises designed to broaden the student's grasp of the mathematical concepts. Students are often asked to provide their own examples to illustrate a point or to design an application with a given solution. We also include "fill in the blank" type exercises and exercises that invite discussion and debate. These often have no single correct answer.

Footnotes We use footnotes throughout the text to provide interesting background, extended discussion, and various asides.

Thorough Integration of Spreadsheet and Graphing Technology Guidance on the use of spreadsheets (we use Microsoft® Excel) and graphing calculators (we use the TI-83) is thoroughly integrated throughout the text. In many examples we include a discussion of the use of spreadsheets or graphing technology to aid in the solution. We use the icon **T** in front of exercises for which we think technology is extremely useful or required.

The Web Site—www.AppliedCalc.com

For the past several years, our site at **www.AppliedCalc.com** has continued to gain recognition for its comprehensiveness and understandability. Students raised in an environment in which computers suffuse both work and play can use their familiar web browsers to engage the material in an active way. At the web site students and faculty can find:

- **Interactive Tutorials** Major topics are presented as the student is guided through exercises that parallel the text.
- **Interactive True–False Chapter Quizzes** True–false quizzes based on the material in each chapter help the students review pertinent concepts and avoid common pitfalls.
- **Chapter Review Exercises** This section provides review questions taken from tests and exams not appearing in the printed text. We are constantly adding topics and new exercises.
- **Detailed Chapter Summaries** Here we review all the basic definitions and problem-solving techniques discussed in the text, often reinforcing the main ideas through the use of interactive elements. Additional examples for review contained in the summaries can be easily printed.

• **Downloadable Excel Tutorials** Students will find detailed Excel tutorials for almost every section of the book. These interactive tutorials expand on the examples given in the text.
• **On-Line Utilities** Our collection of easy-to-use on-line utilities, written in Java™ and Javascript, allows the student to solve many of the technology-based application exercises directly on the web page. These utilities include a function grapher, a function evaluator, regression tools, and a numerical integration tool. All that is required to use these utilities is a standard, Java-capable web browser such as the current versions of Netscape Navigator and Microsoft® Explorer.
• **Downloadable Software** In addition to the web-based utilities, we offer a suite of free and intuitive stand-alone Macintosh® programs, including one for function graphing.
• **Extra Topics** We include complete interactive text and exercise sets for a selection of topics not ordinarily included in printed texts, but often requested by instructors. These extra topics are referred to in the appropriate sections of the text, giving instructors the option to include them in their courses.
• **On-Line Chapters** The web site includes a complete additional chapter on Calculus Applied to Probability and Statistics. This chapter incorporates the same pedagogy as the printed material, and can be readily printed and distributed.

Supplemental Material

For the Instructor
• **Instructor's Solutions Manual** Solutions for even-numbered exercises and model approaches to problem solving make this a key teaching resource.
• **BCA Testing** This revolutionary, Internet-ready testing suite allows instructors to customize exams and track student progress in an accessible format. Algorithmic problems and an extensive concept base make for extraordinarily diverse testing options. ISBN 0534-378536

For the Student
• **Student Solutions Manual** Detailed solutions for odd-numbered exercises are a valuable supplement to the student's classroom learning. ISBN 0534-376541
• **Graphing Calculator Manual** Text-specific graphing calculator manual using the TI 83 Plus and TI 86. ISBN 0534-378994
• **Excel Manual** This distinctive, text-specific manual uses Excel instructions and formulas to reinforce vital concepts in applied calculus. ISBN 0534-381650
• **The Web Site—www.AppliedCalc.com** For further details on this dynamic resource, see page xi.

Other Books in This Series

• ***Finite Mathematics,* 2e** ISBN 0534-366325
• ***Finite Mathematics and Applied Calculus,* 2e** ISBN 0534-366309

Acknowledgments

This project would not have been possible without the contributions and suggestions of numerous colleagues, students, and friends. We are particularly grateful to our colleagues who class tested preliminary versions of the Second Edition, including Frank Anderson and Brad Shelton at the University of Oregon, and Safwan Akbik, Bob Bumcrot, Dan Seabold, and Sylvia Silberger, all at Hofstra. We are also grateful to everyone we worked with at Brooks/Cole for their encouragement and guidance throughout the project. Specifically, we would like to thank Margot Hanis, for her enthusiasm and for believing this would work, Curt Hinrichs for seeing the project through to completion with equal enthusiasm, and Susan Graham, Mary Vezilich, and Karin Sandberg for their ongoing support in bringing this project to fruition.

We would also like to thank the numerous reviewers and accuracy checkers who carefully read our early drafts and provided many helpful suggestions that have shaped the development of this book:

James Ball
Indiana State University

Jon Cole
St. John's University

Matt Coleman
Fairfield University

Casey Cremins
University of Maryland

Julie Daberkow
University of Maryland

Deborah Denvir
Marshall University

Michael Ecker
Pennsylvania State University

Janice Epstein
Texas A&M University

Candy Giovanni
Michigan State University

Jerrold Grossman
Oakland University

Joe Guthrie
University of Texas, El Paso

Thomas Kelley
Metropolitan State College of Denver

Keith Kendig
Cleveland State University

Greg Klein
Texas A&M University

Michael Moses
George Washington University

Ralph Oberste-Vorth
University of South Florida

James Osterburg
University of Cincinnati

James Parks
State University of New York, Potsdam

Joanne Peeples
El Paso Community College

Gordon Savin
University of Utah

Daniel Scanlon
Orange Coast College

Brad Shelton
University of Oregon

Bennette Harris
University of Wisconsin, Whitewater

Loek Helminck
North Carolina State University

Thomas Keller
Southwest Texas State University

Cynthia Siegel
University of Missouri, St Louis

Stephen Stuckwisch
Auburn University

Doug Ulmer
University of Arizona

Will Watkins
University of Texas, Pan American

—Stefan Waner
—Steven R. Costenoble

Applied Calculus

SECOND EDITION

You're the Expert

Modeling Spending on Internet Advertising

You are the new director of Impact Advertising, Inc.'s Internet Division, which has enjoyed a steady 0.25% of the Internet advertising market. You have drawn up an ambitious proposal to expand your division in light of your anticipation that Internet advertising will continue to skyrocket. The VP in charge of financial affairs feels that current projections (based on a linear model) do not warrant the level of expansion you propose. How can you persuade the VP that those projections do not fit the data convincingly?

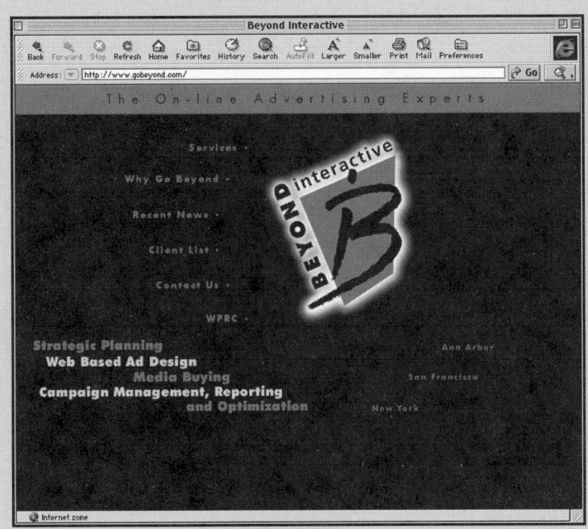

Source: Beyond Interactive. Copyright © 1998 Wolverine Web Productions LLC, Ann Arbor, MI. Reprinted by permission.

Internet Resources for This Chapter

At the web site, follow the path

Web site → Everything for Calculus → Chapter 1

where you will find a detailed chapter summary you can print out, a true–false quiz, and a collection of review exercises. You will also find downloadable Microsoft® Excel tutorials for each section, an on-line grapher, an on-line regression utility, and other resources. In addition, complete text and interactive exercises have been placed on the web site for the following optional topic:

New Functions from Old: Scaled and Shifted Functions

Functions and Linear Models

chapter 1

1.1 Functions from the Numerical and Algebraic Viewpoints
1.2 Functions from the Graphical Viewpoint
1.3 Linear Functions
1.4 Linear Models
1.5 Linear Regression (Optional)

Introduction

To analyze recent trends in spending on Internet advertising and to make reasonable projections, we need a mathematical *model* of this spending. Where do we start? In order to apply mathematics to real-world situations like this, we need a good understanding of basic mathematical concepts. Perhaps the most fundamental of these concepts is that of a *function:* a relationship that shows how one quantity depends on another. Functions may be described numerically and, often, algebraically. They can also be described graphically—a viewpoint that is extremely useful.

The simplest functions—the ones with the simplest formulas and the simplest graphs—are the *linear* functions. Because of their simplicity they are also among the most useful functions, and they can often be used to model real-world situations, at least over short periods of time. In discussing linear functions, we shall introduce the concepts of *slope* and *rate of change,* which are the starting point of calculus.

In the last section of this chapter, we discuss *simple linear regression:* the construction of linear functions that best fit given collections of data. Regression is used extensively in applied mathematics, statistics, and quantitative methods in business. The inclusion of regression utilities in graphing calculators and computer spreadsheets makes this powerful mathematical tool readily available for anyone to use.

1.1 Functions from the Numerical and Algebraic Viewpoints

The following table gives the weights of a particular child at various ages in its first year.

Age (months)	0	2	3	4	5	6	9	12
Weight (pounds)	8	9	13	14	16	17	18	19

Let us write $W(0)$ for the child's weight at birth (in pounds), $W(2)$ for its weight at 2 months, and so on [we read $W(0)$ as "W of 0"]. Thus, $W(0) = 8$, $W(2) = 9$, $W(3) = 13, \ldots, W(12) = 19$. More generally, if we write t for the age of the child (in months) at any time during its first year, then we write $W(t)$ for the the weight of the child at age t. We call W a **function** of the variable t, meaning that for each value of t between 0 and 12, W gives us a single corresponding number $W(t)$ (the weight of the child at that age).

In general, we think of a function as a way of producing new objects from old ones. The functions we deal with in this text produce new numbers from old numbers. The numbers we have in mind are the *real* numbers, including not only positive and negative integers and fractions but also numbers like $\sqrt{2}$ or π (see Appendix A for more on real numbers). For this reason, the functions we use are called **real-valued functions of a real variable.** For example, the function W takes the child's age in months and returns its weight in pounds at that age (Figure 1).

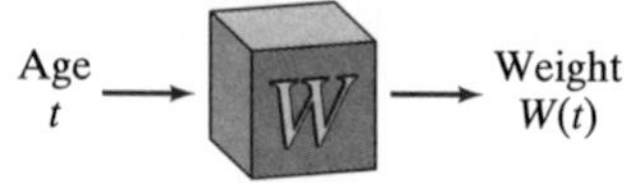

Figure 1

If we introduce another variable—say, y—that stands for the weight of the child, then we can write $y = W(t)$. The function W then tells us exactly how y depends on t. We call y the **dependent variable** because its value depends on the value of t, the **independent variable.**

A function may be specified in several different ways. It may be specified **numerically,** by giving the values of the function for a number of values of the independent variable, as in the table above. It may be specified **verbally,** as in "Let $W(t)$ be the weight of the child at age t months in its first year."[1] In some cases we may be able to use an algebraic formula to calculate the function, and then we say that the function is specified **algebraically.** And in the next section we shall see that a function may also be specified **graphically.**

Question For which values of t does it make sense to ask for $W(t)$? In other words, for which ages t is the function W defined?

Answer Since $W(t)$ refers to the weight of the child at age t months *in its first year,* $W(t)$ is defined when t is any number between 0 and 12—that is, when $0 \le t \le 12$. Using interval notation (see Appendix A), we can say that $W(t)$ is defined when t is in the interval $[0, 12]$. The set of values of the independent variable for which a function is defined is called its **domain** and is a necessary part of the definition of the function. Notice that the table above gives the value of $W(t)$ at only some of the infinitely many possible values in the domain $[0, 12]$.

Here is a summary of the terms we've just introduced.

Functions

A **real-valued function f of a real-valued variable x** assigns to each real number x in a specified set of numbers, called the **domain** of f, a unique real number $f(x)$, read "f of x."

The variable x is called the **independent variable.** If $y = f(x)$, we call y the **dependent variable.**

Note on Domains

The domain of a function is not always specified explicitly; if no domain is specified for the function f, we take the domain to be the largest set of numbers x for which $f(x)$ makes sense. This "largest possible domain" is sometimes called the **natural domain.**

[1]Specifying a function verbally in this way is useful for understanding what the function is doing, but it gives no numerical information.

Quick Examples

1. Let $W(t)$ be the weight (in pounds) at age t months of a particular child during its first year. The independent variable is t. If we write $y = W(t)$, then the dependent variable is y, the child's weight. The domain of W is $[0, 12]$ because it was specified that W gives the child's weight during its first year.
2. Let $f(x) = 1/x$. The function f is specified algebraically. Some specific values of f are

$$f(2) = \frac{1}{2} \qquad f(3) = \frac{1}{3} \qquad f(-1) = \frac{1}{-1} = -1$$

Here, $f(0)$ is not defined because there is no such number as $1/0$. The natural domain of f consists all real numbers except zero because $f(x)$ makes sense for all values of x other than $x = 0$.

example 1

Numerically Specified Function

The following table shows the approximate number of U.S. citizens who visited Europe each year from 1984 to 1994, with $t = 0$ representing 1984:

Years since 1984, t	0	1	2	3	4	5	6	7	8	9	10
Visitors (millions), N	5.5	6.3	6.4	5.0	6.0	6.5	6.8	6.3	7.3	7.5	8.0

Source: Department of Transportation/*New York Times,* May 9, 1995, p. D4.

Viewing N as a function of t, give its domain and the values $N(0)$, $N(3)$, and $N(10)$. Estimate and interpret the value $N(3.5)$.

Solution

The domain of N is the set of numbers t with $0 \le t \le 10$; that is, $[0, 10]$. From the table, we have

$$N(0) = 5.5 \qquad \text{5.5 million visitors in 1984}$$

$$N(3) = 5.0 \qquad \text{5 million visitors in 1987}$$

$$N(10) = 8.0 \qquad \text{8 million visitors in 1994}$$

What about $N(3.5)$? Since $N(3) = 5.0$ and $N(4) = 6.0$, we estimate that

$$N(3.5) \approx 5.5$$

We call the process of estimating values for a function between points where it is already known **interpolation.**

Question How do we interpret $N(3.5)$?

Answer Since $N(3)$ represents the number of U.S. visitors to Europe during the year ending December 1987, and $N(4)$ represents the number during the year ending December 1988, the most logical interpretation of $N(3.5)$ is the number of U.S. visitors to Europe during the year ending June 1988—that is, the number during the one-year period July 1, 1987–June 30, 1988.

Before we go on . . .
Question Can we use the table to estimate $N(t)$ for values of t *outside* the domain—say, $t = 11$?
Answer Strictly speaking, $N(t)$ is defined only when $0 \le t \le 10$. However, we could consider a function with a larger domain, like the function that gives the number of U.S. visitors to Europe each year from 1984 to 2000. Estimating values for a function outside a range where it is already known is called **extrapolation.** As a general rule, extrapolation is far less reliable than interpolation; it is hard to predict the future from current data.

The two functions we have looked at so far were both specified **numerically,** meaning that we were given numerical values of the function evaluated at *certain* values of the independent variable. It would be more useful if we had a formula that allows us to calculate the value of the function for *any* value of the independent variable we wished. If a function is specified by a formula, we say that it is specified **algebraically.**

example 2

Algebraically Defined Function
Let f be the function specified by

$$f(x) = 0.03x^2 - 0.06x + 6, \qquad \text{with domain}^1\ (-2, 10].$$

(This formula gives an approximation of the tourism function N in Example 1.) Use the formula to calculate $f(0), f(10), f(-1), f(a)$, and $f(x + h)$. Is $f(-2)$ defined?

Solution

Let us check first that the values we are asked to calculate are all defined. Since the domain is stated to be $(-2, 10]$, the quantities $f(0)$, $f(10)$, and $f(-1)$ are all defined. The quantities $f(a)$ and $f(x + h)$ will also be defined if a and $x + h$ are understood to be in $(-2, 10]$. However, $f(-2)$ is not defined because -2 is not in the domain $(-2, 10]$.

If we take the formula for $f(x)$ and substitute 0 for x (replace x everywhere it occurs by 0), we get

$$f(0) = 0.03(0)^2 - 0.06(0) + 6 = 6$$

so $f(0) = 6$. Similarly,

$$f(10) = 0.03(10)^2 - 0.06(10) + 6 = 3 - 0.6 + 6 = 8.4$$

$$f(-1) = 0.03(-1)^2 - 0.06(-1) + 6 = 0.03 + 0.06 + 6 = 6.09$$

$$f(a) = 0.03a^2 - 0.06a + 6 \qquad \text{Substitute } a \text{ for } x.$$

$$f(x + h) = 0.03(x + h)^2 - 0.06(x + h) + 6. \qquad \text{Substitute } (x + h) \text{ for } x.$$

$$= 0.03x^2 + 0.06xh + 0.03h^2 - 0.06x - 0.06h + 6$$

---

[1]See Appendix A for a discussion of interval notation.

Before we go on . . . If we had not specified anything about the domain of f, we would have used the natural domain of f. In this case, the natural domain is the set of all real numbers because the quantity $0.03x^2 - 0.06x + 6$ is defined for every real number x.

We said that the function f given in this example is actually an approximation of the tourism function N of Example 1, so $f(x)$ is the approximate number of U.S. visitors to Europe during year x. The following table compares some of their values:

x	0	5	10
$N(x)$	5.5	6.5	8.0
$f(x)$	6.0	6.45	8.4

The function f is a best-fit quadratic curve (with coefficients rounded) based on the data in Example 1.

We call the algebraic function f an **algebraic model** of U.S. tourism in Europe because it models, or represents, the tourism data using an algebraic formula. The particular kind of algebraic model we used is called a **quadratic model** (see the end of this section for the names of some common types of algebraic functions).

Note that it is important to place parentheses around the quantity for which you are evaluating a function. For instance, if $g(x) = x^2 + 1$, then

$$g(-2) = (-2)^2 + 1 = 4 + 1 = 5 \qquad \text{Not } -2^2 + 1 = -4 + 1 = -3 ✗$$

$$g(x + h) = (x + h)^2 + 1 \qquad \text{Not } x^2 + h + 1 \text{ or } x + h^2 + 1 ✗$$

Instead of using "function notation"

$$f(x) = 0.03x^2 - 0.06x + 6 \qquad \text{Function notation}$$

we could use "equation notation":

$$y = 0.03x^2 - 0.06x + 6 \qquad \text{Equation notation}$$

(the choice of the letter y is conventional) and we say that "y is a function of x." This is exactly the way you enter a function in a graphing calculator or spreadsheet (see below).

Note also that there is nothing magical about the letter x. We might just as well write

$$f(t) = 0.03t^2 - 0.06t + 6$$

which defines *exactly the same function* as $f(x) = 0.03x^2 - 0.06x + 6$. For example, to calculate $f(10)$ from the formula for $f(t)$, we would substitute 10 for t, getting $f(10) = 8.4$, just as we did using the formula for $f(x)$.

example 3

T Using Technology to Evaluate a Function

Evaluate the function $f(x) = 0.03x^2 - 0.06x + 6$ of Example 2 for $x = 0, 1, 2, \ldots, 10$.

Solution

This task is tedious to do by hand, but various technologies can make evaluating a function easier.

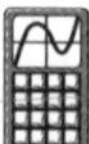

Graphing Calculator

There are several ways to evaluate an algebraically defined function on a graphing calculator such as the TI-83. First, enter the function in the "Y=" screen as

```
Y1 = 0.03*X^2-0.06*X+6
```

or

Y_1 = 0.03X²-0.06X+6

Then, to evaluate $f(0)$, for example, enter the following in the home screen:

Y_1(0)   This evaluates the function Y_1 at 0.

Alternatively, you can use the table feature: After entering the function under Y_1, press [2nd] [TblSet] and set Indpnt to "Ask." (You do this once and for all; it will permit you to specify values for x in the table screen.) Then, press [2nd] [TABLE], and you will be able to evaluate the function at several values of x. Whichever method we use, we obtain the following set of values:

x	0	1	2	3	4	5	6	7	8	9	10
$f(x)$	6	5.97	6	6.09	6.24	6.45	6.72	7.05	7.44	7.89	8.4

Spreadsheet

A spreadsheet program like Excel is ideal for this purpose. In Excel, for example, enter the numbers 0 through 10 in cells A1 through A11.[2] Then enter the formula

```
=0.03*A1^2-0.06*A1+6
```

A1 refers to the x value in cell A1.

in cell B1. Copy cell B1 and paste it into cells B2 through B11. These cells will now show the values of $f(x)$ for the corresponding values of x.

	A	B
1	0	=0.03*A1^2-0.06*A1+6
2	1	
3	2	
4	3	
5	4	
6	5	
7	6	
8	7	
9	8	
10	9	
11	10	

Web Site

At the web site, follow the path

Web site → On-Line Utilities → Function Evaluator & Grapher

Enter

```
0.03*x^2-0.06*x+6
```

or

```
0.03x^2-0.06x+6
```

in the "f(x) =" box, enter the values of x in the "Values of x" boxes (you can use the tab key to move from one box to the next), and press "Evaluate."

[2] A shortcut is available in Excel: Enter 0 in cell A1 and then select A1. Drag the "fill handle" in the corner of the cell down, while holding down the Control key (PC) or Option key (Macintosh), to fill in cells A1 through A11.

Sometimes, as in the following example, we need to use several formulas to specify a single function.

example 4

Piecewise-Defined Function
The percentage $U(t)$ of semiconductor equipment manufactured in the United States during the period 1981–1994 can be approximated by the following function of time t in years ($t = 0$ represents 1980):

$$U(t) = \begin{cases} 76 - 3.3t & \text{if } 1 \le t \le 10 \\ 18 + 2.5t & \text{if } 10 < t \le 14 \end{cases}$$

What percentage of semiconductor equipment was manufactured in the United States in the years 1982, 1990, and 1992?

Source: VLSI Research/*New York Times*, October 9, 1994, Section 3, p. 2.

Solution
The years 1982, 1990, and 1992 correspond, respectively, to $t = 2, 10$, and 12.

$t = 2$: $U(2) = 76 - 3.3(2) = 69.4$ We used the first formula because $1 \le t \le 10$.

$t = 10$: $U(10) = 76 - 3.3(10) = 43$ We used the first formula because $1 \le t \le 10$.

$t = 12$: $U(12) = 18 + 2.5(12) = 48$ We used the second formula because $10 < t \le 14$.

Thus, the percentage of semiconductor equipment that was manufactured in the United States was 69.4% in 1982, 43% in 1990, and 48% in 1992.

Graphing Calculator

The following expression defines the function U on a TI-83:

```
(X≤10)*(76-3.3X) + (X>10)*(18+2.5X)
```

When `X` is less than or equal to 10, the logical expression (`X≤10`) will evaluate to 1 because it is true, whereas the expression (`X>10`) will evaluate to 0 because it is false. The value of the function will be given by the expression (`76-3.3X`). When `X` is greater than 10, the expression (`X≤10`) will evaluate to 0, whereas the expression (`X>10`) will evaluate to 1, so the value of the function will be given by the expression (`18+2.5X`).

The logical operators (≤ and >, for example) are found in the TEST menu.

Web Site

The function utility available on the web site at

Web site → On-Line Utilities → Function Evaluator & Grapher

allows you to use logical expressions similar to those allowed by graphing calculators as described above. The following expression defines the function U:

```
(x<=10)*(76-3.3x) + (x>10)*(18+2.5x)
```

Spreadsheet

In Excel we would specify U by the following formula, assuming that the value of t is in cell A1:

```
=IF(A1<=10, 76-3.3*A1, 18+2.5*A1)
```

The `IF` function evaluates its first argument, which tests to see if the value of t is in the range $t \le 10$. If true, `IF` returns the result of evaluating its second argument; if not, it returns the result of evaluating its third argument. If we set up the spreadsheet as shown, we get a table of values for the function.

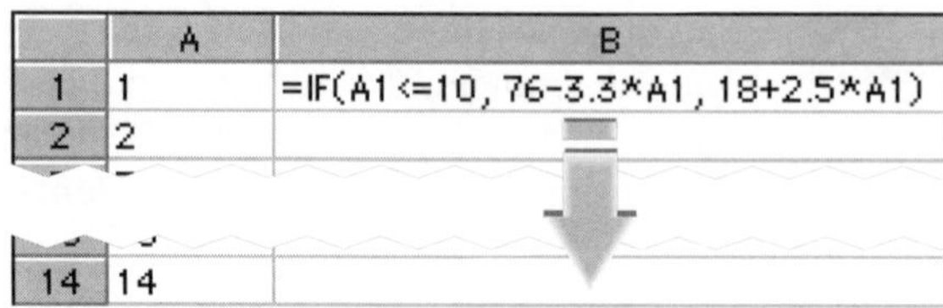

The functions we used in the examples above are **mathematical models** of real-life situations because they model, or represent, situations in mathematical terms.

Mathematical Modeling

To **mathematically model** a situation means to represent it in mathematical terms. The particular representation used is called a **mathematical model** of the situation. Mathematical models do not always represent a situation perfectly or completely. Some (like Example 2) represent a situation only approximately, whereas others represent only some aspects of the situation.

Quick Examples

Situation	Model
1. Albano's bank balance is twice Bravo's.	$a = 2b$ (a = Albano's balance, b = Bravo's)
2. The temperature is now 10° and increasing by 20° per hour.	$T(t) = 10 + 20t$ (t = time in hours, T = temperature)
3. The volume of a rectangular solid with a square base is obtained by multiplying the area of its base by its height.	$V = x^2h$ (h = height, x = length of a side of the base)
4. U.S. tourism in Europe	The table in Example 1 is a **numerical model** of U.S. tourism in Europe. The function in Example 2 is an **algebraic model** of U.S. tourism in Europe.
5. Semiconductor manufacture in the United States	Example 4 gives a **piecewise algebraic model** of the percentage of semiconductor equipment manufactured in the United States.

The following table lists some common types of functions that are often used to model real-world situations.

Common Types of Algebraic Functions

	Type of function	Example
Linear:	$f(x) = mx + b$ m, b constant	$f(x) = 3x - 2$
Quadratic:	$f(x) = ax^2 + bx + c$ a, b, c constant $(a \neq 0)$	$f(x) = -3x^2 + x - 1$
Cubic:	$f(x) = ax^3 + bx^2 + cx + d$ a, b, c, d constant $(a \neq 0)$	$f(x) = 2x^3 - 3x^2 + x - 1$
Polynomial:	$f(x) = ax^n + bx^{n-1} + \cdots + rx + s$ $a, b, \ldots, r, s$ constant (includes all the above functions)	All the above, and $f(x) = x^6 - x^4 + x - 3$
Exponential:	$f(x) = Ab^x$ A, b constant (b positive)	$f(x) = 3(2^x)$
Rational:	$f(x) = \dfrac{P(x)}{Q(x)}$ $P(x)$ and $Q(x)$ polynomials	$f(x) = \dfrac{x^2 - 1}{2x + 5}$

Functions and models other than linear ones are called **nonlinear.**

1.1 exercises

In Exercises 1–4, evaluate or estimate each expression based on the following table:

x	−3	−2	−1	0	1	2	3
$f(x)$	1	2	4	2	1	0.5	0.25

1. **a.** $f(0)$ **b.** $f(2)$

2. **a.** $f(-1)$ **b.** $f(1)$

3. **a.** $f(2) - f(-2)$ **b.** $f(-1)f(-2)$ **c.** $-2f(-1)$

4. **a.** $f(1) - f(-1)$ **b.** $f(1)f(-2)$ **c.** $3f(-2)$

5. Given $f(x) = 4x - 3$, find **a.** $f(-1)$, **b.** $f(0)$, **c.** $f(1)$, **d.** $f(y)$, **e.** $f(a + b)$.

6. Given $f(x) = -3x + 4$, find **a.** $f(-1)$, **b.** $f(0)$, **c.** $f(1)$, **d.** $f(y)$, **e.** $f(a + b)$.

7. Given $f(x) = x^2 + 2x + 3$, find **a.** $f(0)$, **b.** $f(1)$, **c.** $f(-1)$, **d.** $f(-3)$, **e.** $f(a)$, **f.** $f(x + h)$.

8. Given $g(x) = 2x^2 - x + 1$, find **a.** $g(0)$, **b.** $g(-1)$, **c.** $g(r)$, **d.** $g(x + h)$.

9. Given $g(s) = s^2 + \dfrac{1}{s}$, find **a.** $g(1)$, **b.** $g(-1)$, **c.** $g(4)$, **d.** $g(x)$, **e.** $g(s + h)$, **f.** $g(s + h) - g(s)$.

10. Given $h(r) = \dfrac{1}{r + 4}$, find **a.** $h(0)$, **b.** $h(-3)$, **c.** $h(-5)$, **d.** $h(x^2)$, **e.** $h(x^2 + 1)$, **f.** $h(x^2) + 1$.

11. Given

$$f(t) = \begin{cases} -t & \text{if } t < 0 \\ t^2 & \text{if } 0 \le t < 4 \\ t & \text{if } t \ge 4 \end{cases}$$

find **a.** $f(-1)$, **b.** $f(1)$, **c.** $f(4) - f(2)$, **d.** $f(3)f(-3)$.

12. Given

$$f(t) = \begin{cases} t - 1 & \text{if } t \le 1 \\ 2t & \text{if } 1 < t < 5 \\ t^3 & \text{if } t \ge 5 \end{cases}$$

find **a.** $f(0)$, **b.** $f(1)$, **c.** $f(4) - f(2)$, **d.** $f(5) + f(-5)$.

In Exercises 13–16, say whether f(x) is defined for the given values of x. If it is defined, give its value.

13. $f(x) = x - \frac{1}{x^2}$, with domain $(0, +\infty)$

a. $x = 4$ **b.** $x = 0$ **c.** $x = -1$

14. $f(x) = \frac{2}{x} - x^2$, with domain $[2, +\infty)$

a. $x = 4$ **b.** $x = 0$ **c.** $x = 1$

15. $f(x) = \sqrt{x + 10}$, with domain $[-10, 0)$

a. $x = 0$ **b.** $x = 9$ **c.** $x = -10$

16. $f(x) = \sqrt{9 - x^2}$, with domain $(-3, 3)$

a. $x = 0$ **b.** $x = 3$ **c.** $x = -3$

In Exercises 17–20, use technology (such as a spreadsheet, web site utilities, or a graphing calculator) to evaluate each function for the given values of x (when defined there).

T **17.** $f(x) = 0.1x^2 - 4x + 5; x = 0, 1, \ldots, 10$

T **18.** $g(x) = 0.4x^2 - 6x - 0.1; x = -5, -4, \ldots, 4, 5$

T **19.** $h(x) = \frac{x^2 - 1}{x^2 + 1}; x = 0.5, 1.5, 2.5, \ldots, 10.5$ (Round all answers to four decimal places.)

T **20.** $r(x) = \frac{2x^2 + 1}{2x^2 - 1}; x = -1, 0, 1, \ldots, 9$ (Round all answers to four decimal places.)

Applications

21. ***Profit*** The annual profits of Lotus Development Corp. for the years 1990–1994 are shown in the following table. ($t = 0$ represents the year beginning January 1990. Losses are shown as negative.)

Year, t	Profit (millions of dollars), $P(t)$
0	23
1	32
2	80
3	55
4	−20

Source: Datastream: company reports/*New York Times,* June 6, 1995, p. D8.

a. Find or estimate $P(0)$, $P(3)$, and $P(1.5)$. Interpret your answers. **b.** What is the domain of P?

22. ***Net Sales*** Net sales (after-tax revenue[3]) of Lotus Development Corp. for the years 1990–1994 are shown in the following table. ($t = 0$ represents the year beginning January 1990.)

Year, t	Net sales (millions of dollars), $S(t)$
0	700
1	810
2	900
3	990
4	980

Source: Datastream: company reports/*New York Times,* June 6, 1995, p. D8.

a. Find or estimate $S(2)$, $S(4)$, and $S(2.5)$. Interpret your answers. **b.** What is the domain of S?

23. ***Coffee Shops*** The number $C(t)$ of coffee shops and related enterprises in the United States can be approximated by the following function of time t in years since 1990:

$$C(t) = \begin{cases} 500t + 800 & \text{if } 0 \le t \le 4 \\ 1300t - 2400 & \text{if } 4 < t \le 10 \end{cases}$$

a. Evaluate $C(0)$, $C(4)$, and $C(5)$, and interpret the results.

b. Use the model to estimate when there were 5400 coffee shops in the United States.

T **c.** Use technology to generate a table of values for $C(t)$ with $t = 0, 1, \ldots, 10$.

Source for data: Specialty Coffee Association of America/*New York Times*, August 13, 1995, p. F10.

24. ***Semiconductors*** The percentage $J(t)$ of semiconductor equipment manufactured in Japan during the period 1981–1994 can be approximated by the following function of time t in years since 1980:

$$J(t) = \begin{cases} 17 + 3.1t & \text{if } 1 \le t \le 10 \\ 68 - 2t & \text{if } 10 < t \le 14 \end{cases}$$

a. Evaluate $J(1)$, $J(10)$, and $J(12)$, and interpret the results.

b. Use the model to estimate, to the closest year, when Japan manufactured 40% of all semiconductor equipment.

T **c.** Use technology to generate a table of values for $J(t)$ with $t = 1, 2, \ldots, 14$.

Source for data: VLSI Research/*New York Times*, October 9, 1994, Section 3, p. 2.

25. ***Income Taxes*** The U.S. federal income tax is a function of taxable income. Write T for the tax owed on a taxable income of I dollars. In 1997 the function T for a single taxpayer was specified as given in the table.

[3]The **revenue** that results from one or more business transactions is the total payment received, sometimes called the gross proceeds.

If your taxable income was over—	But not over—	Your tax is	Of the amount over—
\$0	\$24,650	15%	\$0
24,650	59,750	\$3697.50 + 28%	24,650
59,750	124,650	13,525.50 + 31%	59,750
124,650	271,050	33,644.50 + 36%	124,650
271,050	—	86,348.50 + 39.6%	271,050

What tax was owed by a single taxpayer on a taxable income of \$26,000? On a taxable income of \$65,000?

26. ***Income Taxes*** The income tax function T in Exercise 25 can also be written in the following form:

$$T(I) = \begin{cases} 0.15I & \text{if } 0 < I \le 24{,}650 \\ 3697.50 + 0.28(I - 24{,}650) & \text{if } 24{,}650 < I \le 59{,}750 \\ 13{,}525.50 + 0.31(I - 59{,}750) & \text{if } 59{,}750 < I \le 124{,}650 \\ 33{,}644.50 + 0.36(I - 124{,}650) & \text{if } 124{,}650 < I \le 271{,}050 \\ 86{,}348.50 + 0.396(I - 271{,}050) & \text{if } I > 271{,}050 \end{cases}$$

What tax was owed by a single taxpayer on a taxable income of \$25,000? On a taxable income of \$125,000?

27. ***Demand*** The demand for Sigma Mu Fraternity plastic brownie dishes is given by

$$q(p) = 361{,}201 - (p + 1)^2$$

where q represents the number of brownie dishes Sigma Mu can sell per month at a price of p¢ each. Use this function to determine the following values:

a. The number of brownie dishes Sigma Mu can sell per month if the price is set at 50¢

b. The number of brownie dishes they can unload per month if they give them away

c. The lowest price at which Sigma Mu will be unable to sell any dishes

28. ***Revenue*** The total weekly revenue earned at Royal Ruby Retailers (RRR) is given by

$$R(p) = -\frac{4}{3}p^2 + 80p$$

where p is the price (in dollars) RRR charges per ruby. Use the function to determine the following values:

a. The weekly revenue, to the nearest dollar, when the price is set at \$20 per ruby

b. The weekly revenue, to the nearest dollar, when the price is set at \$200 per ruby (Interpret your result.)

c. The price RRR should charge to obtain a weekly revenue of \$1200

29. ***Investments in South Africa*** The number of U.S. companies that invested in South Africa from 1986 through 1994 closely followed the function

$$n(t) = 5t^2 - 49t + 232$$

Here, t is the number of years since 1986, and $n(t)$ is the number of U.S. companies that own at least 50% of their South African subsidiaries and employ 1000 or more people.

a. Find the appropriate domain of n.

b. Is $t \ge 0$ an appropriate domain? Give reasons for your answer.

The model is the authors' (least-squares quadratic regression with coefficients rounded to nearest integer). *Source for raw data:* Investor Responsibility Research Center, Inc., Fleming Martin/*New York Times*, June 7, 1994, p. D1.

30. ***Sony Net Income*** The annual net income for Sony Corporation from 1989 through 1994 can be approximated by the function

$$I(t) = -77t^2 + 301t + 524$$

Here, t is the number of years since 1989, and $I(t)$ is Sony's net income in millions of dollars for the corresponding fiscal year.

a. Find the appropriate domain of t.

b. Is $[0, +\infty)$ an appropriate domain? Give reasons for your answer.

The model is the authors' (least-squares quadratic regression with coefficients rounded to nearest integer). *Source for raw data:* Sony Corporation/*New York Times*, May 20, 1994, p. D1.

31. ***Spending on Corrections*** The following table shows the annual spending on corrections by all states in the United States ($t = 0$ represents the year 1990):

Year, t	Spending (millions of dollars), $S(t)$
0	16
1	18
2	18
3	20
4	22
5	26
6	28
7	30

Source: National Association of State Budget Officers/*New York Times*, February 28, 1999, p. A1. Data are rounded.

a. Which of the following functions best fits the given data? (*Warning*: None of them fits exactly, but one fits more closely than the others.)

(1) $S(t) = -0.2t^2 + t + 16$

(2) $S(t) = 0.2t^2 + t + 16$

(3) $S(t) = t + 16$

b. Use your answer to part (a) to "predict" spending on corrections in 1998, assuming that the trend continued.

32. ***Spending on Corrections*** Repeat Exercise 31, this time choosing from the following functions:

(1) $S(t) = 16 + 2t$

(2) $S(t) = 16 + t + 0.5t^2$

(3) $S(t) = 16 + t - 0.5t^2$

33. ***Toxic Waste Treatment*** The cost of treating waste by removing PCPs increases rapidly as the quantity of PCPs removed goes up. Here is a possible model:

$$C(q) = 2000 + 100q^2$$

where q is the reduction in toxicity (in pounds of PCPs removed per day) and $C(q)$ is the daily cost (in dollars) of this reduction.

a. Find the cost of removing 10 pounds of PCPs per day.

b. Government subsidies for toxic waste cleanup amount to

$$S(q) = 500q$$

where q is as above and $S(q)$ is the daily dollar subsidy. Calculate the net cost of removing q pounds of PCPs per day (the cost after the subsidy is taken into account) $N(q)$. Then, find the net cost of removing 20 pounds of PCPs per day.

34. ***Dental Plans*** A company pays for its employees' dental coverage at an annual cost C given by

$$C(q) = 1000 + 100\sqrt{q}$$

where q is the number of employees covered and $C(q)$ is the annual cost in dollars.

a. If the company has 100 employees, find its annual outlay for dental coverage.

b. Assuming that the government subsidizes coverage by an annual dollar amount of

$$S(q) = 200q$$

calculate the net cost function $N(q)$ to the company, and calculate the net cost of subsidizing its 100 employees. Comment on your answer.

T 35. ***Acquisition of Language*** The percentage $p(t)$ of children who are able to speak in at least single words by the age of t months can be approximated by the equation

$$p(t) = 100\left(1 - \frac{12{,}196}{t^{4.478}}\right) \quad \text{where } t \geq 8.5$$

a. Create a table of values of p for $t = 9, 10, \ldots, 20$ (rounding answers to one decimal place).

b. What percentage of children are able to speak in at least single words by the age of 12 months?

c. By what age are 90% or more children speaking in at least single words?

The model is the authors' and is based on data presented in the article *The Emergence of Intelligence* by William H. Calvin, *Scientific American*, October 1994, pp. 101–107.

T 36. ***Acquisition of Language*** The percentage $p(t)$ of children who are able to speak in sentences of five or more words by the age of t months can be approximated by the equation

$$p(t) = 100\left(1 - \frac{5.2665 \times 10^{17}}{t^{12}}\right) \quad \text{where } t \geq 30$$

a. Create a table of values of p for $t = 30, 31, \ldots, 40$ (rounding answers to one decimal place).

b. What percentage of children are able to speak in sentences of five or more words by the age of 36 months?

c. By what age are 75% or more children speaking in sentences of five or more words?

The model is the authors' and is based on data presented in the article *The Emergence of Intelligence* by William H. Calvin, *Scientific American*, October 1994, pp. 101–107.

Communication and Reasoning Exercises

37. Complete the following: If weekly profit P is specified as a function of selling price s, then the independent variable is ______ and the dependent variable is ______.

38. Complete the following: The function notation for the equation $y = 4x^2 - 2$ is ______.

39. Why is this assertion false? If $f(x) = x^2 - 1$, then $f(x + h) = x^2 + h - 1$.

40. Why is this assertion false? If $f(x) = \sqrt{x}$, then $f(x + h) - f(x) = \sqrt{x + h} - \sqrt{x} = \sqrt{h}$.

41. True or false? If a function f is specified numerically, we can compute $f(x)$ for every x in the domain of f.

42. True or false? Every single-valued[4] algebraic expression involving one unknown specifies a function.

[4]By specifying "single-valued," we wish to exclude expressions such as $\pm\sqrt{x}$.

1.2 Functions from the Graphical Viewpoint

Consider again the function W discussed in Section 1.1, which gives a child's weight during its first year. If we represent the data given in Section 1.1 graphically by plotting the given pairs of numbers $(t, W(t))$, we get Figure 2. (We have connected successive points by line segments.)

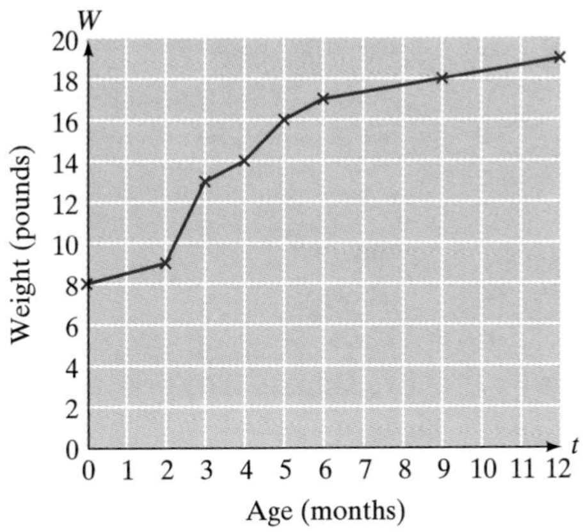

Figure 2

Suppose now that we had only the graph and not the table of data given in Section 1.1. We could use the graph to find values of W. For instance, to find $W(9)$ from the graph we do the following:

1. Scan the t-axis at the bottom of the graph until we reach the desired value ($t = 9$ in this case).
2. Estimate[1] the height (W-coordinate) of the corresponding point on the graph (18 in this case.)

Thus, $W(9) \approx 18$ pounds.

We say that Figure 2 specifies the function W **graphically.** The graph in Figure 2 is not a very accurate specification of W; the actual weights of the child would tend to follow a smooth curve rather than a jagged line. However, the jagged line is useful because it permits us to interpolate; for instance, we can estimate that $W(1) \approx 8.5$ pounds.

example 1

Graphically Specified Function

Figure 3 shows the annual circulation of *Outside* magazine for the period 1993–1997. Here, $t = 0$ represents December 31, 1993, and $C(t)$ represents the circulation in the year ending at time t. Estimate and interpret $C(0)$, $C(2)$, and $C(2.5)$. What is the domain of C?

[1] In a graphically defined function, we can never know the y-coordinates of points exactly; no matter how accurately a graph is drawn, we can obtain only *approximate* values of the coordinates of points. That is why we use the word *estimate* rather than *calculate* and why we use "$W(9) \approx 18$" rather than "$W(9) = 18$."

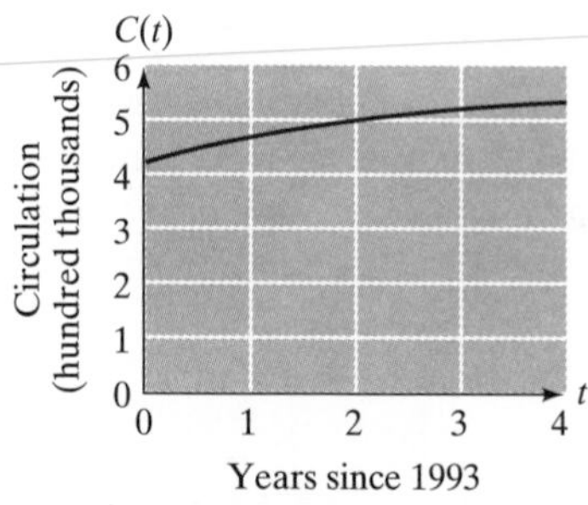

Source: Audit Bureau of Circulations/Publishers Information Bureau/*New York Times,* October 27, 1997, p. D13.

Figure 3

Solution

We carefully estimate the C-coordinates of the points with t-coordinates 0, 2, and 2.5. We have

$$C(0) \approx 4.2$$

which means that the circulation in the year ending December 1993 ($t = 0$) was approximately 420,000 copies. We see that

$$C(2) \approx 5.0$$

so the circulation in the year ending December 1995 ($t = 2$) was approximately 500,000 copies. Also,

$$C(2.5) \approx 5.1$$

meaning that the circulation in the year ending June 1996 ($t = 2.5$) was approximately 510,000 copies.

The domain of C is the set of all values of t for which $C(t)$ is defined: $0 \leq t \leq 4$, or $[0, 4]$.

Sometimes we are interested in drawing the graph of a function that has been specified in some other way—perhaps numerically or algebraically. We do this by plotting points[2] with coordinates $(x, f(x))$. Here is the formal definition of a graph.

Graph of a Function

The **graph of the function** f is the set of all points $(x, f(x))$ in the xy-plane, where we *restrict the values of x to lie in the domain of f.*

To obtain the graph of a function, we plot points of the form $(x, f(x))$ for several values of x in the domain of f. The shape of the entire graph can usually be inferred from sufficiently many points.

[2]Graphing utilities typically draw graphs by plotting a large number of points and joining them.

Quick Example
To sketch the graph of the function

$$f(x) = x^2 \quad \text{Function notation}$$

or

$$y = x^2 \quad \text{Equation notation}$$

with domain the set of all real numbers, we first choose some values of x in the domain and then compute the corresponding y-coordinates.

x	−3	−2	−1	0	1	2	3
$y = x^2$	9	4	1	0	1	4	9

Plotting these points gives the picture on the left, which suggests the graph shown on the right.[3] (This particular curve happens to be called a **parabola**, and its lowest point, at the origin, is called its **vertex.**)

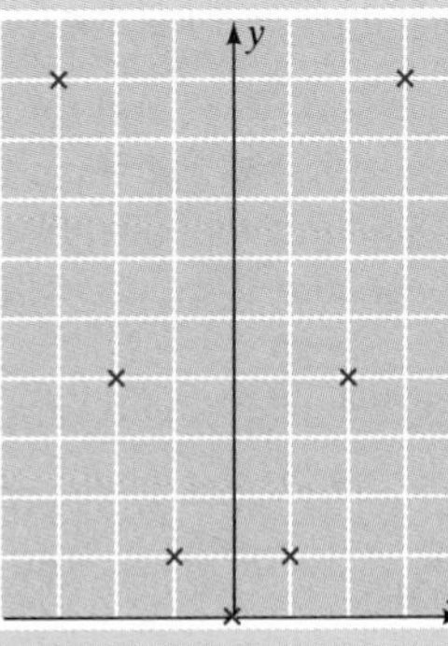

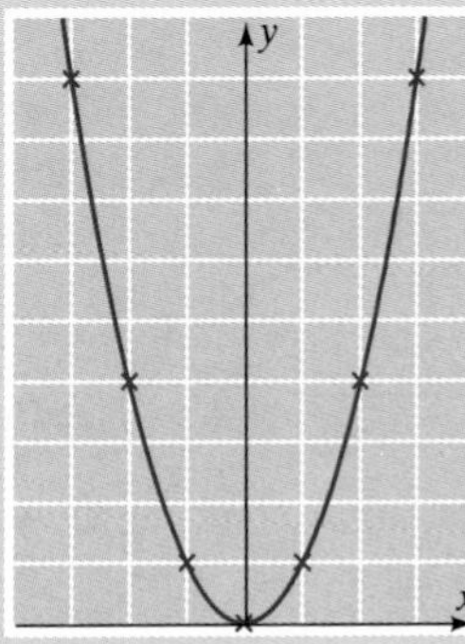

example 2

Graphing a Function

The monthly revenue[4] R from users logging on to your computer game web site depends on the monthly access fee p you charge according to the formula

$$R(p) = -5600p^2 + 14{,}000p \qquad (0 \le p \le 2.5)$$

(R and p are in dollars.) Sketch the graph of R. Find the access fee that will result in the highest monthly revenue.

Solution

To sketch the graph of R, we plot points of the form $(x, R(x))$ for several values of x in the domain of R. First, we calculate several points:

p	0	0.5	1	1.5	2	2.5
$R(p) = -5600p^2 + 14{,}000p$	0	5600	8400	8400	5600	0

[3]If you plot more points, you will find that they lie on a smooth curve as shown. That is why we did not use line segments to connect the points.

[4]The **revenue** that results from one or more business transactions is the total payment received, sometimes called the gross proceeds.

Plotting these points gives the graph on the left in Figure 4, which suggests the parabola shown on the right. The revenue graph appears to reach its highest point when $p = 1.25$, so setting the access fee at $1.25 appears to result in the highest monthly revenue.

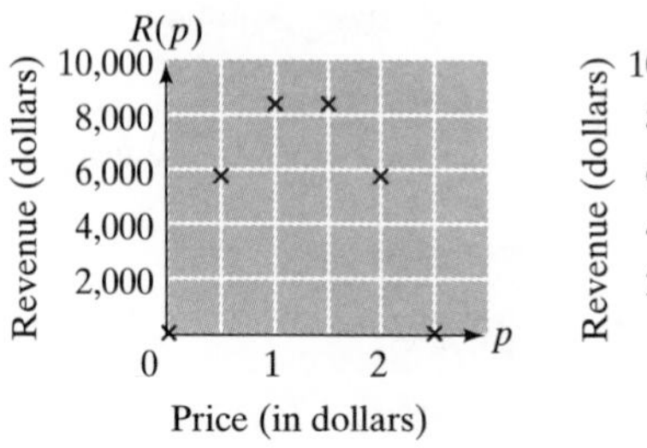

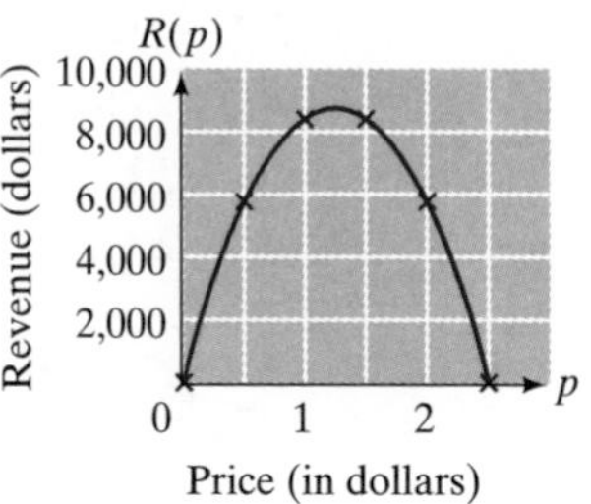

Figure 4

Before we go on . . . As we discussed after Example 2 in Section 1.1, we could write the function R in equation notation as

$$R = -5600p^2 + 14{,}000p \qquad \text{Equation notation}$$

The independent variable is p, and the dependent variable is R. Function notation and equation notation, using the same letter for the function name and the dependent variable, are used interchangeably throughout the literature. It is important to be able to switch back and forth easily from function notation to equation notation.

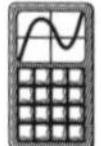

Graphing Calculator

As the name suggests, graphing calculators are designed for graphing functions. If you enter

```
Y1 = -5600*X^2+14000*X
```

in the "Y=" screen, you can reproduce the graph shown above by setting the window coordinates to match Figure 4: Xmin = 0, Xmax = 2.5, Ymin = 0, Ymax = 10000, and then pressing [GRAPH].

If you want to plot individual points (as in the graph on the left in Figure 4) on the TI-83, enter the data in the [STAT] EDIT mode with the values of p in L_1 and the value of $R(p)$ in L_2. Then go to the "Y=" window and turn Plot1 on by selecting it and pressing [ENTER]. Now press ZOOM STAT ([ZOOM] [9]) to obtain the scatter plot.

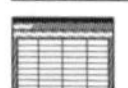

Spreadsheet

A spreadsheet program such as Excel can be used to plot individual data points, as in the graph on the left of Figure 4, and also a continuous curve, as in the graph on the right.

Suppose that you wish to graph a set of data points. As in Example 3 of Section 1.1, first create a table of values for the function by entering the values of the independent variable p in column A and the formula for the function in cell B1, copying it into the cells beneath as shown.

	A	B
1	0	= -5600*A1^2 + 14000*A1
2	0.5	
3	1	
4	1.5	
5	2	
6	2.5	

To draw the graph shown on the left in Figure 4, select both rows of data and then ask the program to insert a chart. When it asks you to specify what type of chart, specify an "XY" or "scatter plot." This tells the program that your data specify the x- and y-coordinates of a sequence of points. You can then set various other options, add labels, and otherwise fiddle with the chart until it looks nice.

Question How do we get a curve similar to the one on the right in Figure 4?
Answer If you chose the option that connects points by lines when setting up the above graph, you should see a jagged-looking version of Figure 4. To get a smoother curve, you will need to plot many points. Following is a method of plotting 100 points, based on the "Graphing Calculator" Excel worksheet posted on the web site at

Web Site

Web site → Everything for Calculus → Chapter 1 → Excel Tutorials →
Section 1.2: Functions from the Graphical Viewpoint

Set up your spreadsheet as follows:

	A	B	C	D
1	=$D$1	= -5600*A1^2 + 14000*A1	Xmin	0
2	=A1+$D$4		Xmax	2.5
3			Number of points	100
4			Delta X	=(D2-D1)/D3
5				
101				

The 100 values of x, the price, will appear in column A. The corresponding values of y, the revenue, will appear in column B. In column D you will see some settings: Xmin = 0 and Xmax = 2.5 (see Figure 4). Delta X (in cell D4) is the amount by which x is increased as you go from one x value in column A to the next, starting with Xmin in A1. Enter the formula for $R(p)$ in cell B1 and copy it into the other cells in column B as shown.

When you are done, graph the data in columns A and B, choosing the scatter plot option with subtype "points connected by lines with no markers." You should see a curve similar to that in Figure 4.

Save this spreadsheet with its graph and you can use it to graph new functions as follows:

1. Enter the new values for Xmin and Xmax in column D.
2. Enter the new function in cell B1 (using "A1" in place of x).
3. Copy the contents of cell B1 to cells B2–B101.

The graph will be updated automatically.

Web Site

At the web site, follow either

Web site → On-Line Utilities → Function Evaluator & Grapher

or Web site → On-Line Utilities → Graphing Functions

Instructions are available on the web pages.

Vertical Line Test

Every point on the graph of a function has the form $(x, f(x))$ for some x in the domain of f. Since f assigns a *single* value $f(x)$ to each value of x in the domain, it follows that, in the graph of f, there should be only one y corresponding to any such value of x—namely, $y = f(x)$. In other words, *the graph of a function cannot contain two or more points with the same x-coordinate—that is, two or more points on the same vertical line.* On the other hand, a vertical line at a value of x not in the domain will not contain any points in the graph. This gives us the following rule.

Vertical Line Test

For a graph to be the graph of a function, every vertical line must intersect the graph in *at most* one point.

Quick Examples

In the following graphs, only graph B passes the vertical line test, so only graph B is the graph of a function.

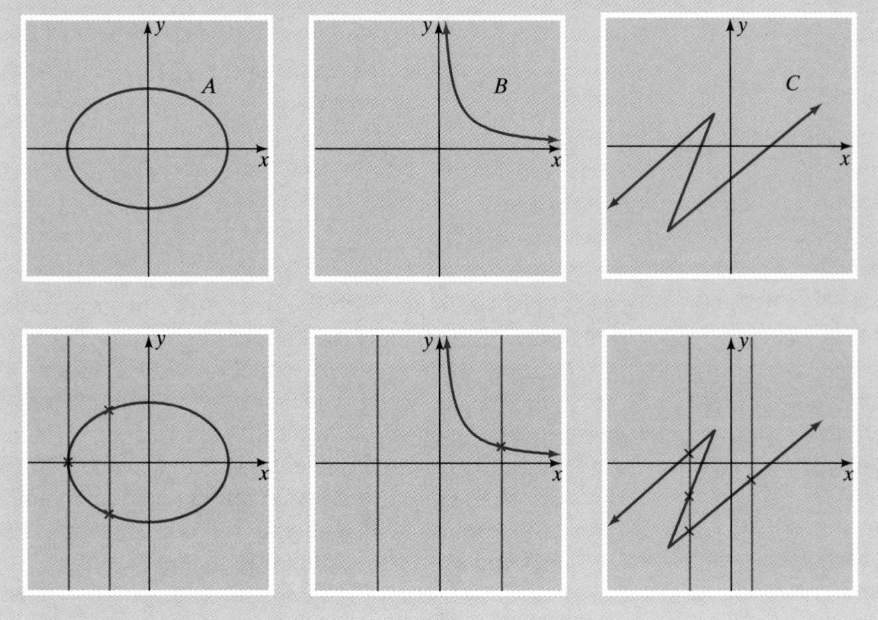

Graphing Piecewise-Defined Functions

Let us revisit the semiconductor example from Section 1.1.

example 3

Graphing a Piecewise-Defined Function

The percentage $U(t)$ of semiconductor equipment manufactured in the United States during the period 1981–1994 can be approximated by the following function of time t in years ($t = 0$ represents 1980):

$$U(t) = \begin{cases} 76 - 3.3t & \text{if } 1 \le t \le 10 \\ 18 + 2.5t & \text{if } 10 < t \le 14 \end{cases}$$

Sketch the graph of U.

Source: VLSI Research/*New York Times,* October 9, 1994, Section 3, p. 2.

Solution

As in Example 2, we compute $U(t)$ for several values of t, plot these points on the graph, and then connect them.

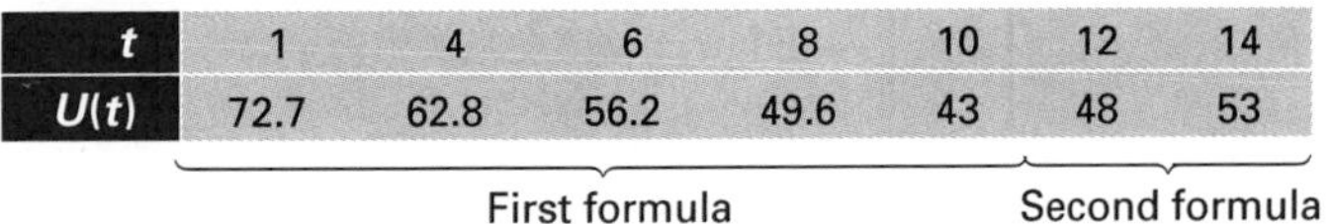

t	1	4	6	8	10	12	14
$U(t)$	72.7	62.8	56.2	49.6	43	48	53

First formula (t = 1 to 10) — Second formula (t = 12, 14)

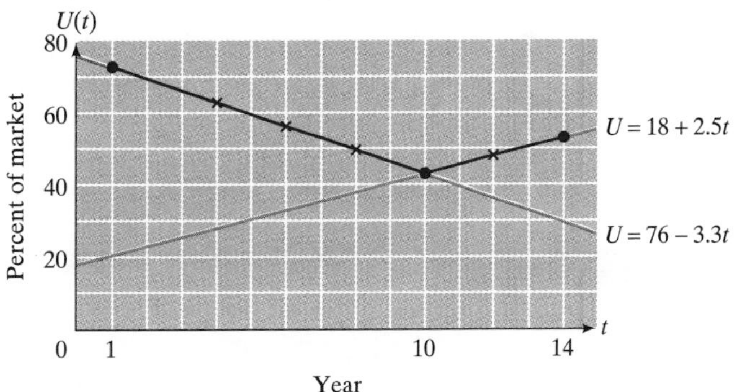

Figure 5

The graph in Figure 5 has the following features:

1. The first formula (the descending line) is used for $1 \le t \le 10$.
2. The second formula (ascending line) is used for $10 < t \le 14$.
3. The domain is $[1, 14]$, so the graph is cut off at $t = 1$ and $t = 14$.
4. The heavy dots at the ends indicate the endpoints of the domain.

T Graphing Technology

Example 4 of Section 1.1 shows how to enter this piecewise-defined function into various technologies. Example 2 of this section shows how to draw the graphs.

example 4

Graphing a Piecewise-Defined Function

Sketch the graph of

$$f(x) = \begin{cases} -1 & \text{if } -4 \le x < -1 \\ x & \text{if } -1 \le x \le 1 \\ x^2 - 1 & \text{if } 1 < x \le 2 \end{cases}$$

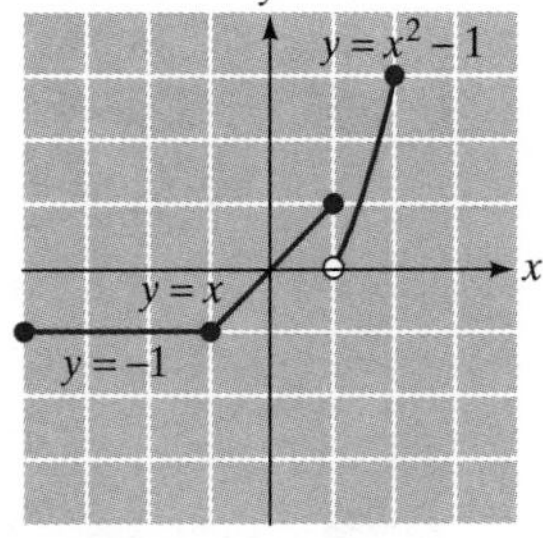

Figure 6

Solution

Using the techniques we've discussed, we sketch the three graphs $y = -1$, $y = x$, and $y = x^2 - 1$ and then use the appropriate portion of each (Figure 6).

Question What are the solid dots and open dots in Figure 6?

Answer The solid dots indicate points on the graph; the open dots indicate points not on the graph. For example, when $x = 1$, the inequalities in the formula tell us that we are to use the middle formula (x) rather than the bottom one ($x^2 - 1$). Thus, $f(1) = 1$, not 0, so we place a solid dot at $(1, 1)$ and an open dot at $(1, 0)$.

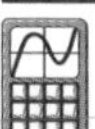

Graphing Calculator

On the TI-83 we enter this function as

```
(X<-1)*(-1)+(-1≤X and X≤1)*X+(1<X)*(X²-1)
```

The logical operator and is found in the TEST LOGIC menu. The TI-83 will not handle the transition at $x = 1$ correctly; it will connect the two parts of the graph with a spurious line segment.

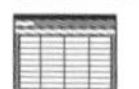

Spreadsheet

In Excel we enter this function as follows, assuming that the value of x is in A1:

```
=IF(A1<-1, -1, IF(A1<=1, X, X^2-1))
```

If we draw a graph with points connected by lines, we will have the same problem mentioned above for graphing calculators.

We end this section with a list of some useful types of functions and their graphs.

Type of function	Examples	
Linear $f(x) = mx + b$ m, b constant Graphs of linear functions are straight lines.	$y = x$	$y = -2x + 2$
Quadratic $f(x) = ax^2 + bx + c$ a, b, c constant ($a \neq 0$) Graphs of quadratic functions are called **parabolas.**	$y = x^2$	$y = -2x^2 + 2x + 4$
Technology formulas:	`x^2`	`-2*x^2+2*x+4`
Cubic $f(x) = ax^3 + bx^2 + cx + d$ a, b, c, d constant ($a \neq 0$)	$y = x^3$	$y = -x^3 + 3x^2 + 1$
Technology formulas:	`x^3`	`-x^3+3*x^2+1`

Type of function	Examples	
Exponential $f(x) = Ab^x$ A, b constant (b positive)	 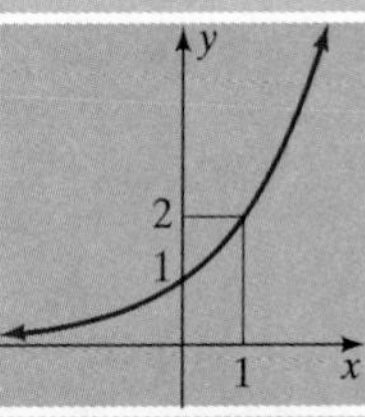	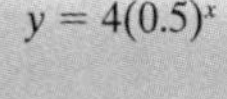 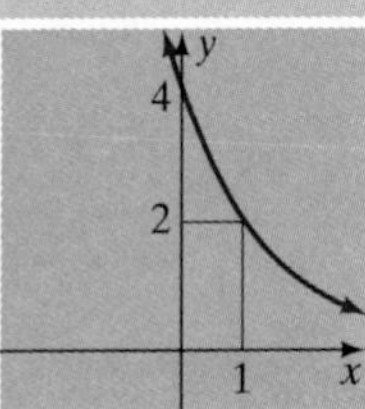
Technology formulas:	`2^x`	`4*0.5^x`
Rational $f(x) = \dfrac{P(x)}{Q(x)}$ $P(x)$ and $Q(x)$ polynomials The graph of $y = 1/x$ is a **hyperbola.** The domain excludes zero because 1/0 is not defined.	$y = \dfrac{1}{x}$ 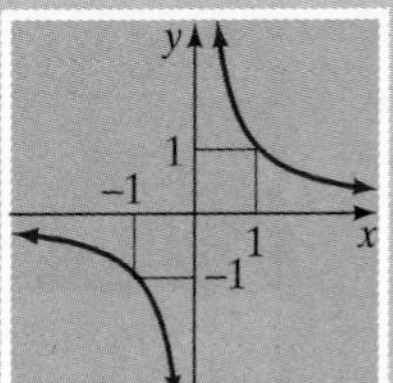	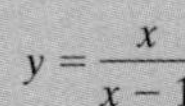 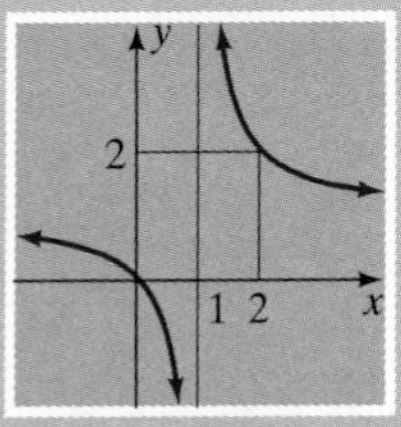
Technology formulas:	`1/x`	`x/(x-1)`
Absolute value For x positive or zero, $y = \|x\|$ agrees with $y = x$. For x negative or zero, $y = 1 \times 1$ agrees with $y = -x$.	$y = \|x\|$ 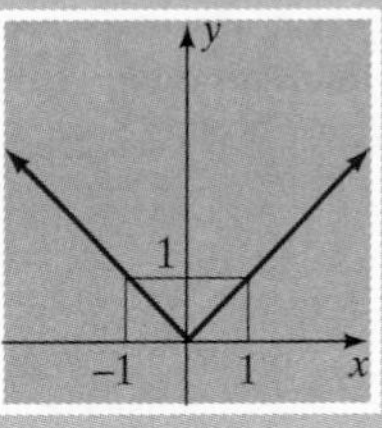	$y = \|2x + 2\|$ 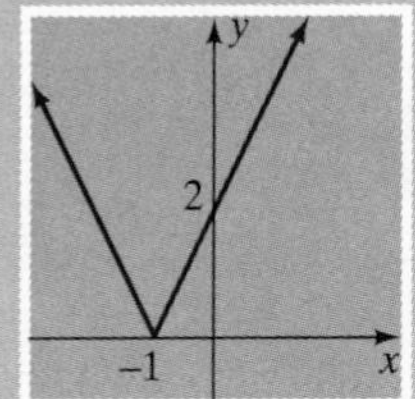
Technology formulas:	`abs(x)`	`abs(2*x+2)`
Square root The domain of $y = \sqrt{x}$ must be restricted to the nonnegative numbers because the square root of a negative number is not real. Its graph is the top half of a horizontally oriented parabola.	$y = \sqrt{x}$ 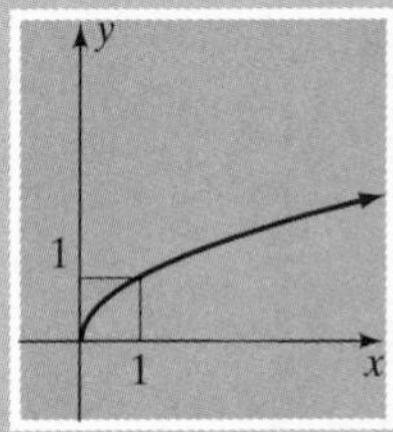	$y = \sqrt{4x - 2}$ 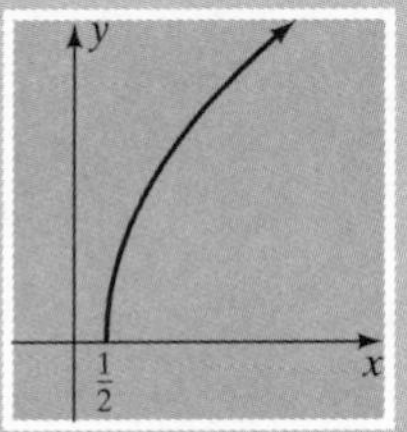
Technology formulas:	`x^0.5` or `√(x)`	`(4*x-2)^0.5` or `√(4*x-2)`

On-Line Optional Section: New Functions from Old

If you follow the path

Web site → Everything for Calculus → Chapter 1

you will find on-line text, interactive examples, and exercises on scaling and translating the graph of a function by changing the formula.

1.2 exercises

In Exercises 1–4, use the graph of the function f to find approximations of the given values.

1.

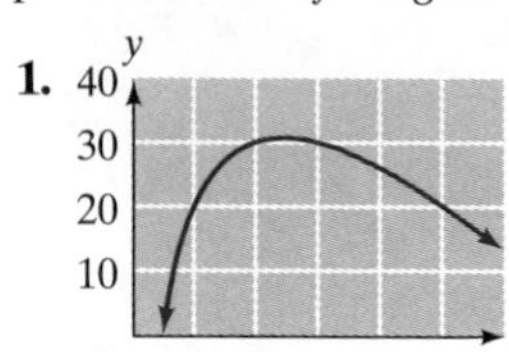

a. $f(1)$
b. $f(2)$
c. $f(3)$
d. $f(5)$
e. $f(3) - f(2)$

2.

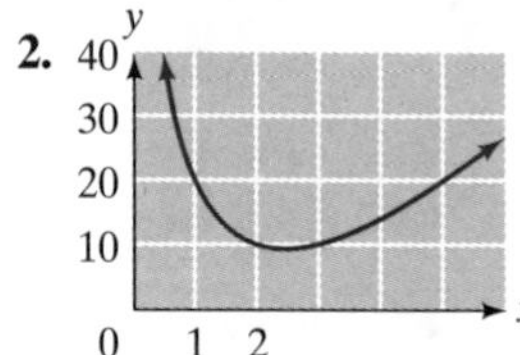

a. $f(1)$
b. $f(2)$
c. $f(3)$
d. $f(5)$
e. $f(3) - f(2)$

3.

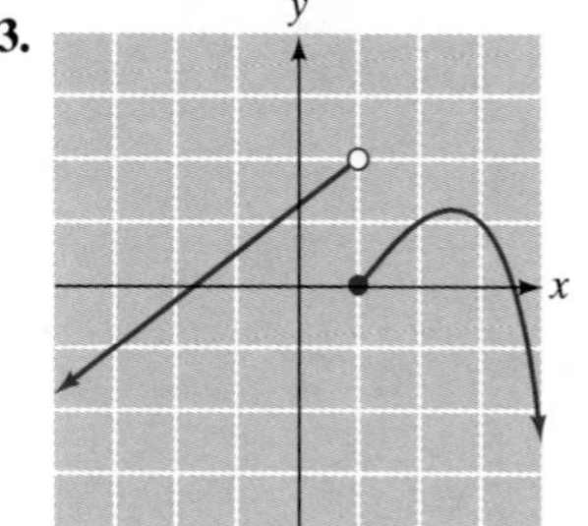

a. $f(-3)$
b. $f(0)$
c. $f(1)$
d. $f(2)$
e. $\dfrac{f(3) - f(2)}{3 - 2}$

4.

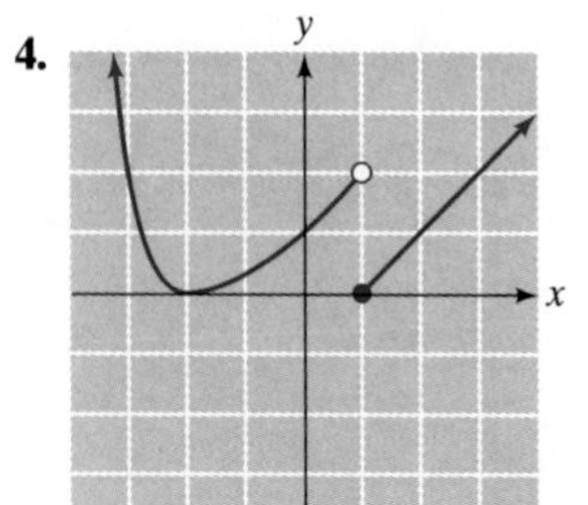

a. $f(-2)$
b. $f(0)$
c. $f(1)$
d. $f(3)$
e. $\dfrac{f(3) - f(1)}{3 - 1}$

In Exercises 5 and 6, match the functions to the graphs. Using technology to draw the graphs is suggested but not required.

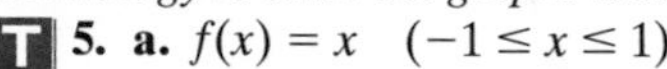

T **5.** **a.** $f(x) = x \quad (-1 \le x \le 1)$
b. $f(x) = -x \quad (-1 \le x \le 1)$
c. $f(x) = \sqrt{x} \quad (0 < x < 4)$
d. $f(x) = x + \dfrac{1}{x} - 2 \quad (0 < x < 4)$
e. $f(x) = |x| \quad (-1 \le x \le 1)$
f. $f(x) = x - 1 \quad (-1 \le x \le 1)$

(I)

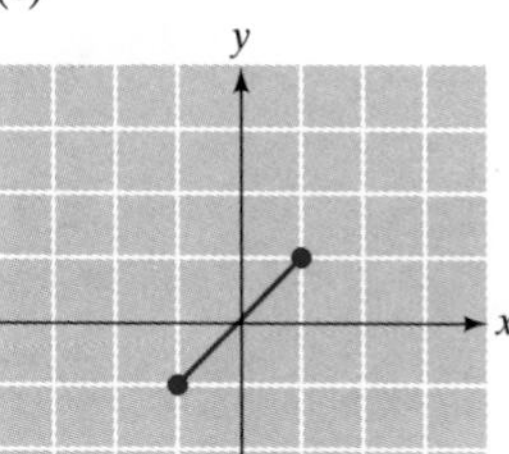

(II)

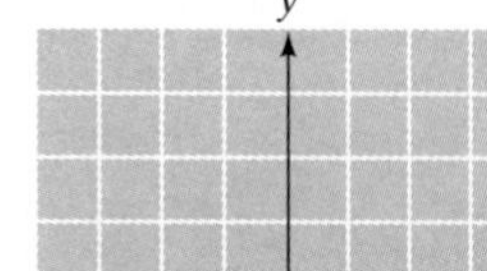

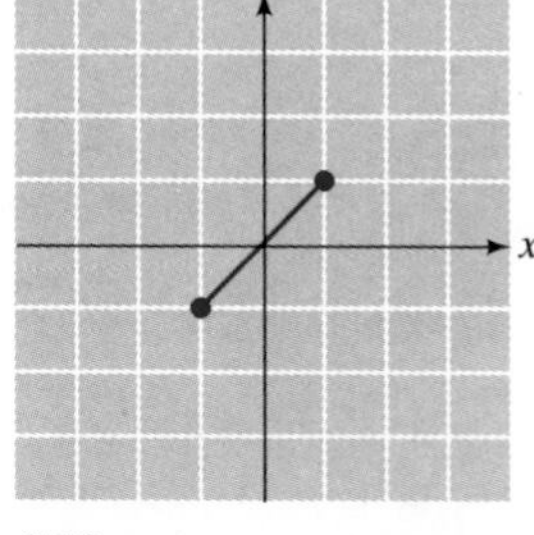

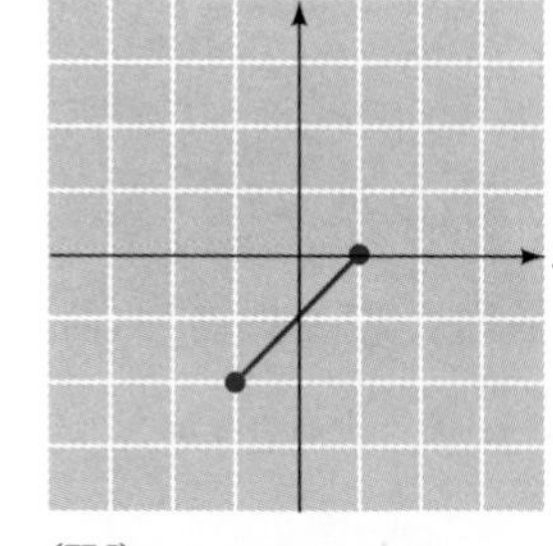

(III)

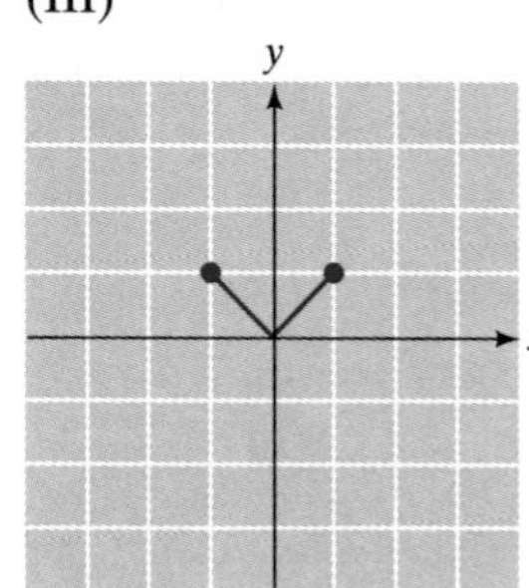

(IV)

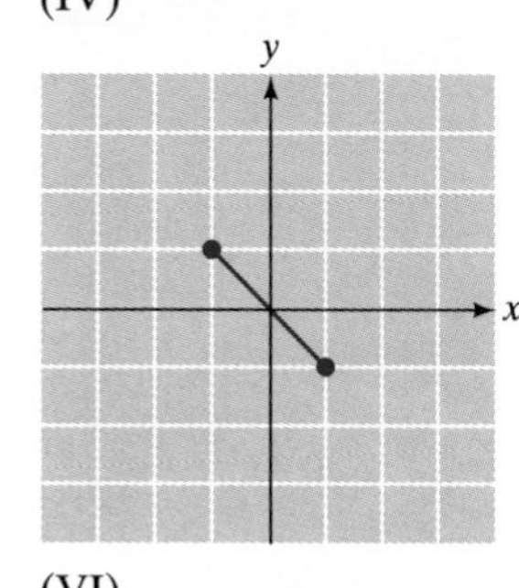

(V)

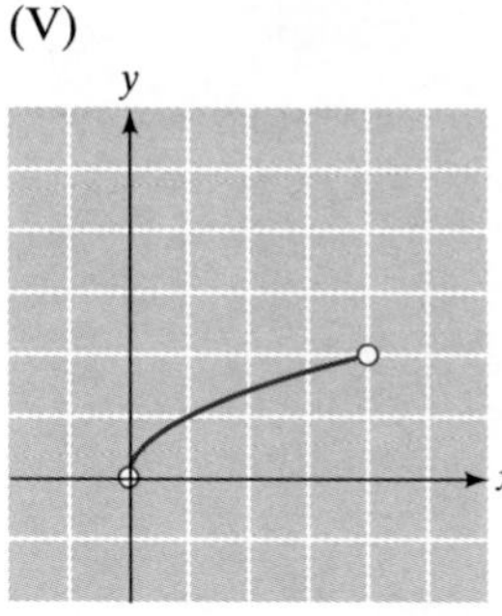

(VI)

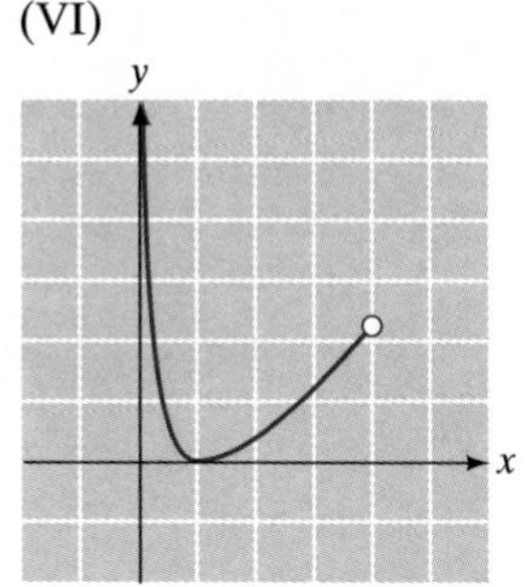

T 6. **a.** $f(x) = -x + 4 \quad (0 < x \le 4)$
b. $f(x) = 2 - |x| \quad (-2 < x \le 2)$
c. $f(x) = \sqrt{x + 2} \quad (-2 < x \le 2)$
d. $f(x) = -x^2 + 2 \quad (-2 < x \le 2)$
e. $f(x) = \frac{1}{x} - 1 \quad (0 < x \le 4)$
f. $f(x) = x^2 - 1 \quad (-2 < x \le 2)$

(I)

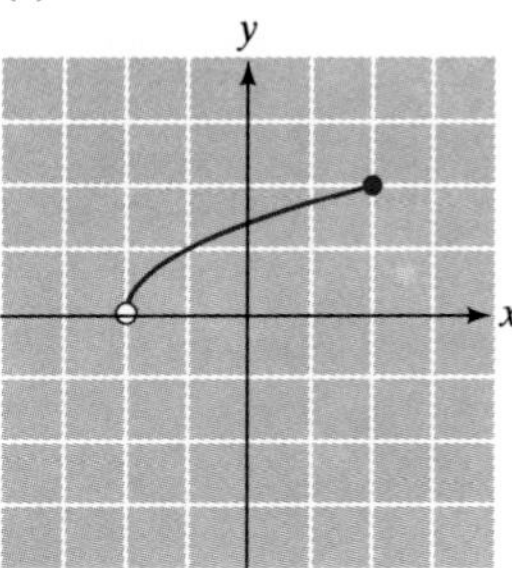

(II)

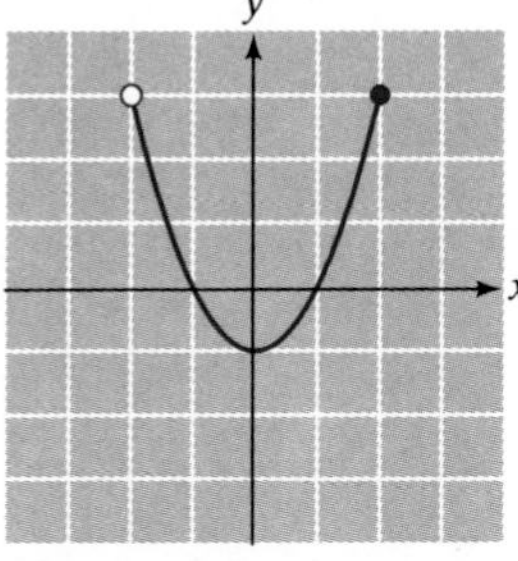

(III)

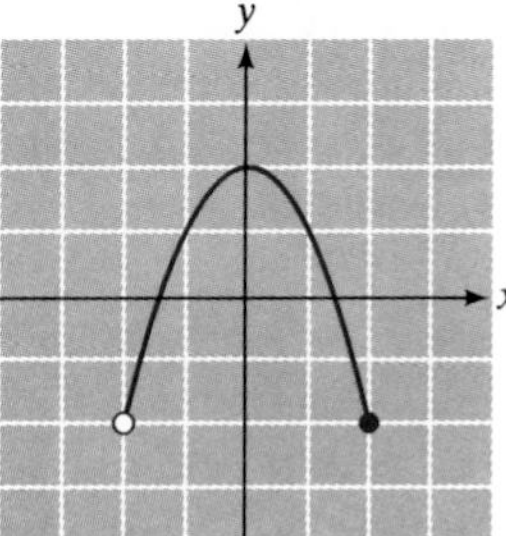

(IV)

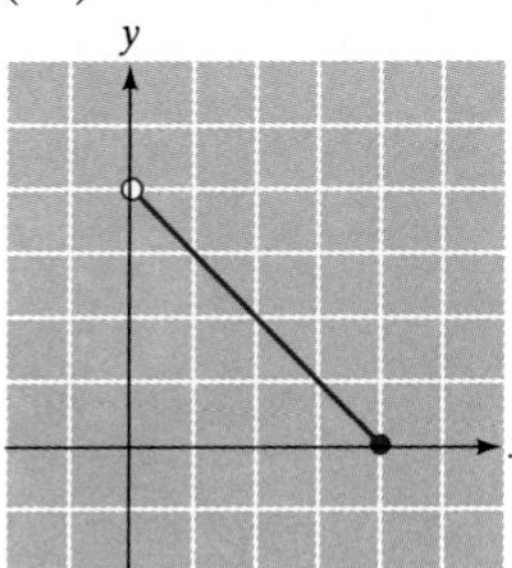

(V)

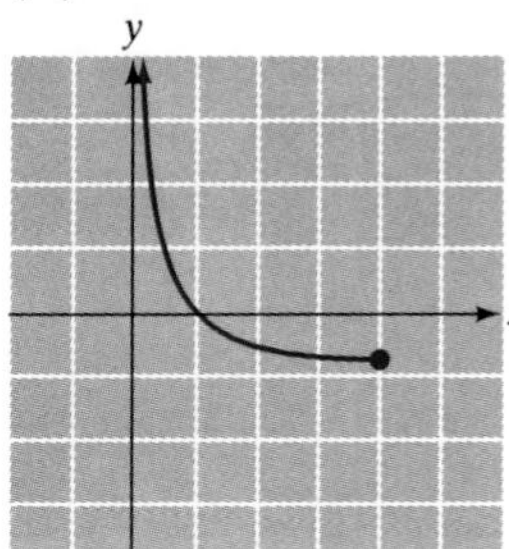

(VI)

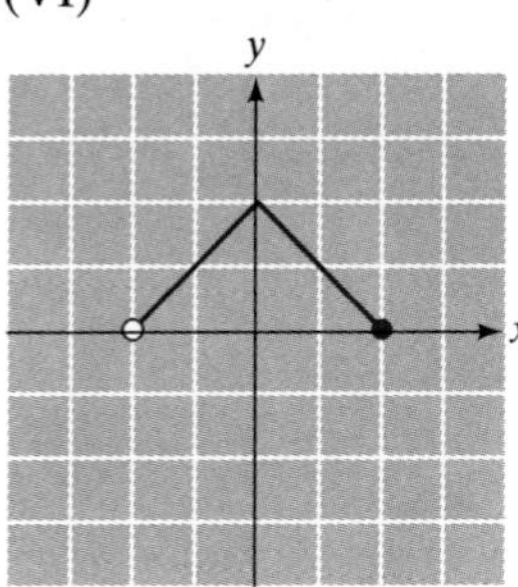

Graph the functions given in Exercises 7–12. We suggest that you become familiar with these graphs in addition to those in the chart at the end of the section.

7. $f(x) = -x^3$, domain $(-\infty, +\infty)$

8. $f(x) = x^3$, domain $[0, +\infty)$

9. $f(x) = x^4$, domain $(-\infty, +\infty)$

10. $f(x) = \sqrt[3]{x}$, domain $(-\infty, +\infty)$

11. $f(x) = \frac{1}{x^2} \quad (x \neq 0)$ **12.** $f(x) = x + \frac{1}{x} \quad (x \neq 0)$

In Exercises 13–18, sketch the graph of the function and evaluate the given expressions.

13. $f(x) = \begin{cases} x & \text{if } -4 \le x < 0 \\ 2 & \text{if } 0 \le x < 4 \end{cases}$ **a.** $f(-1)$ **b.** $f(0)$ **c.** $f(1)$

14. $f(x) = \begin{cases} -1 & \text{if } -4 \le x \le 0 \\ x & \text{if } 0 < x \le 4 \end{cases}$ **a.** $f(-1)$ **b.** $f(0)$ **c.** $f(1)$

15. $f(x) = \begin{cases} x & \text{if } -1 < x \le 0 \\ x + 1 & \text{if } 0 < x \le 2 \\ x & \text{if } 2 < x \le 4 \end{cases}$ **a.** $f(0)$ **b.** $f(1)$ **c.** $f(2)$ **d.** $f(3)$

16. $f(x) = \begin{cases} -x & \text{if } -1 < x < 0 \\ x - 2 & \text{if } 0 \le x \le 2 \\ -x & \text{if } 2 < x \le 4 \end{cases}$ **a.** $f(0)$ **b.** $f(1)$ **c.** $f(2)$ **d.** $f(3)$

T **17.** $f(x) = \begin{cases} x^2 & \text{if } -2 < x \le 0 \\ 1/x & \text{if } 0 < x \le 4 \end{cases}$ **a.** $f(-1)$ **b.** $f(0)$ **c.** $f(1)$

T **18.** $f(x) = \begin{cases} -x^2 & \text{if } -2 < x \le 0 \\ \sqrt{x} & \text{if } 0 < x < 4 \end{cases}$ **a.** $f(-1)$ **b.** $f(0)$ **c.** $f(1)$

Applications

Sales of Sport Utility Vehicles *Exercises 19–24 refer to the graph at the top of page 26, which shows the number of sports utility vehicles $f(t)$ sold in the United States each year from 1980 through 1994. Here, $t = 0$ represents the year 1980, and $f(t)$ represents sales in year t in thousands of vehicles.*

19. Estimate $f(2)$, $f(6)$, and $f(11.5)$. Interpret your answers.

20. Estimate $f(3)$, $f(12)$, and $f(10.5)$. Interpret your answers.

21. Estimate the largest value of $f(t)$ for $6 \le t \le 10$. Interpret the result.

22. Estimate the smallest value of $f(t)$ for $6 \le t \le 10$. Interpret the result.

23. During which year(s) were approximately 900,000 vehicles sold?

24. During which year(s) were more than 1 million vehicles sold?

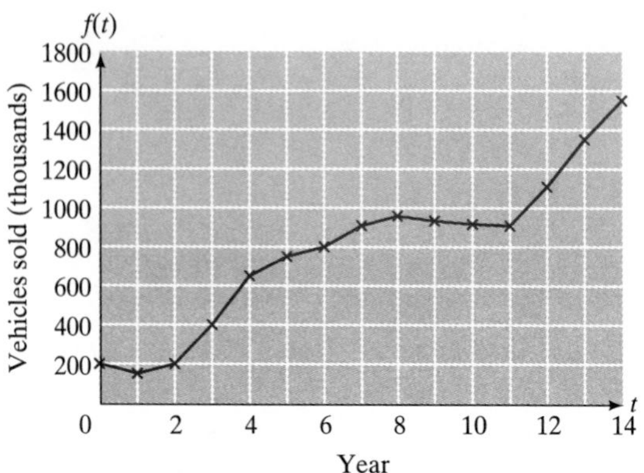

Source: Ford Motor Company/*New York Times*, February 9, 1995, p. D17.

T **25.** ***Acquisition of Language*** The percentage $p(t)$ of children who are able to speak in at least single words by the age of t months can be approximated by the equation[5]

$$p(t) = 100\left(1 - \frac{12{,}196}{t^{4.478}}\right) \quad \text{where } t \geq 8.5$$

a. Graph p for $9 \leq t \leq 20$ and $0 \leq p \leq 100$. Use your graph to answer parts (b) and (c).

b. What percentage of children are able to speak in at least single words by the age of 12 months? (Round your answer to the nearest percentage point.)

c. By what age are 90% of children speaking in at least single words? (Round your answer to the nearest month.)

T **26.** ***Acquisition of Language*** The percentage $p(t)$ of children who are able to speak in sentences of five or more words by the age of t months can be approximated by the equation[6]

$$p(t) = 100\left(1 - \frac{5.2665 \times 10^{17}}{t^{12}}\right) \quad \text{where } t \geq 30$$

a. Graph p for $30 \leq t \leq 45$ and $0 \leq p \leq 100$. Use your graph to answer parts (b) and (c).

b. What percentage of children are able to speak in sentences of five or more words by the age of 36 months? (Round your answer to the nearest percentage point.)

c. By what age are 75% of children speaking in sentences of five or more words? (Round your answer to the nearest percentage point.)

27. ***Hawaiian Tourism*** The following table shows the numbers of visitors to Hawaii in three different years.

Year	1985	1991	1994
Visitors (millions)	4.9	7.0	6.5

Source: Hawaii Visitors Bureau/*New York Times,* September 5, 1995, p. A12.

Which of the following kind of models would best fit the given data? Explain your choice. (a, b, c, and m are constants.)

(A) $n(t) = at^2 + bt + c$

(B) $n(t) = mt + b$

(C) $n(t) = a + \dfrac{b}{t}$

28. ***Magazine Advertising*** The following table shows the numbers of advertising pages in *Outside* magazine for the period 1993–1995.

Year	1993	1994	1995
Advertising pages	1000	1040	1080

Source: Audit Bureau of Circulations/Publishers Information Bureau/*New York Times,* October 27, 1997, p. D13.

Which of the following models would best fit the given data? Explain your choice. (a, b, c, and m are constants.)

(A) $n(t) = at^2 + bt + c$

(B) $n(t) = mt + b$

(C) $n(t) = a + \dfrac{b}{t}$

29. ***Coffee Shops*** (Compare Exercise 23 in Section 1.1.) The number $C(t)$ of coffee shops and related enterprises in the United States can be approximated by the following function of time t in years since 1990.

$$C(t) = \begin{cases} 500t + 800 & \text{if } 0 \leq t \leq 4 \\ 1300t - 2400 & \text{if } 4 < t \leq 10 \end{cases}$$

Sketch the graph of C and use your graph to estimate, to the nearest year, when there were 9000 coffee shops in the United States.

Source for data: Specialty Coffee Association of America/*New York Times*, August 13, 1995, p. F10.

30. ***Semiconductors*** (Compare Exercise 24 in Section 1.1.) The percentage $J(t)$ of semiconductor equipment manufactured in Japan during the period 1981–1994 can be approximated by the following function of time t in years since 1980.

$$J(t) = \begin{cases} 17 + 3.1t & \text{if } 1 \leq t \leq 10 \\ 68 - 2t & \text{if } 10 < t \leq 14 \end{cases}$$

Sketch the graph of J and use your graph to estimate, to the nearest year, when Japan manufactured 45% of all semiconductor equipment.

Source for data: VLSI Research/*New York Times*, October 9, 1994, Section 3, p. 2.

---

[5]The model is the authors' and is based on data presented in the article *The Emergence of Intelligence* by William H. Calvin, *Scientific American,* October 1994, pp. 101–107.

[6]See footnote 5.

Communication and Reasoning Exercises

31. True or false? Every graphically specified function can also be specified numerically. Explain.

32. True or false? Every algebraically specified function can also be specified graphically. Explain.

33. True or false? Every numerically specified function can also be specified graphically. Explain.

34. True or false? Every graphically specified function can also be specified algebraically. Explain.

35. How do the graphs of two functions differ if they are specified by the same formula but have different domains?

36. How do the graphs of two functions $f(x)$ and $g(x)$ differ if $g(x) = f(x) + 10$? (Try an example.)

37. How do the graphs of two functions $f(x)$ and $g(x)$ differ if $g(x) = f(x - 5)$? (Try an example.)

38. How do the graphs of two functions $f(x)$ and $g(x)$ differ if $g(x) = f(-x)$? (Try an example.)

1.3 Linear Functions

Linear functions are among the simplest functions and are perhaps the most useful of all mathematical functions.

Linear Function

A **linear function** is one that can be written in the form

			Quick Example
	$f(x) = mx + b$	Function notation	$f(x) = 3x - 1$
or	$y = mx + b$	Equation notation	$y = 3x - 1$

where m and b are fixed numbers. (The names m and b are traditional.[1])

Linear Functions from the Numerical Point of View

The following table shows values of $y = 3x - 1$ ($m = 3, b = -1$) for various values of x:

x	−4	−3	−2	−1	0	1	2	3	4
y	−13	−10	−7	−4	−1	2	5	8	11

First, note that setting $x = 0$ gives $y = -1$, the value of b.

Thus, b is the value of y when x = 0.

What about m? Notice that the value of y increases by $m = 3$ for every increase of 1 in x. This is caused by the term $3x$ in the formula: For every increase of 1 in x we get an increase of $3 \times 1 = 3$ in y.

Thus, y increases by m units for every one-unit increase in x.

Likewise, for every increase of 2 in x we get an increase of $3 \times 2 = 6$ in y. In general, if x changes by some amount, y will change by three times that amount.

Mathematicians traditionally use Δ (delta, the Greek equivalent of the Roman letter D) to stand for "difference," or "change in." For example, we write Δx to stand for the change in x. With this notation, we can write

[1]Actually, c is sometimes used instead of b. As for m, there has been some research lately into the question of its origin, but still no one knows exactly why the letter m is used.

$$\Delta y = 3\,\Delta x \quad \text{Change in } y = 3 \times \text{Change in } x$$

or

$$\frac{\Delta y}{\Delta x} = 3$$

In general (replace the number 3 by a general number m), we have the following guidelines:

The Numerical Roles of m and b in the Linear Function $f(x) = mx + b$

Role of m

Equation form
If $y = mx + b$, then y changes by m units for every one-unit change in x. A change of Δx units in x results in a change of $\Delta y = m\,\Delta x$ units in y. Thus,

$$m = \frac{\Delta y}{\Delta x} = \frac{\text{Change in } y}{\text{Change in } x}$$

Function form
If $f(x) = mx + b$, then f changes by m units for every one-unit change in x. A change of Δx units in x results in a change of $\Delta f = m\,\Delta x$ units in f. Thus,

$$m = \frac{\Delta f}{\Delta x} = \frac{\text{Change in } f}{\text{Change in } x}$$

Role of b

When $x = 0$, $y = b$ Equation form

$f(0) = b$ Function form

example 1

Recognizing Linear Data Numerically
Which of the following two tables gives the values of a linear function? What is the formula for that function?

x	0	2	4	6	8	10	12
$f(x)$	1	2	4	8	16	32	64

x	0	2	4	6	8	10	12
$g(x)$	−1	3	7	11	15	19	23

Solution
The function f cannot be linear. If it were, we would have $\Delta f = m\,\Delta x$ for some fixed number m. However, although the change in x between successive entries in the table is $\Delta x = 2$ each time, the change in f is not the same each time. On the other hand, the change in g is the same each time—namely, $\Delta g = 4$—so g is linear. Since we know Δg and Δx, we can compute

$$m = \frac{\Delta g}{\Delta x} = \frac{4}{2} = 2$$

From the table, $g(0) = -1$; hence, $b = -1$. Thus,

$$g(x) = 2x - 1 \quad \text{Check that this formula gives the values in the table.}$$

Spreadsheet

The following layout shows how you can use your spreadsheet to compute the successive quotients $m = \Delta y/\Delta x$ and hence check whether a given set of data shows a linear relationship (in which case all the quotients will be the same).

	A	B	C
1	0	-1	
2	2	3	=(B2-B1)/(A2-A1)
3	4	7	
4	6	11	
5	8	15	
6	10	19	
7	12	23	

Linear Functions from the Graphical Point of View

If we graph the linear function $g(x) = 2x - 1$ ($m = 2$, $b = -1$) using some of the points from Example 1, we get the graph shown in Figure 7. The quantities $m = 2$ and $b = -1$ are reflected in the graph of g as follows. First, since $g(0) = -1$, the graph passes through $(0, -1)$. Thus, $b = -1$ is the y-coordinate of the point where the graph crosses the y-axis. We call $b = -1$ the **y-intercept** of the graph.

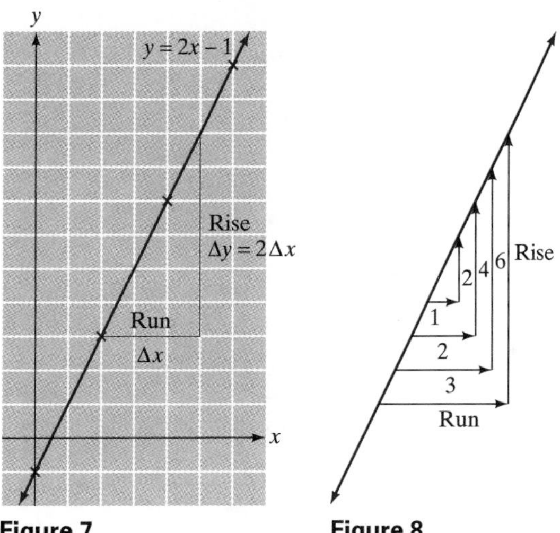

Figure 7 **Figure 8**

The number $m = 2$ shows up in the figure in a different way. Because the value of y increases by exactly 2 for every increase of 1 in x, the graph of g is a straight line rising by 2 for every 1 we go to the right. We say that we have a **rise** of 2 units for each **run** of 1 unit. Because the value of y changes by $\Delta y = 2\,\Delta x$ units for every change of Δx units in x, in general we have a rise of $\Delta y = 2\,\Delta x$ units for each run of Δx units. Thus, we have a rise of 4 for a run of 2, a rise of 6 for a run of 3, and so on (Figure 8).

We see that $m = 2$ is a measure of the steepness of the line; we call m the **slope of the line.**

$$\text{Slope} = m = \frac{\Delta y}{\Delta x} = \frac{\text{Rise}}{\text{Run}}$$

For a general linear function $f(x) = mx + b$, we can draw the graph using the slope and y-intercept.

The Graph of a Linear Function: Slope and y-Intercept

The graph of the linear function

$$f(x) = mx + b \quad \text{Function form}$$

or

$$y = mx + b \quad \text{Equation form}$$

is a straight line with slope m and y-intercept b. The slope m is given by

$$\text{Slope} = m = \frac{\Delta y}{\Delta x} = \frac{\text{Rise}}{\text{Run}}$$

For positive m, the graph rises m units for every one-unit move to the right, and the graph rises $\Delta y = m\,\Delta x$ units for every Δx units moved to the right. For negative m, the graph drops $|m|$ units for every one-unit move to the right, and the graph drops $|m|\,\Delta x$ units for every Δx units moved to the right.

Graph of $y = mx + b$

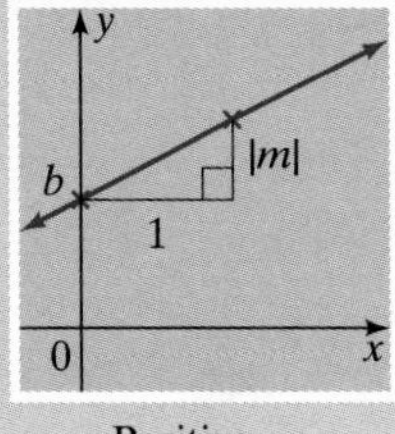

Positive m

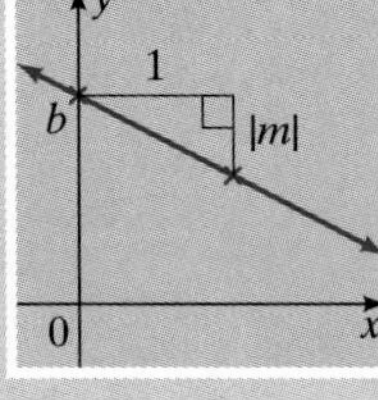

Negative m

Quick Examples

1. The function $f(x) = 2x + 1$ has slope $m = 2$ and y-intercept $b = 1$. To sketch the graph, we start at the y-intercept $b = 1$ on the y-axis and then move one unit to the right and up $m = 2$ units to arrive at a second point on the graph. Now we connect the two points to obtain the graph on the left.

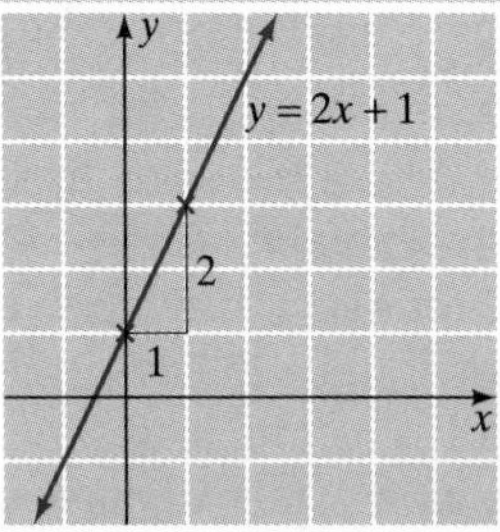

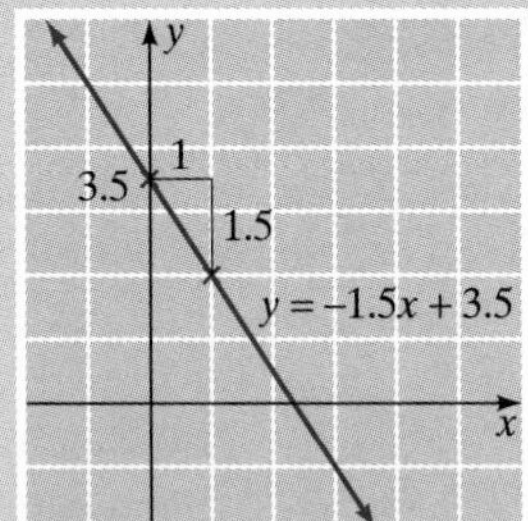

2. The line $y = -1.5x + 3.5$ has slope $m = -1.5$ and y-intercept $b = 3.5$. Since the slope is negative, the graph on the right goes *down* 1.5 units for every one unit it moves to the right.

3. The following graphs illustrate lines with different positive and negative slopes. Notice that the larger the absolute value of the slope, the steeper the line.

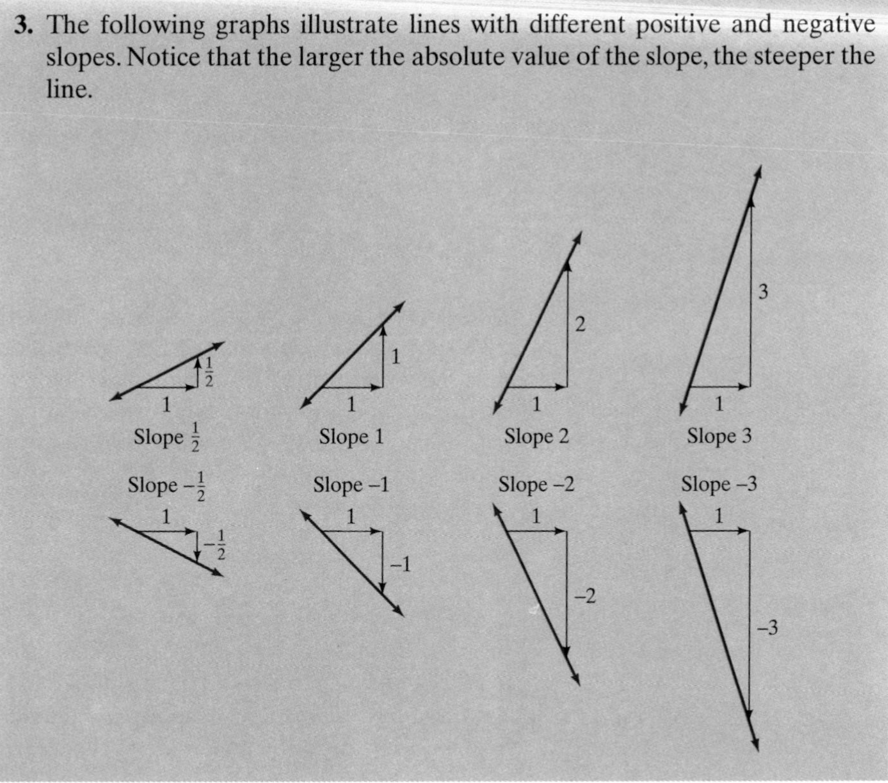

T **Graphing with Technology**

If you plot $y = 2x + 1$ using any of the technologies we have discussed, the line may appear to have the wrong slope. Figure 9 shows the graph of $y = 2x + 1$ twice: once on the xy-plane, with axes scaled the same, and once using technology.

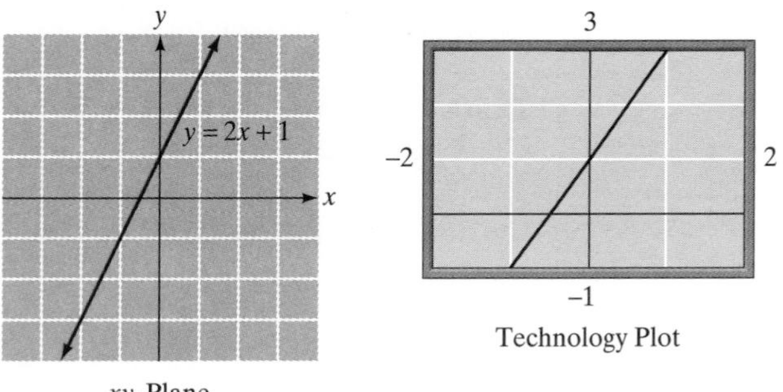

Figure 9

The slope in the graphing calculator window looks smaller (the line is less steep) than the slope in the xy-plane. This distortion occurs because (as with most calculators) the screen is not square. Most graphing calculators have an option to "square up" the graph to remove this distortion. Similar comments hold for other graphing technologies.

example 2 **Graphing a Linear Equation; Intercepts**

Graph the equation $x + 2y = 4$. Where does the line cross the x- and y-axes?

Solution

We first write y as a linear function of x by solving the equation for y:

$$2y = -x + 4$$

so

$$y = -\tfrac{1}{2}x + 2$$

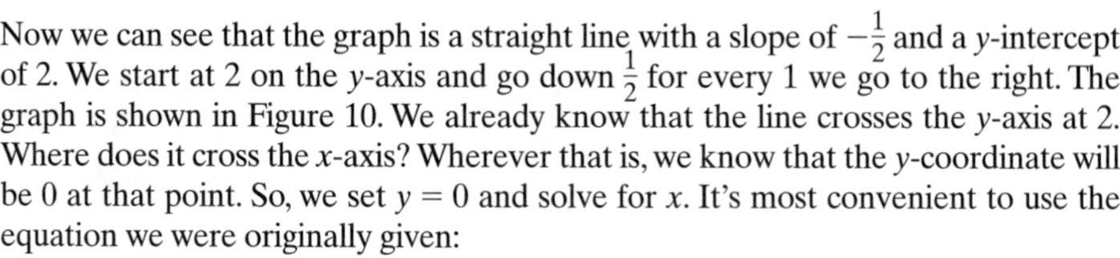

Now we can see that the graph is a straight line with a slope of $-\frac{1}{2}$ and a y-intercept of 2. We start at 2 on the y-axis and go down $\frac{1}{2}$ for every 1 we go to the right. The graph is shown in Figure 10. We already know that the line crosses the y-axis at 2. Where does it cross the x-axis? Wherever that is, we know that the y-coordinate will be 0 at that point. So, we set $y = 0$ and solve for x. It's most convenient to use the equation we were originally given:

Figure 10

$$x + 2(0) = 4$$

so

$$x = 4$$

Thus, the line crosses the x-axis at 4.

Before we go on . . . We could have graphed the equation in another way by first finding the intercepts. Once we know that the line crosses the y-axis at 2 and the x-axis at 4, we can draw those two points and then draw the line that connects them.

It's worth noting what we did to find the intercepts in the preceding example.

Finding the Intercepts

The ***x*-intercept** of a line is where it crosses the x-axis. To find it, we set $y = 0$ and solve for x. The ***y*-intercept** is where the line crosses the y-axis. If the equation of the line is written as $y = mx + b$, then b is the y-intercept. Otherwise, we set $x = 0$ and solve for y.

Quick Example

Consider the equation $3x - 2y = 6$. To find its x-intercept, we set $y = 0$ to find $x = 6/3 = 2$. To find its y-intercept, we set $x = 0$ to find $y = 6/(-2) = -3$. Thus, the line crosses the x-axis at 2 and the y-axis at -3.

Computing the Slope of a Line

We know that the slope of a line is given by

$$\text{Slope} = m = \frac{\text{Rise}}{\text{Run}} = \frac{\Delta y}{\Delta x}$$

Question Two points—say, (x_1, y_1) and (x_2, y_2)—determine a line in the xy-plane. How do we find its slope?

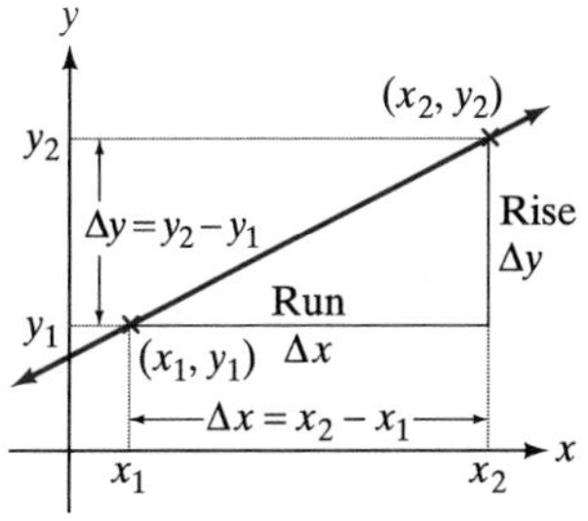

y, y_2, (x_2, y_2), $\Delta y = y_2 - y_1$, Rise Δy, y_1, Run Δx, (x_1, y_1), $\Delta x = x_2 - x_1$, *x*, x_1, x_2

Figure 11

Answer As you can see in Figure 11, the rise is $\Delta y = y_2 - y_1$, the change in the *y*-coordinate from the first point to the second, while the run is $\Delta x = x_2 - x_1$, the change in the *x*-coordinate.

Computing the Slope of a Line

We can compute the slope *m* of the line through the points (x_1, y_1) and (x_2, y_2) with the formula

$$m = \frac{\Delta y}{\Delta x} = \frac{y_2 - y_1}{x_2 - x_1}$$

Quick Examples

1. The slope of the line through $(x_1, y_1) = (1, 3)$ and $(x_2, y_2) = (5, 11)$ is

$$m = \frac{\Delta y}{\Delta x} = \frac{y_2 - y_1}{x_2 - x_1} = \frac{11 - 3}{5 - 1} = \frac{8}{4} = 2$$

2. The slope of the line through $(x_1, y_1) = (1, 2)$ and $(x_2, y_2) = (2, 1)$ is

$$m = \frac{\Delta y}{\Delta x} = \frac{y_2 - y_1}{x_2 - x_1} = \frac{1 - 2}{2 - 1} = \frac{-1}{1} = -1$$

Question What if we had chosen to list the two points in Quick Example 1 in reverse order? That is, suppose we had taken $(x_1, y_1) = (5, 11)$ and $(x_2, y_2) = (1, 3)$.

Answer Then we would have found

$$m = \frac{\Delta y}{\Delta x} = \frac{y_2 - y_1}{x_2 - x_1} = \frac{3 - 11}{1 - 5} = \frac{-8}{-4} = 2$$

the same answer. The order in which we take the points is not important, *as long as we use the same order in both the numerator and the denominator.*

example 3

Zero Slope; Undefined Slope

Find the slope of each of the following lines:

a. Through (2, 3) and (−1, 3) **b.** Through $\left(\frac{1}{2}, 1\right)$ and $\left(\frac{1}{2}, 3\right)$

Solution

a. From the formula,

$$m = \frac{\Delta y}{\Delta x} = \frac{3 - 3}{-1 - 2} = \frac{0}{-3} = 0$$

A line of slope 0 has a 0 rise, so it is a *horizontal* line (Figure 12).

b. $m = \dfrac{\Delta y}{\Delta x} = \dfrac{3 - 1}{\frac{1}{2} - \frac{1}{2}} = \dfrac{2}{0}$ is *undefined.*

If we plot the two points in question, we see that the line that passes through them is *vertical* (Figure 13). Vertical lines are the only ones that have undefined slope.

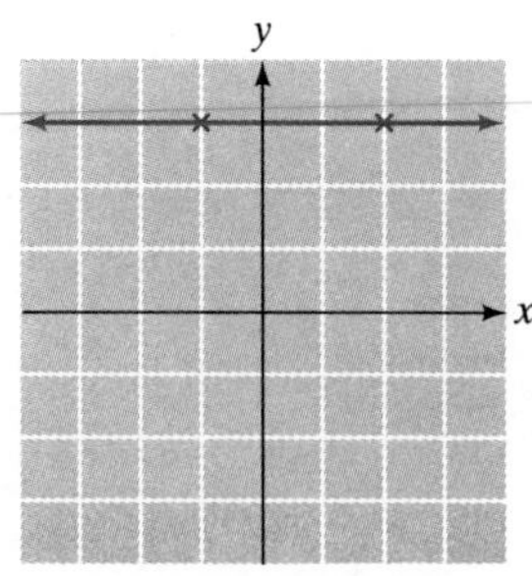

The line through (2, 3) and (–1, 3)

Figure 12

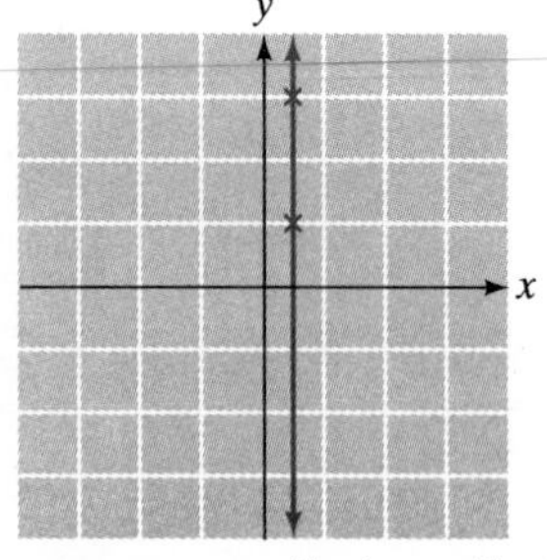

The line through $\left(\frac{1}{2}, 1\right)$ and $\left(\frac{1}{2}, 3\right)$

Figure 13

Finding a Linear Equation from Data: How to Make a Linear Model

If we happen to know the slope and y-intercept of a line, writing down its equation is straightforward. For example, if we know that the slope is 3 and the y-intercept is -1, then the equation is $y = 3x - 1$. Sadly, we seldom have such convenient information. For instance, we may know the slope and a point other than the y-intercept, two points on the line, or other information.

We describe the most straightforward method for finding the equation of a line: the **point-slope method.** As the name suggests, we need two pieces of information:

- The *slope m* (which specifies the direction of the line)
- A *point* (x_0, y_0) on the line (which pins down its location in the plane)

Question What is the equation of the line through the point (x_0, y_0) with slope m?

Answer Before we state the answer in general, let us look at a specific example. Suppose we want to find the equation of the line through (1, 2) with slope 3. If (x, y) is a point on this line other than (1, 2), then, since the line through the points (1, 2) and (x, y) has slope 3, the slope formula tells us that

$$m = 3 = \frac{y - 2}{x - 1}$$

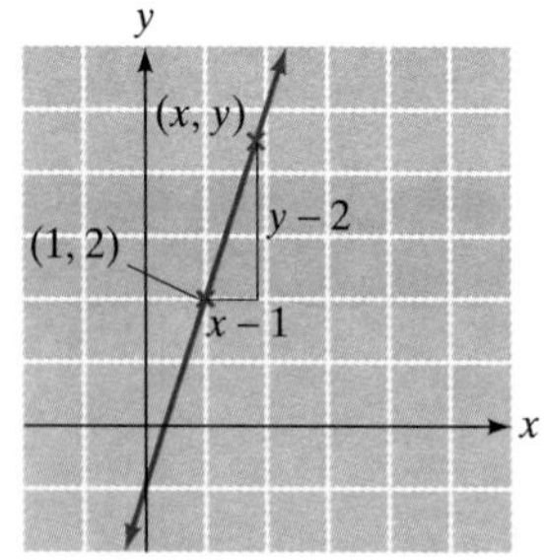

Figure 14

(Figure 14). Multiplying both sides by $x - 1$ gives

$$y - 2 = 3(x - 1)$$

which is an equation of this line because it must be satisfied by any point (x, y) on the line. If we add 2 to both sides, we get y as a linear function of x:

$$y = 2 + 3(x - 1) = 3x - 1$$

There was nothing special in this example about the numbers 1, 2, and 3. We can easily generalize to get the following formula.

Point-Slope Formula

An equation of the line through the point (x_0, y_0) with slope m is given by

$$y = y_0 + m(x - x_0) \quad \text{Equation form}$$

or $$f(x) = y_0 + m(x - x_0) \quad \text{Function form}$$

When to Apply the Point-Slope Formula

We apply the point-slope formula to find the equation of a line whenever we are given information about a point and the slope of the line. The formula does not apply if the slope is undefined, as in a vertical line [see Example 4(e) below].

Quick Example
The line through (2, 3) with slope 4 has equation

$$y = 3 + 4(x - 2) \quad \text{or} \quad f(x) = 3 + 4(x - 2)$$

Caution The point (x_0, y_0) represents a given point, so the terms x_0 and y_0 are always replaced by actual numbers. The terms x and y, on the other hand, remain as x and y because they are the variables in the equation of the line.

example 4

Using the Point-Slope Formula
Find equations for the following straight lines.

a. Through (1, 3) with slope 2
b. Through the points (1, 2) and (3, −1)
c. Through (2, −2) and parallel to the line $3x + 4y = 5$
d. Horizontal and through (−9, 5)
e. Vertical and through (−9, 5)

Solution
In each case other than (e), we apply the point-slope formula.
a. To apply the point-slope formula, we need a point on the line and the slope:

- ***Point*** Given here as $(x_0, y_0) = (1, 3)$
- ***Slope*** Given here as $m = 2$

We can therefore write

$$\begin{aligned} y &= y_0 + m(x - x_0) \\ &= 3 + 2(x - 1) && \text{Substitute for } m, x_0, \text{ and } y_0. \\ &= 3 + 2x - 2 && \text{Use distributive law.} \\ &= 2x + 1 \end{aligned}$$

b. Again, we need a point and a slope:

- ***Point*** We have two to choose from, so we take the first: $(x_0, y_0) = (1, 2)$.

- ***Slope*** Not given *directly,* but we do have enough information to calculate it. Since we are given two points on the line, we can use the slope formula:

$$m = \frac{y_2 - y_1}{x_2 - x_1} = \frac{-1 - 2}{3 - 1} = -\frac{3}{2}$$

An equation of the line is therefore

$$\begin{aligned} y &= y_0 + m(x - x_0) \\ &= 2 - \frac{3}{2}(x - 1) && \text{Substitute for } m, x_0, \text{ and } y_0. \\ &= 2 - \frac{3}{2}x + \frac{3}{2} && \text{Use distributive law.} \\ &= -\frac{3}{2}x + \frac{7}{2} \end{aligned}$$

c. Proceeding as before, we need:

- ***Point*** Given here as $(2, -2)$
- ***Slope*** We use the fact that *parallel lines have the same slope.* (Why?) We can find the slope of $3x + 4y = 5$ by solving for y and then looking at the coefficient of x:

$$y = -\frac{3}{4}x + \frac{5}{4}$$

so the slope is $-\frac{3}{4}$.

Thus, an equation for the desired line is

$$\begin{aligned} y &= y_0 + m(x - x_0) \\ &= -2 - \frac{3}{4}(x - 2) \\ &= -2 - \frac{3}{4}x + \frac{3}{2} \\ &= -\frac{3}{4}x - \frac{1}{2} \end{aligned}$$

d. We are given a point: $(-9, 5)$. Furthermore, we are told that the line is horizontal, which tells us that the slope is 0. Therefore, we get

$$\begin{aligned} y &= y_0 + m(x - x_0) \\ &= 5 + 0(x + 9) \\ &= 5 \end{aligned}$$

e. We are given a point: $(-9, 5)$. This time, we are told that the line is vertical, which means that the slope is undefined. Thus, we can't use the point-slope formula. (That formula makes sense only when the slope of the line is defined.) What can we do? Well, here are some points on the desired line:

$$(-9, 1), (-9, 2), (-9, 3), \ldots$$

so $x = -9$ and $y =$ *anything.* If we simply say that $x = -9$, then these points are all solutions, so the equation is $x = -9$. [See also Example 3(b).]

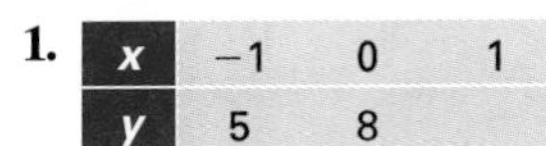

In Exercises 1–6, a table of values for a linear function is given. Fill in the missing value and calculate m in each case.

1.

x	−1	0	1
y	5	8	

2.

x	−1	0	1
y	−1	−3	

3.

x	2	3	5
$f(x)$	−1	−2	

4.

x	2	4	5
$f(x)$	−1	−2	

5.

x	−2	0	2
$f(x)$	4		10

6.

x	0	3	6
$f(x)$	−1		−5

In Exercises 7–10, first find f(0) if it is not supplied and then find the equation of the given linear function.

7.

x	−2	0	2	4
$f(x)$	−1	−2	−3	−4

8.

x	−6	−3	0	3
$f(x)$	1	2	3	4

9.

x	−4	−3	−2	−1
$f(x)$	−1	−2	−3	−4

10.

x	1	2	3	4
$f(x)$	4	6	8	10

In Exercises 11–14, decide which of the two given functions is linear and find its equation.

11.

x	0	1	2	3	4
$f(x)$	6	10	14	18	22
$g(x)$	8	10	12	16	22

12.

x	−10	0	10	20	30
$f(x)$	−1.5	0	1.5	2.5	3.5
$g(x)$	−9	−4	1	6	11

13.

x	0	3	6	10	15
$f(x)$	0	3	5	7	9
$g(x)$	−1	5	11	19	29

14.

x	0	3	5	6	9
$f(x)$	2	6	9	12	15
$g(x)$	−1	8	14	17	26

Sketch the straight lines with the equations given in Exercises 15–28.

15. $y = 2x - 1$

16. $y = x - 3$

17. $y = -\frac{2}{3}x + 2$

18. $y = -\frac{1}{2}x + 3$

19. $y + \frac{1}{4}x = -4$

20. $y - \frac{1}{4}x = -2$

21. $7x - 2y = 7$

22. $2x - 3y = 1$

23. $3x = 8$

24. $2x = -7$

25. $6y = 9$

26. $3y = 4$

27. $2x = 3y$

28. $3x = -2y$

In Exercises 29–44, calculate the slope, if defined, of the straight line through the given pair of points. Try to do as many as you can without writing anything down except the answer.

29. $(0, 0)$ and $(1, 2)$

30. $(0, 0)$ and $(-1, 2)$

31. $(-1, -2)$ and $(0, 0)$

32. $(2, 1)$ and $(0, 0)$

33. $(4, 3)$ and $(5, 1)$

34. $(4, 3)$ and $(4, 1)$

35. $(1, -1)$ and $(1, -2)$

36. $(-2, 2)$ and $(-1, -1)$

37. $(2, 3.5)$ and $(4, 6.5)$

38. $(10, -3.5)$ and $(0, -1.5)$

39. $(300, 20.2)$ and $(400, 11.2)$

40. $(1, -20.2)$ and $(2, 3.2)$

41. $(0, 1)$ and $\left(-\frac{1}{2}, \frac{3}{4}\right)$

42. $\left(\frac{1}{2}, 1\right)$ and $\left(-\frac{1}{2}, \frac{3}{4}\right)$

43. (a, b) and (c, d), $(a \neq c)$

44. (a, b) and (c, b), $(a \neq c)$

45. In the following figures, estimate the slope of each line segment.

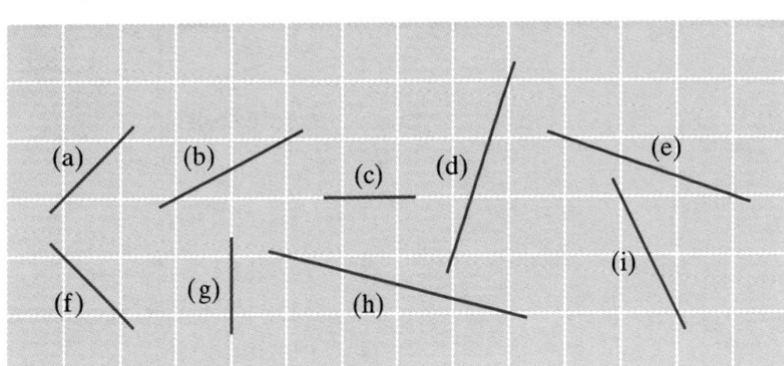

46. In the following figures, estimate the slope of each line segment.

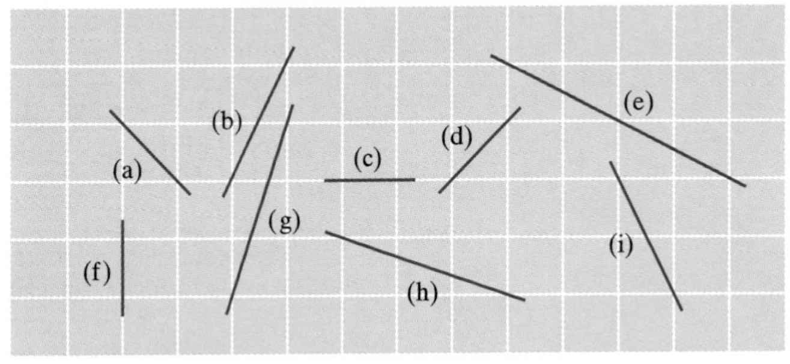

Find the linear equations with graphs that are the straight lines in Exercises 47–60.

47. Through (1, 3) with slope 3

48. Through (2, 1) with slope 2

49. Through $\left(1, -\frac{3}{4}\right)$ with slope $\frac{1}{4}$

50. Through $\left(0, -\frac{1}{3}\right)$ with slope $\frac{1}{3}$

51. Through (20, −3.5) and increasing at a rate of 10 units of y per unit of x

52. Through (3.5, −10) and increasing at a rate of 1 unit of y per 2 units of x

53. Through (2, −4) and (1, 1)

54. Through (1, −4) and (−1, −1)

55. Through (1, −0.75) and (0.5, 0.75)

56. Through (0.5, −0.75) and (1, −3.75)

57. Through (6, 6) and parallel to the line $x + y = 4$

58. Through $\left(\frac{1}{3}, -1\right)$ and parallel to the line $3x - 4y = 8$

59. Through (0.5, 5) and parallel to the line $4x - 2y = 11$

60. Through $\left(\frac{1}{3}, 0\right)$ and parallel to the line $6x - 2y = 11$

Communication and Reasoning Exercises

61. To what linear function of x does the linear equation $ax + by = c$ $(b \neq 0)$ correspond? Why did we specify $b \neq 0$?

62. Complete the following: The slope of the line with equation $y = mx + b$ is the number of units that ____ increases per unit increase in ____.

63. Complete the following: If, in a straight line, y is increasing three times as fast as x, then its ____ is ____.

64. Suppose that y is decreasing at a rate of four units per three-unit increase in x. What can we say about the slope of the linear relationship between x and y? What can we say about the intercept?

65. If y and x are related by the linear expression $y = mx + b$, how will y change as x changes if m is positive? Negative? Zero?

1.4 Linear Models

Using linear functions to describe or approximate relationships in the real world is called **linear modeling.** We start with some examples involving cost, revenue, and profit.

Cost, Revenue, and Profit Functions

example 1

Linear Cost Function

The Yellow Cab Company charges \$1 on entering the cab plus an additional \$2 per mile.[1]

a. Find the cost C of an x-mile trip.
b. Use your answer to calculate the cost of a 40-mile trip.
c. What is the cost of the second mile? What is the cost of the tenth mile?
d. Graph C as a function of x.

Solution

a. We are being asked to find how the cost C depends on the length x of the trip, or to find C as a function of x. Here is the cost in a few cases.

[1]This is equivalent to charging \$3 for the first mile and \$2 for each subsequent mile. Some cab companies charge using a combination of miles and minutes, but we have chosen our example for simplicity.

Cost of a 1-mile trip: $C = 2(1) + 1 = 3$ 1 mile @ \$2 per mile plus \$1 to enter

Cost of a 2-mile trip: $C = 2(2) + 1 = 5$ 2 miles @ \$2 per mile plus \$1 to enter

Cost of a 3-mile trip: $C = 2(3) + 1 = 7$ 3 miles @ \$2 per mile plus \$1 to enter

Do you see the pattern? The cost of an x-mile trip is given by the linear function

$$C = 2x + 1 \quad \text{Equation form}$$

or

$$C(x) = 2x + 1 \quad \text{Function form}$$

Notice that the slope 2 is the incremental cost per mile. In this context we call 2 the **marginal cost;** the varying quantity $2x$ is called the **variable cost.** The C-intercept 1 is the cost to enter the cab, which we call the **fixed cost.** In general, a linear cost function has the following form:

$$C(x) = \overbrace{mx}^{\text{Variable cost}} + b$$

mx: Marginal cost (slope m), b: Fixed cost

b. We can use the formula for the cost function to calculate the cost of a 40-mile trip as

$$C(40) = 2(40) + 1 = \$81$$

c. To calculate the cost of the second mile, we *could* proceed as follows:

Find the cost of a 1-mile trip: $C(1) = 2(1) + 1 = \$3.$

Find the cost of a 2-mile trip: $C(2) = 2(2) + 1 = \$5.$

Therefore, the cost of the second mile is $\$5 - \$3 = \$2$.

But notice that this is just the marginal cost. In fact, the marginal cost is the cost of each additional mile, so we could do this more simply as follows.

Cost of second mile = Cost of tenth mile = Marginal cost = \$2

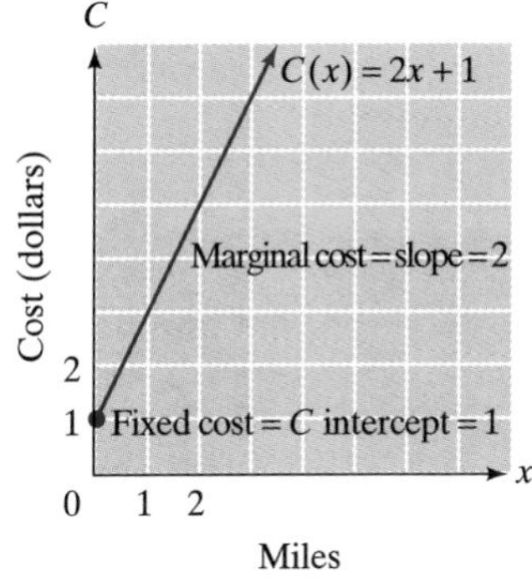

Figure 15

d. Figure 15 shows the graph of the cost function, which we can interpret as a *cost vs. miles* graph. The fixed cost is the starting height on the left, and the marginal cost is the slope of the line.

Here is a summary of the terms used in this example along with an introduction to some new terms.

Cost, Revenue, and Profit Functions

A **cost function** specifies the cost C as a function of the number of items x. Thus, $C(x)$ is the cost of x items. A cost function of the form

$$C(x) = mx + b$$

is called a **linear cost function.** The quantity mx is called the **variable cost,** and the intercept b is called the **fixed cost.** The slope m, the **marginal cost,** measures the incremental cost per item.

The **revenue** that results from one or more business transactions is the total payment received, sometimes called the gross proceeds. If $R(x)$ is the revenue from selling x items at a price of m each, then R is the linear function $R(x) = mx$ and the selling price m can also be called the **marginal revenue.**

The **profit,** on the other hand, is the *net* proceeds, or what remains of the revenue after costs are subtracted. If the profit depends linearly on the number of items, the slope m is called the **marginal profit.** Profit, revenue, and cost are related by the following formula:

$$\text{Profit} = \text{Revenue} - \text{Cost}$$

$$P = R - C$$

If the profit is negative—say, –\$500—we refer to a **loss** (of \$500 in this case). To **break even** means to make neither a profit nor a loss. Thus, break-even occurs when $P = 0$, or

$$R = C \qquad \text{Break-even}$$

The **break-even point** is the number of items x at which break-even occurs.

Quick Example

If the daily cost (including operating costs) of manufacturing x T-shirts is $C(x) = 8x + 100$ and the revenue obtained by selling x T-shirts is $R(x) = 10x$, then the daily profit that results from the manufacture and sale of x T-shirts is

$$\begin{aligned} P(x) &= R(x) - C(x) \\ &= 10x - (8x + 100) = 2x - 100 \end{aligned}$$

Break-even occurs when $P(x) = 0$, or $x = 50$.

example 2

Cost, Revenue, and Profit

The manager of the FrozenAir Refrigerator factory notices that on Monday it cost the company \$25,000 to build 30 refrigerators, and on Tuesday it cost \$30,000 to build 40 refrigerators.

a. Find a linear cost function based on this information. What is the daily fixed cost, and what is the marginal cost?

b. FrozenAir sells its refrigerators for \$1500 per unit. What is the revenue function?

c. What is the profit function? How many refrigerators must FrozenAir sell per day in order to break even?

Solution

a. We are seeking C as a linear function of x, the number of refrigerators sold:

$$C = mx + b$$

We are told that $C = 25{,}000$ when $x = 30$, and this amounts to being told that (30, 25,000) is a point on the graph of the cost function. Similarly, (40, 30,000) is another point on the line. The graph of C as a function of x is shown in Figure 16.

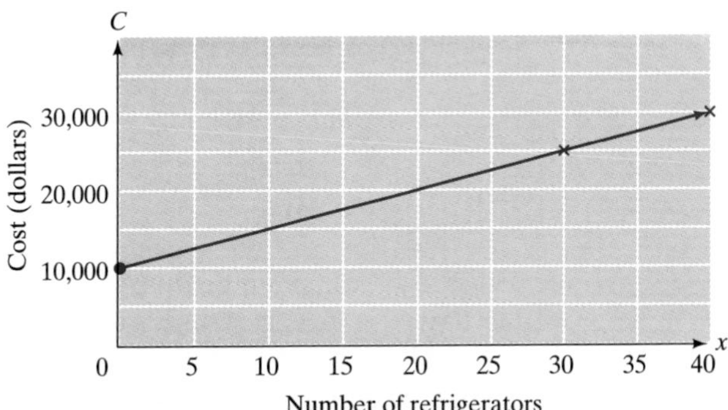

Figure 16

To find the equation of this line, recall that we need two items of information: a point on the line and the slope.

- ***Point*** We are given two points; let us use the first: (30, 25,000).
- ***Slope*** We use the slope equation (with C playing the role of y):

$$m = \frac{C_2 - C_1}{x_2 - x_1} = \frac{30{,}000 - 25{,}000}{40 - 30} = 500$$

In other words, the marginal cost is \$500 per refrigerator. To complete the problem, we use the point-slope formula:

$$\begin{aligned} C &= C_0 + m(x - x_0) \\ &= 25{,}000 + 500(x - 30) \\ &= 500x + 10{,}000 \qquad \text{Cost function—equation notation} \end{aligned}$$

or $\quad C(x) = 500x + 10{,}000 \qquad$ Function notation

Since $b = 10{,}000$, the factory's fixed cost is \$10,000 each day.

b. The revenue FrozenAir obtains from the sale of a single refrigerator is \$1500. So, if they sell x refrigerators, they earn a revenue of

$$R(x) = 1500x$$

c. For the profit, we use the formula

$$\text{Profit} = \text{Revenue} - \text{Cost}$$

For the cost and revenue, we can substitute the answers from parts (a) and (b) and obtain

$$\begin{aligned} P(x) &= R(x) - C(x) && \text{Formula for profit} \\ &= 1500x - (500x + 10{,}000) && \text{Substitute } R(x) \text{ and } C(x). \\ &= 1000x - 10{,}000 \end{aligned}$$

Here, $P(x)$ is the daily profit FrozenAir makes by making and selling x refrigerators per day.

Finally, to break even means to make zero profit. Thus, we need to find x such that $P(x) = 0$. So, all we have to do is set $P(x) = 0$ and solve for x:

$$1000x - 10{,}000 = 0$$

gives $$x = \frac{10{,}000}{1000} = 10$$

Thus, to break even, FrozenAir needs to manufacture and sell 10 refrigerators per day.

Before we go on . . . For values of x less than the break-even point, 10, $P(x)$ is negative, so the company will have a loss. For values of x greater than the break-even point, $P(x)$ is positive, so the company will make a profit. This is the reason we are interested in the point where $P(x) = 0$. Since $P(x) = R(x) - C(x)$, we can also look at the break-even point as the point where revenue = cost: $R(x) = C(x)$ (Figure 17).

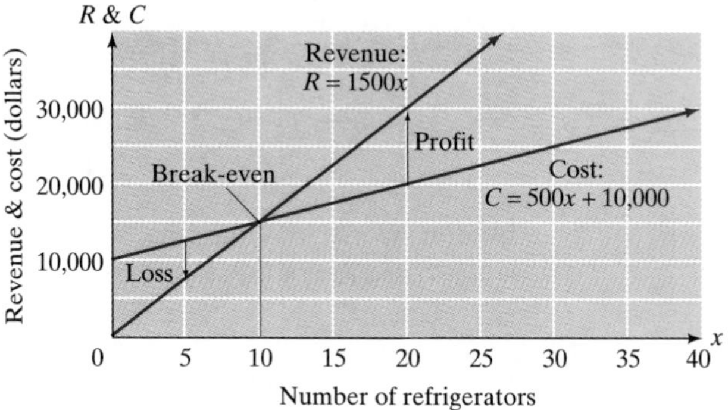

Figure 17

Spreadsheet

Excel has an interesting feature called "Goal Seek" that can be used to find the point of intersection of two lines. The downloadable Excel tutorial for this section contains detailed instructions on using Goal Seek to find break-even points.

Demand and Supply Functions

It is commonplace to observe that demand for a commodity goes down as its price goes up. It is traditional to use the letter q for the (quantity of) demand. Consider the following example.

example 3

Demand Function

You are the owner of the upscale Workout Fever Health Club and have been charging an annual membership fee of \$600. You are disappointed with the response: The club has been averaging only 10 new members per month. To remedy this, you decide to lower the fee to \$500, and you notice that this boosts new membership to an average of 16 per month.

a. Taking the demand q to be the average number of new members per month, express q as a linear function of the annual membership fee p.

b. Use the demand equation to predict monthly new membership if you lower the membership fee to \$350.

Solution

a. A **demand equation** or **demand function** expresses demand q (in this case, the number of new memberships sold per month) as a function of the unit price p (in this case, membership fee). Since we are asking for a *linear* demand function, we are looking for an equation of the form

$$q = mp + b$$

(We take p to play the role of x and q to play the role of y.) Thus, the graph of q versus p will be a straight line. To find m and b, notice that, just as in Example 2(a), we are given two points on the graph: (600, 10) and (500, 16) (Figure 18). We use this information to obtain the equation of the line.

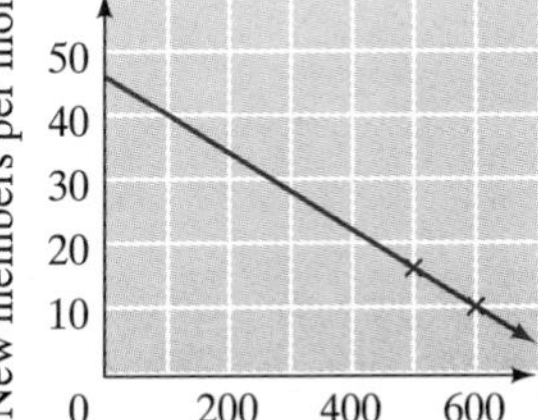

Figure 18

- ***Point*** (600, 10)
- ***Slope*** $m = \dfrac{q_2 - q_1}{p_2 - p_1} = \dfrac{16 - 10}{500 - 600} = -\dfrac{6}{100} = -0.06$

Thus, the equation we want is

$$\begin{aligned} q &= q_0 + m(p - p_0) \\ &= 10 - 0.06(p - 600) \\ &= -0.06p + 46 \qquad \text{Demand function—equation notation} \\ q(p) &= -0.06p + 46 \qquad \text{Function notation} \end{aligned}$$

b. If you lower the annual fee to \$350, then $p = 350$, and so

$$\begin{aligned} q &= -0.06p + 46 \\ &= -0.06(350) + 46 = 25 \text{ new members per month} \end{aligned}$$

Before we go on . . . Suppose, in a fit of greed, you decided to *increase* the price to \$2000. Then

$$q(2000) = -0.06(2000) + 46 = -74$$

The linear model seems to be indicating that -74 new members will join each month! In other words, the model has failed to give a meaningful answer.

Question Just how reliable *is* the linear model?
Answer Look at it this way: The *actual* demand graph could, in principle, be obtained by tabulating new membership figures for a large number of different annual fees. If the resulting points were plotted on the pq-plane, they would probably suggest a curve and not a straight line. However, if you looked at a small enough portion of this curve, you could closely *approximate* it by a straight line. In other words, *over a small range of values of p, the linear model is accurate.* Linear models of real-world situations are generally reliable for only small ranges of the variables. (This idea will come up again in some of the exercises.)

Demand Function

A **demand equation** or **demand function** expresses demand q (the number of items demanded) as a function of the unit price p (the price per item). A **linear demand function** has the form

$$q(p) = mp + b$$

Interpretation of *m*

The (usually negative) slope m measures the change in demand per unit change in price. Thus, for instance, if p is measured in dollars and q in monthly sales, and $m = -400$, then each \$1 increase in the price per item will result in a drop in sales of 400 items per month.

Interpretation of *b*

The y-intercept b gives the demand if the items were given away.[3]

Quick Example

If the demand for T-shirts, measured in daily sales, is given by $q(p) = -4p + 90$, where p is the sale price in dollars, then daily sales drop by four T-shirts for every \$1 increase in price. If the T-shirts were given away, the demand would be 90 T-shirts per day.

We have seen that a demand function gives us the number of items consumers are willing to buy at a given price, and a higher price generally results in a lower demand. As the price rises, however, suppliers will be more inclined to produce these items (as opposed to spending their time and money on other products), so the supply will generally rise. A **supply** function gives q, the number of items suppliers are willing to make available for sale,[4] as a function of p, the price per item.

example 4

Demand, Supply, and Equilibrium Price

You run a small supermarket and must determine how much to charge for Hot 'n Spicy brand baked beans. The following table shows weekly sales figures for Hot 'n Spicy at two different prices (the demand) as well as the number of cans per week you are prepared to place on sale at these prices (the supply).

Price per can	\$0.50	\$0.75
Demand (cans sold per week)	400	350
Supply (cans placed on sale per week)	300	500

a. Model these data with linear demand and supply functions.

b. How much should you charge per can of Hot 'n Spicy beans if you want the demand to equal the supply? How many cans will you sell at that price, known as the **equilibrium price?** What happens if you charge more than the equilibrium price? What happens if you charge less?

---

[3]This demand would not be unlimited if items were given away. For instance, campus newspapers are sometimes given away, and yet piles of them are often left untouched.

[4]Although a bit confusing at first, it is traditional to use the same letter q for the quantity of supply and the quantity of demand, particularly when we want to compare them, as in Example 4.

Solution

a. To model the demand, we use the first two rows of the table, which give us two points: (0.50, 400) and (0.75, 350).

- ***Point*** (0.50, 400)
- ***Slope*** $m = \dfrac{q_2 - q_1}{p_2 - p_1} = \dfrac{350 - 400}{0.75 - 0.50} = \dfrac{-50}{0.25} = -200$

Thus, the demand equation is

$$\begin{aligned} q &= q_0 + m(p - p_0) \\ &= 400 - 200(p - 0.50) = 400 - 200p + 100 \\ &= -200p + 500 \end{aligned}$$

To model the supply, we use the first and third rows of the table, which give us two more points: (0.50, 300) and (0.75, 500).

- ***Point*** (0.50, 300)
- ***Slope*** $m = \dfrac{q_2 - q_1}{p_2 - p_1} = \dfrac{500 - 300}{0.75 - 0.50} = \dfrac{200}{0.25} = 800$

So, the supply equation is

$$\begin{aligned} q &= q_0 + m(p - p_0) \\ &= 300 + 800(p - 0.50) = 300 + 800p - 400 \\ &= 800p - 100 \end{aligned}$$

In function notation, we can write the demand and supply functions as follows:

- ***Demand*** $q(p) = -200p + 500$
- ***Supply*** $q(p) = 800p - 100$

b. To find where the demand equals the supply, we equate the two functions:

$$\begin{aligned} \text{Demand} &= \text{Supply} \\ -200p + 500 &= 800p - 100 \\ -1000p &= -600 \end{aligned}$$

so

$$p = \frac{-600}{-1000} = \$0.60$$

This is the equilibrium price. We can find the corresponding demand by substituting 0.60 for p in the demand (or supply) equation.

$$\text{Equilibrium demand} = -200(0.60) + 500 = 380 \text{ cans per week}$$

Thus, in order to balance supply and demand, you should charge 60¢ per can of Hot 'n Spicy beans and you should place 380 cans on sale per week.

To see what happens if you charge more than 60¢ per can, look at Figure 19, which shows the graphs of demand and supply on the same set of axes. If you charge 80¢ per can ($p = 0.80$), then the supply will be greater than the demand and there will be a weekly surplus. Similarly, if you charge less—say 40¢ per can—then the supply will be less than the demand and there will be a shortage of Hot 'n Spicy beans.

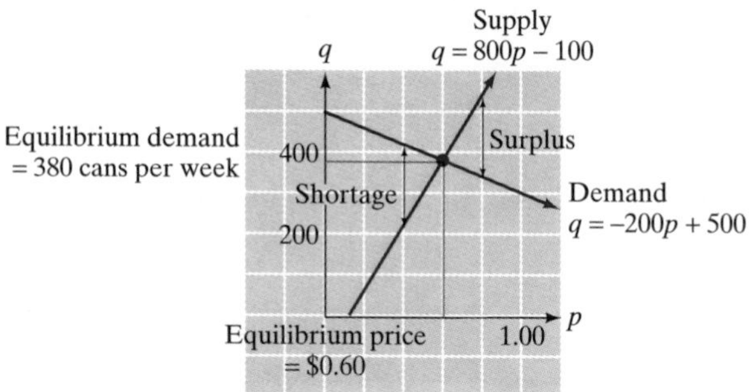

Figure 19

Before we go on . . . We saw that if you charge less than the equilibrium price, there will be a shortage. If you raise your price toward the equilibrium, you will be able to sell more items and hence increase revenue because it is the supply equation—and not the demand equation—that determines what you can sell below the equilibrium price. On the other hand, if you charge more than the equilibrium price, you will be left with a possibly costly surplus of unsold items. Prices tend to move toward the equilibrium, so supply tends to equal demand. When supply equals demand, we say that the market **clears.**

Supply Function and Equilibrium Price

A **supply equation** or **supply function** expresses the supply q (the number of items a supplier is willing to make available) as a function of the unit price p (the price per item). A **linear supply function** has the form

$$q(p) = mp + b$$

It is usually the case that supply increases as the unit price increases, so m is usually positive.

Demand and supply are said to be in **equilibrium** when demand equals supply. The corresponding values of p and q are called the **equilibrium price** and **equilibrium demand,** respectively. To obtain the equilibrium price, set demand equal to supply and solve for the unit price p. To obtain the equilibrium demand, evaluate the demand (or supply) function at the equilibrium price.

Change over Time

Things all around us change with time. Thus, it is natural to think of many quantities, such as your income or the temperature in Honolulu, as functions of time.

example 5

Growth of Sales

The U.S. Air Force's satellite-based Global Positioning System (GPS) allows people with radio receivers to determine their exact location anywhere on earth. The following table shows the estimated total sales of U.S.-made products that use the Global Positioning System.

Year	1994	2000
Sales (billions)	$0.8	$8.3

Source: United States Global Positioning System Industry Council/*New York Times,* March 5, 1996, p. D1. Data estimated from published graph.

a. Use these data to model total sales of GPS–based products as a linear function of time t measured in years since 1994. What is the significance of the slope?

b. Use the model to predict when sales of GPS–based products will reach $13.3 billion, assuming they continue to increase at a steady rate.

Solution

a. First, notice that 1994 corresponds to $t = 0$ and 2000 to $t = 6$. Thus, we are given the coordinates of two points on the graph of sales s as a function of time t: (0, 0.8) and (6, 8.3). The slope is

$$m = \frac{s_2 - s_1}{t_2 - t_1} = \frac{8.3 - 0.8}{6 - 0} = \frac{7.5}{6} = 1.25$$

Using the point (0, 0.8), we get

$$\begin{aligned} s &= s_0 + m(t - t_0) \\ &= 0.8 + 1.25(t - 0) \\ &= 1.25t + 0.8 \end{aligned}$$

Notice that we calculated the slope as the ratio (Change in sales)/(Change in time). Thus, m is the *rate of change* of sales and is measured in units of sales per unit of time, or billions of dollars per year. In other words, to say that $m = 1.25$ is to say that sales are increasing by $1.25 billion per year.

b. Our model of sales as a function of time is

$$s = 1.25t + 0.8$$

Sales of GPS–based products will reach $13.3 billion when $s = 13.3$, or

$$13.3 = 1.25t + 0.8$$

Solving for t, we have

$$1.25t = 13.3 - 0.8 = 12.5$$

$$t = \frac{12.5}{1.25} = 10$$

Thus, in 2004 ($t = 10$) sales are predicted to reach $13.3 billion.

example 6

Velocity

You are driving down the Ohio Turnpike watching the mileage markers to stay awake. Measuring time in hours after you see the 20-mile marker, you see the following markers each half-hour:

Time (hours)	0	0.5	1	1.5	2
Marker (miles)	20	47	74	101	128

Find your location s as a function of t, the number of hours you have been driving. (The number s is called your **position.**)

Solution

If we plot the position s versus the time t, the five markers listed give us the graph in Figure 20. The points on the graph appear to lie along a straight line. We can verify this by calculating how far you traveled in each half-hour. In the first half-hour you traveled $47 - 20 = 27$ miles. In the second half-hour you traveled $74 - 47 = 27$ miles also. In fact, you traveled exactly 27 miles each half-hour. The points we plotted lie on a straight line that rises 27 for every 0.5 we go to the right, for a slope of $27/0.5 = 54$.

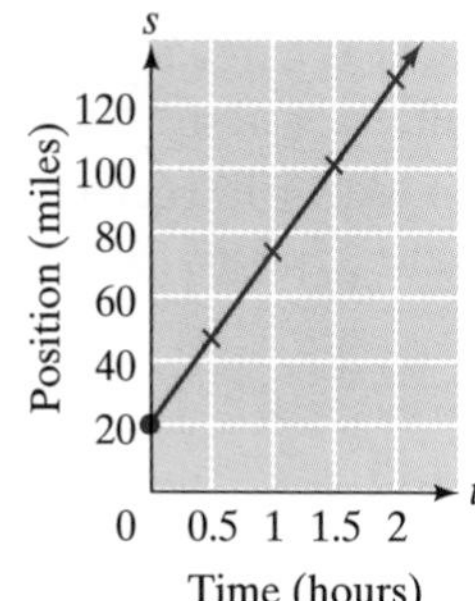

Figure 20

To get the equation of that line, we notice that we have the s-intercept, which is the starting marker of 20. From the slope-intercept formula (using s in place of y and t in place of x) we get

$$s(t) = 54t + 20$$

Before we go on . . . Notice the significance of the slope: For every hour you travel, you drive a distance of 54 miles. In other words, you are traveling at a constant velocity of 54 miles per hour. We have uncovered a very important principle:

In the graph of position vs. time, velocity is given by the slope.

Linear Change over Time

If a quantity q is a linear function of time t, so that

$$q(t) = mt + b$$

then the slope m measures the **rate of change** of q, and b is the quantity at time $t = 0$, the **initial quantity.** If q represents the position of a moving object, then the rate of change is also called the **velocity.**

Units of m

The units of measurement of m are the units of q per unit of time; for instance, if q is income in dollars and t is time in years, then the rate of change m is measured in dollars per year.

Quick Example

If annual Internet-based sales of your video game software are given by $s(t) = 200t + 50$ units, where t is time in years from now, then you are presently selling 50 units per year and your annual sales are increasing at a rate of 200 units per year.

All of the preceding examples share the following common theme.

General Linear Models

If $y = mx + b$ is a linear model of changing quantities x and y, then the slope m is the rate at which y is increasing per unit increase in x, and the y-intercept b is the value of y that corresponds to $x = 0$. The slope m is measured in units of y per unit of x, and the intercept b is measured in units of y.

1.4 exercises

Applications

1. ***Cost*** A piano manufacturer has a daily fixed cost of \$1200 and a marginal cost of \$1500 per piano. Find the cost $C(x)$ of manufacturing x pianos in one day. Use your function to answer the questions.
   a. On a given day, what is the cost of manufacturing three pianos?
   b. What is the cost of manufacturing the third piano that day?
   c. What is the cost of manufacturing the 11th piano that day?

2. ***Cost*** The cost of renting tuxes for the Choral Society's formal is \$20 down plus \$88 per tux. Express the cost C as a function of x, the number of tuxedos rented. Use your function to answer the questions.
   a. What is the cost of renting two tuxes?
   b. What is the cost of the second tux?
   c. What is the cost of the 4098th tux?
   d. What is the marginal cost per tux?

3. ***Cost*** The RideEm Bicycles factory can produce 100 bicycles in a day at a total cost of \$10,500, and it can produce 120 bicycles in a day at a total cost of \$11,000. What are the company's daily fixed costs, and what is the marginal cost per bicycle?

4. ***Cost*** A soft-drink manufacturer can produce 1000 cases of soda in a week at a total cost of \$6000, and 1500 cases of soda at a total cost of \$8500. Find the manufacturer's weekly fixed costs and the marginal cost per case of soda.

5. ***Break-Even Analysis*** Your college newspaper, *The Collegiate Investigator*, has fixed production costs of \$70 per edition and marginal printing and distribution costs of 40¢ per copy. *The Collegiate Investigator* sells for 50¢ per copy.
   a. Write the associated cost, revenue, and profit functions.
   b. What profit (or loss) results from the sale of 500 copies of *The Collegiate Investigator*?
   c. How many copies should be sold in order to break even?

6. ***Break-Even Analysis*** The Audubon Society at Enormous State University (ESU) is planning its annual fund-raising "Eatathon." The society will charge students 50¢ per serving of pasta. The only expenses the society will incur are the cost of the pasta, estimated at 15¢ per serving, and the \$350 cost of renting the facility for the evening.
   a. Write the associated cost, revenue, and profit functions.
   b. How many servings of pasta must the Audubon Society sell to break even?
   c. What profit (or loss) results from the sale of 1500 servings of pasta?

7. ***Demand*** Sales figures show that your company sold 1960 pen sets per week when they were priced at \$1 per pen set, and 1800 pen sets per week when they were priced at \$5 per pen set. What is the linear demand function for your pen sets?

8. ***Demand*** A large department store is prepared to buy 3950 of your neon-colored shower curtains per month for \$5 each, but only 3700 shower curtains per month for \$10 each. What is the linear demand function for your neon-colored shower curtains?

9. ***Demand for Microprocessors*** The following table shows the worldwide quarterly sales of microprocessors and the average wholesale price each quarter.

	1997 2nd quarter	1997 3rd quarter	1998 1st quarter
Wholesale price	\$235	\$215	\$210
Sales (millions)	21	24	23

Source: International Data Corporation/*New York Times*, "Computer Economics 101," June 9, 1998, p. D6. Data are rounded.

Use the 1997 second- and third-quarter data to obtain a linear demand function for microprocessors. Are the 1998 first-quarter data consistent with the demand equation? Explain.

10. ***Demand for Microprocessors*** Refer back to the data in Exercise 9. Use the 1997 second-quarter data and the 1998 first-quarter data to obtain a linear demand function for microprocessors. Use your model to give an estimate of worldwide quarterly sales if the price changes to

\$215. Can all three data points be represented by the same linear demand function?

11. ***Equilibrium Price*** You can sell 90 pet chias per week if they are marked at \$1 each, but only 30 per week if they are marked at \$2 per chia. Your chia supplier is prepared to sell you 20 chias per week if they are marked at \$1 per chia, and 100 per week if they are marked at \$2 per chia.
   a. Write the associated linear demand and supply functions.
   b. At what price should the chias be marked so that there is neither a surplus nor a shortage of chias?

12. ***Equilibrium Price*** The demand for your college newspaper is 2000 copies per week if the paper is given away free of charge, and the demand drops to 1000 per week if the charge is 10¢ per copy. However, the university is prepared to supply only 600 copies per week free of charge but will supply 1400 per week at 20¢ per copy.
   a. Write the associated linear demand and supply functions.
   b. At what price should the college newspapers be sold so that there is neither a surplus nor a shortage of papers?

13. ***Revenue*** Quarterly sales revenues of Creative Labs, Inc., increased in the years 1993–1995 as shown in the following graph, where x is quarter with $x = 1$ being the first quarter of 1993:

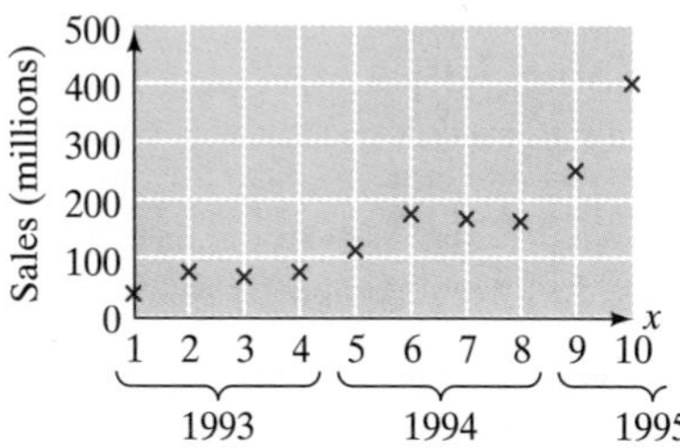

Source: Creative Labs/*New York Times*, February 23, 1995, p. D6.

   a. Find two points on the graph such that the slope of the line segment joining them is the largest possible.
   b. What does your answer tell you about Creative Labs?

14. ***Profit*** Repeat Exercise 13, but base your answers on the following graph of Creative Labs' quarterly profit:

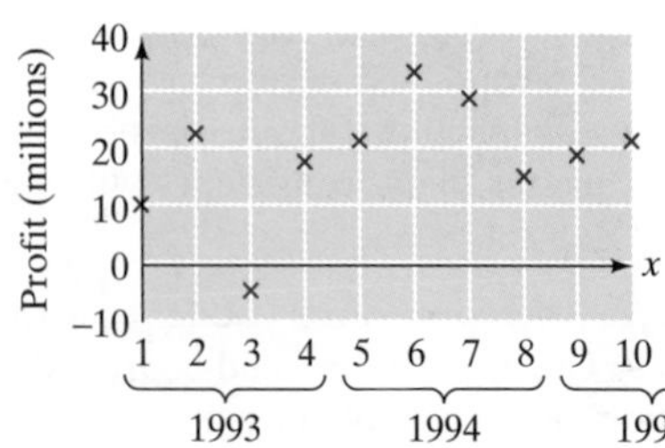

Source: Creative Labs/*New York Times*, February 23, 1995, p. D6.

15. ***Certified Financial Planners*** The number of people who pass the exam to become a Certified Financial Planner (CFP) rose from 400 in 1992 to 2000 in 1995. Find a linear model for the number of people who pass the exam in year t, with $t = 0$ corresponding to 1992. What is the significance of the slope? What are its units?
Source: Cerulli Associates/*New York Times*, June 25, 1995, p. F6. Data are approximate.

16. ***Investment Firms*** The number of firms registered as investment advisers with the Securities and Exchange Commission (SEC) increased significantly in the early 1990s, from 17,400 in 1990 to 21,600 in 1994. Find a linear model for the number n of firms registered as investment advisers in year t, with $t = 0$ corresponding to 1990. What is the significance of the slope? What are its units?
Source: Cerulli Associates/*New York Times*, June 25, 1995, p. F6. Data are approximate.

17. ***Medicare Spending*** Annual federal spending on Medicare (in constant 1994 dollars) increased more or less linearly from \$50 billion in 1973 to \$155 billion in 1994.
   a. Use these data to express s, the annual spending on Medicare (in billions of dollars), as a linear function of t, the number of years since 1973. What was the rate of increase of Medicare spending?
   b. Use your model to predict Medicare spending in 2010, assuming the spending trend continues.
Source: Health Care Financing Administration, Economics Report of the President/*New York Times*, July 23, 1995, p. 1. Figures are rounded to the nearest \$5 billion.

18. ***Pasta Imports*** U.S. imports of pasta increased from 290 million pounds in 1990 ($t = 0$) by an average of 52 million pounds per year.
   a. Use these data to express q, the annual U.S. imports of pasta (in millions of pounds), as a linear function of t, the number of years since 1990.
   b. Use your model to estimate U.S. pasta imports in the year 2000, assuming the import trend continued.
Source: Department of Commerce/*New York Times*, September 5, 1995, p. D4. Data are rounded and are for the period 1990 through 1994.

19. ***Velocity*** The position of a model train, in feet along a railroad track, is given by
$$s(t) = 2.5t + 10$$
after t seconds.
   a. How fast is the train moving?
   b. Where is the train after 4 seconds?
   c. When will the train have moved a distance of 25 feet?

20. ***Velocity*** The height of a falling sheet of paper, in feet from the ground, is given by
$$s(t) = -1.8t + 9$$
after t seconds.
   a. What is the velocity of the sheet of paper?
   b. How high is the paper after 4 seconds?
   c. When will the sheet of paper reach the ground?

21. ***Fast Cars*** A police car was traveling down Ocean Parkway in a high-speed chase from Jones Beach. The car was at Jones Beach at exactly 10:00 P.M. ($t = 10$) and was at Oak Beach, 13 miles from Jones Beach, at exactly 10:06 P.M.
   a. How fast was the police car traveling?
   b. How far was the police car from Jones Beach at time t?

22. ***Fast Cars*** The car that was being pursued by the police in Exercise 21 was at Jones Beach at exactly 9:54 P.M. ($t = 9.9$) and passed Oak Beach (13 miles from Jones Beach) at exactly 10:06 P.M., where it was overtaken by the police.
   a. How fast was the car traveling?
   b. How far was the car from Jones Beach at time t?

23. ***Fahrenheit and Celsius*** In the Fahrenheit temperature scale, water freezes at 32°F and boils at 212°F. In the Celsius (or centigrade) scale, water freezes at 0°C and boils at 100°C. Assuming that the Fahrenheit temperature F and the Celsius temperature C are related by a linear equation, find F in terms of C. Use your equation to find the Fahrenheit temperatures that correspond to 30°C, 22°C, –10°C, and –14°C, to the nearest degree.

24. ***Fahrenheit and Celsius*** Use the information about Celsius and Fahrenheit given in Exercise 23 to obtain a linear equation for C in terms of F, and use your equation to find the Celsius temperatures that correspond to 104°F, 77°F, 14°F, and –40°F, to the nearest degree.

25. ***Income*** The well-known romance novelist Celestine A. Lafleur (a.k.a. Bertha Snodgrass) has decided to sell the screen rights to her latest book, *Henrietta's Heaving Heart*, to Boxoffice Success Productions, Inc., for \$50,000. In addition, the contract assures Ms. Lafleur royalties of 5% of the net profits.[5] Express her income I as a function of the net profit N, and determine the net profit necessary to bring her an income of \$100,000. What is her marginal income (share of each dollar of net profit)?

26. ***Income*** Due to the enormous success of the movie *Henrietta's Heaving Heart* based on a novel by Celestine A. Lafleur (see Exercise 25), Boxoffice Success Productions, Inc., decides to film the sequel, *Henrietta, Oh Henrietta*. At this point, Bertha Snodgrass (whose novels now top the bestseller lists) feels she is in a position to demand \$100,000 for the screen rights and royalties of 8% of the net profits. Express her income I as a function of the net profit N, and determine the net profit necessary to bring her an income of \$1,000,000. What is her marginal income (share of each dollar of net profit)?

27. ***Newspaper Circulation*** The following table gives the average daily circulations of all the newspapers of the Gannett Company and Newhouse Newspapers, Inc., two major newspaper companies.

	Gannett	Newhouse
Number of newspapers	82	26
Daily circulation (millions)	5.8	3.0

Source: Newspaper Association of America/*New York Times*, July 30, 1995, p. E6. Circulation figures are rounded to the nearest 0.1 million and reflect average daily circulation as of September 30, 1994.

   a. Use these data to express a company's daily circulation in millions, c, as a linear function of the number, n, of newspapers it publishes.
   b. What information does the slope give about newspaper publishers?
   c. Gannett Company was planning to add 11 new newspapers in 1995. According to your model, how would this affect daily circulation?

28. ***Newspaper Circulation*** The following table gives the average daily circulations of all the newspapers of Knight Ridder, Inc., and Dow Jones & Co., two major publishers of newspapers:

	Knight Ridder	Dow Jones
Number of newspapers	27	22
Daily circulation (millions)	3.6	2.4

Source: Newspaper Association of America/*New York Times*, July 30, 1995, p. E6. Circulation figures are rounded to the nearest 0.1 million and reflect average daily circulation as of September 30, 1994.

   a. Use these data to express a company's daily circulation, c, as a linear function of the number, n, of newspapers it publishes.
   b. The Times Mirror Company publishes 11 newspapers and has a daily circulation of 2.6 million. To what extent is this consistent with the model?
   c. Find a domain for your model that is consistent with a circulation of between 0 and 4.8 million.

29. ***Profit and Revenue*** Luxotica Group S.p.A. of Italy, a prominent eyeglass frame manufacturer, enjoyed healthy profits from 1991 to 1994. The following table shows Luxotica's annual revenue and profit in each of these years:

	1991	1992	1993	1994
Revenue (millions)	\$400	\$350	\$380	\$500
Profit (millions)	\$55	\$45	\$51	\$75

Source: Datastream: company reports/*New York Times*, April 1, 1995, p. L38. Figures are approximate.

[5] Percentages of net profit are commonly called "monkey points." Few movies ever make a net profit on paper, and anyone with any clout in the business gets a share of the gross, not the net.

a. Verify that these data show a linear relationship between annual revenue and profit, and write Luxotica's annual profit P as a linear function of its annual revenue R.
b. Find an equation that gives Luxotica's annual costs C as a function of its annual revenue.
c. Estimate Luxotica's annual fixed costs.

30. ***Profit and Revenue*** A more accurate reading of Luxotica Group S.p.A.'s annual profits P and revenue R using data from 1990 to 1994 gives the following function, which is a linear regression based on the given data.

$$P = 0.187R - 19.4$$

where P and R are in millions of dollars.
a. What does the slope tell you about the company?
b. Use this model to find an equation that gives Luxotica's annual costs C as a linear function of its annual revenue.
c. What does the function in part (b) tell you about the company?
Source: Datastream: company reports/*New York Times*, April 1, 1995, p. L38.

31. ***Biology*** The Snowtree cricket behaves in a rather interesting way: The rate at which it chirps depends linearly on the temperature. One summer evening you hear a cricket chirping at a rate of 140 chirps per minute, and you notice that the temperature is 80°F. Later in the evening the cricket has slowed down to 120 chirps per minute, and you notice that the temperature has dropped to 75°F. Express the temperature T as a function of the cricket's rate of chirping r. What is the temperature if the cricket is chirping at a rate of 100 chirps per minute?

32. ***Muscle Recovery Time*** Most workout enthusiasts will tell you that muscle recovery time is about 48 hours. But it is not quite as simple as that; the recovery time ought to depend on the number of sets you do involving the muscle group in question. For example, if you do no biceps exercises, then the recovery time for your biceps is (of course) zero. To take a compromise position, let us assume that if you do three sets of exercises on a muscle group, then its recovery time is 48 hours. Use these data to write a linear function that gives the recovery time (in hours) in terms of the number of sets that affect a particular muscle. Use this model to calculate how long it will take your biceps to recover if you do 15 sets of curls. Comment on your answer with reference to the usefulness of a linear model.

33. ***Profit Analysis—Aviation*** The operating cost of a Boeing 747-100, which seats up to 405 passengers, is estimated to be \$5132 per hour. If an airline charges each passenger a fare of \$100 per hour of flight, find the hourly profit P it earns operating a 747-100 as a function of the number of passengers x. (Be sure to specify the domain.) What is the smallest number of passengers it must carry in order to make a profit?
Source: Air Transportation Association of America, 1992.

34. ***Profit Analysis—Aviation*** The operating cost of a McDonnell Douglas DC 10-10, which seats up to 295 passengers, is estimated to be \$3885 per hour. If an airline charges each passenger a fare of \$100 per hour of flight, find the hourly profit P it earns operating a DC 10-10 as a function of the number of passengers x. (Be sure to specify the domain.) What is the smallest number of passengers it must carry in order to make a profit?
Source: Air Transportation Association of America, 1992.

35. ***Break-Even Analysis*** *(based on a question from a CPA exam)* The Oliver Company plans to market a new product. Based on its market studies, Oliver estimates that it can sell up to 5500 units in 2005. The selling price will be \$2 per unit. Variable costs are estimated to be 40% of total revenue. Fixed costs are estimated to be \$6000 for 2005. How many units should the company sell to break even?

36. ***Break-Even Analysis*** *(based on a question from a CPA exam)* The Metropolitan Company sells its latest product at a unit price of \$5. Variable costs are estimated to be 30% of the total revenue, while fixed costs amount to \$7000 per month. How many units should the company sell per month in order to break even, assuming that it can sell up to 5000 units per month at the planned price?

37. ***Break-Even Analysis*** *(from a CPA exam)* Given the following notations, write a formula for the break-even sales level:

SP = Selling price per unit

FC = Total fixed cost

VC = Variable cost per unit

38. ***Break-Even Analysis*** *(based on a question from a CPA exam)* Given the following notation, give a formula for the total fixed cost:

SP = Selling price per unit

VC = Variable cost per unit

BE = Break-even sales level in units

39. ***Break-Even Analysis—Organized Crime*** The organized crime boss and perfume king Butch (Stinky) Rose has overhead costs (bribes to corrupt officials, motel photographers, wages for hitmen, explosives, etc.) amounting to \$20,000 per day. On the other hand, he has a substantial income from his counterfeit perfume racket. He buys imitation French perfume (Chanel № 22.5) at \$20 per

gram, pays an additional \$30 per 100 grams for transportation, and sells the perfume via his street thugs for \$600 per gram. Specify Stinky's profit function $P(x)$, where x is the quantity (in grams) of perfume he buys and sells, and use your answer to calculate how much perfume should pass through his hands per day so that he breaks even.

40. ***Break-Even Analysis—Disorganized Crime*** Butch (Stinky) Rose's counterfeit Chanel № 22.5 racket has run into difficulties. It seems that the *authentic* Chanel № 22.5 perfume is selling at only \$500 per gram, whereas the street thugs have been selling the counterfeit perfume for \$600 per gram, and Stinky's costs are \$400 per gram plus \$30 per gram for transportation and commission. (The perfume's smell is easily detected by specially trained Chanel Hounds, and this necessitates elaborate packaging measures.) Stinky therefore decides to price the perfume at \$420 per gram to undercut the competition. Specify Stinky's profit function $P(x)$, where x is the quantity (in grams) of perfume he buys and sells. Use your answer to calculate how much perfume should pass through his hands per day so that he breaks even. Interpret your answer.

41. ***Venture Capital*** Venture capital firms raise money for promising entrepreneurs to assist them until they can sell shares to the public. The following graphs show the total funds controlled by venture capital firms as well as the number of venture capital firms.

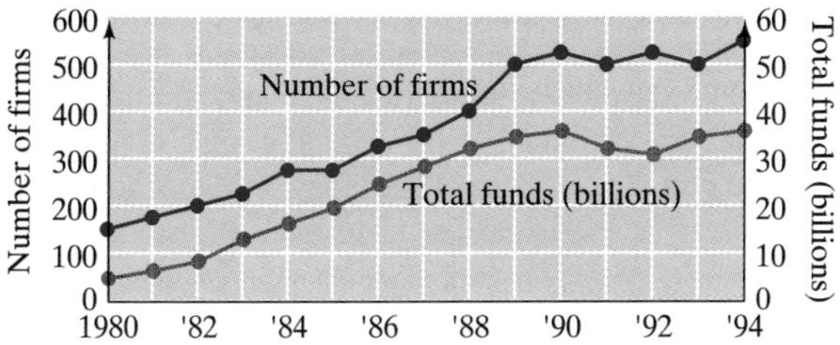

Let $N(t)$ be the number of firms at time t years since 1980, and let $R(t)$ be the total amount of money (in billions of dollars) these firms control. These functions can be approximated by the following linear functions.

$$N(t) = 31t + 145$$

$$R(t) = 2.4t + 7.3$$

a. What do the slopes in these equations represent?

b. What do the intercepts represent?

c. According to the models, the slope of $N(t)$ is more than ten times the slope of $R(t)$, whereas the graphs are roughly parallel. Explain this apparent discrepancy.

Source for data: Venture Economics Information Services, Boston/*New York Times*, May 25, p. D5. Linear models are rounded.

42. ***Airline Capacity*** In airline industry terminology, one A.S.M. represents one passenger seat (whether sold or vacant) flown 1 mile. The following chart shows Kiwi Airlines' growing capacity, measured in A.S.M.'s per month, for the period from January 1994 through March 1995.

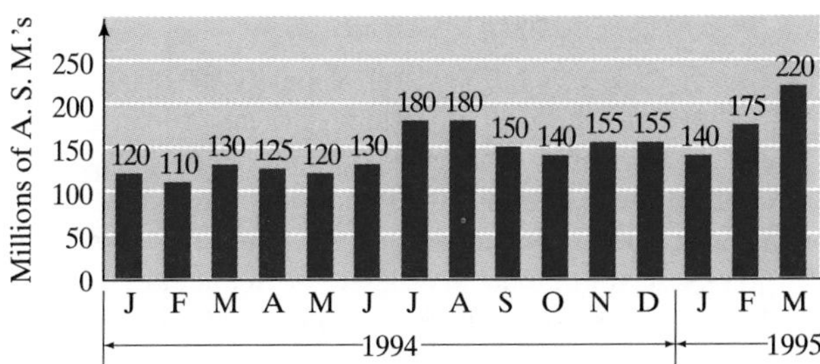

Source: Company reports/*New York Times*, March 25, 1995, p. D5. The figures are rounded.

a. Use the January 1994 and 1995 figures to give Kiwi's capacity C, in millions of A.S.M.'s per month, as a linear function of time t, where t is time in months since January 1994.

b. If Kiwi had used this model in January 1995 to predict its capacity in March 1995, how accurate would its prediction have been?

43. ***Semiconductors*** The percentage $P(t)$ of semiconductor equipment manufactured in the United States during the period 1981–1994 can be approximated by the following function of time t in years since 1980:

$$P(t) = \begin{cases} 76 - 3.3t & \text{if } 1 \le t \le 10 \\ 18 + 2.5t & \text{if } 10 < t \le 14 \end{cases}$$

How fast and in what direction was the percentage of the United States' contribution to the semiconductor market changing in 1985?

Source for data: VLSI Research/*New York Times*, October 9, 1994, Section 3, p. 2

44. ***Semiconductors*** The percentage $J(t)$ of semiconductor equipment manufactured in Japan during the period 1981–1994 can be approximated by the following function of time t in years since 1980:

$$J(t) = \begin{cases} 17 + 3.1t & \text{if } 1 \le t \le 10 \\ 68 - 2t & \text{if } 10 < t \le 14 \end{cases}$$

How fast and in what direction was the percentage of Japan's contribution to the semiconductor market changing in 1986?

Source for data: VLSI Research/*New York Times*, October 9, 1994, Section 3, p. 2

45. ***Career Choices*** In 1989 approximately 30,000 college-bound high school seniors intended to major in computer and information sciences. This number decreased to approximately 23,000 in 1994 and rose to 60,000 in 1999. Model this number C as a piecewise linear function of the time t in years since 1989. Use your model to estimate the number of college-bound high school seniors who intended to major in computer and information sciences in 1992.

Source: The College Board; National Science Foundation/*New York Times*, September 2, 1999, p. C1.

46. ***Career Choices*** In 1989 approximately 100,000 college-bound high school seniors intended to major in engineering. This number decreased to approximately 85,000 in 1995 and rose to 88,000 in 1999. Model this number E as a piecewise linear function of the time t in years since 1989. Use your model to estimate the number of college-bound high school seniors who intended to major in engineering in 1995.

Source: The College Board; National Science Foundation/*New York Times*, September 2, 1999, p. C1.

47. ***Divorce Rates*** A study found that the divorce rate d appears to depend on the ratio r of available men to available women. When the ratio was 1.3 (130 available men per 100 available women), the divorce rate was 22%. The divorce rate rose to 35% when the ratio grew to 1.6 and rose to 30% when the ratio dropped to 1.1. Model these data by expressing d as a piecewise linear function of r. Use your model to estimate the divorce rate when the numbers of available men and women are the same.

Source: New York Times, February 19, 1995, p. 40. The cited study, by Scott J. South and associates, appeared in the *American Sociological Review* (February 1995). Figures are rounded.

48. ***Retirement*** In 1950 the number N of retirees was approximately 150 per thousand people aged 20–64. In 1990 this number rose to approximately 200, and it is projected to rise to 275 in 2020. Model N as a piecewise linear function of the time t in years since 1950. Use your model to project the number of retirees per thousand people aged 20–64 in 2010.

Source: Social Security Adminstration/*New York Times*, April 4, 1999, p. WK3.

Communication and Reasoning Exercises

49. If a quantity is changing linearly with time and it increases by ten units in the first day, what can you say about its behavior in the third day?

50. The quantities Q and T are related by a linear equation of the form

$$Q = mT + b$$

When $T = 0$, Q is positive, but Q decreases to a negative quantity when T is 10. What are the signs of m and b? Explain your answers.

51. Suppose the cost function is $C(x) = mx + b$ (with m and b positive), the revenue function is $R(x) = kx$ $(k > m)$, and the number of items is increased from the break-even quantity. Does this result in a loss or a profit, or is it impossible to say? Explain your answer.

52. You have been constructing a demand equation, and you obtained a (correct) expression of the form $p = mq + b$, but you would have preferred one of the form $q = mp + b$. Should you simply switch p and q in the answer, should you start again from scratch using p in the role of x and q in the role of y, or should you solve your demand equation for q? Give reasons for your decision.

1.5 Linear Regression

We have seen how to find a linear model given two data points: We find the equation of the line that passes through them. However, we often have more than two data points. The points rarely all lie on one line, but they often come close to doing so. The problem is to find the line that comes *closest* to passing through all of the points.

Suppose, for example, that we have the following data about new home sales in a region over one year:

Selling price (thousands)	$150–169	170–189	190–209	210–229	230–249	250–269	270–289
Houses sold	126	103	82	75	82	40	20

Let us simplify by replacing each price range with a single price in the middle of the range.

Selling price (thousands)	$160	180	200	220	240	260	280
Houses sold	126	103	82	75	82	40	20

We would like to use these data to construct a linear demand function for new homes.[1] Recall that a demand function gives the demand (here the annual sales) as a function of the price. Figure 21 shows a plot of sales versus price. These points suggest a descending line, although they clearly do not all lie on a single straight line.

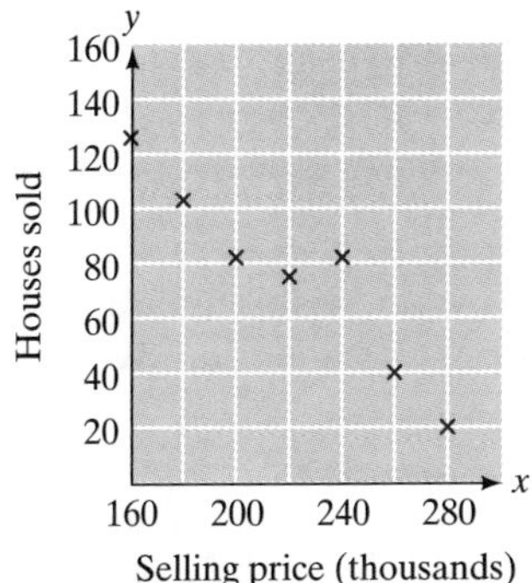

Figure 21

Question What line comes *closest* to passing through these data points?
Answer To answer this question we need to know what is meant by *closest.* We would like the sales figures predicted by the line (the **predicted values**) to be as close as possible to the actual sales figures (the **observed values**). Figure 22 shows one possible line; the differences between the predicted and observed values (the **errors**) are the vertical distances marked.

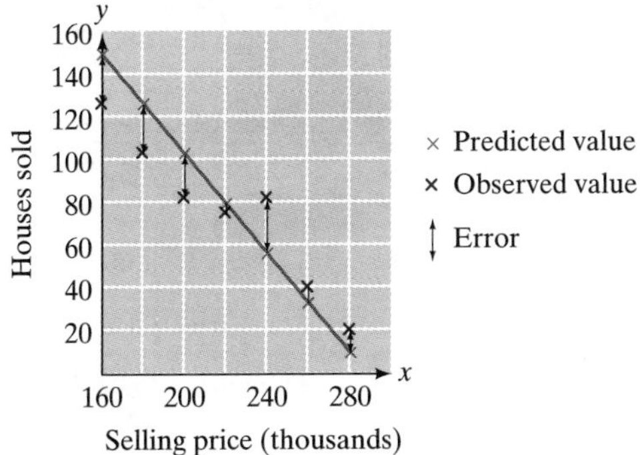

Figure 22

Question So why don't we just find the line that makes all of these errors zero?
Answer We could do that only if there were a line that passed through all of the points. Clearly, there is no such line in this case.

Question So, can we find a line that makes each of these errors as small as possible?
Answer We can't do that either. The line that minimizes the first two errors is the line that passes through the first two data points because it makes each of those errors zero. But that line certainly does not minimize the other errors. There is a tradeoff: Making some errors smaller makes others larger.

[1]We are using the term *demand function* loosely here. Strictly speaking, a demand function models the demand for the same item at different prices, whereas here we are modeling the demand for different homes at different prices.

Question OK, so what can we do?
Answer Since we cannot make all of the errors zero, and we cannot make them each as small as possible, we make some reasonable combination of them as small as possible. One possibility is to minimize the sum of the errors. This turns out to be difficult (largely because distances are measured using absolute values). What is technically easier to deal with is the sum of the *squares* of the errors. The line that minimizes the sum of the squares of the errors is called the **regression line,** the **least-squares line,** or the **best-fit line.**

Question How do we find the regression line?
Answer Here is its equation. We will justify this equation in Chapter 8, using calculus.

Regression Line (Best-Fit Line; Least-Squares Line)

The line that best fits the n data points $(x_1, y_1), (x_2, y_2), \ldots, (x_n, y_n)$ has the form

$$y = mx + b$$

where

$$m = \frac{n(\Sigma xy) - (\Sigma x)(\Sigma y)}{n(\Sigma x)^2 - (\Sigma x)^2}$$

$$b = \frac{\Sigma y - m(\Sigma x)}{n}$$

$$n = \text{number of data points}$$

Here, Σ means "the sum of." Thus, for example,

$$\Sigma x = \textit{The sum of the x values} = x_1 + x_2 + \cdots + x_n$$

and $$\Sigma xy = \textit{The sum of products} = x_1y_1 + x_2y_2 + \cdots + x_ny_n$$

example 1

Computing a Regression Line
Find the line that best fits the following data:

x	1	2	3	4
y	1.5	1.6	2.1	3.0

Solution
Let us organize our work in the form of a table, where the original data are entered in the first two columns and the bottom row contains the column sums.

	x	y	xy	x^2
	1	1.5	1.5	1
	2	1.6	3.2	4
	3	2.1	6.3	9
	4	3.0	12.0	16
Σ (sum)	10	8.2	23.0	30

Thus, $\Sigma x = 10$, $\Sigma y = 8.2$, $\Sigma xy = 23.0$, and $\Sigma x^2 = 30$. Since there are $n = 4$ data points, we get

$$m = \frac{n(\Sigma xy) - (\Sigma x)(\Sigma y)}{n(\Sigma x^2) - (\Sigma x)^2} = \frac{4(23) - (10)(8.2)}{4(30) - (10)^2} = 0.5$$

$$b = \frac{\Sigma y - m(\Sigma x)}{n} = \frac{8.2 - (0.5)(10)}{4} = 0.8$$

So, the best-fit line is

$$y = 0.5x + 0.8$$

The data points and the best-fit line are shown in Figure 23.

$y = 0.5x + 0.8$

Figure 23

Spreadsheet

Of course, the table in Example 1 strongly suggests a spreadsheet. The following layout duplicates the earlier table and also includes the formulas for m and b in row 6.

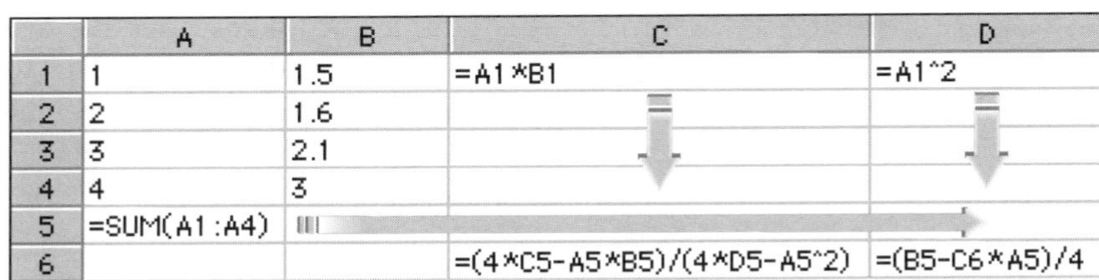

	A	B	C	D
1	1	1.5	=A1*B1	=A1^2
2	2	1.6		
3	3	2.1		
4	4	3		
5	=SUM(A1:A4)			
6			=(4*C5-A5*B5)/(4*D5-A5^2)	=(B5-C6*A5)/4
			formula for m	formula for b

(Most spreadsheets, including Excel, have built-in regression formulas. See the next example for details on using a spreadsheet regression formula.)

Now let us return to the example with which we began this section.

example 2

New Home Sales

Find the linear model that best fits the following data on new home sales. Then extrapolate from the model to predict the sales of homes priced at around \$140,000.

Selling price (thousands)	\$160	180	200	220	240	260	280
Houses sold	126	103	82	75	82	40	20

Solution

Here is a table with the necessary calculations.

	x	y	xy	x^2
	160	126	20,160	25,600
	180	103	18,540	32,400
	200	82	16,400	40,000
	220	75	16,500	48,400
	240	82	19,680	57,600
	260	40	10,400	67,600
	280	20	5,600	78,400
Σ (sum)	1540	528	107,280	350,000

With $n = 7$, the formulas give

$$m = \frac{7(107{,}280) - (1540)(528)}{7(350{,}000) - 1540^2} \approx -0.793$$

$$b = \frac{528 - (-0.7928571429)(1540)}{7} \approx 250$$

Notice that, when computing b, we used the most accurate value for m that we could obtain, $m = -0.7928571429$, rather than the rounded value -0.793. This illustrates a general rule:

When calculating, don't round intermediate results. Rather, use the most accurate results obtainable or have your calculator or computer store them for you.

The least-squares line is

$$y = -0.793x + 250$$

To predict the sales of homes priced at around \$140,000, we substitute 140 for x and get $y \approx 139$. Recall that the data were originally in price ranges of \$20,000. Thus, our model predicts that 139 homes will be sold in the range of \$130,000–\$149,000 each.

Figure 24 shows the original data and the best-fit line.

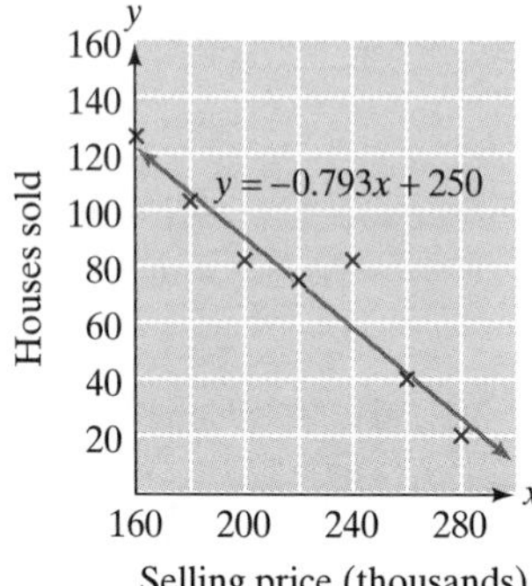

Figure 24

Graphing Calculator

Now that we know how to calculate the best-fit line by hand, we'll let you in on a little secret: Most graphing calculators and computer spreadsheets have the linear regression formulas built in. To find the best-fit line using a TI-83, enter the coordinates of the data points using [STAT] `EDIT` (x values go in L_1 and corresponding y values go in L_2), go to [STAT] `CALC`, and select `LinReg(ax+b)`.

Spreadsheet

In Excel there are at least two ways of finding the best-fit line. The simplest way is to ask Excel to add a **trendline** to a graph of the data points. Once you have created a chart with the data points, choose "Add Trendline . . ." from the Chart menu. Make sure you select the "linear" type of trendline. If you want to see the equation of the line, select the "Display equation on chart" option.

Alternatively, you can use the "LINEST" function (for "linear estimate"). For example, suppose that you have the data entered in the spreadsheet in the following way:

	A	B	C	D
1	160	126		
2	180	103		
3	200	82		
4	220	75		
5	240	82		
6	260	40		
7	280	20		

Select two unused adjacent cells in the same row—say, C1 and D1. Then enter the formula

```
=LINEST(B1:B7,A1:A7)
```

and press Control-Shift-Enter. (This key combination is necessary when using a formula like LINEST that returns values in multiple cells.) Note that the first argument to LINEST is the range of y values, while the second argument is the range of x values. The slope and y-intercept will be calculated and inserted in the two cells you selected.

Web Site

You can use the On-Line Utilities at the web site to give you the equation of the best-fit line. Follow

Web site → On-Line Utilities → Simple Regression

Then enter the coordinates of the data points and press the `mx+b` button to obtain the equation.

Question If my data points happen to lie on a straight line, will that be the best-fit line?
Answer Certainly. If the data points do lie on a straight line, then the smallest possible value for the sum of the squares of the errors is zero, which occurs when we use the line that passes through all of the points.

Question If my data points do not all lie on one straight line, how can I tell how close they are to lying on a straight line?
Answer There is a number that measures the "goodness of fit" of the least-squares line, called the **coefficient of correlation** or **correlation coefficient.** This number, usually denoted r, is between -1 and 1. The closer r is to -1 or 1, the better the fit. For an *exact* fit, $r = -1$ (for a line with negative slope) or $r = 1$ (for a line with positive slope). For a bad fit, r is close to zero. Figure 25 shows several collections of data points with the least-squares lines and the corresponding values of r.

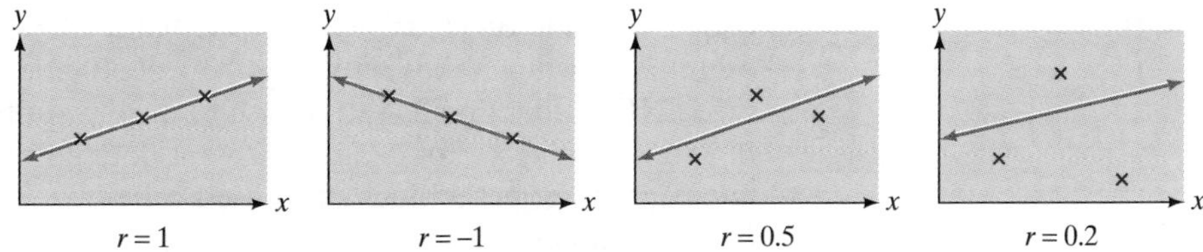

Figure 25

The correlation coefficient is calculated using the following formula. (To obtain this formula requires a fair knowledge of statistics.)

Coefficient of Correlation

The coefficient of correlation of the line that best fits the n data points (x_1, y_1), $(x_2, y_2), \ldots, (x_n, y_n)$ is

$$r = \frac{n(\Sigma xy) - (\Sigma x)(\Sigma y)}{\sqrt{n(\Sigma x^2) - (\Sigma x)^2} \cdot \sqrt{n(\Sigma y^2) - (\Sigma y)^2}}$$

example 3 **Correlation Coefficient**

Find the correlation coefficient for the best-fit line found in Example 2.

Solution

The formula for r requires one calculation that we have not yet done:

$$\Sigma y^2 = 47{,}558$$

Substituting into the formula, we get

$$\begin{aligned} r &= \frac{n(\Sigma xy) - (\Sigma x)(\Sigma y)}{\sqrt{n(\Sigma x^2) - (\Sigma x)^2} \cdot \sqrt{n(\Sigma y^2) - (\Sigma y)^2}} \\ &= \frac{7(107{,}280) - (1540)(528)}{\sqrt{7(350{,}000) - 1540^2} \cdot \sqrt{7(47{,}558) - 528^2}} \\ &\approx -0.9543 \end{aligned}$$

Thus, the fit is a pretty good one, as confirmed by Figure 24.

Before we go on . . . Why is r negative in this case?

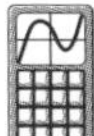

Graphing Calculator

Of course, a utility that will calculate linear regression lines will also calculate the correlation coefficient. To find the correlation coefficient using a TI-83 you need to tell the calculator to show you the coefficient at the same time it shows you the regression line. To do this, press [CATALOG] and select `DiagnosticOn` from the list. The command will be pasted to the home screen, and you should then press [ENTER] to execute the command. Once you have done this, the `LinReg(ax+b)` command will show you not only a and b but also r and r^2.

Spreadsheet

In Excel, when you add a trendline to a chart, you can select the option "Display r-squared value on chart" to show the value of r^2 on the chart. (It is common to examine r^2, which takes on values between 0 and 1, instead of r.) Alternatively, the LINEST function has an option to display quite a few statistics about a best-fit line, including r^2. For example, if your data points are entered as in the preceding example, select a block of unused cells two wide and five tall—for example, C1 through D5.

	A	B	C	D
1	160	126		
2	180	103		
3	200	82		
4	220	75		
5	240	82		
6	260	40		
7	280	20		

Enter the formula

```
=LINEST(B1:B7,A1:A7,TRUE,TRUE)
```

and press Control-Shift-Enter. The result should look something like this:

	A	B	C	D
1	160	126	-0.792857143	249.8571429
2	180	103	0.11109382	24.84133323
3	200	82	0.910609364	11.75706475
4	220	75	50.93427036	5
5	240	82	7040.571429	691.1428571
6	260	40		
7	280	20		

The value of r^2 is the number 0.910609364 in cell C3. For the meanings of the other numbers shown, see the on-line help for LINEST in Excel (a good course in statistics wouldn't hurt, either).

1.5 exercises

Find the best-fit line associated with each set of points in Exercises 1–4. Graph the data and the best-fit line.

1. $(1, 1), (2, 2), (3, 4)$
2. $(0, 1), (1, 1), (2, 2)$
3. $(0, -1), (1, 3), (4, 6), (5, 0)$
4. $(2, 4), (6, 8), (8, 12), (10, 0)$

In Exercises 5 and 6, use correlation coefficients to determine which of the given sets of data is best fit by its associated regression line and which is fit worst. Is it a perfect fit for any of the data sets?

5. **a.** $\{(1, 3), (2, 4), (5, 6)\}$ **b.** $\{(0, -1), (2, 1), (3, 4)\}$
   c. $\{(4, -3), (5, 5), (0, 0)\}$
6. **a.** $\{(1, 3), (-2, 9), (2, 1)\}$ **b.** $\{(0, 1), (1, 0), (2, 1)\}$
   c. $\{(0, 0), (5, -5), (2, -2.1)\}$

Applications

7. ***Pollution Control*** According to recent surveys, the percentage of new plant and equipment expenditures by U.S. manufacturing companies spent on pollution control is as shown ($t = 0$ represents 1975):

Year	0	5	6	9	12
Percent spent	9.3	4.8	4.3	3.3	4.3

Source: Survey of Current Business, 58, 62, 66, 68.

Use a best-fit line to estimate the percentage for 1985. (Round regression coefficients to three significant digits and the final answer to two significant digits.)

8. ***Pollution Control*** The percentage of new plant and equipment expenditures by U.S. public utility companies spent on pollution control is as shown ($t = 0$ represents 1975):

Year	0	5	6	9	12
Percent spent	8.4	8.1	7.3	6.8	5.1

Source: Survey of Current Business, 58, 62, 66, 68.

Use a best-fit line to estimate the percentage for 1985. (Round regression coefficients to three significant digits and the final answer to two significant digits.)

9. ***Oil Recovery*** In 1978 Congress conducted a study of the amount of additional oil that can be extracted from existing oil wells by "enhanced recovery techniques" (such as injecting solvents or steam into an oil well to lower the density of the oil). As the price of oil increases, the amount of oil that can be recovered economically in this manner also increases. The following table gives the study's estimates of recoverable oil based on the price per barrel:

Price per barrel	\$12	\$14	\$22	\$30
Recovery (billions of barrels)	21	30	42	49

Source: U.S. Congress, Office of Technology Assessment, *Enhanced Oil Recovery Potential in the U.S.* (Washington, DC: OTA, 1978):7. The recovery figures are based on a 10% minimum rate of return, and the prices are in constant 1976 dollars and rounded to the nearest dollar.

Use a best-fit line to estimate the additional amount of oil that could be economically recovered if the price of oil dropped to \$10 per barrel. (Round regression coefficients to three significant digits and the final answer to two significant digits.)

10. ***Depletion of Natural Gas Reserves*** The following table shows the estimated U.S. natural gas reserves, in trillions of cubic feet ($t = 0$ represents 1980):

Year	0	2	4	6	8
Reserves	194	200	197	190	183

Source: U.S. Energy Information Administration, *International Energy Annual* (Washington, DC: Government Printing Office): annual.

Use a best-fit linear model to estimate the rate at which natural gas was depleted during the given period.

The use of a graphing calculator or computer spreadsheet is recommended for Exercises 11–16.

T 11. ***Big Brother*** In 1995 the FBI was seeking the ability to monitor 74,250 phone lines simultaneously. The following chart shows the number of phone lines monitored from 1987 through 1993.

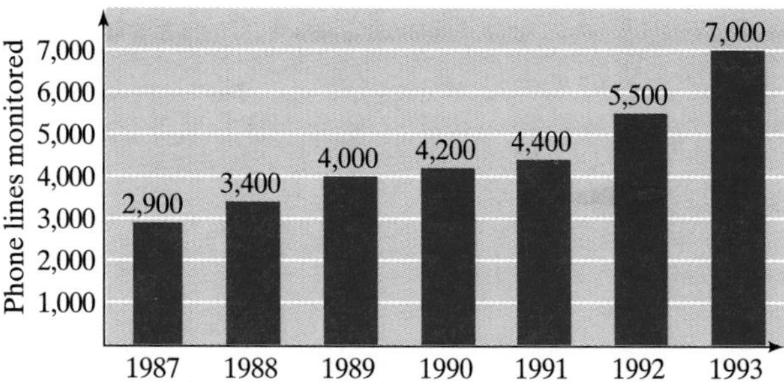

Source: Electronic Privacy Information Center, Justice Department, Administrative Office of the United States Courts/*New York Times*, November 2, 1995, p. D5.

Use a linear model of these data to project the number of phone lines monitored by the FBI in 1999. (Round your answer to the nearest 100 phone lines.)

T 12. ***SAT Scores by Income*** The following chart shows 1994 verbal SAT scores in the United States as a function of parents' income level:

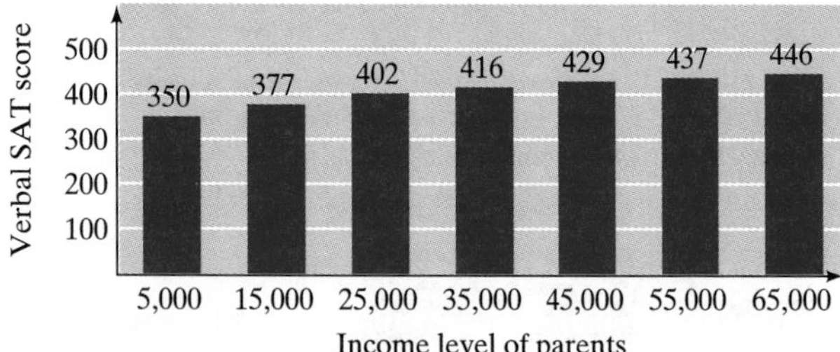

Source: The College Board/*New York Times*, March 5, 1995, p. E16.

Use a linear model of these data to estimate the verbal SAT score of a student whose parents have an income of $70,000.

T 13. ***Life Expectancy*** Life expectancies at birth in the United States for people born in various years are given in the following table:

Year	1920	1930	1940	1950
Life expectancy	54.1	59.7	62.9	68.2

Year	1960	1970	1980	1990
Life expectancy	69.7	70.8	73.7	75.4

Using a least-squares line, estimate the rate at which life expectancy was changing during the given period.

T 14. ***Infant Mortality*** The infant mortality rates (deaths per 1000 live births) in the United States for various years are given in the following table:

Year	1960	1970	1980	1985	1986	1987	1988
Mortality	26	20	12.6	10.6	10.4	10.1	10

Use a least-squares line to predict the infant mortality rate in the year 2010. Is this model realistic?

T 15. ***Profit*** The following chart shows the net income (estimated to the nearest 15 million) of the Walt Disney Company for the years 1984–1992.

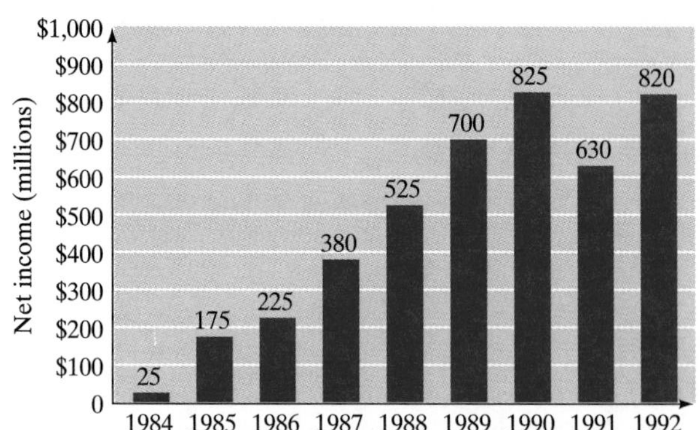

Source: Company reports/*New York Times*, December 1992, p. D1.

Find a best-fit linear model for these data. (Find profit P as a function of the year t, with $t = 0$ corresponding to 1980.) Use your model to predict Disney's net income, to the nearest million dollars, in 1993.

T 16. ***Stock Prices*** Repeat Exercise 15, but model Walt Disney's stock prices in the following chart (these stock prices are rough estimates):

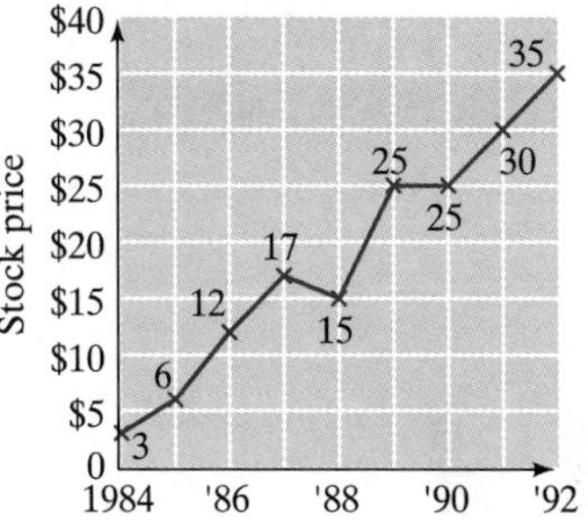

Source: Company reports/*New York Times*, December 1992, p. D1.

Communication and Reasoning Exercises

17. Verify from the formula that the best-fit line associated with the two points (a, b) and (c, d) is the line that passes through both. (Assume that $a \neq c$.)

18. What is the slope of the best-fit line for the points $(0, 0)$, $(-a, a)$, and (a, a)? (Assume that $a \neq 0$.)

19. If the points $(x_1, y_1), (x_2, y_2), \ldots, (x_n, y_n)$ lie on a straight line, what can you say about the regression line associated with these points?

20. If all but one of the points $(x_1, y_1), (x_2, y_2), \ldots, (x_n, y_n)$ lie on a straight line, does the least-squares line pass through all but one of these points?

21. Must the least-squares line pass through at least one of the data points? Illustrate your answer with an example.

22. Why must care be taken when using mathematical models to extrapolate?

You're the Expert

Modeling Spending on Internet Advertising

You are the new director of Impact Advertising, Inc.'s, Internet Division, which has enjoyed a steady 0.25% of the Internet advertising market. You have drawn up an ambitious proposal to expand your division in light of your anticipation that Internet advertising will continue to skyrocket. However, upper management sees things differently and, based on the following e-mail, does not seem likely to approve the budget for your proposal.

TO: JRutah@impact.com (J. R. Utah)
CC: CVodoylePres@impact.com. (C. V. O'Doyle, CEO)
FROM: SGlombardoVP@impact.com (S. G. Lombardo, VP Financial Affairs)
SUBJECT: Your Expansion Proposal
DATE: August 3, 2001

Hi John:

Your proposal reflects exactly the kind of ambitious planning and optimism we like to see in our new upper management personnel. Your presentation last week was most impressive and obviously reflected a great deal of hard work and preparation.

I am in full agreement with you that Internet advertising is on the increase. Indeed, our Market Research Department informs me that, based on a regression of the most recently available data, Internet advertising revenue in the United States will continue to grow by approximately $1 billion per year. This translates into approximately $2.5 million in increased revenues per year for Impact, given our 0.25% market share. This rate of expansion is exactly what our planned 2002 budget anticipates. Your proposal, on the other hand, would require a budget of approximately *twice* the 2002 budget allocation, even though your proposal provides no hard evidence to justify this degree of financial backing.

At this stage, therefore, I am sorry to say that I am inclined not to approve the funding for your project, although I would be happy to discuss this further with you. I plan to present my final decision on the 2002 budget at next week's divisional meeting.

Regards, Sylvia

Refusing to admit defeat, you contact the Market Research Department and request the details of their projections on Internet advertising. They fax you the following information:

Year	1995	1996	1997	1998	1999	2000	2001
Spending on advertising (billions)	$0	$0.3	$0.8	$1.9	$3	$4.3	$5.8

Source: Nielsen NetRatings/*New York Times,* June 7, 1999, p. C17. Figures are approximate, and the 1999–2001 figures are projected.

Regression model $y = 0.9857x - 0.6571$ (x = years since 1995)

Correlation coefficient $r = 0.9781$

Now you see where the VP got that \$1-billion figure: The slope of the regression equation is close to 1, which indicates an increase of just under \$1 billion per year. Also, the correlation coefficient is very high—an indication that the linear model fits the data well. In view of this strong evidence, it seems difficult to argue that revenues will increase by significantly more than the projected \$1 billion per year.

To get a better picture of what's going on, you decide to graph the data together with the regression line in your spreadsheet program. What you get is shown in Figure 26. You immediately notice that the data points seem to suggest a curve and not a straight line. Then again, perhaps the suggestion of a curve is an illusion. There are, you surmise, two possible interpretations of the data:

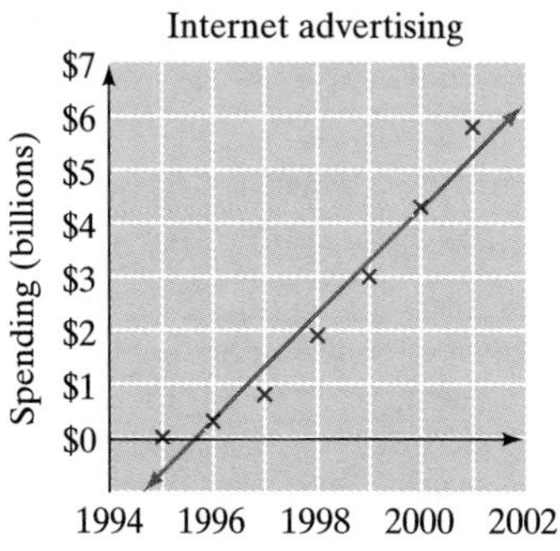

Figure 26

1. *(Your first impression)* As a function of time, spending on Internet advertising is nonlinear and is in fact accelerating (the rate of change is increasing), so a linear model is inappropriate.
2. *(Devil's advocate)* Spending on Internet advertising *is* a linear function of time; the fact that the points do not lie on the regression line is simply a consequence of random factors, such as the state of the economy, the stock market performance, and so on.

You suspect that the VP will probably opt for the second interpretation and discount the graphical evidence of accelerating growth by claiming that it is an illusion: a "statistical fluctuation." That is, of course, a possibility, but you wonder how likely it really is.

For the sake of comparison, you decide to try a regression based on the simplest nonlinear model you can think of—a quadratic function:

$$y = ax^2 + bx + c$$

Your spreadsheet allows you to fit such a function with a click of the mouse. The result is the following:

$$y = 0.1190x^2 + 0.2714x - 0.0619 \qquad (x = \text{number of years since 1995})$$

$$r = 0.9992$$

Figure 27 shows the graph of the regression function together with the original data.

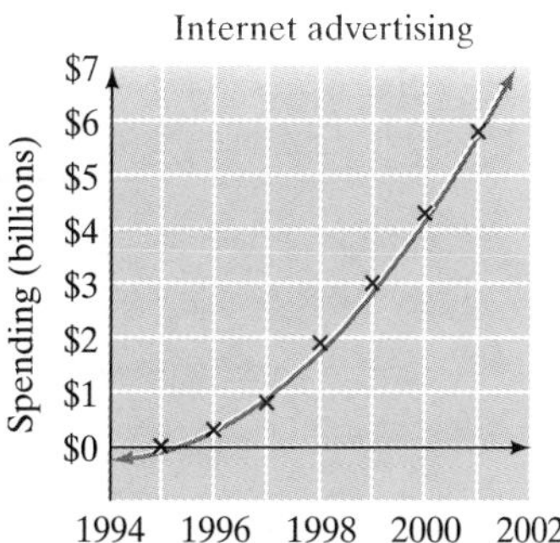

Figure 27

Aha! The fit is visually far better, and the correlation coefficient is even higher! Furthermore, the quadratic model predicts 2002 spending as

$$y = 0.1190(7)^2 + 0.2714(7) - 0.0619 \approx \$7.67 \text{ billion}$$

which is $1.87 billion higher than the 2001 spending figure in the table above. Given Impact Advertising's 0.25% market share, this translates to an increase in revenues of $4.7 million, which is almost double the estimate predicted by the linear model!

You quickly draft an e-mail to Lombardo and are about to press "Send" when you decide, as a precaution, to check with a statistician. He tells you to be cautious: The value of r always tends to increase when you pass from a linear model to a quadratic one due to an increase in "degrees of freedom."[2] A good way to test whether a quadratic model is more appropriate than a linear one is to compute a statistic called the "p-value" associated with the coefficient of x^2. A low value of p indicates a high probability that the coefficient of x^2 cannot be zero (see below). Notice that if the coefficient of x^2 is zero, then you have a linear model.

You can, your friend explains, obtain the p-value using a spreadsheet such as Excel as follows: First, set up the data in columns, with an extra column for the values of x^2.

	A	B	C
1	x	x^2	y
2	0	0	0
3	1	1	0.3
4	2	4	0.8
5	3	9	1.9
6	4	16	3
7	5	25	4.3
8	6	36	5.8

Next, choose "Data analysis" from the Tools menu, and choose "Regression." In the dialog box, give the location of the data as shown in Figure 28. Clicking "OK" then gives you a large chart of statistics. The p-value you want is in the last row of the data: $p = 0.000478$.

[2]The number of degrees of freedom in a regression model is 1 less than the number of coefficients. For a linear model it is 1 (there are two coefficients: the slope m and the intercept b), and for a quadratic model it is 2. For a detailed discussion, consult a text on regression analysis.

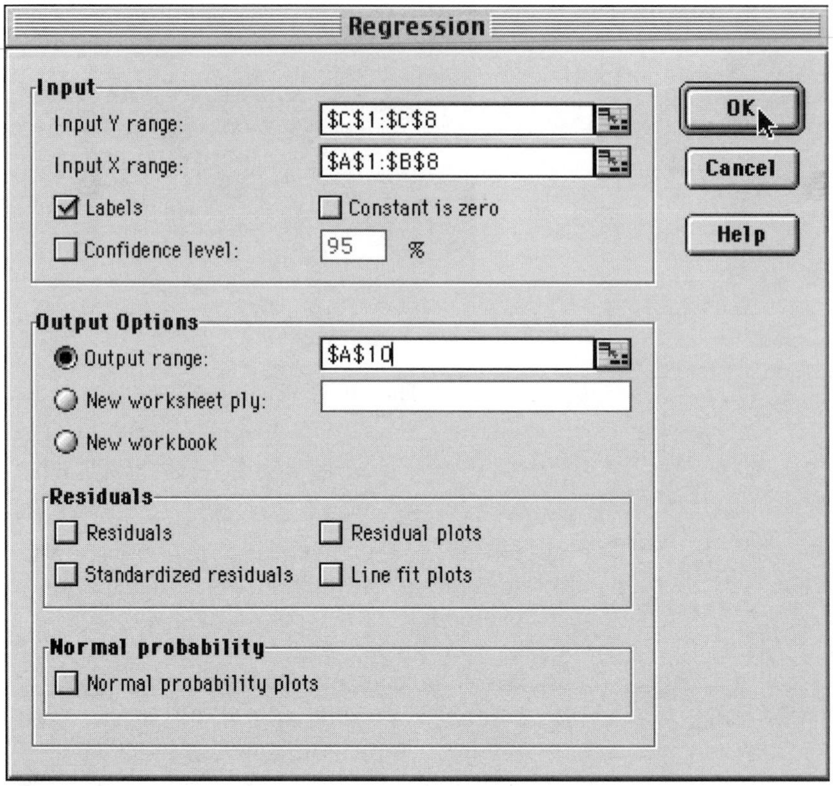

Regression

Input
Input Y range: $C$1:$C$8
Input X range: $A$1:$B$8
☑ Labels ☐ Constant is zero
☐ Confidence level: 95 %

Output Options
◉ Output range: $A$10
○ New worksheet ply:
○ New workbook

Residuals
☐ Residuals ☐ Residual plots
☐ Standardized residuals ☐ Line fit plots

Normal probability
☐ Normal probability plots

OK
Cancel
Help

Figure 28

Question What does p measure?
Answer Roughly speaking, p is the probability, with allowance for random fluctuation in the data, that you are wrong in claiming that the coefficient of x^2 is not zero. In other words, you can be 99.9522% certain that you are not making an error.

Thus, you can go ahead and send your e-mail with 99% confidence!

exercises

Suppose you are given the following data:

Year	1995	1996	1997	1998	1999	2000	2001
Spending on advertising (billions)	$0	$0.3	$1.5	$2.6	$3.4	$4.3	$5.0

T **1.** Obtain a linear regression model and the correlation coefficient r. According to the model, at what rate is spending on Internet advertising increasing in the United States? How does this translate to annual revenues for Impact Advertising?

T **2.** Use a spreadsheet or other technology to graph the data together with the best-fit line. Does the graph suggest a quadratic model (parabola)?

T **3.** Test your impression in the preceding exercise by using technology to fit a quadratic function and graphing the resulting curve together with the data. Does the graph suggest that the quadratic model is appropriate?

T **4.** Perform a regression analysis and find the associated p-value. What does it tell you about the appropriateness of a quadratic model?

chapter 1 review test

1. Graph the following functions and equations.

   a. $y = -2x + 5$ **b.** $2x - 3y = 12$ **c.** $y = \begin{cases} \frac{1}{2}x & \text{if } -1 \le x \le 1 \\ x - 1 & \text{if } 1 < x \le 3 \end{cases}$

   d. $f(x) = 4x - x^2$, domain $[0, 4]$

2. Find the equation of each line.
   a. Through $(3, 2)$ with slope -3
   b. Through $(-1, 2)$ and $(1, 0)$
   c. Through $(1, 2)$ parallel to $x - 2y = 2$
   d. With slope $\frac{1}{2}$ crossing $3x + y = 6$ at its x-intercept

Applications

3. As your on-line bookstore, OHaganBooks.com, has grown in popularity, you have been monitoring book sales as a function of the traffic at your site (measured in "hits" per day) and have obtained the following model:

$$n(x) = \begin{cases} 0.02x & \text{if } 0 \le x \le 1000 \\ 0.025x - 5 & \text{if } 1000 < x \le 2000 \end{cases}$$

   where $n(x)$ is the average number of books sold in a day in which there are x hits at the site.
   a. On average, how many books per day does your model predict you will sell when you have 500 hits in a day? 1000 hits in a day? 1500 hits in a day?
   b. What does the coefficient 0.025 tell you about your book sales?
   c. According to the model, how many hits per day are needed to sell an average of 30 books per day?

4. To increase business at OHaganBooks.com, you plan to place more banner ads at well-known Internet portals. So far, you have the following data on the average number of hits per day at OHaganBooks.com versus your monthly advertising expenditures:

Advertising expenditure per month	\$2000	\$5000
Web site hits per day	1900	2050

   You decide to construct a linear model giving the average number of hits h per day as a function of the advertising expenditure c.
   a. What is the model you obtain?
   b. Based on your model, how much traffic can you anticipate if you budget \$6000 per month for banner ads?
   c. Your goal is eventually to increase traffic at your site to an average of 2500 hits per day. Based on your model, how much do you anticipate you will need to spend on banner ads to accomplish this?

5. A month ago you increased the expenditure on banner ads to \$6000 per month, and you have noticed that the traffic at OHaganBooks.com has not increased

to the level predicted by the linear model in Question 4. Fitting a quadratic function to the data you have gives the model

$$h = -0.000005c^2 + 0.085c + 1750$$

where h is the daily traffic (hits) at your web site and c is the monthly advertising expenditure.

a. According to this model, what is the current traffic at your site?
b. Does this model give a reasonable prediction of traffic at expenditures greater than \$8500 per month? Why?

6. Besides selling books, you are generating additional revenue at OHaganBooks.com through your new on-line publishing service. Readers pay a fee to download the entire text of a novel. Author royalties and copyright fees cost you an average of \$4 per novel, and the monthly cost of operating and maintaining the service amounts to \$900 per month. You are currently charging readers \$5.50 per novel.
a. What are the associated cost, revenue, and profit functions?
b. How many novels must you sell per month in order to break even?
c. If you lower the charge to \$5.00 per novel, how many books will you need to sell in order to break even?

7. To generate a profit from your on-line publishing service, you need to know how the demand for novels depends on the price you charge. During the first month of the service, you were charging \$10 per novel and sold 350. Lowering the price to \$5.50 per novel had the effect of increasing demand to 620 novels per month.
a. Use the given data to construct a linear demand equation.
b. Use the demand equation you constructed in part (a) to estimate the demand if you raised the price to \$15 per novel.
c. Using the information on cost given in Question 6, determine which of the three prices (\$5.50, \$10, and \$15) would result in the largest profit and determine the size of that profit.

8. It is now several months later and you have tried selling your on-line novels at a variety of prices, with the following results:

Price	\$5.50	\$10	\$12	\$15
Demand (monthly sales)	620	350	300	100

a. Use the given data to obtain a linear regression model of demand. (Round coefficients to four decimal places.)
b. Use the demand model you constructed in part (a) to estimate the demand if you charged \$8 per novel. (Round the answer to the nearest novel.)

Additional On-Line Review

If you follow the path

Web site → Everything for Calculus → Chapter 1

you will find the following additional resources to help you review:

- A comprehensive chapter summary (including examples and interactive features)
- Additional review exercises (including interactive exercises and many with help)
- A true–false chapter quiz
- Several useful utilities, including graphers and a regression tool

You're the Expert

Epidemics

A flu epidemic is spreading through the population of the United States. An estimated 150,000,000 people are susceptible to this particular strain. There are 10,000 people already ill, and that number is doubling every 2 weeks. When will the flu have infected 1,000,000 people; when will it have infected 10,000,000; and when will it have infected 100,000,000?

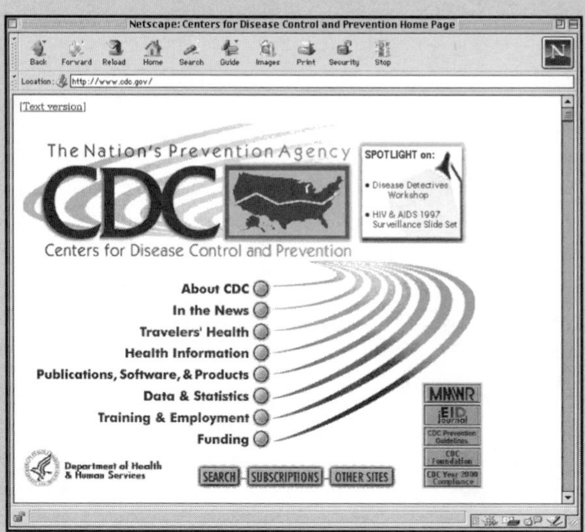

Internet Resources for This Chapter

At the web site, follow the path

Web site → Everything for Calculus → Chapter 2

where you will find a detailed chapter summary you can print out, a true–false quiz, and a collection of review questions. You will also find downloadable Excel tutorials for each section, an on-line regression utility you can use to model with many of the functions we discuss in this chapter, an on-line grapher, and other resources. In addition, complete text and interactive exercises have been placed on the web site covering several optional topics:

- Linear and exponential regression
- Using and deriving algebraic properties of logarithms
- More on trigonometric functions

Nonlinear Models

chapter 2

2.1 Quadratic Functions and Models
2.2 Exponential Functions and Models
2.3 Logarithmic Functions and Models
2.4 Trigonometric Functions and Models (Optional)

Introduction

To predict the course of a flu epidemic we need a model of how the epidemic will progress. Specifically, we need to find the number of infected people as a function of time. Common sense tells us that the number of infected people does not increase at a constant rate during the entire course of the epidemic. We expect the number of infected people to grow comparatively slowly at first, accelerate as the epidemic takes hold, and finally slow as saturation is reached. Thus, we need to find a suitable nonlinear function to model the epidemic.

The nonlinear functions we consider in this chapter are the *quadratic* functions, the simplest nonlinear functions; the *exponential* functions, essential for discussing many kinds of growth and decay, including the growth (and decay) of money in finance and the initial growth of an epidemic; the *logarithmic* functions, needed to fully understand the exponential functions; and the *trigonometric* functions, which describe periodic (repeating) behavior.

The course of an epidemic is often modeled by a *logistic* function. We shall see how to use this kind of model, which is based on the exponential function, in *You're the Expert* at the end of this chapter.

2.1 Quadratic Functions and Models

In Chapter 1 we studied linear functions. Linear functions are useful, but in real-life applications they are often accurate for only a limited range of values of the variables. The relationship between two quantities is often best modeled by a curved line rather than a straight line. The simplest function with a graph that is not straight is a **quadratic** function.

Quadratic Function

A **quadratic function** of the variable x is a function that can be written in the form

$$f(x) = ax^2 + bx + c \qquad \text{Function form}$$

or

$$y = ax^2 + bx + c \qquad \text{Equation form}$$

where a, b, and c are fixed numbers (with $a \neq 0$).

Quick Examples

1. $f(x) = 3x^2 - 2x + 1$ — $a = 3, b = -2, c = 1$
2. $g(x) = -x^2$ — $a = -1, b = 0, c = 0$
3. $R(p) = -5600p^2 + 14{,}000p$ — $a = -5600, b = 14{,}000, c = 0$

Every quadratic function $f(x) = ax^2 + bx + c$ $(a \neq 0)$ has a **parabola** as its graph. Figure 1 shows the two possible orientations of the graph.

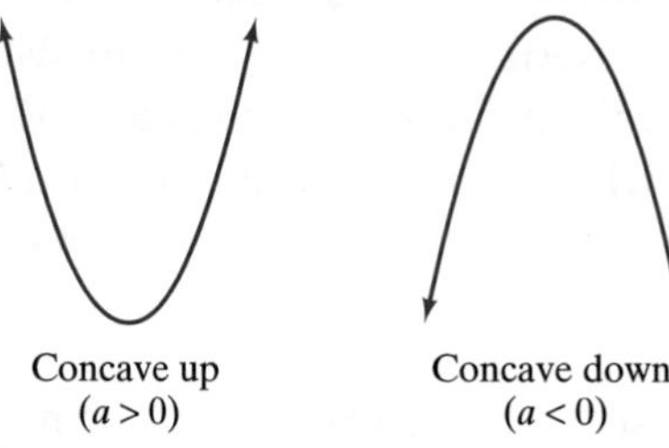

Figure 1

Following is a summary of some features of parabolas that we can use to sketch the graph of any quadratic function.[1]

Features of a Parabola

The graph of $f(x) = ax^2 + bx + c$ is a parabola with the following features:

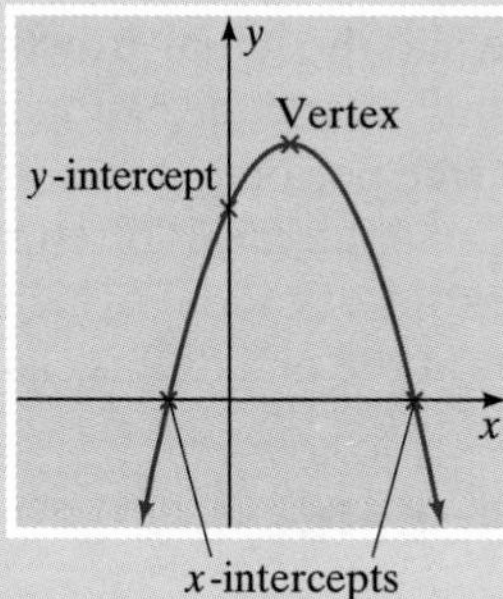

Vertex

The x-coordinate of the vertex is $-b/(2a)$. The y-coordinate is $f(-b/(2a))$.

***x*-Intercepts**

(if any) These occur when $f(x) = 0$; that is, when

$$ax^2 + bx + c = 0$$

[1] We shall not fully justify the formula for the vertex and the axis of symmetry until we have studied some calculus, although it is possible to do so with just algebra.

We can solve this equation for x using the quadratic formula. Thus, the x-intercepts are

$$x = \frac{-b \pm \sqrt{b^2 - 4ac}}{2a}$$

If the **discriminant** $(b^2 - 4ac)$ is positive, there are two x-intercepts. If it is zero, there is a single x-intercept (at the vertex). If it is negative, there are no x-intercepts (so the parabola doesn't touch the x-axis at all).

***y*-Intercept**
This occurs when $x = 0$, so

$$y = a \cdot 0^2 + b \cdot 0 + c = c$$

Symmetry
The parabola is symmetric with respect to the vertical line through the vertex, which is the line $x = -b/(2a)$.

example 1

Sketch the graph of $f(x) = x^2 + 2x - 8$.

Solution
Here, $a = 1, b = 2$, and $c = -8$. Since $a > 0$, the parabola is concave up (Figure 2).

Figure 2

- *Vertex:* The x-coordinate of the vertex is

$$x = -\frac{b}{2a} = -\frac{2}{2} = -1$$

To get its y-coordinate, we substitute the value of x back into $f(x)$ to get

$$y = f(-1) = (-1)^2 + 2(-1) - 8 = 1 - 2 - 8 = -9$$

Thus, the coordinates of the vertex are $(-1, -9)$.

- *x-Intercepts:* To calculate the x-intercepts (if any), we solve the equation

$$x^2 + 2x - 8 = 0$$

Luckily, this equation factors as $(x + 4)(x - 2) = 0$. Thus, the solutions are $x = -4$ and $x = 2$, so these values are the x-intercepts. (We could also have used the quadratic formula here.)

- *y-Intercept:* The y-intercept is given by $c = -8$.
- *Symmetry:* The graph is symmetric around the vertical line $x = -1$. Now we can sketch the curve as in Figure 3.

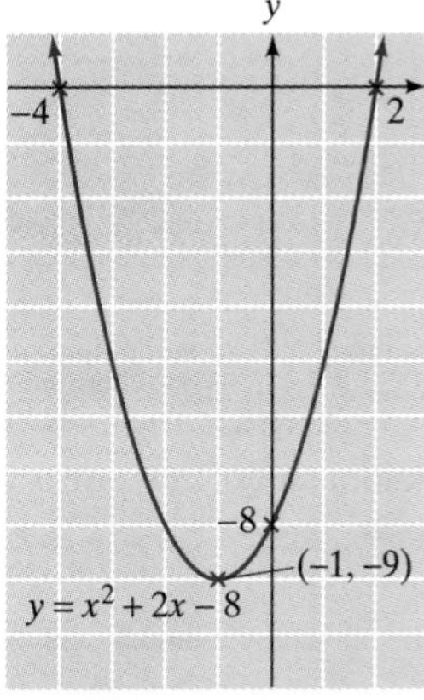

Figure 3

---

example 2

Sketch the graph of $f(x) = 4x^2 - 12x + 9$.

Solution
We have $a = 4, b = -12$, and $c = 9$. Since $a > 0$, this parabola is concave up.

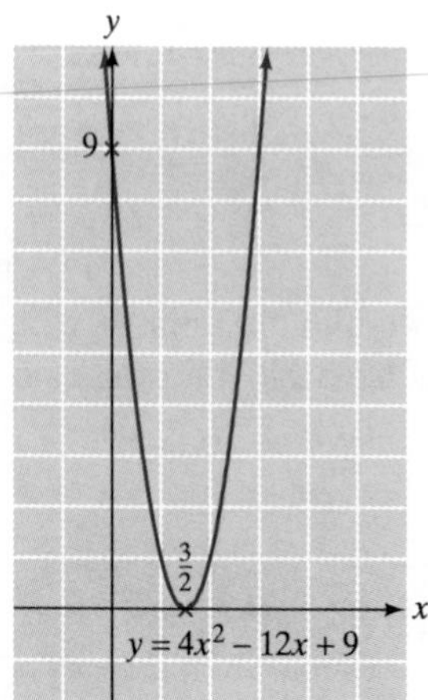

Figure 4

- *Vertex:* $x = -\dfrac{b}{2a} = \dfrac{12}{8} = \dfrac{3}{2}$ x-coordinate of vertex

$$y = f\left(\frac{3}{2}\right) = 4\left(\frac{3}{2}\right)^2 - 12\left(\frac{3}{2}\right) + 9 = 0 \quad \text{y-coordinate of vertex}$$

  Thus, the vertex is at the point $\left(\frac{3}{2}, 0\right)$.

- *x-Intercepts:* $4x^2 - 12x + 9 = 0$

$$(2x - 3)^2 = 0$$

  The only solution is $2x - 3 = 0$, or $x = \frac{3}{2}$. Note that this coincides with the vertex, which lies on the x-axis.

- *y-Intercept:* $c = 9$
- *Symmetry:* The graph is symmetric around the vertical line $x = \frac{3}{2}$. The graph is the narrow parabola shown in Figure 4.

---

example 3

Sketch the graph of $f(x) = -\frac{1}{2}x^2 + 4x - 12$.

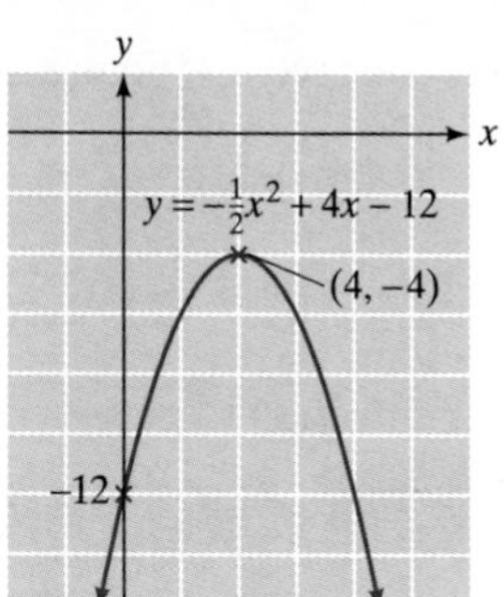

Figure 5

Solution

Here, $a = -\frac{1}{2}$, $b = 4$, and $c = -12$. Since $a < 0$, the parabola is concave down. The vertex has x-coordinate $-b/(2a) = 4$, with corresponding y-coordinate $f(4) = -\frac{1}{2}(4)^2 + 4(4) - 12 = -4$. Thus, the vertex is at $(4, -4)$. For the x-intercepts, we must solve $-\frac{1}{2}x^2 + 4x - 12 = 0$. If we try to use the quadratic formula, we discover that the discriminant is $b^2 - 4ac = 16 - 24 = -8$. Since the result is negative, there are no solutions of the equation, so there are no x-intercepts. The y-intercept is given by $c = -12$, and the graph is symmetric around the vertical line $x = 4$. Since there are no x-intercepts, the graph lies entirely below the x-axis, as shown in Figure 5.

Applications

Recall that the **revenue** that results from one or more business transactions is the total payment received. Thus, if q units of some item are sold at p dollars per unit, the revenue resulting from the sale is

$$\text{Revenue} = \text{Price} \times \text{Quantity}$$

$$R = pq$$

example 4

Demand and Revenue

A publishing company predicts that the demand equation for the sale of its latest sci-fi novel is

$$q = -2000p + 150{,}000$$

where q is the number of books it can sell per year at a price of \$$p$ per book. What price should the company charge to obtain the maximum annual revenue?

Solution

The total revenue depends on the price, as follows:

$$R = pq \qquad \text{Formula for revenue}$$

$$= p(-2000p + 150{,}000) \qquad \text{Substitute for } q \text{ from demand equation.}$$

$$= -2000p^2 + 150{,}000p \qquad \text{Simplify.}$$

We are looking for the price p that gives the maximum possible revenue. Notice that what we have is a quadratic function of the form $R(p) = ap^2 + bp + c$, where $a = -2{,}000$, $b = 150{,}000$, and $c = 0$. Since a is negative, the graph of the function is a parabola, concave down, so its vertex is its highest point (Figure 6). The p-coordinate of the vertex is

$$p = -\frac{b}{2a} = -\frac{150{,}000}{-4000} = 37.5$$

This value of p gives the highest point on the graph and thus gives the largest value of $R(p)$. We may conclude that the company should charge \$37.50 per book to maximize its annual revenue.

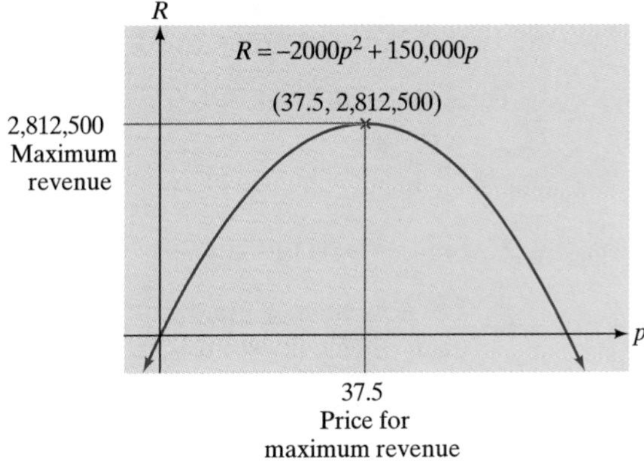

Figure 6

Before we go on . . . You might ask what the maximum annual revenue is. The answer is supplied for us by the revenue function $R(p) = -2000p^2 + 150{,}000p$. Since we have $p = 37.5$, we can substitute this value into the function and obtain $R(37.5) = -2000(37.5)^2 + 150{,}000(37.5) = 2{,}812{,}500$. In other words, the company will earn total annual revenues from this book amounting to \$2,812,500.

---

example 5

Demand, Revenue, and Profit

As the operator of Workout Fever Health Club (see Example 3 in Section 1.4), you calculate that your demand equation is

$$q = -0.06p + 46$$

where q is the number of new members who join the club per month and p is the annual membership fee you charge.

a. Since you are running a shoestring operation, your annual operating costs amount to only \$5,000 per year. Find the revenue and profit as functions of the membership price p.

b. At what price should you set annual memberships in order to break even? Could you still break even if your operating costs went up to \$10,000 per year?

c. At what price should you set annual memberships to obtain the maximum profit?

Solution

a. The annual revenue is given by

$$R = pq \qquad \text{Formula for revenue}$$
$$= p(-0.06p + 46) \qquad \text{Substitute for } q \text{ from demand equation.}$$
$$= -0.06p^2 + 46p \qquad \text{Simplify.}$$

while the annual cost C is fixed at \$5000. Thus, the profit function is

$$P = R - C \qquad \text{Formula for profit}$$
$$= -0.06p^2 + 46p - 5000 \qquad \text{Substitute for revenue and cost.}$$

b. First recall from Section 1.4 that break-even occurs when the profit equals zero, or

$$P = 0$$

or $\quad -0.06p^2 + 46p - 5000 = 0$

We now have a quadratic equation in p, and we must solve it for the break-even price p. To do this, we solve using the quadratic formula to obtain:

$$p \approx 635.55 \text{ or } 131.12$$

Although you may be tempted to choose the larger figure, thinking that you'll get a larger total fee, remember that both these options will result in your operation breaking even—there is no advantage to one or the other. Thus, to break even, you should charge either \$635.55 or \$131.12 per year.

If the operating costs went up to \$10,000 per year, then the break-even equation would become

$$-0.06p^2 + 46p - 10{,}000 = 0$$

This equation has no real solutions because the discriminant, $b^2 - 4ac$, is negative. You simply cannot break even with these costs, and you might as well close down the operation because the profit is always negative (the graph of profit vs. price lies below the p-axis).

c. We saw in part (a) that the profit function is

$$P = -0.06p^2 + 46p - 5000.$$

The graph of P is a parabola whose vertex has p-coordinate $-b/(2a) = 46/0.12 \approx 383.33$. Thus, for a maximum profit, you should charge an annual membership fee of \$383.33.

T We can answer parts (b) and (c) of the example by graphing the revenue and cost functions. We had $R = -0.06p^2 + 46p$ and $C = 5000$. If we plot these functions on the same coordinate system, as well as the hypothetical greater cost of \$10,000, we get the picture shown in Figure 7.

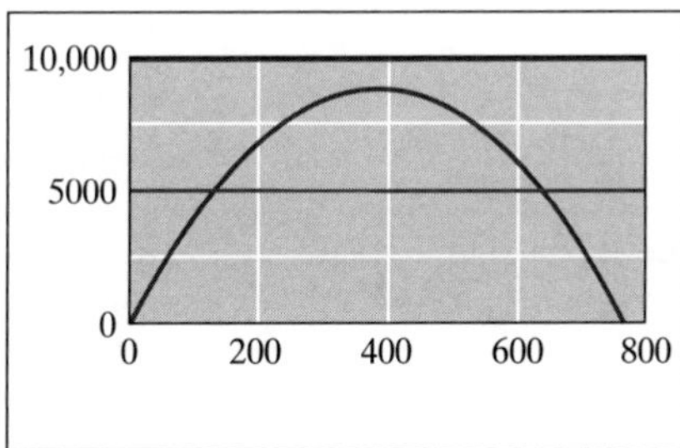

Figure 7

The horizontal lines at heights of \$5000 and \$10,000 represent the two fixed annual costs, while the parabola is the graph of the revenue function. The two intersection points are the break-even points we calculated in part (b). We can zoom and trace to obtain the coordinates of the break-even point and confirm the calculations. The higher-cost line fails to touch the revenue curve, which confirms our conclusion that the health club cannot break even with annual costs of \$10,000.

The graph also shows the price you should charge to obtain the maximum profit. Trace to the highest point of the parabola (which has a p-coordinate midway between the break-even points). You can also find the annual revenue predicted by the demand equation as the y-coordinate.

2.1 exercises

Sketch the graphs of the quadratic functions in Exercises 1–10, indicating the coordinates of the vertex, the y-intercept, and the x-intercepts (if any).

1. $f(x) = x^2 + 3x + 2$ **2.** $f(x) = -x^2 - x$

3. $f(x) = -x^2 + 4x - 4$ **4.** $f(x) = x^2 + 2x + 1$

5. $f(x) = -x^2 - 40x + 500$ **6.** $f(x) = x^2 - 10x - 600$

7. $f(x) = x^2 + x - 1$ **8.** $f(x) = x^2 + \sqrt{2}x + 1$

9. $f(x) = x^2 + 1$ **10.** $f(x) = -x^2 + 5$

For each demand equation in Exercises 11–14, express the total revenue R as a function of the price p per item, sketch the graph of the resulting function, and determine the price p that maximizes total revenue.

11. $q = -4p + 100$ **12.** $q = -3p + 300$

13. $q = -2p + 400$ **14.** $q = -5p + 1200$

Applications

15. *Sport Utility Vehicles* The average weight of a sport utility vehicle (SUV) can be approximated by

$$W = 3t^2 - 90t + 4200 \qquad (5 \le t \le 27)$$

where t is time in years ($t = 0$ represents 1970) and W is the average weight of an SUV in pounds.[2] Sketch the graph of W as a function of t. According to the model, in what year were SUVs the lightest? What was their average weight in that year?

16. *Sedans* The average weight of a sedan can be approximated by

$$W = 6t^2 - 240t + 4800 \qquad (5 \le t \le 27)$$

where t is time in years ($t = 0$ represents 1970) and W is the average weight of a sedan in pounds. Sketch the graph of W as a function of t. According to the model, in

[2]The quadratic models used in Exercises 15 and 16 are based on data published in the *New York Times,* November 30, 1997, p. 43.

what year were sedans the lightest? What was their average weight in that year?

17. ***Fuel Efficiency*** The fuel efficiency (in miles per gallon) of a sport utility vehicle (SUV) depends on its weight according to the formula[3]

$$E = 0.0000016x^2 - 0.016x + 54 \qquad (1800 \le x \le 5400)$$

where x is the weight of an SUV in pounds. According to the model, what is the weight of the least fuel-efficient SUV? Would you trust the model for weights greater than the answer you obtained? Explain.

Source: Environmental Protection Agency, National Highway Traffic Safety Administration, American Automobile Manufacturers' Association, Ford Motor Company/*New York Times*, November 30, 1997, p. 43.

18. ***Global Warming*** The amount of carbon dioxide (in pounds per 15,000 miles) released by a typical sport utility vehicle (SUV) depends on its fuel efficiency according to the formula

$$W = 32x^2 - 2080x + 44{,}000 \qquad (12 \le x \le 33)$$

where x is the fuel efficiency of an SUV in miles per gallon. According to the model, what is the fuel efficiency of the SUV that has the least carbon dioxide pollution? Comment on the reliability of the model for fuel efficiencies that exceed your answer.

Source: See the source in Exercise 17.

19. ***Revenue*** The market research department of the Better Baby Buggy Co. notices that when its buggies are priced at \$80 each, it can sell 100 each month. However, when the price is raised to \$100, it can only sell 90 each month. Assuming that the demand is linear, at what price should the company sell the buggies to get the largest revenue? What is the largest monthly revenue?

20. ***Revenue*** The Better Baby Buggy Co. has just come out with a new model, the Turbo. The market research department now estimates that the company can sell 200 Turbos per month at \$60, but only 120 per month at \$100. If we assume that the demand is linear, at what price should the company sell its buggies to get the largest revenue? What is the largest monthly revenue?

21. ***Revenue*** Pack-Em-In Real Estate is building a new housing development. The more houses it builds, the less people will be willing to pay, due to the crowding and smaller lot sizes. In fact, if the company builds 40 houses in this particular development, it can sell them for \$200,000 each, but if it builds 60 houses, it will be able to get only \$160,000 each. How many houses should the company build to get the largest revenue? What is the largest possible revenue?

22. ***Revenue*** Pack-Em-In has another development in the works. If it builds 50 houses in this development it will be able to sell them at \$190,000 each, but if it builds 70 houses, it will get only \$170,000 each. How many houses should the company build to get the largest revenue? What is the largest possible revenue?

23. ***Web Site Profit*** Encouraged by the popularity of your Dungeons and Dragons® web site, www.mudbeast.net, you have decided to charge users who log on to the site. When you charged a \$2 access fee, your web counter showed a demand of 280 "hits" per month. After you lowered the price to \$1.50, activity increased to 560 hits per month.
   a. Construct a linear demand function for your web site and hence obtain the monthly revenue R as a function of the access fee x.
   b. Your Internet provider charges you \$30 per month to maintain your site. Write down the monthly profit P as a function of the access fee x, and hence determine the access fee you should charge to obtain the largest possible monthly profit. What is the largest possible monthly profit?

24. ***T-Shirt Profit*** The two fraternities Sigma Alpha Mu and Ep Sig plan to raise money jointly to benefit homeless people on Long Island. They will sell Starship Troopers T-shirts in the Student Center, but they are not sure how much to charge. Sigma Alpha Mu treasurer Solo recalls that they once sold 400 shirts in a week at \$8 each, but Ep Sig treasurer Justino claims that, based on past experience, they can sell 600 per week if they charge \$4 each.
   a. Based on this anecdotal information, construct a linear demand equation for Starship Troopers T-shirts, and hence obtain the weekly revenue R as a function of the unit price x.
   b. The university administration charges the fraternities \$500 per week for use of the Student Center. Write down the monthly profit P as a function of the unit price x, and hence determine how much the fraternities should charge to obtain the largest possible weekly profit. What is the largest possible weekly profit?

25. ***Web Site Profit*** The latest demand equation for your Dungeons and Dragons® web site, www.mudbeast.net, is given by

$$q = -400x + 1200$$

where q is the number of "hits" per month and x is the access fee you charge. Your Internet provider bills you as follows:

Site maintenance fee: \$20 per month
High-volume access fee: 50¢ per hit

Find the monthly cost as a function of the access fee x. Hence, find the monthly profit as a function of x, and

[3]Fuel efficiency assumes 50% city driving and 50% highway driving. The quadratic models in Exercises 17 and 18 are based on a quadratic regression using data from 18 models of SUV.

determine the access fee you should charge to obtain the largest possible monthly profit. What is the largest possible monthly profit?

26. ***T-Shirt Profit*** The latest demand equation for your Starship Troopers T-shirts is given by

$$q = -40x + 600$$

where q is the number of shirts you can sell in 1 week if you charge $\$x$ per shirt. The Student Council charges you \$400 per week for use of their facilities, and the T-shirts cost you \$5 each. Find the weekly cost as a function of the unit price x. Hence, find the weekly profit as a function of x, and determine the unit price you should charge to obtain the largest possible weekly profit. What is the largest possible weekly profit?

27. ***Nightclub Management*** You have just opened a new nightclub, Russ's Techno Pitstop, but you are unsure how much to charge for the cover charge (entrance fee). One week you charged \$10 cover per guest and averaged 300 guests per night. The next week you charged \$15 per guest and averaged 250 guests per night.
   a. Find a linear demand equation showing the number of guests q per night as a function of the cover charge p.
   b. Find the nightly revenue R as a function of the cover charge p.
   c. The club will provide two free nonalcoholic drinks for each guest, costing \$3 per head. In addition, the nightly overheads (rent, salaries, dancers, DJ, etc.) amount to \$3000. Obtain the cost C as a function of the cover charge p.
   d. Now obtain the profit in terms of the cover charge p, and hence determine the cover charge you should charge for a maximum profit.

28. ***Television Advertising*** You are the sales manager for Montevideo Productions, Inc., and you are planning to review the prices you charge clients for television advertisement development. You currently charge each client an hourly development fee of \$2500. With this pricing structure, the demand, measured by the number of contracts Montevideo signs per month, is 15 contracts. This is down 5 contracts from the figure last year, when your company charged only \$2000.
   a. Construct a linear demand equation giving the number of contracts q as a function of the hourly fee p Montevideo Productions, Inc. charges for development.
   b. On average, Montevideo Productions, Inc. bills for 50 hours of production time on each contract. Give a formula for the total revenue obtained by charging $\$p$ per hour.
   c. The costs to Montevideo Productions, Inc. are estimated as follows:

   Fixed costs: \$120,000 per month
   Variable costs: \$80,000 per contract

   Express Montevideo Productions' monthly cost **(i)** as a function of the number q of contracts and **(ii)** as a function of the hourly production charge p.
   d. Express Montevideo Productions' monthly profit as a function of the hourly development fee p and hence determine the price it should charge to maximize the profit.

Communication and Reasoning Exercises

29. Suppose the graph of revenue as a function of unit price is a parabola that is concave down. What is the significance of the coordinates of the vertex, the x-intercepts, and the y-intercept?

30. Suppose the height of a stone thrown vertically upward is given by a quadratic function of time. What is the significance of the coordinates of the vertex, the (possible) x-intercepts, and the y-intercept?

31. Explain why, if demand is a linear function of unit price p (with negative slope), there must be a *single value of* p that results in a maximum revenue.

32. Explain why, if the average cost of a commodity is given by $y = 0.1x^2 - 4x - 2$, where x is the number of units sold, there is a single choice of x that results in the lowest possible average cost.

33. If the revenue function for a particular commodity is $R(p) = -50p^2 + 60p$, what is the (linear) demand function? Give a reason for your answer.

34. If the revenue function for a particular commodity is $R(p) = -50p^2 + 60p + 50$, can the demand function be linear? What is the associated demand function?

2.2 Exponential Functions and Models

Although the quadratic models we studied in the preceding section can often be used to model nonlinear behavior, there are three common nonlinear phenomena that are best modeled by a different kind of function. The phenomena are population growth, the value of an investment with interest reinvested, and depreciation or decay; the model is the exponential model.

Exponential Function

An **exponential function** has the form

$$f(x) = Ab^x$$

where A and b are constants with b positive. We call b the **base** of the exponential function.

Quick Examples

1. $f(x) = 2^x$ $\quad A = 1, b = 2$
2. $g(x) = 3 \cdot 2^{-4x} = 3(2^{-4})^x$ $\quad A = 3, b = 2^{-4}$

For reference, we repeat below the list of laws of exponents from the algebra review in Appendix A. Which of the laws did we use in reformulating $g(x)$ above?

The Laws of Exponents

If b and c are positive and x and y are any real numbers, then the following laws hold:

Law	Quick example
1. $b^x b^y = b^{x+y}$	$2^3 2^2 = 2^5 = 32$
2. $\frac{b^x}{b^y} = b^{x-y}$	$\frac{4^3}{4^2} = 4^{3-2} = 4^1 = 4$
3. $\frac{1}{b^x} = b^{-x}$	$9^{-0.5} = \frac{1}{9^{0.5}} = \frac{1}{3}$
4. $b^0 = 1$	$(3.3)^0 = 1$
5. $(b^x)^y = b^{xy}$	$(3^2)^2 = 3^4 = 81$
6. $(bc)^x = b^x c^x$	$(4 \cdot 2)^2 = 4^2 2^2 = 64$
7. $\left(\frac{b}{c}\right)^x = \frac{b^x}{c^x}$	$\left(\frac{4}{3}\right)^2 = \frac{4^2}{3^2} = \frac{16}{9}$

example 1

Exponential Function

The function $f(x) = 20(2^x)$ has $A = 20$ and $b = 2$. We can find its value for several values of x as follows:

$$f(3) = 20(2^3) = 20(8) = 160$$

$$f(-3) = 20(2^{-3}) = 20\left(\frac{1}{8}\right) = 2.5 \qquad 2^{-3} = \frac{1}{2^3} = \frac{1}{8}$$

$$f(0) = 20(2^0) = 20(1) = 20 \qquad 2^0 = 1$$

Technology

You can use a spreadsheet, the function evaluator at the web site, or a graphing calculator to automate these calculations. The function evaluator is available via the path

Web site → On-Line Utilities → Function Evaluator & Grapher

The caret ^ is the standard symbol for exponents in most technology platforms. Thus, to compute $f(x)$, type `20*2^x`.

Exponential Functions from the Numerical Point of View

Suppose you invest \$5000 in municipal bonds that yield 10% per year, with the interest reinvested **(compounded)** at the end of each year. The following table tracks the value of your investment over a 5-year period:

Year, t	0	1	2	3	4	5
Value, V	5000	5500	6050	6655	7320.50	8052.55

Notice that the value of your investment each year is 1.1 times its value the previous year. Thus, for every increase of 1 in t, V is multiplied by 1.1.

Question Why is this?
Answer The value of your investment increases by 10% each year. Thus, at the end of each year, your investment is worth 110% of (or 1.10 times) its value at the start of the year.

Question What has this got to do with exponential models?
Answer Let us look once again at your investment: It starts as \$5000 and is multiplied by 1.1 each year. Thus, after t years, the \$5000 has been multiplied by 1.1 a total of t times; that is,

$$V(t) = 5000(1.1)^t \quad \text{This is } Ab^t \text{ with } A = 5000 \text{ and } b = 1.1.$$

Thus, the value of your investment follows an exponential model.

In general, if $f(x) = Ab^x$, then the value of $f(x)$ is *multiplied* by b every time x increases by 1, as shown in the following table:

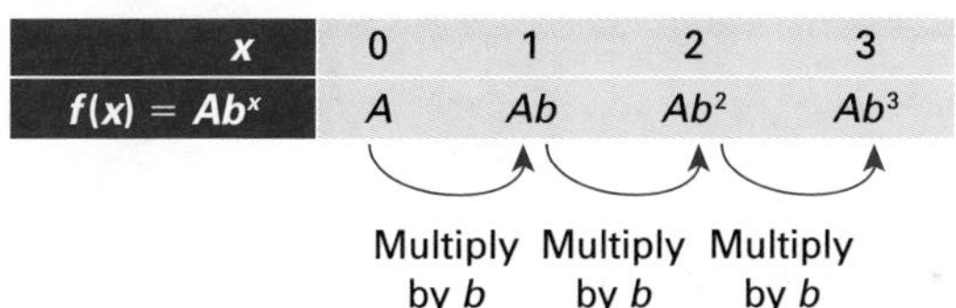

x	0	1	2	3
$f(x) = Ab^x$	A	Ab	Ab^2	Ab^3

example 2

Recognizing Exponential Data Numerically
Some of the values of two functions, f and g, are given in the following table:

x	−2	−1	0	1	2
$f(x)$	−7	−3	1	5	9
$g(x)$	$\frac{2}{9}$	$\frac{2}{3}$	2	6	18

One of these functions is linear and the other is exponential. Which is which?

Solution

Remember that a linear function increases (or decreases) by the same amount every time x increases by 1. The values of f behave this way: Every time x increases by 1, the value of $f(x)$ increases by 4. Therefore, f is a linear function with a *slope* of 4. Since $f(0) = 1$, we see that

$$f(x) = 4x + 1$$

is a linear formula that fits the data.

On the other hand, every time x increases by 1, the value of $g(x)$ is *multiplied* by 3. Since $g(0) = 2$, we find that

$$g(x) = 2(3^x)$$

is an exponential function that fits the data.

Exponential Functions from the Graphical Point of View

We can visualize the data for the two functions f and g given in Example 2 by plotting the points as in Figure 8.

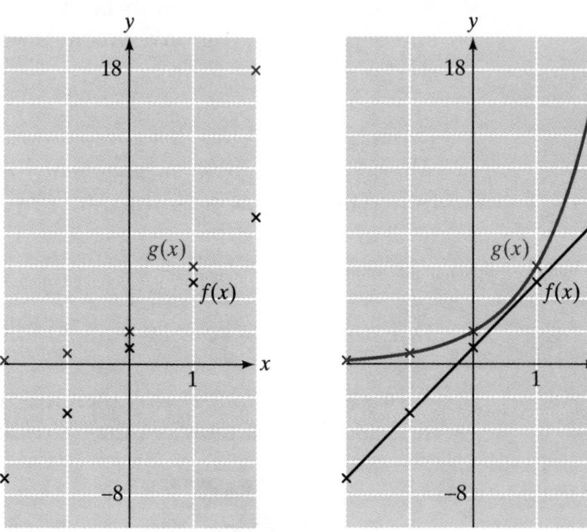

Figure 8

The given values of $f(x)$ clearly lie along a straight line, whereas the values of $g(x)$ lie along a curve, as shown in the graph on the right. Notice that the curve becomes steeper and steeper as the value of x increases. (Also notice how the value of your municipal bond investment increases by larger amounts as the years go by.) This phenomenon is called **exponential growth.**

example 3

Graphing Exponential Functions
On the same set of axes, graph the functions $f(x) = 2^x$ and $g(x) = 2^{-x}$.

Solution

Although we can graph these functions easily using technology, we can also graph them easily by hand. Here is a table of values to start with:

x	-3	-2	-1	0	1	2	3
$f(x) = 2^x$	$\frac{1}{8}$	$\frac{1}{4}$	$\frac{1}{2}$	1	2	4	8
$g(x) = 2^{-x}$	8	4	2	1	$\frac{1}{2}$	$\frac{1}{4}$	$\frac{1}{8}$

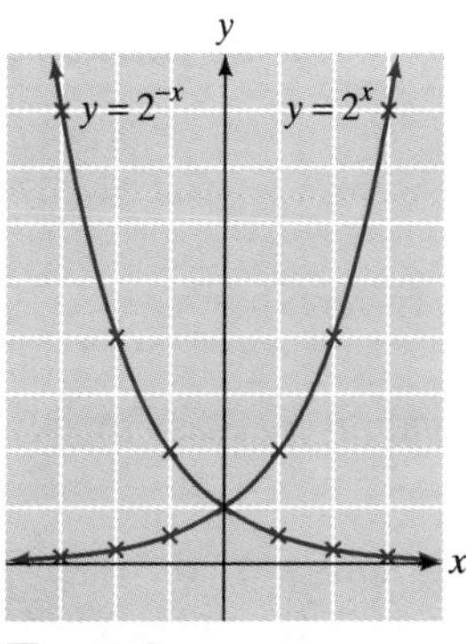

Figure 9

Notice the symmetry in the table: Values of $f(-x)$ correspond to values of $g(x)$. This is a consequence of the fact that

$$f(-x) = 2^{-x} = g(x)$$

and is reflected by a symmetry between their graphs (see Figure 9). Notice also how the curve $y = 2^x$ goes "shooting up" for larger values of x. In fact, the y-coordinate doubles for each increase of 1 in x. [For example, if we had plotted an extra point for $x = 4$, the coordinates would be (4, 16), which is much higher than we've allowed for in the picture.] On the other hand, notice how the curve "levels off" for larger and larger negative values of x as the sequence $2^{-1}, 2^{-2}, 2^{-3}, \ldots$ gets closer and closer to zero. The curve $y = 2^{-x}$ is the mirror image of $y = 2^x$. The curves meet on the y-axis at $y = 1$ because $2^0 = 2^{-0} = 1$.

Graphing Calculator

If you enter

```
Y1 = 2^X
Y2 = 2^(-X)
```

in the "Y=" screen, you can reproduce the graph by setting the window coordinates to match Figure 9: `Xmin` = −3, `Xmax` = 3, `Ymin` = −1, `Ymax` = 8.

Spreadsheet

In Section 1.2 you saw how to set up Excel for graphing a continuous curve. By inserting an extra column (column C below), you can include a second function.

	A	B	C	D	E
1	=$D$1	=2^A1	=2^(-A1)	Xmin	-3
2	=A1+$D$4			Xmax	3
3				Number of points	100
4				Delta x	=(E2-E1)/E3
5					
101					

Graph the data in columns A–C by choosing the scatter plot option with subtype "points connected by lines with no markers."

Web Site

A grapher can be found by following the path

Web site → On-Line Utilities → Function Evaluation & Grapher

More Exponential Graphs

On the same set of axes, sketch the functions $f_1(x) = \left(\frac{1}{2}\right)^x$, $f_2(x) = 1^x$, $f_3(x) = 2^x$, $f_4(x) = 3^x$, and $f_5(x) = 4^x$.

Solution

We can save ourselves some work by noting the following things about these functions:

$y = \left(\frac{1}{2}\right)^x$ is the same as $y = 2^{-x}$, by the laws of exponents, and we have already drawn that curve.

$y = 1^x$ is the same as $y = 1$, a horizontal line with y-intercept 1.

$y = 2^x$ we have already drawn.

$y = 3^x$ looks like $y = 2^x$, except that the y-values triple, rather than double, for each increase of 1 in x.

$y = 4^x$ behaves like $y = 2^x$ and $y = 3^x$, except that the y-values increase by a factor of 4 for each increase of 1 in x.

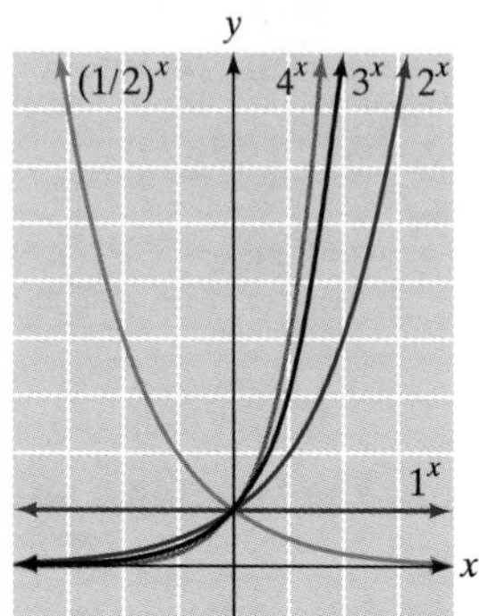

Figure 10

The graphs are shown in Figure 10. Notice that all the graphs pass through the point (0, 1) because of the identity $b^0 = 1$. If we were to sketch the curve $y = \left(\frac{3}{2}\right)^x$, it would have the same general shape as $y = 2^x$, but it would lie between the line $y = 1^x$ and the curve $y = 2^x$ because $\frac{3}{2}$ lies between 1 and 2. Notice also that the graphs *cross* at (0, 1). For example, the curve $y = 3^x$ lies above $y = 2^x$ on the right of the y-axis, but it lies below it to the left of the y-axis. (Why?)

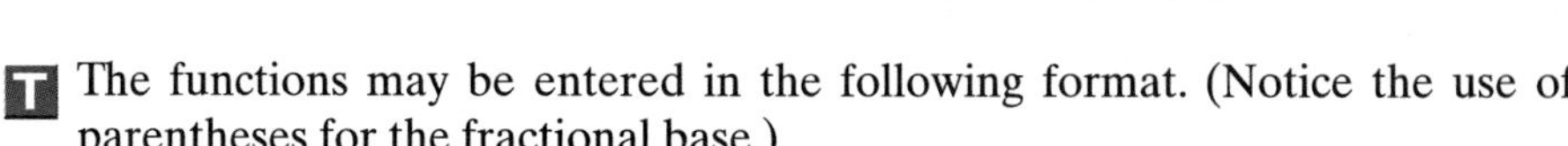

T The functions may be entered in the following format. (Notice the use of parentheses for the fractional base.)

```
(1/2)^X, 4^X, 3^X, 2^X, 1^X
```

Before we go on . . . Also try graphing the function $g(x) = 3 \cdot 2^x$. You can use the format `3*2^X` or `3*(2^X)`. The parentheses are not necessary because of the standard convention on the order of operations.

Finding an Exponential Equation from Data: How to Make an Exponential Model

When we discussed linear functions in Section 1.3, we gave a method for calculating the equation of the line that passes through two given points. In the case of exponential functions, we *could* give a formula for the exponential curve that passes through two given points, but the formula is a little complicated. Instead, we suggest using the method shown in the next example.

example 5

Finding the Exponential Curve Through Two Points

Find an equation of the exponential curve that passes through (1, 6) and (3, 24).

Solution

We are looking for an equation of the form

$$y = Ab^x \qquad (b > 0)$$

that passes through the two points. Substituting the coordinates of the given points, we get

$$6 = Ab^1 \qquad \text{Substituting (1, 6)}$$

$$24 = Ab^3 \qquad \text{Substituting (3, 24)}$$

If we now divide the second equation by the first, the As cancel and we get

$$\frac{24}{6} = \frac{Ab^3}{Ab} = b^2.$$

Thus, $b^2 = 4$

and so $b = 2$ Since $b > 0$

Now that we have b, we can substitute its value into the first equation to obtain

$6 = 2A$ Substitute $b = 2$ in $6 = Ab$.

$A = 3$

We have both constants, $A = 3$ and $b = 2$, so the model is

$$y = 3(2^x)$$

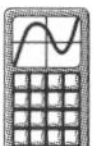

Graphing Calculator

To find the equation of the exponential curve that passes through two given points using a TI-83, enter the coordinates of the data points using STAT EDIT (x values go in L_1, y values in L_2), go to STAT CALC, and select ExpReg. This command can also be used to find the exponential function that best fits a set of data points, similar to the linear regression discussed in Chapter 1.

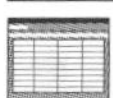

Spreadsheet

In Excel you can find the equation of an exponential curve by adding a "trendline" to a graph of the data points. Once you have created a chart with the data points, choose "Add Trendline . . ." from the Chart menu. Make sure that you select the "exponential" type of trendline. If you want to see the equation of the line, select the "Display equation on chart" option. If you have more than two data points, the trendline shown is the one that best fits the data.

Web Site

At the web site, follow the path

Web site → On-Line Utilities → Simple Regression

Enter the x- and y-coordinates of the two points, and press the "y=a(b^x)" button to find the equation. You can then graph the curve and data points by pressing "graph." This utility can also be used to find the exponential function that best fits a set of data points.

Applications

example 6

Epidemics

In the early stages of the AIDS epidemic, the number of people infected was doubling every 6 months, with an estimated 1.3 million persons infected by January 1985.

a. Assuming an exponential growth model, find an exponential model that predicts the number of people infected t years later.

b. Use the model to estimate the number of people infected by October 1985.

Solution

a. We could use the method in Example 5 or reason as follows: At time $t = 0$ (January 1985) the number of infected people was 1.3 million. Since the number doubled every 6 months, it quadrupled every year. After t years, we therefore need to multiply the original 1.3 million by 4^t, so the model is

$$y = 1.3(4^t) \text{ million people}$$

b. October 1985 corresponds to $t = \frac{9}{12} = 0.75$ (since October is 9 months after January). Substituting this value of t in the model gives

$$y = 1.3(4^t) = 1.3(4^{0.75}) \approx 3.68 \text{ million people infected}$$

Before we go on . . . Doubling every 6 months couldn't continue for very long, and this is borne out by observations. If doubling every 6 months did continue, then in 20 years the number of infected people would be

$$(1.3)4^{20} \text{ million} \approx 1{,}429{,}365{,}116{,}000{,}000{,}000$$

or more than 200 million times greater than the (human) population of the earth! Thus, the exponential model is unreliable for predicting long-term trends.

Epidemiologists use more sophisticated models to measure the spread of epidemics, and these models predict a "leveling-off" phenomenon as the number of cases becomes a significant part of the total population (see *You're the Expert* at the end of this chapter). However, the exponential growth model *is* fairly reliable *in the early stages* of an epidemic.

Exponential functions arise in finance and economics mainly through the idea of **compound interest.** Suppose you invest $500 (the **present** value) in a bank account with an annual yield of 15%, and the interest is reinvested at the end of every year. Let t represent the number of years since you made the initial $500 investment. Each year, the investment is worth 115% (or 1.15 times) its value the previous year. The **future value** A of your investment after t years surely depends on the time t, so we think of A as a function of t. The following table illustrates how we can calculate the future value for several values of t:

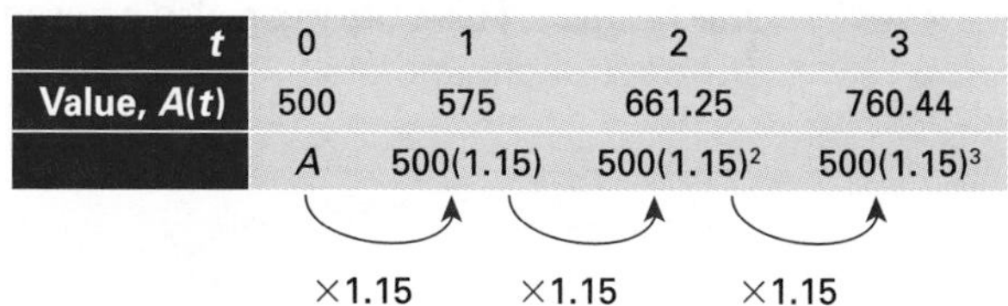

t	0	1	2	3
Value, $A(t)$	500	575	661.25	760.44
	A	$500(1.15)$	$500(1.15)^2$	$500(1.15)^3$

Thus, $A(t) = 500(1.15)^t$. The traditional way to write this formula is

$$A(t) = P(1 + r)^t$$

where P is the present value ($P = 500$) and r is the annual interest rate ($r = 0.15$). If, instead of compounding the interest once a year, we compound it every 3 months (four times a year), the formula becomes

$$A(t) = P\left(1 + \frac{r}{4}\right)^{4t}$$

since we receive a total of $4t$ interest payments at an interest rate of one-fourth the annual rate, $r/4$.

Compound Interest

If an amount **(present value)** P is invested for t years at an annual rate of r, and if the interest is compounded (reinvested) m times per year, then the **future value** A is

$$A(t) = P\left(1 + \frac{r}{m}\right)^{mt}$$

A special case is **interest compounded once a year:**

$$A(t) = P(1 + r)^t$$

Quick Example

If \$2000 is invested for two and a half years in a mutual fund with an annual yield of 12.6% and the earnings are reinvested each month, then $P = 2000$, $r = 0.126$, $m = 12$, and $t = 2.5$, which gives

$$A(2.5) = 2000\left(1 + \frac{0.126}{12}\right)^{12t} = 2000(1.0105)^{12(2.5)} = 2000(1.0105)^{30} = \$2736.02$$

example 7

Investments (Compound Interest)

Consider the scenario in the Quick Example above: you invest \$2000 in a mutual fund with an annual yield of 12.6% and the interest is reinvested each month.

a. Find the associated exponential model.

b. Use the model to estimate the year when the value of your investment will reach \$5000.

Solution

a. We apply the formula

$$A(t) = P\left(1 + \frac{r}{m}\right)^{mt}$$

with $P = 2000$, $r = 0.126$, and $m = 12$. We get

$$A(t) = 2000\left(1 + \frac{0.126}{12}\right)^{12t} = 2000(1.0105)^{12t}$$

This is the exponential model.

Technology

The format for the exponential model $2000(1.0105)^{12t}$ is

```
2000*(1+0.126/12)^(12*X)
```

X represents the variable *t*.

(What would happen if we left out the last set of parentheses?) This can be used as a spreadsheet formula or placed in the "Y=" menu of the TI-83 or the "f(x)=" box in the Function Evaluator & Grapher web page (refer to the discussion in Example 1).

b. We need to find the value of t for which $A(t) = \$5000$, so we need to solve the equation

$$5000 = 2000(1.0105)^{12t}$$

In the next section we will use logarithms to do this algebraically, but we can answer the question now using a spreadsheet, the Function Evaluator & Grapher tool at the web site, or a graphing calculator. Just enter the model and use the table feature to look up the balance at the end of several years. Here is a table you might obtain:

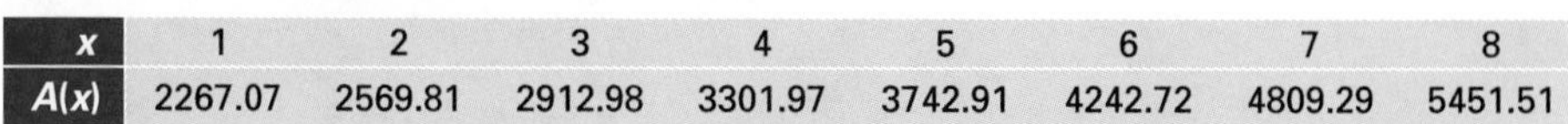

x	1	2	3	4	5	6	7	8
$A(x)$	2267.07	2569.81	2912.98	3301.97	3742.91	4242.72	4809.29	5451.51

Since the balance first exceeds \$5000 in year 8, the answer is $t = 8$ years.

example 8

Radioactive Decay
Carbon-14, an unstable isotope of carbon, decays continuously to nitrogen. The amount of carbon-14 remaining in a sample that originally contained A grams of carbon-14 is given by

$$C(t) = A(0.999879)^t$$

where t is time in years. A recently discovered plant contains 0.50 gram of carbon-14 and is known to be 50,000 years old. How much carbon-14 did the plant originally contain?

Solution
We are given the following information: $C = 0.50$, $A =$ the unknown, and $t = 50{,}000$. Substituting gives

$$0.50 = A(0.999879)^{50{,}000}$$

Solving for A gives

$$A = \frac{0.50}{0.999879^{50{,}000}} \approx 212 \text{ grams}$$

Before we go on . . . The formula we used for A has the form

$$A(t) = \frac{C}{0.999879^t}$$

which gives the original amount of carbon-14 t years ago in terms of the amount C that is left now. A similar formula can be used in finance to find the present value, given the future value.

The Number e and More Applications

In nature we find examples of growth that occurs *continuously,* as though "interest" is being added more often than every second or fraction of a second. To model this, we need to see what happens to our compound interest formula as we let m (the number of times interest is added per year) become extremely large. Something very interesting does happen: We end up with a more compact and elegant formula than we began with. To see why, let's look at a very simple situation.

Suppose we invest \$1 in the bank for 1 year at 100% interest, compounded m times per year. If $m = 1$, then 100% interest is added every year, and so our money doubles at the end of the year. In general, the accumulated capital at the end of the year is

$$A = 1\left(1 + \frac{1}{m}\right)^m = \left(1 + \frac{1}{m}\right)^m$$

Now, we are interested in what A becomes for large values of m. We can make a chart that shows how this quantity behaves as m increases.

m	1	10	100	1000	10,000	100,000	1,000,000	10,000,000
$\left(1 + \frac{1}{m}\right)^m$	2	2.59374246	2.70481383	2.71692393	2.71814593	2.71826824	2.71828047	2.71828169

Something interesting *does* seem to be happening! The numbers appear to be getting closer and closer to a specific value. In mathematical terminology, we say that the numbers *converge* to a fixed number, 2.71828 . . . . This number is one of the most important in mathematics and is referred to as e. The number e is irrational, just as the more familiar number π is, so we cannot write down its exact numerical value. To 20 decimal places, $e = 2.71828182845904523536 \ldots$.

We now say that, if \$1 is invested for 1 year at 100% interest **compounded continuously,** the accumulated money at the end of that year will amount to $\$e = \$2.72$ (to the nearest cent). But what about the following more general question?

Question Suppose we invest an amount P for t years at an interest rate of r, compounded continuously. What will be the accumulated amount A at the end of that period?

Answer In the special case above, we took the compound interest formula and let m get larger and larger. We do the same again more generally, after a little preliminary work with the algebra of exponentials.

$$A = P\left(1 + \frac{r}{m}\right)^{mt}$$

$$= P\left(1 + \frac{1}{(m/r)}\right)^{mt} \qquad \text{Substituting } \frac{r}{m} = \frac{1}{(m/r)}$$

$$= P\left(1 + \frac{1}{(m/r)}\right)^{(m/r)rt} \qquad \text{Substituting } m = \left(\frac{m}{r}\right)r$$

$$= P\left[\left(1 + \frac{1}{(m/r)}\right)^{(m/r)}\right]^{rt} \qquad \text{Using the rule } a^{bc} = (a^b)^c$$

For continuous compounding of interest, we let m, and hence m/r, get very large. This affects only the term in brackets, which converges to e, and we get the formula shown next.

Continuous Compounding and Continuous Growth

If $\$P$ is invested at an annual interest rate r compounded continuously, then the accumulated amount after t years is

$$A(t) = Pe^{rt}$$

Quick Examples

1. If \$100 is invested in an account that bears 15% interest compounded continuously, then, at the end of 10 years, the investment will be worth

$$A(10) = 100e^{(0.15)(10)} = \$448.17$$

2. If \$1 is invested in an account that bears 100% interest compounded continuously, then, at the end of x years, the investment will be worth

$$A(x) = e^x$$

The graph of $A(x) = e^x$ is shown in the figure.

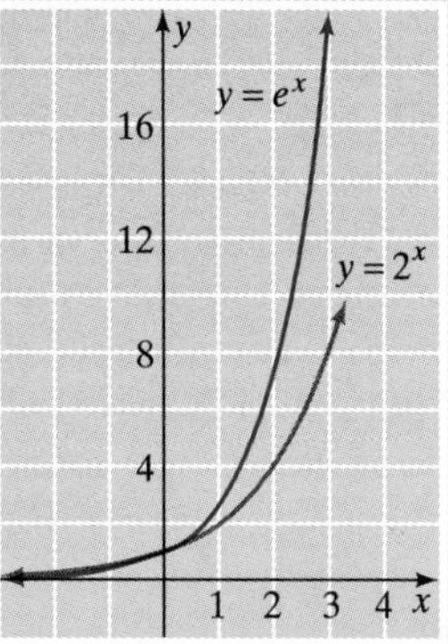

Technology format: `e^(X)`

(Note that the scales on the x- and y-axes are not the same. If they were, the graph would appear much steeper.)

Note If we write $A(t) = P(e^r)^t$ we see that $A(t)$ is an exponential function of t, where the base is $b = e^r$, so we have really not introduced a new kind of function. As we shall see, the exponential function with base e is the simplest from the point of view of calculus, and this is the real importance of e.

example 9

Continuous Compounding

a. You invest \$10,000 at Fastrack Savings & Loan, which pays 6% compounded continuously. Express the balance in your account as a function of the number of years n, and calculate the amount of money you will have after 5 years.

b. Your friend has just invested \$10,000 in Constant Growth Funds, which contains stocks that are continuously declining at a rate of 6% per year. How much will her investment be worth in 5 years?

Solution

a. We use the continuous growth formula with $P = 10{,}000$, $r = 0.06$, and t variable, getting

$$A(t) = Pe^{rt}$$

$$= 10{,}000e^{0.06t} \qquad \text{Technology format: } \texttt{10000e^(0.06*X)}$$

In 5 years,

$$A(5) = 10{,}000e^{0.06(5)} \qquad \text{Technology format: } \texttt{10000e^(0.06*5)}$$

$$= 10{,}000e^{0.3} = \$13{,}498.59$$

b. Since the investment is depreciating, we use a negative value for r and take $P = 10{,}000$, $r = -0.06$, and $t = 5$, getting

$$A(t) = Pe^{rt}$$

$$A(5) = 10{,}000e^{-0.06(5)} = 10{,}000e^{-0.3} = \$7{,}408.18$$

Before we go on . . .

Question How does continuous compounding compare with monthly compounding?

Answer To repeat the calculation in part (a) using monthly compounding instead of continuous compounding, we use the compound interest formula with $P = 10{,}000$, $r = 0.06$, $m = 12$, and $t = 5$, and find

$$A(5) = 10{,}000\left(1 + \frac{0.06}{12}\right)^{60} = \$13{,}488.50$$

Thus, continuous compounding earns approximately \$10 more than monthly compounding. On a \$10,000 investment after 5 years, this is little to get excited about.

2.2 exercises

For each function in Exercises 1–12, complete a table like the following:

x	−3	−2	−1	0	1	2	3
$f(x)$							

1. $f(x) = 4^x$ **2.** $f(x) = 3^x$

3. $f(x) = 3^{-x}$ **4.** $f(x) = 4^{-x}$

5. $g(x) = 2(2^x)$ **6.** $g(x) = 2(3^x)$

7. $h(x) = -3(2^{-x})$ **8.** $h(x) = -2(3^{-x})$

9. $r(x) = 2^x - 1$ **10.** $r(x) = 2^{-x} + 1$

11. $s(x) = 2^{x-1}$ **12.** $s(x) = 2^{1-x}$

Using a chart of values, graph each of the functions in Exercises 13–18 by hand. (Use $-3 \le x \le 3$.)

13. $f(x) = 3^{-x}$ **14.** $f(x) = 4^{-x}$

15. $g(x) = 2(2^x)$ **16.** $g(x) = 2(3^x)$

17. $h(x) = -3(2^{-x})$ **18.** $h(x) = -2(3^{-x})$

T *For each function in Exercises 19–24, complete a table like the following (round each answer to four significant digits):*

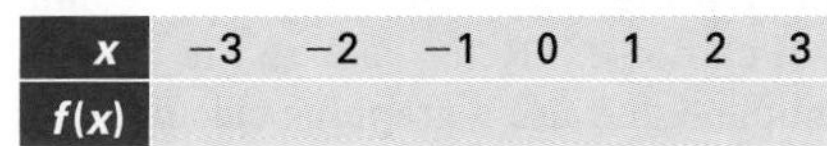

x	−3	−2	−1	0	1	2	3
$f(x)$							

19. $f(x) = e^{-2x}$ **20.** $g(x) = e^{x/5}$

21. $h(x) = 1.01(2.02^{-4x})$ **22.** $h(x) = 3.42(3^{-x/5})$

23. $r(x) = 50\left(1 + \frac{1}{3.2}\right)^{2x}$ **24.** $r(x) = 0.043\left(4.5 - \frac{5}{1.2}\right)^{-x}$

For Exercises 25–32, model the data using an exponential function $f(x) = Ab^x$.

25.

x	0	1	2
$f(x)$	500	250	125

26.

x	0	1	2
$f(x)$	500	1000	2000

27.

x	0	1	2
$f(x)$	10	30	90

28.

x	0	1	2
$f(x)$	90	30	10

29.

x	0	1	2
$f(x)$	500	225	101.25

30.

x	0	1	2
$f(x)$	5	3	1.8

31.

x	1	2
$f(x)$	−110	−121

32.

x	1	2
$f(x)$	−41	−42.025

T *In Exercises 33–40, supply a valid input format for the given function. For example,* $3\dfrac{1-x}{2^x}$ *can be represented by either* `3*(1-x)/2^x` *or* `(3*(1-x))/2^x`.

33. 2^{x-1} **34.** 2^{-4x} **35.** $\dfrac{2}{1-2^{-4x}}$ **36.** $\dfrac{2^{3-x}}{1-2^x}$

37. $\dfrac{(3+x)^{3x}}{x+1}$ **38.** $\dfrac{20.3^{3x}}{1+20.3^{2x}}$ **39.** $2e^{(1+x)/x}$ **40.** $\dfrac{2e^{2/x}}{x}$

T *On the same set of axes, graph the pairs of functions in Exercises 41–48 with $-3 \le x \le 3$. Identify which graph corresponds to which function.*

41. $f_1(x) = 1.6^x, f_2(x) = 1.8^x$ **42.** $f_1(x) = 2.2^x, f_2(x) = 2.5^x$

43. $f_1(x) = 300(1.1^x), f_2(x) = 300(1.1^{2x})$

44. $f_1(x) = 100(1.01^{2x}), f_2(x) = 100(1.01^{3x})$

45. $f_1(x) = 2.5^{1.02x}, f_2(x) = e^{1.02x}$

46. $f_1(x) = 2.5^{-1.02x}, f_2(x) = e^{-1.02x}$

47. $f_1(x) = 1000(1.045^{-3x}), f_2(x) = 1000(1.045^{3x})$

48. $f_1(x) = 1202(1.034^{-3x}), f_2(x) = 1202(1.034^{3x})$

Find equations for exponential functions that pass through the pairs of points given in Exercises 49–54.

49. (1, 3) and (3, 6) **50.** (1, 2) and (4, 6)

51. (−1, 5) and (4, 20) **52.** (−1, 7) and (3, 14)

53. (2, 3) and (6, 2) **54.** (−1, 2) and (3, 1)

Applications

55. *Investments* As of October 18, 1993, the annual percentage yield on government bonds in Germany was 5.97%. Assuming that this rate of return continued until 1998 and that interest was reinvested annually, how much would a \$5000 bond purchased at the start of 1993 be worth at the end of 1998? (Answer to the nearest \$1.)
Source: S. G. Wartburg & Company, as quoted in the *New York Times*, October 18, 1993.

56. *Investments* Refer back to the yield on government bonds in Germany in Exercise 55. How much should one have invested at the start of 1993 to accumulate a total of \$500,000 by the end of 1998? (Answer to the nearest \$1.)

57. *Carbon Dating* A fossil originally contained 104 grams of carbon-14. Refer back to the formula for $C(t)$ in Example 8 and use a graphing calculator or the web tools to estimate the amount of carbon-14 left in the sample after 5000, 10,000, 15,000, . . . , and 35,000 years.

58. *Carbon Dating* A fossil presently contains 4.06 grams of carbon-14. Refer back to the formula for $A(t)$ in Example 8 and use a graphing calculator or the web tools to estimate the amount of carbon-14 left in the sample 5000, 10,000, 15,000, . . . , and 35,000 years ago.

59. *Carbon Dating* A fossil presently contains 4.06 grams of carbon-14. It is estimated that the fossil originally contained 46 grams of carbon-14. Use the method in Exercise 58 to estimate the age of the sample to the nearest 5000 years.

60. *Carbon Dating* A fossil presently contains 2.8 grams of carbon-14. It is estimated that the fossil originally contained 104 grams of carbon-14. Use the method in Exercise 57 to estimate the age of the sample to the nearest 5000 years.

61. *Bacteria* A bacteria culture starts with 1000 bacteria and doubles in size every 3 hours. Find an exponential model for the size of the culture as a function of time t in hours, and use the model to predict how many bacteria there will be after 2 days.

62. *Bacteria* A bacteria culture starts with 1000 bacteria. Two hours later there are 1500 bacteria. Find an exponential model for the size of the culture as a function of time t in hours, and use the model to predict how many bacteria there will be after 2 days.

63. *Aspirin* Soon after taking an aspirin, a patient has absorbed 300 mg of the drug. If the amount of aspirin in the bloodstream decays exponentially, with half being removed every 2 hours, find the amount of aspirin in the bloodstream after 5 hours.

64. ***Alcohol*** After several drinks a person has a blood alcohol level of 0.20 mg/dL. If the amount of alcohol in the blood decays exponentially, with one-fourth being removed every hour, find the person's blood alcohol level after 4 hours.

65. ***Profit*** After-tax profits of South African Breweries (SAB) were \$350 million in 1991 and \$600 million in 1997. Find **a.** a linear model and **b.** an exponential model for SAB's profit P as a function of time t since 1991. Which model better fits the data given in the table?

t	0 (1991)	1	2	3
Profit (\$ millions)	350	360	375	400
t	4	5	6 (1997)	
Profit (\$ millions)	500	580	600	

Source: Company reports/Bloomberg Financial Markets/*New York Times,* August 27, 1997, p. D4.

66. ***Revenue*** Revenues of South African Breweries (SAB) were \$6 billion in 1992 and \$8.4 billion in 1997. Find **a.** a linear model and **b.** an exponential model for SAB's revenue R as a function of time t since 1992. Which model better fits the data given in the table?

t	0 (1992)	1	2	3	4	5 (1997)
Revenue (\$ billions)	6	7	7	7.8	8	8.4

Source: Company reports/Bloomberg Financial Markets/*New York Times,* August 27, 1997, p. D4.

67. ***Frogs*** Frogs in Nassau County have been breeding like flies! Each year the pledge class of Epsilon Delta is instructed by the brothers to tag all the frogs residing on the ESU campus (Nassau County Branch) as an educational exercise. Two years ago the students managed to tag all 50,000 of them (with little Epsilon Delta Fraternity tags). This year's pledge class discovered that last year's tags had all fallen off, and they wound up tagging a total of 75,000 frogs.

a. Find an exponential model for the frog population.

b. Assuming exponential population growth, and that all this year's tags have fallen off, how many tags should Epsilon Delta order for next year's pledge class?

68. ***Flies*** Flies in Suffolk County have been breeding like frogs! Three years ago the Health Commission caught 4000 flies in 1 hour in a trap. This year it caught 7000 flies in 1 hour.

a. Find an exponential model for the fly population.

b. Assuming exponential population growth, how many flies should the commission expect to catch next year?

69. ***U.S. Population*** The U.S. population was 180 million in 1960 and 250 million in 1990. Assuming exponential population growth, what will the population be in the year 2010?

Source: The World Almanac and Book of Facts 1992, Pharos Books, New York.

70. ***World Population*** World population was estimated at 1.6 billion people in 1900 and 5.3 billion people in 1990. Assuming exponential growth, when were there only two people in the world? Comment on your answer.

Source: The World Almanac and Book of Facts 1992, Pharos Books, New York.

71. ***Investments***

a. You have \$100 invested in Quarterly Savings and Loan Company, which pays 5% interest compounded quarterly. By how much will your investment have grown after 4 years?

b. Continuity Continental Corp. advertises that it can improve on Quarterly's offer by giving the same interest rate but compounded continuously. How much more would you have earned had you invested with Continuity?

72. ***Investments*** Rock Solid Bank & Trust is offering a CD that pays 4% compounded continuously. How much interest would a \$1000 deposit earn over 10 years?

73. ***Savings*** SemiSolid Savings & Loan is offering a savings account that pays $3\frac{1}{2}$% interest compounded continuously. How much interest would a deposit of \$2000 earn over 10 years?

74. ***Global Warming*** The most abundant greenhouse gas is carbon dioxide. According to the United Nations "worst-case scenario" prediction, the amount of carbon dioxide in the atmosphere (in parts of volume per million) can be approximated by

$$C(t) \approx 277e^{0.00353t} \qquad (0 \le t \le 350)$$

where t is time in years since 1750.[1]

a. Use the model to estimate the amount of carbon dioxide in the atmosphere in 1950, 2000, 2050, and 2100.

b. According to the model, when, to the nearest decade, will the level surpass 700 parts per million?

Source: Tom Boden/Oak Ridge National Laboratory, Scripps Institute of Oceanography/University of California, International Panel on Climate Change/*New York Times,* December 1, 1997, p. F1.

75. ***Global Warming*** Repeat Exercise 74 using the United Nations "midrange" scenario prediction:

$$C(t) \approx 277e^{0.00267t} \qquad (0 \le t \le 350)$$

where t is time in years since 1750.

[1] Obtained using exponential regression based on the 1750 figure and the 2100 UN prediction.

T ***Employment*** *Exercises 76 and 77 are based on the following chart, which shows the number of people employed by Northrop Grumman in Long Island, New York, from 1987 to 1995.*[2]

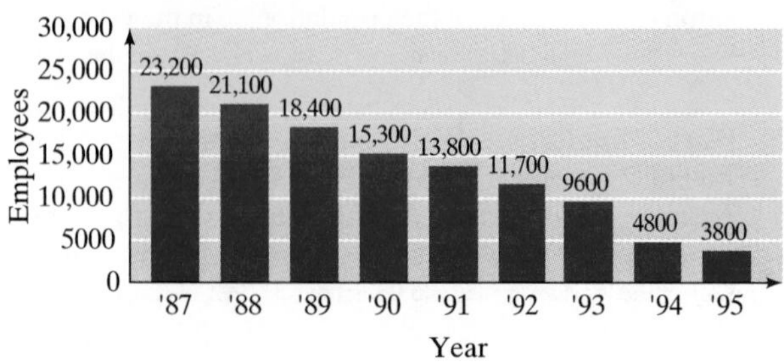

Source: Grumman Corp., Northrop Grumman Corp./*Newsday*, September 23, 1994, p. A5.

[2]The 1987–1994 figures show employment at Grumman Corp. before it was acquired by Northrop Corp. in 1994. The 1994 and 1995 figures are company projections.

The method of exponential regression gives the following model:[3]

$$P = 28{,}131.7e^{-0.2208t}$$

where t is the number of years since 1987.

76. According to the regression model, when will employment by Grumman Northrop Long Island drop to 1000 employees? (Give the answer to the nearest year.)

77. According to the regression model, when will employment by Grumman Northrop Long Island drop to 500 employees? (Give the answer to the nearest year.)

[3]This method is described at the web site and also in the chapter on the calculus of several variables.

Communication and Reasoning Exercises

78. Which of the following three functions will be largest for large values of x?
a. $f(x) = x^2$ **b.** $r(x) = 2^x$ **c.** $h(x) = x^{10}$

79. Which of the following three functions will be smallest for large values of x?
a. $f(x) = x^{-2}$ **b.** $r(x) = 2^{-x}$ **c.** $h(x) = x^{-10}$

80. What limitations apply to using an exponential function to model growth in real-life situations? Illustrate your answer with an example.

81. Describe two real-life situations in which a linear model would be more appropriate than an exponential model and two situations in which an exponential model would be more appropriate than a linear model.

82. Explain in words why 5% per year compounded continuously yields more interest than 5% per year compounded monthly.

83. Your local banker tells you that the reason his bank doesn't compound interest continuously is that it would be too demanding of computer resources because the computer would need to spend a great deal of time keeping all accounts updated. Comment on his reasoning.

84. Your other local banker tells you that the reason *her* bank doesn't offer continuously compounded interest is that it is equivalent to offering a fractionally higher interest rate compounded daily. Comment on her reasoning.

2.3 Logarithmic Functions and Models

Logarithms were invented by John Napier (1550–1617) in the late 16th century as a means of aiding calculation. His invention made possible the prodigious hand calculations of astronomer Johannes Kepler (1571–1630), who was the first to describe accurately the orbits and motions of the planets. Today computers and electronic calculators have done away with that use of logarithms, but many other uses remain. In particular, the logarithm is a key tool for manipulating exponential models.

The equation

$$2^3 = 8$$

tells us that the power to which we need to raise 2 in order to get 8 is 3. We abbreviate the phrase "the power to which we need to raise 2 in order to get 8" as "$\log_2 8$." Thus, another way of writing the equation $2^3 = 8$ is

$$\log_2 8 = 3 \qquad \text{The power to which we need to raise 2 in order to get 8 is 3.}$$

This is read "the base 2 logarithm of 8 is 3" or "the log, base 2, of 8 is 3."

Here is the general definition.

Base *b* Logarithm

The **base *b* logarithm of *x***, $\log_b x$, is the power to which we need to raise b in order to get x. Symbolically,

$$\log_b x = y \qquad \text{means} \qquad b^y = x$$

Logarithmic form — *Exponential form*

The number $\log_b x$ is defined only if b and x are both positive, and $b \neq 1$.

Quick Examples

The following table lists some exponential equations and their equivalent logarithmic forms:

Exponential form	$10^3 = 1000$	$4^2 = 16$	$3^3 = 27$	$5^1 = 5$
Logarithmic form	$\log_{10}1000 = 3$	$\log_4 16 = 2$	$\log_3 27 = 3$	$\log_5 5 = 1$
Exponential form	$7^0 = 1$	$4^{-2} = \frac{1}{16}$	$25^{1/2} = 5$	
Logarithmic form	$\log_7 1 = 0$	$\log_4 \frac{1}{16} = -2$	$\log_{25} 5 = \frac{1}{2}$	

example 1

Calculating Logarithms by Hand

Calculate the following logarithms:

a. $\log_3 9$ **b.** $\log_{10} 10{,}000$ **c.** $\log_9 1$ **d.** $\log_3 \frac{1}{27}$

Solution

a. In words:

$$\log_3 9 = \text{the power to which we need to raise 3 in order to get 9}$$

Since $3^2 = 9$, this power is 2, so

$$\log_3 9 = 2$$

Algebraically:

$\log_3 9 = \square$ — Problem in logarithmic form

$3^{\square} = 9$ — Problem in exponential form

$3^{\boxed{2}} = 9$ — Solve the problem.

$\log_3 9 = 2$ — Answer in logarithmic form

b. In words:

$$\log_{10} 10{,}000 = \text{the power to which we need to raise 10 in order to get 10,000}$$

Since $10^4 = 10{,}000$, this power is 4, so

$$\log_{10} 10{,}000 = 4$$

c. To what power must we raise 9 to get 1? We know that $9^0 = 1$, so

$$\log_9 1 = 0$$

Similarly,

$$\log_b 1 = 0$$

for every positive number b other than 1. ($\log_1$ is not defined; why?)

d. To what power must we raise 3 to get $\frac{1}{27}$? Since $3^3 = 27$, we see that $3^{-3} = \frac{1}{27}$, so

$$\log_3 \frac{1}{27} = -3$$

Logarithms cannot always be calculated by hand. For instance, to calculate $\log_{10} 2.452$, we need to answer the question, To what power must we raise 10 to get 2.452? The answer, 0.38952 . . . , cannot be found easily by hand but can be obtained easily (if approximately) on a calculator.

Logarithms on a Calculator

The following are standard abbreviations:

Base 10: $\log_{10} x = \log x$ Common logarithm

Base *e*: $\log_e x = \ln x$ Natural logarithm

Thus, to obtain, say, $\log_{10} 5$ on a traditional scientific calculator, press [5] followed by [LOG]. (You should get 0.6989 . . . .) On a graphing calculator like the TI-83 or for the Function Evaluator & Grapher at the web site, use

`log(5)` On the TI-83, use the "log" key.

In Excel, use the formula

`=LOG(5)`

For bases other than 10, we use the next formula.

Change-of-Base Formula

$$\log_b a = \frac{\log a}{\log b} = \frac{\ln a}{\ln b}$$

For example, to find $\log_{3.45} 2.261$, we divide log 2.261 by log 3.45, getting 0.6588 (to four significant digits). We get the same answer by dividing ln 2.261 by ln 3.45. (Try it.)[1]

---

[1]Here is a quick explanation of why this formula works: To calculate $\log_a b$, we ask, To what power must we raise a to get b? To check the formula, we try using $\log b/\log a$ as the exponent:

$$a^{\log b/\log a} = (10^{\log a})^{\log b/\log a} \quad \text{since } a = 10^{\log a}$$
$$= 10^{\log b} = b$$

so this exponent works!

Quick Examples

1. $\log_{11}9 = \dfrac{\log 9}{\log 11} \approx 0.91631$ Technology: `log(9)/log(11)` or `ln(9)/ln(11)`

2. $\log_{3.2}\left(\dfrac{1.42}{3.4}\right) \approx -0.75065$ Technology: `log(1.42/3.4)/log(3.2)`

example 2

Graphs of Logarithmic Functions

a. Sketch the graph of $f(x) = \log_2 x$ by hand.

b. Use graphing technology to compare the graph in part (a) with the graphs of $\log_b x$ for $b = \frac{1}{4}, \frac{1}{2}$, and 4.

Solution

a. To sketch the graph of $f(x) = \log_2 x$ by hand, we begin, as usual, with a table of values. Since $\log_2 x$ is not defined when $x = 0$, we choose several values of x close to zero and also some larger values, all chosen so that their logarithms are easy to compute:

x	$\frac{1}{8}$	$\frac{1}{4}$	$\frac{1}{2}$	1	2	4	8
$f(x) = \log_2 x$	−3	−2	−1	0	1	2	3

Graphing these points and joining them by a smooth curve give us Figure 11.

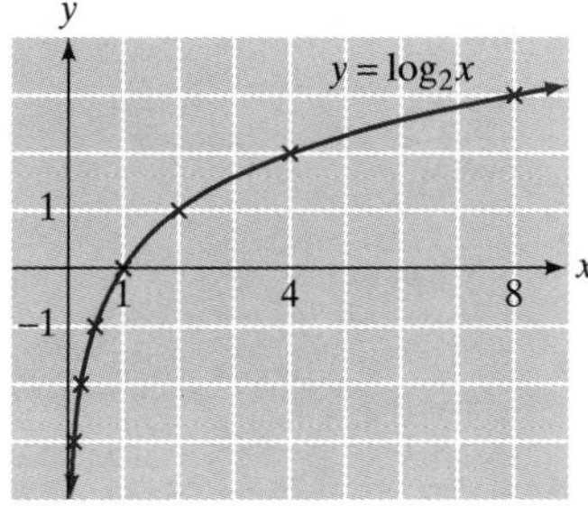

Figure 11

b. To enter the logarithmic functions in a graphing utility, we use the change-of-base formula:

$$\log_b x = \frac{\log x}{\log b}$$

For example, in a TI-83 we would enter the functions

```
Y1 = log(X)/log(0.25)
Y2 = log(X)/log(0.5)
Y3 = log(X)/log(2)
Y4 = log(X)/log(4)
```

Figure 12 shows the resulting graphs.

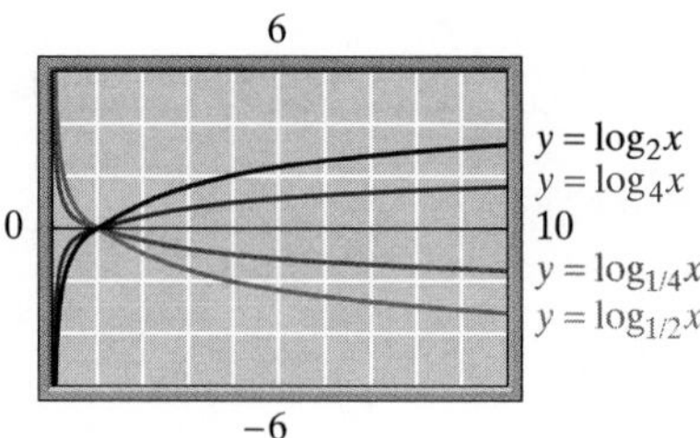

Figure 12

Before we go on . . . Notice that all the graphs in Figure 12 pass through the point (1, 0). (Why?) Notice further that the graphs of the logarithmic functions with bases less than 1 are upside-down versions of the others. Finally, how are these graphs related to the graphs of exponential functions?

The next table lists some important properties of logarithms. These properties follow from the laws from exponents.

Algebraic Properties of Logarithms

The following identities hold for all positive bases $a \neq 1$ and $b \neq 1$, all positive numbers x and y, and every real number r.

Identity	Quick example
1. $\log_b(xy) = \log_b x + \log_b y$	$\log_2 16 = \log_2 8 + \log_2 2$
2. $\log_b\left(\frac{x}{y}\right) = \log_b x - \log_b y$	$\log_2\left(\frac{5}{3}\right) = \log_2 5 - \log_2 3$
3. $\log_b(x^r) = r \log_b x$	$\log_2(6^5) = 5 \log_2 6$
4. $\log_b b = 1$; $\log_b 1 = 0$	$\log_2 2 = 1$; $\log_{11} 1 = 0$
5. $\log_b\left(\frac{1}{x}\right) = -\log_b x$	$\log_2\left(\frac{1}{3}\right) = -\log_2 3$
6. $\log_b x = \frac{\log_a x}{\log_a b}$	$\log_2 5 = \frac{\log_{10} 5}{\log_{10} 2} = \frac{\log 5}{\log 2}$

Optional Internet Topic: Using and Deriving Algebraic Properties of Logarithms

At the web site you will find a list of these properties you can print out, some exercises on using the properties, and a discussion on where they come from.

Applications

example 3

Investments: How Long?

Thirty-year U.S. Treasury bonds are yielding an average of 5.2% per year.[2] At that interest rate, how long will it take a \$1000 investment to be worth \$1500 if the interest is compounded monthly?

[2]Bond rates for the week ending September 18, 1998, are from *Federal Reserve Statistical Release H.15: Selected Interest Rates,* available on the World Wide Web through the White House's Economics Statistics Briefing Room, http://www.whitehouse.gov/fsbr/esbr.html.

Solution

Substituting $A = 1500, P = 1000, r = 0.052$, and $m = 12$ in the compound interest equation gives

$$1500 = 1000\left(1 + \frac{0.052}{12}\right)^{12t}$$

or $$1500 = 1000(1.004333)^{12t}$$

and we must solve for t. We first divide both sides by 1000, getting an equation in exponential form:

$$1.5 = 1.004333^{12t}$$

In logarithmic form, this becomes

$$12t = \log_{1.004333}(1.5)$$

We can now solve for t:

$$t = \frac{\log_{1.004333}(1.5)}{12} \approx 7.8 \text{ years}$$

Technology: `(log(1.5)/log(1+0.052/12))/12`

Thus, it will take approximately 7.8 years for a \$1000 investment to be worth \$1500.

---

example 4

Half-Life

a. The weight of carbon-14 that remains in a sample that originally weighed A grams is given by

$$C(t) = A(0.999879)^t$$

where t is time in years. Find the **half-life,** the time it takes half of the carbon-14 in a sample to decay.

b. Another radioactive material has a half-life of 7000 years. How long will it take for 99.95% of the substance in a sample to decay?

Solution

a. We want to find the value of t for which $C(t)$ = the weight of undecayed carbon-14 left = half the original weight = $0.5A$. Substituting, we get

$$0.5A = A(0.999879)^t$$

Dividing both sides by A gives

$$0.5 = 0.999879^t$$ Exponential form

Thus, $$t = \log_{0.999879}0.5 \approx 5728 \text{ years}$$ Logarithmic form

Technology: `log(0.5)/log(0.999879)`

b. This time we are given the half-life, which we can use to find the exponential model. Since half of the sample decays in 7000 years, a sample of A grams now ($t = 0$) will decay to $0.5A$ grams in 7000 years ($t = 7000$). This gives two data points: $(0, A)$ and $(7000, 0.5A)$. We can use the method in Example 5 of the preceding section to find

$$y = A(0.5)^{t/7000}$$

Now that we have the model, we can answer the question by substituting

y = amount of *undecayed* material left = 0.05% of the original amount = $0.0005A$.

We have

$$0.0005A = A(0.5)^{t/7000}$$

Dividing both sides by A gives

$$0.0005 = (0.5)^{t/7000}$$

so

$$\frac{t}{7000} = \log_{0.5}(0.0005)$$

$$t = 7000 \log_{0.5}(0.0005) \approx 76{,}760 \text{ years}$$

2.3 exercises

In Exercises 1–4, complete the following tables.

1.

Exponential form	$10^4 = 10{,}000$	$4^2 = 16$	$3^3 = 27$
Logarithmic form			
Exponential form	$5^1 = 5$	$7^0 = 1$	$4^{-2} = \frac{1}{16}$
Logarithmic form			

2.

Exponential form	$4^3 = 64$	$10^{-1} = 0.1$	$2^8 = 256$
Logarithmic form			
Exponential form	$5^0 = 1$	$(0.5)^2 = 0.25$	$6^{-2} = \frac{1}{36}$
Logarithmic form			

3.

Exponential form			
Logarithmic form	$\log_{0.5}0.25 = 2$	$\log_5 1 = 0$	$\log_{10}0.1 = -1$
Exponential form			
Logarithmic form	$\log_4 64 = 3$	$\log_2 256 = 8$	$\log_2 \frac{1}{4} = -2$

4.

Exponential form			
Logarithmic form	$\log_5 5 = 1$	$\log_4 \frac{1}{16} = -2$	$\log_4 16 = 2$
Exponential form			
Logarithmic form	$\log_{10}10{,}000 = 4$	$\log_3 27 = 3$	$\log_7 1 = 0$

Applications

5. ***Investments*** How long will it take a \$500 investment to be worth \$700 if it is continuously compounded at 10% per year? (Give the answer to two decimal places.)

6. ***Investments*** How long will it take a \$500 investment to be worth \$700 if it is continuously compounded at 15% per year? (Give the answer to two decimal places.)

7. ***Investments*** How long, to the nearest year, will it take an investment to triple if it is continuously compounded at 10% per year?

8. ***Investments*** How long, to the nearest year, will it take me to become a millionaire if I invest \$1000 at 10% interest compounded continuously?

9. ***Stocks*** Professor Stefan Schwartzenegger invests $1000 in Tarnished Trade (TT) stock, which is depreciating continuously at a rate of 20% per year. Luckily for Professor Schwartzenegger, TT declares bankruptcy at the instant his investment has fallen to $666. How long after the initial investment did this occur?

10. ***Investments***

a. The Hofstra Choral Society's investment of $100,000 in Tarnished Teak Enterprises is depreciating continuously at 16% per year. At this rate, how long, to the nearest year, will it take for the investment to be worth $1?

b. After six agonizing months watching its savings dwindle, the Choral Society pulls out what is left of its investment and buys Sammy Solid Trust Bonds, which earn a steady 6% interest, compounded semiannually. How long will it take the Choral Society to recover its losses?

11. ***Carbon Dating*** The amount of carbon-14 remaining in a sample that weighs A grams is given by

$$C(t) = A(0.999879)^t$$

where t is time in years. If tests on a fossilized skull reveal that 99.95% of the carbon-14 has decayed, how old is the skull?

12. ***Carbon Dating*** Refer back to Exercise 11. How old is a fossil in which only 30% of the carbon-14 has decayed?

Government Bonds *Exercises 13–22 are based on the following table, which lists annual percentage yields on government bonds in several countries as of October 18, 1993:*

Country	U.S.	Japan	Germany
Yield	5.16%	3.86%	5.97%
Country	Britain	Canada	Mexico
Yield	6.81%	6.58%	13.7%

Source: Salamon Brothers, Mexican Government, S. G. Wartburg & Company, J. P. Morgan Global Research, as quoted in the *New York Times,* October 18, 1993.

13. Assuming that you invest $10,000 in U.S. government bonds, how long (to the nearest year) must you wait before your investment is worth $15,000 if the interest is compounded annually?

14. Assuming that you invest $10,000 in Japanese government bonds, how long (to the nearest year) must you wait before your investment is worth $15,000 if the interest is compounded annually?

15. If you invest $10,400 in German government bonds and the interest is compounded monthly, how many months will it take for your investment to grow to $20,000?

16. If you invest $10,400 in British government bonds and the interest is compounded monthly, how many months will it take for your investment to grow to $20,000?

17. How long, to the nearest year, will it take an investment in Mexico to double its value if the interest is compounded every 6 months?

18. How long, to the nearest year, will it take an investment in Canada to double its value if the interest is compounded every 6 months?

19. How long will it take a $1000 investment in Canadian government bonds to be worth the same as an $800 investment in Mexican government bonds? (Assume all interest is compounded annually and give the answer to the nearest year.)

20. How long will it take a $1000 investment in Japanese government bonds to be worth the same as an $800 investment in U.S. government bonds? (Assume all interest is compounded annually and give the answer to the nearest year.)

21. If the interest on U.S. government bonds is compounded continuously, how long will it take the value of an investment to double? (Give the answer correct to two decimal places.)

22. If the interest on Canadian government bonds is compounded continuously, how long will it take the value of an investment to double? (Give the answer correct to two decimal places.)

23. ***Carbon Dating*** The half-life of carbon-14 is 5730 years. It is found that in a fossilized math professor's skull, 99.875% of the carbon-14 has decayed since the time it was part of a living math professor (if you call that living, which we do). How old is the skull? (Round your answer to three significant digits.)

24. ***Carbon Dating*** A fossilized math book has just been dug up, and archaeologists find that 99.5% of the carbon-14 in the paper has decayed. How long ago were the trees cut down to make the paper for the book? (Round your answer to three significant digits.)

25. ***Automobiles*** The rate of auto thefts is tripling every 6 months. Find the doubling time.

26. ***Televisions*** The rate of television thefts is doubling every 4 months. Find the tripling time.

27. ***Radioactive Decay*** Uranium-235 is used as fuel for some nuclear reactors. It has a half-life of 710 million years. How long will it take 10 grams of uranium-235 to decay down to 1 gram? (Round your answer to three significant digits.)

28. ***Radioactive Decay*** Plutonium-239 is used as fuel for some nuclear reactors and also as the fissionable material in atomic bombs. It has a half-life of 24,400 years. How long will it take 10 grams of plutonium-239 to decay to 1 gram? (Round your answer to three significant digits.)

29. ***Radioactive Decay*** You are trying to determine the half-life of a new radioactive element you have isolated. You start with 1 gram, and 2 days later you determine that it has decayed down to 0.7 gram. What is its half-life? (Round your answer to three significant digits.)

30. ***Radioactive Decay*** You have just isolated a new radioactive element. If you can determine its half-life, you will win the Nobel Prize in physics. You purify a sample of 2 grams. One of your colleagues steals half of it, and 3 days later you find that 0.1 gram of the radioactive material is still left. What is the half-life? (Round your answer to three significant digits.)

31. ***Aspirin*** Soon after taking an aspirin, a patient has absorbed 300 mg of the drug. If the amount of aspirin in the bloodstream decays exponentially, with half being removed every 2 hours, find the time it will take for the amount of aspirin in the bloodstream to decrease to 100 mg.

32. ***Alcohol*** After several drinks a person has a blood alcohol level of 0.20 mg/dL. If the amount of alcohol in the blood decays exponentially, with one-fourth being removed every hour, find the time it will take for the person's blood alcohol level to decrease to 0.08 mg/dL.

Employment *Exercises 33–38 are based on the following chart, which shows the number of people employed by Northrop Grumman in Long Island, New York, for the period from 1987 to 1995.*[3]

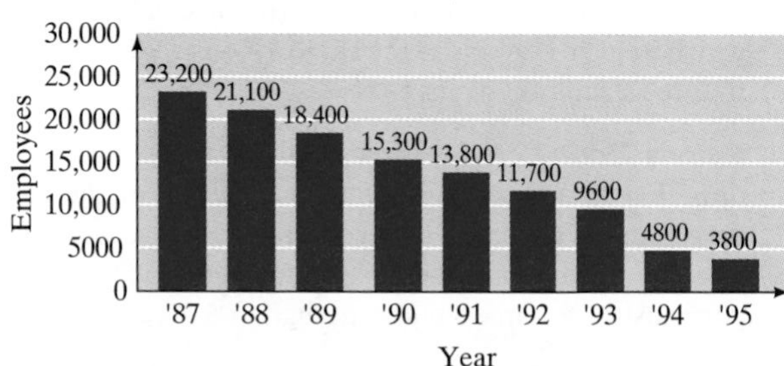

Source: Grumman Corp., Northrop Grumman Corp./*Newsday*, September 23, 1994, p. A5.

33. Use the 1987 and 1995 figures to construct an exponential decay model for employment at Northrop Grumman/Long Island, and use your model to predict the number of people employed there in 1997. Your model should have the form

$$P = P_0e^{-kt}$$

where t is the number of years since 1987. (Round your answer to the nearest 100 people.)

34. Repeat Exercise 33 using the 1987 and 1994 figures.

35. The method of least-squares exponential regression gives the model[4]

$$P = 28{,}131.7e^{-0.220800t}$$

where t is the number of years since 1987. Refer back to the model you obtained in Exercise 33, and tell which model predicts a higher employment figure for 1997.

36. Repeat Exercise 35 using your model from Exercise 34.

37. According to the least-squares model in Exercise 35, when will employment by Northrop Grumman/Long Island drop to 1000 employees? (Give the answer to the nearest year.)

38. According to the least-squares model in Exercise 35, when will employment by Northrop Grumman/Long Island drop to 500 employees? (Give the answer to the nearest year.)

39. ***Richter Scale*** The **Richter scale** is used to measure the intensity of earthquakes. The Richter scale rating of an earthquake is given by the formula

$$R = \frac{2}{3}(\log E - 4.4)$$

where E is the energy released by the earthquake (measured in joules[5]).

a. The San Francisco earthquake of 1906 registered $R = 8.2$ on the Richter scale. How many joules of energy were released?

b. In 1989 another San Francisco earthquake registered 7.1 on the Richter scale. Compare the two: The energy released in the 1989 earthquake was what percentage of the energy released in the 1906 quake?

c. Show that if two earthquakes registering R_1 and R_2 on the Richter scale release E_1 and E_2 joules of energy, respectively, then

$$\frac{E_2}{E_1} = 10^{1.5(R_2-R_1)}$$

d. Fill in the blank: If one earthquake registers one point more on the Richter scale than another, then it releases ________ times the amount of energy.

---

[3]The 1987–1994 figures show employment at Grumman Corp. before it was acquired by Northrop Corp. in 1994. The 1994 and 1995 figures are company projections.

[4]This method is described in the chapter on functions of several variables. It is built into many calculators and spreadsheets, as described in Example 5 of the preceding section.

[5]A joule is a unit of energy; 100 joules of energy light up a 100-watt lightbulb for 1 second.

40. ***Sound Intensity*** The loudness of a sound is measured in **decibels.** The decibel level of a sound is given by the formula

$$D = 10 \log \frac{I}{I_0}$$

where D is the decibel level (dB), I is its intensity in watts per square meter (W/m²),and $I_0 = 10^{-12}$ W/m² is the intensity of a barely audible "threshold" sound. A sound intensity of 90 dB or greater causes damage to the average human ear.

a. Find the decibel levels of each of the following, rounding to the nearest decibel:

Whisper: 115×10^{-12} W/m²

TV (average volume from 10 feet): 320×10^{-7} W/m²

Loud music: 900×10^{-3} W/m²

Jet aircraft (from 500 feet): 100 W/m²

b. Which of the above sounds damages the average human ear?

c. Show that if two sounds of intensity I_1 and I_2 register decibel levels of D_1 and D_2, respectively, then

$$\frac{I_2}{I_1} = 10^{0.1(D_2 - D_1)}$$

d. Fill in the blank: If one sound registers one decibel more than another, then it is ________ times as intense.

41. ***Sound Intensity*** The decibel level of a TV set decreases with the distance from the set according to the formula

$$D = 10 \log\left(\frac{320 \times 10^7}{r^2}\right)$$

where D is the decibel level and r is the distance from the TV set in feet.

a. Find the decibel level (to the nearest decibel) at distances of 10, 20, and 50 feet.

b. Express D in the form $D = A + B \log r$ for suitable constants A and B. (Round A and B to two significant digits.)

c. How far must a listener be from a TV so that the decibel level drops to zero? (Round the answer to two significant digits.)

42. ***Acidity*** The acidity of a solution is measured by its pH, which is given by the formula

$$pH = -\log(H^+)$$

where H^+ measures the concentration of hydrogen ions in moles per liter.[6] The pH of pure water is 7. A solution is referred to as *acidic* if its pH is below 7 and as *basic* if its pH is above 7.

a. Calculate the pH of each of the following substances:

Blood: 3.9×10^{-8} moles/liter

Milk: 4.0×10^{-7} moles/liter

Soap solution: 1.0×10^{-11} moles/liter

Black coffee: 1.2×10^{-7} moles/liter

b. How many moles of hydrogen ions are contained in a liter of acid rain that has a pH of 5.0?

c. Complete the following: If the pH of a solution increases by 1.0, then the concentration of hydrogen ions ________.

[6] A mole corresponds to about 6.0×10^{23} hydrogen ions. (This number is known as Avogadro's number.)

Communication and Reasoning Exercises

43. Why is the logarithm of a negative number not defined?

44. Of what use are logarithms, now that they are no longer needed to perform complex calculations?

45. If $y = 4^x$, then $x =$ _____

46. If $y = \log_6 x$, then $x =$ _____

47. Evaluate $2^{\log_2 8}$.

48. Evaluate $e^{\ln x}$.

49. Evaluate $\ln(e^x)$.

50. How is $\ln \sqrt{a}$ related to $\ln a$?

2.4 Trigonometric Functions and Models

The Sine Function

Figure 13 shows the approximate average daily high temperatures in New York's Central Park. If we draw the graph for several years, we get the repeating pattern shown in Figure 14, in which the x-coordinate represents time in years, with $x = 0$

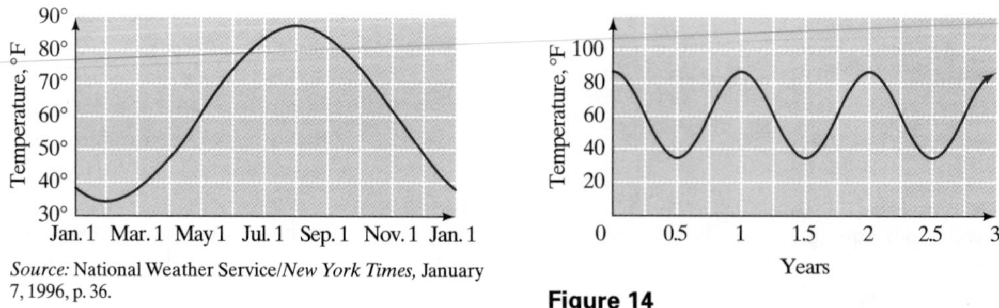

Source: National Weather Service/*New York Times,* January 7, 1996, p. 36.

Figure 13

Figure 14

representing August 1, and the y-coordinate represents the temperature in °F. This is an example of **cyclical** or **periodic** behavior.

Cyclical behavior is common in the business world. Just as there are seasonal fluctuations in the temperature in Central Park, there are seasonal fluctuations in the demand for surfing equipment, swim wear, snow shovels, and many other items. The graph in Figure 15 even suggests cyclical behavior in employment at securities firms in the United States.

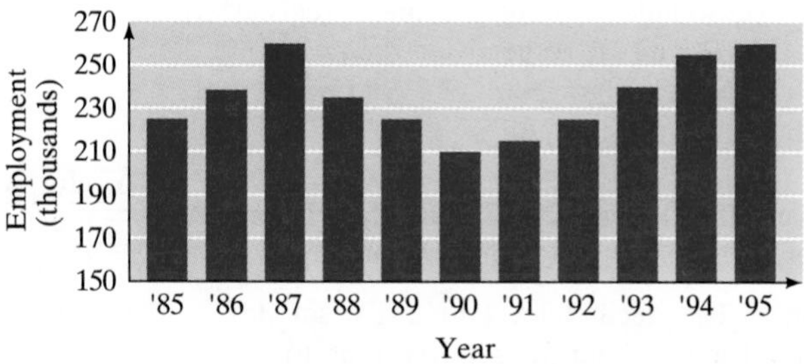

Source: Securities Industry Association/*New York Times,* September 1, 1996, p. F9.

Figure 15

From a mathematical point of view, the simplest models of cyclical behavior are the **sine** and **cosine** functions. An easy way to describe these functions is as follows: Imagine a bicycle wheel that has a radius of one unit, with a marker attached to the rim of the rear wheel, as shown in Figure 16.

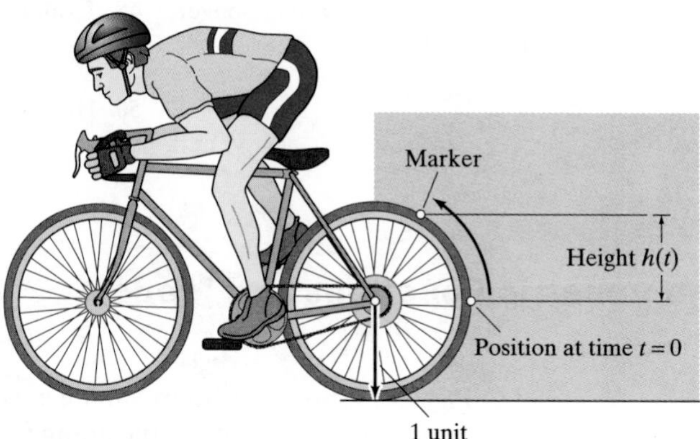

Figure 16

Now, we can measure the height $h(t)$ of the marker above the center of the wheel. As the wheel rotates, $h(t)$ fluctuates between -1 and $+1$. Suppose that, at time $t = 0$, the marker was at height zero as shown in the diagram, so $h(0) = 0$. Since the wheel has a radius of one unit, its circumference (the distance all around) is 2π, where $\pi = 3.14159265....$ If the cyclist happens to be moving at a speed of one unit per second, it will take the bicycle wheel 2π seconds to make one complete revolution. During the time interval $[0, 2\pi]$, the marker will first rise to a maximum height of $+1$, drop to a low point of -1, and then return to the starting position of 0 at $t = 2\pi$. This function $h(t)$ is called the **sine function,** denoted by $\sin(t)$. Figure 17 shows its graph.

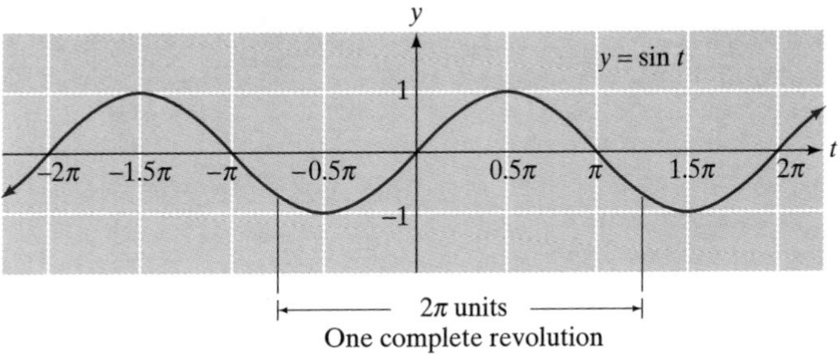

Figure 17

The Sine Function

"Bicycle Wheel" Definition
If a wheel with radius one unit rolls forward at a speed of one unit per second, then $\sin(t)$ is the height after t seconds of a marker on the rim of the wheel, starting in the position shown in Figure 16.

Geometric Definition
The **sine** of a real number t is the y-coordinate (height) of the point P in the following diagram, where $|t|$ is the length of the arc shown.

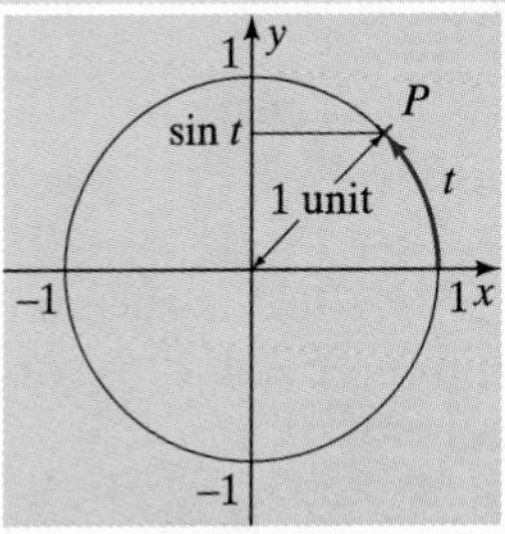

sin(*t*) = *y*-coordinate of the point *P*

example 1

T Trigonometric Function
Use technology to plot the following pairs of graphs on the same set of axes.

a. $f(t) = \sin(t); g(t) = 2\sin(t)$

b. $f(t) = \sin(t); g(t) = \sin(t + \pi/4)$

c. $f(t) = \sin(t); g(t) = \sin(3t)$

Solution

a. (Important note: If you are using a calculator, make sure it is set to *radians mode*, not degree mode.) We enter these functions as `sin(X)` and `2*sin(X)`, respectively. We use window coordinates suggested by the graph of sin(t) shown in Figure 17 but with larger y-coordinates (why?):

$$-6.5 \le x \le 6.5 \quad \text{and} \quad -2.5 \le y \le 2.5$$

The graphs are shown in Figure 18.

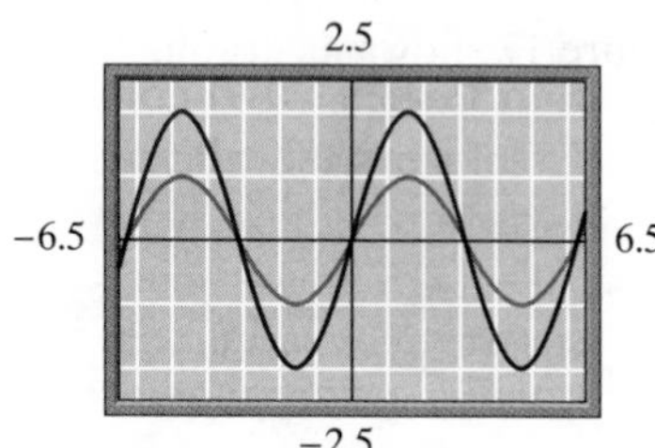

Figure 18

Here, $f(t) = \sin(t)$ is shown in gray and $g(t) = 2\sin(t)$ is shown in color. Notice that multiplication by 2 has doubled the **amplitude,** or *the distance it oscillates up and down.* Whereas the original sine curve oscillates between −1 and 1, the new curve oscillates between −2 and 2. In general:

A sin(x) has amplitude A.

b. We enter these functions as `sin(X)` and `sin(X+π/4)`, respectively. Using the window coordinates $-6.5 \le x \le 6.5$ and $-1.25 \le y \le 1.25$, we get Figure 19.

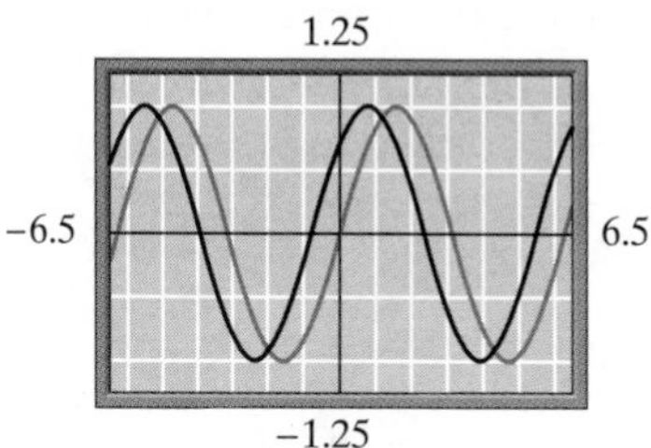

Figure 19

Once again, $f(t) = \sin(t)$ is shown in gray and $g(t) = \sin(t + \pi/4)$ is shown in color. The addition of $\pi/4$ to the argument has shifted the graph to the left by $\pi/4$ units. In general:

Replacing x by $x + c$ shifts the graph to the left c units.

How would we shift the graph to the *right* $\pi/4$ units?

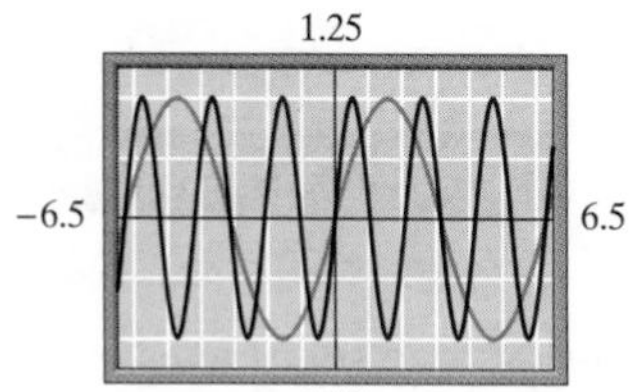

Figure 20

c. We enter these functions as `sin(X)` and `sin(3*X)`, respectively. Using the same window coordinates and color coding as in part (b), we obtain the graph in Figure 20. The graph of sin($3t$) oscillates three times as fast as the graph of sin(t). In general:

Replacing x by bx multiplies the rate of oscillation by b.

Web Site

At the web site, follow the path

Web site → On-Line Text → New Functions from Old: Scaled and Shifted Functions

for more discussion of the operations we just used to modify the graph of the sine function.

We can combine the operations in Example 1 to obtain the general case.

The General Sine Function

The general sine function is

$$f(x) = A \sin[\omega(x - \alpha)] + C$$

Its graph is shown here.

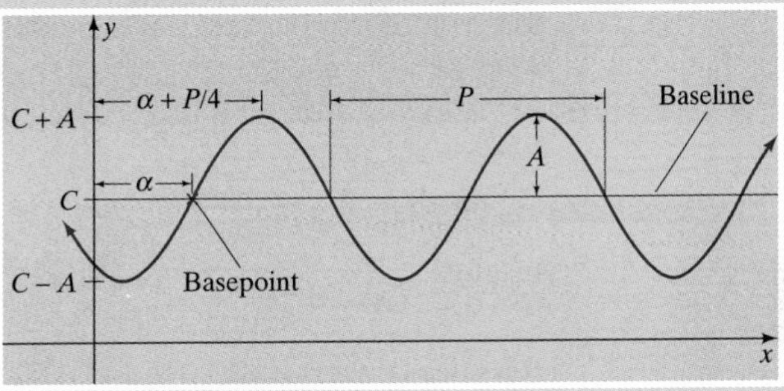

- A is the **amplitude** (the height of each peak above the baseline).
- C is the **vertical offset** (height of the baseline).
- P is the **period** or **wavelength** (the length of each cycle).
- ω is the **angular frequency,** and is given by

$$\omega = 2\pi/P \text{ or } P = 2\pi/\omega.$$

- α is the **phase shift.**

example 2

Electrical Current

The typical voltage V supplied by an electrical outlet in the United States is a sinusoidal function that oscillates between −165 volts and +165 volts with a frequency of 60 cycles per second. Find an equation for the voltage as a function of time t.

Solution

What we are looking for is a function of the form

$$V(t) = A \sin[\omega(t - \alpha)] + C$$

Referring to the above figure, we can look at the constants one at a time.

- *Amplitude A and vertical offset C:* Since the voltage oscillates between −165 volts and +165 volts, we see that $A = 165$ and $C = 0$.
- *Period P:* Since the electric current oscillates 60 times in 1 second, the length of time it takes to oscillate once is $\frac{1}{60}$ second. Thus, the period is $P = \frac{1}{60}$.

- *Angular Frequency ω:* This is given by the formula

$$\omega = \frac{2\pi}{P} = 2\pi(60) = 120\pi$$

- *Phase Shift α:* The phase shift α tells us when the curve first crosses the t-axis as it ascends. Since we are free to specify what time $t = 0$ represents, let us say that the curve crosses 0 when $t = 0$, so $\alpha = 0$. Thus,

$$V(t) = A\sin[\omega(t - \alpha)] + C = 165\sin(120\pi t)$$

where t is time in seconds.

---

example 3

Cyclical Employment Patterns

An economist consulted by your employment agency indicates that the demand for temporary employment (measured in thousands of job applications per week) in your county can be roughly approximated by the function

$$d = 4.3\sin(0.82t - 0.3) + 7.3$$

where t is time in years since January 2000. Calculate the amplitude, the vertical offset, the phase shift, the angular frequency, and the period, and interpret the results.

Solution

To calculate these constants, we write

$$d = A\sin[\omega(t - \alpha)] + C = A\sin[\omega t - \omega\alpha] + C$$
$$= 4.3\sin(0.82t - 0.3) + 7.3$$

and we see right away that $A = 4.3$ (the amplitude), $C = 7.3$ (vertical offset), and $\omega = 0.82$ (angular frequency). We also have

$$\omega\alpha = 0.3$$

so that $$\alpha = \frac{0.3}{\omega} = \frac{0.3}{0.82} \approx 0.37$$

(rounded to two significant digits; notice that all the terms are given to two digits). Finally, we get the period using the formula

$$P = \frac{2\pi}{\omega} = \frac{2\pi}{0.82} \approx 7.7$$

We can interpret these numbers as follows: The demand for temporary employment fluctuates in cycles of 7.7 years about a baseline of 7300 job applications per week. Every cycle, the demand peaks at 11,600 applications per week (4300 above the baseline) and dips to a low of 3000. In April 2000 ($t = 0.37$) the demand for employment was at the baseline level and rising.

The Cosine Function

Closely related to the sine function is the cosine function, defined as follows (refer to the definition of the sine function for comparison).

The Cosine Function

Geometric Definition

The **cosine** of a real number t is the x-coordinate of the point P in the following diagram, in which $|t|$ is the length of the arc shown.

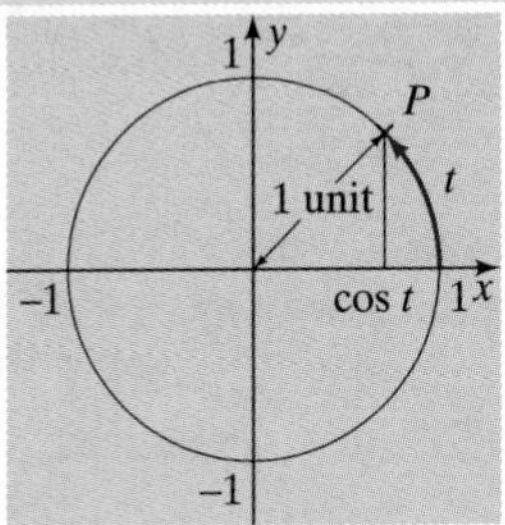

cos(t) = x-coordinate of the point P

Graph of the Cosine Function

The graph of the cosine function is identical to the graph of the sine function, except that it is shifted $\pi/2$ units to the left.

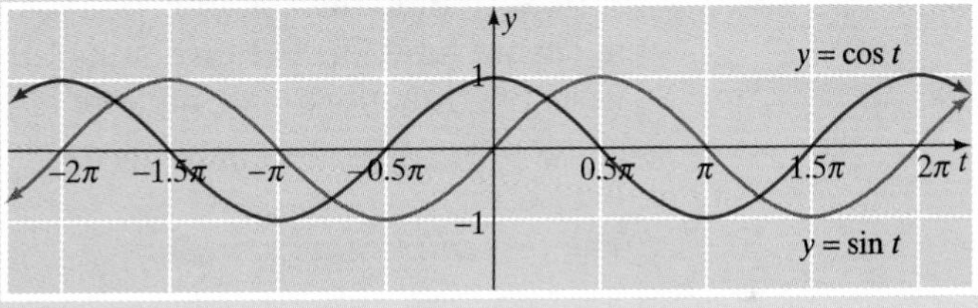

Notice that the coordinates of the point P in the above diagram are $(\cos t, \sin t)$ and that the distance from P to the origin is one unit. However, the Pythagorean theorem tells us that the distance from a point (x, y) to the origin is $\sqrt{x^2 + y^2}$. Thus,

$$\text{Square of the distance from } P \text{ to } (0, 0) = 1$$

$$(\sin t)^2 + (\cos t)^2 = 1$$

We often write $(\sin t)^2$ as $\sin^2 t$ and similarly for the cosine, so we can rewrite the equation as

$$\sin^2 t + \cos^2 t = 1$$

This equation is one of the important relationships between the sine and cosine functions.

Fundamental Trigonometric Identities

Relationships Between Sine and Cosine

The sine and cosine of a number t are related by

$$\sin^2 t + \cos^2 t = 1$$

We can obtain the cosine curve by shifting the sine curve to the left a distance of $\pi/2$. [See Example 1(b) for a shifted sine function.] Conversely, we can obtain the sine curve from the cosine curve by shifting it $\pi/2$ units to the right. These facts can be expressed as

$$\cos t = \sin(t + \pi/2)$$

$$\sin t = \cos(t - \pi/2)$$

Alternative Formulation

We can also obtain the cosine curve by first inverting the sine curve vertically (replace t by $-t$) and then shifting to the *right* a distance of $\pi/2$. This gives us two alternative formulas (which are easier to remember):

$$\cos t = \sin(\pi/2 - t)$$ Cosine is the sine of the complement.

$$\sin t = \cos(\pi/2 - t)$$

Question Since we can rewrite the cosine function in terms of the sine function, who needs the cosine function anyway?
Answer Technically, nobody. We don't need the cosine function and we can get by with only the sine function. On the other hand, it is convenient to have the cosine function because it starts at its highest point rather than at zero. These two functions and their relationship play important roles throughout mathematics.

The General Cosine Function

The general cosine function is

$$f(x) = A\cos\left[\omega(x - \alpha)\right] + C$$

Its graph is shown here.

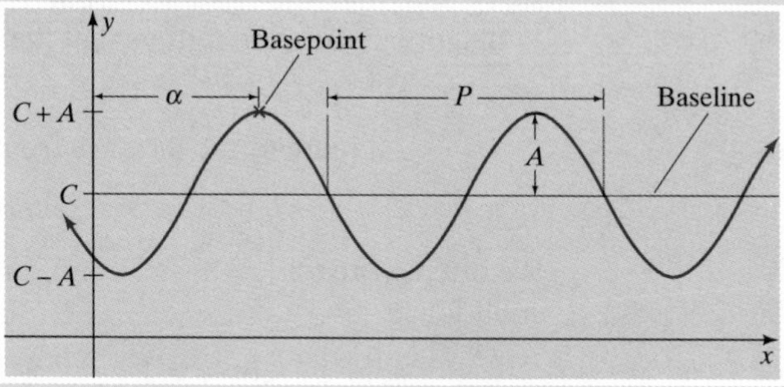

Note that the basepoint is at the highest point of the curve. All the constants have the same meaning as for the general sine curve:

- A is the amplitude (the height of each peak above the baseline).
- C is the vertical offset (height of the baseline).
- P is the period or wavelength (the length of each cycle).
- ω is the angular frequency, and is given by

$$\omega = 2\pi/P \text{ or } P = 2\pi/\omega.$$

- α is the phase shift.

example 4

Cash Flows into Stock Funds

The annual cash flow into stock funds (measured as a percentage of total assets) has fluctuated in cycles of approximately 40 years since 1955, when it was at a high point. The highs were roughly +15% of total assets, whereas the lows were roughly −10% of total assets.

a. Model this cash flow with a cosine function of the time t in years, with $t = 0$ representing 1955.

b. Convert the answer in part (a) to a sine function model.

Source: Investment Company Institute/*New York Times,* February 2, 1997, p. F8.

Solution

a. Cosine modeling is similar to sine modeling; we are seeking a function of the form

$$P(t) = A\cos\left[\omega(t - \alpha)\right] + C$$

- *Amplitude A and Vertical Offset C:* The cash flow fluctuates between −10% and +15%. We can express this as a fluctuation of $A = 12.5$ about the average $C = 2.5$.
- *Period P:* This is given as $P = 40$.
- *Angular frequency ω:* We find ω from the formula

$$\omega = \frac{2\pi}{P} = \frac{2\pi}{40} = \frac{\pi}{20} \approx 0.157$$

- *Phase Shift α:* The basepoint is at the high point of the curve, and we are told that cash flow was at its high point at $t = 0$. Therefore, the basepoint occurs at $t = 0$, and so $\alpha = 0$.

Putting the model together gives

$$\begin{aligned} P(t) &= A\cos\left[\omega(t - \alpha)\right] + C \\ &= 12.5\cos(0.157t) + 2.5 \end{aligned}$$

where t is time in years.

b. To convert between a sine and cosine model, we can use one of the relationships given earlier. Let's use the formula

$$\cos x = \sin(x + \pi/2)$$

Therefore,

$$P(t) = 12.5\cos(0.157t) + 2.5 = 12.5\sin(0.157t + \pi/2) + 2.5$$

The Other Trigonometric Functions

The ratios and reciprocals of sine and cosine are given their own names.

Tangent, Cotangent, Secant, Cosecant

Tangent: $\tan x = \dfrac{\sin x}{\cos x}$

Cotangent: $\cot x = \text{cotan}\, x = \dfrac{\cos x}{\sin x} = \dfrac{1}{\tan x}$

Secant: $\sec x = \dfrac{1}{\cos x}$

Cosecant: $\csc x = \text{cosec}\, x = \dfrac{1}{\sin x}$

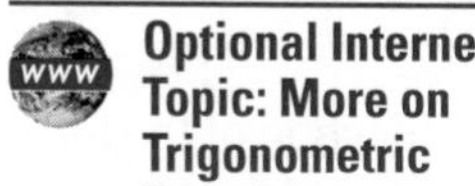

Optional Internet Topic: More on Trigonometric Functions

At the web site you will find more discussion of the graphs of the trigonometric functions, their relationship to right triangles, and some exercises. Follow the path

Web site → On-Line Text → Trigonometric Functions and Calculus → The Six Trigonometric Functions

2.4 exercises

In Exercises 1–12, graph the given functions or pairs of functions on the same set of axes.

a. *Sketch the curve without any technological help by consulting the discussion in Example 1.*

T **b.** *Use technology to check your sketches.*

1. $f(t) = \sin(t);\, g(t) = 3\sin(t)$

2. $f(t) = \sin(t);\, g(t) = 2.2\sin(t)$

3. $f(t) = \sin(t);\, g(t) = \sin(t - \pi/4)$

4. $f(t) = \sin(t);\, g(t) = \sin(t + \pi)$

5. $f(t) = \sin(t);\, g(t) = \sin(2t)$

6. $f(t) = \sin(t);\, g(t) = \sin(-t)$

7. $f(t) = 2\sin[3\pi(t - 0.5)] - 3$

8. $f(t) = 2\sin[3\pi(t + 1.5)] + 1.5$

9. $f(t) = \cos(t);\, g(t) = 5\cos[3(t - 1.5\pi)]$

10. $f(t) = \cos(t);\, g(t) = 3.1\cos(t)$

11. $f(t) = \cos(t);\, g(t) = -2.5\cos(t)$

12. $f(t) = \cos(t);\, g(t) = 2\cos(t - \pi)$

In Exercises 13–18, model each curve with a sine function. (Note that not all are drawn with the same scale on the two axes.)

13.

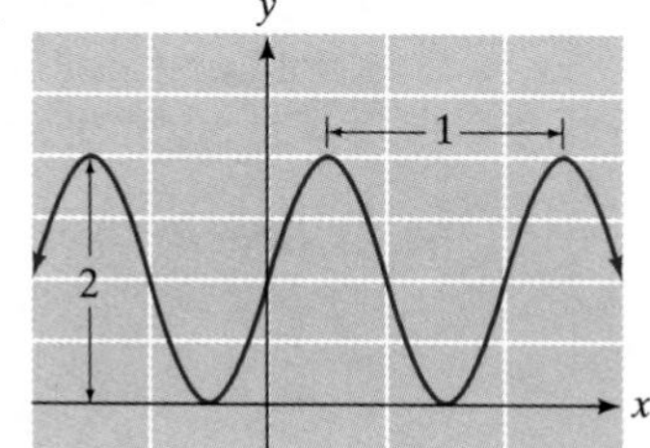

14.

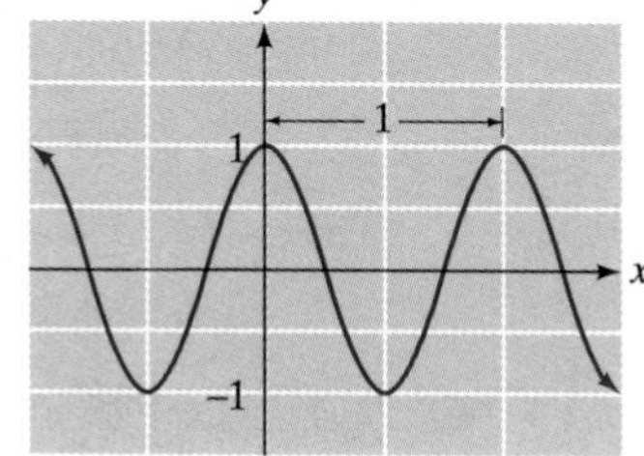

15.

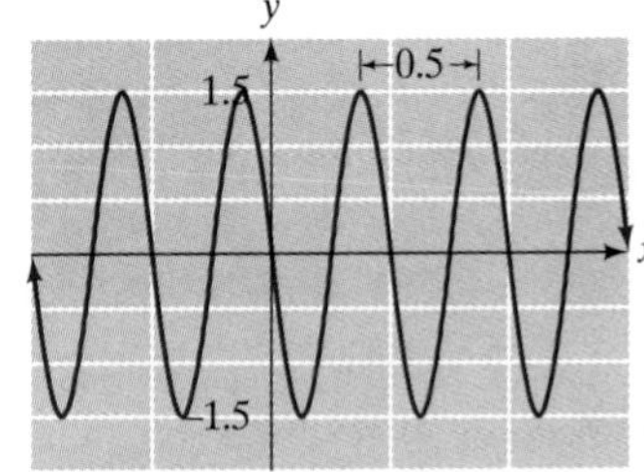

16.

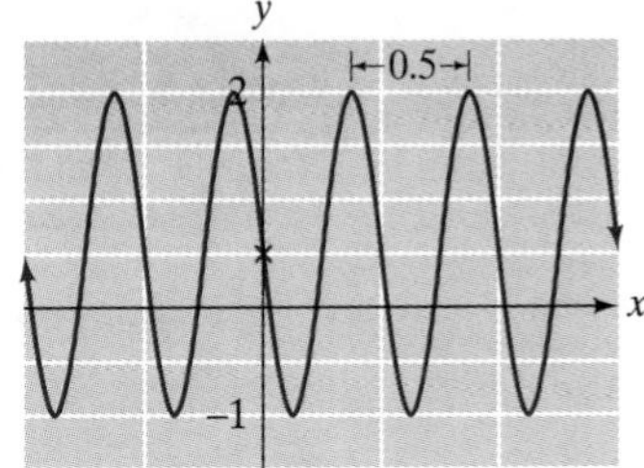

17.

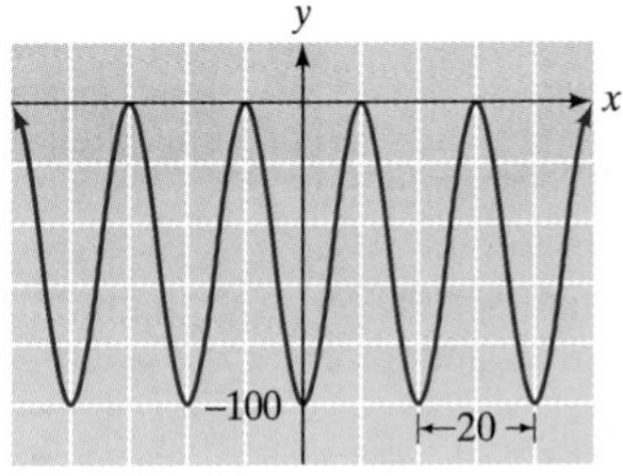

18.

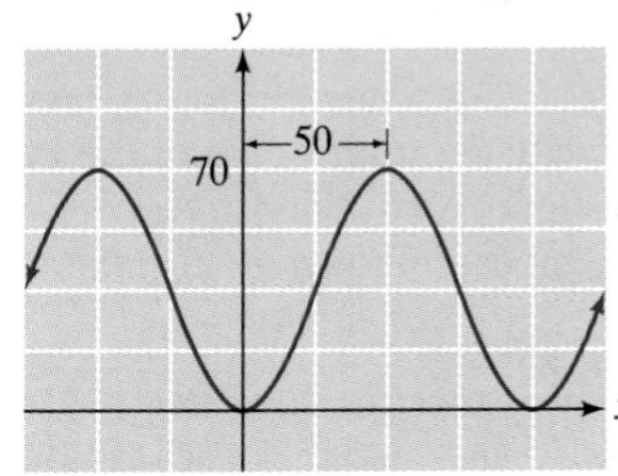

In Exercises 19–24 model each curve with a cosine function. (Note that not all are drawn with the same scale on the two axes.)

19.

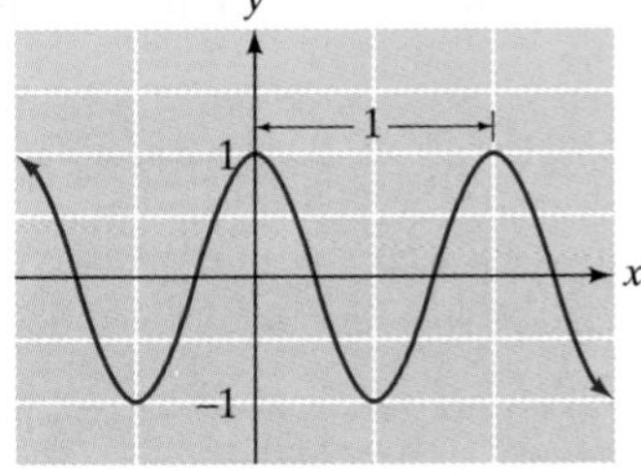

20.

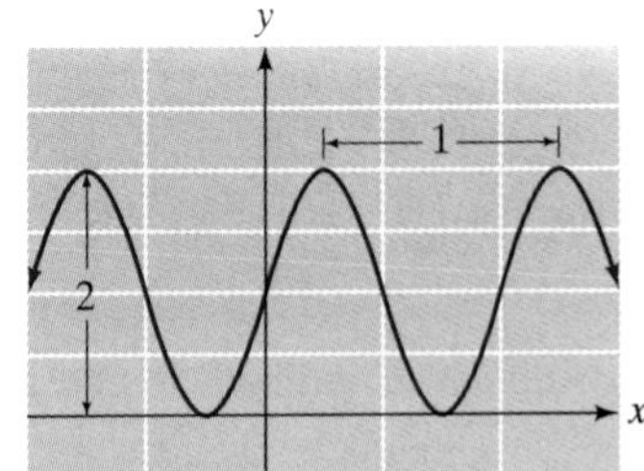

21.

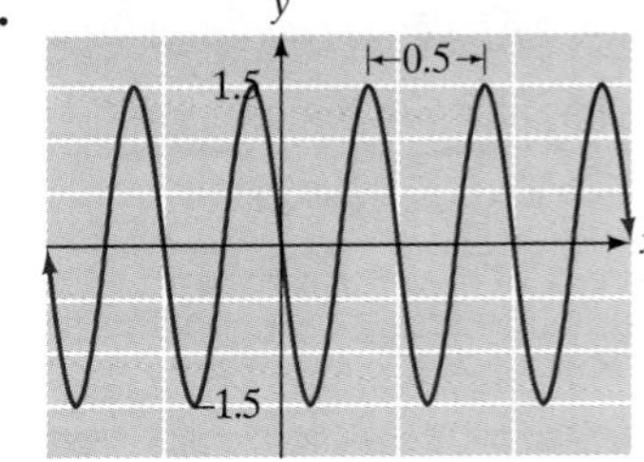

22.

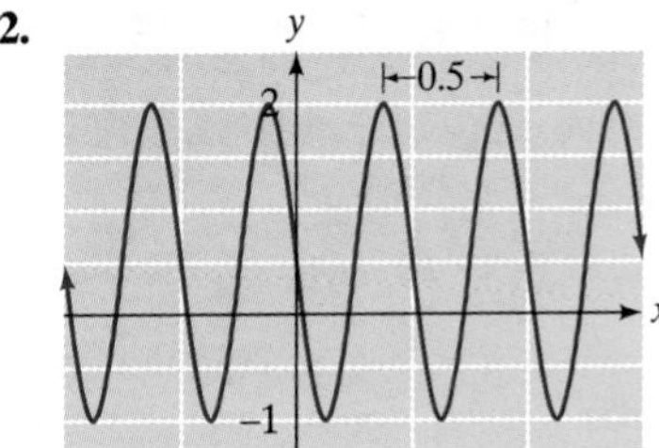

23.

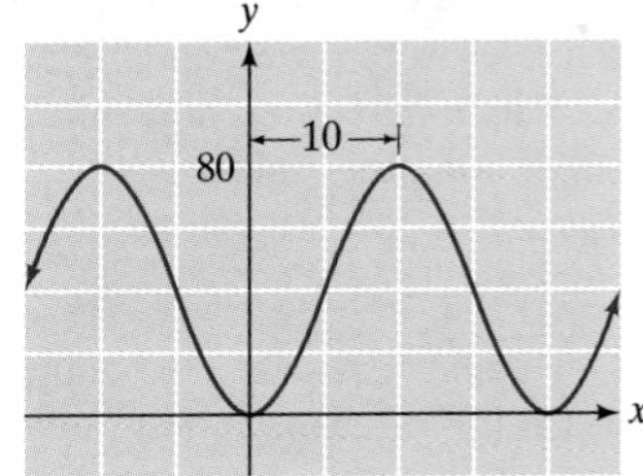

24.

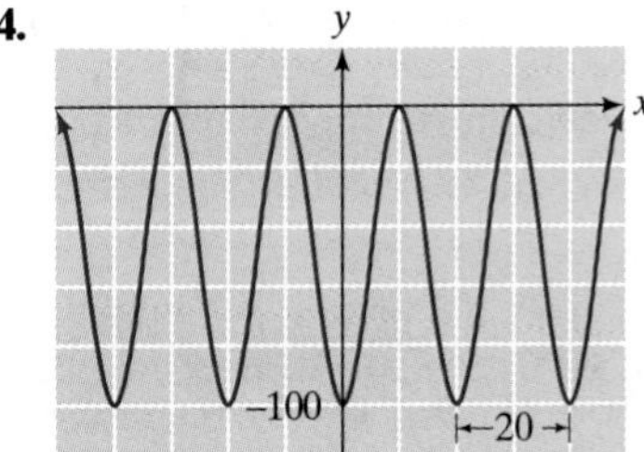

In Exercises 25–28, use the conversion formula $\cos x = \sin(\pi/2 - x)$ *to replace each expression by a sine function.*

25. $f(t) = 4.2\cos(2\pi t) + 3$ **26.** $f(t) = 3 - \cos(t - 4)$

27. $g(x) = 4 - 1.3\cos[2.3(x - 4)]$

28. $g(x) = 4.5\cos[2\pi(3x - 1)] + 7$

Some Identities *Starting with the identity* $\sin^2 x + \cos^2 x = 1$ *and then dividing both sides of the equation by a suitable trigonometric function, derive the trigonometric identities in Exercises 29 and 30.*

29. $\sec^2 x = 1 + \tan^2 x$ **30.** $\csc^2 x = 1 + \cot^2 x$

Exercises 31–38 are based on the following ***addition formulas:***

$$\sin(x + y) = \sin x \cos y + \cos x \sin y$$

$$\sin(x - y) = \sin x \cos y - \cos x \sin y$$

$$\cos(x + y) = \cos x \cos y - \sin x \sin y$$

$$\cos(x - y) = \cos x \cos y + \sin x \sin y$$

31. Calculate $\sin(\pi/3)$, given that $\sin(\pi/6) = 1/2$ and $\cos(\pi/6) = \sqrt{3}/2$.

32. Calculate $\cos(\pi/3)$, given that $\sin(\pi/6) = 1/2$ and $\cos(\pi/6) = \sqrt{3}/2$.

33. Use the formula for $\sin(x + y)$ to obtain the identity $\sin(t + \pi/2) = \cos t$.

34. Use the formula for $\cos(x + y)$ to obtain the identity $\cos(t - \pi/2) = \sin t$.

35. Show that $\sin(\pi - x) = \sin x$.

36. Show that $\cos(\pi - x) = -\cos x$.

37. Use the addition formulas to express $\tan(x + \pi)$ in terms of $\tan(x)$.

38. Use the addition formulas to express $\operatorname{cotan}(x + \pi)$ in terms of $\operatorname{cotan}(x)$.

Applications

T **39.** ***Computer Sales*** Sales of computers are subject to seasonal fluctuations. Computer City's sales of computers in 1995 and 1996 can be approximated by the function

$$s(t) = 0.106\sin(1.39t + 1.61) + 0.455 \qquad (1 \le t \le 8)$$

where t is time in quarters ($t = 1$ represents the end of the first quarter of 1995) and $s(t)$ is computer sales (quarterly revenue) in billions of dollars.[1]

a. Use graphing technology to plot sales versus time for the 2-year period January 1995 through January 1997. Then use your graph to estimate the value of t and the quarter during which sales were lowest and highest.

b. Estimate Computer City's maximum and minimum quarterly revenue from computer sales.

c. Indicate how the answer to part (b) can be obtained directly from the equation for $s(t)$.

T **40.** ***Computer Sales*** Repeat Exercise 39 using the following model for CompUSA's quarterly sales of computers:

$$s(t) = 0.0778\sin(1.52t + 1.06) + 0.591$$

41. ***Computer Sales*** (based on Exercise 39, but no graphing calculator required) Computer City's sales of computers in 1995 and 1996 can be approximated by the function

$$s(t) = 0.106\sin(1.39t + 1.61) + 0.455 \qquad (1 \le t \le 8)$$

where t is time in quarters ($t = 1$ represents the end of the first quarter of 1995) and $s(t)$ is computer sales (quarterly revenue) in billions of dollars. Calculate the amplitude, the vertical offset, the phase shift, the angular frequency, and the period, and interpret the results.

42. ***Computer Sales*** Repeat Exercise 41 using the following model for CompUSA's quarterly sales of computers:

$$s(t) = 0.0778\sin(1.52t + 1.06) + 0.591$$

43. ***Sales Fluctuations*** Sales of General Motors cars and light trucks in 1996 fluctuated from a high of \$95 billion in October ($t = 0$) to a low of \$80 billion in April ($t = 6$).[2] Construct a sinusoidal model for the monthly sales $s(t)$ of General Motors.

44. ***Sales Fluctuations*** Sales of Ocean King Boogie Boards fluctuate from a low of 50 units per week each February 1 ($t = 1$) to a high of 350 units per week each August 1 ($t = 7$). Use a sine function to model the weekly sales $s(t)$ of Ocean King Boogie Boards, where t is time in months.

45. ***Sales Fluctuations*** Repeat Exercise 43, but this time use a cosine function for your model.

46. ***Sales Fluctuations*** Repeat Exercise 44, but this time use a cosine function for your model.

47. ***Tides*** The depth of water at my favorite surfing spot varies from 5 to 15 feet, depending on the time. Last Sunday high tide occurred at 5:00 A.M. and the next high tide occurred at 6:30 P.M. Use a sine function model to

[1] The models in Exercises 39–42 are based on a regression of data that appeared in the *New York Times,* January 8, 1997, p. D1. Constants are rounded to three significant digits.

[2] These are rough figures based on the percentage of the market held by GM as published in the *New York Times,* January 9, 1997, p. D4.

describe the depth of water as a function of time t in hours since midnight on Sunday morning.

48. ***Tides*** Repeat Exercise 47 using data from the depth of water at my second favorite surfing spot, where the tide last Sunday varied from a low of 6 feet at 4:00 A.M. to a high of 10 feet at noon.

49. ***Inflation*** The uninflated cost of Dugout brand snow shovels currently varies from a high of \$10 on January 1 ($t = 0$) to a low of \$5 on June 1 ($t = 0.5$).

a. Assuming this trend were to continue indefinitely, calculate the uninflated cost $u(t)$ of Dugout snow shovels as a function of time t in years. (Use a sine function.)

b. If we assume a 4% annual rate of inflation in the cost of snow shovels, the cost of a snow shovel t years from now, adjusted for inflation, will be 1.04^t times the uninflated cost. Find the cost $c(t)$ of Dugout snow shovels as a function of time t.

50. ***Deflation*** Sales of my exclusive 1997 vintage Chateau Petit Mont Blanc vary from a high of ten bottles per day on April 1 ($t = 0.25$) to a low of four bottles per day on October 1.

a. Assuming this trend were to continue indefinitely, find the undeflated sales $u(t)$ of Chateau Petit Mont Blanc as a function of time t in years. (Use a sine function.)

b. Regrettably, ever since that undercover exposé of my wine-making process, sales of Chateau Petit Mont Blanc have been declining at an annual rate of 12%. Using the preceding exercise as a guide, write down a model for the deflated sales $s(t)$ of Chateau Petit Mont Blanc t years from now.

Music *Musical sounds exhibit the same kind of periodic behavior as the trigonometric functions. High-pitched notes have short periods (less than 1/1000 second), whereas the lowest audible notes have periods of about 1/100 second. Some electronic synthesizers work by superimposing (adding) sinusoidal functions of different frequencies to create different textures. Exercises 51–54 show some examples of how superposition can be used to create interesting periodic functions.*

T **51.** ***Sawtooth Wave***

a. Graph the following functions in a window with $-7 \le x \le 7$ and $-1.5 \le y \le 1.5$:

$$y_1 = \frac{2}{\pi}\cos x$$

$$y_3 = \frac{2}{\pi}\cos x + \frac{2}{3\pi}\cos 3x$$

$$y_5 = \frac{2}{\pi}\cos x + \frac{2}{3\pi}\cos 3x + \frac{2}{5\pi}\cos 5x$$

b. Following the pattern established above, give a formula for y_{11} and graph it in the same window.

c. How would you modify y_{11} to approximate a sawtooth wave with an amplitude of 3 and a period of 4π?

T **52.** ***Square Wave*** Repeat Exercise 51 using sine functions in place of cosine functions to approximate a square wave.

T **53.** ***Harmony*** If we add two sinusoidal functions with frequencies that are exact ratios of each other, the result is a pleasing sound. The following function models two notes an octave apart together with the intermediate fifth:

$$y = \cos(x) + \cos(1.5x) + \cos(2x)$$

Graph this function in the window $0 \le x \le 20$ and $-3 \le y \le 3$ and estimate the period of the resulting wave.

T **54.** ***Discord*** If we add two sinusoidal functions with similar, but unequal, frequencies, the result is a function that "pulsates," or exhibits "beats." (Piano tuners and guitar players use this phenomenon to help them tune an instrument.) Graph the function

$$y = \cos(x) + \cos(0.9x)$$

in the window $-50 \le x \le 50$ and $-2 \le y \le 2$ and estimate the period of the resulting wave.

Communication and Reasoning Exercises

55. When are the seasonal highs and lows for sales of a commodity modeled by a function of the form $s(t) = A\sin(2\pi t) + B$ (A and B constants)?

56. Your friend has come up with the following model for choral society Tupperware stock inventory: $r(t) = 4\sin[2\pi(t-2)/3] + 2.3$, where t is time in weeks and $r(t)$ is the number of items in stock. Comment on the model.

57. Your friend is telling everybody that all six trigonometric functions can be obtained from the single function $\sin x$. Is he correct? Explain your answer.

58. Another friend claims that all six trigonometric functions can be obtained from the single function $\cos x$. Is she correct? Explain your answer.

59. If weekly sales of sodas at a movie theater are given by $s(t) = A + B\cos(\omega t)$, what is the largest B can be? Explain your answer.

60. Complete the following: If the cost of an item is given by $c(t) = A + B\cos[\omega(t - \alpha)]$, then the cost fluctuates by ________ with a period of ________ about a base of ________, peaking at time $t =$ ________.

You're the Expert

Epidemics

A flu epidemic is spreading through the population of the United States. An estimated 150,000,000 people are susceptible to this particular strain. There are 10,000 people already ill, and that number is doubling every 2 weeks. As adviser to the Surgeon General, you have the job of predicting the course of the epidemic. In particular, the Surgeon General needs to know when the flu will have infected 1,000,000 people; when it will have infected 10,000,000 people; and when it will have infected 100,000,000 people.

Although the initial spread of an epidemic appears to be exponential, it cannot continue to be so because the size of the susceptible population is limited. A commonly used model for epidemics is the **logistic curve,** given by the function

$$A(t) = \frac{NP_0}{P_0 + (N - P_0)k^{-t}}$$

where $A(t)$ = the infected population at time t, P_0 = the population initially infected, and N = the total susceptible population. The number k is a constant that governs the rate of spread of the epidemic (its exact meaning will be made clear in a moment). The graph of this function is shown in Figure 21.

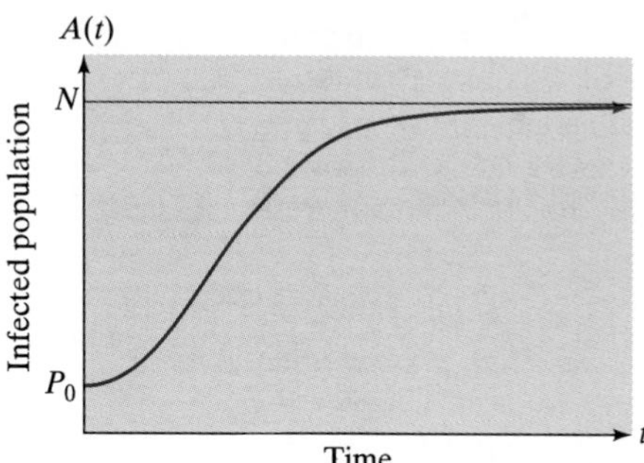

Figure 21

The initial part of this graph shows roughly exponential growth. To see why, multiply the top and bottom of the formula for $A(t)$ by k^t, which gives

$$A(t) = \frac{NP_0k^t}{P_0k^t + (N - P_0)}$$

For small t, the quantity k^t is close to $k^0 = 1$, and so the denominator of this expression is approximately

$$P_0(1) + (N - P_0) = N$$

This gives

$$A(t) \approx P_0k^t$$

for small t; that is, $A(t)$ grows approximately exponentially in the early part of the epidemic. This also tells us that the number k is the base for the exponential growth that governs the early stages of the epidemic. Thus, in the early stages of the epidemic, the number of infected people increases by a factor of k per unit of time.

On the other hand, as t gets large, the term k^{-t} in the original formula gets very small, and thus $A(t)$ gets close to the number N. You can see this in the right-most part of the graph as $A(t)$ levels off under the line at height N.

The problems you now face are (1) to determine the constants to use in the logistic curve, and (2) to use the formula to predict the course of the epidemic. From the data you have, you know that $P_0 = 10{,}000$ and $N = 150{,}000{,}000$. The main problem is to find k. But remember that the initial spread of the flu epidemic is given by $A \approx P_0 k^t$. You know that initially the infected population is doubling every 2 weeks. Thus, if we take t to be time in weeks, then the curve resembles an exponential curve that passes through the points (0, 10,000) and (2, 20,000). Using the technique of Section 2.2, you find that

$$A(t) \approx 10{,}000(2^{t/2}) = 10{,}000(2^{(1/2)})^t = 10{,}000(\sqrt{2})^t$$

Thus, we take $k = \sqrt{2}$. (Do you see why the number of infected people increases by a factor of $\sqrt{2}$ per unit of time?)

Now we can write the logistic curve that governs this particular epidemic:

$$A(t) = \frac{1{,}500{,}000{,}000{,}000}{10{,}000 + 149{,}990{,}000(\sqrt{2})^{-t}} = \frac{150{,}000{,}000}{1 + 14{,}999(\sqrt{2})^{-t}}$$

The graph of this function is shown in Figure 22.

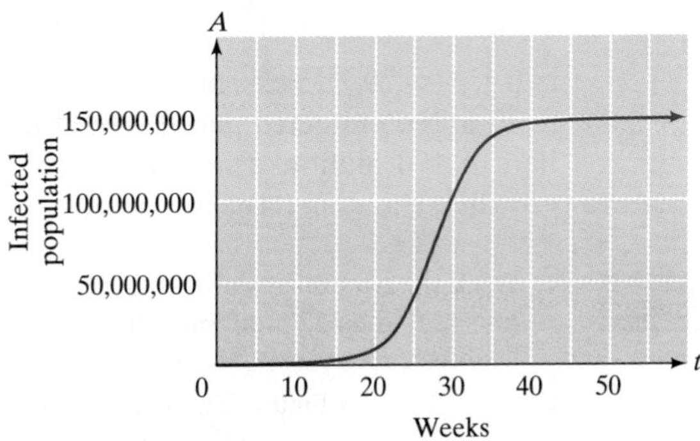

Figure 22

To check your calculation, calculate $A(2) \approx 20{,}000$, showing the expected doubling after 2 weeks.

Now you can tackle the question of prediction: When will the disease infect 1,000,000 people? This is asking: When is $A(t) = 1{,}000{,}000$? Setting $A(t)$ equal to 1,000,000, you get the equation

$$1{,}000{,}000 = \frac{150{,}000{,}000}{1 + 14{,}999(\sqrt{2})^{-t}}$$

Now you solve for t. First, cross-multiply:

$$1{,}000{,}000(1 + 14{,}999(\sqrt{2})^{-t}) = 150{,}000{,}000$$

$$1 + 14{,}999(\sqrt{2})^{-t} = 150 \qquad \text{Divide both sides by 1,000,000.}$$

$$14{,}999(\sqrt{2})^{-t} = 149 \qquad \text{Subtract 1 from both sides.}$$

$$(\sqrt{2})^{-t} = 149/14{,}999 \qquad \text{Divide by 14,999.}$$

To solve this equation, you rewrite it in logarithmic form:

$$-t = \log_{\sqrt{2}}\left(\frac{149}{14{,}999}\right) = \frac{\log(149/14{,}999)}{\log\sqrt{2}} \approx -13.3$$

so $\qquad t \approx 13.3$ weeks

So, in just over 13 weeks, 1,000,000 people are expected to be infected.

When will 10,000,000 people be infected? Set $A(t) = 10{,}000{,}000$ and solve for t as before:

$$10{,}000{,}000 = \frac{150{,}000{,}000}{1 + 14{,}999(\sqrt{2})^{-t}}$$

$$1 + 14{,}999(\sqrt{2})^{-t} = 15$$

$$(\sqrt{2})^{-t} = 14/14{,}999$$

$$-t = \log_{\sqrt{2}}(14/14{,}999) \approx -20.1$$

so $\qquad t \approx 20.1$ weeks

The epidemic will reach 10,000,000 people in just over 20 weeks.

Finally, to determine when 100,000,000 people will be infected, you solve

$$100{,}000{,}000 = \frac{150{,}000{,}000}{1 + 14{,}999(\sqrt{2})^{-t}}$$

and get $t \approx 29.7$ weeks. Notice that it takes less than 7 weeks to go from 1 million to 10 million infected people, but it takes more than 9 weeks to go from 10 million to 100 million. The epidemic slows down after it passes the exponential growth it shows in its early stages.

exercises

1. Track the later stages of the epidemic by determining when 110 million people will be infected, when 120 million will be, when 130 million will be, and when 140 million will be infected.
2. Refer back to Figure 22. When, to the nearest 5 weeks, would you estimate the epidemic to be spreading the fastest?
3. How many people would you estimate to be infected at the time the epidemic is spreading the fastest?
4. Give a logistic model for the following: You have sold 100 "I ♥ Calculus" T-shirts, and sales are increasing continuously at a rate of 30% per day. You estimate the total market for "I ♥ Calculus" T-shirts to be 3000. Use your model to predict when you will have sold 2000 T-shirts.
5. In Russia the average consumer drank two servings of Coca-Cola® in 1993. This amount appeared to be increasing exponentially with a doubling time of 2 years.[1] Given a long-range market saturation estimate of 100 servings per year, find a logistic model for the consumption of Coca-Cola in Russia, and use your model to predict when the average consumption will be 50 servings per year.
Source: New York Times, September 26, 1994, p. D2.

[1]The doubling time is based on retail sales of Coca-Cola products in Russia. Sales in 1993 were double those in 1991 and were expected to double again by 1995.

6. The following graph shows the monthly total U.S. personal income from April 1, 1992, to April 1, 1993, in trillions of dollars.

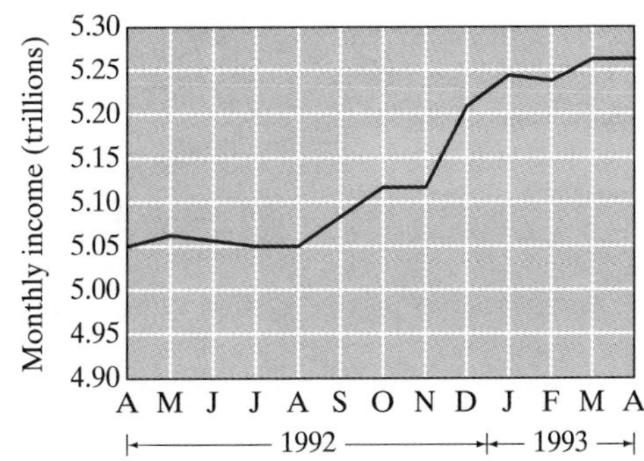

Source: National Association of Purchasing Management, Department of Commerce/*New York Times,* June 2, 1993, section 3, p. 1. Our graph is a facsimile of the published graph (though accuracy may have suffered slightly).

Find a rough logistic model for the income by estimating P_0 and N from the graph and experimenting with several values of k. You may also want to add a constant to the function.

chapter 2 review test

1. Sketch the graph of each of the following quadratic functions, indicating the coordinates of the vertex, the y-intercept, and the x-intercepts (if any).
   a. $f(x) = x^2 + 2x - 3$ **b.** $f(x) = -x^2 - x - 1$

2. Find a formula of the form $f(x) = Ab^x$ for each of the following functions.
   a. $f(1) = 2, f(3) = 18$ **b.** $f(1) = 10, f(3) = 5$

3. Model each of the following curves with a sine function (the scales on the two axes may not be the same).

a.

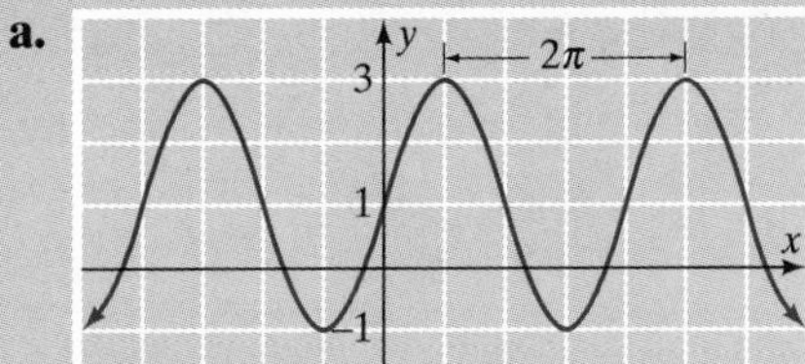

b.

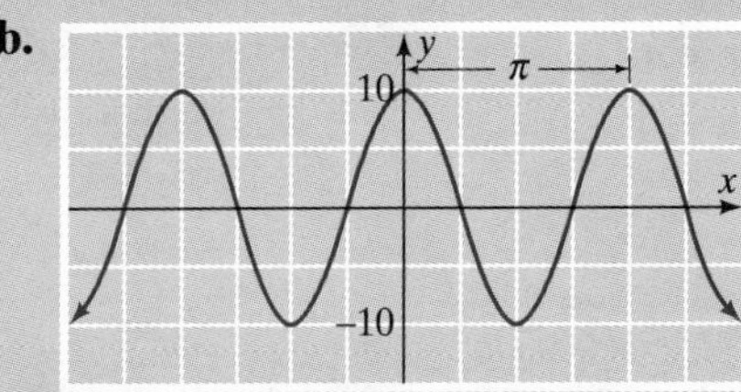

Applications

4. The daily traffic ("hits per day") at OHaganBooks.com seems to depend on the monthly expenditure on advertising through banner ads on well-known Internet portals. The following model, based on information you have collected over the past few months, shows the approximate relationship:

$$h = -0.000005c^2 + 0.085c + 1750$$

where h is the average number of hits per day at the web site and c is the monthly advertising expenditure.

a. According to the model, what monthly advertising expenditure will result in the largest volume of traffic at the site? What is that volume?

b. In addition to predicting a maximum volume of traffic, the model predicts that the traffic will eventually drop to zero if the advertising expenditure is increased too much. What expenditure (to the nearest dollar) results in no web site traffic?

c. What feature of the formula for this quadratic model indicates that it will predict an eventual decline in traffic as advertising expenditure increases?

5. Some time ago you formulated the following linear model of demand:

$$q = -60p + 950$$

where q is the monthly demand for OHaganBooks.com's on-line novels at a price of p dollars per novel.

a. Use this model to express the monthly revenue as a function of the unit price p, and hence determine the price you should charge for a maximum monthly revenue.

b. Author royalties and copyright fees cost you an average of $4 per novel, and the monthly cost of operating and maintaining the on-line publishing service amounts to $900 per month. Express the monthly profit P as a function of the unit price p, and hence determine the unit price you should charge for a maximum monthly profit. What is the resulting profit (or loss)?

6. In the period immediately following its initial public offering (IPO), OHaganBooks.com's stock is doubling in value every 3 hours.

a. If you bought $10,000 worth of the stock when it was first offered, how much was your stock worth after 8 hours?

b. How long from the initial offering did it take your investment to reach $50,000?

c. After 10 hours of trading, the stock turns around and starts losing one-third of its value every 4 hours. How long (from the initial offering) will it be before your stock is once again worth $10,000?

7. OHaganBooks.com modeled its weekly sales over a period of time with the function

$$s(t) = 6053 + \frac{4474}{1 + e^{-0.55(t-4.8)}}$$

as shown in the following graph (t is measured in weeks).

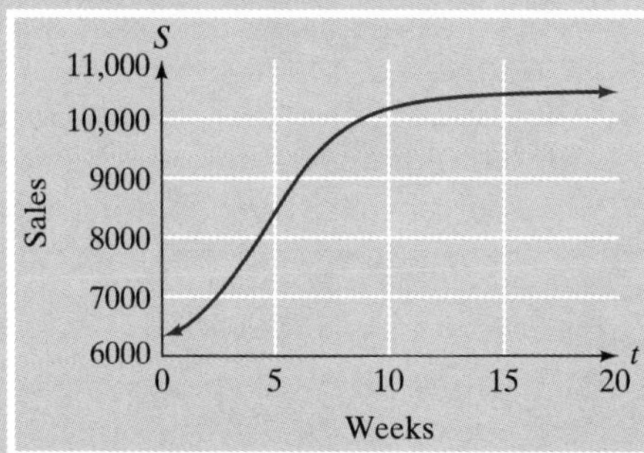

a. As time goes on, it appears that weekly sales are leveling off. At what value are they leveling off?
b. When did weekly sales rise above 10,000?

8. After several years OHaganBooks.com noticed that its sales showed seasonal fluctuations, so that weekly sales oscillated in a sine wave from a low of 9000 books per week to a high of 12,000 books per week, with the high point of the year being three-quarters of the way through the year, in October. Model OHaganBooks.com's weekly sales as a function of t, the number of weeks into the year.

Additional On-Line Review

If you follow the path

Web site → Everything for Calculus → Chapter 2

you will find the following additional resources to help you review:

- A comprehensive chapter summary (including examples and interactive features)
- Additional review exercises (including interactive exercises and many with help)
- A true–false chapter quiz
- Several useful graphing utilities

You're the Expert

Reducing Sulfur Emissions

The Environmental Protection Agency (EPA) wants to formulate a policy that will encourage utilities to reduce sulfur emissions. Its goal is to reduce annual emissions of sulfur dioxide by a total of 10 million tons from the current level of 25 million tons by imposing a fixed charge for every ton of sulfur released into the environment per year. The EPA has some data showing the marginal cost to utilities of reducing sulfur emissions. As a consultant to the EPA, you must determine the amount to be charged per ton of sulfur emissions in light of these data.

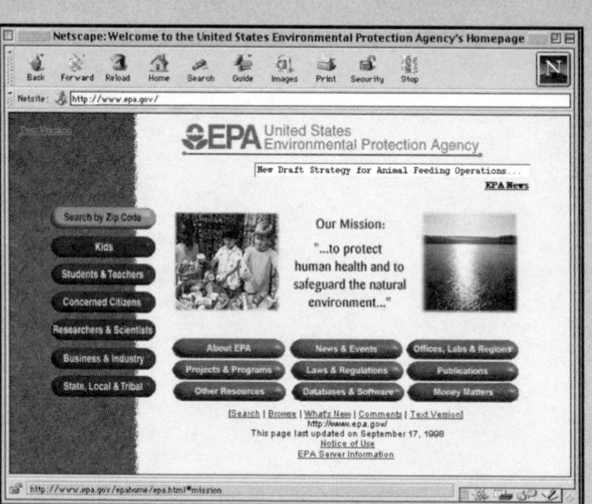

Internet Resources for This Chapter

At the web site, follow the path

Web site → Everything for Calculus → Chapter 3

where you will find links to step-by-step tutorials for the main topics in this chapter, a detailed chapter summary, a true–false quiz, and a collection of sample test questions. You will also find an on-line grapher and other useful resources. Complete text and interactive exercises have been placed on the web site covering the following optional topics:

- Sketching the graph of the derivative
- Proof of the power rule
- Continuity and differentiability

Introduction to the Derivative

chapter 3

3.1 Average Rate of Change
3.2 The Derivative As Rate of Change: A Numerical Approach
3.3 The Derivative As Slope: A Geometric Approach
3.4 The Derivative As a Function: An Algebraic Approach
3.5 A First Application: Marginal Analysis
3.6 Limits and Continuity: Numerical and Graphical Approaches (Optional)
3.7 Limits and Continuity: Algebraic Approach (Optional)

Introduction

With this chapter we begin our study of calculus—one of the most important, most useful, most used parts of mathematics—with the derivative. In the world around us, everything is changing, and the derivative tells us the rate of change. How fast and in which direction is the change occurring? Is the Dow Jones average going up, and if so, how fast? If I raise my prices, how many customers will I lose? If I launch this missile, how high will it go, and where will it come down?

We have discussed the concept of rate of change for linear functions (straight lines), where it is measured by a single number, the slope. But this works only because a straight line maintains a constant rate of change along its whole length. Other functions rise faster here than there—or rise in one place and fall in another—so that the rate of change varies along the graph. The first achievement of calculus is to provide a systematic and straightforward way of calculating (hence the name) these rates of change. To describe a changing world, we need a language of change, and that is what calculus is.

The history of calculus is an interesting story of personalities, intellectual movements, and controversy. Credit for its invention is given to two mathematicians: Isaac Newton (1642–1727) and Gottfried Leibniz (1646–1716). Newton, an English mathematician and scientist, developed calculus first, probably in the 1660s. We say "probably" because, for various reasons, he did not publish his ideas until much later. This allowed Leibniz to publish his own version of calculus first, in 1684. Fifteen years later, stirred up by nationalist fervor in England and on the Continent, controversy erupted over who should get the credit for the invention of calculus. The debate got so heated that the Royal Society (of which Newton and Leibniz were both members) set up a commission to investigate the question. The commission decided in favor of Newton, who happened to be president of the society at the time. The consensus today is that both mathematicians deserve credit because they came to the same conclusions working independently. This is not really surprising: Both built on well-known work of other people, and it was almost inevitable that someone would put it all together at about that time.

3.1 Average Rate of Change

In applications we often have a changing quantity and want to know how *fast* it is changing. Up to this point, we have talked about straight lines and linear functions, where the rate of change is measured by the slope of the line. Let us see what we can use in place of the slope when we have a function that is not linear.

Numerical Point of View

example 1

Average Rate of Change from a Table

The following table lists the approximate price of Boston Chicken stock each year during the 5-year period following its initial public offering[1] ($t = 3$ represents 1993, when the company first went public):

Year, t	3	4	5	6	7	8
Stock price, $S(t)$	\$20	20	20	30	20	10

Source: Bloomberg Financial Markets/*New York Times,* May 2, 1998, p. D1.

a. What was the average rate of change in the value of Boston Chicken stock over its first 3 years (the period $3 \le t \le 6$ or $[3, 6]$ in interval notation); over its first 5 years (the period $3 \le t \le 8$ or $[3, 8]$); and over the period $[4, 7]$?

b. Graph the values shown in the table. How are the rates of change reflected in the graph?

Solution

a. During the 3-year period $[3, 6]$, the value of the stock changed as follows:

Start of the period ($t = 3$):	$S(3) = 20$
End of the period ($t = 6$):	$S(6) = 30$
Change during the period $[3, 6]$:	$S(6) - S(3) = 10$

Thus, the value of Boston Chicken stock increased by \$10 in 3 years, giving an average rate of change of $\frac{10}{3} \approx \$3.33$ per year. We can write the calculation this way:

$$\text{Average rate of change of } S = \frac{\text{Change in } S}{\text{Change in } t}$$

$$= \frac{\Delta S}{\Delta t} \qquad \begin{array}{l}\Delta S \text{ means "change in } S\text{."}\\ \Delta t \text{ means "change in } t\text{."}\end{array}$$

$$= \frac{S(6) - S(3)}{6 - 3} = \frac{10}{3} \approx \$3.33 \text{ per year}$$

Interpreting the result: During the period $[3, 6]$ (the first 3 years), the value of Boston Chicken stock increased at an average rate of \$3.33 per year.

Similarly, the average rate of change during the period $[3, 8]$ was

$$\text{Average rate of change of } S = \frac{\Delta S}{\Delta t} = \frac{S(8) - S(3)}{8 - 3} = \frac{10 - 20}{8 - 3} = -\frac{10}{5} = -\$2.00 \text{ per year}$$

[1]The given stock values are rounded and reflect the average stock performance during the given year.

Interpreting the result: During the period [3, 8] (the first 5 years), the value of Boston Chicken stock *decreased* at an average rate of \$2.00 per year.

Finally, during the period [4, 7], the average rate of change was

$$\text{Average rate of change of } S = \frac{\Delta S}{\Delta t} = \frac{S(7) - S(4)}{7 - 4} = \frac{20 - 20}{7 - 4} = \frac{0}{3} = \$0 \text{ per year}$$

Interpreting the result: During the period [4, 7], the rate of change of the Boston Chicken stock price was zero (even though its price did fluctuate during that period).

b. In Chapter 1 we saw that the rate of change of a quantity that changes linearly with time is measured by the slope of its graph versus time t. However, Boston Chicken's stock price does not change linearly with time. Figure 1 shows the data plotted in two different ways: as a bar chart and as a piecewise linear graph. Although bar charts are used more commonly in the media, the graph on the right illustrates the changing stock price more clearly.

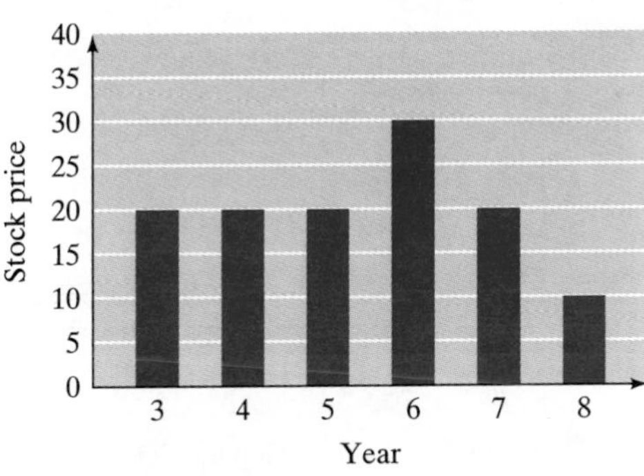

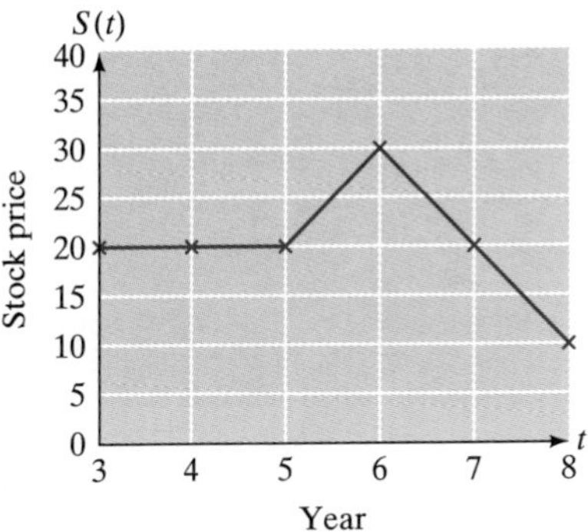

Figure 1

We saw in part (a) that the average rate of change over the interval [3, 6] is the ratio

$$\text{Rate of change of } S = \frac{\Delta S}{\Delta t} = \frac{S(6) - S(3)}{6 - 3} = \frac{\text{Change in } S}{\text{Change in } t}$$

That is also the slope of the line through P and Q shown in Figure 2.

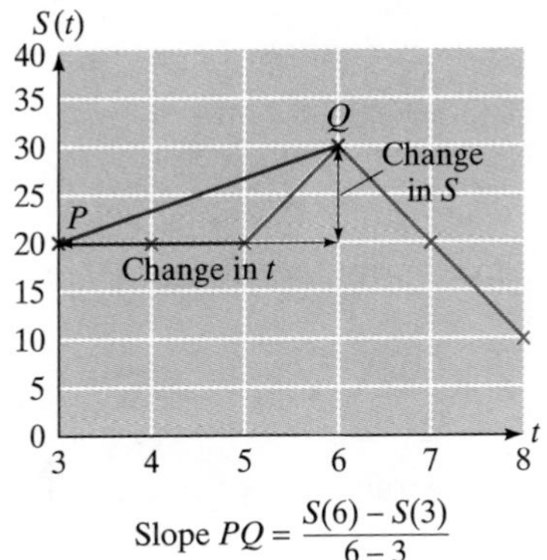

Figure 2

Interpreting the result: The average rate of change of the Boston Chicken stock price over the interval [3, 6] is the slope of the line passing through the points on the graph where $t = 3$ and $t = 6$.

Similarly, the average rates of change of the stock price over the intervals [3, 8] and [4, 7] are the slopes of the lines that pass through pairs of corresponding points.

Here is the formal definition of the rate of change of a function over an interval.

Average Rate of Change of f over $[a, b]$: Difference Quotient

The **average rate of change** of the function f over the interval $[a, b]$ is

$$\text{Average rate of change of } f = \frac{\Delta f}{\Delta x} = \frac{f(b) - f(a)}{b - a}$$

$$= \text{Slope of line through points } P \text{ and } Q \text{ (see figure)}$$

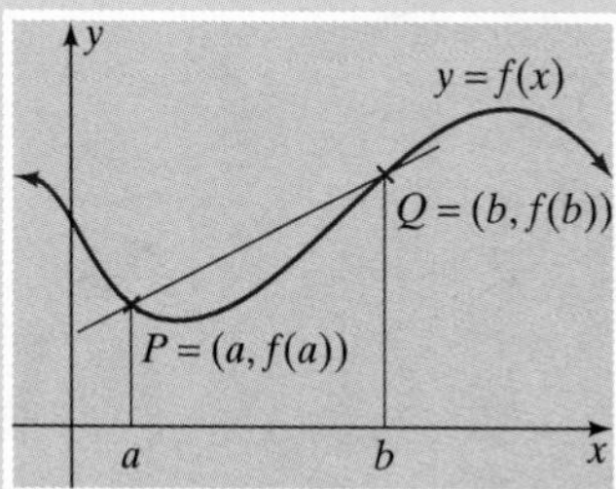

Average rate of change = Slope of PQ

We also call this average rate of change the **difference quotient** of f over the interval $[a, b]$. [Notice that it is the *quotient* of the *differences* $f(b) - f(a)$ and $b - a$.]

Units

The units of the average rate of change are units of $f(x)$ per unit of x.

Quick Example

If $f(3) = -1$ dollar and $f(5) = 0.5$ dollar, and if x is measured in years, then the average rate of change of f over the interval [3, 5] is given by

$$\text{Average rate of change} = \frac{f(5) - f(3)}{5 - 3} = \frac{0.5 - (-1)}{2} = 0.75 \text{ dollars per year}$$

Alternative Formula: Average Rate of Change of f over $[a, a + h]$

(Replace b above by $a + h$.) The average rate of change of f over the interval $[a, a + h]$ is

$$\text{Average rate of change of } f = \frac{f(a + h) - f(a)}{h}$$

Graphical Point of View

In Example 1 we saw that the average rate of change of a quantity can be determined directly from a graph. Here is an example that further illustrates the graphical approach.

Example 2

Average Rate of Change from a Graph

Figure 3 shows the number of sport utility vehicles (SUVs) sold in the United States each year from 1980 through 1994.

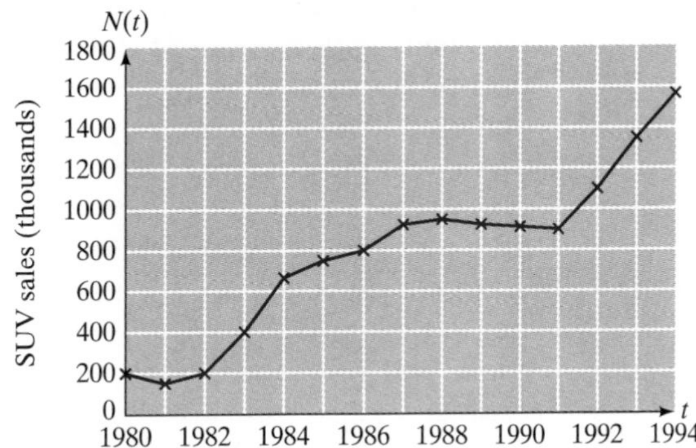

Source: Ford Motor Company/*New York Times,* February 9, 1995, p. D17.

Figure 3

a. Use the graph to estimate the average rate of change of $N(t)$ with respect to time t over the interval [1983, 1991] and interpret the result.

b. Over which 1-year period(s) was the average rate of change of $N(t)$ the greatest?

Solution

a. The average rate of change of N over the interval [1983, 1991] is given by the slope of the line through the points P and Q shown in Figure 4.

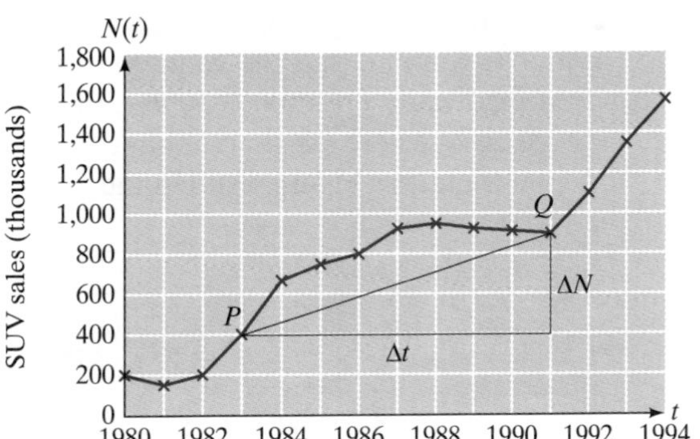

Figure 4

From the figure, we see

$$\text{Average rate of change of } N = \frac{\Delta N}{\Delta t} = \text{Slope } PQ \approx \frac{900 - 400}{1991 - 1983} = \frac{500}{8} = 62.5$$

Thus, the rate of change of N over the interval [1983, 1991] is approximately 62.5.

Question How do we interpret the result?
Answer A clue is given by the units of the average rate of change: units of N per unit of t. The units of N are thousands of SUVs and the units of t are years. Thus, the average rate of change of N is measured in thousands of SUVs per year, and we can now interpret the result as follows: Annual sales of SUVs were increasing at an average rate of 62,500 SUVs per year from 1983 to 1991.

b. The rates of change of annual sales over successive 1-year periods are given by the slopes of the individual line segments that make up the graph in Figure 3. Thus, the greatest average rate of change over a single year corresponds to the segments with the largest slope. If you look carefully at the figure, two segments, corresponding to [1983, 1984] and [1992, 1993], have slope approximately 250. Thus, the average rate of change of annual sales was largest over the 1-year periods from 1983 to 1984 and from 1992 to 1993.

Before we go on . . . Notice that we do not get exact answers from a graph; the best we can do is *estimate* the rates of change: Was the exact answer to part (a) closer to 62.5 or 63? Two people can reasonably disagree about results read from a graph, and you should bear this in mind when you check the answers to the exercises.

Perhaps the most sophisticated way to compute the average rate of change of a quantity is by using a mathematical formula or model for the quantity in question.

Algebraic Point of View

example 3

Average Rate of Change from a Formula
You are based in Indonesia, and you monitor the value of the U.S. dollar on the foreign exchange market very closely during a rather active 5-day period. Suppose you find that the value of one U.S. dollar can be well approximated by the function $R(t) = 7500 + 500t - 100t^2$ rupiahs (the rupiah is the Indonesian currency[2]), where t is time in days ($t = 0$ represents the value of the dollar at noon on Monday). Figure 5 shows the graph of R as a function of time t for $0 \le t \le 5$.

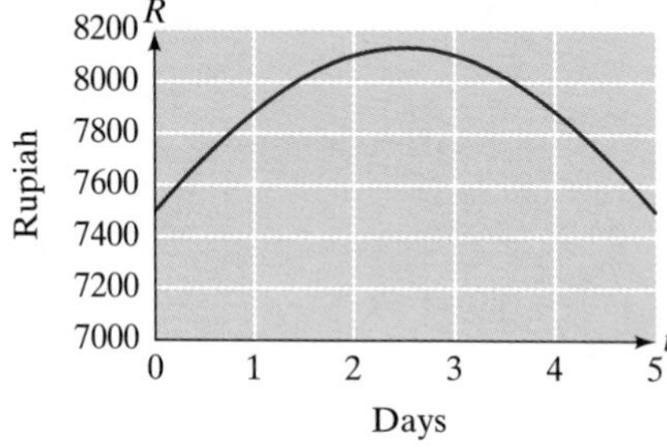

Figure 5

[2] The U.S. dollar was trading at around 8100 rupiahs in May 1999, according to the *New York Times,* May 28, 1999, p. C12.

Looking at the graph, we can see that the value of the U.S. dollar rose rather rapidly at the beginning of the week, but by the middle of the week the rise had slowed, until the market faltered and the value of the dollar began to fall more and more rapidly toward the end of the week. What was the average rate of change of the dollar's value over the 2-day period starting at noon on Tuesday (the interval [1, 3] on the t-axis)?

Solution

We have

$$\text{Average rate of change of } R \text{ over } [1, 3] = \frac{\Delta R}{\Delta t} = \frac{R(3) - R(1)}{3 - 1}$$

From the formula for $R(t)$, we find

$$R(3) = 7500 + 500(3) - 100(3)^2 = 8100$$

$$R(1) = 7500 + 500(1) - 100(1)^2 = 7900$$

Thus, the average rate of change of R is given by

$$\frac{R(3) - R(1)}{3 - 1} = \frac{8100 - 7900}{2} = \frac{200}{2} = 100 \text{ rupiahs per day}$$

In other words, the value of the U.S. dollar was increasing at an average rate of 100 rupiahs per day over the given 2-day period.

Before we go on . . . We saw that the average rate of change corresponds to the slope of a line segment through two points on the graph. Which two points?

example 4

T Difference Quotients with Technology

Continuing with Example 3, use technology to compute the average rate of change of

$$R(t) = 7500 + 500t - 100t^2$$

over the intervals $[3, 3 + h]$, where $h = 1, 0.1, 0.01, 0.001$, and 0.0001. What do the answers tell us about the value of the dollar?

Solution

We use the alternative formula

$$\text{Average rate of change of } R \text{ over } [a, a + h] = \frac{R(a + h) - R(a)}{h}$$

so

$$\text{Average rate of change over } [3, 3 + h] = \frac{R(3 + h) - R(3)}{h}$$

Graphing Calculator

If we use a graphing calculator like the TI-83, we can enter the function R as Y_1 (using X for t):

```
Y1 = 7500+500*X-100*X^2
```

We can now find the average rate of change for $h = 1$ by evaluating, on the home screen,

$(Y_1(3+1)-Y_1(3))/1$

which gives −200. To evaluate for $h = 0.1$, recall the expression using 2nd ENTER and then change the 1, both places it occurs, to 0.1, getting

$(Y_1(3+0.1)-Y_1(3))/0.1$

which gives −110. Continuing, we can evaluate the average rate of change for all the desired values of *h*. We get the values in the following table:

h	1	0.1	0.01	0.001	0.0001
Average rate of change $= \frac{R(3+h)-R(3)}{h}$	−200	−110	−101	−100.1	−100.01

Each of these values is an average rate of change of *R*. For example, the value corresponding to $h = 0.01$ is −101, which tells us: *Over the interval* [3, 3.01], *the value of the U.S. dollar was decreasing at an average rate of 101 rupiahs per day.* In other words, during the first one-hundredth of a day starting at $t = 3$ (approximately the first 15 minutes of Thursday afternoon), the value of the dollar was decreasing at an average rate of 101 rupiahs per day. Put another way, in those 15 minutes, the value of the dollar decreased at a rate that, if continued, would have resulted in a drop of 101 rupiahs over the course of a day.

We'll return to this example at the beginning of the next section.

Spreadsheet

To compute the average rate of change in a spreadsheet, like Excel, first enter the values of *h* in row 1.

	A	B	C	D	E
1	1	0.1	0.01	0.001	0.0001
2					

In row 2, we can now compute the difference quotients. In cell A2, enter the formula for the difference quotient, using the value in cell A1 for *h:*

```
                 R(3 + h)                           R(3)
=((7500+500*(3+A1)-100*(3+A1)^2) - (7500+500*3-100*3^2))/A1
```

Next, copy the contents of cell A2 to cells B2 through E2. This gives the following table of difference quotients.

	A	B	C	D	E
1	1	0.1	0.01	0.001	0.0001
2	-200	-110	-101	-100.1	-100.01

Web Site

We can use the function evaluator found by following the path

Web site → On-Line Utilities → Function Evaluator & Grapher

to do these calculations. Enter

```
((7500+500*(3+x)-100*(3+x)^2) - (7500+500*3-100*3^2))/x
```

in the "f(x)=" box (thinking of x as *h,* the value we would like to change), enter the values of *h* in the "Values of x" boxes, and press "Evaluate."

3.1 exercises

In Exercises 1–18, calculate the average rate of change of the given function over the given interval. Where appropriate, specify the units of measurement.

1.

x	0	1	2	3
$f(x)$	3	5	2	−1

Interval: [1, 3]

2.

x	0	1	2	3
$f(x)$	−1	3	2	1

Interval: [0, 2]

3.

x	−3	−2	−1	0
$f(x)$	−2.1	0	−1.5	0

Interval: [−3, −1]

4.

x	−2	−1	0	1
$f(x)$	−1.5	−0.5	4	6.5

Interval: [−1, 1]

5.

t (months)	2	4	6
$R(t)$ ($ millions)	20.2	24.3	20.1

Interval: [2, 6]

6.

x (kilos)	1	2	3
$C(x)$ (£)	2.20	3.30	4.00

Interval: [1, 3]

7.

p ($)	5.00	5.50	6.00
$q(p)$ (items)	400	300	150

Interval: [5, 5.5]

8.

t (hours)	0	0.1	0.2
$D(t)$ (miles)	0	3	6

Interval: [0.1, 0.2]

9.

Inflation (%)
7 6 5 4 3 2 1 0
1993 1994 1995 1996 1997 1998
Year

Interval: [94, 97]

10.

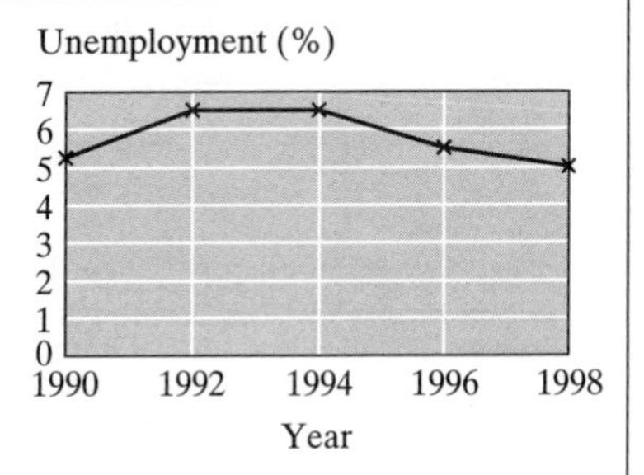

Interval: [92, 98]

11.

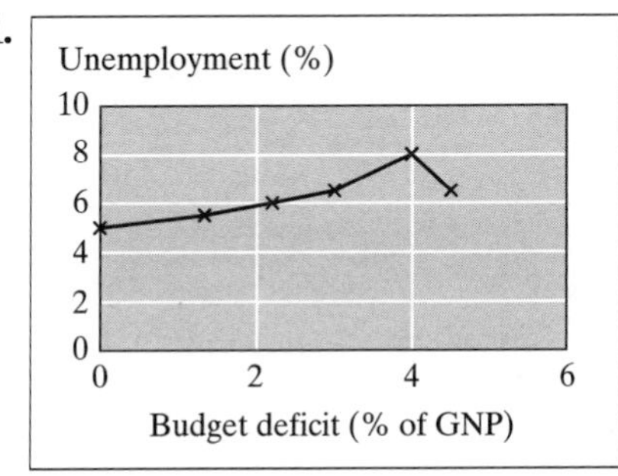

Interval: [0, 4]

12.

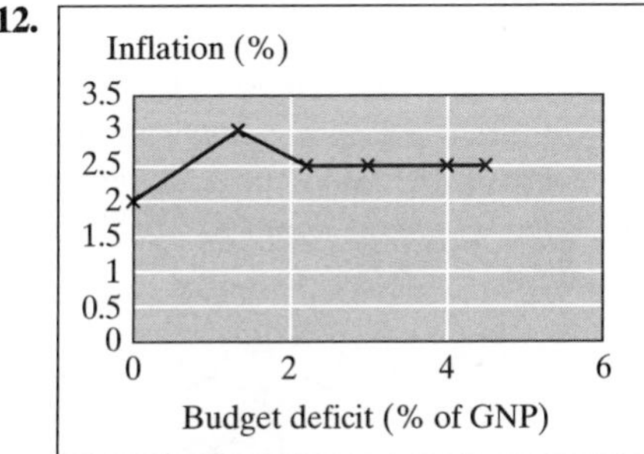

Interval: [0, 4]

13. $f(x) = x^2 - 3; [1, 3]$

14. $f(x) = 2x^2 + 4; [-1, 2]$

15. $f(x) = 2x + 4; [-2, 0]$

16. $f(x) = \frac{1}{x}; [1, 4]$

17. $f(x) = \frac{x^2}{2} + \frac{1}{x}; [2, 3]$

18. $f(x) = 3x^2 - \frac{x}{2}; [3, 4]$

T *In Exercises 19–24, use technology to calculate the average rate of change of the given function f over the intervals $[a, a+h]$, where $h = 1, 0.1, 0.01, 0.001$, and 0.0001. (It will be easier to do this if you first simplify the difference quotient as much as possible.)*

19. $f(x) = 2x^2; a = 0$

20. $f(x) = \frac{x^2}{2}; a = 1$

21. $f(x) = \frac{1}{x}; a = 2$

22. $f(x) = \frac{2}{x}; a = 1$

23. $f(x) = x^2 + 2x; a = 3$

24. $f(x) = 3x^2 - 2x; a = 0$

Applications

25. ***Profit*** The annual profits of Lotus Development Corp. for the years 1990–1994 are listed in the following table. (Losses are shown as negative.)

Year	1990	1991	1992	1993	1994
Profit (millions)	$23	$32	$80	$55	−$20

Source: Datastream: Company reports/*New York Times*, June 6, 1995, p. D8. Figures are approximate.

During which 2-year interval was the average rate of change of annual profits **a.** largest **b.** smallest? Interpret your answers by referring to the rates of change.

26. ***Net Sales*** Net sales (after-tax revenues) of Lotus Development Corp. for the years 1990–1994 are listed in the following table:

Year	1990	1991	1992	1993	1994
Net sales (millions)	$700	$810	$900	$990	$980

Source: Datastream: Company reports/*New York Times*, June 6, 1995, p. D8. Figures are approximate.

During which 2-year interval was the average rate of change of annual net sales **a.** greatest **b.** smallest? Interpret your answers by referring to the rates of change.

Venture Capital *Venture capital firms raise money to assist promising entrepreneurs until they can sell shares to the public. Exercises 27–34 are based on the following pair of graphs and the table below, which show the total amount of money controlled by venture capital firms as well as the number of venture capital firms.*

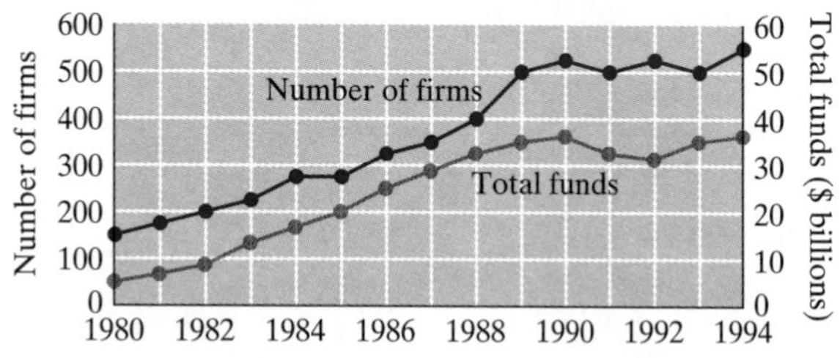

27. What was the average rate of increase in the number of venture capital firms in the 10-year period 1982–1992? (Remember to give the units of measurement.)

28. What was the average rate of increase in the total funds controlled by venture capital firms in the 5-year period 1982–1987? (Remember to give the units of measurement.)

29. During which 1-year period(s) was the number of venture capital firms increasing fastest? What was this rate of increase?

30. During which 1-year period(s) was the total funds controlled by venture capital firms increasing fastest? What was this rate of increase?

31. What is the longest period over which the average rate of change in the number of venture capital firms was zero? (There may be more than one answer.)

32. What is the longest period over which the average rate of change in the total funds controlled by venture capital firms was zero? (There may be more than one answer.)

33. If $N(t)$ is the number of firms at time t years since 1980 and $R(t)$ is the total amount of money these firms control, what does the ratio $R(t)/N(t)$ measure?

34. Refer back to Exercise 33. How fast was $R(t)/N(t)$ changing during the 1-year period 1993–1994? Interpret the result.

35. ***Market Index*** Joe Downs runs a small investment company from his basement. Every week he publishes a report on the success of his investments, including the progress of the "Joe Downs Index." At the end of one particularly memorable week, he reported that the index for that week had the value $I(t) = 1000 + 1500t - 800t^2 + 100t^3$ points, where t represents the number of business days into the week; t ranges from 0 at the beginning of the week to 5 at end of the week. The graph of I is shown below.

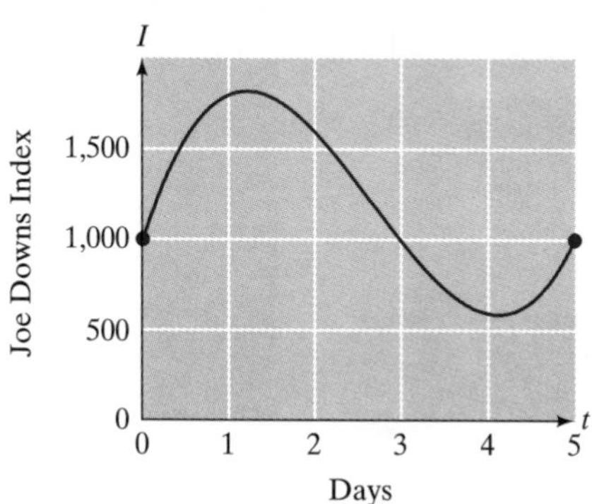

On average, how fast, and in what direction, was the index changing over the first two business days (the interval [0, 2])?

Year	'80	'81	'82	'83	'84	'85	'86	'87	'88	'89	'90	'91	'92	'93	'94
Number of firms	150	175	200	225	275	275	325	350	400	500	525	500	525	500	550
Total funds (billions)	$5	6	9	14	16	20	25	29	32	35	36	32	31	35	36

Source: Venture Economics Information Services, Boston/*New York Times*, May 25, 1995, p. D5. Data are rounded.

36. ***Market Index*** Refer to the Joe Downs Index in Exercise 35. On average, how fast, and in which direction, was the index changing over the last three business days (the interval [2, 5])?

37. ***Revenue*** Annual revenues earned by South African Breweries (SAB) from 1991 through 1997 can be modeled by the function

$$R(t) = -0.13t^2 + 0.54t + 5.22 \qquad (1 \le t \le 7)$$

where $R(t)$ is revenue in billions of dollars and t is time in years since January 1990.[3] (Thus, $t = 1$ represents January 1991.) Estimate the average rate of change of $R(t)$ over the period January 1991 through January 1995, and interpret your answer.

38. ***Profit*** Annual after-tax profits earned by South African Breweries (SAB) from 1991 through 1997 can be modeled by the function

$$P(t) = 7.0t^2 - 9.5t + 350 \qquad (1 \le t \le 7)$$

where $P(t)$ is profit in millions of dollars and t is time in years since January 1990. (Thus, $t = 1$ represents January 1991.) Estimate the average rate of change of $P(t)$ over the period January 1991 through January 1995, and interpret your answer.

[3] The models in Exercises 37 and 38 are based on a quadratic regression of graphical data published in the *New York Times,* August 27, 1997, p. D4.

Communication and Reasoning Exercises

39. Describe three ways of determining the average rate of change of f over an interval $[a, b]$. Which of the three methods is *least* precise? Explain.

40. If f is a linear function of x with slope m, what is its average rate of change over any interval $[a, b]$?

41. If the rate of change of quantity A is 2 units of quantity A per unit of quantity B, and the rate of change of quantity B is 3 units of quantity B per unit of quantity C, what is the rate of change of quantity A with respect to quantity C?

42. If the rate of change of quantity A is 2 units of quantity A per unit of quantity B, what is the rate of change of quantity B with respect to quantity A?

3.2 The Derivative As Rate of Change: A Numerical Approach

In the preceding section (Example 4) we looked at the average rate of change of the function $R(t) = 7500 + 500t - 100t^2$, which represented the value of the U.S. dollar in Indonesian currency (rupiah) during a hypothetical 5-day period, over the intervals $[3, 3 + h]$ for successively smaller values of h. Here are the values we got:

h getting smaller; interval $[3, 3 + h]$ getting smaller →

h	1	0.1	0.01	0.001	0.0001
Average rate of change over $[3, 3 + h]$	−200	−110	−101	−100.1	−100.01

Rate of change approaching −100 points per day →

The average rate of change of the value of the U.S. dollar over smaller and smaller periods of time starting at the instant $t = 3$ (noon on Thursday) appears to be getting closer and closer to −100 rupiahs per day. As we look at these shrinking periods of time, we are getting closer to looking at what happens at the *instant* $t = 3$. So it seems reasonable to say that the average rates of change are approaching the **instantaneous rate of change** at $t = 3$, which the table suggests was −100 rupiahs per day. This is how fast the value of the U.S. dollar was changing at $t = 3$ (noon on Thursday).

The process of letting h get smaller and smaller is called taking the **limit** as h approaches 0. We write $h \to 0$ as shorthand for "h approaches 0." Taking the limit of the average rates of change gives us the instantaneous rate of change. Here is our notation for this limit.

Instantaneous Rate of Change of $f(x)$ at $x = a$: Derivative

The **instantaneous rate of change** of $f(x)$ at $x = a$ is defined by taking the limit of the average rates of change of f over the intervals $[a, a + h]$, as h approaches 0.[1] We write:

$$\text{Instantaneous rate of change} = \lim_{h \to 0} \frac{f(a + h) - f(a)}{h}$$

(We read $\lim_{h \to 0}$ as "the limit as h approaches zero of".) We also call the instantaneous rate of change the **derivative** of $f(x)$ at $x = a$, which we write as $f'(a)$ (read "f prime of a"). Thus,

$$f'(a) = \lim_{h \to 0} \frac{f(a + h) - f(a)}{h}$$

The units of f' are units of f per unit of x.

Quick Example
If $f(x) = 7500 + 500x - 100x^2$, then the table above suggests (correctly) that

$$f'(3) = \lim_{h \to 0} \frac{f(3 + h) - f(3)}{h} = -100$$

Notes

1. For now, we shall trust our intuition when it comes to limits. We discuss limits in detail in the last two sections of this chapter.
2. $f'(a)$ is a number we can calculate, or at least approximate, for various values of a, as we did in the example above. Since $f'(a)$ depends on the value of a, we can think of f' as a function of a. An old name for f' is "the function *derived from f*," which has been shortened to the *derivative* of f.
3. It is because f' is a function that we sometimes refer to $f'(a)$ as "the derivative of f *evaluated* at a" or "the derivative of $f(x)$ evaluated at $x = a$."
4. Finding the derivative of f is called **differentiating f.**

Question We obtained the instantaneous rate of change above by observing that the average rates of change over the smaller and smaller intervals $[3, 3.1]$, $[3, 3.01]$, $[3, 3.001]$, . . . approach a fixed number (-100). What happens if the average rates of change do not approach any fixed number at all?
Answer That can certainly happen. When, as in the example above, the average rates of change $[f(a + h) - f(a)]/h$ *do* approach a fixed value, we say that f is

[1] We also allow $h < 0$, in which case the interval is $[a + h, a]$.

differentiable at $x = a$. Thus, $f(x) = 7500 + 500x - 100x^2$ is differentiable at $x = 3$. If, on the other hand, the average rates of change $[f(a + h) - f(a)]/h$ do *not* approach some fixed number as h approaches zero, then f is **not differentiable** at $x = a$. An example is $f(x) = x^{1/3}$, which is not differentiable at $x = 0$. (This example is discussed in Example 7 in Section 3.4.)

example 1

Instantaneous Rate of Change

The air temperature one spring morning was given by the function $f(t) = 50 + 2t^2$ °F, t hours after 7:00. How fast was the temperature rising at 9:00?

Solution

We are being asked to find the instantaneous rate of change of the temperature at $t = 2$, so we need to find $f'(2)$. To do this we examine the average rates of change

$$\frac{f(2 + h) - f(2)}{h}$$

for values of h approaching zero. Calculating the difference quotient for $h = 1$, 0.1, 0.01, 0.001, and 0.0001, we get the following values:

h	1	0.1	0.01	0.001	0.0001
Average rate of change over $[2, 2 + h]$	10	8.2	8.02	8.002	8.0002

The average rates of change are clearly approaching the number 8, so we can say that $f'(2) = 8$. Thus, at 9:00 in the morning the temperature was rising at the rate of 8°F per hour.

Before we go on . . . Actually, the table above shows only what happens in the periods of time immediately following 9:00. When examining the limit, we should also look at the periods of time immediately before 9:00, which correspond to negative values of h. If we calculate the difference quotient for several negative values of h approaching zero, we get the following table:

h	−1	−0.1	−0.01	−0.001	−0.0001
Average rate of change over $[2 + h, 2]$	6	7.8	7.98	7.998	7.9998

From this table we can see that the average rates of change in the periods just prior to 9:00 are also approaching 8°F per hour.

example 2

T Instantaneous Rate of Change Using Technology

The economist Henry Schultz calculated the following demand function for corn:

$$q(p) = \frac{176{,}000}{p^{0.77}}$$

where p is the price (in dollars) per bushel of corn and q is the number of bushels of corn that could be sold at the price p in 1 year.[2] Estimate $q'(2)$, and interpret the result.

Source: Henry Schultz, *The Theory and Measurement of Demand* [as cited in *Introduction to Mathematical Economics* by A. L. Ostrosky, Jr., and J. V. Koch (Prospect Heights, IL: Waveland Press, 1979)].

Solution

First, we calculate the average rates of change over the interval $[2, 2 + h]$ using the difference quotient:

$$\frac{q(2+h) - q(2)}{h} = \frac{\left[\dfrac{176{,}000}{(2+h)^{0.77}} - \dfrac{176{,}000}{2^{0.77}}\right]}{h}$$

Using technology, as in Example 4 of the preceding section, we can calculate the average rate of change for various values of h. We get the following table:

h	1	0.1	0.01	0.001	0.0001
Average rate of change	−27,678	−38,055	−39,561	−39,718	−39,734

These numbers show what will happen if the price rises. We are also interested in what will happen if the price *decreases*. This means that we should also look at negative h.

h	−1	−0.1	−0.01	−0.001	−0.0001
Average rate of change	−72,791	−41,579	−39,912	−39,753	−39,737

Although it is not obvious from these tables *exactly* what number is being approached, we can say what it is *approximately*. In fact, with some certainty, we can say that, to three significant digits, $q'(2) \approx -39{,}700$ bushels per dollar. Thus, at a price of \$2 per bushel, the demand for corn is falling at a rate of approximately 39,700 bushels per dollar increase in price.

Before we go on . . .

Question What's the difference between $q(2)$ and $q'(2)$? What do they mean?

Answer Briefly, $q(2)$ is the *value of* q when $p = 2$, whereas $q'(2)$ is the *rate at which* q *is changing* when $p = 2$. Here, $q(2) = 176{,}000/2^{0.77} \approx 103{,}209$ bushels. Thus, at a price of \$2 per bushel, the demand for corn (measured by annual sales) is 103,209 bushels. On the other hand, $q'(2) \approx -39{,}700$ bushels per dollar. This means that, at a price of \$2 per bushel, the demand *is falling at a rate of 39,700 bushels per \$1 increase in price.*

[2] This demand function is based on data for the period 1915–1929.

Question Do we always need to make tables of difference quotients to calculate an approximate value for the derivative?
Answer We can usually *approximate* the value of the derivative by using a single, small value of h. In the example above, the value $h = 0.0001$ would have given a pretty good approximation. The problems with using a fixed value of h are that (1) we do not get an *exact* answer, only an *approximation* of the derivative, and (2) how good an approximation it is depends on the function we are differentiating. With most of the functions we consider, it is a good enough approximation.

Calculating a Quick Approximation of the Derivative

We can calculate an approximate value of $f'(a)$ by using the formula

$$f'(a) \approx \frac{f(a+h) - f(a)}{h}$$

with a small value of h. The value $h = 0.0001$ often works (but see the following section for a graphical way of determining a good value to use).

Alternative Formula: The "Balanced Difference Quotient"

The following alternative formula often gives a more accurate result and is the one used in many calculators (the `nDeriv` function of the TI-83 does this; by default it uses $h = 0.001$, but this may be changed via an optional argument):

$$f'(a) \approx \frac{f(a+h) - f(a-h)}{2h}$$

(See Exercise 50 in this section and Exercise 48 in the following section for more explanation of the formula for the balanced difference quotient.)

example 3

Quick Approximation of the Derivative
Calculate an approximate value of $f'(5)$ if $f(x) = x^{0.4}$.

Solution
If we use $h = 0.0001$, the approximation given above is

$$f'(5) \approx \frac{f(5 + 0.0001) - f(5)}{0.0001} = \frac{f(5.0001) - f(5)}{0.0001} = \frac{5.0001^{0.4} - 5^{0.4}}{0.0001} \approx 0.1522914$$

This answer is accurate to five decimal places; in fact, $f'(5) = 0.152292315097\ldots$.

Spreadsheet

You can compute both the difference quotient and the balanced difference quotient approximations in a spreadsheet as follows. (We mentioned that the TI-83 calculator has the balanced difference quotient built in.)

	A	B	C	D	E
1	5	=A1-A2	=B1^0.4		
2	0.0001	=A1		=(C2-C1)/$A$2	
3		=A1+A2			=0.5*(D2+D3)
	a and *h*	*a* − *h*, *a*, and *a* + *h*	*f*(*x*)	difference quotients	balanced quotient

The function we are using is in cell C1 (and copied to the cells below as shown). The two difference quotients in column D use $h = -0.0001$ and $h = 0.0001$, respectively. The balanced quotient is their average (column E).

Before we go on... For comparison, the "balanced" approximation with $h = 0.0001$ is

$$f'(5) \approx \frac{f(5+0.0001)-f(5-0.0001)}{0.0002} = \frac{f(5.0001)-f(4.9999)}{0.0002} = \frac{5.0001^{0.4}-4.9999^{0.4}}{0.0002} \approx 0.15229231500$$

which is accurate to nine decimal places!

Delta Notation

We introduced the notation $f'(x)$ for the derivative of f at x, but there is another interesting notation. We have written the difference quotient as

$$\text{Average rate of change} = \frac{\Delta f}{\Delta x} \qquad \frac{\text{Change in } f}{\text{Change in } x}$$

which leads to the notation

$$\text{Instantaneous rate of change} = \lim_{\Delta x \to 0} \frac{\Delta f}{\Delta x} = \frac{df}{dx}$$

for the derivative; that is, df/dx is just another notation for $f'(x)$. Do not think of df/dx as a real quotient; it is only a symbolic form for the derivative. We read df/dx as "the derivative of f with respect to x." It is necessary to include the phrase "with respect to x" when other variables are around, and in any case it reminds us which variable is being used as the independent variable. This is Leibniz's original notation for the derivative. We say more about Leibniz notation in Section 3.4.

We finish this section with another important application.

example 4

Velocity

My friend Eric, an enthusiastic baseball player, claims he can "probably" throw a ball upward at a speed of 100 feet/second.[3] Our physicist friends tell us that the ball's height s (in feet) t seconds later would be $s = 100t - 16t^2$. Find its average velocity over the interval [2, 3] and its instantaneous velocity exactly 2 seconds after Eric throws it.

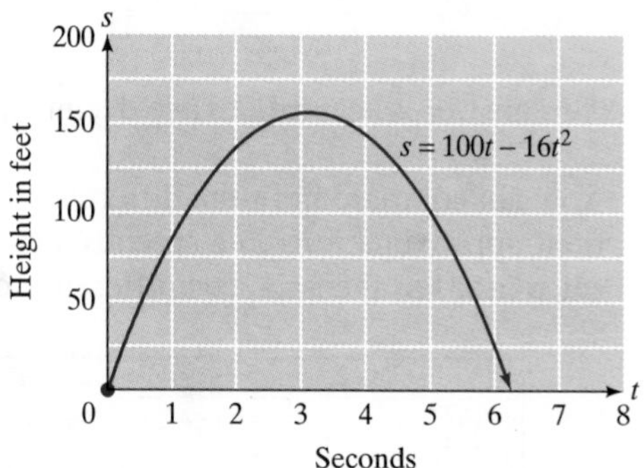

Figure 6

[3] Eric's claim is difficult to believe; 100 feet/second corresponds to around 68 miles per hour (mph), and professional pitchers can throw *forward* at about 100 mph.

Solution

The graph of the ball's height as a function of time is shown in Figure 6. Asking for the velocity is really asking for the rate of change of height with respect to time. Why? Consider average velocity first. To compute the **average velocity** of the ball from time 2 to time 3, we first compute the change in height:

$$\Delta s = s(3) - s(2) = 156 - 136 = 20 \text{ ft}$$

Since the ball rises 20 feet in $\Delta t = 1$ second, we use the defining formula *speed = distance/time* to get an average velocity:

$$\text{Average velocity} = \frac{\Delta s}{\Delta t} = \frac{20}{1} = 20 \text{ ft/sec}$$

from time $t = 2$ to $t = 3$. This is just the difference quotient, so

The average velocity is the average rate of change of height.

To get the **instantaneous velocity** at $t = 2$, we find the instantaneous rate of change of height. In other words, we need to calculate the derivative ds/dt at $t = 2$. Using the quick approximation described above, we get

$$\frac{ds}{dt} \approx \frac{s(2 + 0.0001) - s(2)}{0.0001} = \frac{s(2.0001) - s(2)}{0.001}$$

$$= \frac{\left[100(2.0001) - 16(2.0001)^2\right] - \left[100(2) - 16(2)^2\right]}{0.0001} = 35.9984 \text{ ft/sec}$$

In fact, the instantaneous velocity at $t = 2$ is exactly 36 ft/sec. (Try an even smaller value of h to persuade yourself.)

Before we go on . . . If we repeat the calculation at time $t = 5$, we get

$$\frac{ds}{dt} = \lim_{h\to 0} \frac{s(5 + h) - s(5)}{h} = -60 \text{ ft/sec}$$

The negative sign tells that the ball is *falling* at a rate of 60 feet per second at time $t = 5$. (How does the fact that it is falling at $t = 5$ show up on the graph?)

The preceding example gives another interpretation of the derivative.

Velocity

For an object that is moving in a straight line with position $s(t)$ at time t, the **average velocity** from time t to time $t + h$ is the average rate of change of position with respect to time:

$$v_{\text{average}} = \frac{s(t + h) - s(t)}{h} = \frac{\Delta s}{\Delta t}$$

The **instantaneous velocity** at time t is

$$v = \lim_{h\to 0} \frac{s(t + h) - s(t)}{h} = \frac{ds}{dt}$$

In other words, velocity is the rate of change of position with respect to time, or velocity is the derivative of position with respect to time.

Here is one last bit of notation. In the last example, we could have written the velocity either as s' or as ds/dt, as we chose to do. To write the answer to the question, that the velocity at $t = 2$ sec was 36 ft/sec, we can write either

$$s'(2) = 36$$

or

$$\left.\frac{ds}{dt}\right|_{t=2} = 36$$

The notation $|_{t=2}$ is read "evaluated at $t = 2$." Similarly, if $y = f(x)$, we can write the instantaneous rate of change of f at $x = 5$ either in functional notation as

$$f'(5) \qquad \text{The derivative of } f \text{ evaluated at } x = 5$$

or in Leibniz notation as

$$\left.\frac{dy}{dx}\right|_{x=5} \qquad \text{The derivative of } y \text{ evaluated at } x = 5$$

The latter notation is obviously more cumbersome than the functional notation $f'(5)$, but the notation dy/dx has compensating advantages, as we shall see. You should feel free to use whichever notation is more convenient in each case. In fact, you should practice using both notations.

3.2 exercises

Consider the functions in Exercises 1–4 as representing the value of the U.S. dollar in Indian rupees as a function of the time t in days.[4] *Find the average rates of change of R(t) over the time intervals [t, t + h], where t is as indicated and h = 1, 0.1, and 0.01 days. Hence, estimate the instantaneous rate of change of R at time t, specifying the units of measurement. (Use smaller values of h to check your estimates.)*

1. $R(t) = 50t - t^2; t = 5$

2. $R(t) = 60t - 2t^2; t = 3$

3. $R(t) = 100 + 20t^3; t = 1$

4. $R(t) = 1000 + 50t - t^3; t = 2$

Each of the functions in Exercises 5–8 gives the cost to manufacture x items. Find the average cost per unit of manufacturing h more items (i.e., the average rate of change of the total cost) at a production level of x, where x is as indicated and h = 10 and 1. Hence, estimate the instantaneous rate of change of the total cost at the given production level x, specifying the units of measurement. (Use smaller values of h to check your estimates.)

5. $C(x) = 10{,}000 + 5x - \dfrac{x^2}{10{,}000}; x = 1000$

6. $C(x) = 20{,}000 + 7x - \dfrac{x^2}{20{,}000}; x = 10{,}000$

7. $C(x) = 15{,}000 + 100x + \dfrac{1000}{x}; x = 100$

8. $C(x) = 20{,}000 + 50x + \dfrac{10{,}000}{x}; x = 100$

In Exercises 9–12, estimate the derivative of the given function at the indicated point.

9. $f(x) = 1 - 2x; x = 2$

10. $f(x) = \dfrac{x}{3} - 1; x = -3$

11. $f(x) = \dfrac{x^2}{4} - \dfrac{x^3}{3}; x = -1$

12. $f(x) = \dfrac{x^2}{2} + \dfrac{x}{4}; x = 2$

In Exercises 13–22, estimate the given quantity.

13. $f(x) = x^3$; estimate $f'(-1)$

14. $g(x) = x^4$; estimate $g'(-2)$

15. $g(t) = \dfrac{1}{t^5}$; estimate $g'(1)$

16. $s(t) = \dfrac{1}{t^3}$; estimate $s'(-2)$

17. $y = 4x^2$; estimate $\left.\dfrac{dy}{dx}\right|_{x=2}$

18. $y = 1 - x^2$; estimate $\left.\dfrac{dy}{dx}\right|_{x=-1}$

19. $s = 4t + t^2$; estimate $\left.\dfrac{ds}{dt}\right|_{t=-2}$

20. $s = t - t^2$; estimate $\left.\dfrac{ds}{dt}\right|_{t=2}$

21. $R = \dfrac{1}{p}$; estimate $\left.\dfrac{dR}{dp}\right|_{p=20}$

22. $R = \sqrt{p}$; estimate $\left.\dfrac{dR}{dp}\right|_{p=400}$

[4] The U.S. dollar was trading at around 43 rupees in May 1999, according to the *New York Times*, May 28, 1999, p. C12.

E[5] *Exercises 23–26 are based on logarithmic and exponential functions. Estimate the given quantity.*

E **23.** $f(x) = e^x$; estimate $f'(0)$

E **24.** $f(x) = 2e^x$; estimate $f'(1)$

E **25.** $f(x) = \ln x$; estimate $f'(1)$

E **26.** $f(x) = \ln x$; estimate $f'(2)$

[5]E indicates exercises using exponential functions.

Δ[6] *Exercises 27–30 are based on trigonometric functions. Estimate the given quantity.*

Δ **27.** $f(x) = \sin x$; estimate $f'(0)$

Δ **28.** $f(x) = \cos x$; estimate $f'(\pi/2)$

Δ **29.** $f(x) = \tan x$; estimate $f'(\pi/4)$

Δ **30.** $f(x) = \sec x$; estimate $f'(\pi/4)$

[6]Δ indicates exercises using trigonometric functions.

Applications

31. ***Revenue*** Annual revenues earned by South African Breweries (SAB) from 1991 through 1997 can be modeled by the function

$$R(t) = -0.13t^2 + 0.54t + 5.22 \qquad (1 \le t \le 7)$$

where $R(t)$ is revenue in billions of dollars and t is time in years since January 1990. (Thus, $t = 1$ represents January 1991.)[7]

a. Estimate the average rate of change of $R(t)$ over the period January 1991 through January 1995, and interpret your answer.

b. Estimate the instantaneous rate of change of $R(t)$ in January 1991, and interpret your answer.

c. The answers to parts (a) and (b) have opposite signs. What does this indicate about SAB's revenues?

32. ***Profit*** Annual after-tax profits earned by South African Breweries (SAB) from 1991 through 1997 can be modeled by the function

$$P(t) = 7.0t^2 - 9.5t + 350 \qquad (1 \le t \le 7)$$

where $P(t)$ is profit in millions of dollars and t is time in years since January 1990. (Thus, $t = 1$ represents January 1991.)

a. Estimate the average rate of change of $P(t)$ over the period January 1991 through January 1995, and interpret your answer.

b. Estimate the instantaneous rate of change of $P(t)$ in January 1991, and interpret your answer.

c. The answer to part (b) is less than the answer to part (a). What does this indicate about SAB's after-tax profits?

33. ***Demand*** Suppose the demand for a new brand of sneakers is given by

$$q = \frac{5{,}000{,}000}{p}$$

where p is the price per pair of sneakers, in dollars, and q is the number of pairs of sneakers that can be sold at price p. Find $q(100)$ and estimate $\left.\frac{dq}{dp}\right|_{p=100}$. Interpret your answers.

34. ***Demand*** Suppose the demand for an old brand of TV is given by

$$q = \frac{100{,}000}{p + 10}$$

where p is the price per TV set, in dollars, and q is the number of TV sets that can be sold at price p. Find $q(190)$ and estimate $\left.\frac{dq}{dp}\right|_{p=190}$. Interpret your answers.

35. ***Velocity*** If a stone is dropped from a height of 100 feet, its height after t seconds is given by $s = 100 - 16t^2$.

a. Find the stone's average velocity over the period $[2, 4]$.

b. Estimate its instantaneous velocity at time $t = 4$.

36. ***Velocity*** If a stone is thrown down at a speed of 120 ft/sec from a height of 1000 feet, its height after t seconds is given by $s = 1000 - 120t - 16t^2$.

a. Find the stone's average velocity over the period $[1, 3]$.

b. Estimate its instantaneous velocity at time $t = 3$.

Exercises 37 and 38 are applications of Einstein's Special Theory of Relativity and relate to objects that are moving extremely fast. In science fiction terminology, a speed of warp 1 *is the speed of light—about 3×10^8 meters per second. (Thus, for instance, a speed of warp 0.8 corresponds to 80% of the speed of light—about 2.4×10^8 meters per second.)*

37. ***Lorentz Contraction*** According to Einstein's Special Theory of Relativity, a moving object appears to get shorter to a stationary observer as its speed approaches the speed of light. If a spaceship that has a length of 100 meters at rest travels at a speed of warp p, its length in meters, as measured by a stationary observer, is given by

$$L(p) = 100\sqrt{1 - p^2} \qquad \text{with domain } [0, 1]$$

[7]The models in Exercises 31 and 32 are based on a quadratic regression of graphical data published in the *New York Times*, August 27, 1997, p. D4.

Estimate $L(0.95)$ and $L'(0.95)$. What do these figures tell you?

38. ***Time Dilation*** Another prediction of Einstein's Special Theory of Relativity is that, to a stationary observer, clocks (as well as all biological processes) in a moving object appear to go more and more slowly as the speed of the object approaches the speed of light. If a spaceship travels at a speed of warp p, then the time it takes for an on-board clock to register 1 second, as measured by a stationary observer, is given by

$$T(p) = \frac{1}{\sqrt{1-p^2}} \text{ seconds} \qquad \text{with domain } [0, 1)$$

Estimate $T(0.95)$ and $T'(0.95)$. What do these figures tell you?

39. ***Learning to Speak*** Let $p(t)$ represent the percentage of children who are able to speak at the age of t months.

a. It is found that $p(10) = 60$ and $\left.\frac{dp}{dt}\right|_{t=10} = 18.2$. What does this mean?

b. As t increases, what happens to p and $\frac{dp}{dt}$?

Source: Data presented in "The Emergence of Intelligence" by William H. Calvin in *Scientific American,* October 1994, pp. 101–107.

40. ***Learning to Read*** Let $p(t)$ represent the number of children in your class who learned to read at the age of t years.

a. Assuming that everyone in your class could read by the age of 7, what does this tell you about $p(7)$ and $\left.\frac{dp}{dt}\right|_{t=7}$?

b. Assuming that 25.0% of the people in your class could read by the age of 5, and that 25.3% of them could read by the age of 5 years and 1 month, estimate $\left.\frac{dp}{dt}\right|_{t=5}$. Remember to give its units.

41. ***On-Line Services*** On January 1, 1996, America Online was the largest on-line service provider, with 4.5 million subscribers, and it was adding new subscribers at a rate of 60,000 per week. If $A(t)$ is the number of America Online subscribers t weeks after January 1, 1996, what do the given data tell you about values of the function A and its derivative?

Source: Information and Interactive Services Report/*New York Times,* January 2, 1996, p. C14.

42. ***On-Line Services*** On January 1, 1996, Prodigy was the third-largest on-line service provider, with 1.6 million subscribers, but it was losing subscribers. If $P(t)$ is the number of Prodigy subscribers t weeks after January 1, 1996, what do the given data tell you about values of the function P and its derivative?

Source: See the source in Exercise 41.

Exercises marked with **E** *are based on logarithmic and exponential functions.*

E **43.** ***Sales*** Weekly sales of a new brand of sneakers are given by

$$S(t) = 200 - 150e^{-t/10}$$

pairs per week, where t is the number of weeks since the introduction of the brand. Estimate $S(5)$ and $\left.\frac{dS}{dt}\right|_{t=5}$, and interpret your answers.

E **44.** ***Sales*** Weekly sales of an old brand of TV are given by

$$S(t) = 100e^{-t/5}$$

sets per week, where t is the number of weeks after the introduction of a competing brand. Estimate $S(5)$ and $\left.\frac{dS}{dt}\right|_{t=5}$, and interpret your answers.

E **45.** ***Logistic Growth in Demand*** The demand for a new product can be modeled by a **logistic** curve of the form

$$q(t) = \frac{N}{1 + ke^{-rt}}$$

where $q(t)$ is the total number of units sold t months after the introduction of the new product, and N, k, and r are constants that depend on the product and the market. Assume that the demand for video game units is determined by this formula, with $N = 10{,}000$, $k = 0.5$, and $r = 0.4$. Estimate $q(2)$ and $\left.\frac{dq}{dt}\right|_{t=2}$, and interpret the results.

E **46.** ***Information Highway*** When the National Science Foundation's Internet network was in its infancy, the amount of information transmitted each month could be modeled by the equation

$$q(t) = \frac{2e^{0.69t}}{3 + 1.5e^{-0.4t}} \qquad (0 \le t \le 6)$$

where q is the amount of information transmitted each month, in billions of data packets, and t is the number of years since the start of 1988.[8] Estimate the number of data packets transmitted during the first month of 1990, and also the rate at which this number was increasing.

[8]This is the authors' model, based on figures published in the *New York Times,* November 3, 1993.

Communication and Reasoning Exercises

47. Use the difference quotient to explain the fact that if f is a linear function, then the average rate of change over any interval equals the instantaneous rate of change at any point.

48. Give a numerical explanation of the fact that if f is a linear function, then the average rate of change over any interval equals the instantaneous rate of change at any point.

49. Explain why we cannot put $h = 0$ in the formula

$$f'(x) = \lim_{h \to 0} \frac{f(x+h) - f(x)}{h}$$

for the derivative of f.

50. The balanced difference quotient

$$f'(a) \approx \frac{f(a + 0.0001) - f(a - 0.0001)}{0.0002}$$

is the average rate of change of f over what interval?

3.3 The Derivative As Slope: A Geometric Approach

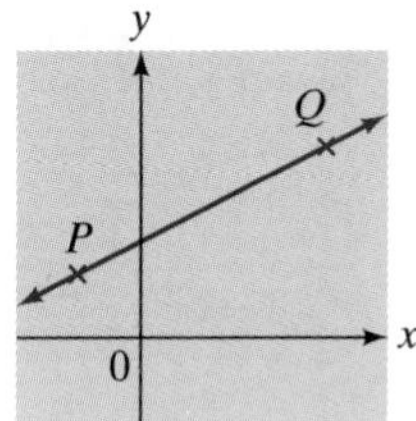

Figure 7

We saw in Chapter 1 that the rate of change of a linear function is the slope of the corresponding line. For a function whose graph is not a straight line, we have no notion of *the* slope of the graph. Look at the line in Figure 7. This line is just as steep at the point P as it is at the point Q. Put another way, the line is rising just as fast at P as it is at Q. Now look at the curve shown in Figure 8. This graph is steeper at Q than it is at P. This suggests that the "slope" of the curve increases from left to right, so no single number can measure the steepness of the whole graph. Instead, we have to assign a number *to each point of the graph* to measure its steepness *at that point.*

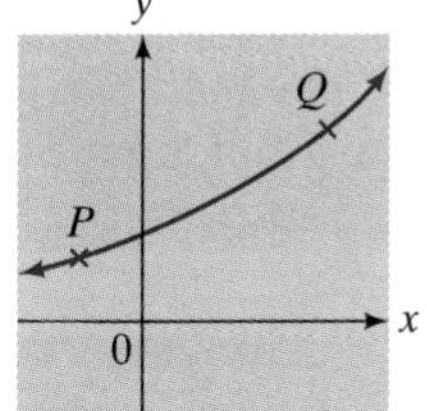

Figure 8

Question How can we measure the steepness of a graph at a specified point?
Answer Building on what we already know, the steepness of a line, we say that the steepness of a graph at a specified point is *the slope of the tangent line to the graph at that point.*

Question What is the tangent line to the graph at a point?
Answer A tangent line to a *circle* is a line that touches the circle in just one point. A tangent line gives the circle "a glancing blow," as shown in Figure 9. In the case of smooth curves other than circles, a tangent line may touch the given curve at more than one point or pass through it (Figure 10).

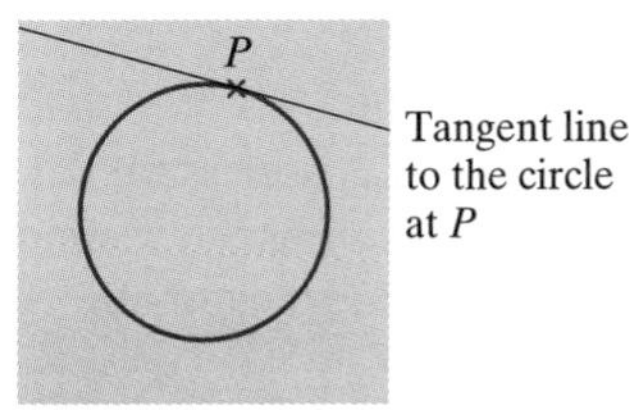

Figure 9

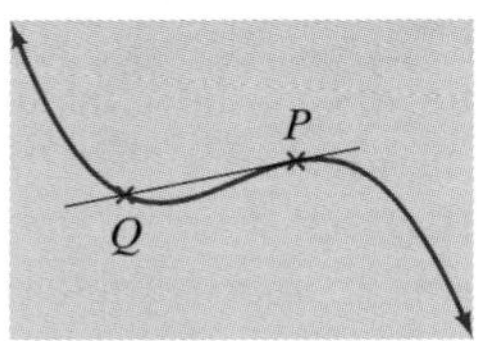

Tangent line at P intersects graph at Q

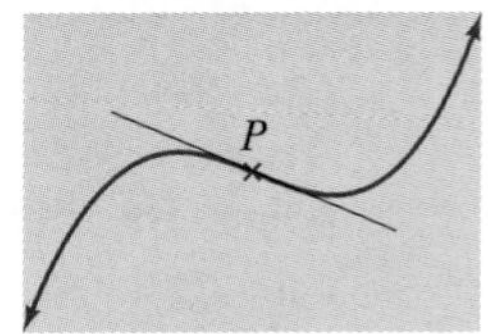

Tangent line at P passes through curve at P

Figure 10

However, all the tangent lines we have sketched have the following property in common: If we focus on a small portion of the curve very close to the point P—in other words, if we "zoom in" to the graph near the point P—the curve will appear almost straight and almost indistinguishable from the tangent line (Figure 11).

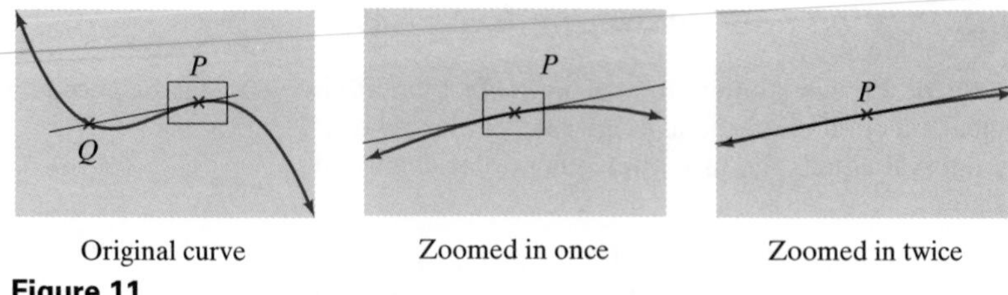

Figure 11

This observation leads to the following definition.

Preliminary Definition of Tangent Line

The tangent line to a curve at a point P is the (one and only) line through P that becomes indistinguishable from the curve itself if we zoom in sufficiently close to the graph near P.

Note
This definition is imprecise: Exactly what do we mean by "sufficiently close" and "indistinguishable"? We will remedy this by giving a more precise definition below.

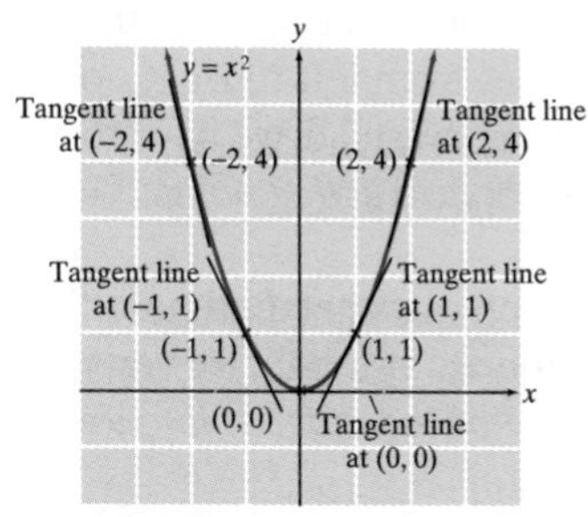

Figure 12

Figure 12 shows lines tangent to the parabola $y = x^2$ at the five points $(-2, 4)$, $(-1, 1)$, $(0, 0)$, $(1, 1)$, and $(2, 4)$.

The steepness of the parabola at the point $(-2, 4)$ is the slope of the tangent line at $(-2, 4)$. This slope is negative and of large magnitude. Similarly, the steepness of the parabola at $(-1, 1)$ is the slope of the tangent line at $(-1, 1)$. This slope is also negative but of smaller magnitude. The steepness of the graph at $(0, 0)$ is 0 because the tangent line there is horizontal (it happens to be the x-axis). The steepness of the graph at $(1, 1)$ is positive because the tangent line there has a positive slope, and the steepness of the graph at $(2, 4)$ is also positive, but larger.

In summary:

Steepness of a Graph

The steepness of a graph at a point P is measured by the slope of the tangent to the curve at P.

Question How do we calculate the slope of a tangent to a curve?
Answer We saw in Chapter 1 that to calculate the slope of a line we need two points. The difficulty is that we have only one point, which by itself is useless for calculating the slope. We can approximate the slope, however, by choosing as our second point another point on the graph "close" to the first, as we do in the next example.

example 1

Slope of the Tangent Equals Rate of Change
Estimate the steepness of the curve $y = x^2 - 3x + 3$ at the point on the graph where $x = 2$.

Solution
Figure 13 shows the graph of $f(x) = x^2 - 3x + 3$ with the tangent line whose slope we would like to calculate.

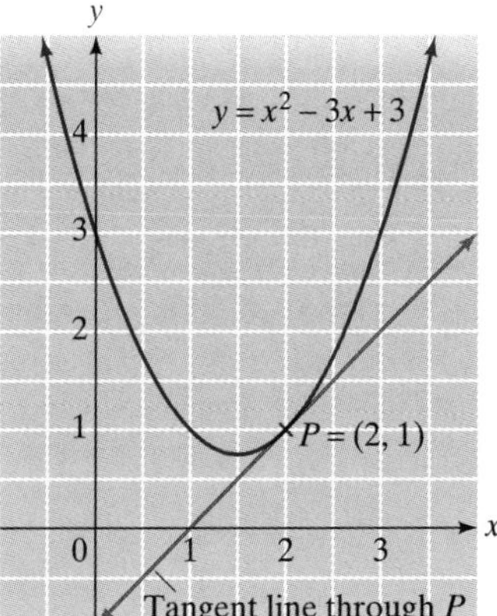

Figure 13

To estimate the line's slope, what if we choose a nearby point Q on the graph, as shown in Figure 14?

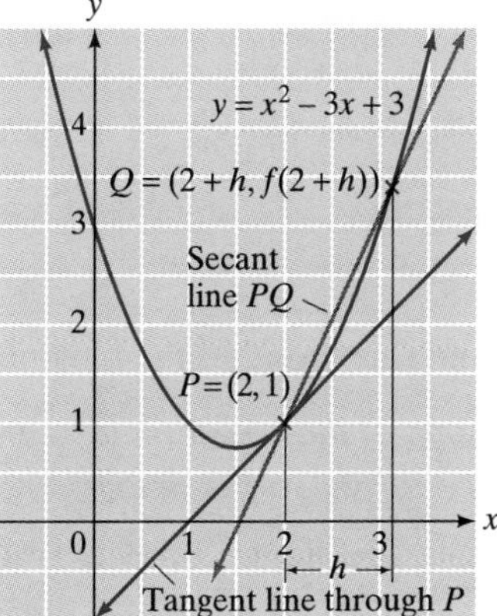

Figure 14

We get the point Q as follows: Since the original point P has x-coordinate 2, we choose a nearby value by adding a small quantity h, getting $2 + h$. The point Q is then taken to be the point on the graph with x-coordinate $2 + h$ ($h > 0$ in Figure 14, but h may also be negative, which would put Q to the left of P). The y-coordinate of Q is $f(2 + h)$. Thus, Q is the point $(2 + h, f(2 + h))$.

The line through P and Q, or any line through two points on the graph, is called a **secant line** of the graph. Its slope, which we denote by m_{sec}, can be calculated using the usual slope formula:

$$m_{\text{sec}} = \frac{y_2 - y_1}{x_2 - x_1} = \frac{f(2 + h) - f(2)}{(2 + h) - 2} = \frac{f(2 + h) - f(2)}{h}$$

Do you recognize this? It is the formula for the average rate of change that we saw in Section 3.1. In other words, *the slope of the secant line PQ is the average rate of change of f(x) over the interval* $[2, 2 + h]$.

As Figure 15 shows, the smaller we choose h, the more accurately the secant line approximates the tangent line, so the more accurately its slope approximates that of the tangent.

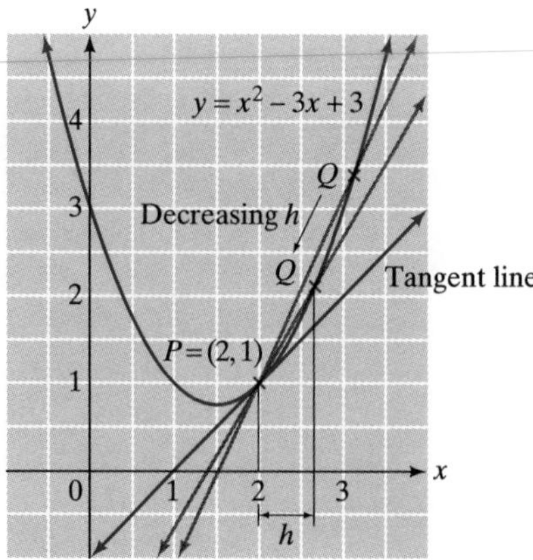

Figure 15

In other words,

$$\begin{aligned}
\text{Slope of tangent line at } (2,1) &= \lim_{h\to 0} (\text{slope of secant line over } [2, 2+h]) \\
&= \lim_{h\to 0} \frac{f(2+h) - f(2)}{h} \\
&= \text{Instantaneous rate of change of } f(x) \text{ at } x = 2 \\
&= \text{Derivative of } f(x) \text{ at } x = 2 \\
&= f'(2)
\end{aligned}$$

In short—and this is one of the most important facts in this book:

The slope of the tangent line to the graph of the function f, at the point where $x = a$, is the derivative at a, $f'(a)$.

To estimate the slope in this example, let us try the balanced approximation we used in Example 3 in the preceding section:

$$\text{Slope of tangent} \approx \frac{f(2+h) - f(2-h)}{2h} = \frac{f(2.0001) - f(1.9999)}{0.0002} = \frac{1.0001 - 0.9999}{0.0002} = 1$$

In fact, this happens to be the exact answer![1]

Before we go on . . . Did this example seem a bit long? Stripped of the explanations, here is all we did: The slope of the tangent line to the graph of $f(x) = x^2 - 3x + 3$ at the point where $x = 2$ is $f'(2)$:

$$\text{Slope of tangent} = f'(2) \approx \frac{f(2+h) - f(2-h)}{2h} = \frac{f(2.0001) - f(1.9999)}{0.0002} = 1$$

We can summarize the ideas presented above.

[1] The balanced difference quotient *always* gives the exact answer for quadratic functions.

Secant and Tangent Lines

The *slope of the secant line* through the points on the graph of f where $x = a$ and $x = a + h$ is given by the average rate of change, or the difference quotient:

$$m_{\text{sec}} = \text{Slope of secant} = \text{Average rate of change} = \frac{f(a+h) - f(a)}{h}$$

The *slope of the tangent line* through the point on the graph of f where $x = a$ is given by the instantaneous rate of change, or the derivative:

$$m_{\text{tan}} = \text{Slope of tangent} = \text{Derivative} = f'(a) = \lim_{h \to 0} \frac{f(a+h) - f(a)}{h}$$

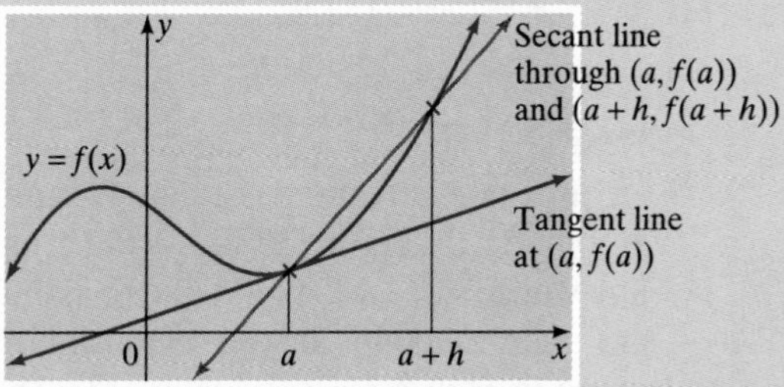

We can approximate the slope of the tangent through the point where $x = a$ using the balanced difference quotient:

$$m_{\text{tan}} \approx \frac{f(a + 0.0001) - f(a - 0.0001)}{0.0002}$$

Note

We can now give a more precise definition of what we mean by the tangent line to a point P on the graph of f at a given point. First recall from Section 1.3 that we can completely describe a line by specifying a point on the line and its slope. This is how we will specify the tangent line (refer to the diagram above).

Definition of Tangent Line

The **tangent line** to the graph of f at the point $P(a, f(a))$ is the straight line that passes through P with slope $f'(a)$.

Estimating the Derivative Graphically

Take another look at Figure 13. As we saw, the tangent line and the curve itself almost coincide if we focus on a small portion of the graph very close to the point $P(2, 1)$. Thus, if we zoomed in to the graph near the point $(2, 1)$, the actual graph would appear almost straight—and almost indistinguishable from the tangent line. The next example illustrates how to use this fact to get a good approximation of the derivative.

example 2

T Estimating the Slope by Zooming In

Use graphing technology to estimate the derivative of $f(x) = x^3 - 2x^2 + x$ at the point on the graph where $x = 3$.

Solution

Here is a step-by-step approach to estimate the derivative of $f(x)$ at the point $x = a$.

Step 1 *Graph the curve $y = f(x)$ so that the point where $x = a$ is centered in the x-range.* First, enter the function

```
X^3-2*X^2+X
```

Here, $a = 3$ and we wish to set the window so that the point on the graph where $x = 3$ is centered in the x-range. To do this, we take $h = 1$ and set

$$X_{min} = a - h \qquad \text{and} \qquad X_{max} = a + h$$

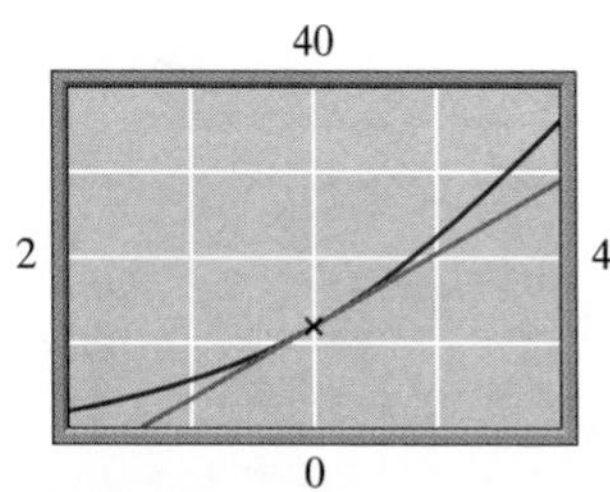

Figure 16

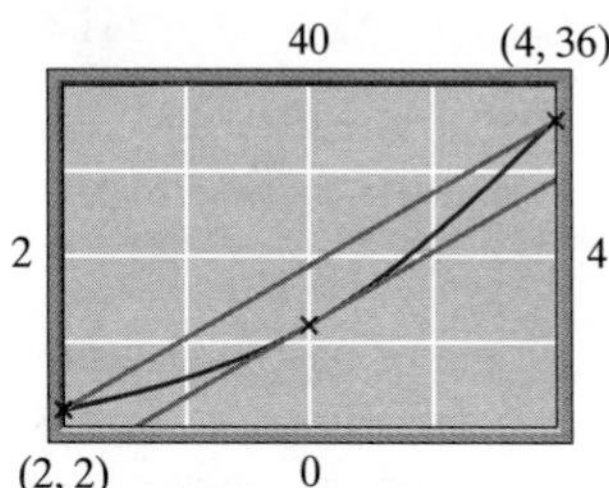

Figure 17

Here, $X_{min} = 3 - 1 = 2$ and $X_{max} = 3 + 1 = 4$. We can set the y-range by trial and error or take advantage of many graphing technologies' abilities to set the y-range automatically. Figure 16 shows the graph with $0 \le y \le 40$ together with the tangent line whose slope we would like to estimate. Here is a crude estimate of the slope we are after: the slope of the secant line through the left and right end points of the graph in Figure 16 (see Figure 17). These are the points $(3 - 1, f(3 - 1)) = (2, 2)$ and $(3 + 1, f(3 + 1)) = (4, 36)$. The slope of the line through these two points is

$$m = \frac{f(4) - f(2)}{4 - 2} = \frac{36 - 2}{4 - 2} = 17$$

What we have just calculated is, in fact, the balanced difference quotient with $h = 1$:

$$m = \frac{f(3 + 1) - f(3 - 1)}{2 \times 1} = 17$$

(We can calculate the balanced difference quotient using technology, as we did with the ordinary difference quotient in Section 3.1. In particular, if we use a TI-83, we can take advantage of the fact that we will already have the function f entered as Y_1.) Since the graph in Figure 16 curves well away from the tangent line, the slope of this secant line may not be a good approximation to the slope of the tangent line. To obtain a more accurate estimate, we need to zoom in.

Step 2 *Zoom in, leaving the point $x = a$ centered in the x-range.* We do this by choosing a smaller value of h. If we choose a much smaller value—say, $h = 0.0001$, we will set $X_{min} = 3 - 0.0001 = 2.9999$ and $X_{max} = 3 + 0.0001 = 3.0001$. Figure 18 shows the resulting graph with $11.998 \le y \le 12.002$.

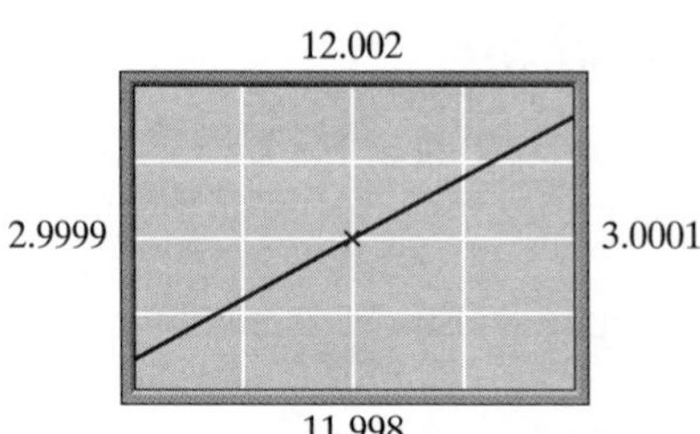

Figure 18

The curve is now indistinguishable from a straight line; the graph, the tangent line, and the secant line all look the same. Therefore, the slope of the tangent line will be well approximated by the slope of the secant line. If we had not chosen so small a value of h at first, we would have to go on to the next step.

Step 3 *Repeat Step 2 using smaller and smaller values of h until the graph appears to be a straight line; then measure the slope of the secant line that passes through the endpoints.* We have already chosen a value of h small enough to do the job, as we see in Figure 18. The slope of the secant line is the balanced difference quotient with $h = 0.0001$, so we calculate

$$m_{\tan} \approx m_{\sec} = \frac{f(3 + 0.0001) - f(3 - 0.0001)}{2 \times 0.0001} \approx \frac{12.0016 - 11.9984}{0.0002} = 16$$

Thus, the slope of the tangent is approximately 16.

Optional Internet Topic: Sketching the Graph of the Derivative

If you follow the path

Web site → On-Line Text → Sketching the Graph of the Derivative

you will find on-line text, examples, and exercises on making a rough sketch of the graph of the derivative of a function based on a knowledge of the graph of the function itself.

3.3 exercises

In each of the graphs in Exercises 1–6, say at which labeled point the slope of the tangent is ***a.*** *greatest* ***b.*** *least (in the sense that −7 is less than 1).*

1.

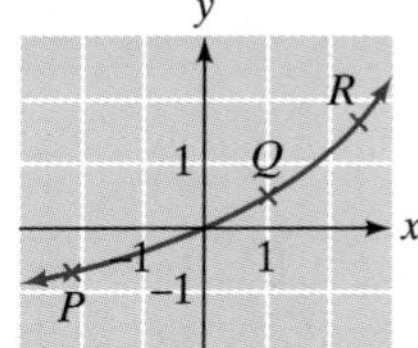

2.

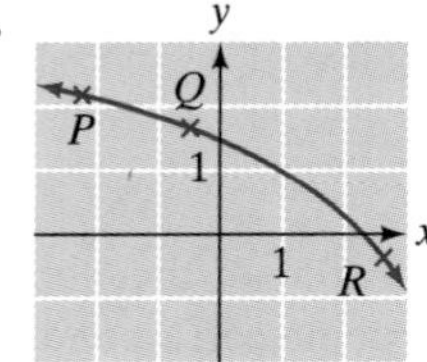

3.

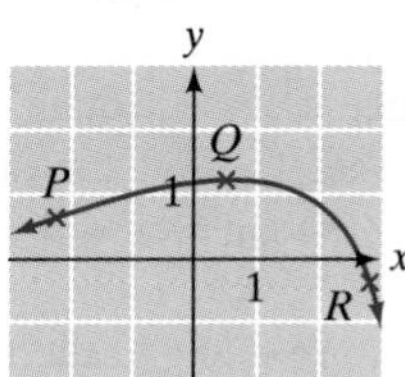

4.

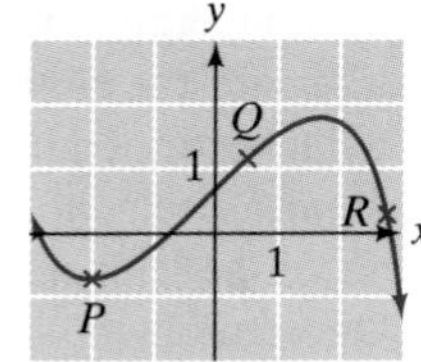

5.

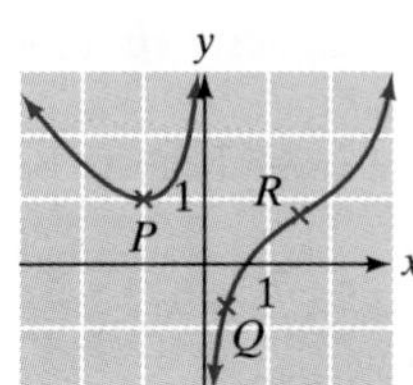

6.

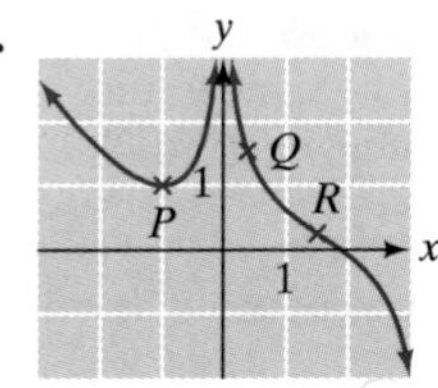

In each of Exercises 7–10, three slopes are given. For each slope, determine at which of the labeled points on the graph the tangent line has that slope.

7. a. 0 **b.** 4 **c.** −1

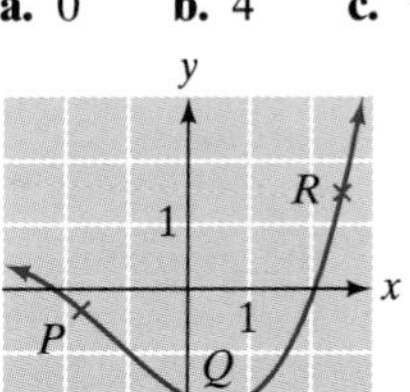

8. a. 0 **b.** 1 **c.** −1

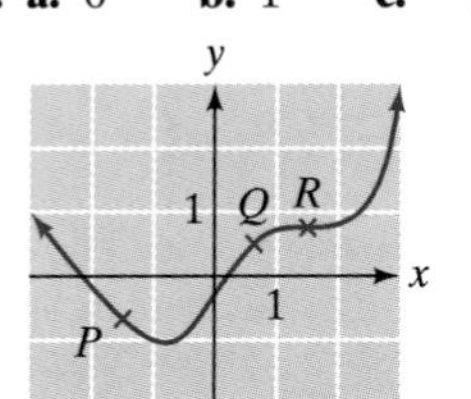

9. a. 0 **b.** 3 **c.** −3

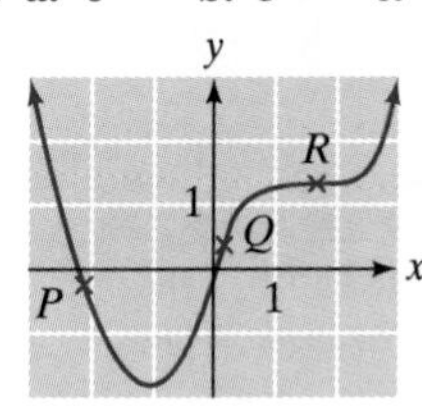

10. a. 0 **b.** 3 **c.** 1

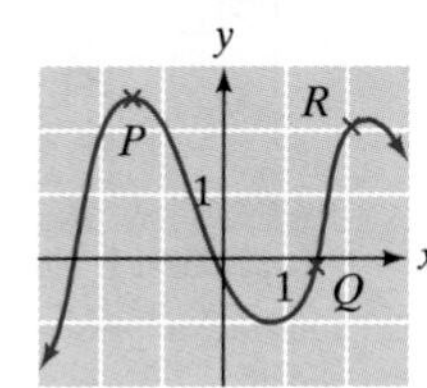

In each of Exercises 11–14, find the approximate coordinates of all points (if any) where the slope of the tangent is **a.** *0,* **b.** *1,* **c.** *−1.*

11.

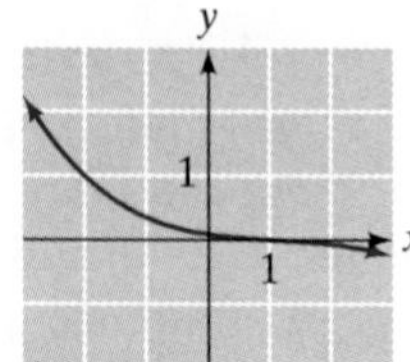

12.

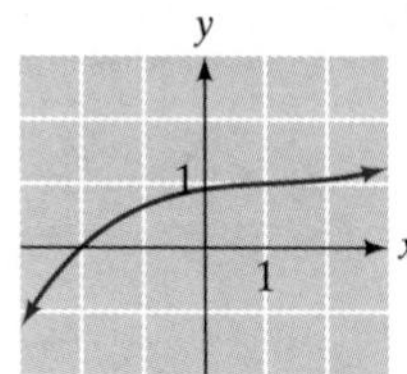

13.

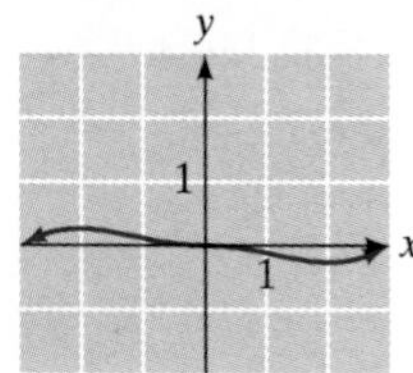

14.

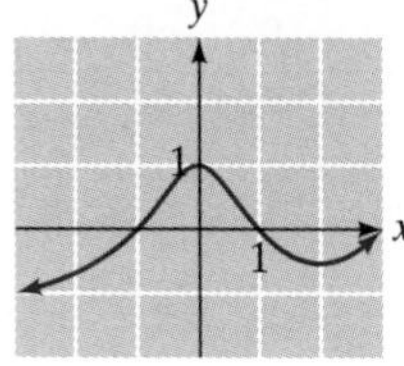

15. Complete the following: The tangent to the graph of the function f at the point where $x = a$ is the line that passes through ______ with slope ______.

16. Complete the following: The difference quotient for f at the point where $x = a$ gives the slope of the ______ that passes through ______.

In Exercises 17–22, **a.** *find the slope of the secant line through the points where $x = a$ and $x = a + h$,* **b.** *use the answer to part (a) to estimate the slope of the tangent to the graph of f at the point where $x = a$.*

17. $f(x) = x^2 + 1\ (a = 2, h = 3)$

18. $f(x) = x^2 - 3\ (a = 1, h = -3)$

19. $f(x) = 2 - x^2\ (a = -1, h = 2)$

20. $f(x) = -1 - x^2\ (a = 0, h = -1)$

21. $f(x) = 1 - 2x\ (a = 2, h = -5)$

22. $f(x) = \frac{x}{3} - 1\ (a = -3, h = 1)$

In Exercises 23–28, **a.** *use any method to find the slope of the tangent to the graph of the given function at the point that has the given x-coordinate;* **b.** *find an equation of the tangent line in part (a) In each case, sketch the curve together with the appropriate tangent line.* [Hint for part (b): *To find the equation of the tangent line, use the point-slope formula and the result of part (a)*]

23. $f(x) = x^3; x = -1$

24. $f(x) = x^2; x = 0$

25. $f(x) = x + \frac{1}{x}; x = 2$

26. $f(x) = \frac{1}{x^2}; x = 1$

27. $f(x) = \sqrt{x}; x = 4$

28. $f(x) = 2x + 4; x = -1$

T *In Exercises 29–32, use graphing technology to estimate the derivative of the given function at the given point.*

29. $f(x) = \frac{1}{x^{1.1} - 4}; x = 1$

30. $f(x) = \frac{1}{x^{1.1} - 4}; x = 2$

31. $f(x) = \sqrt{1 - x^2}; x = 0.5$

32. $f(x) = \sqrt{1 - x^2}; x = 0.75$

T *In Exercises 33–36, use a graph of the given function $f(x)$ over the given range of values of x to determine the approximate values of x in that range for which $f'(x) = 0$ (if any).*

33. $f(x) = x^{3.4} - x^{1.2}; 0 \le x \le 5$

34. $f(x) = 2x^{4.1} - x^{1.3}; 0 \le x \le 5$

35. $f(x) = (x - 1)(x - 2)(x - 3); -1 \le x \le 5$

36. $f(x) = (x - 1)^2(x - 2)^2(x - 3); -1 \le x \le 5$

Applications

37. ***Weather*** The following graph shows the approximate average daily high temperatures in New York's Central Park.

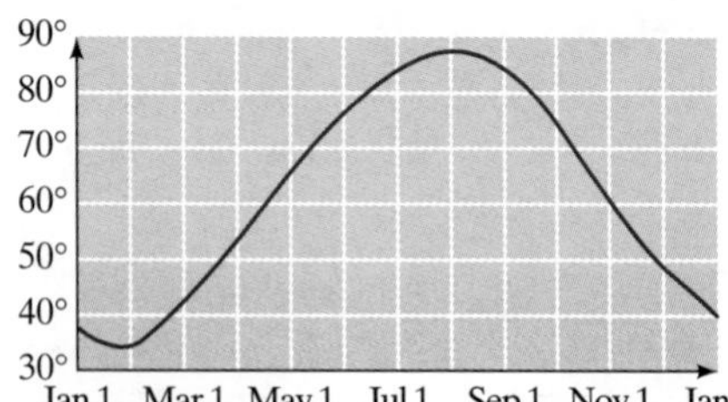

Source: National Weather Service/*New York Times,* January 7, 1996, p. 36.

a. Estimate the slope of the secant line through the points corresponding to April 1 and September 1. (Give the units of measurement.) What does the answer tell you about the temperature in Central Park?

b. Estimate the slope of the tangent line through the point corresponding to April 1. (Give the units of measurement.) What does the answer tell you about the temperature in Central Park?

c. Estimate the slope of the tangent line through the point corresponding to September 1. (Give the units of measurement.) What does the answer tell you about the temperature in Central Park?

d. On what date(s) is the temperature rising at the rate of 5° per month?

e. On what date is the temperature highest? How fast is the temperature changing on that date?

f. On what date is the temperature rising fastest?

38. ***Daylight*** The following graph shows the number of hours of daylight each day at a location 70° north of the equator (which is above the Arctic Circle).

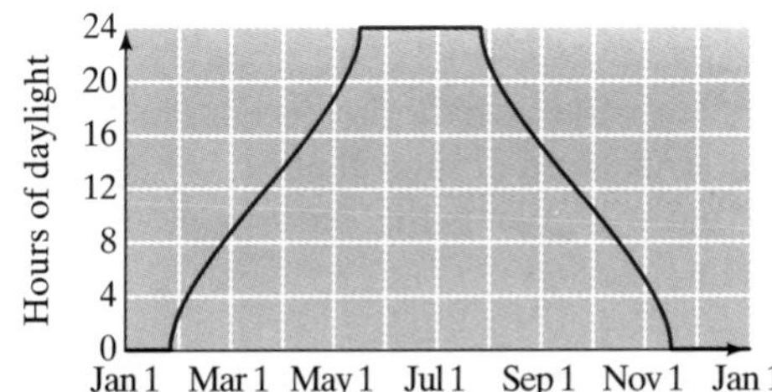

a. Estimate the slope of the secant line through the points corresponding to October 1 and November 1. (Give the units of measurement.) What does the answer tell you about the hours of daylight at this latitude?

b. Estimate the slope of the tangent line through the point corresponding to August 1. (Give the units of measurement.) What does the answer tell you about the hours of daylight at this latitude?

c. Estimate the slope of the tangent line through the point corresponding to March 1. (Give the units of measurement.) What does the answer tell you about the hours of daylight at this latitude?

d. On what date(s) is the number of hours of daylight increasing at the rate of 8 hours per month?

e. When is the number of hours of daylight the greatest? How fast is the number of hours of daylight changing then?

f. When is the number of hours of daylight changing most quickly?

T *Exercises 39–42 were given in the preceding section, but this time we ask you to use technology to solve them graphically.*

39. ***Demand*** Suppose the demand for a new brand of sneakers is given by

$$q = \frac{5{,}000{,}000}{p}$$

where p is the price per pair of sneakers, in dollars, and q is the number of pairs of sneakers that can be sold at price p. Graph q for $50 \le p \le 200$, use your graph to find $q(100)$ and estimate $\left.\frac{dq}{dp}\right|_{p=100}$, and interpret your answers.

40. ***Demand*** Suppose the demand for an old brand of TV is given by

$$q = \frac{100{,}000}{p + 10}$$

where p is the price per TV set, in dollars, and q is the number of TV sets that can be sold at price p. Graph q for $50 \le p \le 400$, use your graph to find $q(190)$ and estimate $\left.\frac{dq}{dp}\right|_{p=190}$, and interpret your answers.

41. ***Lorentz Contraction*** According to Einstein's Special Theory of Relativity, a moving object appears to get shorter to a stationary observer as its speed approaches the speed of light. If a spaceship that has a length of 100 meters at rest travels at a speed of warp p, its length in meters, as measured by a stationary observer, is given by

$$L(p) = 100\sqrt{1 - p^2} \qquad \text{with domain } [0, 1]$$

Graph L as a function of p, and use the graph to estimate $L(0.95)$ and $L'(0.95)$. What do these figures tell you?

42. ***Time Dilation*** Another prediction of Einstein's Special Theory of Relativity is that, to a stationary observer, clocks (as well as all biological processes) in a moving object appear to go more and more slowly as the speed of the object approaches the speed of light. If a spaceship travels at a speed of warp p, the time it takes for an on-board clock to register 1 second, as measured by a stationary observer, is given by

$$T(p) = \frac{1}{\sqrt{1 - p^2}} \text{ seconds} \qquad \text{with domain } [0, 1)$$

Graph T as a function of p, and use the graph to estimate $T(0.95)$ and $T'(0.95)$. What do these figures tell you?

Communication and Reasoning Exercises

43. If the derivative of f is zero at a point, what do you know about the graph of f near that point?

44. Sketch the graph of a function whose derivative never exceeds 1.

45. Sketch the graph of a function whose derivative exceeds 1 at every point.

46. Sketch the graph of a function whose derivative is exactly 1 at every point.

47. If the derivative of f is always positive, what do you know about the graph of f?

48. Draw a sketch comparing the secant line through $(a - h, f(a - h))$ and $(a + h, f(a + h))$ to the secant line through $(a, f(a))$ and $(a + h, f(a + h))$ and to the tangent line at $(a, f(a))$. Which secant line's slope appears to be a more accurate approximation of the derivative? How does this relate to the balanced difference quotient used in the preceding section?

3.4 The Derivative As a Function: An Algebraic Approach

In the preceding two sections we saw how to estimate the derivative of a function $f(x)$ at a specific value of x using numerical and graphical approaches. In this section, we use an algebraic approach that will give us the *exact value* of the derivative rather than just an approximation.

As we have observed, the derivative $f'(x)$ is a number we can calculate, or at least approximate, for various values of x. Since $f'(x)$ depends on the value of x, we may think of f' as a function of x. This function is the **derivative function.** The algebraic approach in this section gives us a method of calculating the derivative function—or, as is often said, "taking the derivative"—exactly. Another word for a method of calculating is a **calculus.**

Computing the Derivative from the Difference Quotient

example 1

Calculating the Derivative Algebraically

Let $f(x) = x^2$.

a. Use the definition of the derivative to compute $f'(x)$ algebraically.

b. Use the result of part (a) to compute $f'(-3)$.

Solution

a. We start with the definition:

$$f'(x) = \lim_{h\to 0} \frac{f(x+h) - f(x)}{h} \qquad \text{Formula for the derivative}$$

$$= \lim_{h\to 0} \frac{\overbrace{(x+h)^2}^{f(x+h)} - \overbrace{x^2}^{f(x)}}{h} \qquad \text{Substitute for } f(x) \text{ and } f(x+h).$$

$$= \lim_{h\to 0} \frac{(x^2 + 2xh + h^2) - x^2}{h} \qquad \text{Expand } (x+h)^2.$$

$$= \lim_{h\to 0} \frac{2xh + h^2}{h} \qquad \text{Cancel the } x^2.$$

$$= \lim_{h\to 0} \frac{h(2x + h)}{h} \qquad \text{Factor out } h.$$

$$= \lim_{h\to 0} (2x + h) \qquad \text{Cancel the } h.$$

Now we let h approach zero. As h gets closer and closer to zero, the sum $2x + h$ clearly gets closer and closer to $2x + 0 = 2x$. Thus,

$$f'(x) = \lim_{h\to 0} (2x + h) = 2x$$

This is the derivative function.

b. Since $f'(x) = 2x$, we have

$$f'(-3) = 2(-3) = -6$$

Before we go on . . . We did the following calculation in part (a): If $f(x) = x^2$, then $f'(x) = 2x$. This is our first complete calculation of a derivative function.

We can now quantify what we saw in Figure 12 in the last section. For instance, the tangent line at the point where $x = -2$ has slope $f'(-2) = 2(-2) = -4$; the tangent line at the point where $x = -1$ has slope $f'(-1) = -2$.

---

example 2

More Computations of $f'(x)$

Compute $f'(x)$ in each of the following cases.

a. $f(x) = x^3$ **b.** $f(x) = 2x^2 - x$

Solution

a. This is similar to the first example:

$$f'(x) = \lim_{h\to 0} \frac{f(x+h) - f(x)}{h}$$ Derivative formula

$$= \lim_{h\to 0} \frac{\overbrace{(x+h)^3}^{f(x+h)} - \overbrace{x^3}^{f(x)}}{h}$$ Substitute for $f(x)$ and $f(x+h)$.

$$= \lim_{h\to 0} \frac{(x^3 + 3x^2h + 3xh^2 + h^3) - x^3}{h}$$ Expand $(x+h)^3$.

$$= \lim_{h\to 0} \frac{3x^2h + 3xh^2 + h^3}{h}$$ Cancel the x^3.

$$= \lim_{h\to 0} \frac{h(3x^2 + 3xh + h^2)}{h}$$ Factor out h.

$$= \lim_{h\to 0} (3x^2 + 3xh + h^2)$$ Cancel the h.

$$= 3x^2$$ Let h approach zero.

b.

$$f'(x) = \lim_{h\to 0} \frac{f(x+h) - f(x)}{h}$$ Derivative formula

$$= \lim_{h\to 0} \frac{\overbrace{\left[2(x+h)^2 - (x+h)\right]}^{f(x+h)} - \overbrace{(2x^2 - x)}^{f(x)}}{h}$$ Substitute for $f(x)$ and $f(x+h)$.

$$= \lim_{h\to 0} \frac{(2x^2 + 4xh + 2h^2 - x - h) - (2x^2 - x)}{h}$$ Expand.

$$= \lim_{h\to 0} \frac{4xh + 2h^2 - h}{h}$$ Cancel the $2x^2$ and x.

$$= \lim_{h\to 0} \frac{h(4x + 2h - 1)}{h}$$ Factor out h.

$$= \lim_{h\to 0} (4x + 2h - 1)$$ Cancel the h.

$$= 4x - 1$$ Let h approach zero.

---

example 3

T Comparison of Algebraic and Numerical Approaches

Use graphing technology to compare the derivative of $f(x) = x^3$ found algebraically as above to an approximate derivative found numerically.

Solution

Here is how we can use the TI-83 to do this comparison. We begin by entering $f(x)$ as Y_1:

```
Y1=X^3
```

We let Y_2 be the balanced difference quotient with $h = 0.0001$:

```
Y2=(Y1(X+0.0001)-Y1(X-0.0001))/0.0002
```

Finally, we enter the derivative of f, which we found algebraically in the preceding example, as Y_3:

```
Y3=3*X²
```

We can now graph Y_2 and Y_3 to see that they are essentially the same. We can use TRACE to see that the values of the two functions are almost identical.

Before we go on . . . Experiment with other values of h to see how they affect the accuracy of the estimate given by the balanced difference quotient. You can also use the technique of this example to check other derivatives that we find algebraically. Try it.

Shortcut Formula: The Power Rule

We have now calculated the derivatives of several functions using the definition of the derivative as the limit of a difference quotient. The calculations of derivatives as limits are often tedious, so it would be nice to have a quicker method. By the end of the next chapter, we shall be able to find fairly quickly the derivative of almost any function we can write.

If you look at Examples 1 and 2 again, you may notice a pattern:

$$f(x) = x^2 \quad \Rightarrow \quad f'(x) = 2x$$

$$f(x) = x^3 \quad \Rightarrow \quad f'(x) = 3x^2$$

Shortcut: The Power Rule

If n is any constant and $f(x) = x^n$, then

$$f'(x) = nx^{n-1}$$

Quick Examples

1. If $f(x) = x^2$, then $f'(x) = 2x^1 = 2x$.
2. If $f(x) = x^3$, then $f'(x) = 3x^2$.
3. If $f(x) = x$, rewrite as $f(x) = x^1$, so $f'(x) = 1x^0 = 1$.
4. If $f(x) = 1$, rewrite as $f(x) = x^0$, so $f'(x) = 0x^{-1} = 0$.

WWW **Optional Internet Topic: Proof of the Power Rule**

At the web site, follow the path

Web site → Everything for Calculus → Chapter 3 → Proof of the Power Rule

for a proof of the power rule.

example 4

Using the Power Rule for Negative and Fractional Exponents
Calculate the derivatives of the following.

a. $f(x) = \frac{1}{x}$ **b.** $f(x) = \frac{1}{x^2}$ **c.** $f(x) = \sqrt{x}$

Solution

a. Rewrite as $f(x) = x^{-1}$. Then $f'(x) = (-1)x^{-2} = -\frac{1}{x^2}$

b. Rewrite as $f(x) = x^{-2}$. Then $f'(x) = (-2)x^{-3} = -\frac{2}{x^3}$

c. Rewrite as $f(x) = x^{0.5}$. Then $f'(x) = 0.5x^{-0.5} = \frac{0.5}{x^{0.5}}$

Alternatively, rewrite $f(x)$ as $x^{1/2}$, so that $f'(x) = \frac{1}{2}x^{-1/2} = \frac{1}{2x^{1/2}} = \frac{1}{2\sqrt{x}}$

Before we go on . . . We can now evaluate the derivative at any value of x just as we evaluate any other function. For example, if $f(x) = x^{100}$, then $f'(x) = 100x^{99}$, and so

$$f'(-1) = 100(-1)^{99} = -100$$

Some of the derivatives in Example 4 are very useful to remember, so we summarize them in a table. We suggest that you add to this table as you learn more derivatives. It is *extremely* helpful to remember the derivatives of common functions such as $1/x$ and $\sqrt{x}$, even though they can be obtained using the power rule as in the above example.

Table of Derivative Formulas

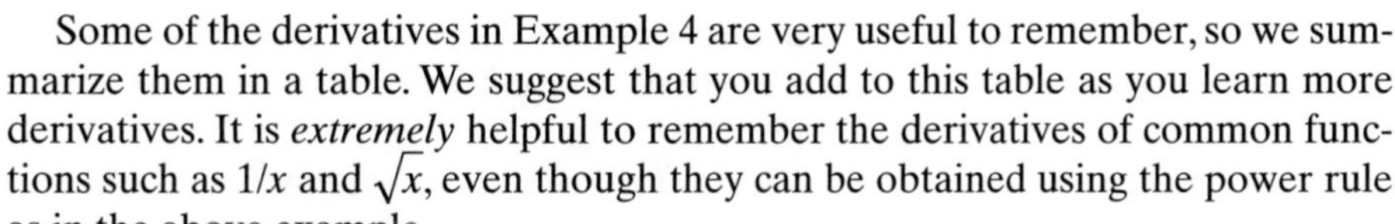

$f(x)$	x^n	x	1	$\frac{1}{x}$	$\frac{1}{x^2}$	$\sqrt{x}$
$f'(x)$	nx^{n-1}	1	0	$-\frac{1}{x^2}$	$-\frac{2}{x^3}$	$\frac{1}{2\sqrt{x}}$

Another Notation: Differential Notation

Here is a useful notation based on the "d-" notation we discussed in Section 3.2. **Differential notation** is based on an abbreviation for the phrase "the derivative with respect to x." For example, we learned in Example 2 that if $f(x) = x^3$, then $f'(x) = 3x^2$. When we say "$f'(x) = 3x^2$," we mean the following:

The derivative of x^3 with respect to x equals $3x^2$

You may wonder why we sneaked in the words "with respect to x." All this means is that the variable of the function is x and not any other variable.[1] Since we use the phrase "the derivative with respect to x" often, we use the abbreviation given next.

[1] This may seem odd in the case of $f(x) = x^3$ because there are no other variables to worry about. But in expressions like st^3 that involve variables other than x, it is necessary to specify just what the variable of the function is. This is the same reason we write "$f(x) = x^3$" rather than just "$f = x^3$."

Differential Notation; Differentiation

$\frac{d}{dx}$ means "the derivative with respect to x."

Thus, $\frac{d}{dx}[f(x)]$ is the same thing as $f'(x)$, the derivative of $f(x)$ with respect to x. If y is a function of x, then the derivative of y with respect to x is

$$\frac{d}{dx}(y) \qquad \text{or more compactly} \qquad \frac{dy}{dx}$$

To **differentiate** a function $f(x)$ with respect to x means to take its derivative with respect to x.

Notes

1. $\frac{dy}{dx}$ is Leibniz's notation for the derivative we discussed in Section 3.2 (see Example 4 there).

2. Leibniz's notation illustrates units nicely: units of $\frac{dy}{dx}$ are units of y per unit of x.

Quick Examples
Differential Notation

1. The derivative with respect to x of x^3 is $3x^2$. $\quad \frac{d}{dx}(x^3) = 3x^2$

2. The derivative with respect to t of $\frac{1}{t}$ is $-\frac{1}{t^2}$. $\quad \frac{d}{dt}\left(\frac{1}{t}\right) = -\frac{1}{t^2}$

Leibniz Notation

1. If $y = x^4$, then $\frac{dy}{dx} = 4x^3$.

2. If $u = \frac{1}{t^2}$, then $\frac{du}{dt} = -\frac{2}{t^3}$.

The Rules for Sums and Constant Multiples

We can now find the derivatives of more complicated functions, such as polynomials, using the following rules.

Derivatives of Sums, Differences, and Constant Multiples

If $f(x)$ and $g(x)$ are any two functions with derivatives $f'(x)$ and $g'(x)$, respectively, and if c is any constant, then

$$[f(x) \pm g(x)]' = f'(x) \pm g'(x)$$
$$[cf(x)]' = cf'(x)$$

In Words

- The derivative of a sum is the sum of the derivatives, and the derivative of a difference is the difference of the derivatives.
- The derivative of c times a function is c times the derivative of the function.

In Differential Notation

$$\frac{d}{dx}\left[f(x) \pm g(x)\right] = \frac{d}{dx}f(x) \pm \frac{d}{dx}g(x)$$

$$\frac{d}{dx}\left[cf(x)\right] = c\,\frac{d}{dx}f(x)$$

A proof of the sum rule appears at the end of this section. The proof of the rule for constant multiples is similar.

example 5

Using the Rules for Sums and Constant Multiples

a. Let $f(x) = 5x^3$. Then

Derivative of $5x^3$ = 5 times derivative of x^3

$$\frac{d}{dx}(5x^3) = 5\left(\frac{d}{dx}x^3\right) \qquad \text{Rule for constant multiples}$$

$$= 5(3x^2) = 15x^2 \qquad \text{Power rule}$$

In other words, we multiply the coefficient (5) by the exponent (3), and then decrease the exponent by 1.

b. Let $f(x) = 12x$. Then

$$\frac{d}{dx}(12x) = 12\,\frac{d}{dx}(x) \qquad \text{Rule for constant multiples}$$

$$= 12(1) = 12 \qquad \text{Power rule}$$

In other words, the derivative of a constant coefficient times x is just the coefficient.

c. Let $f(x) = 3$. Then think of $f(x)$ as $3(1)$, so that

$$\frac{d}{dx}(3(1)) = 3\,\frac{d}{dx}(1) \qquad \text{Rule for constant multiples}$$

$$= 3(0) = 0 \qquad \text{Power rule}$$

In other words, the derivative of a constant by itself is zero.

d. Let $f(x) = 3x^2 + 2x - 4$. Then

$$\frac{d}{dx}(3x^2 + 2x - 4) = \frac{d}{dx}(3x^2) + \frac{d}{dx}(2x - 4) \qquad \text{Rule for sums}$$

$$= \frac{d}{dx}(3x^2) + \frac{d}{dx}(2x) - \frac{d}{dx}(4) \qquad \text{Rule for differences}$$

$$= 3(2x) + 2(1) - 0 = 6x + 2 \qquad \text{See parts (a), (b), (c).}$$

Before we go on . . . Notice that in part (d) we had three terms in the expression for $f(x)$, not just two. By applying the rule for sums and differences twice, we see that the derivative of a sum or difference of three terms is the sum or difference of the derivatives of the terms.

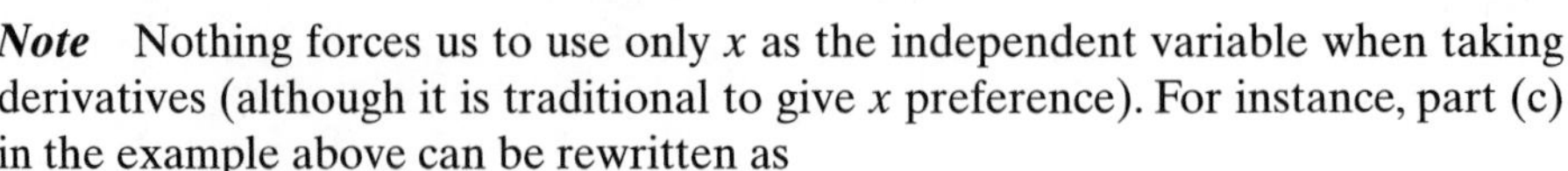

Note Nothing forces us to use only x as the independent variable when taking derivatives (although it is traditional to give x preference). For instance, part (c) in the example above can be rewritten as

$$\frac{d}{dt}(3t^2 + 2t - 4) = 6t + 2 \qquad \frac{d}{dt} \text{ means “derivative with respect to } t\text{”}$$

or

$$\frac{d}{du}(3u^2 + 2u - 4) = 6u + 2 \qquad \frac{d}{du} \text{ means “derivative with respect to } u\text{”}$$

In the preceding example, we saw instances of the following important facts. (Think about these graphically to see why they must be true.)

Derivative of a Constant Times x; Derivative of a Constant

If c is any constant, then

$$\frac{d}{dx}(cx) = c$$

$$\frac{d}{dx}(c) = 0$$

Quick Examples

$$\frac{d}{dx}(6x) = 6 \qquad \frac{d}{dx}(-x) = -1$$

$$\frac{d}{dx}(5) = 0 \qquad \frac{d}{dx}(\pi) = 0$$

Sometimes we need to do a little rewriting before we are ready to take the derivative of a function, as the following example shows.

example 6

Converting to Exponent Form

Find the derivative of $f(x) = \dfrac{2x}{3} - \dfrac{6}{x} + \dfrac{2}{3x^{0.2}} - \dfrac{x^4}{2}$.

Solution

The rules we have apply to only sums or differences of functions of the form

Constant $\times$ x raised to a power

So, we must first rewrite each of the terms in $f(x)$ in this “exponent form”:

$$f(x) = \frac{2x}{3} - \frac{6}{x} + \frac{2}{3x^{0.2}} - \frac{x^4}{2} \qquad \text{Rational form}$$

$$= \frac{2}{3}x - 6x^{-1} + \frac{2}{3}x^{-0.2} - \frac{1}{2}x^4 \qquad \text{Exponent form}^2$$

---

[2] See the section on exponents in Appendix A to brush up.

We are now ready to take the derivative:

$$f'(x) = \frac{2}{3}(1) - 6(-1)x^{-2} + \frac{2}{3}(-0.2)x^{-1.2} - \frac{1}{2}(4x^3)$$

$$= \frac{2}{3} + 6x^{-2} - \frac{0.4}{3}x^{-1.2} - 2x^3 \qquad \text{Exponent form}$$

$$= \frac{2}{3} + \frac{6}{x^2} - \frac{0.4}{3x^{1.2}} - 2x^3 \qquad \text{Rational form}$$

---

example 7

Functions Not Differentiable at a Point
Find the natural domains of the derivatives of $f(x) = x^{1/3}$ and $g(x) = x^{2/3}$.

Solution
By the power rule,

$$f'(x) = \frac{1}{3}x^{-2/3} = \frac{1}{3x^{2/3}}$$

and

$$g'(x) = \frac{2}{3}x^{-1/3} = \frac{2}{3x^{1/3}}$$

Since $f'(x)$ and $g'(x)$ are defined for only nonzero values of x, their natural domains consist of all real numbers except zero. Thus, the derivatives f' and g' do not exist at $x = 0$. As in Section 3.2, we say that f and g are **not differentiable** at $x = 0$, or that $f'(0)$ and $g'(0)$ **do not exist.** If we look at Figure 19, we notice why these functions fail to be differentiable at $x = 0$: The graph of g comes to a sharp point (called a **cusp**) at 0, so it is not meaningful to speak about a tangent line at that point. The graph of f has a vertical tangent line at 0. Since such a line has undefined slope, the derivative is undefined at that point.

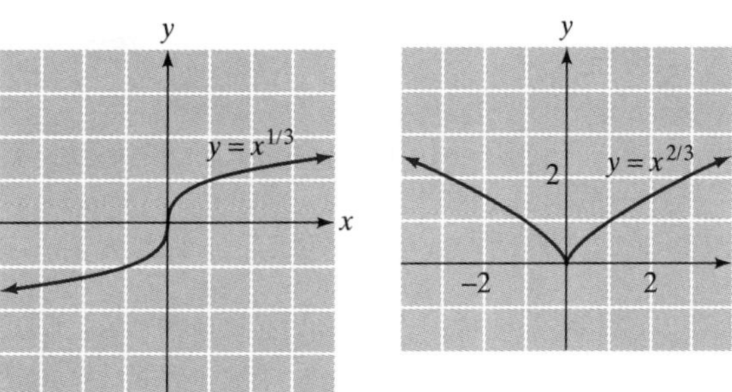

Figure 19

We can also detect this nondifferentiability by computing some difference quotients numerically. In the case of $f(x) = x^{1/3}$, we get the following table:

h	± 1	± 0.1	± 0.01	± 0.001	± 0.0001
$\dfrac{f(0+h) - f(0)}{h}$	1	4.6416	21.544	100	464.16

The values suggest that the difference quotients $[f(0 + h) - f(0)]/h$ grow large without bound rather than approach any fixed number as h approaches zero. (Can you see how the behavior of the difference quotients in the table is reflected in the graph?)

T Using Technology

If you try to graph the function $f(x) = x^{2/3}$ using the format

```
X^(2/3)
```

you may get only the right-hand portion of Figure 19 because graphing utilities are (often) not programmed to raise negative numbers to fractional exponents. (However, most will handle `X^(1/3)` correctly, as a special case they recognize.) To avoid this difficulty, you can take advantage of the identity

$$x^{2/3} = (x^2)^{1/3}$$

so that it is always a nonnegative number that is being raised to a fractional exponent. Thus, use the format

```
(X^2)^(1/3)
```

to obtain both portions of the graph.

Proof of the Sum Rule

Here is a quick proof of the sum rule:

$$\begin{aligned}
\frac{d}{dx}\left[f(x) + g(x)\right] &= \lim_{h\to 0} \frac{\left[f(x+h) + g(x+h)\right] - \left[f(x) + g(x)\right]}{h} \\
&= \lim_{h\to 0} \frac{\left[f(x+h) - f(x)\right] + \left[g(x+h) - g(x)\right]}{h} \\
&= \lim_{h\to 0} \left(\frac{f(x+h) - f(x)}{h} + \frac{g(x+h) - g(x)}{h}\right) \\
&= \lim_{h\to 0} \frac{f(x+h) - f(x)}{h} + \lim_{h\to 0} \frac{g(x+h) - g(x)}{h} \\
&= \frac{d}{dx}\left[f(x)\right] + \frac{d}{dx}\left[g(x)\right]
\end{aligned}$$

The next-to-last step uses a property of limits: The limit of a sum is the sum of the limits. Think about why this should be true. The last step uses the definition of the derivative again.

3.4 exercises

In Exercises 1–10, **a.** *use the definition of the derivative as a limit to compute $f'(x)$ (no shortcuts, please)* **b.** *find the slope of the tangent to the graph of f at the indicated point.*

1. $f(x) = x^2 + 1; (2, 5)$

2. $f(x) = x^2 - 3; (1, -2)$

3. $f(x) = 3x^2 + x; (1, 4)$

4. $f(x) = 2x^2 + x; (-2, 6)$

5. $f(x) = 2x - x^2; (-1, -3)$

6. $f(x) = -x - x^2; (0, 0)$

7. $f(x) = x^3 + 2x; (2, 12)$

8. $f(x) = x - 2x^3; (1, -1)$

9. $f(x) = \frac{1}{x}; (1, 1)$

10. $f(x) = \frac{1}{x^2}; (1, 1)$

In Exercises 11–20, use the shortcut rules to calculate the derivative of the given function mentally.

11. $f(x) = x^5$

12. $f(x) = x^4$

13. $f(x) = 2x^{-2}$

14. $f(x) = 3x^{-1}$

15. $f(x) = -x^{0.25}$

16. $f(x) = -x^{-0.5}$

17. $f(x) = 2x^4 + 3x^3 - 1$

18. $f(x) = -x^3 - 3x^2 - 1$

19. $f(x) = -x + \frac{1}{x} + 1$

20. $f(x) = \frac{1}{x} + \frac{1}{x^2}$

In Exercises 21–28, obtain the derivative dy/dx and state the rules you use.

21. $y = 10$

22. $y = x^3$

23. $y = x^2 + x$

24. $y = x - 5$

25. $y = 4x^3 + 2x - 1$

26. $y = 4x^{-1} - 2x - 10$

27. $y = x^{104} - 99x^2 + x$

28. $y = x^{3/2} + x^{5/2}$

In Exercises 29–50, find the derivative of each function.

29. $f(x) = x^2 - 3x + 5$

30. $f(x) = 3x^3 - 2x^2 + x$

31. $f(x) = x + x^{0.5}$

32. $f(x) = x^{0.5} + 2x^{-0.5}$

33. $g(x) = x^{-2} - 3x^{-1} - 2$

34. $g(x) = 2x^{-1} + 4x^{-2}$

35. $g(x) = \frac{1}{x} - \frac{1}{x^2}$

36. $g(x) = \frac{1}{x^2} + \frac{1}{x^3}$

37. $h(x) = \frac{2}{x^{0.4}}$

38. $h(x) = -\frac{1}{2x^{0.2}}$

39. $h(x) = \frac{1}{x^2} + \frac{2}{x^3}$

40. $h(x) = \frac{2}{x} - \frac{2}{x^3} + \frac{1}{x^4}$

41. $r(x) = \frac{2}{3x} - \frac{1}{2x^{0.1}} + \frac{4x^{1.1}}{3} + 2$

42. $r(x) = \frac{4}{3x^2} + \frac{1}{x^{3.2}} - \frac{2x^2}{3} - 4$

43. $r(x) = \frac{2x}{3} - \frac{x^{0.1}}{2} + \frac{4}{3x^{1.1}} - 2$

44. $r(x) = \frac{4x^2}{3} + \frac{x^{3.2}}{6} - \frac{2}{3x^2} + 4$

45. $s(x) = \sqrt{x} + \frac{1}{\sqrt{x}}$

46. $s(x) = x + \frac{7}{\sqrt{x}}$

[Hint for Exercises 47–50: *First expand the given function.*]

47. $s(x) = x\left(x^2 - \frac{1}{x}\right)$

48. $s(x) = x^{-1}\left(x - \frac{2}{x}\right)$

49. $t(x) = \frac{x^2 - 2x^3}{x}$

50. $t(x) = \frac{2x + x^2}{x}$

In Exercises 51–56, evaluate each expression.

51. $\frac{d}{dx}\left(x + \frac{1}{x^2}\right)$

52. $\frac{d}{dx}\left(2x - \frac{1}{x}\right)$

53. $\frac{d}{dx}(2x^{1.3} - x^{-1.2})$

54. $\frac{d}{dx}(2x^{4.3} + x^{0.6})$

55. $\frac{d}{dt}(at^3 - 4at)$; (a constant)

56. $\frac{d}{dt}(at^2 + bt + c)$; ($a, b, c$ constant)

In Exercises 57–62, find the indicated derivative.

57. $y = \frac{x^{10.3}}{2} + 99x^{-1}; \frac{dy}{dx}$

58. $y = \frac{x^{1.2}}{3} - \frac{x^{0.9}}{2}; \frac{dy}{dx}$

59. $s = 2.3 + \frac{2.1}{t^{1.1}} - \frac{t^{0.6}}{2}; \frac{ds}{dt}$

60. $s = \frac{2}{t^{1.1}} + t^{-1.2}; \frac{ds}{dt}$

61. $V = \frac{4}{3}\pi r^3; \frac{dV}{dr}$

62. $A = 4\pi r^2; \frac{dA}{dr}$

In Exercises 63–70, find the slope of the tangent to the graph of the given function at the indicated point.

63. $f(x) = x^3; (-1, -1)$

64. $g(x) = x^4; (-2, 16)$

65. $f(x) = 1 - 2x; (2, -3)$

66. $f(x) = \frac{x}{3} - 1; (-3, -2)$

67. $h(x) = x^{0.25}; (16, 2)$

68. $s(x) = x^{-0.5}; (1, 1)$

69. $g(t) = \frac{1}{t^5}; (1, 1)$

70. $s(t) = \frac{1}{t^3}; \left(-2, -\frac{1}{8}\right)$

In Exercises 71–76, find the equation of the tangent line to the graph of the given function at the point with the indicated x-coordinate. In each case, sketch the curve together with the appropriate tangent line.

71. $f(x) = x^3; x = -1$

72. $f(x) = x^2; x = 0$

73. $f(x) = x + \frac{1}{x}; x = 2$

74. $f(x) = \frac{1}{x^2}; x = 1$

75. $f(x) = \sqrt{x}; x = 4$

76. $f(x) = 2x + 4; x = -1$

In Exercises 77–82, find all the values of x (if any) where the tangent line to the graph of the given equation is horizontal.

77. $y = 2x^2 + 3x - 1$

78. $y = -3x^2 - x$

79. $y = 2x + 8$

80. $y = -x + 1$

81. $y = x + \frac{1}{x}$

82. $y = x - \sqrt{x}$

83. Write out the proof that $\frac{d}{dx}(x^4) = 4x^3$.

84. Write out the proof that $\frac{d}{dx}(x^5) = 5x^4$.

Applications

85. ***Cost*** Consider the two cost functions $C_1(x) = 10{,}000 + 5x - 0.1x^2$ and $C_2(x) = 20{,}000 + 10x - 0.2x^2$. How do the rates of change of these cost functions at the same production levels compare?

86. ***Cost*** The cost of making x teddy bears at the Cuddly Companion Company used to be $C_1(x) = 100 + 40x - 0.001x^2$. Due to rising health costs, it now is $C_2(x) = 1000 + 40x - 0.001x^2$. How does the rate of change of cost at a production level of x teddy bears compare to what it used to be?

87. ***Profit*** The cost to manufacture x cases of beer per week is given by $C(x)$, while the revenue from selling x cases is given by $R(x)$. How must the rate of change of cost and the rate of change of revenue be related when the rate of change of profit is zero?

88. ***Profit*** The cost to manufacture x cases of beer per week is given by $C(x)$, while the revenue from selling x cases is given by $R(x)$. How must the rate of change of cost and the rate of change of revenue be related when the rate of change of profit is positive?

89. ***Embryo Development*** The oxygen consumption of a bird embryo increases from the time the egg is laid through the time the chick hatches. In a typical galliform bird, the oxygen consumption (in milliliters per hour) can be approximated by

$$c(t) = -0.00271t^3 + 0.137t^2 - 0.892t + 0.149 \qquad (8 \le t \le 30)$$

where t is the time (in days) since the egg was laid.[3] (An egg will typically hatch at around $t = 28$.) Find $c'(15)$ and $c'(30)$. What do these results tell about the embryo's oxygen consumption just prior to hatching?

90. ***Embryo Development*** The oxygen consumption of a turkey embryo increases from the time the egg is laid through the time the turkey chick hatches. In a brush turkey, the oxygen consumption (in milliliters per hour) can be approximated by

$$c(t) = -0.00118t^3 + 0.119t^2 - 1.83t + 3.972 \qquad (20 \le t \le 50)$$

where t is the time (in days) since the egg was laid. (An egg will typically hatch at around $t = 50$.) Find $c'(30)$ and $c'(50)$. What do these results tell about the embryo's oxygen consumption just prior to hatching?

91. ***Velocity*** If a stone is dropped from a height of 100 feet, its height s after t seconds is given by $s = 100 - 16t^2$, with s in feet.

a. Find the stone's velocity at times $t = 0, 1, 2, 3$, and 4 seconds.

b. When does it reach the ground, and how fast is it traveling when it hits the ground?

92. ***Velocity*** If a stone is thrown down at 120 feet/second from a height of 1000 feet, its height s after t seconds is given by $s = 1000 - 120t - 16t^2$, with s in feet.

a. Find the stone's velocity at times $t = 0, 1, 2, 3$, and 4 seconds.

b. When does it reach the ground, and how fast is it traveling when it hits the ground?

[3] The models in Exercises 89 and 90 approximate graphical data published in the article "The Brush Turkey" by Roger S. Seymour in *Scientific American,* December, 1991, pp. 108–114.

Communication and Reasoning Exercises

93. What instructions would you give to a fellow student who wants to accurately graph the tangent line to the curve $y = 3x^2$ at the point $(-1, 3)$?

94. What instructions would you give to a fellow student who wants to accurately graph a line at a right angle to the curve $y = 4/x$ at the point where $x = 0.5$?

95. Consider $f(x) = x^2$ and $g(x) = 2x^2$. How do the slopes of the tangent lines of f and g at the same x compare?

96. Consider $f(x) = x^3$ and $g(x) = x^3 + 3$. How do the slopes of the tangent lines of f and g compare?

97. Of the three methods (numerical, graphical, and algebraic) we can use to estimate the derivative of a function at a given value of x, which is always the most accurate? Explain.

98. How would you respond to an acquaintance who says, "I finally understand what the derivative is: It is nx^{n-1}! Why weren't we taught that in the first place instead of the difficult way using limits?"

99. Following is an excerpt from your friend's graded homework:

$$3x^4 + 11x^5 = 12x^3 + 55x^4$$ ✗ WRONG (−8)

Why was it marked wrong?

3.5 A First Application: Marginal Analysis

In Chapter 1 we considered linear *cost functions* of the form $C(x) = mx + b$, where C is the total cost, x is the number of items, and m and b are constants. The slope m is the *marginal cost.* It measures the *cost of producing one more item.* Notice that the derivative of $C(x) = mx + b$ is $C'(x) = m$. In other words, for a linear cost function, *the marginal cost is the derivative of the cost function.*

In general, we have the following definition.

Marginal Cost

A **cost function** specifies the total cost C as a function of the number of items x. In other words, $C(x)$ is the total cost of x items. The **marginal cost function** is the derivative $C'(x)$ of the cost function $C(x)$. It measures the rate of change of cost with respect to x.

Units

The units of marginal cost are units of cost (dollars, say) per item. *We interpret $C'(x)$ as the approximate cost of producing one more item.*[1]

Quick Example

If $C(x) = 400x + 1000$ dollars, then the marginal cost function is $C'(x) = \$400$ per item (a constant).

example 1

Marginal Cost

Suppose that the cost in dollars to manufacture portable CD players is given by

$$C(x) = 150{,}000 + 20x - 0.0001x^2$$

where x is the number of CD players manufactured.[2] Find the marginal cost function $C'(x)$ and use it to estimate the cost of manufacturing the 50,001st CD player.

Solution

Since $C(x) = 150{,}000 + 20x - 0.0001x^2$, the marginal cost function is

$$C'(x) = 20 - 0.0002x$$

The units of $C'(x)$ are units of C (dollars) per unit of x (CD players). Thus, $C'(x)$ is measured in dollars per CD player. The cost of the 50,001st CD player is the amount by which the total cost would rise if we increased production from 50,000 CD players to 50,001. Thus, we need to know the rate at which the total cost rises

---

[1] See Example 1.

[2] You might well ask where on earth this formula came from. There are two approaches to obtaining cost functions in real life: analytical and numerical. The *analytical* approach is to calculate the cost function from scratch. For example, in this situation, we might have fixed costs of \$150,000 plus a production cost of \$20 per CD player. The term $0.0001x^2$ may reflect a cost saving for high levels of production, such as a bulk discount in the cost of electronic components. In the *numerical* approach, we first obtain the cost at several different production levels by direct observation. This gives several points on the (as yet unknown) cost versus production level graph. Then we find the equation of the curve that best fits these points. This is called *curve-fitting* or *regression* and uses techniques of calculus. The resulting equation gives the required cost function.

as we increase production. This rate of change is measured by the derivative, or marginal cost, which we just computed. At $x = 50{,}000$, we get

$$C'(50{,}000) = 20 - 0.0002(50{,}000) = \$10 \text{ per CD player}$$

We estimate that the 50,001st CD player will cost approximately $10.

Before we go on . . . The marginal cost is really only an approximation to the cost of the 50,001st CD player:

$$\begin{aligned} C'(50{,}000) &\approx \frac{C(50{,}001) - C(50{,}000)}{1} \quad \text{Set } h = 1 \text{ in the definition of the derivative.} \\ &= C(50{,}001) - C(50{,}000) \\ &= \text{Cost of the 50,001st CD player} \end{aligned}$$

The exact cost of the 50,001st CD player is

$$\begin{aligned} &C(50{,}001) - C(50{,}000) \\ &= \left[150{,}000 + 20(50{,}001) - 0.0001(50{,}001)^2\right] - \left[150{,}000 + 20(50{,}000) - 0.0001(50{,}000)^2\right] \\ &= \$9.9999 \end{aligned}$$

So the marginal cost is a good approximation to the actual cost.

Graphically, we are using the tangent line to approximate the cost function near a production level of 50,000. Figure 20 shows the graph of the cost function together with the tangent line at $x = 50{,}000$. Notice that the tangent line is essentially indistinguishable from the graph of the function for some distance on either side of 50,000.

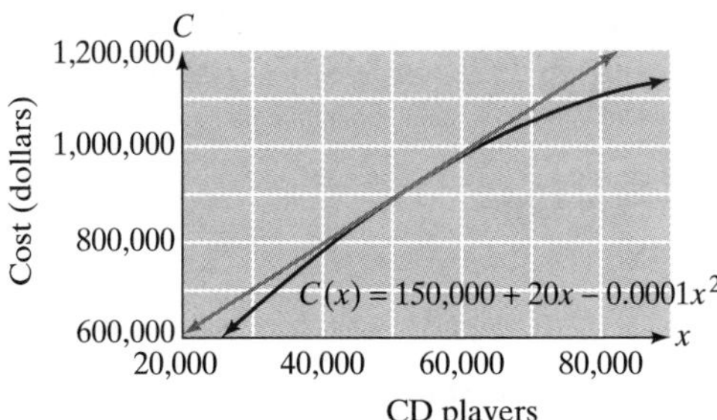

Figure 20

Notes

1. In general, the difference quotient $[C(x + h) - C(x)]/h$ gives the **average cost per item** to produce h more items at a current production level of x items. (Why?)
2. Notice that $C'(x)$ is much easier to calculate than $[C(x + h) - C(x)]/h$. (Try it.)

As the following examples show, the term *marginal* can also be applied to quantities other than cost.

example 2

Marginal Revenue and Profit

In Chapter 1 we obtained the following demand equation for the Workout Fever Health Club.

$$q = -0.06p + 46 = -\frac{3}{50}p + 46$$

where p is the membership fee and q is the average number of new members per month. Assume that each new member costs Workout Fever Health Club \$100 for medical evaluation and personal training.

a. Calculate the monthly revenue and profit as functions of q.

b. Calculate the marginal revenue and profit functions, dR/dq and dP/dq.

c. Calculate the marginal profit for $q = 15$, $q = 20$, and $q = 25$, and interpret the results.

Solution

a. The monthly revenue is given by $R = pq$. Since we need to express R as a function of q only, we must replace p in this equation by a function of q. The relationship between p and q is given in the demand equation above, and we want p as a function of q, so we first need to rewrite the demand equation by solving for p. This gives

$$p = \frac{50}{3}(46 - q)$$ Solve the demand equation for p.

We can substitute this expression for p in $R = pq$ to obtain

$$R(q) = \frac{50}{3}(46 - q)q$$ Substitute for p in the revenue equation.

$$= \frac{2300}{3}q - \frac{50}{3}q^2$$

This is the monthly revenue as a function of q. For the profit function, recall that

$$P = R - C$$ Revenue − Cost

Since it costs Workout Fever \$100 per new member, the cost of q new members per month is $100q$. Thus,

$$P(q) = \frac{2300}{3}q - \frac{50}{3}q^2 - 100q = \frac{2000}{3}q - \frac{50}{3}q^2$$

b. The marginal revenue and profit functions are the derivatives:

$$\text{Marginal revenue} = \frac{dR}{dq} = \frac{2300}{3} - \frac{100}{3}q$$

$$\text{Marginal profit} = \frac{dP}{dq} = \frac{2000}{3} - \frac{100}{3}q$$

and the units for both are dollars per new member. Think of these functions as approximating the *additional* revenue and profit the club receives for each new member.

c. Substituting the given values of q gives

$$\left.\frac{dP}{dq}\right|_{q=15} = \frac{2000}{3} - \frac{100}{3}(15) = \$166.67 \text{ per new member}$$

This means that, at 15 new members per month, the profit would increase by about $166.67 with the addition of one more new member. It would therefore pay Workout Fever to increase its monthly new membership by lowering the membership fee. We find

$$\left.\frac{dP}{dq}\right|_{q=20} = \frac{2000}{3} - \frac{100}{3}(20) = \$0 \text{ per new member}$$

This means that at 20 new members per month, the profit would remain approximately the same with the addition of one more new member. Also,

$$\left.\frac{dP}{dq}\right|_{q=25} = \frac{2000}{3} - \frac{100}{3}(25) = -\$166.67 \text{ per new member}$$

This means that at 25 new members per month, the profit would decrease by about $166.67 with the addition of one more new member. Workout Fever should therefore decrease its membership by raising the membership fee.

Before we go on . . . This analysis shows that Workout Fever should sign more than 15 new members per month but less than 25. The fact that $P'(20) = 0$ tells us that the health club should adjust the price to attract exactly 20 new members per month. If it attracts slightly fewer—say, 19—then the marginal profit will be positive [as you can check by calculating $P'(19)$], indicating that it should increase monthly membership. Similarly, $P'(21)$ is negative, indicating that the club should decrease monthly membership. In general, *for maximum profit, $P'(q)$ must be 0.*

Figure 21 shows the graph of $P(q)$ for this example. We can see clearly in the graph that $P'(15) > 0$, $P'(20) = 0$, $P'(25) < 0$, and the largest value of $P(q)$ occurs at $q = 20$.

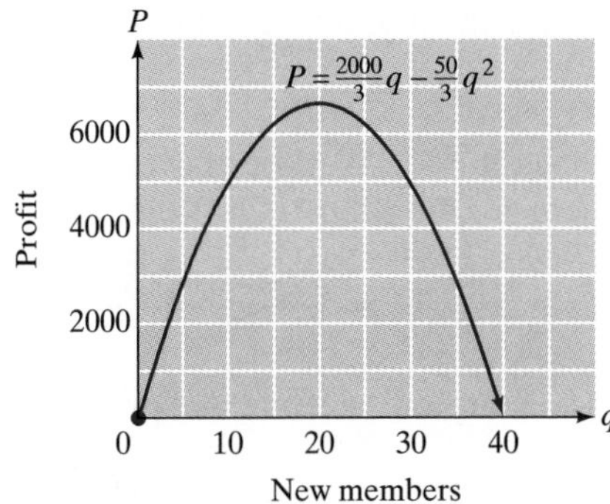

Figure 21

---

example 3

Marginal Product

A consultant determines that Precision Manufacturers, Inc.'s annual profit (in dollars) is given by

$$P(n) = -200{,}000 + 400{,}000n - 4600n^2 - 10n^3$$

where n is the number of assembly line workers it employs.

a. Calculate dP/dn, which is called the **marginal product** at the employment level of n assembly line workers. What are its units?

b. Calculate $P(10)$ and $dP/dn\,|_{n=10}$, and interpret the results.

c. Precision Manufacturers currently employs ten assembly line workers and is considering laying off some of them. What advice would you give the company's management?

Solution

a. Taking the derivative gives

$$\frac{dP}{dn} = 400{,}000 - 9200n - 30n^2$$

The units of dP/dn are profit (in dollars) per worker.

b. Substituting into the formula for $P(n)$, we get

$$P(10) = -200{,}000 + 400{,}000(10) - 4600(10)^2 - 10(10)^3 = \$3{,}330{,}000$$

Thus, Precision Manufacturers will make an annual profit of \$3,330,000 if it employs ten assembly line workers. On the other hand,

$$\left.\frac{dP}{dn}\right|_{n=10} = 400{,}000 - 9200(10) - 30(10)^2 = \$305{,}000 \text{ per worker}$$

Thus, at an employment level of ten assembly line workers, annual profit is increasing at a rate of \$305,000 per additional worker. In other words, if the company were to employ one more assembly line worker, its annual profit would increase by approximately \$305,000.

c. Since the marginal product is positive, profits will increase if the company increases the number of workers and will decrease if it decreases the number of workers, so your advice would be to hire additional assembly line workers. Downsizing the assembly line work force would reduce annual profits.

Before we go on . . . The following question might have occurred to you:

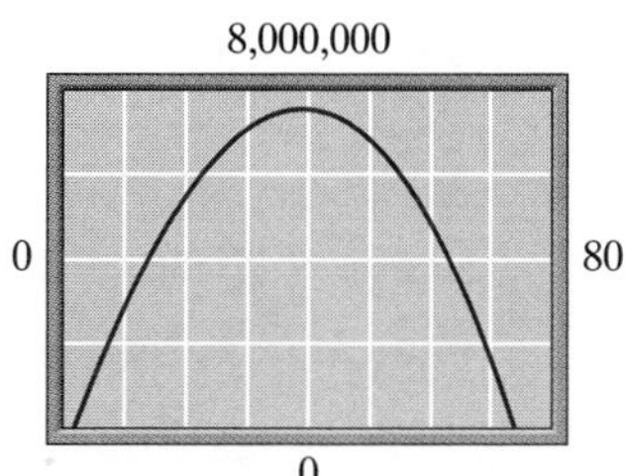

Figure 22

Question How many additional assembly line workers should the company hire to obtain the maximum annual profit?

Answer Since we have profit as a function of the number of workers n, we can graph the function $P(n)$. Figure 22 shows the graph of $P(n)$ for $0 \le n \le 80$, as drawn by a graphing calculator. (For the y-axis scale, we used $0 \le P \le 8{,}000{,}000$.) The horizontal axis represents the number of assembly line workers, and the vertical axis represents annual profit. At the point on the graph where $n = 10$, the slope is positive [as confirmed by our calculation of $P'(10)$]. At approximately $n = 40$, the slope is zero (so the marginal product is zero) and the profit is largest. This tells us that the company should employ approximately 40 assembly line workers for a maximum profit. Thus, the company should hire approximately 30 additional assembly line workers.

Notice that Figure 22 also shows that the company begins to show a loss at an employment level of about 75 assembly line workers.

example 4

Average Cost

Suppose that the cost in dollars to manufacture portable CD players is given by

$$C(x) = 150{,}000 + 20x - \frac{x^2}{10{,}000}$$

where x is the number of CD players manufactured. Find the average cost per CD player if 50,000 CD players are manufactured.

Solution

The total cost of manufacturing 50,000 CD players is given by

$$C(50{,}000) = 150{,}000 + 20(50{,}000) - \frac{50{,}000^2}{10{,}000} = \$900{,}000$$

Since 50,000 CD players cost a total of \$900,000 to manufacture, the average cost of manufacturing one CD player is this total cost divided by 50,000:

$$\overline{C}(50{,}000) = \frac{900{,}000}{50{,}000} = \$18 \text{ per CD player}$$

Thus, if 50,000 CD players are manufactured, each CD player costs the manufacturer an average of \$18 to make.

There is nothing special about the number 50,000. If we replace it by x, we get the average cost of manufacturing x CD players:

$$\overline{C}(x) = \frac{C(x)}{x}$$

$$= \frac{1}{x}\left(150{,}000 + 20x - \frac{x^2}{10{,}000}\right) = \frac{150{,}000}{x} + 20 - \frac{x}{10{,}000}$$

The function $\overline{C}$ is called the **average cost function.**

Question In Example 1 we calculated the *marginal cost* at a production level of 50,000 CD players as

$$C'(50{,}000) = \$10 \text{ per CD player}$$

What, if anything, does this have to do with the average cost?

Answer Average cost and marginal cost convey different, but related, information. The average cost $\overline{C}(50{,}000)$ is the cost per item of manufacturing the first 50,000 CD players, whereas the marginal cost $C'(50{,}000)$ gives the (approximate) cost of manufacturing the *next* CD player. Thus, according to our calculations, the first 50,000 CD players cost an average of \$18, but it costs only about \$10 to manufacture the next one.

Question The marginal cost at the production level of 50,000 CD players is lower than the average cost. Does this mean that the average cost to manufacture CDs is going up or down with increasing volume?

Answer Down. Think about why.

Average Cost

Given a cost function C, the **average cost** of the first x items is given by

$$\bar{C}(x) = \frac{C(x)}{x}$$

The average cost is different from the marginal cost $C'(x)$, which tells the approximate cost of the *next* item.

3.5 exercises

Think of each function in Exercises 1–4 as representing the cost to manufacture x items. Find the marginal cost at the given production level x and state the units of measurement.

1. $C(x) = 10{,}000 + 5x - \dfrac{x^2}{10{,}000}$; $x = 1000$

2. $C(x) = 20{,}000 + 7x - \dfrac{x^2}{20{,}000}$; $x = 10{,}000$

3. $C(x) = 15{,}000 + 100x + \dfrac{1000}{x}$; $x = 100$

4. $C(x) = 20{,}000 + 50x + \dfrac{10{,}000}{x}$; $x = 100$

In Exercises 5 and 6, find the marginal cost, marginal revenue, and marginal profit functions, and find all the values of x for which the marginal profit is zero. Interpret your answer.

5. $C(x) = 4x$; $R(x) = 8x - \dfrac{x^2}{1000}$

6. $C(x) = 5x^2$; $R(x) = x^3 + 7x + 10$

Applications

7. ***Advertising Costs*** The cost, in thousands of dollars, of airing x television commercials during a Super Bowl game is given by[3]

$$C(x) = 150 + 1200x - 0.002x^2$$

a. Find the marginal cost function and use it to estimate how fast the cost is going up when $x = 4$. Compare this with the exact cost of airing the fifth commercial.

b. Find the average cost function $\bar{C}$, and evaluate $\bar{C}(4)$. What does the answer tell you?

Source: *New York Times*, May 5, 1995, p. D5.

8. ***Marginal Cost and Average Cost*** The cost of producing x teddy bears per day at the Cuddly Companion Company is calculated by their marketing staff to be given by the formula

$$C(x) = 100 + 40x - 0.001x^2$$

a. Find the marginal cost function and use it to estimate how fast the cost is going up at a production level of 100 teddy bears. Compare this with the exact cost of producing the 101st teddy bear.

b. Find the average cost function $\bar{C}$, and evaluate $\bar{C}(100)$. What does the answer tell you?

9. ***Marginal Profit*** Suppose that $P(x)$ represents the profit on the sale of x videocassettes. If $P(1000) = 3000$ and $P'(1000) = -3$, what do these values tell you about the profit?

10. ***Marginal Loss*** An automobile retailer calculates that its loss on the sale of Type M cars is given by $L(50) = 5000$ and $L'(50) = -200$, where $L(x)$ represents the loss on the sale of x Type M cars. What do these values tell you about losses?

11. ***Marginal Profit*** Your monthly profit (in dollars) from selling magazines is given by

$$P = 5x + \sqrt{x}$$

where x is the number of magazines you sell in a month. If you are currently selling $x = 50$ magazines per month, find your profit and your marginal profit. Interpret your answers.

12. ***Marginal Profit*** Your monthly profit (in dollars) from your newspaper route is given by

$$P = 2n - \sqrt{n}$$

where n is the number of subscribers on your route. If you currently have 100 subscribers, find your profit and your marginal profit. Interpret your answers.

[3] The NBC television network planned to charge approximately \$1,200,000 per 30-second television spot during the 1996 Super Bowl XXX game. This explains the coefficient of x in the cost function.

13. ***Marginal Product*** A car wash firm calculates that its daily profit (in dollars) depends on the number n of workers it employs according to the formula

$$P = 400n - 0.5n^2$$

Calculate the marginal product at an employment level of 50 workers, and interpret the result.

14. ***Marginal Product*** Repeat Exercise 13 using the formula

$$P = -100n + 25n^2 - 0.005n^4$$

15. ***Marginal Revenue: Pricing Tuna*** Assume that the demand function for tuna in a small coastal town is given by

$$p = \frac{20{,}000}{q^{1.5}} \qquad (200 \le q \le 800)$$

where p is the price (in dollars) per pound of tuna and q is the number of pounds of tuna that can be sold at the price p in 1 month.

a. Calculate the price that the town's fishery should charge for tuna in order to produce a demand of 400 pounds of tuna per month.

b. Calculate the monthly revenue R as a function of the number of pounds of tuna q.

c. Calculate the revenue and marginal revenue (derivative of the revenue with respect to q) at a demand level of 400 pounds per month, and interpret the results.

d. If the town fishery's monthly tuna catch amounts to 400 pounds of tuna, and the price is at the level in part (a), would you recommend that the fishery raise or lower the price of tuna in order to increase its revenue?

16. ***Marginal Revenue: Pricing Tuna*** Repeat Exercise 15 assuming a demand equation of

$$p = \frac{60}{q^{0.5}} \qquad (200 \le q \le 800)$$

17. ***Demand for Poultry*** The demand for poultry can be modeled as

$$q = 63.15 - 0.45p + 0.12b$$

where q is the per capita demand for poultry in pounds per year, p is the wholesale price of poultry in cents per pound, and b is the wholesale price of beef in cents per pound.[4] Assume that the wholesale price of beef is fixed at 45¢ per pound.

a. Find the revenue as a function of q, and hence obtain the marginal revenue as a function of q. (Round constants to two decimal places.)

b. Find the annual per capita revenue that will result at a demand level of 50 pounds of poultry per year, and estimate the change in revenue that will result if the price is raised to yield a demand level of 49 pounds of poultry per year.

c. If a farmer breeds chickens at an average cost of 10¢ per pound, find the annual profit function P in terms of the per capita demand for poultry, evaluate $P'(50)$, and interpret the result.

Source: A. H. Studenmund, *Using Econometrics,* 2d ed. (New York: HarperCollins, 1992), pp. 180–181.

18. ***Demand for Beef*** Refer back to the model for demand for poultry in Exercise 17. Express the annual per capita revenue R as a function of the price of beef if the wholesale price of poultry is fixed at 40¢ per pound. Calculate $R(45)$ and $R'(45)$, and interpret the results.

19. ***Advertising Cost*** Your company is planning to air a number of television commercials during the ABC television network's presentation of the Academy Awards. ABC is charging your company $685,000 per 30-second spot.[5] Additional fixed costs (development and personnel costs) amount to $500,000, and the network has agreed to provide a discount of

$$D(x) = \$10{,}000\sqrt{x}$$

for x television spots.

a. Write down the cost function C, marginal cost function C', and average cost function $\overline{C}$.

b. Compute $C'(3)$ and $\overline{C}(3)$. (Round all answers to three significant digits.) Use these two answers to determine whether the average cost is increasing or decreasing as x increases.

Source: New York Times, May 5, 1995, p. D5.

20. ***Housing Costs*** The cost C of building a house is related to the number k of carpenters used and the number e of electricians used by the formula

$$C = 15{,}000 + 50k^2 + 60e^2$$

a. Assuming that ten carpenters are currently being used, find the marginal cost as a function of e.

b. If ten carpenters and ten electricians are currently being used, use your answer to part (a) to estimate the cost of hiring an additional electrician.

c. If ten carpenters and ten electricians are currently being used, what is the cost of hiring an additional carpenter?

Source: Based on an exercise in *Introduction to Mathematical Economics* by A. L. Ostrosky, Jr., and J. V. Koch (Prospect Heights, IL: Waveland Press, 1979).

21. ***Emission Control*** The cost of controlling emissions at a firm goes up rapidly as the amount of emissions reduced goes up. Here is a possible model:

$$C(q) = 4000 + 100q^2$$

---

[4]This equation is based on data from poultry sales in the period from 1950 to 1984.

[5]This is what ABC actually charged for a 30-second spot during the 1995 Academy Awards presentation.

where q is the reduction in emissions (in pounds of pollutant per day) and C is the daily cost (in dollars) of this reduction.

a. If a firm is currently reducing its emissions by 10 pounds each day, what is the marginal cost of reducing emissions even further?

b. Government clean-air subsidies to the firm are based on the formula

$$S(q) = 500q$$

where q is again the reduction in emissions (in pounds per day) and S is the subsidy (in dollars). At what reduction level does the marginal cost surpass the marginal subsidy?

c. Calculate the net cost function, $N(q) = C(q) - S(q)$, given the cost function and subsidy above, and find the value of q that gives the lowest net cost. What is this lowest net cost? Compare your answer to that in part (b), and comment on what you find.

22. ***Taxation Schemes*** Here is a curious proposal for taxation rates based on income:

$$T(i) = 0.001i^{0.5}$$

where i represents total annual income and $T(i)$ is the income tax rate as a percentage of total annual income. (Thus, for example, an income of \$50,000 per year would be taxed at about 22%, while an income of double that amount would be taxed at about 32%.)[6]

a. Calculate the after-tax (net) income $N(i)$ an individual can expect to earn as a function of income i.

b. Calculate an individual's marginal after-tax income at income levels of \$100,000 and \$500,000.

c. At what income does an individual's marginal after-tax income become negative? What is the after-tax income at that level, and what happens at higher income levels?

d. What do you suspect is the most anyone can earn after taxes? (See footnote 6.)

23. ***Fuel Economy*** Your Porsche's gas mileage (in miles per gallon) is given as a function $M(x)$ of speed x in miles per hour. It is found that

$$M'(x) = \frac{3600x^{-2} - 1}{(3600x^{-1} + x)^2}$$

Find $M'(10)$, $M'(60)$, and $M'(70)$. What do the answers tell you about your car?

24. ***Marginal Revenue*** The estimated marginal revenue for sales of ESU soccer team T-shirts is given by

$$R'(p) = \frac{(8 - 2p)e^{-p^2+8p}}{10{,}000{,}000}$$

where p is the price (in dollars) the soccer players charge for each shirt. Find $R'(3)$, $R'(4)$, and $R'(5)$. What do the answers tell you?

25. ***Transportation Costs*** Before the Alaskan pipeline was built, there was speculation as to whether it might be more economical to transport the oil by large tankers. The following cost equation was estimated by the National Academy of Sciences:

$$C = 0.03 + \frac{10}{T} - \frac{200}{T^2}$$

where C is the cost in dollars of transporting a barrel of oil 1000 nautical miles, and T is the size of an oil tanker in deadweight tons.

a. How much would it cost to transport a barrel of oil 1000 nautical miles in a tanker that weighs 1000 tons?

b. By how much is this cost increasing or decreasing as the weight of the tanker increases beyond 1000 tons?

Source: *Use of Satellite Data on the Alaskan Oil Marine Link, Practical Applications of Space Systems: Cost and Benefits* (Washington, DC: National Academy of Sciences, 1975), p. B-23.

26. ***Transportation Costs*** Refer back to the cost equation in Exercise 25. Find the value of T so that $C'(T) = 0$. Interpret the result. By calculating values of $C(T)$ for T close to and on either side of this amount, what more can you say?

27. ***Marginal Cost*** *(from the GRE economics test)* In a multiple-plant firm in which the different plants have different and continuous cost schedules, if the costs of production for a given output level are to be minimized, which of the following is essential?

a. Marginal costs must equal marginal revenue.

b. Average variable costs must be the same in all plants.

c. Marginal costs must be the same in all plants.

d. Total costs must be the same in all plants.

e. Output per worker-hour must be the same in all plants.

28. ***Study Time*** *(from the GRE economics test)* A student has a fixed number of hours to devote to study and is certain of the relationship between hours of study and the final grade for each course. Grades are given on a numerical scale (e.g., 0 to 100), and each course is counted equally in computing the grade average. To maximize his or her grade average, the student should allocate these hours to different courses to meet which criterion?

a. The grade in each course is the same.

b. The marginal product of an hour's study (in terms of final grade) in each course is zero.

c. The marginal product of an hour's study (in terms of final grade) in each course is equal, though not necessarily equal to zero.

d. The average product of an hour's study (in terms of final grade) in each course is equal.

e. The numbers of hours spent in study for each course are equal.

---

[6]This model has the following interesting feature: An income of \$1 million per year would be taxed at 100%, leaving the individual penniless!

29. ***Marginal Product*** *(from the GRE economics test)* Assume that the marginal product of an additional senior professor is 50% higher than the marginal product of an additional junior professor and that junior professors are paid one-half the amount that senior professors receive. With a fixed overall budget, a university that wishes to maximize its quantity of output from professors should do which of the following?
   a. Hire equal numbers of senior professors and junior professors.
   b. Hire more senior professors and junior professors.
   c. Hire more senior professors and discharge junior professors.
   d. Discharge senior professors and hire more junior professors.
   e. Discharge all senior professors and half of the junior professors.

30. ***Marginal Product*** *(based on a question from the GRE economics test)* Assume that the marginal product of an additional senior professor is twice the marginal product of an additional junior professor and that junior professors are paid two-thirds the amount that senior professors receive. With a fixed overall budget, a university that wishes to maximize its quantity of output from professors should do which of the following?
   a. Hire equal numbers of senior professors and junior professors.
   b. Hire more senior professors and junior professors.
   c. Hire more senior professors and discharge junior professors.
   d. Discharge senior professors and hire more junior professors.
   e. Discharge all senior professors and half of the junior professors.

Communication and Reasoning Exercises

31. What is the cost function? Carefully explain the difference between *average cost* and *marginal cost* **a.** in terms of their mathematical definition, **b.** in terms of graphs, and **c.** in terms of interpretation.

32. The cost function for your grand piano manufacturing plant has the property that $\bar{C}(1000) = \$3000$ per unit and $C'(1000) = \$2500$ per unit. Will the average cost increase or decrease if your company manufactures a slightly larger number of pianos? Explain your reasoning.

33. If the average cost to manufacture one grand piano increases as the production level increases, which is greater, the marginal cost or the average cost?

34. If your analysis of a manufacturing company yielded positive marginal profit but negative profit at the company's current production levels, what would you advise the company to do?

35. If a company's marginal average cost is zero at the current production level, positive for a slightly higher production level, and negative for a slightly lower production level, what should you advise the company to do?

36. The **acceleration** of cost is defined as the derivative of the marginal cost function; that is, the derivative of the derivative—or *second derivative*—of the cost function. What are the units of acceleration of cost, and how does one interpret this measure?

3.6 Limits and Continuity: Numerical and Graphical Approaches

The derivative is defined using a limit, and it is now time to say more precisely what that means. It is possible to speak of limits by themselves, rather than in the context of the derivative. The story of limits is a long one that we will try to make as concise as possible.

Evaluating Limits Numerically

Let's start with a simple example. Look at the function $f(x) = 2 + x$ and ask: What happens to $f(x)$ as x approaches 3? The following table shows the value of $f(x)$ for values of x close to, and on either side of, 3:

	x approaching 3 from the left →					← x approaching 3 from the right			
x	2.9	2.99	2.999	2.9999	3	3.0001	3.001	3.01	3.1
$f(x) = 2 + x$	4.9	4.99	4.999	4.9999		5.0001	5.001	5.01	5.1

We have left the entry under 3 blank to emphasize that, when calculating the limit of $f(x)$ as x *approaches* 3, we are not interested in its value when x *equals* 3. Notice from the table that the closer x gets to 3 from either side, the closer $f(x)$ gets to 5. We write this as

$$\lim_{x\to 3} f(x) = 5 \qquad \text{The limit of } f(x)\text{, as } x \text{ approaches 3, equals 5.}$$

Question Why all the fuss? Can't we simply put $x = 3$ and avoid having to use a table?

Answer This method happens to work for *some* functions, but not for *all* functions. The following example illustrates this point.

example 1

Estimating a Limit Numerically

Use a table to estimate the following limits.

a. $\lim_{x\to 2} \dfrac{x^3 - 8}{x - 2}$ **b.** $\lim_{x\to 0} \dfrac{e^{2x} - 1}{x}$

Solution

a. We cannot simply substitute $x = 2$ because the function $f(x) = (x^3 - 8)/(x - 2)$ is not defined at $x = 2$. (Why?)[1] We can set up a table of values with x approaching 2 from both sides:

	x approaching 2 from the left →					← x approaching 2 from the right			
x	1.9	1.99	1.999	1.9999	**2**	2.0001	2.001	2.01	2.1
$f(x) = \dfrac{x^3 - 8}{x - 2}$	11.41	11.9401	11.9940	11.9994		12.0006	12.0060	12.0601	12.61

We notice that as x approaches 2 from either side, $f(x)$ approaches 12. This suggests that the limit is 12, and we write

$$\lim_{x\to 2} \frac{x^3 - 8}{x - 2} = 12$$

b. The function $g(x) = (e^{2x} - 1)/x$ is not defined at $x = 0$ (nor can it even be simplified to one that *is* defined at $x = 0$). In the following table, we allow x to approach zero from both sides:

	x approaching 0 from the left →					← x approaching 0 from the right			
x	−0.1	−0.01	−0.001	−0.0001	**0**	0.0001	0.001	0.01	0.1
$g(x) = \dfrac{e^{2x} - 1}{x}$	1.8127	1.9801	1.9980	1.9998		2.0002	2.0020	2.0201	2.2140

The table suggests that $\lim_{x\to 0} \dfrac{e^{2x} - 1}{x} = 2$.

[1]However, if you factor $x^3 - 8$, you will find that $f(x)$ can be simplified to a function that *is* defined at $x = 2$. This point will be discussed (and this example redone) in the next section. The function in part (b) cannot be simplified in this way.

Spreadsheet

Spreadsheets are ideal for calculating limits numerically. You can set up your spreadsheet to duplicate the table in part (a) as follows:

	A	B	C	D
1	1.9	=(A1^3-8)/(A1-2)	2.1	
2	1.99		2.01	
3	1.999		2.001	
4	1.9999		2.0001	

(The formula in cell B1 is copied to columns B and D.) The values of $f(x)$ will be calculated in the B and D columns. For part (b), use the formula `=(EXP(2*A1)-1)/A1` in cell B1, and in columns A and C use the values of x shown in the table for part (b).

Before we go on . . . Although the table *suggests* that the limit in part (b) is 2, it by no means establishes that fact conclusively. It is *conceivable* (though not in fact the case here) that putting $x = 0.000000087$ will result in $g(x) = 426$. Using a table can only suggest a value for the limit. In the next section we discuss algebraic techniques for finding limits.

Before we continue, let us give a more formal definition.

Definition of a Limit

If $f(x)$ approaches the number L as x approaches (but is not equal to) a from both sides, then we say that the **limit** of $f(x)$ as $x \to a$ ("x approaches a") is L. We write

$$\lim_{x \to a} f(x) = L$$

or $\quad f(x) \to L$ as $x \to a$

If $f(x)$ *fails* to approach *a single fixed number* as x approaches a from both sides, then we say that $f(x)$ **has no limit** as $x \to a$, or

$$\lim_{x \to a} f(x) \textbf{ does not exist}$$

Notes

1. It is important that $f(x)$ approach the same number as x approaches a from either side. For instance, if $f(x)$ approaches 5 for $x = 1.9, 1.99, 1.999, \ldots$ but approaches 4 for $x = 2.1, 2.01, 2.001, \ldots$, then the limit as $x \to 2$ does not exist. (See the next example for such a situation.)
2. It may happen that $f(x)$ does not approach any fixed number at all as $x \to a$ from either side. In this case, we also say that the limit does not exist.
3. We are deliberately suppressing the exact definition of "approaches"; instead, we trust your intuition. The following phrasing of the definition of the limit is closer to the one used by mathematicians: *We can make $f(x)$ be as close to L as we like by making x be sufficiently close to a.*

example 2

Nonexistent Limit

Let $f(x) = \dfrac{|x|}{x}$. Does $\lim_{x\to 0} f(x)$ exist?

Solution

Here is a table of values with x approaching zero from both sides:

	x approaching 0 from the left →					← x approaching 0 from the right			
x	−0.1	−0.01	−0.001	−0.0001	0	0.0001	0.001	0.01	0.1
$f(x) = \dfrac{\lvert x\rvert}{x}$	−1	−1	−1	−1		1	1	1	1

The table shows that $f(x)$ does not approach the same limit as x approaches zero from both sides. There appear to be two *different* limits: the limit as we approach zero from the left and the limit as we approach zero from the right. We write

$$\lim_{x\to 0^-} f(x) = -1$$

read "the limit as x approaches 0 from the left (or from below) is -1" and

$$\lim_{x\to 0^+} f(x) = 1$$

read "the limit as x approaches 0 from the right (or from above) is 1." These are called the **one-sided limits** of $f(x)$. In order for the **two-sided limit** to exist (the one we are asked to compute), the two one-sided limits must be equal. Since they are not, we conclude that $\lim_{x\to 0} f(x)$ does not exist.

In another useful kind of limit we let x approach either $+\infty$ or $-\infty$, by which we mean that we let x get arbitrarily large or let x become an arbitrarily large negative number. The next example illustrates this.

example 3

Limits at Infinity

Use a table to estimate **a.** $\lim_{x\to+\infty} \dfrac{2x^2-4x}{x^2-1}$ and **b.** $\lim_{x\to-\infty} \dfrac{2x^2-4x}{x^2-1}$.

Solution

a. By saying that x is "approaching $+\infty$," we mean that x is getting larger and larger without bound, so we make the following table:

				x approaching $+\infty$ →	
x	10	100	1000	10,000	100,000
$f(x) = \dfrac{2x^2-4x}{x^2-1}$	1.6162	1.9602	1.9960	1.9996	2.0000

(Note that we are approaching $+\infty$ only from the left because we can hardly approach it from the right!) What seems to be happening is that $f(x)$ is approaching 2. Thus, we write

$$\lim_{x\to+\infty} f(x) = 2$$

b. Here, x is approaching $-\infty$, so we make a similar table, this time with x assuming negative values of greater and greater magnitude (read this table from right to left):

← x approaching $-\infty$

x	−100,000	−10,000	−1000	−100	−10
$f(x) = \dfrac{2x^2 - 4x}{x^2 - 1}$	2.0000	2.0004	2.0040	2.0402	2.4242

Once again, $f(x)$ is approaching 2. Thus,

$$\lim_{x \to -\infty} f(x) = 2$$

Estimating Limits Graphically

Estimating Limits Graphically

The graph of a function f is shown in Figure 23. (Recall that the solid dots indicate points on the graph, whereas the open dots indicate points not on the graph.) From the graph, analyze the following limits.

a. $\lim_{x \to -2} f(x)$ **b.** $\lim_{x \to 0} f(x)$ **c.** $\lim_{x \to 1} f(x)$ **d.** $\lim_{x \to +\infty} f(x)$

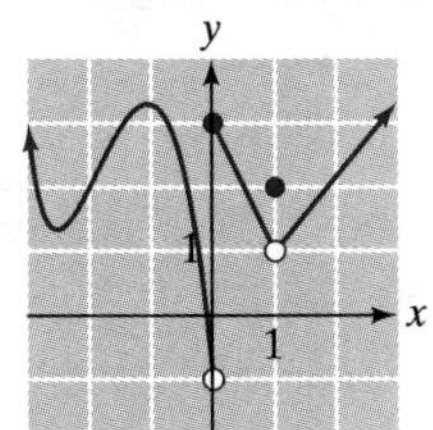

Figure 23

Solution

Since we are given only a graph of f, we must analyze these limits graphically.

a. Imagine that Figure 23 was drawn on a graphing calculator equipped with a trace feature that allows us to move a cursor along the graph and see the coordinates as we go. To simulate this, place a pencil point on the graph to the left of $x = -2$, and move it along the curve so that the x-coordinate approaches -2. (See Figure 24.) We evaluate the limit by observing the behavior of the y-coordinates.[2] We can see directly from the graph that the y-coordinate approaches 2. Similarly, if we place our pencil point to the right of $x = -2$ and move it to the left, the y-coordinate will approach 2 from that side as well (Figure 25). Therefore, as x approaches -2 from either side, $f(x)$ approaches 2, so

$$\lim_{x \to -2} f(x) = 2$$

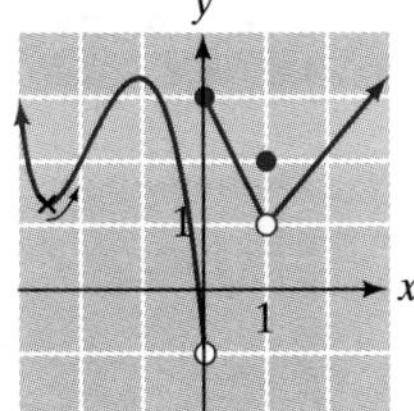

Figure 24

b. This time we move our pencil point toward $x = 0$. If we start from the left of $x = 0$ and approach 0 (by moving right), the y-coordinate approaches -1 (Figure 26). However, if we start from the right of $x = 0$ and approach 0 (by moving left), the y-coordinate approaches 3. Thus (see Example 2),

$$\lim_{x \to 0^-} f(x) = -1$$

and $$\lim_{x \to 0^+} f(x) = 3$$

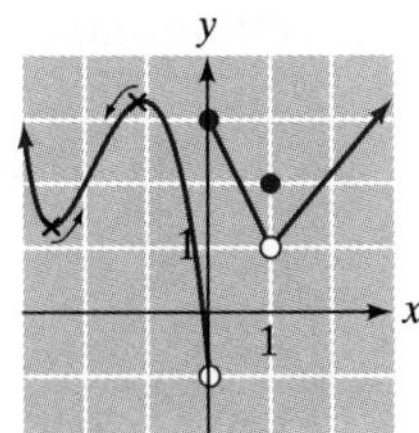

Figure 25

Since these limits are not equal, we conclude that

$$\lim_{x \to 0} f(x) \text{ does not exist}$$

[2]For a visual animation of this process, look at the on-line tutorial for this section at the web site.

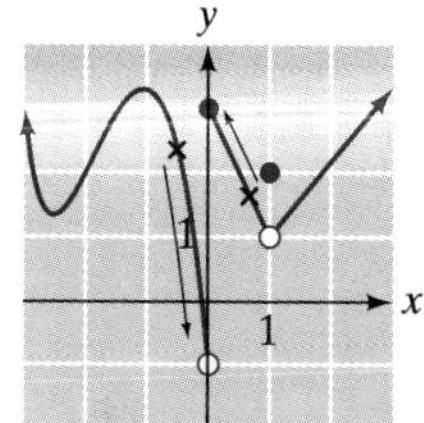

Figure 26

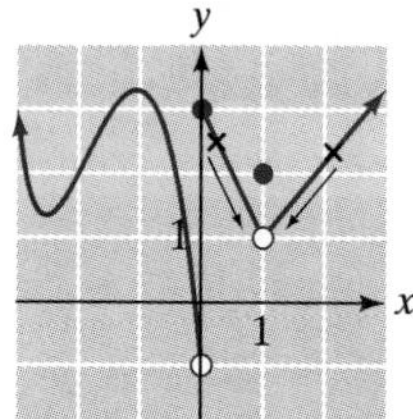

Figure 27

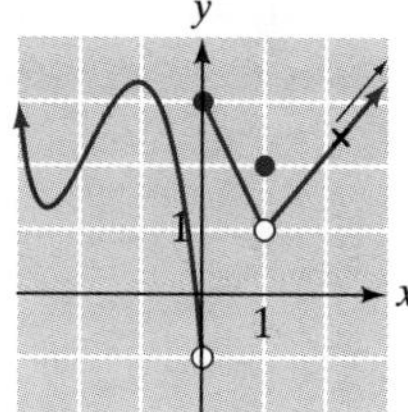

Figure 28

In this case there is a "break" in the graph at $x = 0$, and we say that the function is **discontinuous** at $x = 0$ (see below).

c. Once more we think about a pencil point moving along the graph with the x-coordinate this time approaching $x = 1$ from the left and from the right (Figure 27). As the x-coordinate of the point approaches 1 from either side, the y-coordinate approaches 1 also. Therefore,

$$\lim_{x \to 1} f(x) = 1$$

d. For this limit, x is supposed to approach infinity. We think about a pencil point moving along the graph farther and farther to the right as shown in Figure 28. As the x-coordinate gets larger, the y-coordinate also gets larger and larger **without bound;** that is, if you name any number, no matter how large, $f(x)$ will be even larger than that if x is sufficiently large. Since $f(x)$ is not approaching a specific real number, the limit does not exist. Since $f(x)$ is becoming arbitrarily large, we also say that the limit **diverges to $+\infty$,** and we write

$$\lim_{x \to +\infty} f(x) = +\infty$$

Similarly,

$$\lim_{x \to -\infty} f(x) = +\infty$$

Before we go on . . . In part (c) $\lim_{x \to 1} f(x) = 1$ but $f(1) = 2$. (Why?) Thus, $\lim_{x \to 1} f(x) \neq f(1)$. In other words, the limit of $f(x)$ as x *approaches* 1 is not the same as the value of f *at* $x = 1$. Always keep in mind that when we evaluate a limit as $x \to a$, *we do not care about the value of the function at* $x = a$. We care only about the value of $f(x)$ *as* x *approaches* a. In other words, $f(a)$ may or may not equal $\lim_{x \to a} f(x)$. As in part (b), there is a break in the graph of the function and we say that f is **discontinuous** at $x = 1$.

Part (a) was a case in which the limit and the value of the function are equal:

$$\lim_{x \to -2} f(x) = 2 = f(-2)$$

We say that the function is **continuous** at $x = -2$; its graph has no break at $x = -2$.

We can summarize the graphical method we used in Example 4 as follows.

Evaluating Limits Graphically

To decide whether $\lim_{x \to a} f(x)$ exists, and to find its value if it does:

1. Draw the graph of $f(x)$ by hand or with graphing technology.
2. Position your pencil point (or the "trace" cursor) on a point of the graph to the right of $x = a$.
3. Move the point *along the graph* toward $x = a$ from the right and read the y-coordinate as you go. The value the y-coordinate approaches (if any) is the limit $\lim_{x \to a^+} f(x)$.

4. Repeat Steps 2 and 3, this time starting from a point on the graph to the left of $x = a$ and approaching $x = a$ along the graph from the left. The value the y-coordinate approaches (if any) is $\lim_{x \to a^-} f(x)$.
5. If the left and right limits both exist and have the same value L, then $\lim_{x \to a} f(x)$ exists and equals L.
6. To evaluate $\lim_{x \to +\infty} f(x)$, move the pencil point toward the far right of the graph and estimate the value the y-coordinate approaches (if any). For $\lim_{x \to -\infty} f(x)$, move the pencil point toward the far left.

In the example above we spoke of points at which the function is continuous and discontinuous. We use this concept to evaluate limits in the next section. Here is the mathematical definition.

Continuous Function

Let f be a function, and let a be a number in the domain of f. Then f is **continuous at a** if $\lim_{x \to a} f(x)$ exists and $\lim_{x \to a} f(x) = f(a)$. The function f is said to be **continuous on its domain** if it is continuous at each point in its domain.

If f is not continuous at a particular a in its domain, then we say that f is **discontinuous** at a or that f has a **discontinuity** at a. Thus, a discontinuity can occur at $x = a$ if either:

a. $\lim_{x \to a} f(x)$ does not exist [as in part (b) of Example 4]
b. $\lim_{x \to a} f(x)$ exists but is not equal to $f(a)$ [as in part (c) of Example 4]

Note
If the number a is not in the domain of f—that is, if $f(a)$ is not defined—then f is neither continuous nor discontinuous at a. (See the discussion at the end of the next example.)

In the next example we use both the numerical and graphical approaches.

example 5

Infinite Limit

Does $\lim_{x \to 0^+} \frac{1}{x}$ exist?

Solution

Numerical Method Since we are asked for only the right-hand limit, we need only list values of x approaching zero from the right:

← x approaching 0 from the right

x	0	0.0001	0.001	0.01	0.1
$f(x) = \frac{1}{x}$		10,000	1000	100	10

What seems to be happening as x approaches zero from the right is that $f(x)$ is increasing without bound, as in Example 4(d). That is, if you name any number, no matter how large, $f(x)$ will be even larger than that if x is sufficiently close to zero. Thus, the limit diverges to $+\infty$, so

$$\lim_{x\to 0^+} \frac{1}{x} = +\infty$$

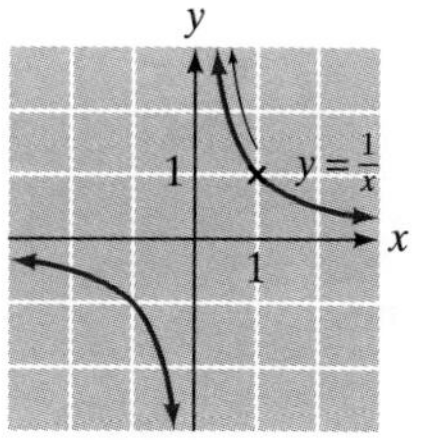

Graphical Method Recall that the graph of $f(x) = 1/x$ is the standard hyperbola shown in Figure 29. The figure also shows the pencil point moving so that its x-coordinate approaches zero from the right. As the point moves along the graph, it is forced to go higher and higher. In other words, its y-coordinate becomes larger and larger, approaching $+\infty$. Thus, we conclude again that

$$\lim_{x\to 0^+} \frac{1}{x} = +\infty$$

Figure 29

Before we go on . . . You should also check that

$$\lim_{x\to 0^-} \frac{1}{x} = -\infty$$

We say that as x approaches zero from the left, $1/x$ diverges to $-\infty$. Also, check that

$$\lim_{x\to +\infty} \frac{1}{x} = \lim_{x\to -\infty} \frac{1}{x} = 0$$

Question At what point (or points) of its domain is the function f discontinuous?
Answer The only point at which there is a break in the graph is $x = 0$. However, 0 is not in the domain of f. At every point that *is* in the domain of f, the function is continuous. For example, at $x = -1$, we find that

$$\lim_{x\to -1} \frac{1}{x} = -1 = f(-1)$$

so the function is continuous there.

Application

example 6

Demand Equation

Economist Henry Schultz calculated the following demand function for corn:

$$q = \frac{130{,}000}{p^{0.75}}$$

where p is the price in dollars per bushel and q is the number of bushels of corn that could be sold at the price p in 1 year.[3]

a. Estimate $\lim_{p\to +\infty} q(p)$ and interpret the answer.

b. Estimate $\lim_{p\to 0^+} q(p)$ and interpret the answer.

Source: Henry Schultz, *The Theory and Measurement of Demand* [as cited in *Introduction to Mathematical Economics* by A. L. Ostrosky, Jr., and J. V. Koch (Prospect Heights, IL: Waveland Press, 1979)].

[3] This demand function is based on data for the period 1915–1929.

Solution

a. Figure 30 shows a plot of $q(p)$ for $0 < p < 100$ and $0 \le q \le 40{,}000$.

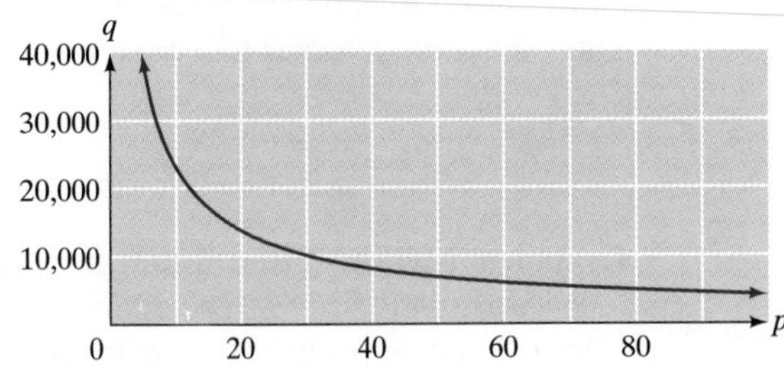

Figure 30

Using either the numerical or graphical approach, we find

$$\lim_{p \to +\infty} q(p) = \lim_{p \to +\infty} \frac{130{,}000}{p^{0.75}} = 0$$

Thus, as the price per bushel becomes higher and higher, the demand drops toward zero. This is what we reasonably expect of a demand equation.

b. The limit here is

$$\lim_{p \to 0^+} q(p) = \lim_{p \to 0^+} \frac{130{,}000}{p^{0.75}} = +\infty$$

So, as the price per bushel decreases toward \$0, the demand soars without bound. We do not reasonably expect this of a demand equation; even if corn was given away free, there would be a finite demand. (Think of what happens when food or other items are freely available at campus events.) We conclude that this demand equation cannot possibly be valid near $p = 0$, and so we need to restrict its domain. We would need more data to determine the lowest price for which the equation is valid.

3.6 exercises

Estimate the limits in Exercises 1–20 numerically.

1. $\lim_{x \to 0} \frac{x^2}{x+1}$

2. $\lim_{x \to 0} \frac{x-3}{x-1}$

3. $\lim_{x \to 2} \frac{x^2-4}{x-2}$

4. $\lim_{x \to 2} \frac{x^2-1}{x-2}$

5. $\lim_{x \to -1} \frac{x^2+1}{x+1}$

6. $\lim_{x \to -1} \frac{x^2+2x+1}{x+1}$

7. $\lim_{x \to +\infty} \frac{3x^2+10x-1}{2x^2-5x}$

8. $\lim_{x \to +\infty} \frac{6x^2+5x+100}{3x^2-9}$

9. $\lim_{x \to -\infty} \frac{x^5-1000x^4}{2x^5+10{,}000}$

10. $\lim_{x \to -\infty} \frac{x^6+3000x^3+1{,}000{,}000}{2x^6+1000x^3}$

11. $\lim_{x \to +\infty} \frac{10x^2+300x+1}{5x+2}$

12. $\lim_{x \to +\infty} \frac{2x^4+20x^3}{1000x^6+6}$

13. $\lim_{x \to +\infty} \frac{10x^2+300x+1}{5x^3+2}$

14. $\lim_{x \to +\infty} \frac{2x^4+20x^3}{1000x^3+6}$

15. $\lim_{x \to 2} e^{x-2}$

16. $\lim_{x \to +\infty} e^{-x}$

17. $\lim_{x \to +\infty} xe^{-x}$

18. $\lim_{x \to -\infty} xe^{x}$

19. $\lim_{x \to 0} \frac{\sin x}{x}$

20. $\lim_{x \to 0} \frac{\cos x - 1}{x^2}$

In Exercises 21–32, a graph of f is given. Use the graph to compute the quantities asked for.

21. **a.** $\lim_{x\to 1} f(x)$ **b.** $\lim_{x\to -1} f(x)$

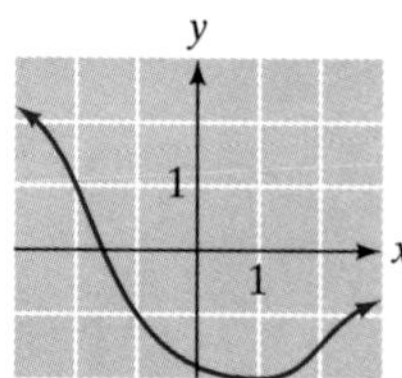

22. **a.** $\lim_{x\to -1} f(x)$ **b.** $\lim_{x\to 1} f(x)$

23. **a.** $\lim_{x\to 0} f(x)$ **b.** $\lim_{x\to 2} f(x)$
c. $\lim_{x\to -\infty} f(x)$ **d.** $\lim_{x\to +\infty} f(x)$

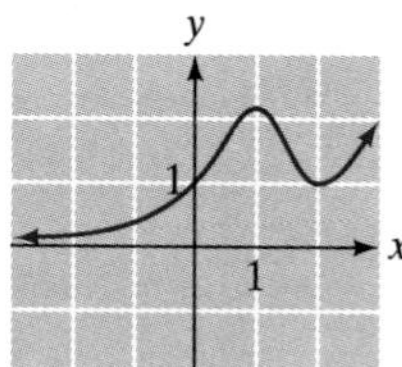

24. **a.** $\lim_{x\to -1} f(x)$ **b.** $\lim_{x\to 1} f(x)$
c. $\lim_{x\to +\infty} f(x)$ **d.** $\lim_{x\to -\infty} f(x)$

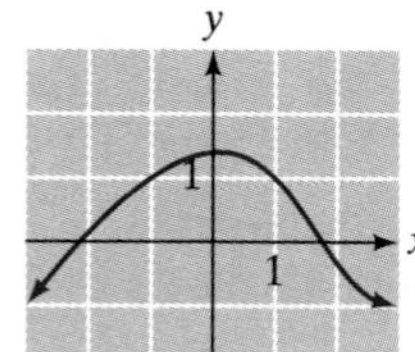

25. **a.** $\lim_{x\to 2} f(x)$ **b.** $\lim_{x\to 0^+} f(x)$
c. $\lim_{x\to 0^-} f(x)$ **d.** $\lim_{x\to 0} f(x)$
e. $f(0)$ **f.** $\lim_{x\to -\infty} f(x)$

26. **a.** $\lim_{x\to 3} f(x)$ **b.** $\lim_{x\to 1^+} f(x)$
c. $\lim_{x\to 1^-} f(x)$ **d.** $\lim_{x\to 1} f(x)$
e. $f(1)$ **f.** $\lim_{x\to +\infty} f(x)$

27. **a.** $\lim_{x\to -2} f(x)$ **b.** $\lim_{x\to -1^+} f(x)$
c. $\lim_{x\to -1^-} f(x)$ **d.** $\lim_{x\to -1} f(x)$
e. $f(-1)$ **f.** $\lim_{x\to +\infty} f(x)$

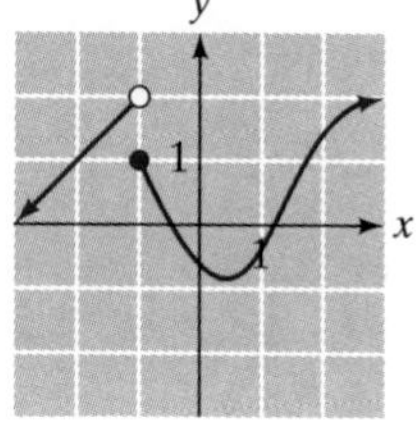

28. **a.** $\lim_{x\to -1} f(x)$ **b.** $\lim_{x\to 0^+} f(x)$
c. $\lim_{x\to 0^-} f(x)$ **d.** $\lim_{x\to 0} f(x)$
e. $f(0)$ **f.** $\lim_{x\to -\infty} f(x)$

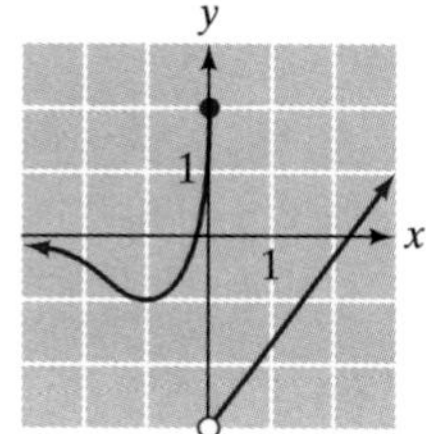

29. **a.** $\lim_{x\to -1} f(x)$ **b.** $\lim_{x\to 0^+} f(x)$
c. $\lim_{x\to 0^-} f(x)$ **d.** $\lim_{x\to 0} f(x)$
e. $f(0)$ **f.** $\lim_{x\to +\infty} f(x)$

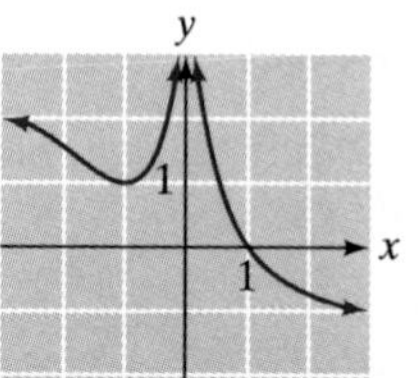

30. **a.** $\lim_{x\to 1} f(x)$ **b.** $\lim_{x\to 0^+} f(x)$
c. $\lim_{x\to 0^-} f(x)$ **d.** $\lim_{x\to 0} f(x)$
e. $f(0)$ **f.** $\lim_{x\to -\infty} f(x)$

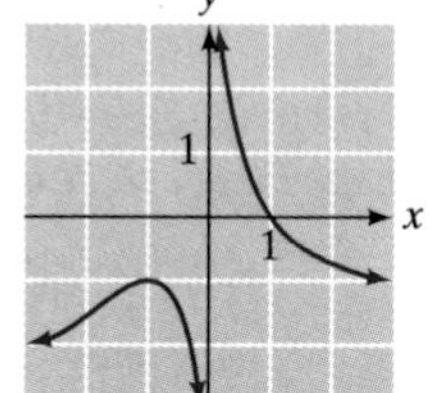

31. **a.** $\lim_{x\to -1} f(x)$ **b.** $\lim_{x\to 0^+} f(x)$
c. $\lim_{x\to 0^-} f(x)$ **d.** $\lim_{x\to 0} f(x)$
e. $f(0)$ **f.** $f(-1)$

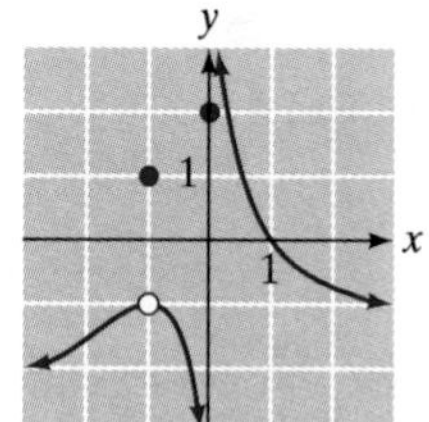

32. **a.** $\lim_{x\to 0^-} f(x)$ **b.** $\lim_{x\to 1^+} f(x)$
c. $\lim_{x\to 0} f(x)$ **d.** $\lim_{x\to 1} f(x)$
e. $f(0)$ **f.** $f(1)$

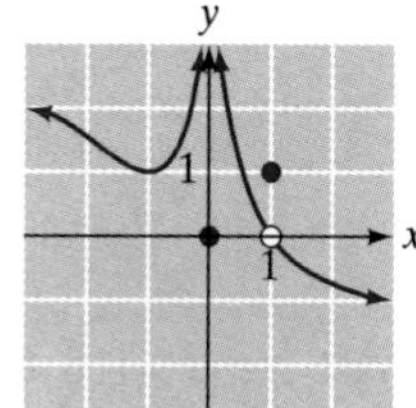

In Exercises 33–44, determine whether each function given in Exercises 21–32 is continuous on its domain. If a particular function is not continuous on its domain, say why.

33. Graph from Exercise 21

34. Graph from Exercise 22

35. Graph from Exercise 23

36. Graph from Exercise 24

37. Graph from Exercise 25

38. Graph from Exercise 26

39. Graph from Exercise 27

40. Graph from Exercise 28

41. Graph from Exercise 29

42. Graph from Exercise 30

43. Graph from Exercise 31

44. Graph from Exercise 32

Applications

Hawaiian Tourism *The number of visitors, in millions of visitors per year, to Hawaii in the years 1985 to 1993 can be approximated by*

$$n(t) = -0.067t^2 + 0.706t + 4.87$$

where $t = 0$ represents June 30, 1985. During the same period, visitor spending can be approximated by

$$r(t) = -0.164t^2 + 1.60t + 6.71$$

where $t = 0$ represents June 30, 1985.[4] *Exercises 45 and 46 are based on these models.*

Source: Hawaii Visitors Bureau/*New York Times*, September 5, 1995, p. A12.

45. Assuming the trends in the models above continue indefinitely, numerically estimate

$$\lim_{t\to+\infty} r(t) \quad \text{and} \quad \lim_{t\to+\infty} \frac{r(t)}{n(t)}$$

Then interpret your answers and comment on the results.

46. Repeat Exercise 45, this time calculating

$$\lim_{t\to+\infty} n(t) \quad \text{and} \quad \lim_{t\to+\infty} \frac{n(t)}{r(t)}$$

47. *Book Sales* The following graph shows the approximate number of books sold annually in the United States in year t, where $t = 0$ represents 1990 and $n(t)$ is the number of books sold, in billions, in the tth year since 1990:[5]

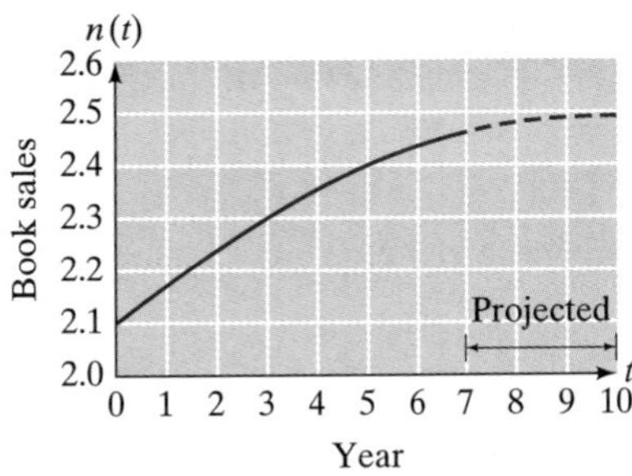

Source: Book Industry Study Group/*New York Times*, August 13, 1995, p. F2.

a. Estimate $\lim_{t\to+\infty} n(t)$ and interpret your answer.
b. Estimate $\lim_{t\to+\infty} n'(t)$ and interpret your answer.

48. *Employment* The following graph shows the number of new employees per year at Amerada Hess Corp. from 1984 ($t = 0$):[6]

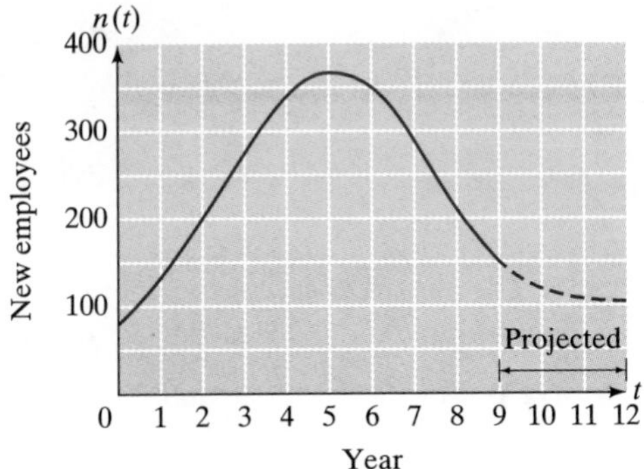

Source: Hoover's Handbook Database (World Wide Web site), The Reference Press, Inc., Austin, TX, 1995.

a. Estimate $\lim_{t\to+\infty} n(t)$ and interpret your answer.
b. Estimate $\lim_{t\to+\infty} n'(t)$ and interpret your answer.

E **49. *SAT Scores by Income*** The following bar graph shows average 1994 U.S. verbal SAT scores as a function of parents' income level:

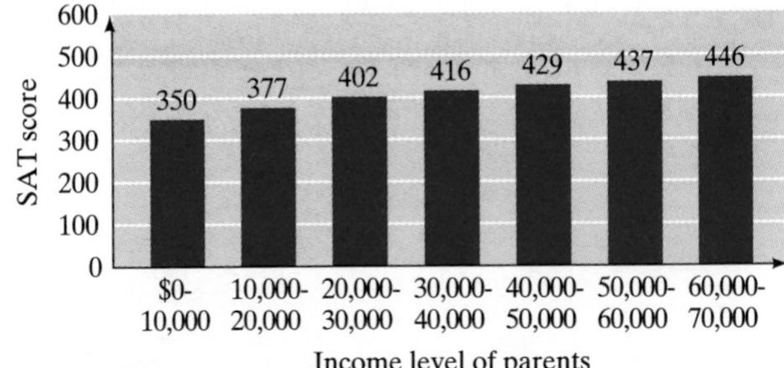

Source: The College Board/*New York Times*, March 5, 1995, p. E16.

These data can be modeled by

$$S(x) = 470 - 136e^{-0.00002645x}$$

where $S(x)$ is the average verbal SAT score of a student whose parents earn \$$x$ per year. Evaluate $\lim_{x\to+\infty} S(x)$ and interpret the result.

---

[4] The models are based on a best-fit quadratic for (approximate) tourism data, as measured in constant 1993 dollars.

[5] The model is the authors'.

[6] The projected part of the curve (from $t = 9$ on) is fictitious. The model is based on a best-fit logistic curve.

E **50.** ***SAT Scores by Income*** The following bar graph shows average 1994 U.S. math SAT scores as a function of parents' income level.

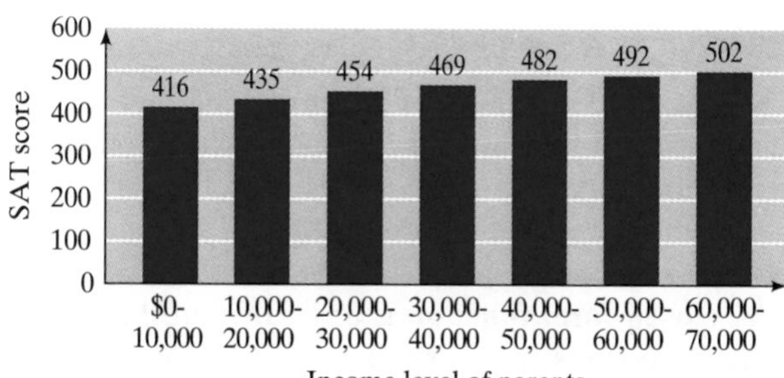

Source: The College Board/*New York Times,* March 5, 1995, p. E16.

These data can be modeled by

$$S(x) = 535 - 136e^{-0.000021286x}$$

where $S(x)$ is the average math SAT score of a student whose parents earn \$$x$ per year. Evaluate $\lim_{x\to+\infty} S(x)$ and interpret the result.

Communication and Reasoning Exercises

51. Describe the two methods of evaluating limits discussed in this section. Give at least one disadvantage of each.

52. What is wrong with the following statement? If $f(a)$ is defined, then $\lim_{x\to a} f(x)$ exists and equals $f(a)$.

53. Give an example of a function f with $\lim_{x\to 1} f(x) = f(2)$.

54. If $D(t)$ is the Dow Jones average at time t and $\lim_{t\to+\infty} D(t) = +\infty$, is it possible that the Dow will fluctuate indefinitely into the future?

55. If $S(t)$ represents the size of the universe in billions of light years at time t years since the Big Bang, and $\lim_{t\to+\infty} S(t) = 130{,}000$, is it possible that the universe will continue to expand forever?

3.7 Limits and Continuity: Algebraic Approach

Although numerical and graphical estimation of limits is effective, the estimates these methods yield may not be perfectly accurate. The algebraic method, when it can be used, always yields an exact answer. Moreover, algebraic analysis of a function often enables us to take a function apart and see "what makes it tick."

Let us start with the same example we used to begin the preceding section. Consider the function $f(x) = 2 + x$ and ask: What happens to $f(x)$ as x approaches 3? To answer this algebraically, notice that, as x gets closer and closer to 3, the quantity $2 + x$ must get closer and closer to $2 + 3 = 5$. Hence,

$$\lim_{x\to 3} f(x) = \lim_{x\to 3} (2 + x)$$

$$= 2 + 3 = 5$$

Question Is that all there is to the algebraic method? Just substitute $x = a$?
Answer Notice that by substituting $x = 3$, we *evaluated the function at* $x = 3$. In other words, we relied on the fact that

$$\lim_{x\to 3} f(x) = f(3)$$

In the last section we said that a function that satisfies this equation is said to be *continuous* at $x = 3$. Thus:

If we know that the function f is continuous at a point a, then we can compute $\lim_{x \to a} f(x)$ by simply substituting $x = a$ into $f(x)$.

In order to use this fact, we need to know how to recognize continuous functions when we see them. Geometrically, they are easy to spot: A function is continuous at $x = a$ if its graph has no break at $x = a$. Algebraically, there is a large class of functions that are known to be continuous on their domains—those, roughly speaking, that are *specified by a single formula.*

We can be more precise. A **closed-form function** is any function that can be obtained by combining constants, powers of x, exponential functions, radicals, logarithms, and trigonometric functions (and some other functions we shall not encounter in this text) into a *single* mathematical formula by means of the usual arithmetic operations and composition of functions. Examples of closed-form functions are

$$3x^2 - x + 1, \qquad \frac{\sqrt{x^2 - 1}}{6x - 1}, \qquad e^{-(4x^2 - 1)/x}, \qquad \sqrt{\log_3(x^2 - 1)}$$

They can be as complicated as we like. The following is *not* a closed-form function:

$$f(x) = \begin{cases} -1 & \text{if } x \le -1 \\ x^2 + x & \text{if } -1 < x \le 1 \\ 2 - x & \text{if } 1 < x \le 2 \end{cases}$$

The reason is that $f(x)$ is not specified by a *single* mathematical formula. What is useful about closed-form functions is the following general rule.

Continuity of Closed-Form Functions

Every closed-form function is continuous on its domain. Thus, if f is a closed-form function and $f(a)$ is defined, then we have $\lim_{x \to a} f(x) = f(a)$.

Mathematics majors spend a great deal of time studying the proof of this result. We ask you to accept it without proof.

example 1

Limit of a Closed-Form Function

Evaluate $\lim_{x \to 1} \dfrac{x^3 - 8}{x - 2}$ algebraically.

Solution

First, notice that $(x^3 - 8)/(x - 2)$ is a closed-form function because it is specified by a single algebraic formula. Also, $x = 1$ is in the domain of this function. Therefore,

$$\lim_{x \to 1} \frac{x^3 - 8}{x - 2} = \frac{1^3 - 8}{1 - 2} = 7$$

In the definition of the derivative, we must take the limit of the difference quotient $[f(x + h) - f(x)]/h$ as $h \to 0$. Although the difference quotients we encounter are usually closed-form functions, we cannot evaluate them by substitution: Since h appears in the denominator, $h = 0$ is not in the domain of the difference quotient. However—and this is the key to finding such limits—some preliminary algebraic simplification may allow us to obtain a closed-form function with $h = 0$ in its domain. We can find the limit by substituting $h = 0$ in the new function. Many limits can be computed by this technique.

example 2

Simplifying to Obtain the Limit

Evaluate $\lim_{x\to 2} \dfrac{x^3 - 8}{x - 2}$ algebraically.

Solution

Although $(x^3 - 8)/(x - 2)$ is a closed-form function, $x = 2$ is not in its domain. Thus, we cannot obtain the limit by substitution. Instead, we first simplify $f(x)$ to obtain a new function with $x = 2$ in its domain. To do this, notice first that the numerator can be factored as

$$x^3 - 8 = (x - 2)(x^2 + 2x + 4)$$

Thus,

$$\frac{x^3 - 8}{x - 2} = \frac{(x - 2)(x^2 + 2x + 4)}{x - 2} = x^2 + 2x + 4$$

After we have canceled the offending $(x - 2)$ in the denominator, we are left with a closed-form function *with 2 in its domain*. Thus,

$$\lim_{x\to 2} \frac{x^3 - 8}{x - 2} = \lim_{x\to 2} (x^2 + 2x + 4)$$

$$= 2^2 + 2(2) + 4 = 12 \qquad \text{Substitute } x = 2.$$

This confirms the answer we found numerically in Example 1 in the last section.

Before we go on . . . If the given function fails to simplify, we can always evaluate the limit numerically. It may very well be that the limit does not exist in that case.

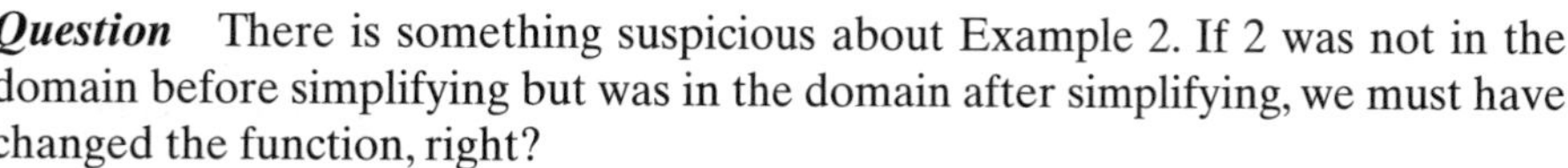

Question There is something suspicious about Example 2. If 2 was not in the domain before simplifying but was in the domain after simplifying, we must have changed the function, right?

Answer Correct. In fact, when we said that

$$\underset{\text{Domain excludes 2.}}{\frac{x^3 - 8}{x - 2}} = \underset{\text{Domain includes 2.}}{x^2 + 2x + 4}$$

we were lying a little bit. What we really meant is that these two expressions are equal *where both are defined.* The functions $(x^3 - 8)/(x - 2)$ and $x^2 + 2x + 4$ are different functions. The difference is that $x = 2$ is not in the domain of $(x^3 - 8)/(x - 2)$ and is in the domain of $x^2 + 2x + 4$. Since $\lim_{x\to 2} f(x)$ explicitly *ignores* any value that f may have at 2, this does not matter. From the point of view of the limit at 2, these functions *are* equal. In general we have the following rule.

Functions with Equal Limits

If $f(x) = g(x)$ for all x except possibly $x = a$, then

$$\lim_{x\to a} f(x) = \lim_{x\to a} g(x)$$

The next example gives an application of the simplifying technique in the calculation of a derivative. (See Examples 1 and 2 in Section 3.3 for other such calculations.)

example 3

Calculating Derivatives from Scratch

Verify the following result directly from the definition of the derivative. If $g(x) = 1/x$, then $g'(x) = -1/x^2$.

Solution

$$f'(x) = \lim_{h\to 0} \frac{f(x+h) - f(x)}{h} \qquad \text{Definition of derivative}$$

$$= \lim_{h\to 0} \frac{\left[\dfrac{1}{x+h} - \dfrac{1}{x}\right]}{h} \qquad \text{Definition of } f(x)$$

$$= \lim_{h\to 0} \frac{\left[\dfrac{x-(x+h)}{(x+h)x}\right]}{h} \qquad \text{Addition of fractions}$$

$$= \lim_{h\to 0} \frac{x - x - h}{h(x+h)x} \qquad \text{Simplify the numerator.}$$

$$= \lim_{h\to 0} \frac{-h}{h(x+h)x} \qquad \text{Simplify some more.}$$

$$= \lim_{h\to 0} \frac{-1}{(x+h)x} \qquad \text{Cancel } h.$$

$$= -\frac{1}{x^2} \qquad \text{Set } h = 0.$$

Note that we couldn't set $h = 0$ until we canceled the h. (Why?)

We can also use algebraic techniques to analyze functions that are not given in closed form.

example 4

Non-Closed-Form Function

For which values of x are the following piecewise defined functions continuous?

a. $f(x) = \begin{cases} x^2 + 2 & \text{if } x < 1 \\ 2x - 1 & \text{if } x \geq 1 \end{cases}$ **b.** $g(x) = \begin{cases} x^2 - x + 1 & \text{if } x \leq 0 \\ 1 - x & \text{if } 0 < x \leq 1 \\ x - 3 & \text{if } x > 1 \end{cases}$

Solution

a. The function $f(x)$ is given in closed form over the intervals $(-\infty, 1)$ and $[1, +\infty)$. At $x = 1$, $f(x)$ suddenly switches from one closed-form function to another, so $x = 1$ is the only point where there is a potential problem with continuity. To investigate the continuity of $f(x)$ at $x = 1$, let us calculate the limit there:

$$\lim_{x\to1^-} f(x) = \lim_{x\to1^-} (x^2 + 2) \quad \text{Since } f(x) = x^2 + 2 \text{ for } x < 1$$

$$= (1)^2 + 2 = 3 \quad x^2 + 2 \text{ is closed form.}$$

$$\lim_{x\to1^+} f(x) = \lim_{x\to1^+} (2x - 1) \quad \text{Since } f(x) = 2x - 1 \text{ for } x > 1$$

$$= 2(1) - 1 = 1 \quad 2x - 1 \text{ is closed form.}$$

Since the left and right limits are different, $\lim_{x\to1} f(x)$ does not exist, and so $f(x)$ is discontinuous at $x = 1$.

b. The only potential points of discontinuity for $g(x)$ occur at $x = 0$ and $x = 1$:

$$\lim_{x\to0^-} g(x) = \lim_{x\to0^-} x^2 - x + 1 = 1$$

$$\lim_{x\to0^+} g(x) = \lim_{x\to0^+} 1 - x = 1$$

Thus, $\lim_{x\to0} g(x) = 1$. Furthermore, $g(0) = 1 - (0) = 1$ from the formula, and so

$$\lim_{x\to0} g(x) = g(0)$$

which shows that $g(x)$ is continuous at $x = 0$. At $x = 1$, we have

$$\lim_{x\to1^-} g(x) = \lim_{x\to1^-} 1 - x = 0$$

$$\lim_{x\to1^+} g(x) = \lim_{x\to1^+} x - 3 = -2$$

so that $\lim_{x\to1} g(x)$ does not exist. Thus, $g(x)$ is discontinuous at $x = 1$. We conclude that $g(x)$ is continuous at every real number x except $x = 1$.

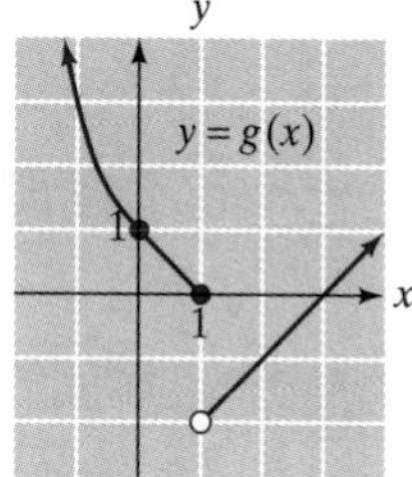

Figure 31

Before we go on . . . Figure 31 shows the graph of g. Notice how the discontinuity at $x = 1$ shows up as a break in the graph, whereas at $x = 0$ the two pieces "fit together" at the point $(0, 1)$.

Optional Internet Topic: Continuity and Differentiability

If you follow the path

Web site → On-Line Text → Continuity and Differentiability

you will find on-line text, interactive examples, and exercises on the relationship between continuity and differentiability.

Limits at Infinity

Let us look once again at Example 3 in the preceding section.

example 5

Limits at Infinity

Compute the following limits, if they exist.

a. $\lim_{x\to+\infty} \dfrac{2x^2 - 4x}{x^2 - 1}$ **b.** $\lim_{x\to-\infty} \dfrac{2x^2 - 4x}{x^2 - 1}$

Solution

While calculating the values for the tables used in Example 3 in the preceding section, you might have noticed that the highest power of x in both the numerator and the denominator dominated the calculations. For instance, when $x = 100{,}000$, the term $2x^2$ in the numerator has the value of 20,000,000,000, whereas the term $4x$ has the comparatively insignificant value of 400,000. Similarly, the term x^2 in the denominator overwhelms the term -1. In other words, for large values of x (or negative values with large magnitude),

$$\frac{2x^2 - 4x}{x^2 - 1} \approx \frac{2x^2}{x^2} \qquad \text{Use only highest powers on top and bottom.}$$

$$= 2$$

Therefore,

$$\lim_{x\to\pm\infty} \frac{2x^2 - 4x}{x^2 - 1} = 2$$

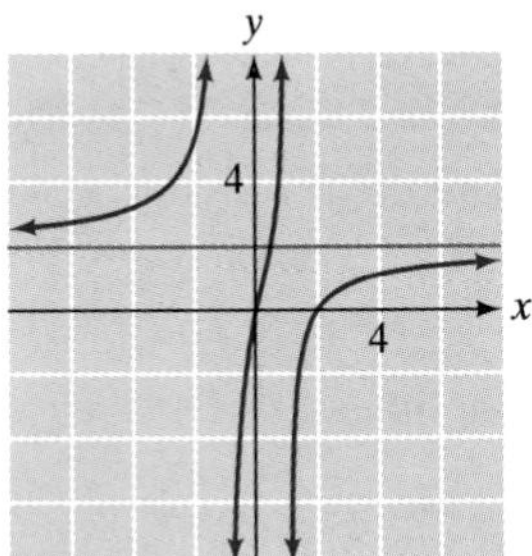

Figure 32

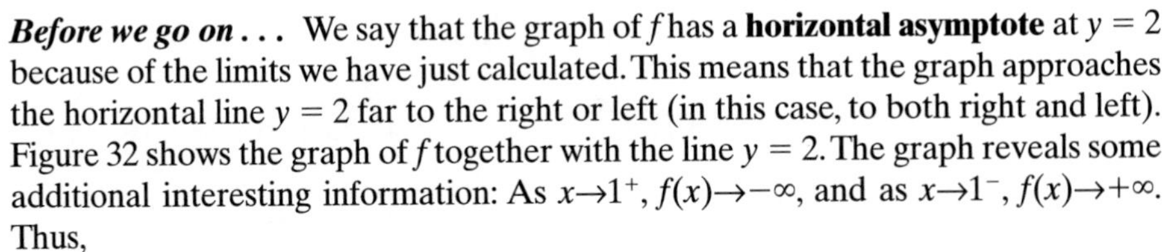
Before we go on . . . We say that the graph of f has a **horizontal asymptote** at $y = 2$ because of the limits we have just calculated. This means that the graph approaches the horizontal line $y = 2$ far to the right or left (in this case, to both right and left). Figure 32 shows the graph of f together with the line $y = 2$. The graph reveals some additional interesting information: As $x\to 1^+$, $f(x)\to -\infty$, and as $x\to 1^-$, $f(x)\to +\infty$. Thus,

$$\lim_{x\to 1} f(x) \text{ does not exist}$$

Can you determine what happens as $x\to -1$?

In the example above, $f(x)$ was a **rational function:** a quotient of polynomial functions. We calculated the limit of $f(x)$ at $\pm\infty$ by ignoring all powers of x in both the numerator and denominator except for the largest. It is possible to prove that this procedure is valid for any rational function (by dividing both top and bottom by the highest power of x present).

Evaluating the Limit of a Rational Function at $\pm\infty$

If $f(x)$ has the form

$$f(x) = \frac{c_n x^n + \cdots + c_2 x^2 + c_1 x + c_0}{d_m x^m + \cdots + d_2 x^2 + d_1 x + d_0}$$

with the c_i and d_i constants ($c_n \neq 0$ and $d_m \neq 0$), then we can calculate the limit of $f(x)$ as $x\to\pm\infty$ by ignoring all powers of x except the highest in both the numerator and the denominator. Thus,

$$\lim_{x\to\pm\infty} f(x) = \lim_{x\to\pm\infty} \frac{c_n x^n}{d_m x^m}$$

example 6

Limits at Infinity
Calculate the limits.

a. $\lim_{x\to+\infty} \frac{3x^4 - x^3 + 1}{x^3 + 40x^2}$ **b.** $\lim_{x\to-\infty} \frac{x^3 + 40x^2}{10x^4}$

Solution

a. Ignoring all but the highest powers of x, we have

$$\lim_{x\to+\infty} \frac{3x^4 - x^3 + 1}{x^3 + 40x^2} = \lim_{x\to+\infty} \frac{3x^4}{x^3} = \lim_{x\to+\infty} 3x$$

Now we have a far simpler limit to evaluate. We can even say what the limit is without a table: As $x\to+\infty$, $3x\to+\infty$ as well. Thus,

$$\lim_{x\to+\infty} \frac{3x^4 - x^3 + 1}{x^3 + 40x^2} = +\infty$$

b. We have

$$\lim_{x\to-\infty} \frac{x^3 + 40x^2}{10x^4} = \lim_{x\to-\infty} \frac{x^3}{10x^4} = \lim_{x\to-\infty} \frac{1}{10x}$$

At this stage, a table would be helpful, but once again we can manage without it. If x is, say, $-10{,}000$, then $1/10x = -1/100{,}000 = -0.00001$, extremely close to zero. In fact, the larger x gets in magnitude, the smaller $1/10x$ must get. Thus,

$$\lim_{x\to-\infty} \frac{x^3 + 40x^2}{10x^4} = 0$$

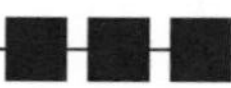

3.7 exercises

Calculate the limits in Exercises 1–38 algebraically. If a limit does not exist, say why.

1. $\lim_{x\to0} (x + 1)$

2. $\lim_{x\to0} (2x - 4)$

3. $\lim_{x\to2} \frac{2 + x}{x}$

4. $\lim_{x\to-1} \frac{4x^2 + 1}{x}$

5. $\lim_{x\to-1} \frac{x + 1}{x}$

6. $\lim_{x\to4} (x + \sqrt{x})$

7. $\lim_{x\to8} (x - \sqrt[3]{x})$

8. $\lim_{x\to1} \frac{x - 2}{x + 1}$

9. $\lim_{h\to1} (h^2 + 2h + 1)$

10. $\lim_{h\to0} (h^3 - 4)$

11. $\lim_{h\to3} 2$

12. $\lim_{h\to0} -5$

13. $\lim_{h\to0} \frac{h^2}{h + h^2}$

14. $\lim_{h\to0} \frac{h^2 + h}{h^2 + 2h}$

15. $\lim_{x\to1} \frac{x^2 - 2x + 1}{x^2 - x}$

16. $\lim_{x\to-1} \frac{x^2 + 3x + 2}{x^2 + x}$

17. $\lim_{x\to2} \frac{x^3 - 8}{x - 2}$

18. $\lim_{x\to-2} \frac{x^3 + 8}{x^2 + 3x + 2}$

19. $\lim_{x\to0^+} \frac{1}{x^2}$

20. $\lim_{x\to0^+} \frac{1}{x^2 - x}$

21. $\lim_{x\to-1} \frac{x^2 + 1}{x + 1}$

22. $\lim_{x\to-1^-} \frac{x^2 + 1}{x + 1}$

23. $\lim_{x\to+\infty} \frac{3x^2 + 10x - 1}{2x^2 - 5x}$

24. $\lim_{x\to+\infty} \frac{6x^2 + 5x + 100}{3x^2 - 9}$

25. $\lim_{x\to+\infty} \frac{x^5 - 1000x^4}{2x^5 + 10{,}000}$

26. $\lim_{x\to+\infty} \frac{x^6 + 3000x^3 + 1{,}000{,}000}{2x^6 + 1000x^3}$

27. $\lim_{x\to+\infty} \frac{10x^2 + 300x + 1}{5x + 2}$

28. $\lim_{x\to+\infty} \frac{2x^4 + 20x^3}{1000x^3 + 6}$

29. $\lim_{x\to+\infty} \frac{10x^2 + 300x + 1}{5x^3 + 2}$

30. $\lim_{x\to+\infty} \frac{2x^4 + 20x^3}{1000x^6 + 6}$

31. $\lim_{x\to-\infty} \dfrac{3x^2 + 10x - 1}{2x^2 - 5x}$

32. $\lim_{x\to-\infty} \dfrac{6x^2 + 5x + 100}{3x^2 - 9}$

33. $\lim_{x\to-\infty} \dfrac{x^5 - 1000x^4}{2x^5 + 10{,}000}$

34. $\lim_{x\to-\infty} \dfrac{x^6 + 3000x^3 + 1{,}000{,}000}{2x^6 + 1000x^3}$

35. $\lim_{x\to-\infty} \dfrac{10x^2 + 300x + 1}{5x + 2}$

36. $\lim_{x\to-\infty} \dfrac{2x^4 + 20x^3}{1000x^3 + 6}$

37. $\lim_{x\to-\infty} \dfrac{10x^2 + 300x + 1}{5x^3 + 2}$

38. $\lim_{x\to-\infty} \dfrac{2x^4 + 20x^3}{1000x^6 + 6}$

In Exercises 39–58, use the definition to calculate the derivative of the given function.

39. $f(x) = -14$

40. $f(x) = 5$

41. $f(x) = 2x - 3$

42. $f(x) = -3x + 5$

43. $g(x) = -4x - 1$

44. $g(x) = 10x - 100$

45. $g(x) = x^2 - 2x$

46. $g(x) = 3x^2 + 1$

47. $h(x) = -5x^2 + 2x - 1$

48. $h(x) = -3x^2 - x + 5$

49. $f(t) = t^3 + t$

50. $f(t) = 2t^3 - t^2$

51. $g(t) = t^4 - t$

52. $g(t) = 3t^4 + 2t^2$

53. $h(t) = 6/t$

54. $h(t) = -1/t$

55. $f(x) = x + \dfrac{1}{x}$

56. $f(x) = 6 - \dfrac{6}{x}$

57. $f(x) = \dfrac{1}{x - 2}$

58. $f(x) = \dfrac{1}{2x + 1}$

In Exercises 59–66, find all points of discontinuity of the given function.

59. $f(x) = \begin{cases} x + 2 & \text{if } x < 0 \\ 2x - 1 & \text{if } x \ge 0 \end{cases}$

60. $g(x) = \begin{cases} 1 - x & \text{if } x \le 1 \\ x - 1 & \text{if } x > 1 \end{cases}$

61. $g(x) = \begin{cases} x + 2 & \text{if } x < 0 \\ 2x + 2 & \text{if } 0 \le x < 2 \\ x^2 + 2 & \text{if } x \ge 2 \end{cases}$

62. $f(x) = \begin{cases} 1 - x & \text{if } x \le 1 \\ x + 2 & \text{if } 1 < x < 3 \\ x^2 - 4 & \text{if } x \ge 3 \end{cases}$

63. $h(x) = \begin{cases} x + 2 & \text{if } x < 0 \\ 0 & \text{if } x = 0 \\ 2x + 2 & \text{if } x > 0 \end{cases}$

64. $h(x) = \begin{cases} 1 - x & \text{if } x < 1 \\ 1 & \text{if } x = 1 \\ x + 2 & \text{if } x > 1 \end{cases}$

65. $f(x) = \begin{cases} 1/x & \text{if } x < 0 \\ x & \text{if } 0 \le x \le 2 \\ 2^{x-1} & \text{if } x > 2 \end{cases}$

66. $f(x) = \begin{cases} x^3 + 2 & \text{if } x \le -1 \\ x^2 & \text{if } -1 < x < 0 \\ x & \text{if } x \ge 0 \end{cases}$

Applications

67. ***Social Ills*** The number of DWI arrests in New Jersey during the period from 1990 to 1993 can be modeled by the equation[1]

$$n(t) = \frac{18{,}000}{(t + 1)^{0.4}}$$

where $n(t)$ is the number of DWI arrests in year t, with $t = 0$ representing 1990. Calculate $\lim_{t\to+\infty} n(t)$ and interpret your answer.

68. ***Social Ills*** Repeat Exercise 67 using the linear model

$$n(t) = -2.4t + 19.5$$

69. ***Acquisition of Language*** The percentage $p(t)$ of children who are able to speak in at least single words by the age of t months can be approximated by the equation[2]

$$p(t) = 100\left(1 - \frac{12{,}196}{t^{4.478}}\right) \qquad (t \ge 8.5)$$

Calculate $\lim_{t\to+\infty} p(t)$ and $\lim_{t\to+\infty} p'(t)$ and interpret the results.

70. ***Acquisition of Language*** The percentage $q(t)$ of children who are able to speak in sentences of five or more words by the age of t months can be approximated by the equation

$$q(t) = 100\left(1 - \frac{5.2665 \times 10^{17}}{t^{12}}\right) \qquad (t \ge 30)$$

If p is the function referred to in Exercise 69, calculate $\lim_{t\to+\infty} [p(t) - q(t)]$ and interpret the result.

[1] This is a regression model based on data gleaned from a graph from the New Jersey Administrative Office of the Courts/*New York Times,* September 26, 1994, p. B1.

[2] The models in Exercises 69 and 70 are the authors' and are based on data presented in the article "The Emergence of Intelligence" by William H. Calvin in *Scientific American,* October 1994, pp. 101–107.

Communication and Reasoning Exercises

71. Describe the three methods of evaluating limits discussed in this section and the last. Give at least one disadvantage of each.

72. Give an example of a function f specified by means of algebraic formulas such that f is not continuous at $x = 2$.

73. What is wrong with the following statement? If $f(x)$ is specified algebraically, and $f(a)$ is defined, then $\lim_{x\to a} f(x)$ exists and equals $f(a)$.

74. What is wrong with the following statement? $\lim_{x\to -2} (x^2 - 4)/(x + 2)$ does not exist, because substituting $x = -2$ yields 0/0, which is undefined.

75. Find a function that is continuous everywhere except at two points.

76. Find a function that is continuous everywhere except at three points.

You're the Expert

Reducing Sulfur Emissions

The Environmental Protection Agency (EPA) wants to formulate a policy that will encourage utilities to reduce sulfur emissions. Its goal is to reduce annual emissions of sulfur dioxide by a total of 10 million tons from the current level of 25 million tons by imposing a fixed charge for every ton of sulfur released into the environment per year. As a consultant to the EPA, you must determine the amount to be charged per ton of sulfur emissions.

You have the following data, which show the marginal cost to the utility industry of reducing sulfur emissions at several levels:[1]

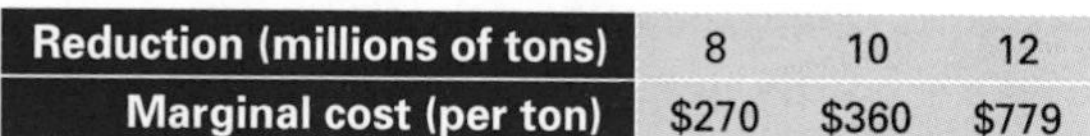

Reduction (millions of tons)	8	10	12
Marginal cost (per ton)	\$270	\$360	\$779

Source: Congress of the United States, Congressional Budget Office, *Curbing Acid Rain: Cost, Budget and Coal Market Effects* (Washington, DC: Government Printing Office, 1986), xx, xxii, 23, 80.

If $C(q)$ is the cost of removing q tons of sulfur dioxide, the table tells you that $C'(8{,}000{,}000) = \$270$ per ton, $C'(10{,}000{,}000) = \$360$ per ton, and $C'(12{,}000{,}000) = \$779$ per ton. Recalling that $C'(q)$ is the slope of the tangent to the graph of the cost function, you can see from the table that this slope is positive and increasing as q increases, so the graph of this cost function has the general shape shown in Figure 33.

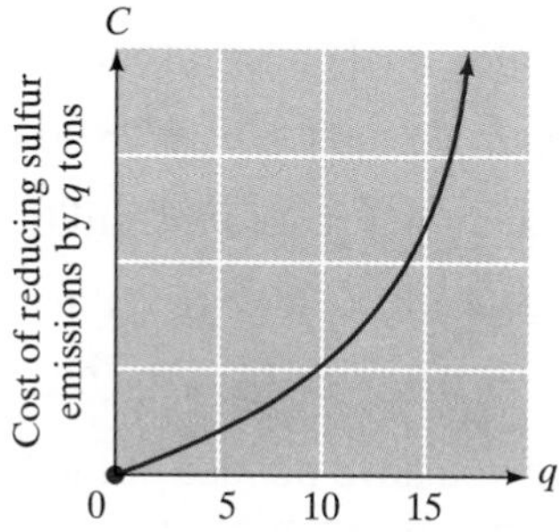

Figure 33

(Notice that the slope is increasing as you move to the right.) Thus, the utility industry has no cost incentive to reduce emissions. What you would like to do—if the goal

[1] These figures were produced in a computerized study of reducing sulfur emissions from the 1980 level by the given amounts.

of reducing total emissions by 10 million tons is to be reached—is alter this cost curve so that it has the general shape shown in Figure 34.

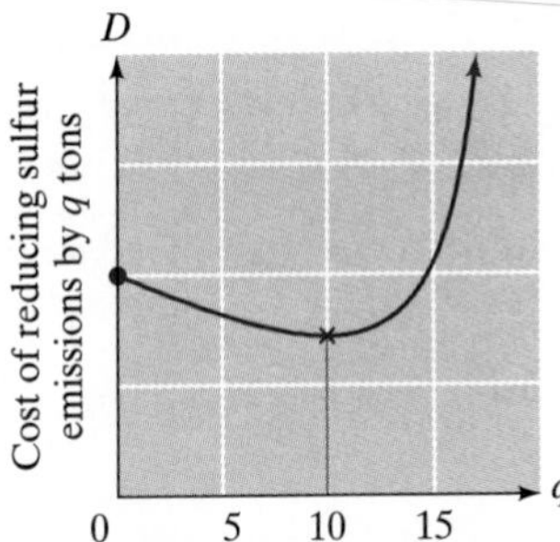

Figure 34

In this curve, the cost D to utilities is lowest at a reduction level of 10 million tons, so if the utilities act to minimize cost, they can be expected to reduce emissions by 10 million tons, which is the EPA goal. From the graph, you can see that $D'(10{,}000{,}000) = \$0$ per ton, whereas $D'(q)$ is negative for $q < 10{,}000{,}000$ and positive for $q > 10{,}000{,}000$.

At first, you are bothered by the fact that you were not given a cost function. Only the marginal costs were supplied, but you decide to work as best you can without knowing the original cost function $C(q)$.

You now assume that the EPA will impose an annual emission charge of $\$k$ per ton of sulfur released into the environment. It is your job to calculate k. Since you are working with q as the independent variable, you decide that it would be best to formulate the emission charge as a function of q, where q represents the amount by which sulfur emissions are *reduced.* The relationship between the annual sulfur emissions and the amount q by which emissions are reduced from the original 25 million tons is given by

$$\begin{aligned}\text{Annual sulfur emissions} &= \text{Original emissions} - \text{Amount of reduction}\\ &= 25{,}000{,}000 - q\end{aligned}$$

Thus, the total annual emission charge to the utilities is

$$k(25{,}000{,}000 - q) = 25{,}000{,}000k - kq$$

This results in a total cost to the utilities of

$$\text{Total cost} = \text{Cost of reducing emissions} + \text{Emission charge}$$

$$D(q) = C(q) + 25{,}000{,}000k - kq$$

Even though you have no idea of the form of $C(q)$, you remember that the derivative of a sum is the sum of the derivatives, and so you differentiate both sides with respect to q:

$$D'(q) = C'(q) + 0 - k = C'(q) - k$$

Remember that you want

$$D'(10{,}000{,}000) = 0$$

Thus, $\quad C'(10{,}000{,}000) - k = 0$

Referring to the table, you see that

$$360 - k = 0$$

so $$k = \$360 \text{ per ton}$$

In other words, all you need to do is set the emission charge at $k = \$360$ per ton of sulfur emitted. Furthermore, to ensure that the resulting curve will have the general shape shown in Figure 34, you would like to have $D'(q)$ negative for $q < 10{,}000{,}000$ and positive for $q > 10{,}000{,}000$. To check this, write

$$D'(q) = C'(q) - k = C'(q) - 360$$

and refer to the table to obtain

$$D'(8{,}000{,}000) = 270 - 360 = -90 < 0 \quad ✔$$

and $$D'(12{,}000{,}000) = 779 - 360 = 419 > 0 \quad ✔$$

Thus, based on the given data, the resulting curve will have the shape you require. You therefore inform the EPA that an annual emissions charge of \$360 per ton of sulfur released into the environment will create the desired incentive: to reduce sulfur emissions by 10 million tons per year.

One week later, you are informed that this charge would be unrealistic because the utilities cannot possibly afford such a cost. You are asked whether there is an alternative plan that accomplishes the 10-million-ton reduction goal and yet is cheaper to the utilities by \$5 billion per year. You then look at your expression for the emission charge

$$25{,}000{,}000k - kq$$

and notice that, if you decrease this amount by \$5 billion, the derivative will not change at all because the derivative of a constant is zero. Thus, you propose the following revised formula for the emission charge:

$$\begin{aligned} 25{,}000{,}000k - kq - 5{,}000{,}000{,}000 &= 25{,}000{,}000(360) - 360q - 5{,}000{,}000{,}000 \\ &= 4{,}000{,}000{,}000 - 360q \end{aligned}$$

At the expected reduction level of 10 million tons, the total amount paid by the utilities will then be

$$4{,}000{,}000{,}000 - 360(10{,}000{,}000) = \$400{,}000{,}000$$

Thus, your revised proposal is the following: Impose an annual emissions charge of \$360 per ton of sulfur released into the environment and hand back \$5 billion in the form of subsidies. The effect of this policy will be to cause the utilities industry to reduce sulfur emissions by 10 million tons per year and will result in \$400 million in annual revenues to the government.

Notice that this policy also provides an incentive for the utilities to search for cheaper ways to reduce emissions. For instance, a reduction level of 12 million tons will result in an additional cost to the industry of

$$4{,}000{,}000{,}000 - 360(12{,}000{,}000) = -\$320{,}000{,}000$$

The fact that this is negative means that the government will be paying the utilities industry \$320 million in annual subsidies.

exercises

1. Excluding subsidies, what should the annual emission charge be if the goal is to reduce sulfur emissions by 8 million tons?
2. Excluding subsidies, what should the annual emission charge be if the goal is to reduce sulfur emissions by 12 million tons?
3. What is the *marginal emission charge* in your revised proposal (as stated before the exercise set)? What is the relationship between the marginal cost of reducing sulfur emissions before emissions charges are implemented and the marginal emission charge, at the optimal reduction under your revised proposal?
4. We said that the revised policy provided an incentive for utilities to find cheaper ways to reduce emissions. How would $C(q)$ have to change to make 12 million tons the optimum reduction?
5. What change in $C(q)$ would make 8 million tons the optimum reduction?
6. If the scenario in Exercise 5 took place, what would the EPA have to do to make 10 million tons the optimum reduction once again?
7. Due to intense lobbying by the utility industry, you are asked to revise the proposed policy so that the utility industry will pay no charge if sulfur emissions are reduced by the desired 10 million tons. How can you accomplish this?
8. Suppose that, instead of imposing a fixed charge per ton of emission, you decide to use a sliding scale, so that the total charge to the industry for annual emissions of x tons will be $\$kx^2$ for some k. What must k be to again make 10 million tons the optimum reduction?

chapter 3 review test

1. For each of the following functions, find the average rate of change over the interval $[a, a + h]$ for $h = 1, 0.01$, and 0.001. (Round answers to four decimal places.) Then estimate the slope of the tangent line to the graph of the function at a.

a. $f(x) = \dfrac{1}{x + 1}; a = 0$ **b.** $f(x) = x^x; a = 2$

c. $f(x) = e^{2x}; a = 0$ **d.** $f(x) = \ln(2x); a = 1$

2. Following are the graphs of functions with four points marked. Determine at which (if any) of these points the derivative of the function is (**i**) -1, (**ii**) 0, (**iii**) 1, and (**iv**) 2.

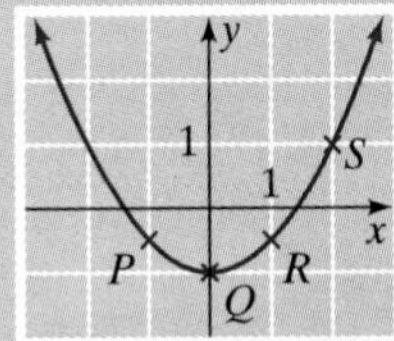

b.
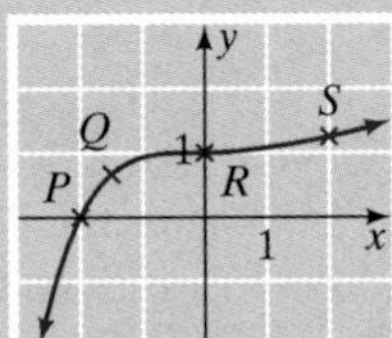

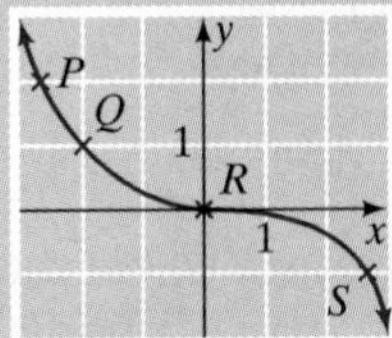

d.
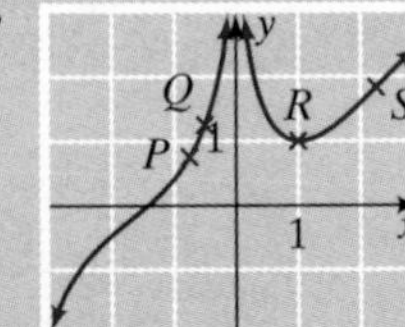

3. Let f have the graph shown.

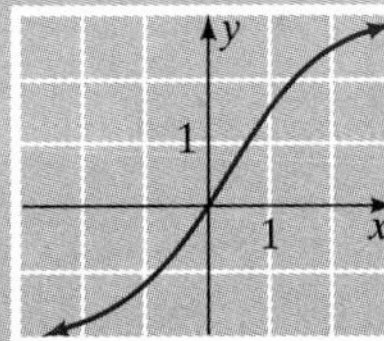

Select the correct answer.

a. The average rate of change of f over the interval $[0, 2]$ is (**A**) greater than, (**B**) less than, (**C**) approximately equal to $f'(0)$.

b. The average rate of change of f over the interval $[-1, 1]$ is (**A**) greater than, (**B**) less than, (**C**) approximately equal to $f'(0)$.

c. Over the interval $[0, 2]$, the instantaneous rate of change of f is (**A**) increasing, (**B**) decreasing, (**C**) neither.

d. Over the interval $[-2, 2]$, the instantaneous rate of change of f is (**A**) increasing and then decreasing, (**B**) decreasing and then increasing, (**C**) approximately constant.

e. When $x = 2$, $f(x)$ is (**A**) approximately 1 and increasing at a rate of about 2.5 units per unit of x, (**B**) approximately 1.2 and increasing at a rate of about 1 unit per unit of x, (**C**) approximately 2.5 and increasing at a rate of about 1.5 units per unit of x, (**D**) approximately 2.5 and increasing at a rate of about 2.5 units per unit of x.

4. Find the derivative of each of the following functions.

a. $f(x) = 10x^5 + \frac{1}{2}x^4 - x + 2$ **b.** $f(x) = \frac{10}{x^5} + \frac{1}{2x^4} - \frac{1}{x} + 2$

c. $f(x) = 3x^3 + 3\sqrt[3]{x}$ **d.** $f(x) = \frac{2}{x^{2.1}} - \frac{x^{0.1}}{2}$

5. Use the definition of the derivative to calculate the derivative of each of the following functions algebraically.

a. $f(x) = x^2 + x$ **b.** $f(x) = \frac{1}{x} + 1$

Applications

6. Since the start of July, OHaganBooks.com has seen its weekly sales increase as shown in the following table:

Week	1	2	3	4	5	6
Sales (books)	6500	7000	7200	7800	8500	9000

a. What was the average rate of increase of weekly sales over this time period?

b. During which 1-week interval did the rate of increase of sales exceed the average rate?

c. During which 2-week interval did the weekly sales increase at the highest average rate, and what was that average rate?

7. OHaganBooks.com fit the curve

$$w(t) = -3.7t^3 + 74.6t^2 + 135.5t + 6333.3$$

to its weekly sales figures from Question 6, as shown in the following graph:

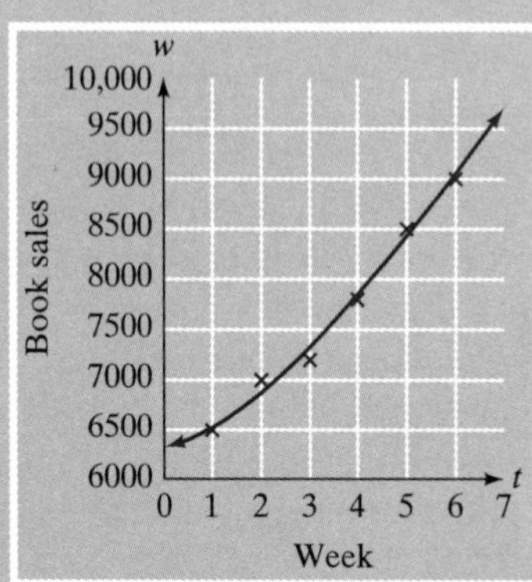

a. According to the model, what was the rate of increase of sales at the beginning of the second week ($t = 1$)? (Round your answer to the nearest unit.)
b. If we extrapolate the model, what would be the rate of increase of weekly sales at the beginning of the eighth week ($t = 7$)?
c. Graph the function w for $0 \le t \le 20$. Would it be realistic to use the function to predict sales through week 20? Why?

8. OHaganBooks.com decided that the curve above was not suitable for extrapolation, so instead it tried

$$s(t) = 6053 + \frac{4474}{1 + e^{-0.55(t-4.8)}}$$

which is shown in the following graph:

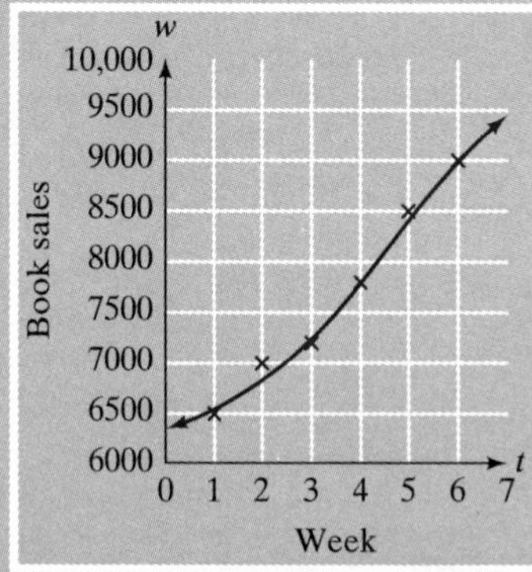

a. Using this function, estimate the rate of increase of weekly sales at the beginning of the seventh week ($t = 6$). (Round your answer to the nearest unit.)
b. If we extrapolate the model, what would be the rate of increase of weekly sales at the beginning of the 15th week ($t = 14$)?

c. Graph the function s for $0 \leq t \leq 20$. What is the long-term prediction for weekly sales? What is the long-term prediction for the rate of change of weekly sales?

9. As OHaganBooks.com's sales increase, so do its costs. If we take into account volume discounts from suppliers and shippers, the weekly cost of selling x books is

$$C(x) = -0.00002x^2 + 3.2x + 5400 \text{ dollars}$$

a. What is the marginal cost at a sales level of 8000 books per week?
b. What is the average cost per book at a sales level of 8000 books per week?
c. What is the marginal average cost at a sales level of 8000 books per week?
d. Interpret the results of parts (a)–(c).

Additional On-Line Review

If you follow the path

Web site → Everything for Calculus → Chapter 3

you will find the following additional resources to help you review:

- A comprehensive chapter summary (including examples and interactive features)
- Additional review exercises (including interactive exercises and many with help)
- A true–false chapter quiz
- Graphing utilities

You're the Expert

Projecting Market Growth

You are on the board of directors at Fullcourt Academic Press. The sales director of the high school division has just burst into your office with a proposal for a major expansion strategy based on the assumption that the number of high school seniors in the United States will be growing at a rate of at least 20,000 per year through the year 2005. Since the figures actually appear to be leveling off, you are suspicious about this estimate. You would like to devise a model that predicts the trend before tomorrow's scheduled board meeting. How do you go about doing this?

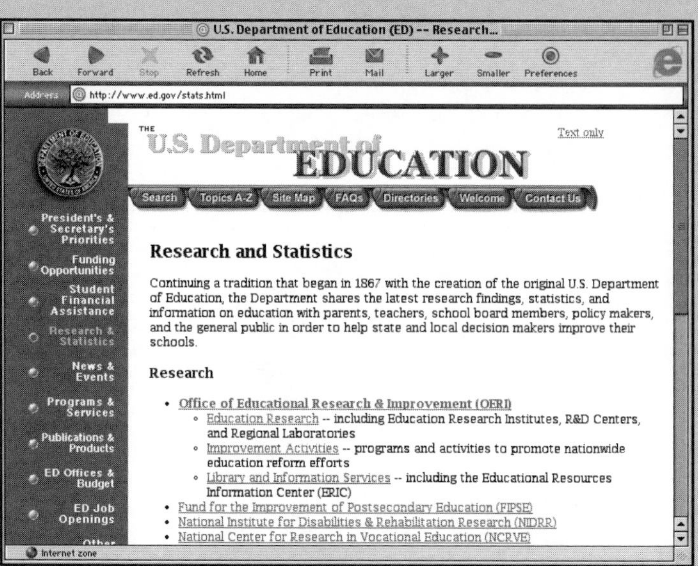

Internet Resources for This Chapter

At the web site, follow the path

Web site → Everything for Calculus → Chapter 4

where you will find links to step-by-step tutorials for the main topics in this chapter, a detailed chapter summary you can print out, a true–false quiz, and a collection of sample test questions. You will also find an on-line grapher and other useful resources. Complete text and interactive exercises have been placed on the web site for the following optional topic:

- Linear approximation and error estimation

Techniques of Differentiation

chapter 4

4.1 The Product and Quotient Rules
4.2 The Chain Rule
4.3 Derivatives of Logarithmic and Exponential Functions
4.4 Derivatives of Trigonometric Functions (Optional)
4.5 Implicit Differentiation

Introduction

In Chapter 3 we studied the concept of the derivative of a function, and we saw some of the applications for which derivatives are useful. However, the only functions we could differentiate easily were sums of terms of the form ax^n, where a and n are constants.

In this chapter we develop techniques that enable us to differentiate any closed-form function—that is, any function, no matter how complicated, that can be specified by a formula involving powers, radicals, exponents, logarithms, and (optionally) trigonometric functions. We also show how to find the derivatives of functions that are specified only *implicitly*—that is, functions for which we are not given an explicit formula for y in terms of x but only an equation relating x and y.

4.1 The Product and Quotient Rules

We know how to find the derivatives of functions that are sums of powers, like polynomials. In general, if a function is a sum or difference of functions whose derivatives we know, then we know how to find its derivative. But what about *products and quotients* of functions whose derivatives we know? For instance, how do we calculate the derivative of an expression like $x^2/(x + 1)$?

Question I don't see what the problem is. Can't we just say that the derivative of $x^2/(x + 1)$ is $2x/1 = 2x$?

Answer No. Your calculation is based on an assumption: that the derivative of a quotient is the quotient of the derivatives. But that assumption leads to inconsistencies. For instance, we know that

$$\frac{d}{dx}\left(\frac{x^3}{x}\right) = \frac{d}{dx}(x^2) = 2x$$

However, if we used the method suggested in your question, we would get

$$\frac{d}{dx}\left(\frac{x^3}{x}\right) = \frac{3x^2}{1} = 3x^2 \qquad ✘ \text{ WRONG}$$

Thus, the derivative of a quotient is *not* the quotient of the derivatives. Similarly, the following calculation is also *wrong:*

$$\frac{d}{dx}(x^3 \cdot x) = 3x^2 \cdot 1 = 3x^2 \qquad \text{✗ WRONG}$$

since $x^3 \cdot x = x^4$, and its derivative is $4x^3$, not $3x^2$. Thus, the derivative of a product is *not* the product of the derivatives.

Question If that is not how we find the derivatives of products and quotients, how *do* we?

Answer To motivate the correct method, let us ask a question: We know that the daily revenue that results from the sale of q items per day at a price of p dollars per item is given by the product, $R = pq$ dollars. Suppose you are currently selling wall posters on campus. At this time your daily sales are 50 posters, and sales are increasing at a rate of 4 per day. Furthermore, you are currently charging \$10 per poster, and you are raising the price at a rate of \$2 per day. Can you estimate how fast your daily revenue is increasing? In other words, can you estimate the rate of change, dR/dt, of the revenue R?

Question Let me see . . . there are two contributions to the rate of change of the daily revenue: the increase in daily sales and the increase in the unit price. We have

$$\frac{dR}{dt} \text{ due to increasing price:} \quad \$2 \text{ per day} \times 50 \text{ posters} = \$100 \text{ per day}$$

$$\frac{dR}{dt} \text{ due to increasing sales:} \quad \$10 \text{ per poster} \times 4 \text{ posters per day} = \$40 \text{ per day}$$

Thus, I estimate the daily revenue to be increasing at a rate of \$100 + \$40 = \$140 per day.

Answer Correct. Now you have just answered your original question about the rate of change of a product. Let us translate what you have said into symbols:

$$\frac{dR}{dt} \text{ due to increasing price:} \quad \frac{dp}{dt} \times q$$

$$\frac{dR}{dt} \text{ due to increasing sales:} \quad p \times \frac{dq}{dt}$$

Thus, the rate of change of the revenue is given by

$$\frac{dR}{dt} = \frac{dp}{dt}q + p\frac{dq}{dt}$$

Since $R = pq$, we have discovered the following rule for differentiating a product:

$$\frac{d(pq)}{dt} = \frac{dp}{dt}q + p\frac{dq}{dt}$$

The derivative of a product is the derivative of the first times the second, plus the first times the derivative of the second.

This rule and a similar rule for differentiating quotients are given next. After a little practice using them, we discuss how these results are proved.

Product Rule

$$\frac{d}{dx}\left[f(x)g(x)\right] = f'(x)g(x) + f(x)g'(x)$$

In Words

The derivative of a product is the derivative of the first times the second, plus the first times the derivative of the second.

Quick Example

$$\frac{d}{dx}\left[x^2(3x-1)\right] = \underset{\text{Derivative of first}}{\underset{\uparrow}{2x}} \cdot \underset{\text{Second}}{\underset{\uparrow}{(3x-1)}} + \underset{\text{First}}{\underset{\uparrow}{x^2}} \cdot \underset{\text{Derivative of second}}{\underset{\uparrow}{(3)}}$$

Quotient Rule

$$\frac{d}{dx}\left(\frac{f(x)}{g(x)}\right) = \frac{f'(x)g(x) - f(x)g'(x)}{\left[g(x)\right]^2}$$

In Words

The derivative of a quotient is the derivative of the top times the bottom, minus the top times the derivative of the bottom, all over the bottom squared.

Quick Example

$$\frac{d}{dx}\left(\frac{x^3}{x^2+1}\right) = \frac{\overset{\text{Derivative of top}}{\overset{\downarrow}{3x^2}}\overset{\text{Bottom}}{\overset{\downarrow}{(x^2+1)}} - \overset{\text{Top}}{\overset{\downarrow}{x^3}} \cdot \overset{\text{Derivative of bottom}}{\overset{\downarrow}{2x}}}{\underset{\text{Bottom squared}}{\underset{\uparrow}{(x^2+1)^2}}}$$

Notes

- Don't try to remember the rules by the symbols we have used, but remember them in words. (The slogans are easy to remember, even if the terms are not precise.)
- One more time: *The derivative of a product is NOT the product of the derivatives, and the derivative of a quotient is NOT the quotient of the derivatives.* To find the derivative of a product, you must use the product rule, and to find the derivative of a quotient, you must use the quotient rule. Forgetting this is a mistake everyone makes from time to time.[1]

Question Wait a minute! The expression $2x^3$ is a product, and we already know that its derivative is $6x^2$. Where did we use the product rule?

Answer To differentiate functions such as $2x^3$, we have used the rule from Section 3.4: *The derivative of c times a function is c times the derivative of the function.* The product rule gives us the same result:

[1]Leibniz made this mistake at first, too, so you are in good company.

Derivative of first, Second, First, Derivative of Second

$$\frac{d}{dx}(2x^3) = (0)(x^3) + (2)(3x^2) = 6x^2 \qquad \text{Product rule}$$

$$\frac{d}{dx}(2x^3) = (2)(3x^2) = 6x^2 \qquad \text{Derivative of a constant times a function}$$

We do not recommend that you use the product rule to differentiate functions like $2x^3$; continue to use the simpler rule when one of the factors is a constant.

example 1

Using the Product Rule

Compute the following derivatives.

a. $\frac{d}{dx}\left[(x^{3.2}+1)(1-x)\right]$ Simplify the answer.

b. $\frac{d}{dx}\left[(x+1)(x^2+1)(x^3+1)\right]$ Do not expand the answer.

Solution

a. We can do the calculation in two ways.

Product rule: Derivative of first, Second, First, Derivative of second

$$\frac{d}{dx}\left[(x^{3.2}+1)(1-x)\right] = (3.2x^{2.2})(1-x) + (x^{3.2}+1)(-1)$$

$$= 3.2x^{2.2} - 3.2x^{3.2} - x^{3.2} - 1 \qquad \text{Expand the answer.}$$

$$= -4.2x^{3.2} + 3.2x^{2.2} - 1$$

Avoiding the product rule: First, expand the given expression.

$$(x^{3.2}+1)(1-x) = -x^{4.2} + x^{3.2} - x + 1$$

Thus,

$$\frac{d}{dx}\left[(x^{3.2}+1)(1-x)\right] = \frac{d}{dx}(-x^{4.2} + x^{3.2} - x + 1)$$

$$= -4.2x^{3.2} + 3.2x^{2.2} - 1$$

In this example the product rule saves us little or no work, but later sections contain examples that can be done in no other way. Learn how to use the product rule now!

b. Here we have a product of *three* functions, not just two. We can find the derivative by using the product rule twice:

$$\frac{d}{dx}\left[(x+1)(x^2+1)(x^3+1)\right] = \frac{d}{dx}(x+1)\cdot\left[(x^2+1)(x^3+1)\right] + (x+1)\cdot\frac{d}{dx}\left[(x^2+1)(x^3+1)\right]$$

$$= (1)(x^2+1)(x^3+1) + (x+1)\left[(2x)(x^3+1) + (x^2+1)(3x^2)\right]$$

$$= (1)(x^2+1)(x^3+1) + (x+1)(2x)(x^3+1) + (x+1)(x^2+1)(3x^2)$$

We can see here a more general product rule:

$$(fgh)' = f'gh + fg'h + fgh'$$

Notice that every factor has a chance to contribute to the rate of change of the product. There are similar formulas for products of four or more functions.

---

example 2

Using the Quotient Rule

Compute the derivatives. **a.** $\dfrac{d}{dx}\left(\dfrac{1-3.2x^{-0.1}}{x+1}\right)$ **b.** $\dfrac{d}{dx}\left(\dfrac{(x+1)(x+2)}{x-1}\right)$

Solution

Derivative of top ↓ Bottom ↓ Top ↓ Derivative of bottom ↓

a. $$\frac{d}{dx}\left(\frac{1-3.2x^{-0.1}}{x+1}\right) = \frac{(0.32x^{-1.1})(x+1) - (1-3.2x^{-0.1})(1)}{(x+1)^2}$$

↑ Bottom squared

$$= \frac{0.32x^{-0.1} + 0.32x^{-1.1} - 1 + 3.2x^{-0.1}}{(x+1)^2}$$ Expand the numerator.

$$= \frac{3.52x^{-0.1} + 0.32x^{-1.1} - 1}{(x+1)^2}$$

b. Here we have both a product and a quotient. Which rule do we use, the product or the quotient rule? Here is a way to decide. Think about how we would calculate, step by step, the value of $(x + 1)(x + 2)/(x - 1)$ for a specific value of x—say, $x = 11$. Here is how we would probably do it:

1. Calculate $(x + 1)(x + 2) = (11 + 1)(11 + 2) = 156$.
2. Calculate $x - 1 = 11 - 1 = 10$.
3. Divide 156 by 10 to get 15.6.

Now ask: What was the last operation we performed? The last operation was division, so we can regard the whole expression as a *quotient*—that is, as $(x + 1)(x + 2)$ *divided by* $(x - 1)$. Therefore, we should use the quotient rule.

The first thing the quotient rule tells us to do is take the derivative of the numerator. Now, the numerator is a product, so we must use the product rule to take its derivative. Here is the calculation.

Derivative of top ↓ Bottom ↓ Top ↓ Derivative of bottom ↓

$$\frac{d}{dx}\left(\frac{(x+1)(x+2)}{x-1}\right) = \frac{\overbrace{[(1)(x+2) + (x+1)(1)]}(x-1) - \overbrace{[(x+1)(x+2)]}(1)}{(x-1)^2}$$

↑ Bottom squared

$$= \frac{(2x+3)(x-1) - (x+1)(x+2)}{(x-1)^2}$$

$$= \frac{x^2 - 2x - 5}{(x-1)^2}$$

It was important to determine the *order of operations,* and in particular to determine the last operation to be performed. Pretending to do an actual calculation reminds us of the order of operations; we call this technique the **calculation thought experiment.**

Before we go on . . . We used the quotient rule because the function was a quotient; we used the product rule to calculate the derivative of the numerator because the numerator was a product. Get used to this: Differentiation rules usually must be used in combination; once you have determined one rule to use, do not assume that you can forget the others.

Here is another way we could have done this problem: Our calculation thought experiment could have taken the following form:

1. Calculate $(x + 1)/(x - 1) = (11 + 1)/(11 - 1) = 1.2$.
2. Calculate $x + 2 = 11 + 2 = 13$.
3. Multiply 1.2 by 13 to get 15.6.

Then we would have regarded the expression as a *product*—the product of the factors $(x + 1)/(x - 1)$ and $(x + 2)$—and used the product rule instead. We can't escape the quotient rule, however: We would need to use it to take the derivative of the first factor, $(x + 1)/(x - 1)$. Try this approach for practice and check that you get the same answer.

Calculation Thought Experiment

The **calculation thought experiment** is a technique used to determine whether to treat an algebraic expression as a product, quotient, sum, or difference. Given an expression, consider the steps you would use in computing its value. If the last operation is multiplication, treat the expression as a product; if the last operation is division, treat the expression as a quotient; and so on.

Quick Examples

1. $(3x^2 - 4)(2x + 1)$ can be computed by first calculating the expressions in parentheses and then multiplying. Since the last step is multiplication, we can treat the expression as a product.
2. $(2x - 1)/x$ can be computed by first calculating the numerator and denominator and then dividing one by the other. Since the last step is division, we can treat the expression as a quotient.
3. $x^2 + (4x - 1)(x + 2)$ can be computed by first calculating x^2, then calculating the product $(4x - 1)(x + 2)$, and finally adding the two answers. Thus, we can treat the expression as a sum.
4. $(3x^2 - 1)^5$ can be computed by first calculating the expression in parentheses and then raising the answer to the fifth power. Thus, we can treat the expression as a power. (We shall see how to differentiate powers of expressions in the next section.)

It often happens that the same expression can be calculated in different ways; for example, $(x + 1)(x + 2)/(x - 1)$ can be treated as either a quotient or a product (see Example 2b).

example 3

Using the Calculation Thought Experiment

Find $\frac{d}{dx}\left[6x^2 + 5\left(\frac{x}{x-1}\right)\right]$.

Solution

The calculation thought experiment tells us that the expression we are asked to differentiate can be treated as a *sum.* Since the derivative of a sum is the sum of the derivatives, we get

$$\frac{d}{dx}\left[6x^2 + 5\left(\frac{x}{x-1}\right)\right] = \frac{d}{dx}(6x^2) + \frac{d}{dx}\left[5\left(\frac{x}{x-1}\right)\right]$$

In other words, we must take the derivatives of $6x^2$ and $5[x/(x-1)]$ separately and then add the answers. The derivative of $6x^2$ is $12x$. There are two ways of taking the derivative of $5[x/(x-1)]$: We could first multiply the expression $[x/(x-1)]$ by 5 to get $[5x/(x-1)]$ and then take its derivative using the quotient rule, or we could pull the 5 out, as we do next:

$$\frac{d}{dx}\left[6x^2 + 5\left(\frac{x}{x-1}\right)\right] = \frac{d}{dx}(6x^2) + \frac{d}{dx}\left[5\left(\frac{x}{x-1}\right)\right] \qquad \text{Derivative of sum}$$

$$= 12x + 5\frac{d}{dx}\left(\frac{x}{x-1}\right) \qquad \text{Constant} \times \text{Function}$$

$$= 12x + 5\left[\frac{(1)(x-1) - (x)(1)}{(x-1)^2}\right] \qquad \text{Quotient rule}$$

$$= 12x + 5\left[\frac{-1}{(x-1)^2}\right]$$

$$= 12x - \frac{5}{(x-1)^2}$$

Derivation of the Product Rule

To calculate the derivative of the function $f(x)g(x)$, we use the definition of the derivative:

$$\frac{d}{dx}\left[f(x)g(x)\right] = \lim_{h\to 0}\frac{f(x+h)g(x+h) - f(x)g(x)}{h}$$

Question How can we rewrite this expression so that we can evaluate the limit?
Answer There are several ways to do this. Here is one approach: Notice that the numerator reflects a simultaneous change in f [from $f(x)$ to $f(x+h)$] and g [from $g(x)$ to $g(x+h)$]. To separate the two effects, we add and subtract a quantity in the numerator that reflects a change in only one of the functions:

$$\frac{d}{dx}\left[f(x)g(x)\right] = \lim_{h\to 0}\frac{f(x+h)g(x+h) - f(x)g(x)}{h}$$

$$= \lim_{h\to 0}\frac{f(x+h)g(x+h) - f(x)g(x+h) + f(x)g(x+h) - f(x)g(x)}{h}$$

We subtracted and added the quantity[2] $f(x)g(x+h)$.

$$= \lim_{h\to 0}\frac{\left[f(x+h) - f(x)\right]g(x+h) + f(x)\left[g(x+h) - g(x)\right]}{h}$$

Common factors

$$= \lim_{h\to 0}\left(\frac{f(x+h) - f(x)}{h}\right)g(x+h) + \lim_{h\to 0} f(x)\left(\frac{g(x+h) - g(x)}{h}\right)$$

Limit of sum

$$= \lim_{h\to 0}\left(\frac{f(x+h) - f(x)}{h}\right)\lim_{h\to 0} g(x+h) + \lim_{h\to 0} f(x)\lim_{h\to 0}\left(\frac{g(x+h) - g(x)}{h}\right)$$

Limit of product

Now we already know the following four limits:

$$\lim_{h\to 0}\frac{f(x+h) - f(x)}{h} = f'(x)$$ Definition of derivative of f

$$\lim_{h\to 0}\frac{g(x+h - g(x)}{h} = g'(x)$$ Definition of derivative of g

$$\lim_{h\to 0} g(x+h) = g(x)$$ If g is differentiable, it must be continuous.[3]

$$\lim_{h\to 0} f(x) = f(x)$$ Limit of a constant

Putting these limits into the one we're calculating, we get

$$\frac{d}{dx}\left[f(x)g(x)\right] = f'(x)g(x) + f(x)g'(x)$$

which is the product rule.

The quotient rule can be proved in a very similar way. You can find a proof at the web site by following the path

Web site → Everything for Calculus → Chapter 4 → Proof of Quotient Rule

[2] Adding an appropriate form of zero is an age-old math ploy.

[3] To see why differentiable functions are continuous, follow

Web Site → On-Line Text → Continuity & Differentiability

4.1 exercises

*In Exercises 1–12, **a.** calculate the derivative of the given function mentally without using either the product or the quotient rule; then **b.** use the product or quotient rule to find the derivative. Check that you obtain the same answer.*

1. $f(x) = 3x$ **2.** $f(x) = 2x^2$ **3.** $g(x) = x \cdot x^2$

4. $g(x) = x \cdot x$ **5.** $h(x) = x(x + 3)$ **6.** $h(x) = x(1 + 2x)$

7. $r(x) = 100x^{2.1}$ **8.** $r(x) = 0.2x^{-1}$ **9.** $s(x) = \dfrac{2}{x}$

10. $t(x) = \dfrac{x}{3}$ **11.** $u(x) = \dfrac{x^2}{3}$ **12.** $s(x) = \dfrac{3}{x^2}$

Calculate dy/dx in Exercises 13–42. You need not expand your answers.

13. $y = (x + 1)(x^2 - 1)$ **14.** $y = (4x^2 + x)(x - x^2)$

15. $y = (2x^{0.5} + 4x - 5)(x - x^{-1})$

16. $y = (x^{0.7} - 4x - 5)(x^{-1} + x^{-2})$

17. $y = (2x^2 - 4x + 1)^2$ **18.** $y = (2x^{0.5} - x^2)^2$

19. $y = \left(\dfrac{x}{3.2} + \dfrac{3.2}{x}\right)(x^2 + 1)$ **20.** $y = \left(\dfrac{x^{2.1}}{7} + \dfrac{2}{x^{2.1}}\right)(7x - 1)$

21. $y = x^2(2x + 3)(7x + 2)$ **22.** $y = x(x^2 - 3)(2x^2 + 1)$

23. $y = (5.3x - 1)(1 - x^{2.1})(x^{-2.3} - 3.4)$

24. $y = (1.1x + 4)(x^{2.1} - x)(3.4 - x^{-2.1})$

25. $y = \left(\sqrt{x} + 1\right)\left(\sqrt{x} + \dfrac{1}{x^2}\right)$ **26.** $y = \left(4x^2 - \sqrt{x}\right)\left(\sqrt{x} - \dfrac{2}{x^2}\right)$

27. $y = \dfrac{2x + 4}{3x - 1}$ **28.** $y = \dfrac{3x - 9}{2x + 4}$

29. $y = \dfrac{2x^2 + 4x + 1}{3x - 1}$ **30.** $y = \dfrac{3x^2 - 9x + 11}{2x + 4}$

31. $y = \dfrac{x^2 - 4x + 1}{x^2 + x + 1}$ **32.** $y = \dfrac{x^2 + 9x - 1}{x^2 + 2x - 1}$

33. $y = \dfrac{x^{0.23} - 5.7x}{1 - x^{-2.9}}$ **34.** $y = \dfrac{8.43x^{-0.1} - 0.5x^{-1}}{3.2 + x^{2.9}}$

35. $y = \dfrac{\sqrt{x} + 1}{\sqrt{x} - 1}$ **36.** $y = \dfrac{\sqrt{x} - 1}{\sqrt{x} + 1}$

37. $y = \dfrac{\frac{1}{x} + \frac{1}{x^2}}{x + x^2}$ **38.** $y = \dfrac{1 - \frac{1}{x^2}}{x^2 - 1}$

39. $y = \dfrac{(x + 3)(x + 1)}{3x - 1}$ **40.** $y = \dfrac{x}{(x - 5)(x - 4)}$

41. $y = \dfrac{(x + 3)(x + 1)(x + 2)}{3x - 1}$

42. $y = \dfrac{3x - 1}{(x - 5)(x - 4)(x - 1)}$

In Exercises 43–48, compute the derivatives.

43. $\dfrac{d}{dx}\left[(x^2 + x)(x^2 - x)\right]$ **44.** $\dfrac{d}{dx}\left[(x^2 + x^3)(x + 1)\right]$

45. $\dfrac{d}{dx}\left[(x^3 + 2x)(x^2 - x)\right]\Big|_{x=2}$

46. $\dfrac{d}{dx}\left[(x^2 + x)(x^2 - x)\right]\Big|_{x=1}$

47. $\dfrac{d}{dt}\left[(t^2 - t^{0.5})(t^{0.5} + t^{-0.5})\right]\Big|_{t=1}$

48. $\dfrac{d}{dt}\left[(t^2 + t^{0.5})(t^{0.5} - t^{-0.5})\right]\Big|_{t=1}$

In Exercises 49–54, find the equation of the line tangent to the graph of the given function at the point with the indicated x-coordinate.

49. $f(x) = (x^2 + 1)(x^3 + x); x = 1$

50. $f(x) = (x^{0.5} + 1)(x^2 + x); x = 1$

51. $f(x) = \dfrac{x + 1}{x + 2}; x = 0$ **52.** $f(x) = \dfrac{\sqrt{x} + 1}{\sqrt{x} + 2}; x = 4$

53. $f(x) = \dfrac{x^2 + 1}{x}; x = -1$ **54.** $f(x) = \dfrac{x}{x^2 + 1}; x = 1$

Applications

55. *Revenue* The monthly sales of Sunny Electronics' new stereo system are given by $S(x) = 20x - x^2$ hundred units per month, x months after its introduction. The price Sunny charges is $p(x) = 1000 - x^2$ dollars per stereo system, x months after introduction. The revenue Sunny earns must then be $R(x) = 100S(x)p(x)$. Find, 5 months after the introduction, the rate of change of monthly sales, the rate of change of the price, and the rate of change of revenue. Interpret your answers.

56. *Revenue* The monthly sales of Sunny Electronics' new portable tape player are given by $S(x) = 20x - x^2$ hundred units per month, x months after its introduction. The price Sunny charges is $p(x) = 100 - x^2$ dollars per tape player, x months after introduction. The revenue Sunny earns must then be $R(x) = 100S(x)p(x)$. Find, 6 months after the introduction, the rate of change of monthly sales, the rate of change of the price, and the rate of change of revenue. Interpret your answers.

57. ***Revenue*** Dorothy Wagner is currently selling 20 "I ❤ Calculus" T-shirts per day, but sales are dropping at a rate of 3 per day. She is currently charging \$7 per T-shirt, but to compensate for dwindling sales, she is increasing the unit price by \$1 per day. How fast, and in what direction, is her daily revenue currently changing?

58. ***Pricing Policy*** Let us turn Exercise 57 around a little: Dorothy Wagner is currently selling 20 "I ❤ Calculus" T-shirts per day, but sales are dropping at a rate of 3 per day. She is currently charging \$7 per T-shirt, and she wishes to increase her daily revenue by \$10 per day. At what rate should she increase the unit price to accomplish this (assuming that the price increase does not affect sales)?

59. ***Bus Travel*** The Thoroughbred Bus Company finds that its monthly costs for one particular year were given by $C(t) = 10{,}000 + t^2$ dollars after t months. After t months, the company had $P(t) = 1000 + t^2$ passengers per month. How fast was its cost per passenger changing after 6 months?

60. ***Bus Travel*** The Thoroughbred Bus Company finds that its monthly costs for one particular year were given by $C(t) = 100 + t^2$ dollars after t months. After t months, the company had $P(t) = 1000 + t^2$ passengers per month. How fast was its cost per passenger changing after 6 months?

Some of the following exercises are variations of exercises in the preceding chapter.

61. ***Fuel Economy*** Your Porsche's gas mileage (in miles per gallon) is given as a function $M(x)$ of speed x in miles per hour, where

$$M(x) = \frac{1}{x + 3600x^{-1}}$$

Calculate $M'(x)$, and then $M'(10)$, $M'(60)$, and $M'(70)$. What do the answers tell about your car?

62. ***Fuel Economy*** Your used Chevy's gas mileage (in miles per gallon) is given as a function $M(x)$ of speed x in miles per hour, where

$$M(x) = \frac{10}{x + 3025x^{-1}}$$

Calculate $M'(x)$, and hence determine the *sign* of each of the following: $M'(40)$, $M'(55)$, and $M'(60)$. Interpret your results.

63. ***Expansion*** The number of Toys "Я" Us® stores (including Kids "Я" Us stores) worldwide increased from 171 at the start of 1984 to 1032 at the start of 1994. If the annual revenue at each store was \$600,000 in 1984 and was increasing by \$50,000 per year, how fast would Toys "Я" Us's worldwide revenue have been increasing at the start of 1990? (Use a linear model for the number of stores.)
Source: Company reports/Associated Press/*New York Times,* January 12, 1994, p. D4.

64. ***Investments*** The price of GTE® stock rose from about \$22 per share in January 1989 to \$35 per share in January 1994. If you had purchased 100 shares of GTE in January 1989 and steadily purchased additional shares at a rate of 10 shares per month, how fast would the value of your investment have been increasing in January 1994? (Use a linear model for the share price.)
Source: Company reports/Datastream/*New York Times,* January 14, 1994, p. D1.

65. ***Biology—Reproduction*** The "Verhulst model" for population growth specifies the reproductive rate of an organism as a function of the total population according to the following formula:

$$R(p) = \frac{r}{1 + kp}$$

where p is the total population in thousands of organisms, r and k are constants that depend on the particular circumstances and organism being studied, and $R(p)$ is the reproduction rate in thousands of organisms per hour. If $k = 0.125$ and $r = 45$, find $R'(p)$ and then $R'(4)$. Interpret the result.
Source: Mathematics in Medicine and the Life Sciences by F. C. Hoppensteadt and C. S. Peskin (New York: Springer-Verlag, 1992), pp. 20–22.

66. ***Biology—Reproduction*** Another model, the "predator satiation model" for population growth, specifies that the reproductive rate of an organism as a function of the total population varies according to the following formula:

$$R(p) = \frac{rp}{1 + kp}$$

where p is the total population in thousands of organisms, r and k are constants that depend on the particular circumstances and organism being studied, and $R(p)$ is the reproduction rate in new organisms per hour. Given that $k = 0.2$ and $r = 0.08$, find $R'(p)$ and $R'(2)$. Interpret the result.
Source: See the source for Exercise 65.

67. ***Embryo Development*** Bird embryos consume oxygen from the time the egg is laid through the time the chick hatches. For a typical galliform bird, the total oxygen consumption (in milliliters) t days after the egg was laid can be approximated by[3]

$$C(t) = -0.0163t^4 + 1.096t^3 - 10.704t^2 + 3.576t \qquad (t \le 30)$$

(An egg will usually hatch at around $t = 28$.) Suppose that at time $t = 0$ you have a collection of 30 newly laid eggs and that the number of eggs decreases linearly to zero at time $t = 30$ days. How fast is the total oxygen consumption of your collection of embryos changing

[3]The models in Exercises 67 and 68 are derived from graphical data published in the article "The Brush Turkey" by Roger S. Seymour in *Scientific American,* December 1991, pp. 108–114.

after 25 days? (Answer to the nearest whole number.) Interpret the result.

68. ***Embryo Development*** Turkey embryos consume oxygen from the time the egg is laid through the time the chick hatches. For a brush turkey, the total oxygen consumption (in milliliters) t days after the egg was laid can be approximated by

$$C(t) = -0.00708t^4 + 0.952t^3 - 21.96t^2 + 95.328t \qquad (t \le 50)$$

(An egg will typically hatch at around $t = 50$.) Suppose that at time $t = 0$ you have a collection of 100 newly laid eggs and that the number of eggs decreases linearly to zero at time $t = 50$ days. How fast is the total oxygen consumption of your collection of embryos changing after 40 days? (Answer to the nearest whole number.) Interpret the result.

Communication and Reasoning Exercises

69. You have come across the following in a newspaper article: "Revenues of HAL Home Heating Oil, Inc. are rising by \$4.2 million per year. This is due to an annual increase of 70¢ per gallon in the price HAL charges for heating oil and an increase in sales of 6 million gallons of oil per year." Comment on this analysis.

70. Your friend says that since average cost is obtained by dividing the cost function by the number of units x, it follows that the derivative of average cost is the same as marginal cost because the derivative of x is 1. Comment on this analysis.

71. Find a demand function $q(p)$ such that at a price per item of p = \$100, revenue will rise if the price per item is increased.

72. What must be true about a demand function $q(p)$ so that at a price per item of p = \$100, revenue will decrease if the price per item is increased?

73. ***Marginal Product*** *(from the GRE economics test)* Which of the following statements about average product and marginal product is correct?
    a. If average product is decreasing, marginal product must be less than average product.
    b. If average product is increasing, marginal product must be increasing.
    c. If marginal product is decreasing, average product must be less than marginal product.
    d. If marginal product is increasing, average product must be decreasing.
    e. If marginal product is constant over some range, average product must be constant over that range.

74. ***Marginal Cost*** *(based on a question from the GRE economics test)* Which of the following statements about average cost and marginal cost is correct?
    a. If average cost is increasing, marginal cost must be increasing.
    b. If average cost is increasing, marginal cost must be decreasing.
    c. If average cost is increasing, marginal cost must be greater than average cost.
    d. If marginal cost is increasing, average cost must be increasing.
    e. If marginal cost is increasing, average cost must be greater than marginal cost.

4.2 The Chain Rule

We can now find the derivatives of expressions that involve powers of x combined using addition, subtraction, multiplication, and division, but we still cannot take the derivative of an expression like $(3x + 1)^{0.5}$. For this we need one more rule. The function $h(x) = (3x + 1)^{0.5}$ is not a sum, difference, product, or quotient. We can use the calculation thought experiment to find the last operation we would perform in calculating $h(x)$.

1. Calculate $3x + 1$.
2. Take the 0.5 power (square root) of the answer.

Thus, the last operation is "take the 0.5 power." We do not yet have a rule for finding the derivative of the 0.5 power of a quantity other than x.

There is a way to build $h(x)$ out of two simpler functions: $f(x) = x^{0.5}$ (the function that corresponds to the second step in the calculation above) and $u(x) = 3x + 1$ (the first step):

$$h(x) = (3x + 1)^{0.5}$$

$$= f(3x + 1) \qquad f(x) = x^{0.5}$$

$$= f(u(x)) \qquad u(x) = 3x + 1$$

We say that h is the **composite** of f and u. We read $f(u(x))$ as "f of u of x."

In order to compute $h(1)$, say, we first compute $3 \cdot 1 + 1 = 4$ and then take the square root of 4, giving $h(1) = 2$. In order to compute $f(u(1))$, we can follow exactly the same steps: First compute $u(1) = 4$ and then $f(u(1)) = f(4) = 2$. We always compute $f(u(x))$ numerically from the inside out: Given x, compute first $u(x)$ and then $f(u(x))$.

Now, f and u are functions *whose derivatives we know.* The *chain rule* allows us to use our knowledge of the derivatives of f and u to find the derivative of $f(u(x))$. For the purposes of stating the rule, let us avoid some of the nested parentheses by abbreviating $u(x)$ as u. Thus, we write $f(u)$ instead of $f(u(x))$ and remember that u is a function of x.

Chain Rule

If f is a differentiable function of u and u is a differentiable function of x, then the composite $f(u)$ is a differentiable function of x, and

$$\frac{d}{dx}[f(u)] = f'(u)\frac{du}{dx} \qquad \text{Chain rule}$$

In words

The derivative of f(quantity) is the derivative of f, evaluated at that quantity, times the derivative of the quantity.

Quick Examples

1. Take $f(u) = u^2$. Then

$$\frac{d}{dx}[u^2] = 2u\frac{du}{dx} \qquad \text{Since } f'(u) = 2u$$

The derivative of a quantity squared is 2 times the quantity, times the derivative of the quantity.

2. Take $f(u) = u^{0.5}$. Then

$$\frac{d}{dx}(u^{0.5}) = 0.5u^{-0.5}\frac{du}{dx} \qquad \text{Since } f'(u) = 0.5u^{-0.5}$$

The derivative of a quantity raised to the 0.5 is 0.5 times the quantity raised to the -0.5, times the derivative of the quantity.

As the quick examples illustrate, for every power of a function u whose derivative we know, we now get a "generalized" differentiation rule. The table gives more examples.

Original rule	Generalized rule	In words
$\frac{d}{dx} x^2 = 2x$	$\frac{d}{dx} u^2 = 2u \frac{du}{dx}$	The derivative of a quantity squared is twice the quantity, times the derivative of the quantity.
$\frac{d}{dx} x^3 = 3x^2$	$\frac{d}{dx} u^3 = 3u^2 \frac{du}{dx}$	The derivative of a quantity cubed is 3 times the quantity squared, times the derivative of the quantity.
$\frac{d}{dx}\left(\frac{1}{x}\right) = -\frac{1}{x^2}$	$\frac{d}{dx}\left(\frac{1}{u}\right) = -\frac{1}{u^2}\frac{du}{dx}$	The derivative of 1 over a quantity is negative 1 over the quantity squared, times the derivative of the quantity.
Power rule	**Generalized power rule**	
$\frac{d}{dx} x^n = nx^{n-1}$	$\frac{d}{dx} u^n = nu^{n-1} \frac{du}{dx}$	The derivative of a quantity raised to the n is n times the quantity raised to the $n - 1$, times the derivative of the quantity.

Question Why should I accept the chain rule?
Answer To motivate it, let us see why it is true in a few special cases: when $f(u) = u^n$, where $n = \pm1, \pm2, \pm3, \ldots$. In these cases, the chain rule tells us, for example, that

$$\frac{d}{dx}(u^2) = 2u\frac{du}{dx}$$

$$\frac{d}{dx}(u^3) = 3u^2\frac{du}{dx}$$

and so on. But we could have done the first of these another way, using the product rule:

$$\frac{d}{dx}(u^2) = \frac{d}{dx}(u \cdot u) = \frac{du}{dx}u + u\frac{du}{dx} = 2u\frac{du}{dx}$$

which gives us the same result. Similarly, we can use the product rule to verify that

$$\frac{d}{dx}(u^3) = 3u^2\frac{du}{dx}$$

and so on for higher positive powers of u. We can now use the quotient rule and the chain rule for positive powers to verify the chain rule for *negative* powers.

Question The argument that the chain rule works in this special case does not compel me to accept that it works for *all* differentiable functions of u. How can I convince myself of the general case?

Answer Although the proof of the chain rule is beyond the scope of this book, you can find one on the web site by following the path

Web site → Everything for Calculus → Chapter 4 → Proof of Chain Rule

example 1

Using the Chain Rule

Compute the following derivatives.

a. $\frac{d}{dx}(2x^2 + x)^3$ **b.** $\frac{d}{dx}(x^3 + x)^{100}$ **c.** $\frac{d}{dx}\sqrt{3x + 1}$

Solution

a. Using the calculation thought experiment, we see that the last operation we would perform in calculating $(2x^2 + x)^3$ is *cubing*. Thus, we think of $(2x^2 + x)^3$ as *a quantity cubed*. There are two similar methods we can use to calculate its derivative.

Method 1: Using the formula. We think of $(2x^2 + x)^3$ as u^3, where $u = 2x^2 + x$. By the formula,

$$\frac{d}{dx}u^3 = 3u^2\frac{du}{dx} \qquad \text{Generalized power rule}$$

Now we substitute for u:

$$\frac{d}{dx}(2x^2 + x)^3 = 3(2x^2 + x)^2\frac{d}{dx}(2x^2 + x) = 3(2x^2 + x)^2(4x + 1)$$

Method 2: Using the verbal form. If we prefer to use the verbal form, we get: The derivative of $(2x^2 + x)$ cubed is three times $(2x^2 + x)$ squared, times the derivative of $(2x^2 + x)$. In symbols,

$$\frac{d}{dx}(2x^2 + x)^3 = 3(2x^2 + x)^2(4x + 1)$$

as we obtained above.

b. First, the calculation thought experiment: If we were computing $(x^3 + x)^{100}$, the last operation we would perform is *raising a quantity to the power 100.* Thus, we are dealing with *a quantity raised to the power 100,* and so we again use the generalized power rule. According to the verbal form of the generalized power rule, the derivative of a quantity raised to the power 100 is 100 times that quantity raised to the power 99, times the derivative of that quantity. In symbols,

$$\frac{d}{dx}(x^3 + x)^{100} = 100(x^3 + x)^{99}(3x^2 + 1)$$

c. We first rewrite the expression $\sqrt{3x + 1}$ as $(3x + 1)^{0.5}$ and then use the generalized power rule as in parts (a) and (b): The derivative of a quantity raised to the 0.5 is 0.5 times the quantity raised to the -0.5, times the derivative of the quantity. Thus,

$$\frac{d}{dx}(3x + 1)^{0.5} = 0.5(3x + 1)^{-0.5} \cdot 3 = 1.5(3x + 1)^{-0.5}$$

Before we go on . . . The following are examples of common errors:

$$\frac{d}{dx}(x^3 + x)^{100} = 100(3x^2 + 1)^{99} \qquad \text{✗ WRONG}$$

$$\frac{d}{dx}(x^3 + x)^{100} = 100(x^3 + x)^{99} \qquad \text{✗ WRONG}$$

Remember that the generalized power rule says that the derivative of a quantity to the power 100 is 100 times *that same quantity* raised to the power 99, *times the derivative of that quantity.*

Question It seems that there are now two formulas for the derivative of an nth power:

1. $\frac{d}{dx}x^n = nx^{n-1}$
2. $\frac{d}{dx}u^n = nu^{n-1}\frac{du}{dx}$

Which one do I use?

Answer Formula 1 is the original power rule, which applies only to a power of x. Thus, for instance, it applies to x^{10} but it does not apply to $(2x + 1)^{10}$ because the quantity that is being raised to a power is not x. Formula 2 applies to a power of any *function of x*, such as $(2x + 1)^{10}$. It can even be used in place of the original power rule. For example, if we take $u = x$ in formula 2, we obtain

$$\frac{d}{dx}x^n = nx^{n-1}\frac{dx}{dx}$$

$$= nx^{n-1} \qquad \text{The derivative of } x \text{ with respect to } x \text{ is 1.}$$

Thus, the generalized power rule really *is* a generalization of the original power rule, as its name suggests.

example 2

More Examples Using the Chain Rule

Find the derivatives.

a. $\frac{d}{dx}(2x^5 + x^2 - 20)^{-2/3}$ **b.** $\frac{d}{dx}\left(\frac{1}{\sqrt{x+2}}\right)$ **c.** $\frac{d}{dx}\left(\frac{1}{x^2+x}\right)$

Solution

Each of the given functions is, or can be rewritten as, a power of a function whose derivative we know. Thus, we can use the method of Example 1.

a. $\frac{d}{dx}(2x^5 + x^2 - 20)^{-2/3} = -\frac{2}{3}(2x^5 + x^2 - 20)^{-5/3}(10x^4 + 2x)$

b. $\frac{d}{dx}\left(\frac{1}{\sqrt{x+2}}\right) = \frac{d}{dx}(x+2)^{-1/2} = -\frac{1}{2}(x+2)^{-3/2} \cdot 1 = -\frac{1}{2(x+2)^{3/2}}$

c. $\frac{d}{dx}\left(\frac{1}{x^2+x}\right) = \frac{d}{dx}(x^2+x)^{-1} = -(x^2+x)^{-2}(2x+1) = -\frac{2x+1}{(x^2+x)^2}$

Before we go on . . . In part (c) we could have used the quotient rule instead of the generalized power rule. We can think of the quantity $1/(x^2 + x)$ in two different ways using the calculation thought experiment:

1. As 1 divided by something—in other words, as a quotient
2. As something raised to the -1 power

Of course, we get the same derivative using either approach.

We now look at some more complicated examples.

example 3

Harder Examples Using the Chain Rule

Find $\frac{dy}{dx}$ in each case.

a. $y = \left[(x + 1)^{-2.5} + 3x\right]^{-3}$ **b.** $y = (x + 10)^3\sqrt{1 - x^2}$

Solution

a. The calculation thought experiment tells us that the last operation we would perform in calculating y is raising the quantity $[(x + 1)^{-2.5} + 3x]$ to the power -3. Thus, we use the generalized power rule:

$$\frac{dy}{dx} = -3\left[(x + 1)^{-2.5} + 3x\right]^{-4}\frac{d}{dx}\left[(x + 1)^{-2.5} + 3x\right]$$

We are not yet done; we must still find the derivative of $(x + 1)^{-2.5} + 3x$. Finding the derivative of a complicated function in several steps helps to keep the problem manageable. Continuing, we have

$$\frac{dy}{dx} = -3\left[(x + 1)^{-2.5} + 3x\right]^{-4}\frac{d}{dx}\left[(x + 1)^{-2.5} + 3x\right]$$

$$= -3\left[(x + 1)^{-2.5} + 3x\right]^{-4}\left[\frac{d}{dx}(x + 1)^{-2.5} + \frac{d}{dx}(3x)\right] \quad \text{Derivative of a sum}$$

Now we have two derivatives left to calculate. The second of these we know to be 3, and the first is the derivative of a quantity raised to the –2.5 power. Thus,

$$\frac{dy}{dx} = -3\left[(x + 1)^{-2.5} + 3x\right]^{-4}\left[-2.5(x + 1)^{-3.5} \cdot 1 + 3\right]$$

b. The expression $(x + 10)^3\sqrt{1 - x^2}$ is a product, so we use the product rule:

$$\frac{d}{dx}\left[(x + 10)^3\sqrt{1 - x^2}\right] = \left[\frac{d}{dx}(x + 10)^3\right]\sqrt{1 - x^2} + (x + 10)^3\frac{d}{dx}\sqrt{1 - x^2}$$

$$= 3(x + 10)^2\sqrt{1 - x^2} \cdot 1 + (x + 10)^3\frac{1}{2\sqrt{1 - x^2}}(-2x)$$

$$= 3(x + 10)^2\sqrt{1 - x^2} - \frac{x(x + 10)^3}{\sqrt{1 - x^2}}$$

Applications

The next example is a new treatment of Example 3 from Section 3.5.

example 4

Marginal Product

Precision Manufacturers, Inc., is informed by a consultant that its annual profit is given by

$$P = -200{,}000 + 4000q - 0.46q^2 - 0.00001q^3$$

where q is the number of surgical lasers it sells per year. The consultant also informs Precision that the number of surgical lasers it can manufacture per year depends on the number n of assembly line workers it employs according to the equation

$$q = 100n \qquad \text{Each worker contributes 100 lasers per year.}$$

Use the chain rule to find the marginal product dP/dn.

Solution

We could calculate the marginal product by substituting the expression for q in the expression for P to obtain P as a function of n (as given in Chapter 3) and then finding dP/dn. Alternatively—and this will simplify the calculation—we can use the chain rule. To see how the chain rule applies, notice that P is a differentiable function of q, where q in turn is given as a differentiable function of n. Thus, by the chain rule,

$$\frac{dP}{dn} = P'(q)\frac{dq}{dn} \qquad \text{Chain rule}$$

$$= \frac{dP}{dq} \cdot \frac{dq}{dn} \qquad \text{Notice how the "quantities" } dq \text{ appear to cancel.}$$

Now we compute

$$\frac{dP}{dq} = 4000 - 0.92q - 0.00003q^2$$

and

$$\frac{dq}{dn} = 100$$

Substituting into the equation for dP/dn gives

$$\frac{dP}{dn} = (4000 - 0.92q - 0.00003q^2)(100)$$

$$= 400{,}000 - 92q - 0.003q^2$$

Notice that the answer has q as a variable. We can express dP/dn as a function of n by substituting $100n$ for q:

$$\frac{dP}{dn} = 400{,}000 - 92(100n) - 0.003(100n)^2$$

$$= 400{,}000 - 9200n - 30n^2$$

The equation

$$\frac{dP}{dn} = \frac{dP}{dq} \cdot \frac{dq}{dn}$$

in the example above is an appealing way of writing the chain rule because it suggests that the "quantities" dq cancel. In general, we can write the chain rule as shown on the next page.

Chain Rule in Differential Notation

If y is a differentiable function of u, and u is a differentiable function of x, then

$$\frac{dy}{dx} = \frac{dy}{du}\frac{du}{dx}$$

Notice how the units cancel:

$$\frac{\text{Units of } y}{\text{Units of } x} = \frac{\text{Units of } y}{\cancel{\text{Units of } u}}\frac{\cancel{\text{Units of } u}}{\text{Units of } x}$$

Quick Example

If $y = u^3$, where $u = 4x + 1$, then

$$\frac{dy}{dx} = \frac{dy}{du}\frac{du}{dx} = 3u^2 \cdot 4 = 12u^2 = 12(4x + 1)^2$$

You can see one of the reasons we still use Leibniz's differential notation: The chain rule looks like a simple "cancellation" of du terms.

example 5

Marginal Revenue

Suppose that a company's weekly revenue R is given as a function of the unit price p, and that p in turn is given as a function of weekly sales q (by means of a demand equation). If

$$\left.\frac{dR}{dp}\right|_{q=1000} = \$40 \text{ per } \$1 \text{ increase in price}$$

$$\left.\frac{dp}{dq}\right|_{q=1000} = -\$20 \text{ per additional item sold per week}$$

find the marginal revenue when sales are 1000 items per week.

Solution

The marginal revenue is dR/dq. By the chain rule, we have

$$\frac{dR}{dq} = \frac{dR}{dp}\frac{dp}{dq}$$

Units: Revenue per item
= Revenue per \$1 price increase × Price increase per additional item

Since we are interested in the marginal revenue at a demand level of 1000 items per week, we have

$$\left.\frac{dR}{dq}\right|_{q=1000} = (40)(-20) = -\$800 \text{ per additional item sold}$$

Thus, if the price is lowered to increase the demand from 1000 to 1001 items per week, the weekly revenue will drop by approximately \$800.

4.2 exercises

Mentally calculate the derivatives of the functions in Exercises 1–12.

1. $f(x) = (2x + 1)^2$ **2.** $f(x) = (3x - 1)^2$

3. $f(x) = (x - 1)^{-1}$ **4.** $f(x) = (2x - 1)^{-2}$

5. $f(x) = (2 - x)^{-2}$ **6.** $f(x) = (1 - x)^{-1}$

7. $f(x) = (2x + 1)^{0.5}$ **8.** $f(x) = (-x + 2)^{1.5}$

9. $f(x) = (4x - 1)^{-1}$ **10.** $f(x) = (x + 7)^{-2}$

11. $f(x) = \dfrac{1}{3x - 1}$ **12.** $f(x) = \dfrac{1}{(x + 1)^2}$

Calculate the derivatives of the functions in Exercises 13–46.

13. $f(x) = (x^2 + 2x)^4$ **14.** $f(x) = (x^3 - x)^3$

15. $f(x) = (2x^2 - 2)^{-1}$ **16.** $f(x) = (2x^3 + x)^{-2}$

17. $g(x) = (x^2 - 3x - 1)^{-5}$ **18.** $g(x) = (2x^2 + x + 1)^{-3}$

19. $h(x) = \dfrac{1}{(x^2 + 1)^3}$ **20.** $h(x) = \dfrac{1}{(x^2 + x + 1)^2}$

21. $r(x) = (0.1x^2 - 4.2x + 9.5)^{1.5}$

22. $r(x) = (0.1x - 4.2x^{-1})^{0.5}$

23. $r(s) = (s^2 - s^{0.5})^4$ **24.** $r(s) = (2s + s^{0.5})^{-1}$

25. $f(x) = \sqrt{1 - x^2}$ **26.** $f(x) = \sqrt{x + x^2}$

27. $h(x) = 2[(x + 1)(x^2 - 1)]^{-1/2}$

28. $h(x) = 3[(2x - 1)(x - 1)]^{-1/3}$

29. $h(x) = (3.1x - 2)^2 - \dfrac{1}{(3.1x - 2)^2}$

30. $h(x) = \left[(3.1x^2 - 2) - \dfrac{1}{3.1x - 2}\right]^2$

31. $f(x) = [(6.4x - 1)^2 + (5.4x - 2)^3]^2$

32. $f(x) = (6.4x - 3)^{-2} + (4.3x - 1)^{-2}$

33. $f(x) = (x^2 - 3x)^{-2}(1 - x^2)^{0.5}$

34. $f(x) = (3x^2 + x)(1 - x^2)^{0.5}$

35. $s(x) = \left(\dfrac{2x + 4}{3x - 1}\right)^2$ **36.** $s(x) = \left(\dfrac{3x - 9}{2x + 4}\right)^3$

37. $g(z) = \left(\dfrac{z}{1 + z^2}\right)^3$ **38.** $g(z) = \left(\dfrac{z^2}{1 + z}\right)^2$

39. $f(x) = [(1 + 2x)^4 - (1 - x)^2]^3$

40. $f(x) = [(3x - 1)^2 + (1 - x)^5]^2$

41. $t(x) = [2 + (x + 1)^{-0.1}]^{4.3}$

42. $t(x) = [(x + 1)^{0.1} - 4x]^{-5.1}$

43. $r(x) = \left(\sqrt{2x + 1} - x^2\right)^{-1}$ **44.** $r(x) = \left(\sqrt{x + 1} + \sqrt{x}\right)^3$

45. $f(x) = \{1 + [1 + (1 + 2x)^3]^3\}^3$

46. $f(x) = 2x + [2x + (2x + 1)^3]^3$

Find the indicated derivatives in Exercises 47–54. In each case, the independent variable is a (unspecified) function of t.

47. $y = x^{100} + 99x^{-1}; \dfrac{dy}{dt}$ **48.** $y = x^{0.5}(1 + x); \dfrac{dy}{dt}$

49. $s = \dfrac{1}{r^3} + r^{0.5}; \dfrac{ds}{dt}$ **50.** $s = r + r^{-1}; \dfrac{ds}{dt}$

51. $V = \dfrac{4}{3}\pi r^3; \dfrac{dV}{dt}$ **52.** $A = 4\pi r^2; \dfrac{dA}{dt}$

53. $y = x^3 + \dfrac{1}{x}, x = 2$ when $t = 1, \left.\dfrac{dx}{dt}\right|_{t=1} = -1; \left.\dfrac{dy}{dt}\right|_{t=1}$

54. $y = \sqrt{x} + \dfrac{1}{\sqrt{x}}, x = 9$ when $t = 1, \left.\dfrac{dx}{dt}\right|_{t=1} = -1; \left.\dfrac{dy}{dt}\right|_{t=1}$

Applications

55. ***On-Line Trading*** The average commission c (in dollars) per trade earned by the Charles Schwab Corporation decreased as more of its customers traded on-line according to the formula[1]

$c(u) = 100u^2 - 160u + 110$ dollars/trade
(u = fraction of trades done on-line)

During that time, the fraction of trades done on-line increased according to the formula

$u(t) = 0.42 + 0.02t$ (t = months since January 1, 1998)

Use direct substitution to express the average commission per trade c as a function of time (do not simplify the expression), and then use the chain rule to estimate the rate of change of the average commission per trade at the beginning of September 1998. Be sure to specify the units.

56. ***On-Line Trading*** The profitability p (measured in quarterly net income) of the Charles Schwab Corporation increased as more of its customers traded on-line according to the formula

$p(u) = 520u^2 - 300u + 100$ million dollars
(u = fraction of trades done on-line)

[1] The models in Exercises 55 and 56 are based on data for the period September 1997–September 1998. The function c is reliable only in the range $0.4 \le u \le 0.6$. *Source for data:* Charles Schwab/*New York Times*, February 10, 1999, p. C1.

During that time, the fraction of trades done on-line increased according to the formula

$$u(t) = 0.42 + 0.02t \qquad (t = \text{months since January 1, 1998})$$

Use direct substitution to express the quarterly profits p as a function of time (do not simplify the expression), and then use the chain rule to estimate the rate of change of profitability at the beginning of September 1998. Be sure to specify the units.

57. ***Marginal Revenue*** Weekly sales of rubies by Royal Ruby Retailers is given by

$$q = -\frac{4p}{3} + 80$$

where p is the price per ruby (in dollars).

a. Express RRR's weekly revenue as a function of p, and then calculate $\left.\frac{dR}{dp}\right|_{q=60}$.

b. Use the demand equation to calculate $\left.\frac{dp}{dq}\right|_{q=60}$.

c. Use the answers to parts (a) and (b) to find the marginal revenue at a demand level of 60 rubies per week, and interpret the result.

58. ***Marginal Revenue*** Repeat Exercise 57 for a demand level of $q = 52$ rubies per week.

59. ***Marginal Product*** Paramount Electronics, Inc., has an annual profit given by

$$P = -100{,}000 + 5000q - 0.25q^2$$

where q is the number of laptop computers it sells per year. The number of laptop computers it can manufacture per year depends on the number n of electrical engineers Paramount employs, according to the equation

$$q = 30n + 0.01n^2$$

Use the chain rule to find $\left.\frac{dP}{dn}\right|_{n=10}$, and interpret the result.

60. ***Marginal Product*** Refer back to Exercise 59. Give a formula for the average profit per computer

$$\bar{P} = \frac{P}{q}$$

as a function of q, and then determine the **marginal average product,** $d\bar{P}/dn$, at an employee level of ten engineers. Interpret the result.

61. ***Ecology*** Manatees are grazing sea mammals sometimes referred to as "sea sirens." Increasing numbers of manatees have been killed by boats off the Florida coast. Since 1976 the number M of manatees killed by boats each year is roughly linear, with[2]

$$M(t) \approx 2.27t + 11.5 \qquad (0 \le t \le 16)$$

where t is the number of years since 1976. Over the same period, the total number B of boats registered in Florida has also been increasing at a roughly linear rate, given by

$$B(t) \approx 19{,}500t + 436{,}000 \qquad (0 \le t \le 16)$$

Use the chain rule to give an estimate of dM/dB. What does the answer tell about manatee deaths?

62. ***Ecology*** The linear models used in Exercise 61 are rough, as shown in the following graphs of the actual data:

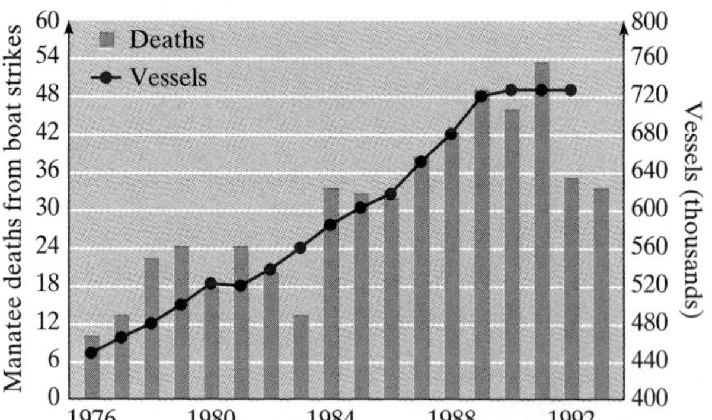

As you can see, the annual number of deaths was declining by 1992, despite an increase in the number of registered boats. This suggests that a linear model is not appropriate for the most recent period. If we consider only the data from 1986 onward, use a quadratic model for manatee deaths, and use a linear model for the number of vessels, we obtain

$$M(t) \approx -1.35t^2 + 9.64t + 30.92$$

$$B(t) \approx 16{,}700t + 641{,}000$$

where t is the number of years since 1986. Use the chain rule to obtain dM/dB. What does the answer tell about manatee deaths? Give a possible explanation of this result.

63. ***Pollution*** An offshore oil well is leaking oil and creating a circular oil slick. If the radius of the slick is growing at a rate of 2 miles per hour, find the rate at which the area is increasing when the radius is 3 miles. (The area of a disc of radius r is $A = \pi r^2$.)

64. ***Mold*** A mold culture in a dorm refrigerator is circular and growing. The radius is increasing at a rate of 0.3 cm/day. How fast is the area growing when the cul-

[2]The models in Exercises 61 and 62 are best-fit linear functions based on graphical data in the article "Manatees" by Thomas J. O'Shea in *Scientific American,* July 1994, pp. 66–72. *Source:* Florida Department of Environmental Protection.

ture is 4 centimeters in radius? (The area of a disc of radius r is $A = \pi r^2$.)

65. ***Budget Overruns*** The Pentagon is planning to build a new satellite that will be spherical. As is typical in these cases, the specifications keep changing, so that the size of the satellite keeps growing. In fact, the radius of the planned satellite is growing 0.5 foot per week. Its cost will be $1000 per cubic foot. At the point when the plans call for a satellite 10 feet in radius, how fast is the cost growing? (The volume of a solid sphere of radius r is $V = \frac{4}{3}\pi r^3$.)

66. ***Soap Bubbles*** The soap bubble I am blowing has a radius that is growing at a rate of 4 cm/s. How fast is the surface area growing when the radius is 10 cm? (The surface area of a sphere of radius r is $S = 4\pi r^2$.)

T E **67.** ***Revenue Growth*** The demand for the Cyberpunk II arcade video game is modeled by the logistic curve

$$q(t) = \frac{10{,}000}{1 + 0.5e^{-0.4t}}$$

where $q(t)$ is the total number of units sold t months after the game's introduction.

a. Use technology to estimate $q'(4)$.

b. Assume that the manufacturers of Cyberpunk II sell each unit for $800. What is the company's marginal revenue, dR/dq?

c. Use the chain rule to estimate the rate at which revenue is growing 4 months after the introduction of the video game.

T E **68.** ***Information Highway*** The amount of information transmitted each month on the Internet for the years from 1988 to 1994 can be modeled by the equation

$$q(t) = \frac{2e^{0.69t}}{3 + 1.5e^{-0.4t}}$$

where q is the amount of information transmitted each month in billions of data packets and t is the number of years since the start of 1988.[3]

a. Use technology to estimate $q'(2)$.

b. Assume that it costs $5 to transmit a million packets of data. What is the marginal cost $C'(q)$?

c. How fast was the cost increasing at the start of 1990?

[3]This is the authors' model, based on figures published in the *New York Times*, November 3, 1993.

Money Stock *Exercises 69–72 are based on the following demand function for money (taken from a question on the GRE economics test):*

$$M_d = (2) \times (y)^{0.6} \times (r)^{-0.3} \times (p)$$

where M_d = demand for nominal money balances (money stock)

y = real income

r = an index of interest rates

p = an index of prices

These exercises also use the idea of **percentage rate of growth:**

$$\text{Percentage rate of growth of } M = \frac{\text{Rate of growth of } M}{M} = \frac{dM/dt}{M}$$

69. *(from the GRE economics test)* If the interest rate and price level are to remain constant while real income grows at 5% per year, the money stock must grow at what percent per year?

70. *(from the GRE economics test)* If real income and the price level are to remain constant while the interest rate grows at 5% per year, the money stock must change by what percent per year?

71. *(from the GRE economics test)* If the interest rate is to remain constant while real income grows at 5% per year and the price level rises at 5% per year, the money stock must grow at what percent per year?

72. *(from the GRE economics test)* If real income grows by 5% per year, the interest rate grows by 2% per year, and the price level drops by 3% per year, the money stock must change by what percent per year?

Communication and Reasoning Exercises

73. Formulate a simple procedure for deciding whether to apply first the chain rule, the product rule, or the quotient rule when finding the derivative of a function.

74. Give an example of a function f with the property that calculating $f'(x)$ requires use of the following rules in the given order: (1) the chain rule, (2) the quotient rule, and (3) the chain rule.

75. Give an example of a function f with the property that calculating $f'(x)$ requires use of the chain rule five times in succession.

76. What can you say about composites of linear functions?

4.3 Derivatives of Logarithmic and Exponential Functions

At this point we know how to take the derivative of any algebraic expression in x (involving powers, radicals, and so on). We now turn to the derivatives of logarithmic and exponential functions.

Derivative of the Natural Logarithm

$$\frac{d}{dx}\ln x = \frac{1}{x}$$

Recall that $\ln x = \log_e x$.

Quick Examples

1. $\frac{d}{dx}(3\ln x) = 3 \cdot \frac{1}{x} = \frac{3}{x}$ — Derivative of a constant times a function

2. $\frac{d}{dx}(x\ln x) = 1 \cdot \ln x + x \cdot \frac{1}{x}$ — Product rule because $x \ln x$ is a product

$$= \ln x + 1$$

The above simple formula works only for the natural logarithm (the logarithm with base e). For logarithms with bases other than e, we have the following formula.

Derivative of the Logarithm with Base b

$$\frac{d}{dx}\log_b x = \frac{1}{x\ln b}$$

Notice that, if $b = e$, we get the same formula as above.

Quick Examples

1. $\frac{d}{dx}\log_3 x = \frac{1}{x\ln 3} \approx \frac{1}{1.0986x}$

2. $\frac{d}{dx}\log_2(x^4) = \frac{d}{dx}(4\log_2 x)$ — We used the logarithm identity $\log_b(x^r) = r\log_b x$.

$$= 4 \cdot \frac{1}{x\ln 2} \approx \frac{4}{0.6931x}$$

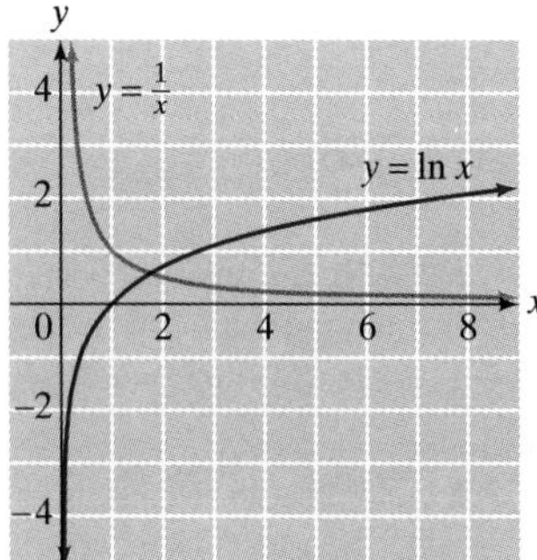

Figure 1

Question Where do these formulas come from?

Answer We will see at the end of this section. For now, let us look at the graphs of $y = \ln x$ and $y = 1/x$ to see whether it is *reasonable* that the derivative of $\ln x$ is $1/x$.

We refer to Figure 1 and notice that if x is close to zero, then the graph of $\ln x$ is rising steeply, so the derivative should be a large positive number. And indeed, $1/x$ is positive and large. As x increases, the graph of $\ln x$ becomes less steep, although it continues to rise. Therefore, the derivative (which gives the slope of the tangent line) should become smaller but remain positive, and this is exactly what happens to $1/x$. So $1/x$ at least behaves the way the derivative of $\ln x$ is supposed to behave.

If we were to take the derivative of the natural logarithm of a *quantity* (a function of x), rather than just x, we would need to use the chain rule, as follows.

Derivatives of Logarithms of Functions

Original rule	Generalized rule	In words
$\frac{d}{dx}\ln x = \frac{1}{x}$	$\frac{d}{dx}\ln u = \frac{1}{u}\frac{du}{dx}$	The derivative of the natural logarithm of a quantity is 1 over that quantity, times the derivative of that quantity.
$\frac{d}{dx}\log_b x = \frac{1}{x\ln b}$	$\frac{d}{dx}\log_b u = \frac{1}{u\ln b}\frac{du}{dx}$	The derivative of the log to base b of a quantity is 1 over the product of ln b and that quantity, times the derivative of that quantity.

Quick Examples

1. $\frac{d}{dx}\ln(x^2+1) = \frac{1}{x^2+1}\frac{d}{dx}(x^2+1)$ $\qquad u = x^2+1$ (see footnote 1)

$$= \frac{1}{x^2+1}(2x) = \frac{2x}{x^2+1}$$

2. $\frac{d}{dx}\log_2(x^3+x) = \frac{1}{(x^3+x)\ln 2}\frac{d}{dx}(x^3+x)$ $\qquad u = x^3+x$

$$= \frac{1}{(x^3+x)\ln 2}(3x^2+1) = \frac{3x^2+1}{(x^3+x)\ln 2}$$

example 1

Derivative of a Logarithmic Function

Find $\frac{d}{dx}\ln\sqrt{x+1}$.

Solution

The calculation thought experiment tells us that we have the natural logarithm of a quantity, so

$$\frac{d}{dx}\ln\sqrt{x+1} = \frac{1}{\sqrt{x+1}}\frac{d}{dx}\sqrt{x+1} \qquad \frac{d}{dx}\ln u = \frac{1}{u}\frac{du}{dx}$$

$$= \frac{1}{\sqrt{x+1}}\cdot\frac{1}{2\sqrt{x+1}} \qquad \frac{d}{dx}\sqrt{u} = \frac{1}{2\sqrt{u}}\frac{du}{dx}$$

$$= \frac{1}{2(x+1)}$$

Before we go on . . . What happened to the square root? As with many problems that involve logarithms, we could have done this one differently and with less

[1]If we were to evaluate ln (x^2+1), the last operation we would perform is to take the natural logarithm of a quantity. Thus, the calculation thought experiment tells us that we are dealing with ln *of a quantity,* and so we need the generalized logarithm rule as stated above.

bother if we had simplified the expression $\ln\sqrt{x+1}$ using the properties of logarithms *before* differentiating. Doing this, we get

$$\ln\sqrt{x+1} = \ln(x+1)^{1/2} = \frac{1}{2}\ln(x+1) \qquad \text{Simplify the logarithm first.}$$

Thus,
$$\frac{d}{dx}\ln\sqrt{x+1} = \frac{d}{dx}\left[\frac{1}{2}\ln(x+1)\right]$$
$$= \frac{1}{2}\left(\frac{1}{x+1}\right)\cdot 1 = \frac{1}{2(x+1)}$$

the same answer as above.

---

example 2

Derivative of a Logarithmic Function

Find $\frac{d}{dx}\ln[(1+x)(2-x)]$.

Solution

This time, we simplify the expression $\ln[(1+x)(2-x)]$ before taking the derivative.

$$\ln[(1+x)(2-x)] = \ln(1+x) + \ln(2-x) \qquad \text{Simplify the logarithm first.}$$

Thus,

$$\frac{d}{dx}\ln[(1+x)(2-x)] = \frac{d}{dx}\ln(1+x) + \frac{d}{dx}\ln(2-x)$$
$$= \frac{1}{1+x} - \frac{1}{2-x} \qquad \text{Because } \frac{d}{dx}\ln(2-x) = -\frac{1}{2-x}$$
$$= \frac{1-2x}{(1+x)(2-x)}$$

Before we go on . . . For practice, try doing this example without simplifying first. What other differentiation rule do you need to use?

---

example 3

Logarithm of an Absolute Value

Find $\frac{d}{dx}\ln|x|$.

Solution

Before we start, you might ask why we are considering the natural log of the absolute value of x to begin with. The reason is this: $\ln x$ is defined for only positive values of x, so its domain is the set of positive real numbers. The domain of $\ln|x|$, on the other hand, is the set of *all* nonzero real numbers; for example, $\ln|-2| = \ln 2 \approx 0.6931$. For this reason, it often turns out to be more useful than the ordinary logarithm function.

Now we'll get to work. The calculation thought experiment tells us that $\ln|x|$ is the natural logarithm of a quantity, so we try the chain rule:

$$\frac{d}{dx}\ln|x| = \frac{1}{|x|}\frac{d}{dx}|x| \qquad u = |x|$$

To go further, we need to know the derivative of $|x|$. As we saw in Chapter 3, the derivative of $|x|$ exists everywhere other than at $x = 0$ and is given by

$$\frac{d}{dx}|x| = \begin{cases} 1 & \text{if } x > 0 \\ -1 & \text{if } x < 0 \end{cases}$$

which we can write in a single formula as

$$\frac{d}{dx}|x| = \frac{|x|}{x} \qquad \frac{|x|}{x} \text{ is also } +1 \text{ for positive } x \text{ and } -1 \text{ for negative } x.$$

Thus, $$\frac{d}{dx}\ln|x| = \frac{1}{|x|}\frac{|x|}{x} = \frac{1}{x}$$

Before we go on . . . Figure 2(a) shows the graphs of $y = \ln|x|$ and $y = 1/x$. Figure 2(b) shows the graphs of $y = \ln|x|$ and $y = 1/|x|$. You should be able to see from these graphs why the derivative of $\ln|x|$ is $1/x$ and not $1/|x|$.

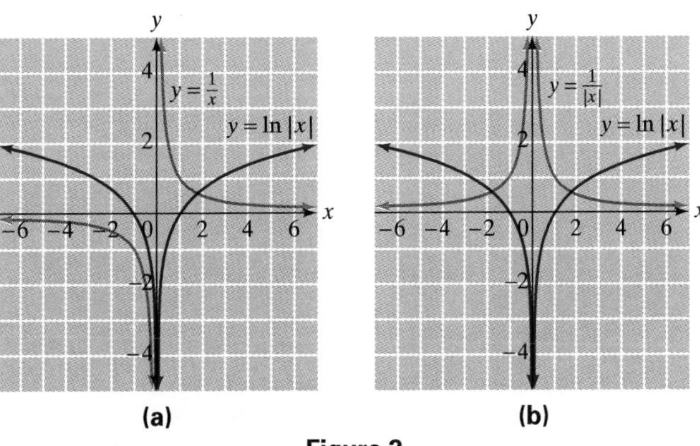

Figure 2

This last example, in conjunction with the chain rule, gives us the following formulas.

Derivative of Logarithms of Absolute Values

Original rule	Generalized rule	In words
$\frac{d}{dx}\ln\|x\| = \frac{1}{x}$	$\frac{d}{dx}\ln\|u\| = \frac{1}{u}\frac{du}{dx}$	The derivative of the natural logarithm of the absolute value of a quantity is 1 over that quantity, times the derivative of that quantity.
$\frac{d}{dx}\log_b\|x\| = \frac{1}{x\ln b}$	$\frac{d}{dx}\log_b\|u\| = \frac{1}{u\ln b}\frac{du}{dx}$	The derivative of the log to base b of the absolute value of a quantity is 1 over the product of ln b and that quantity, times the derivative of that quantity.

Quick Examples

1. $\frac{d}{dx}\ln|x^2+1| = \frac{1}{x^2+1}\frac{d}{dx}(x^2+1)$ $\qquad u = x^2 + 1$

$= \frac{1}{x^2+1}(2x) = \frac{2x}{x^2+1}$

2. $\frac{d}{dx}\log_2|x^3+x| = \frac{1}{(x^3+x)\ln 2}\frac{d}{dx}(x^3+x)$ $\qquad u = x^3 + x$

$= \frac{1}{(x^3+x)\ln 2}(3x^2+1) = \frac{3x^2+1}{(x^3+x)\ln 2}$

In other words, when taking the derivative of the logarithm of the absolute value of a quantity, we can simply ignore the absolute value sign!

We now turn to the derivatives of *exponential* functions—that is, functions of the form $f(x) = a^x$. We begin by showing how *not* to differentiate them.

Caution: The derivative of a^x is *not* xa^{x-1}. The power rule applies only to *constant* exponents. In this case, the exponent is decidedly *not* constant, and so the power rule does not apply.

The following shows the correct way of differentiating a^x. (We justify the formula, as well as the formula for the derivative of $\log_b x$, at the end of the section.)

Derivatives of Exponential Functions

Derivative of e^x

$$\frac{d}{dx}e^x = e^x$$

Thus, e^x has the amazing property that its derivative is itself![2] For bases other than e, we have the following generalization.

Derivative of b^x

If b is any positive number, then

$$\frac{d}{dx}b^x = b^x \ln b$$

Note that if $b = e$, we obtain the previous formula.

Quick Examples

1. $\frac{d}{dx}3^x = 3^x \ln 3$

2. $\frac{d}{dx}3e^x = 3\frac{d}{dx}e^x = 3e^x$

[2]There is another—very simple—function that is its own derivative. What is it?

3. $\frac{d}{dx}\left(\frac{e^x}{x}\right) = \frac{e^x x - e^x(1)}{x^2}$ Quotient rule

$= \frac{e^x(x-1)}{x^2}$

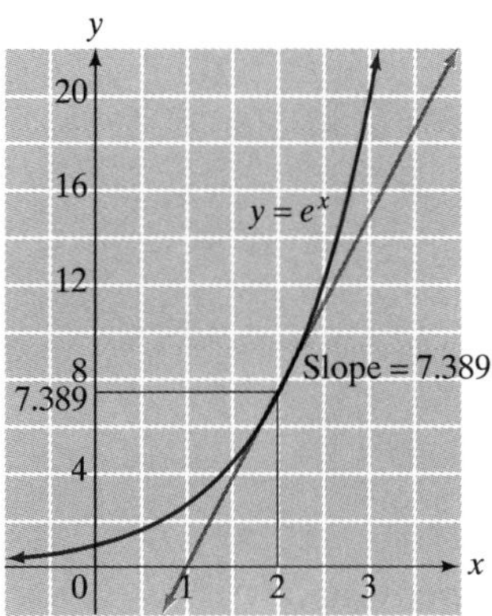

Figure 3

Question Why should I believe this? How can the derivative of the function e^x be the same as the original function?

Answer At the end of this section we give a derivation of this formula. For now, let us consider whether it is *plausible* that e^x equals its own derivative. For a function f to equal its own derivative, its value at every point x must be the same as the slope of the tangent at that point. For instance, if $f(2) = 5$, then the slope of the tangent at the point where $x = 2$ should also be 5. To check whether $f(x) = e^x$ has this property, we take a look at the graph of $f(x) = e^x$ (Figure 3).

Notice that as x increases, both the y-coordinate [the value of $f(x)$] and the slope [the value of $f'(x)$] are increasing. More specifically, if we look at any particular value of x—say, $x = 2$—we notice that $f(x)$ and the slope of the tangent at x appear to be the same. [For instance, $f(2) \approx 7.4$, and the slope at $x = 2$ is also approximately 7.4.]

If we take the derivative of e raised to a *quantity,* not just x, we need to use the chain rule, as follows.

Derivatives of Exponentials of Functions

Original rule	Generalized rule	In words
$\frac{d}{dx}e^x = e^x$	$\frac{d}{dx}e^u = e^u\frac{du}{dx}$	The derivative of e raised to a quantity is e raised to that quantity, times the derivative of that quantity.
$\frac{d}{dx}b^x = b^x \ln b$	$\frac{d}{dx}b^u = b^u \ln b\frac{du}{dx}$	The derivative of b raised to a quantity is b raised to that quantity, times $\ln b$, times the derivative of that quantity.

Quick Examples

1. $\frac{d}{dx}e^{x^2+1} = e^{x^2+1}\frac{d}{dx}(x^2+1)$ $u = x^2 + 1$ (see footnote 3)

$= e^{x^2+1}(2x) = 2x\,e^{x^2+1}$

2. $\frac{d}{dx}2^{3x} = 2^{3x}\ln 2\frac{d}{dx}(3x)$ $u = 3x$

$= 2^{3x}(\ln 2)(3) = (3\ln 2)2^{3x}$

[3]The calculation thought experiment tells us that we have e raised to a quantity.

Applications

example 4

Epidemics

At the start of 1990, the number of U.S. residents infected with HIV was estimated to be 0.4 million. This number was growing exponentially, doubling every 6 months. Had this trend continued, how many new cases per month would have been occurring by the start of 1994?

Solution

To find the answer, we model this exponential growth using the methods of Chapter 2. After t years, the number of cases is

$$A = 0.4(4^t)$$

We are asking for the number of new cases per month. In other words, we want the rate of change, dA/dt:

$$\frac{dA}{dt} = 0.4(\ln 4)4^t \text{ million cases per year}$$

At the start of 1994, $t = 4$, so the number of new cases per year is

$$\left.\frac{dA}{dt}\right|_{t=4} = 0.4(\ln 4)4^4 \approx 141.96 \text{ million cases per year}$$

Thus, the number of new cases *per month* would be $141.96/12 = 11.83$ million!

Before we go on . . . The reason this figure is astronomically large is that we assumed exponential growth—doubling every 6 months—would continue. A more realistic model for the spread of a disease is the logistic model. (See *You're the Expert* in Chapter 2 as well as the next example.)

example 5

Sales Growth

The demand for the Cyberpunk II arcade video game is modeled by the logistic curve

$$q(t) = \frac{10{,}000}{1 + 0.5e^{-0.4t}}$$

where $q(t)$ is the total number of units sold t months after its introduction. How fast is the game selling 2 years after its introduction?

Solution

We are asked for $q'(24)$. We can find the derivative of $q(t)$ using the quotient rule, or we can first write

$$q(t) = 10{,}000(1 + 0.5e^{-0.4t})^{-1}$$

and then use the generalized power rule:

$$q'(t) = -10{,}000(1 + 0.5e^{-0.4t})^{-2}(0.5e^{-0.4t})(-0.4) = \frac{2000e^{-0.4t}}{(1 + 0.5e^{-0.4t})^2}$$

Thus, $$q'(24) = \frac{2000e^{-0.4(24)}}{(1 + 0.5e^{-0.4(24)})^2} \approx 0.135 \text{ unit per month}$$

Thus, after 2 years, sales are quite slow.

We can check this answer graphically. If we plot the total sales curve for $0 \leq t \leq 30$ and $6000 \leq q \leq 10{,}000$, we get the graph shown in Figure 4. Notice that total sales level off at about 10,000 units.[4] We computed $q'(24)$, which is the slope of the curve at the point with t-coordinate 24. If we zoom in to the portion of the curve near $t = 24$, we obtain the graph shown in Figure 5.

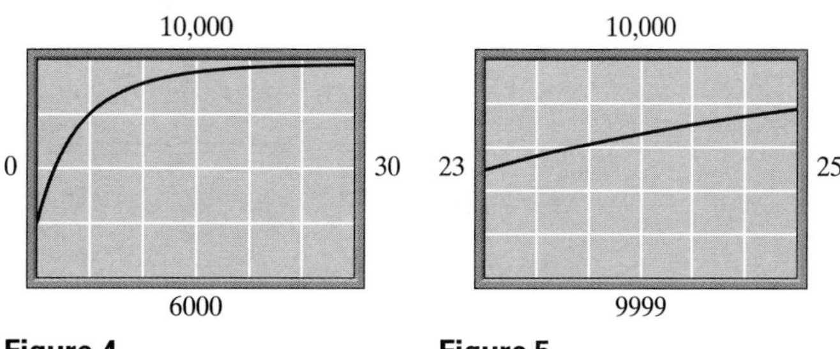

Figure 4 **Figure 5**

The curve is almost linear in this range. If we use the two endpoints of this segment of the curve, (23,9999.4948) and (25,9999.7730), we can approximate the derivative as

$$\frac{9999.7730 - 9999.4948}{25 - 23} = 0.1391$$

which is accurate to two decimal places.

Let us now explain where the formulas for the derivatives of $\ln x$ and e^x come from. We start with $\ln x$.

Question Why is $\frac{d}{dx}\ln x = \frac{1}{x}$?

Answer To compute $d/dx \ln x$, we need to use the definition of the derivative. We also use properties of the logarithm to help evaluate the limit.

$$\frac{d}{dx}\ln x = \lim_{h\to 0}\frac{\ln(x+h) - \ln x}{h} \qquad \text{Definition of the derivative}$$

$$= \lim_{h\to 0}\frac{1}{h}\left[\ln(x+h) - \ln x\right] \qquad \text{Algebra}$$

$$= \lim_{h\to 0}\frac{1}{h}\ln\left(\frac{x+h}{x}\right) \qquad \text{Properties of the logarithm}$$

$$= \lim_{h\to 0}\frac{1}{h}\ln\left(1 + \frac{h}{x}\right) \qquad \text{Algebra}$$

$$= \lim_{h\to 0}\ln\left(1 + \frac{h}{x}\right)^{1/h} \qquad \text{Properties of the logarithm}$$

which we rewrite as

$$\lim_{h\to 0}\ln\left[\left(1 + \frac{1}{(x/h)}\right)^{x/h}\right]^{1/x}$$

[4]We can also say this using limits: $\lim_{t\to+\infty} q(t) = 10{,}000$.

As $h \to 0^+$, the quantity x/h is getting large and positive, and so the quantity in brackets is approaching e (see the definition of e in Section 2.2), which leaves us with

$$\ln(e)^{1/x} = \frac{1}{x} \ln e = \frac{1}{x}$$

which is the derivative we are after.[5] What about the limit as $h \to 0^-$? We will glide over that case and leave it for the interested reader to pursue.[6]

The rule for the derivative of $\log_b x$ follows from the fact that $\log_b x = \ln x \ln b$.

Question Why is $\frac{d}{dx} e^x = e^x$?

Answer To find the derivative of e^x we use a shortcut.[7] We write $g(x) = e^x$. Then

$$\ln g(x) = x$$

We take the derivative of both sides of this equation to get

$$\frac{g'(x)}{g(x)} = 1$$

or $$g'(x) = g(x) = e^x$$

In other words, the exponential function with base e is its own derivative. The rule for exponential functions with other bases follows from the equality $a^x = e^{x \ln a}$ (why?) and the chain rule. (Try it.)

---

[5] We actually used the fact that the logarithm function is continuous when we took the limit.

[6] Here is an outline of the argument for negative h. Since x must be positive for $\ln x$ to be defined, we find that $x/h \to -\infty$ as $h \to 0^-$, and so we must consider the quantity $(1 + 1/m)^m$ for large *negative m*. It turns out the limit is still e (check it numerically!) and so the computation above still works.

[7] This shortcut is an example of a technique called *logarithmic differentiation*, which is occasionally useful. We will see it again in the next section.

4.3 exercises

Find the derivatives of the functions in Exercises 1–14 mentally.

1. $f(x) = \ln(x - 1)$

2. $f(x) = \ln(x + 3)$

3. $f(x) = \log_2 x$

4. $f(x) = \log_3 x$

5. $g(x) = \ln |x^2 + 3|$

6. $g(x) = \ln |2x - 4|$

7. $h(x) = e^{x+3}$

8. $h(x) = e^{x^2}$

9. $f(x) = e^{-x}$

10. $f(x) = e^{1-x}$

11. $g(x) = 4^x$

12. $g(x) = 5^x$

13. $h(x) = 2^{x^2-1}$

14. $h(x) = 3^{x^2-x}$

Find the derivatives of the functions in Exercises 15–70.

15. $f(x) = x \ln x$

16. $f(x) = 3 \ln x$

17. $f(x) = (x^2 + 1) \ln x$

18. $f(x) = (4x^2 - x) \ln x$

19. $f(x) = (x^2 + 1)^5 \ln x$

20. $f(x) = (x + 1)^{0.5} \ln x$

21. $g(x) = \ln |3x - 1|$

22. $g(x) = \ln |5 - 9x|$

23. $g(x) = \ln |2x^2 + 1|$

24. $g(x) = \ln |x^2 - x|$

25. $g(x) = \ln(x^2 - 2.1x^{0.3})$

26. $g(x) = \ln(x - 3.1x^{-1})$

27. $h(x) = \ln [(-2x + 1)(x + 1)]$

28. $h(x) = \ln [(3x + 1)(-x + 1)]$

29. $h(x) = \ln \left(\frac{3x + 1}{4x - 2}\right)$

30. $h(x) = \ln \left(\frac{9x}{4x - 2}\right)$

31. $r(x) = \ln \left|\frac{(x + 1)(x - 3)}{-2x - 9}\right|$

32. $r(x) = \ln \left|\frac{-x + 1}{(3x - 4)(x - 9)}\right|$

33. $s(x) = \ln(4x - 2)^{1.3}$

34. $s(x) = \ln(x - 8)^{-2}$

35. $s(x) = \ln \left| \frac{(x+1)^2}{(3x-4)^3(x-9)} \right|$

36. $s(x) = \ln \left| \frac{(x+1)^2(x-3)^4}{2x+9} \right|$

37. $h(x) = \log_2(x+1)$

38. $h(x) = \log_3(x^2+x)$

39. $r(t) = \log_3(t + 1/t)$

40. $r(t) = \log_3(t + \sqrt{t})$

41. $f(x) = (\ln |x|)^2$

42. $f(x) = \frac{1}{\ln |x|}$

43. $r(x) = \ln(x^2) - [\ln(x-1)]^2$

44. $r(x) = [\ln(x^2)]^2$

45. $f(x) = xe^x$

46. $f(x) = 2e^x - x^2e^x$

47. $r(x) = \ln(x+1) + 3x^3e^x$

48. $r(x) = \ln |x + e^x|$

49. $f(x) = e^x \ln |x|$

50. $f(x) = e^x \log_2 |x|$

51. $f(x) = e^{2x+1}$

52. $f(x) = e^{4x-5}$

53. $h(x) = e^{x^2-x+1}$

54. $h(x) = e^{2x^2-x+1/x}$

55. $s(x) = x^2e^{2x-1}$

56. $s(x) = \frac{e^{4x-1}}{x^3-1}$

57. $r(x) = (e^{2x-1})^2$

58. $r(x) = (e^{2x^2})^3$

59. $g(x) = \frac{e^x + e^{-x}}{e^x - e^{-x}}$

60. $g(x) = \frac{1}{e^x + e^{-x}}$

61. $g(x) = e^{3x-1}e^{x-2}e^x$

62. $g(x) = e^{-x+3}e^{2x-1}e^{-x+11}$

63. $f(x) = \frac{1}{x \ln x}$

64. $f(x) = \frac{e^{-x}}{xe^x}$

65. $f(x) = [\ln(e^x)]^2 - \ln[(e^x)^2]$

66. $f(x) = e^{\ln x} - e^{2\ln(x^2)}$

67. $f(x) = \ln|\ln x|$

68. $f(x) = \ln |\ln|\ln x||$

69. $s(x) = \ln \sqrt{\ln x}$

70. $s(x) = \sqrt{\ln(\ln x)}$

Find the equations of the straight lines described in Exercises 71–76. Use graphing technology to check your answers by plotting the given curve together with the tangent line.

71. Tangent to $y = e^x \log_2 x$ at the point $(1, 0)$

72. Tangent to $y = e^x + e^{-x}$ at the point $(0, 2)$

73. Tangent to $y = \ln \sqrt{2x+1}$ at the point where $x = 0$

74. Tangent to $y = \ln \sqrt{2x^2+1}$ at the point where $x = 1$

75. At right angles to $y = e^{x^2}$ at the point where $x = 1$

76. At right angles to $y = \log_2(3x+1)$ at the point where $x = 1$

Applications

77. *Investments* If \$10,000 is invested in a savings account yielding 4% per year, compounded continuously, how fast is the balance growing after 3 years?

78. *Investments* If \$20,000 is invested in a savings account yielding 3.5% per year, compounded continuously, how fast is the balance growing after 3 years?

79. *Population Growth* The population of Lower Anchovia was 4,000,000 at the start of 1995 and was doubling every 10 years. How fast was it growing per year at the start of 1995? (Round your answer to three significant digits.)

80. *Population Growth* The population of Upper Anchovia was 3,000,000 at the start of 1996 and was doubling every 7 years. How fast was it growing per year at the start of 1996? (Round your answer to three significant digits.)

81. *Radioactive Decay* Plutonium-239 has a half-life of 24,400 years. How fast is a lump of 10 grams decaying after 100 years?

82. *Radioactive Decay* Carbon-14 has a half-life of 5730 years. How fast is a lump of 20 grams decaying after 100 years?

83. *Investments* If \$10,000 is invested in a savings account yielding 4% per year, compounded semiannually, how fast is the balance growing after 3 years?

84. *Investments* If \$20,000 is invested in a savings account yielding 3.5% per year, compounded semiannually, how fast is the balance growing after 3 years?

85. *Life Span in Ancient Rome* The percentage $P(t)$ of people surviving to age t years in ancient Rome can be approximated by[8]

$$P(t) = 92e^{-0.0277t}$$

Calculate $P'(22)$ and explain what the result indicates.

86. *Communication Among Bees* The audible signals honeybees use to communicate are in the frequency range 0–500 hertz and are generated by their wings. The speed with which a honeybee's wings must move the air to generate these signals depends on the frequency of the signal, and this relationship can be approximated by[9]

$$V(f) = 95.6e^{0.0049f}$$

where $V(f)$ is the speed of the air near the honeybee's wings in millimeters per second, and f is the frequency of the communication signal in hertz. Calculate $V'(200)$ and explain what the result indicates.

[8] Based on graphical data in Marvin Minsky's article "Will Robots Inherit the Earth?" in *Scientific American*, October 1994, pp. 109–113.

[9] Based on graphical data in the article "The Sensory Basis of the Honeybee's Dance Language" by Wolfgang H. Kirchner and William F. Towne in *Scientific American*, June 1994, pp. 74–80.

87. ***SAT Scores by Income*** The following chart shows 1994 U.S. average verbal SAT scores as a function of parents' income level:

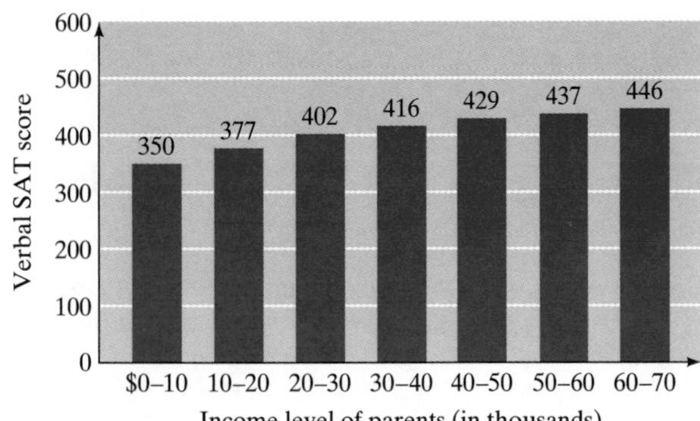

Source: The College Board/*New York Times,* March 5, 1995, p. E16.

a. The data can best be modeled by which equation?

(A) $S(x) = 470 - 136e^{-0.0000264x}$

(B) $S(x) = 136e^{-0.0000264x}$

(C) $S(x) = 355(1.000004^x)$

(D) $S(x) = 470 - 355(1.000004^x)$

($S(x)$ is the average verbal SAT score of students whose parents earn \$$x$ per year.)

b. Use $S'(x)$ to predict how a student's verbal SAT score is affected by a \$1000 increase in parents' income for a student whose parents earn \$45,000.

c. Does $S'(x)$ increase or decrease as x increases? Interpret your answer.

88. ***SAT Scores by Income*** The following chart shows 1994 U.S. average math SAT scores as a function of parents' income level:

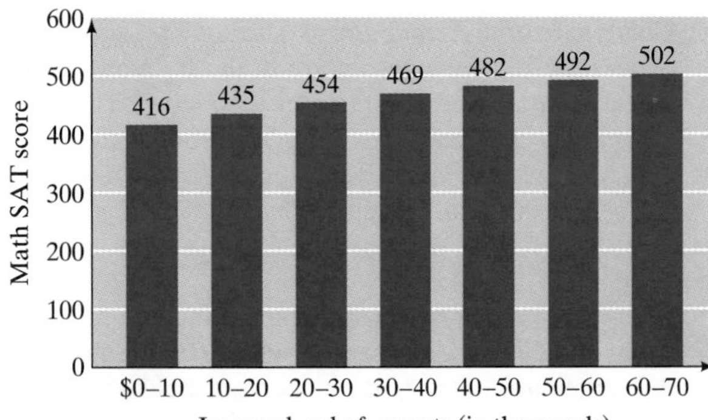

Source: The College Board/*New York Times,* March 5, 1995, p. E16.

a. The data can best be modeled by which function?

(A) $S(x) = 535 - 415(1.000003^x)$

(B) $S(x) = 535 - 136e^{0.0000213x}$

(C) $S(x) = 535 - 136e^{-0.0000213x}$

(D) $S(x) = 415(1.000003^x)$

($S(x)$ is the average math SAT score of students whose parents earn \$$x$ per year.)

b. Use $S'(x)$ to predict how a student's math SAT score is affected by a \$1000 increase in parents' income for a student whose parents earn \$45,000.

c. Does $S'(x)$ increase or decrease as x increases? Interpret your answer.

89. ***Epidemics*** The epidemic described in *You're the Expert* in Chapter 2 followed the curve

$$A = \frac{150{,}000{,}000}{1 + 14{,}999e^{-0.3466t}}$$

where A is the number of people infected and t is the number of months after the start of the disease. How fast is the epidemic growing (i.e., how many new cases are there per month) after 20 months? After 30 months? After 40 months? (Round your answers to three significant digits.)

90. ***Epidemics*** Another epidemic follows the curve

$$A = \frac{200{,}000{,}000}{1 + 20{,}000e^{-0.549t}}$$

where t is in years. How fast is the epidemic growing after 10 years? After 20 years? After 30 years? (Round your answers to three significant digits.)

91. ***Big Brother*** In 1995 the FBI was seeking the ability to monitor 74,250 phone lines at once. The following chart shows the number of phone lines monitored from 1987 through 1993:

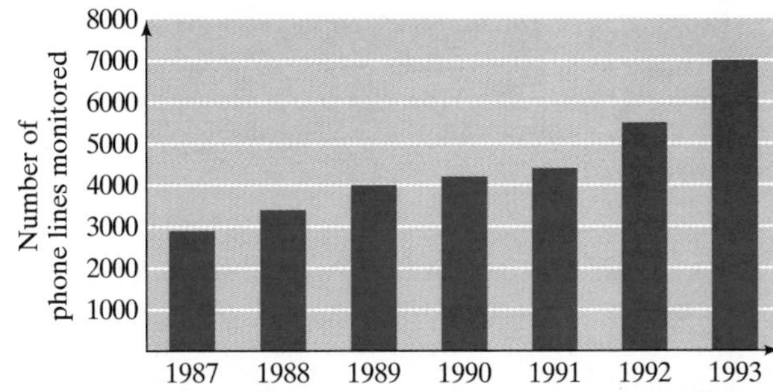

Source: Electronic Privacy Information Center, Justice Department, Administrative Office of the United States Courts/*New York Times,* November 2, 1995, p. D5.

The following logistic model is based on the assumption that, eventually, the number of lines monitored will grow to the proposed capacity of 74,250 phone lines:[10]

$$n(t) = \frac{215{,}000}{2.90 + 74.0e^{-0.141t}}$$

[10] The model is based on a best-fit logistic curve using the given long-term prediction of 74,250.

where $n(t)$ is the number of wiretaps (phone lines tapped) by government agencies during year t, with $t = 0$ representing June 1987.

a. Calculate $n'(t)$ and use it to approximate $n'(13)$. To how many significant digits should we round the answer? Why? What does the answer tell you?

T **b.** Graph the function $n'(t)$ and use your graph to estimate when, to the nearest year, the number of monitored phone lines will be growing most rapidly.

92. ***Employment*** The number of employees at Amerada Hess Corp. for the period 1984 ($t = 0$) through 1993 ($t = 9$) could be approximated by the formula[11]

$$n(t) = 7500 + \frac{23{,}000}{8 + 100e^{-0.5t}}$$

a. Evaluate $n(9)$ and $n'(9)$, round both answers to two significant digits, and interpret the results.

b. Evaluate $\lim_{t\to+\infty} n(t)$ and $\lim_{t\to+\infty} n'(t)$, and interpret the results.

T **93.** ***Diffusion of New Technology*** Numeric control is a technology in which the operation of machines is controlled by numeric instructions on disks, tapes, or cards. In a study, E. Mansfield *et al.* modeled the growth of this technology using the equation

$$p(t) = \frac{0.80}{1 + e^{4.46-0.477t}}$$

where $p(t)$ is the fraction of firms that use numeric control in year t.

a. Graph this function for $0 \le t \le 20$, and estimate $p'(10)$ graphically. Interpret the result.

b. Use your graph to estimate $\lim_{t\to+\infty} p(t)$, and interpret the result.

c. Compute $p'(t)$, graph it, and again find $p'(10)$.

d. Use your graph to estimate $\lim_{t\to+\infty} p'(t)$, and interpret the result.

Source: "The Diffusion of a Major Manufacturing Innovation" in *Research and Innovation in the Modern Corporation* (New York: Norton, 1971), pp. 186–205.

T **94.** ***Diffusion of New Technology*** Repeat Exercise 93 using the revised formula

$$p(t) = \frac{0.90e^{-0.1t}}{1 + e^{4.50-0.477t}}$$

which takes into account that in the long run, this new technology will eventually become outmoded and will be replaced by a newer technology. Draw your graphs using the range $0 \le t \le 40$.

T **95.** ***Growth of HMOs*** The enrollment in health maintenance organizations (HMOs) in the years from 1975 to 1992 can be modeled by the equation[12]

$$n(t) = 5 + \frac{40e^{0.002t}}{1 + 25e^{3-0.5t}}$$

where $n(t)$ represents the total number (in millions) of U.S. residents enrolled in HMOs t years after 1975. Calculate $n'(t)$, graph it, and use your graph to determine the value of t ($0 \le t \le 20$) when $n'(t)$ was a maximum. Interpret your result.

T **96.** ***Demand for Poultry*** The demand for poultry can be modeled as

$$q = -60.5 - 0.45p + 0.12b + 12.21\ln(d)$$

where q is the per capita demand for chicken in pounds per year, p is the wholesale price of chicken in cents per pound, b is the wholesale price of beef in cents per pound, and d is the per capita annual disposable income in dollars per year.[13] Assume that the wholesale prices of chicken and beef are fixed at 25¢ per pound and 50¢ per pound, respectively, and that the mean disposable income in t years' time will be $d = 25{,}000 + 1000t$ dollars. Calculate and graph $q'(t)$ for $0 \le t \le 20$, and use your graph to estimate the value of t for which $q'(t) = 0.30$. Interpret the result.

Source: A. H. Studenmund, *Using Econometrics,* 2d ed. (New York: Harper Collins, 1992), pp. 180–181.

[11] The model is based on a best-fit logistic curve using data from Hoover's Handbook Database (World Wide Web site), The Reference Press, Inc., Austin, Texas, 1995.

[12] This is the authors' model, based on data supplied by the Group Health Association of America published in the *New York Times,* October 18, 1993.

[13] This equation is based on actual data from poultry sales in the period 1950–1984.

Communication and Reasoning Exercises

97. A quantity P is growing exponentially with time. Explain the difference between $P(10)$ and $P'(10)$. If P is measured in kilograms and t is measured in days, what are the units of $P'(10)$?

98. The number N of graphing calculators sold on campus is growing exponentially with time. Can $N'(t)$ grow linearly with time? Explain.

In Exercises 99 and 100, select the correct option.

99. If $Q = 100e^{-0.3t}$, then Q is **(A)** increasing **(B)** decreasing with increasing t, and Q' is **(A)** increasing **(B)** decreasing with increasing t.

100. If $Q = 2000 - e^{0.3t}$, then **(A)** both Q and Q' are increasing with increasing t; **(B)** both Q and Q' are decreasing with increasing t; **(C)** Q is increasing and Q' is decreasing with increasing t; **(D)** Q is decreasing and Q' is increasing with increasing t.

The ***percentage rate of change*** *or* ***fractional rate of change*** *of a function is defined to be the ratio $f'(x)/f(x)$. (It is customary to express this as a percentage when speaking about percentage rate of change.)*

101. Show that the fractional rate of change of the exponential function e^{kx} is equal to k, which is often called its **fractional growth rate.**

102. Show that the fractional rate of change of $f(x)$ is the rate of change of $\ln(f(x))$.

103. Let $A(t)$ represent a quantity growing exponentially. Show that the percentage rate of change, $A'(t)/A(t)$, is constant.

104. Let $A(t)$ be the amount of money in an account paying interest compounded some number of times per year. Show that the percentage rate of growth, $A'(t)/A(t)$, is constant. What might this constant represent?

4.4 Derivatives of Trigonometric Functions

We start with the derivatives of the sine and cosine functions.

Derivatives of the Sine and Cosine Functions

$$\frac{d}{dx}\sin x = \cos x$$

$$\frac{d}{dx}\cos x = -\sin x$$

Notice the sign change.

Quick Examples

1. $\frac{d}{dx}(x\cos x) = 1 \cdot \cos x + x \cdot (-\sin x)$ Product rule: $x \cos x$ is a product.[1]

$$= \cos x - x\sin x$$

2. $\frac{d}{dx}\left(\frac{x^2+x}{\sin x}\right) = \frac{(2x+1)(\sin x) - (x^2+x)(\cos x)}{\sin^2 x}$ Quotient rule

We justify these formulas at the end of the section, but we can see right away that they are plausible by examining Figure 6, which shows the graphs of the sine and cosine functions together with their derivatives.

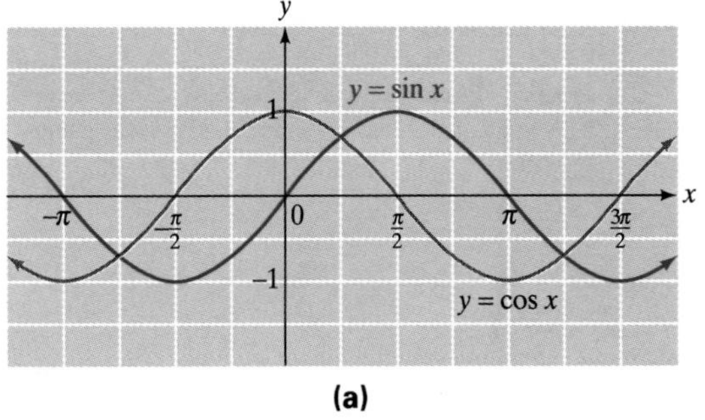

(a)

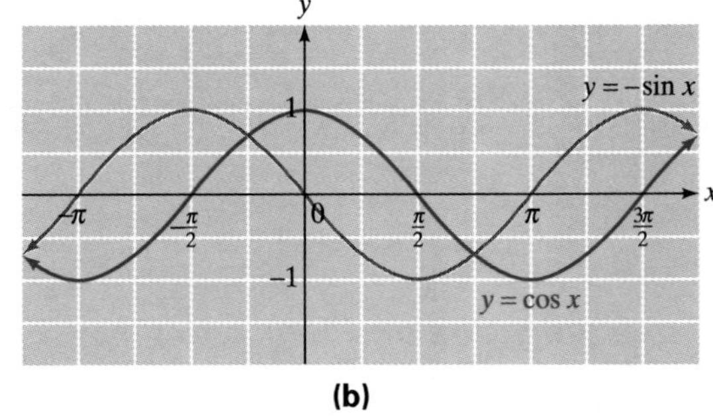

(b)

Figure 6

[1]Apply the calculation thought experiment: If we were to compute $x \cos x$, the last operation we would perform is the multiplication of x and $\cos x$. Hence, $x \cos x$ is a product.

Notice, for instance, that in Figure 6(a) the graph of sin x is rising most rapidly when $x = 0$, corresponding to the maximum value of its derivative, cos x. When $x = \pi/2$, the graph of sin x levels off, so that its derivative, cos x, is 0. Another point to notice: Since periodic functions (such as sine and cosine) repeat their behavior, their derivatives must also be periodic.

Just as with logarithmic and exponential functions, the chain rule can be used to find more general derivatives.

Derivatives of Sines and Cosines of Functions

Original rule	Generalized rule	In words
$\frac{d}{dx}\sin x = \cos x$	$\frac{d}{dx}\sin u = \cos u\frac{du}{dx}$	The derivative of the sine of a quantity is the cosine of that quantity, times the derivative of that quantity.
$\frac{d}{dx}\cos x = -\sin x$	$\frac{d}{dx}\cos u = -\sin u\frac{du}{dx}$	The derivative of the cosine of a quantity is the negative sine of that quantity, times the derivative of that quantity.

Quick Examples

1. $\frac{d}{dx}\sin(3x^2 - 1) = \cos(3x^2 - 1)\frac{d}{dx}(3x^2 - 1)$ $u = 3x^2 - 1$ (see footnote 2)

   $= 6x\cos(3x^2 - 1)$ We placed the $6x$ in front—see below.

2. $\frac{d}{dx}\cos(x^3 + x) = -\sin(x^3 + x)\frac{d}{dx}(x^3 + x)$ $u = x^3 + x$

   $= -(3x^2 + 1)\sin(x^3 + x)$

Note Avoid writing ambiguous expressions like $\cos(3x^2 - 1)(6x)$. Does this mean

$\cos[(3x^2 - 1)(6x)]$ The cosine of the quantity $(3x^2 - 1)(6x)$

or does it mean

$[\cos(3x^2 - 1)](6x)$ The product of $\cos(3x^2 - 1)$ and $6x$

To avoid the ambiguity, place the $6x$ in front of the cosine expression and write

$6x\cos(3x^2 - 1)$ The product of $6x$ and $\cos(3x^2 - 1)$

example 1

Derivatives of Trigonometric Functions

Find the derivatives of the following functions.

a. $f(x) = \sin^2 x$ **b.** $g(x) = \sin^2(x^2)$ **c.** $h(x) = e^{-x}\cos(2x)$

Solution

a. Recall that $\sin^2 x = (\sin x)^2$. The calculation thought experiment tells us that $f(x)$ is the square of a quantity.[3] Therefore, we use the chain rule (or generalized power rule) for differentiating the square of a quantity:

---

[2]If we were to evaluate $\sin(3x^2 - 1)$, the last operation we would perform is taking the sine of a quantity. Thus, the calculation thought experiment tells us that we are dealing with the *sine of a quantity,* and we use the generalized rule.

[3]Notice the difference between $\sin^2 x$ and $\sin(x^2)$. The first is the square of sin x, whereas the second is the sin of the quantity x^2.

$$\frac{d}{dx}(u^2) = 2u\frac{du}{dx}$$

$$\frac{d}{dx}(\sin x)^2 = 2(\sin x)\frac{d(\sin x)}{dx} \qquad u = \sin x$$

$$= 2 \sin x \cos x$$

Thus, $f'(x) = 2 \sin x \cos x$.

b. We rewrite the function $g(x) = \sin^2(x^2)$ as $[\sin(x^2)]^2$. Since $g(x)$ is the square of a quantity, we have

$$\frac{d}{dx}\sin^2(x^2) = \frac{d}{dx}\left[\sin(x^2)\right]^2 \qquad \text{Rewrite } \sin^2(-) \text{ as } [\sin(-)]^2.$$

$$= 2\sin(x^2)\frac{d\left[\sin(x^2)\right]}{dx} \qquad \frac{d}{dx}(u^2) = 2u\frac{du}{dx} \text{ with } u = \sin(x^2)$$

$$= 2\sin(x^2)\cdot\cos(x^2)\cdot 2x \qquad \frac{d}{dx}\sin u = \cos u\frac{du}{dx} \text{ with } u = x^2$$

Thus, $g'(x) = 4x\sin(x^2)\cos(x^2)$.

c. Since $h(x)$ is the product of e^{-x} and $\cos(2x)$, we use the product rule:

$$h'(x) = (-e^{-x})\cos(2x) + e^{-x}\frac{d}{dx}\left[\cos(2x)\right]$$

$$= (-e^{-x})\cos(2x) - e^{-x}\sin(2x)\frac{d}{dx}(2x) \qquad \frac{d}{dx}\cos u = -\sin u\frac{du}{dx}$$

$$= -e^{-x}\cos(2x) - 2e^{-x}\sin(2x)$$

$$= -e^{-x}\left[\cos(2x) + 2\sin(2x)\right]$$

Question What about the derivatives of the other four trigonometric functions?
Answer The remaining trigonometric functions are tan, cotan, sec, and cosec. Each is a ratio of sines and cosines, so we can use the quotient rule to find their derivatives. For example, we can find the derivative of tan as follows:

$$\frac{d}{dx}\tan x = \frac{d}{dx}\left(\frac{\sin x}{\cos x}\right)$$

$$= \frac{(\cos x)(\cos x) - (\sin x)(-\sin x)}{\cos^2 x}$$

$$= \frac{\cos^2 x + \sin^2 x}{\cos^2 x} = \frac{1}{\cos^2 x} = \sec^2 x$$

We ask you to derive the other three derivatives in the exercises. Here is a list of the derivatives of all six trigonometric functions and their chain rule variants.

Derivatives of the Trigonometric Functions

Original rule	Generalized rule
$\frac{d}{dx}\sin x = \cos x$	$\frac{d}{dx}\sin u = \cos u\frac{du}{dx}$
$\frac{d}{dx}\cos x = -\sin x$	$\frac{d}{dx}\cos u = -\sin u\frac{du}{dx}$
$\frac{d}{dx}\tan x = \sec^2 x$	$\frac{d}{dx}\tan u = \sec^2 u\frac{du}{dx}$
$\frac{d}{dx}\text{cotan } x = -\text{cosec}^2 x$	$\frac{d}{dx}\text{cotan } u = -\text{cosec}^2 u\frac{du}{dx}$
$\frac{d}{dx}\sec x = \sec x\tan x$	$\frac{d}{dx}\sec u = \sec u\tan u\frac{du}{dx}$
$\frac{d}{dx}\text{cosec } x = -\text{cosec } x\text{ cotan } x$	$\frac{d}{dx}\text{cosec } u = -\text{cosec } u\text{ cotan } u\frac{du}{dx}$

Quick Examples

1. $\frac{d}{dx}\tan(x^2-1) = \sec^2(x^2-1)\frac{d(x^2-1)}{dx}$ $\quad u = x^2 - 1$

$= 2x\sec^2(x^2-1)$

2. $\frac{d}{dx}\text{cosec}(e^{3x}) = -\text{cosec}(e^{3x})\text{ cotan}(e^{3x})\frac{d(e^{3x})}{dx}$ $\quad u = e^{3x}$

$= -3e^{3x}\text{ cosec}(e^{3x})\text{ cotan}(e^{3x})$ $\quad$ The derivative of e^{3x} is $3e^{3x}$.

example 2

Gas Heating Demand

In Section 2.4 we saw that seasonal fluctuations in temperature suggested a sine function. For instance, we can use the function

$$T = 60 + 25\sin\left[\frac{\pi}{6}(x-4)\right]$$ t = temperature in °F; x = months since January 1

to model a temperature that fluctuates between 35°F on February 1 ($x = 1$) and 85°F on August 1 ($x = 7$) (see Figure 7).

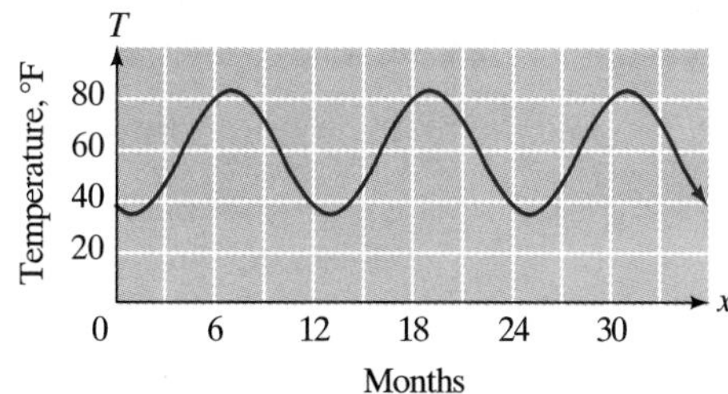

Figure 7

The demand for gas at a utility company can be expected to fluctuate in a similar way because demand grows with increased heating requirements. A reasonable model might therefore be

$$G = 400 - 100 \sin\left[\frac{\pi}{6}(x - 4)\right] \qquad \text{Why did we subtract the sine term?}$$

where G is the demand for gas in cubic yards per day. Find and interpret $G'(10)$.

Solution

First, we take the derivative of G:

$$G'(x) = -100 \cos\left[\frac{\pi}{6}(x - 4)\right] \cdot \frac{\pi}{6}$$

$$= -\frac{50\pi}{3} \cos\left[\frac{\pi}{6}(x - 4)\right] \text{ cubic yards per day, per month}$$

Thus, $$G'(10) = -\frac{50\pi}{3} \cos\left[\frac{\pi}{6}(10 - 4)\right]$$

$$= -\frac{50\pi}{3} \cos(\pi) = \frac{50\pi}{3} \qquad \text{Since } \cos \pi = -1$$

Since the units of $G'(10)$ are cubic yards per day per month, we interpret the result as follows: On November 1 ($x = 10$) the daily demand for gas is increasing at a rate of $50\pi/3 \approx 52.36$ cubic yards per month. This is consistent with Figure 7, which shows the temperature decreasing on that date.

Now, where did these formulas for the derivatives of $\sin x$ and $\cos x$ come from? We first calculate the derivative of $\sin x$ from scratch, using the definition of the derivative:

$$\frac{d}{dx} f(x) = \lim_{h \to 0} \frac{f(x + h) - f(x)}{h}$$

$$\frac{d}{dx} \sin x = \lim_{h \to 0} \frac{\sin(x + h) - \sin x}{h}$$

There is an **addition formula** that we can use to expand $\sin(x + h)$:

$$\sin(x + h) = \sin x \cos h + \cos x \sin h$$

Substituting this expression for $\sin(x + h)$ gives

$$\frac{d}{dx} \sin x = \lim_{h \to 0} \frac{\sin x \cos h + \cos x \sin h - \sin x}{h}$$

Grouping the first and third terms together and factoring out the term $\sin x$ give

$$\frac{d}{dx} \sin x = \lim_{h \to 0} \frac{\sin x\,(\cos h - 1) + \cos x \sin h}{h}$$

$$= \lim_{h \to 0} \frac{\sin x\,(\cos h - 1)}{h} + \lim_{h \to 0} \frac{\cos x \sin h}{h} \qquad \text{Limit of a sum}$$

$$= \sin x \lim_{h \to 0} \frac{\cos h - 1}{h} + \cos x \lim_{h \to 0} \frac{\sin h}{h}$$

and we are left with two limits to evaluate. Calculating these limits analytically requires a little trigonometry.[4] Alternatively, we can get a good idea of what these two limits are by estimating them numerically or graphically. Figures 8 and 9 show the graphs of $(\cos h - 1)/h$ and $(\sin h)/h$, respectively. We find that

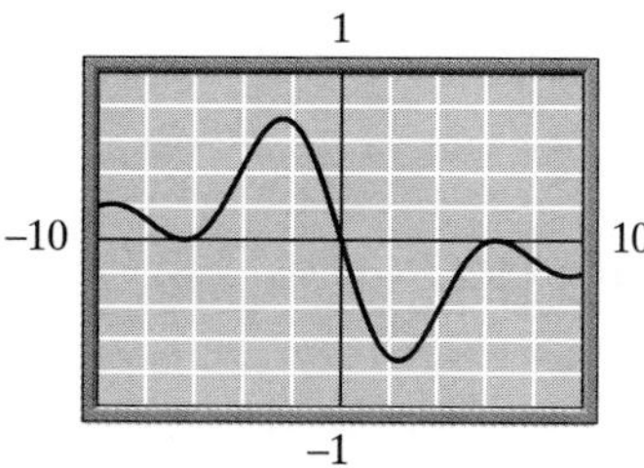

Figure 8

$$\lim_{h\to 0} \frac{\cos h - 1}{h} = 0$$

and

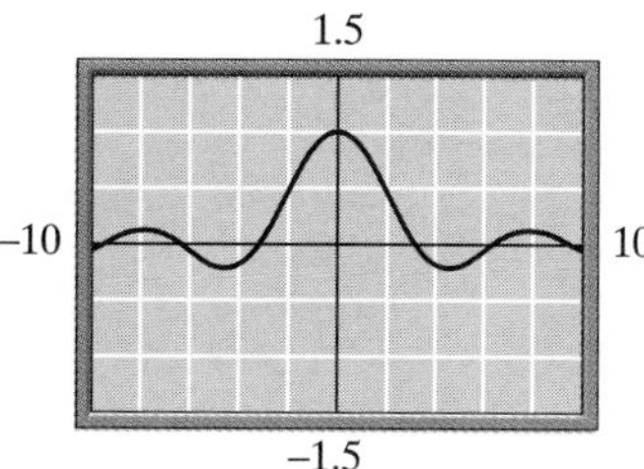

Figure 9

$$\lim_{h\to 0} \frac{\sin h}{h} = 1$$

Therefore,

$$\frac{d}{dx}\sin x = (\sin x)(0) + (\cos x)(1) = \cos x$$

Question What about the derivative of the cosine function?
Answer Let us use the identity

$$\cos x = \sin(\pi/2 - x)$$

from Section 2.6. If $y = \cos x = \sin(\pi/2 - x)$, then, using the chain rule, we have

$$\begin{aligned}\frac{dy}{dx} &= \cos(\pi/2 - x)\frac{d}{dx}(\pi/2 - x)\\ &= (-1)\cos(\pi/2 - x)\\ &= -\sin x \qquad \text{Using the identity } \cos(\pi/2 - x) = \sin x\end{aligned}$$

[4]You can find these calculations on the web site by following the path Web site → Everything for Calculus → Chapter 3 → Proof of Some Trigonometic Limits.

4.4 exercises

In Exercises 1–28, find the derivatives of the given functions.

1. $f(x) = \sin x - \cos x$
2. $f(x) = \tan x - \sin x$
3. $g(x) = (\sin x)(\tan x)$
4. $g(x) = (\cos x)(\cotan x)$
5. $h(x) = 2 \operatorname{cosec} x - \sec x + 3x$
6. $h(x) = 2 \sec x + 3 \tan x + 3x$
7. $r(x) = x \cos x + x^2 + 1$
8. $r(x) = 2x \sin x - x^2$
9. $s(x) = (x^2 - x + 1) \tan x$
10. $s(x) = \dfrac{\tan x}{x^2 - 1}$
11. $t(x) = \dfrac{\cotan x}{1 + \sec x}$
12. $t(x) = (1 + \sec x)(1 - \cos x)$
13. $k(x) = \cos^2 x$
14. $k(x) = \tan^2 x$
15. $j(x) = \sec^2 x$
16. $j(x) = \operatorname{cosec}^2 x$
17. $p(x) = 2 + 5 \sin\left[\frac{\pi}{5}(x - 4)\right]$
18. $p(x) = 10 - 3 \cos\left[\frac{\pi}{6}(x + 3)\right]$
19. $u(x) = \cos(x^2 - x)$
20. $u(x) = \sin(3x^2 + x - 1)$
21. $v(x) = \sec(x^{2.2} + 1.2x - 1)$
22. $v(x) = \tan(x^{2.2} + 1.2x - 1)$
23. $w(x) = \sec x \tan(x^2 - 1)$
24. $w(x) = \cos x \sec(x^2 - 1)$
25. $y(x) = \cos(e^x) + e^x \cos x$
26. $y(x) = \sec(e^x)$
27. $z(x) = \ln|\sec x + \tan x|$
28. $z(x) = \ln|\operatorname{cosec} x + \cotan x|$

In Exercises 29–32, derive the derivatives from the derivatives of sine and cosine.

29. $\dfrac{d}{dx} \sec x = \sec x \tan x$
30. $\dfrac{d}{dx} \cotan x = -\operatorname{cosec}^2 x$
31. $\dfrac{d}{dx} \operatorname{cosec} x = -\operatorname{cosec} x \cotan x$
32. $\dfrac{d}{dx} \ln|\sec x| = \tan x$

Calculate the derivatives in Exercises 33–40.

33. $\dfrac{d}{dx}[e^{-2x} \sin(3\pi x)]$
34. $\dfrac{d}{dx}[e^{5x} \sin(-4\pi x)]$
35. $\dfrac{d}{dx}[\sin(3x)]^{0.5}$
36. $\dfrac{d}{dx} \cos\left(\dfrac{x^2}{x - 1}\right)$
37. $\dfrac{d}{dx} \sec\left(\dfrac{x^3}{x^2 - 1}\right)$
38. $\dfrac{d}{dx}\left(\dfrac{\tan x}{2 + e^x}\right)^2$
39. $\dfrac{d}{dx}\{[\ln|x|][\cotan(2x - 1)]\}$
40. $\dfrac{d}{dx} \ln|\sin x - 2xe^{-x}|$

Applications

41. ***Cost*** The cost of Dig-It brand snow shovels is given by

$$c(t) = 3.5 \sin[2\pi(t - 0.75)]$$

where t is time in years since January 1, 1997. How fast, in dollars per week, is the cost increasing each October 1?

42. ***Sales*** Daily sales of Doggy brand cookies can be modeled by

$$s(t) = 400 \cos[2\pi(t - 2)/7]$$

cartons, where t is time in days since Monday morning. How fast are sales changing on Thursday morning?

43. ***Tides*** The depth of the water at my favorite surfing spot varies from 5 to 15 feet, depending on the time. Last Sunday high tide occurred at 5:00 A.M., and the next high tide occurred at 6:30 P.M.
    a. Obtain a cosine model describing the depth of water as a function of time t in hours since 5:00 A.M. on Sunday morning.
    b. How fast was the tide rising (or falling) at noon on Sunday?

44. ***Tides*** Repeat Exercise 43 using data from the depth of the water at my other favorite surfing spot, where the tide last Sunday varied from a low of 6 feet at 4:00 A.M. to a high of 10 feet at noon. (As in Exercise 43, take t as time in hours since 5:00 A.M.)

45. ***Inflation*** If we take a 3.5% rate of inflation into account, the cost of DigIn brand snow shovels is given by

$$c(t) = 1.035^t[0.8 \sin(2\pi t) + 10.2]$$

where t is time in years since January 1, 1997. How fast, in dollars per week, is the cost of DigIn shovels increasing on January 1, 1998?

46. ***Deflation*** Sales, in bottles per day, of my exclusive mass-produced 1998 vintage Chateau Petit Mont Blanc follow the function

$$s(t) = 4.5e^{-0.2t} \sin(2\pi t)$$

where t is time in years since January 1, 1998. How fast were sales rising or falling on January 1, 1999?

47. ***Tilt of the Earth's Axis*** The tilt of the Earth's axis from its plane of rotation about the sun oscillates between approximately 22.5° and 24.5° with a period of approximately 40,000 years. We know that 500,000 years ago, the tilt of the Earth's axis was 24.5°.

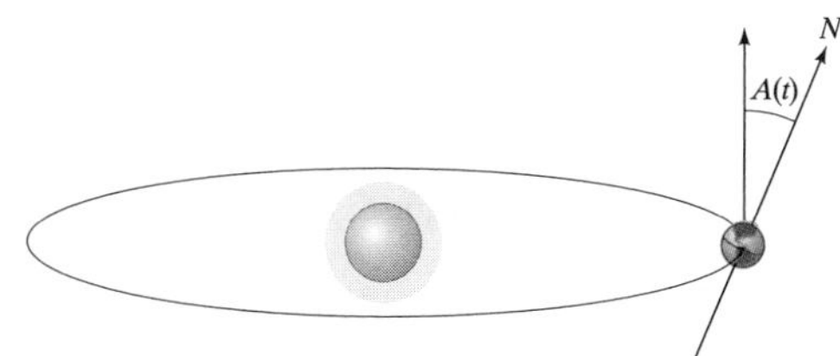

a. Which of the following functions best models the tilt of the Earth's axis?

(A) $A(t) = 23.5 + 2\sin\left(\dfrac{2\pi t + 500}{40}\right)$

(B) $A(t) = 23.5 + \cos\left(\dfrac{t + 500}{80\pi}\right)$

(C) $A(t) = 23.5 + \cos\left(\dfrac{2\pi(t + 500)}{40}\right)$

where $A(t)$ is the tilt in degrees and t is time in thousands of years, with $t = 0$ being the present time.

b. Use the model you selected in part (a) to estimate the rate at which the tilt was changing 150,000 years ago. (Round your answer to three decimal places, and be sure to give the units of measurement.)

Source: Dr. David Hodell, University of Florida/Juan Valesco/*New York Times,* February 16, 1999, p. F1.

48. ***Eccentricity of the Earth's Orbit*** The eccentricity of the Earth's orbit (that is, the deviation of the Earth's orbit from a perfect circle) can be modeled by[5]

$$E(t) = 0.025\left[\cos\left(\frac{2\pi(t + 200)}{400}\right) + \cos\left(\frac{2\pi(t + 200)}{100}\right)\right]$$

where $E(t)$ is the eccentricity and t is time in thousands of years, with $t = 0$ being the present time. What was the value of the eccentricity 200,000 years ago, and how fast was it changing?

Source: See the source in Exercise 47.

[5] This is a rough model based on the actual data.

Communication and Reasoning Exercises

49. Give two examples of a function $f(x)$ with the property that $f''(x) = -f(x)$. ($f''(x)$ is the derivative of $f'(x)$.)

50. Give two examples of a function $f(x)$ with the property that $f''(x) = -4f(x)$. ($f''(x)$ is the derivative of $f'(x)$.)

51. Give two examples of a function $f(x)$ with the property that $f'(x) = -f(x)$.

52. Give four examples of a function $f(x)$ with the property that $f^{(4)}(x) = f(x)$. ($f^{(4)}(x)$ is obtained from $f(x)$ by differentiating 4 times.)

53. By referring to the graph of $f(x) = \cos x$, explain why $f'(x) = -\sin x$ rather than $\sin x$.

54. At what angle does the graph of $f(x) = \sin x$ depart from the origin?

4.5 Implicit Differentiation

Consider the equation $y^5 + y + x = 0$, whose graph is shown in Figure 10.

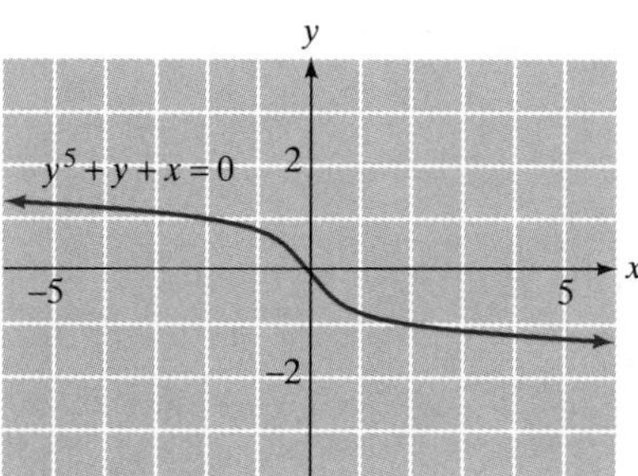

Figure 10

How did we find this graph? We did not solve for y as a function of x; that is impossible. In fact, we solved for x in terms of y to find points to plot. Nonetheless, the graph in Figure 10 is the graph of a function because it passes the vertical line test: Every vertical line crosses the graph no more than once, so for each value of x there is no more than one corresponding value of y. Since we cannot solve for y explicitly in terms of x, we say that the equation determines y as an **implicit function** of x.

Now, suppose we want to find the slope of the tangent line to this curve at, say, the point $(2, -1)$ (which, you should check, is a point on the curve).

Question How would we find dy/dx without first solving for y?
Answer We use the chain rule and a little cleverness. We think of y as a function of x and take the derivative with respect to x of both sides of the equation:

$$y^5 + y + x = 0 \qquad \text{Original equation}$$

$$\frac{d}{dx}(y^5 + y + x) = \frac{d}{dx}(0) \qquad \text{Derivative with respect to } x \text{ of both sides}$$

$$\frac{d}{dx}(y^5) + \frac{d}{dx}(y) + \frac{d}{dx}(x) = 0 \qquad \text{Derivative rules}$$

Now we must be careful. The derivative *with respect to x* of y^5 is *not* $5y^4$. Rather, since y is a function of x, we must use the chain rule, which tells us that

$$\frac{d}{dx}(y^5) = 5y^4\frac{dy}{dx}$$

Thus, we get

$$5y^4\frac{dy}{dx} + \frac{dy}{dx} + 1 = 0$$

We want to find dy/dx, so we *solve for it:*

$$(5y^4 + 1)\frac{dy}{dx} = -1 \qquad \text{Isolate } dy/dx \text{ on one side.}$$

$$\frac{dy}{dx} = -\frac{1}{5y^4 + 1} \qquad \text{Divide both sides by } 5y^4 + 1.$$

Question The formula we just found for dy/dx is not a function of x because there is a y in it. Is this okay?
Answer We should not expect to obtain dy/dx as an explicit function of x if y was not an explicit function of x to begin with. The result is still useful because we can evaluate the derivative at any point on the graph. For instance, at the point $(2, -1)$ on the graph, we get

$$\frac{dy}{dx} = -\frac{1}{5y^4 + 1} = -\frac{1}{5(-1)^4 + 1} = -\frac{1}{6}$$

Thus, the slope of the tangent line to the curve $y^5 + y + x = 0$ at the point $(2, -1)$ is $-\frac{1}{6}$. Figure 11 shows the graph and this tangent line.

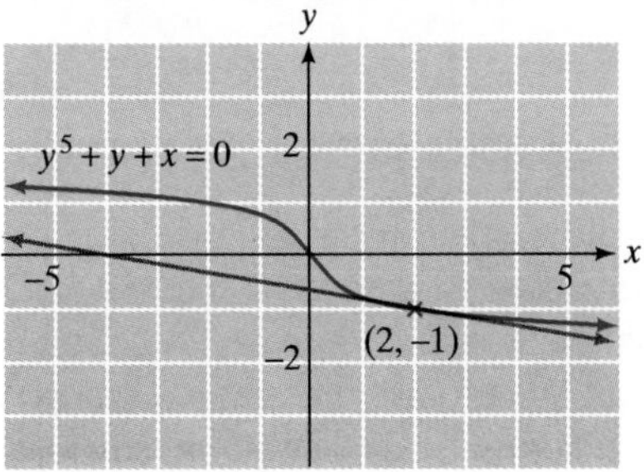

Figure 11

This procedure of differentiating an equation to find dy/dx without first solving the equation for y is called **implicit differentiation.**

An equation in x and y need not determine y as a function of x. Consider, for example, the equation

$$2x^2 + y^2 = 2$$

Solving for y yields $y = \pm\sqrt{2 - 2x^2}$. The $\pm$ sign reminds us that for some values of x there are two corresponding values for y. We can graph this equation by superimposing the graphs of

$$y = \sqrt{2 - 2x^2} \qquad \text{and} \quad y = -\sqrt{2 - 2x^2}$$

The graph, an *ellipse,* is shown in Figure 12.

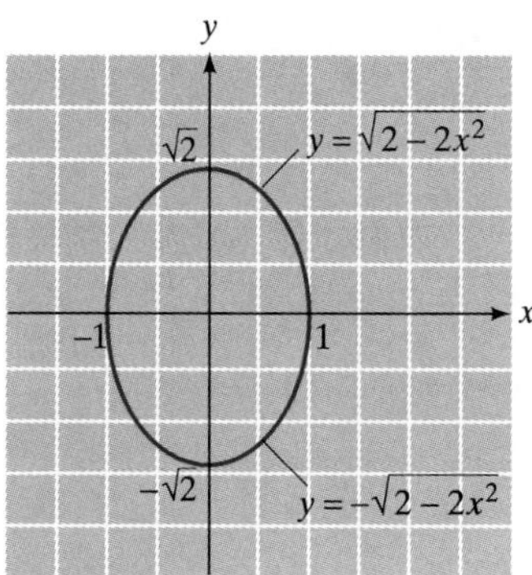

Figure 12

The graph of $y = \sqrt{2 - 2x^2}$ constitutes the top half of the ellipse, and the graph of $y = -\sqrt{2 - 2x^2}$ constitutes the bottom half.

example 1

Implicit Differentiation

Find the slope of the tangent line to the ellipse $2x^2 + y^2 = 2$ at the point $(1/\sqrt{2}, 1)$.

Solution

Since $(1/\sqrt{2}, 1)$ is on the top half of the ellipse in Figure 12, we *could* differentiate the function $y = \sqrt{2 - 2x^2}$ to obtain the result, but it is actually easier to apply implicit differentiation to the original equation.

$$2x^2 + y^2 = 2 \qquad \text{Original equation}$$

$$\frac{d}{dx}(2x^2 + y^2) = \frac{d}{dx}(2) \qquad \text{Derivative with respect to } x \text{ of both sides}$$

$$4x + 2y\frac{dy}{dx} = 0$$

$$2y\frac{dy}{dx} = -4x$$

$$\frac{dy}{dx} = -\frac{4x}{2y} = -\frac{2x}{y}$$

To find the slope at $(1/\sqrt{2}, 1)$ we now substitute for x and y:

$$\left.\frac{dy}{dx}\right|_{(1/\sqrt{2},1)} = -\frac{2/\sqrt{2}}{1} = -\sqrt{2}$$

Thus, the slope of the tangent to the ellipse at the point $(1/\sqrt{2}, 1)$ is $-\sqrt{2} \approx -1.414$.

example 2

Tangent Line to Implicit Curve

Find the equation of the tangent line to the curve $\ln y = xy$ at the point where $y = 1$.

Solution

First, we use implicit differentiation to find dy/dx:

$$\frac{d}{dx}(\ln y) = \frac{d}{dx}(xy)$$ Take d/dx of both sides.

$$\frac{1}{y}\boxed{\frac{dy}{dx}} = (1)y + x\boxed{\frac{dy}{dx}}$$ Chain rule on left; product rule on right

We have boxed the quantity dy/dx to remind us that it is the unknown. To solve for it, we bring all the terms containing dy/dx to the left-hand side and all terms not containing it to the right-hand side:

$$\frac{1}{y}\boxed{\frac{dy}{dx}} - x\boxed{\frac{dy}{dx}} = y$$ Bring the terms with dy/dx to the left.

$$\boxed{\frac{dy}{dx}}\left(\frac{1}{y} - x\right) = y$$ Factor out dy/dx.

$$\boxed{\frac{dy}{dx}}\left(\frac{1 - xy}{y}\right) = y$$

$$\frac{dy}{dx} = y\left(\frac{y}{1 - xy}\right) = \frac{y^2}{1 - xy}$$ Solve for dy/dx.

The derivative gives the slope of the tangent line, so we want to evaluate the derivative at the point where $y = 1$. However, the formula for dy/dx requires values for both x and y. We get the value of x by substituting $y = 1$ in the original equation:

$$\ln y = xy$$

$$\ln(1) = x \cdot 1$$

But $\ln(1) = 0$, and so $x = 0$ for this point. Thus,

$$\left.\frac{dy}{dx}\right|_{(0,1)} = \frac{1^2}{1 - (0)(1)} = 1$$

Therefore, the tangent line is the line through $(x, y) = (0, 1)$ with slope 1, which is

$$y = x + 1$$

Before we go on . . . This is an example of an implicit function for which it is simply not possible to solve for y. Try it.

Sometimes it is easiest to differentiate a complicated function of x by first taking the logarithm and then using implicit differentiation—a technique called **logarithmic differentiation.**

example 3

Logarithmic Differentiation

Find $\frac{d}{dx}\left[\frac{(x+1)^{10}(x^2+1)^{11}}{(x^3+1)^{12}}\right]$ without using the product or quotient rules.

Solution

We write

$$y = \frac{(x+1)^{10}(x^2+1)^{11}}{(x^3+1)^{12}}$$

and then take the natural logarithm of both sides:

$$\ln y = \ln\left[\frac{(x+1)^{10}(x^2+1)^{11}}{(x^3+1)^{12}}\right]$$

We can use properties of the logarithm to simplify the right-hand side:

$$\begin{aligned}\ln y &= \ln(x+1)^{10} + \ln(x^2+1)^{11} - \ln(x^3+1)^{12} \\ &= 10\ln(x+1) + 11\ln(x^2+1) - 12\ln(x^3+1)\end{aligned}$$

Now we can find dy/dx using implicit differentiation:

$$\frac{1}{y}\frac{dy}{dx} = \frac{10}{x+1} + \frac{22x}{x^2+1} - \frac{36x^2}{x^3+1} \quad \text{Take } d/dx \text{ of both sides.}$$

$$\frac{dy}{dx} = y\left(\frac{10}{x+1} + \frac{22x}{x^2+1} - \frac{36x^2}{x^3+1}\right) \quad \text{Solve for } dy/dx.$$

$$= \frac{(x+1)^{10}(x^2+1)^{11}}{(x^3+1)^{12}}\left(\frac{10}{x+1} + \frac{22x}{x^2+1} - \frac{36x^2}{x^3+1}\right) \quad \text{Substitute for } y.$$

Before we go on . . . Redo this example using the product and quotient rules (and the chain rule) instead of logarithmic differentiation, and compare the answers. Compare also the amounts of work involved in both methods.

Application

Productivity usually depends on both labor and capital. Suppose, for example, you are managing an automobile assembly plant. You can measure its productivity by counting the number of automobiles the plant produces each year. As a measure of labor you can use the number of employees, and as a measure of capital you can use its operating budget. The Cobb-Douglas model uses a function of the form

$$P = Kx^a y^{1-a} \quad \text{Cobb-Douglas model for productivity}$$

where P stands for the number of automobiles produced per year, x is the number of employees, and y is the operating budget. The numbers K and a are constants that depend on the particular factory studied, with a between 0 and 1.

example 4

Cobb-Douglas Production Function

The automobile assembly plant you manage has the Cobb-Douglas production function

$$P = x^{0.3}y^{0.7}$$

where P is the number of automobiles it produces per year, x is the number of employees, and y is the daily operating budget (in dollars). Assume a production level of 1000 automobiles per year.

a. Find dy/dx.

b. Evaluate this derivative at $x = 80$ and interpret the answer.

Solution

Since the production level is 1000 automobiles per year, we have $P = 1000$, so the equation is

$$1000 = x^{0.3}y^{0.7}$$

a. We find dy/dx by implicit differentiation:

$$0 = \frac{d}{dx}(x^{0.3}y^{0.7}) \qquad \text{Take } d/dx \text{ of both sides.}$$

$$0 = 0.3x^{-0.7}y^{0.7} + x^{0.3}(0.7)y^{-0.3}\frac{dy}{dx} \qquad \text{Product and chain rules}$$

$$-0.7x^{0.3}y^{-0.3}\frac{dy}{dx} = 0.3x^{-0.7}y^{0.7} \qquad \text{Bring term with } dy/dx \text{ to left.}$$

$$\frac{dy}{dx} = -\frac{0.3x^{-0.7}y^{0.7}}{0.7x^{0.3}y^{-0.3}} \qquad \text{Solve for } dy/dx.$$

$$= -\frac{3y}{7x}$$

b. To evaluate this derivative at $x = 80$, we must first find the corresponding value of y. To obtain y, we substitute $x = 80$ in the original equation $1000 = x^{0.3}y^{0.7}$ and solve for y:

$$1000 = 80^{0.3}y^{0.7}$$

$$y^{0.7} = \frac{1000}{80^{0.3}}$$

To obtain y on its own, we raise both sides to the power 1/0.7 to obtain

$$y = (y^{0.7})^{1/0.7} = \left(\frac{1000}{80^{0.3}}\right)^{1/0.7} \approx 2951.92$$

Now that we have the corresponding value for y, we evaluate the derivative at $x = 80$ and $y = 2951.92$:

$$\left.\frac{dy}{dx}\right|_{x=80} \approx -\frac{3(2951.92)}{7(80)} \approx -15.81 \qquad \text{Using } \frac{dy}{dx} = -\frac{3y}{7x}$$

How do we interpret this result? The first clue is to look at the units of the derivative: We recall that the units of dy/dx are units of y per unit of x. Since y is the daily budget, its units are dollars; since x is the number of employees, its units are employees. Thus,

$$\left.\frac{dy}{dx}\right|_{x=80} = -\$15.81 \text{ per employee}$$

Next, we recall that dy/dx measures the rate of change of y as x changes. Thus (since the answer is negative), the daily budget to maintain the production of 1000 automobiles is decreasing by approximately \$15.81 per additional employee at an employment level of 80 employees. In other words, increasing the work force by one worker will result in a saving of approximately \$15.81 per day. Roughly speaking, *a new employee is worth \$15.81 per day* at the current levels of employment and production.

4.5 exercises

In Exercises 1–10, find dy/dx using implicit differentiation. In each case, compare your answer with the result obtained by first solving for y as a function of x and then taking the derivative.

1. $2x + 3y = 7$ **2.** $4x - 5y = 9$

3. $x^2 - 2y = 6$ **4.** $3y + x^2 = 5$

5. $2x + 3y = xy$ **6.** $x - y = xy$

7. $e^x y = 1$ **8.** $e^x y - y = 2$

9. $y \ln x + y = 2$ **10.** $\frac{\ln x}{y} = 2 - x$

In Exercises 11–34, find the indicated derivative using implicit differentiation.

11. $x^2 + y^2 = 5; dy/dx$ **12.** $2x^2 - y^2 = 4; dy/dx$

13. $x^2y - y^2 = 4; dy/dx$ **14.** $xy^2 - y = x; dy/dx$

15. $3xy - \frac{y}{3} = \frac{2}{x}; dy/dx$ **16.** $\frac{xy}{2} - y^2 = 3; dy/dx$

17. $x^2 - 3y^2 = 8; dx/dy$ **18.** $(xy)^2 + y^2 = 8; dx/dy$

19. $p^2 - pq = 5p^2q^2; dp/dq$ **20.** $q^2 - pq = 5p^2q^2; dp/dq$

21. $xe^y - ye^x = 1; dy/dx$ **22.** $x^2e^y - y^2 = e^x; dy/dx$

23. $e^{st} = s^2; ds/dt$ **24.** $e^{s^2t} - st = 1; ds/dt$

25. $e^x/y^2 = 1 + e^y; dy/dx$ **26.** $x/e^y + xy = 9y; dy/dx$

27. $\ln(y^2 - y) + x = y; dy/dx$

28. $\ln(xy) - x \ln y = y; dy/dx$

29. $\ln(xy + y^2) = e^y; dy/dx$ **30.** $\ln(1 + e^{xy}) = y; dy/dx$

Δ **31.** $x = \tan y$; find dy/dx Δ **32.** $x = \cos y$; find dy/dx

Δ **33.** $x + y + \sin(xy) = 1$; find dy/dx

Δ **34.** $xy + x \cos y = x$; find dy/dx

In Exercises 35–38, use logarithmic differentiation to find dy/dx.

35. $y = (x^3 + x)\sqrt{x^3 + 2}$ **36.** $y = \sqrt{\frac{x-1}{x^2+2}}$

37. $y = x^x$ **38.** $y = x^{-x}$

In Exercises 39–50, use implicit differentiation to evaluate dy/dx at the indicated point on the graph. (If only the x-coordinate is given, you must also find the y-coordinate.)

39. $4x^2 + 2y^2 = 12; (1, -2)$ **40.** $3x^2 - y^2 = 11; (-2, 1)$

41. $2x^2 - y^2 = xy; (-1, 2)$ **42.** $2x^2 + xy = 3y^2; (-1, -1)$

43. $3x^{0.3}y^{0.7} = 10; x = 20$ **44.** $2x^{0.4}y^{0.6} = 10; x = 50$

45. $x^{0.4}y^{0.6} - 0.2x^2 = 100; x = 20$

46. $x^{0.4}y^{0.6} - 0.3x^2 = 10; x = 10$

47. $e^{xy} - x = 4x; x = 3$ **48.** $e^{-xy} + 2x = 1; x = -1$

49. $\ln(x + y) - x = 3x^2; x = 0$

50. $\ln(x - y) + 1 = 3x^2; x = 0$

Applications

51. ***Demand*** The demand equation for soccer tournament T-shirts is

$$pq - 2000 = q$$

where q is the number of T-shirts the Enormous State University soccer team can sell for \$$p$ each.

a. How many T-shirts can the team sell at \$5 each?

b. Find $\left.\frac{dq}{dp}\right|_{p=5}$ and interpret the result.

52. ***Cost Equations*** The cost c (in cents) of producing x gallons of Ectoplasm hair gel is given by the cost equation

$$c^2 - 10cx = 200$$

a. Find the cost of producing 1 gallon and 3.5 gallons.
b. Evaluate dc/dx at $x = 1$ and $x = 3.5$, and interpret the results.

53. ***Housing Costs*** The cost C (in dollars) of building a house is related to the number k of carpenters used and the number e of electricians used by the formula

$$C = 15{,}000 + 50k^2 + 60e^2$$

If the cost of the house comes to \$200,000, find $\left.\dfrac{dk}{de}\right|_{e=15}$ and interpret your result.
Source: Based on an exercise in *Introduction to Mathematical Economics* by A. L. Ostrosky, Jr., and J. V. Koch (Prospect Heights, IL: Waveland Press, 1979).

54. ***Employment*** An employment research company estimates that the value of a recent MBA graduate to an accounting company is

$$V = 3e^2 + 5g^3$$

where V is the value of the graduate, e is the number of years of prior business experience, and g is the graduate school grade point average. If $V = 200$, find de/dg when $g = 3.0$ and interpret the result.

55. ***Grades*** A production formula for a student's performance on a difficult English examination is

$$g = 4tx - 0.2t^2 - 10x^2 \qquad (t < 30)$$

where g is the grade the student can expect to obtain, t is the number of hours of study for the examination, and x is the student's grade point average.

a. For how long should a student with a 3.0 grade point average study to score 80 on the examination?
b. Find dt/dx for a student who earns a score of 80, evaluate it when $x = 3.0$, and interpret the result.
Source: See the source in Exercise 53.

56. ***Grades*** Repeat Exercise 55 using the following production formula for a basket-weaving examination:

$$g = 10tx - 0.2t^2 - 10x^2 \qquad (t < 10)$$

Comment on the result.

57. ***Productivity*** The number p of CDs the Snappy Hardware Co. can manufacture at its plant in one day is given by

$$P = x^{0.6}y^{0.4}$$

where x is the number of workers at the plant and y is the annual expenditure at the plant (in dollars). Compute $\left.\dfrac{dy}{dx}\right|_{x=100}$ at a production level of 20,000 CDs per day and interpret the result.

58. ***Productivity*** Repeat Exercise 57 using a productivity formula of

$$P = x^{0.5}y^{0.5}$$

Exercises 59 and 60 are based on the following demand function for money (taken from a question on the GRE economics test):

$$M_d = (2) \times (y)^{0.6} \times (r)^{-0.3} \times (p)$$

where M_d = demand for nominal money balances (money stock)
y = real income
r = an index of interest rates
p = an index of prices

59. ***Money Stock*** If real income grows while the money stock and the price level remain constant, the interest rate must change at what rate? (Find first dr/dy and then dr/dt; your answers will be expressed in terms of r and y.)

60. ***Money Stock*** If real income grows while the money stock and the interest rate remain constant, the price level must change at what rate?

Communication and Reasoning Exercises

61. Use logarithmic differentiation to give another proof of the product rule.

62. Use logarithmic differentiation to give a proof of the quotient rule.

63. If y is given explicitly as a function of x by an equation $y = f(x)$, compare finding dy/dx by implicit differentiation to finding it explicitly in the usual way.

64. Explain why one should not expect dy/dx to be a function of x if y is not a function of x.

65. True or false? If y is a function of x and $dy/dx \neq 0$ at some point, then, regarding x as an implicit function of y, we have

$$\frac{dx}{dy} = \frac{1}{dy/dx}$$

Explain your answer.

66. If you are given an equation in x and y such that dy/dx is a function of x only, what can you say about the graph of the equation?

You're the Expert

Projecting Market Growth

You are on the board of directors at Fullcourt Academic Press, and TJM, the sales director of the high school division, has just burst into your office with data showing the number of high school graduates over the past decade (Figure 13).

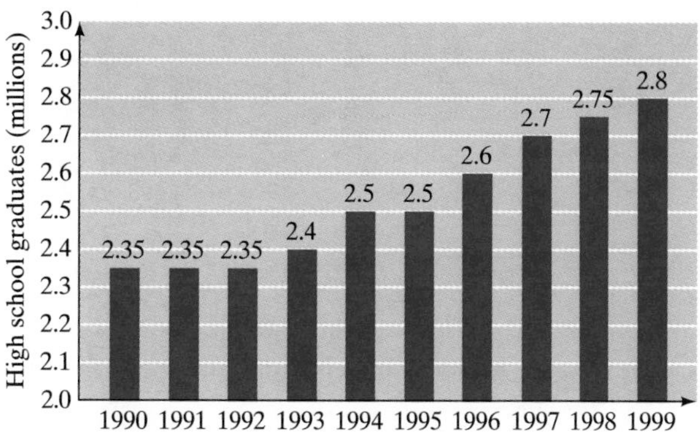

Figure 13

Source: Western Interstate Commission for Higher Education and the College Board/*New York Times,* February 17, 1999, p. B9. Data are rounded.

TJM is pleased that the figures appear to support a basic premise of his recent proposal for a major expansion strategy: that the number of high school seniors in the United States will be growing at a rate of at least 20,000 per year through the year 2005. The rate of increase, as he points out, has far exceeded 20,000 per year since 1993, so it is not overly optimistic to assume that the trend will continue—at least for the next five years.

Although you are tempted to support TJM's proposal at the next board meeting, you would like to estimate first whether the 20,000 figure is a realistic expectation, especially because the graph suggests that the number of graduates began to "level off" (in the language of calculus, the *derivative appears to be decreasing*) during the second half of the decade. Moreover, you recall reading somewhere that the numbers of students in the lower grades have also begun to level off, so it is safe to predict that the slowing of growth in the senior class will continue over the next few years. You really need precise data about numbers in the lower grades to make a meaningful prediction, but TJM's report is scheduled to be presented tomorrow and you would like a quick and easy way of "extending the curve to the right" by then.

It would certainly be helpful if you had a mathematical model of the data in Figure 13 that you could use to project the current trend. But what kind of model should you use? A linear model would be no good because it would not show any change in the derivative (the derivative of a linear function is constant). In addition, best-fit polynomial and exponential functions do not accurately reflect the leveling off, as you realize after trying to fit a few of them (Figure 14). Although logistic curves do level off, the best-fit logistic model for the given data turns out to be almost identical to the best-fit exponential curve shown in the figure, and it begins to level off only around 2025. So you look for an alternative model.

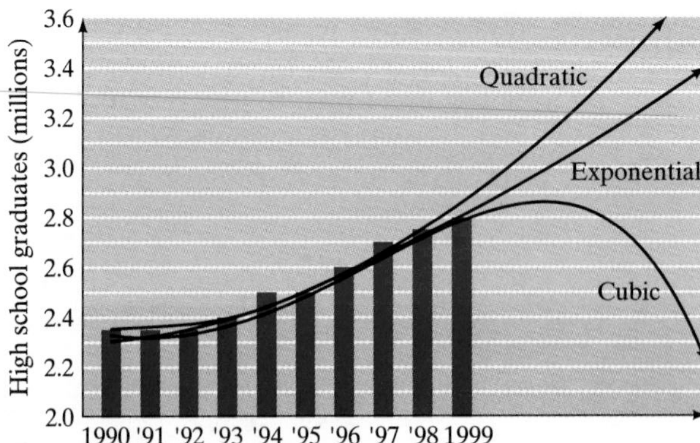

Figure 14

Flipping through a calculus book, you stumble across a function whose graph looks similar to the one you have (Figure 15).

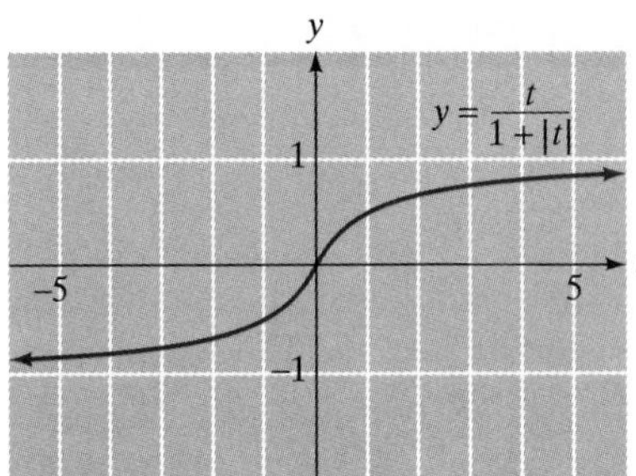

Figure 15

Curves of this form are sometimes called "predator satiation" curves, and they are used to model the population of predators in an environment with limited prey. Although the curve does seem to have the proper shape, you realize that you will need to shift and scale the function in order to fit the actual data. The most general scaled and shifted version of this curve has the form

$$y = c + b\frac{a(t-m)}{1 + a|t-m|}; \qquad a, b, c, m \text{ constant}$$

and you decide to try a model of this form,[1] where y will represent the number of high school seniors (in millions) and t will represent years since 1990.

For the moment, you postpone the question of finding the best values for the constants a, b, c, and m, and you decide first to calculate its derivative in terms of the given constants. The derivative, dy/dt, will represent the rate of increase of high school graduates, which is exactly what you wish to estimate.

[1]The scaled and shifted function is obtained from the original $t/(1 + |t|)$ by scaling by a factor of b in the y direction and a factor of a in the t direction, and then shifting m units in the t direction and c units in the y direction. For a detailed on-line treatment of scaled and shifted functions, follow the path Web Site → Everything for Calculus → Chapter 1 → New Functions from Old: Scaled and Shifted Functions.

The main part of the function is a quotient, so you start with the quotient rule:

$$\frac{dy}{dt} = b\,\frac{a(1 + a|t - m|) - a(t - m)\dfrac{d}{dt}(1 + a|t - m|)}{(1 + a|t - m|)^2}$$

where you suddenly hit a snag: What is the derivative of the absolute value of a quantity? Something your calculus instructor said about the absolute value function then comes back to you:

$$\frac{d}{dt}|t| = \frac{|t|}{t} = \begin{cases} 1 & \text{if } t > 0 \\ -1 & \text{if } t < 0 \end{cases}$$

which, you are disturbed to notice, is not defined at $t = 0$. Undaunted, you make a mental note and press on. Since you need the derivative of the absolute value of a quantity other than t, you use the chain rule, which tells you that

$$\frac{d}{dt}|u| = \frac{|u|}{u}\frac{du}{dt}$$

Thus,

$$\frac{d}{dt}|t - m| = \frac{|t - m|}{t - m} \cdot 1 \qquad u = t - m;\ m = \text{constant}$$

(This is not defined when $t = m$.) Substituting into the formula for dy/dt, you find

$$\begin{aligned}\frac{dy}{dt} &= b\,\frac{a(1 + a|t - m|) - a(t - m) \cdot a\dfrac{|t - m|}{(t - m)}}{(1 + a|t - m|)^2} \\ &= b\,\frac{a(1 + a|t - m|) - a^2|t - m|}{(1 + a|t - m|)^2} && \text{Cancel } (t - m). \\ &= \frac{ab}{(1 + a|t - m|)^2} && a^2|t - m| - a^2|t - m| = 0\end{aligned}$$

It is interesting that, although the derivative of $|t - m|$ is not defined when $t = m$, we see that the offending term $t - m$ was canceled, so that dy/dt seems to be defined[2] at $t = m$.

Now you have a simple-looking expression for dy/dt, which will give you an estimate of the rate of change of the high school senior population. However, you still need values for the constants a, b, c, and m. (You don't really need the value of c to compute the derivative—where has it gone?—but it is a part of the model.) How do you find the values of a, b, c, and m that result in the curve that best fits the given data?

Turning once again to your calculus book, you see that a best-fit curve is one that minimizes the sum of squares of the differences between the actual values for y and the ones predicted by the model. Here is an Excel spreadsheet showing these errors for $a = 2$, $b = 0.5$, $c = 2.5$, and $m = 6$:

[2] In fact, it is defined and has the value given by the formula just derived: ab. To show this takes a bit more work. How might you do it?

	A	B	C	D	E	F
1	T	Y	Y Predicted	Square Error	Constants	
2	0	2.35	2.03846154	0.09705621	a	2
3	1	2.35	2.04545455	0.09274793	b	0.5
4	2	2.35	2.05555556	0.08669753	c	2.5
5	3	2.4	2.07142857	0.10795918	m	6
6	4	2.5	2.1	0.16		
7	5	2.5	2.16666667	0.11111111		
8	6	2.6	2.5	0.01		
9	7	2.7	2.83333333	0.01777778		
10	8	2.75	2.9	0.0225		
11	9	2.8	2.92857143	0.01653061		
12				0.72238036	SSE	

The first two columns show the given data (T = year and Y = number of high school graduates in millions). The formula for Y-Predicted is the model

$$y = c + b\frac{a(t-m)}{1 + a|t-m|}$$

formatted in cell C2 as

```
=$F$4+$F$3*$F$2*(T-$F$5)/(1+$F$2*ABS(T-$F$5))
  c  +  b  * a  * (t - m) / (1 + a  *  |t - m|)
```

and then copied into the cells below it. Since the square error is defined as the square of the difference between Y and Y-Predicted, we enter

```
=(C2-B2)^2
```

in cell D2 and then copy into the cells below it. The sum-of-squares error, SSE (sum of the entries in D2–D11), is then placed in cell D12.

The values of *a, b, c,* and *d* shown are initial values and don't matter too much (but see below); you will have Excel change these values to improve your model. The smaller the SSE is, the better your model. (For a perfect fit, the Y-column would equal the Y-Predicted column and SSE would be zero.) Hence, the goal is now to find values of *a, b, c,* and *m* that make the value of SSE as small as possible. (Compare the section on linear regression in Chapter 1. Finding these values analytically is an extremely difficult mathematical problem. However, there is software, such as Excel's built-in "Solver" routine,[3] that can be used to find *numerical* solutions.

Figure 16 shows how to set up Solver to find the best values for *a, b, c,* and *m* for the setup used in this spreadsheet.

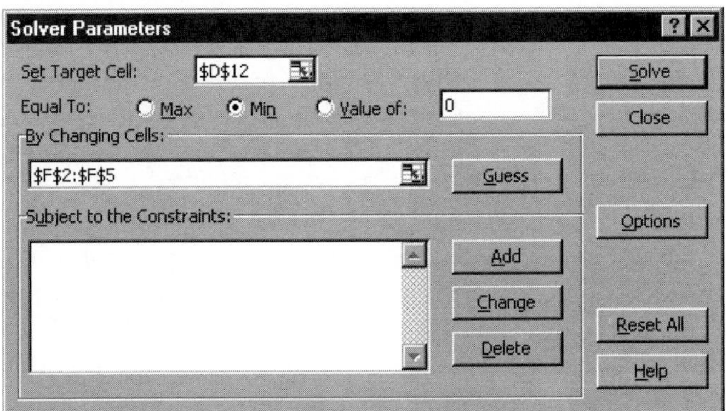

Figure 16

[3]If "Solver" does not appear in the Tools menu, you should first install it using your Excel installation software. (Solver is one of the "Excel Add-Ins.")

The Target Cell, D12, contains the value of SSE, which is to be minimized. The Changing Cells are the cells containing the values of the constants a, b, c, and m that we want to change. That's it. Now press "Solve." After thinking about it for a few seconds, Excel gives the optimal values of a, b, c, and m in cells F2–F5 and the minimum value of SSE in cell D12. You find

$$a = 0.23227749, \qquad b = 0.47131905, \qquad c = 2.60798717, \qquad m = 6.05095439$$

Figure 17 shows that this choice of model and constants gives an excellent fit (with a sum-of-squares error, SSE, of only 0.0037).

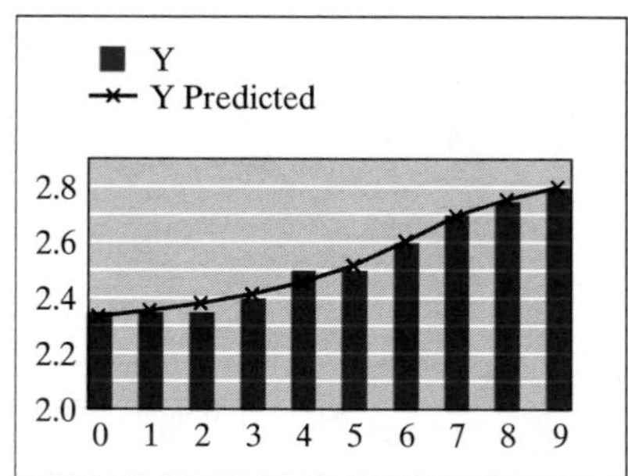

Figure 17

Figure 18 shows how the model predicts the long-term leveling-off phenomenon you were looking for.

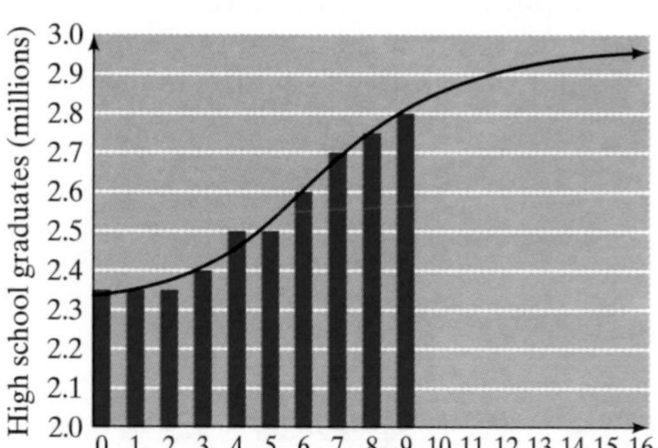

Figure 18

You now turn back to the problem at hand: projecting the rate of increase of the number of high school graduates in 2005. You have the formula

$$\frac{dy}{dt} = \frac{ab}{(1 + a|t - m|)^2}$$

and also values for the constants. So you compute.

$$\frac{dy}{dt} = \frac{(0.23227749)(0.47131905)}{(1 + 0.23227749[15 - 6.05095439])^2} \qquad t = 15 \text{ in } 2005$$

$$\approx 0.01155 \text{ million students per year}$$

or 11,550 students per year—far less than the optimistic estimate of 20,000 in the proposal!

You now conclude that TJM's prediction is suspect and that further research will have to be done before the board can support the proposal.

Question Which values of the constants should I use as starting values when using Excel to find the best-fit curve?

Answer If the starting values of the constants are far from the optimal values, Solver may find a nonoptimal solution. Thus, you need to obtain some rough initial estimate of the constants by examining the graph. Figure 19 shows some important features of the curve that you can use to obtain estimates of a, b, c, and m by inspecting the graph.

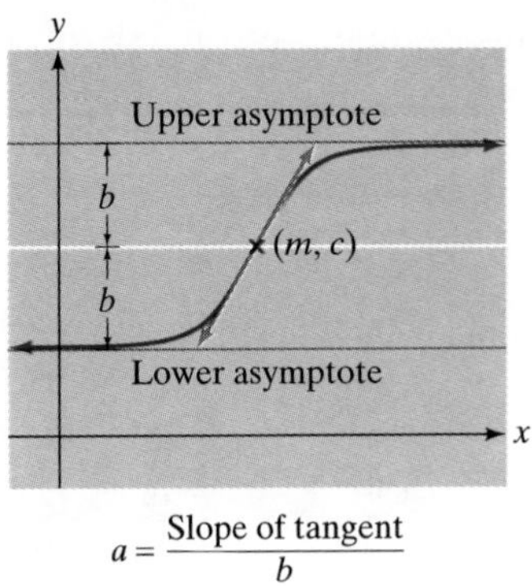

Figure 19

From the graph, m and c are the coordinates of the point on the curve where it is steepest, and b is the vertical distance from that point to the upper or lower asymptote (where the curve "levels off"). To estimate a, first estimate the slope of the tangent at the point of steepest inclination, then divide by b. If b is negative (and a is positive), we obtain an "upside-down" version of the curve (Figure 20).

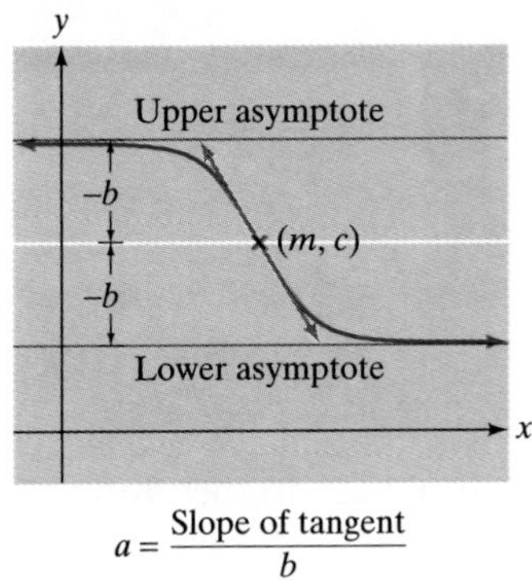

Figure 20

exercises

1. What does the best-fit model predict the high school senior population will be in 2005? (Answer to the nearest 0.1 million.)
2. What is the long-term prediction of the model?
3. Find $\lim_{t\to\infty}(dy/dt)$ and interpret the result.
4. You receive a memo to the effect that the 1994 and 1995 figures are not accurate. Use Excel Solver to re-estimate the best-fit constants a, b, c, and m in the absence of this data, and obtain new estimates for the 1994 and 1995 data. What does the new model predict the rate of change in the number of high school seniors will be in 2005?

5. *Logistic Model* Using the original data, find the best-fit shifted logistic curve of the form

$$f(t) = c + \frac{2b}{1 + e^{-a(t-m)}}$$

What does this model predict will be the growth rate of the number of high school graduates in 2005? (Start with the following values: $c = 2, b = 0.3, a = 1, m = 6$, and round your answers to four decimal places.)

6. *Demand for Freon* The demand for chloroflurocarbon-12 (CFC-12)—the ozone-depleting refrigerant commonly known as freon[4]—has been declining significantly in response to regulation and concern about the ozone layer. The chart below shows the projected demand for CFC-12 for the period 1994–2005:[5]

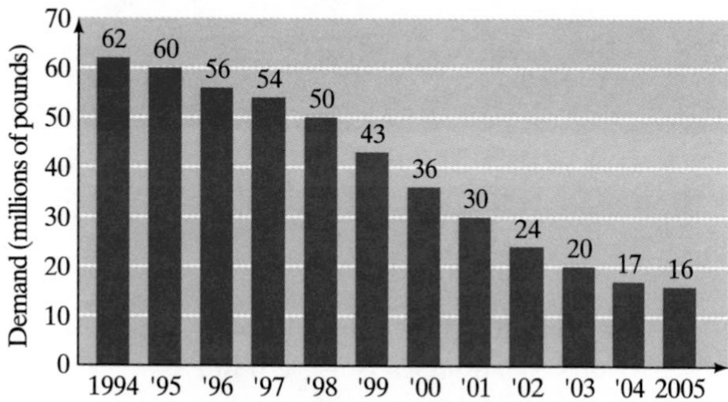

Source: The Automobile Consulting Group/*New York Times,* December 26, 1993, p. F23.

a. Use Excel Solver to obtain the best-fit equation of the form

$$f(t) = c + b\frac{a(t-m)}{1 + a|t-m|}$$

where t = years since 1990. Use your function to estimate the total demand for CFC-12 from the start of the year 2000 to the start of 2010. (Start with the following values: $a = 1$, $b = -25$, $c = 35$, and $m = 10$, and round your answers to four decimal places.)

b. According to your model, how fast is the demand for freon declining in 2000?

[4]The name given to it by Du Pont.

[5]The exact figures were not given, and the chart is a reasonable facsimile of the chart that appeared in the *New York Times.*

chapter 4 review test

1. Find the derivative of each function.

a. $f(x) = e^x(x^2 - 1)$

b. $f(x) = \dfrac{x^2 + 1}{x^2 - 1}$

c. $f(x) = (x^2 - 1)^{10}$

d. $f(x) = \dfrac{1}{(x^2 - 1)^{10}}$

e. $f(x) = e^x(x^2 + 1)^{10}$

f. $f(x) = \left(\dfrac{x - 1}{3x + 1}\right)^3$

g. $f(x) = e^{x^2-1}$

h. $f(x) = (x^2 + 1)e^{x^2-1}$

i. $f(x) = \ln(x^2 - 1)$

j. $f(x) = \dfrac{\ln(x^2 - 1)}{x^2 - 1}$

k. $f(x) = \cos(x^2 - 1)$

l. $f(x) = \sin(x^2 + 1)\cos(x^2 - 1)$

2. Find dy/dx for each of the following equations.

a. $x^2 - y^2 = x$

b. $2xy + y^2 = y$

c. $e^{xy} + xy = 1$

d. $\ln\left(\dfrac{y}{x}\right) = y$

e. $\sin y = x^2 + y^2$

f. $x \cos y = 1$

3. Find all values of x (if any) where the tangent line to the graph of the given equation is horizontal.

a. $y = x - e^{2x-1}$

b. $y = e^{x^2}$

c. $y = \dfrac{x}{x + 1}$

d. $y = \sqrt{x}\,(x - 1)$

Applications

4. At the moment, OHaganBooks.com is selling 1000 books per week and its sales are increasing at a rate of 200 books per week. Also, it is now selling all its books for \$20 each, but its price is dropping at a rate of \$1 per week.

a. At what rate is OHaganBooks.com's revenue rising or falling?

b. John O'Hagan would have preferred to see the company's revenue increase at a rate of \$5000 per week. At what rate would sales have to have been increasing to accomplish that goal, assuming all the other information is as given?

c. The percentage rate of change of a quantity Q is Q'/Q. Why is the percentage rate of change of revenue always equal to the sum of the percentage rates of change of unit price and weekly sales?

5. At the beginning of last week, OHaganBooks.com stock was selling for \$100 per share, rising at a rate of \$50 per year. Its earnings amounted to \$1 per share, rising at a rate of \$0.10 per year.

a. At what rate was its price-to-earnings (P/E) ratio, the ratio of its stock price to its earnings per share, rising or falling?

b. Curt Hinrichs, who recently invested in OHaganBooks.com stock, would have liked to see the P/E ratio increase at a rate of 100 points per year. How fast would the stock have to have been rising, assuming all the other information is as given?

c. The percentage rate of change of a quantity Q is Q'/Q. Why is the percentage rate of change of P/E always equal to the percentage rate of change of unit price minus the percentage rate of change of earnings?

6. OHaganBooks.com modeled its weekly sales over a period of time with the function

$$s(t) = 6053 + \frac{4474}{1 + e^{-0.55(t-4.8)}}$$

as shown in the following graph:

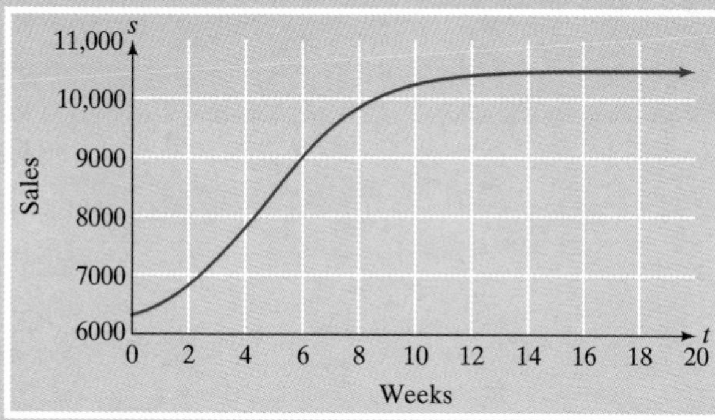

a. Compute $s'(t)$.
b. Use the answer to part (a) to compute the rate of increase of weekly sales at the beginning of the 7th week ($t = 6$). (Round your answer to the nearest unit.)
c. Find the rate of increase of weekly sales at the beginning of the 15th week ($t = 14$).

7. The number of "hits" on OHaganBooks.com's web site was 1000 per day at the beginning of the year, growing at a rate of 5% per week. If this growth rate continued for the whole year (52 weeks), find the rate of increase (in hits per day per week) at the end of the year.

8. The price p that OHaganBooks.com charges for its latest leather-bound gift edition of *Lord of the Rings* is related to the demand q in weekly sales by the equation

$$100pq + q^2 = 5{,}000{,}000$$

Suppose that the price is set at \$40, which would make the demand 1000 copies per week.
a. Using implicit differentiation, compute the rate of change of demand with respect to price, and interpret the result. (Round the answer to two decimal places.)
b. Use the result of part (a) to compute the rate of change of revenue with respect to price. Should the price be raised or lowered to increase revenue?

Additional On-Line Review

If you follow the path

Web site → Everything for Calculus → Chapter 4

you will find the following additional resources to help you review:

- A comprehensive chapter summary (including examples and interactive features)
- Additional review exercises (including interactive exercises and many with help)
- A true–false chapter quiz
- Interactive tutorials

You're the Expert

Production Lot Size Management

Your publishing company is planning the production of its latest bestseller, which it predicts will sell 100,000 copies per month over the coming year. The book will be printed in several batches of the same number, evenly spaced throughout the year. Each print run has a setup cost of \$5000, a single book costs \$1 to produce, and monthly storage costs for books awaiting shipment average 1¢ per book. To meet the anticipated demand at minimum total cost to your company, how many print runs should you plan?

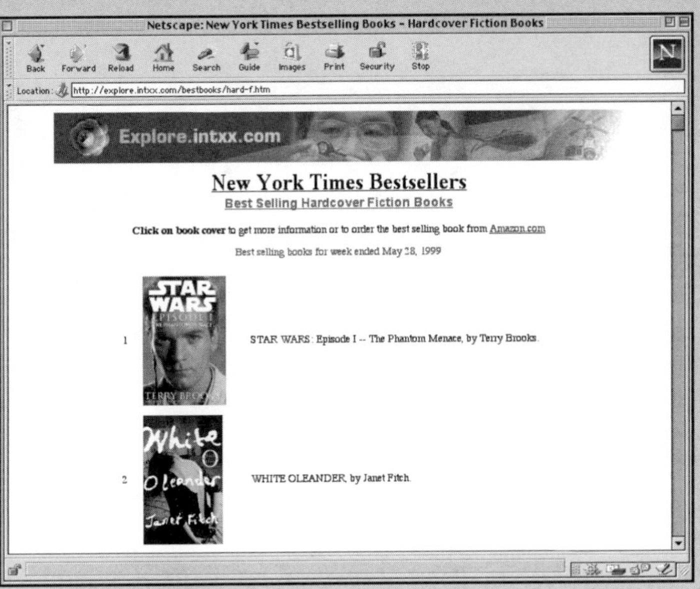

Source: www.explore.intxx. Copyright © 2000 by Integrated Technology, Inc. Reprinted by permission.

Internet Resources for This Chapter

At the web site, follow the path

Web site → Everything for Calculus → Chapter 5

where you will find links to step-by-step tutorials for the main topics in this chapter, a detailed chapter summary you can print out, a true–false quiz, and a collection of sample test questions. You will also find downloadable Excel tutorials for each section, an on-line grapher, and other resources.

Applications of the Derivative

chapter 5

5.1 Maxima and Minima
5.2 Applications of Maxima and Minima
5.3 The Second Derivative and Analyzing Graphs
5.4 Related Rates
5.5 Elasticity of Demand

Introduction

In this chapter we begin to see the power of calculus as an optimization tool. In Chapter 2 we saw how to price an item to get the largest revenue when the demand function is linear. Using calculus, we can handle much more general, nonlinear functions. In the first section of this chapter we show how calculus can be used to solve the problem of finding the values of a variable that lead to a maximum or minimum value of a given function. In the second section we show how this helps us in real-world applications.

Another theme in this chapter is that calculus can help us to understand the graph of a function. By the time you have completed the material in the first section, you will be able to locate and sketch some of the important features of a graph. In the third section we discuss further how to explain what you see in a graph (drawn, for example, using graphing technology) and to locate its most important points.

We also include sections on related rates and elasticity of demand. The first of these examines further the concept of the derivative as a rate of change. The second returns to the problem of optimizing revenue based on the demand equation, looking at it in a new way that leads to an important idea in economics—elasticity.

5.1 Maxima and Minima

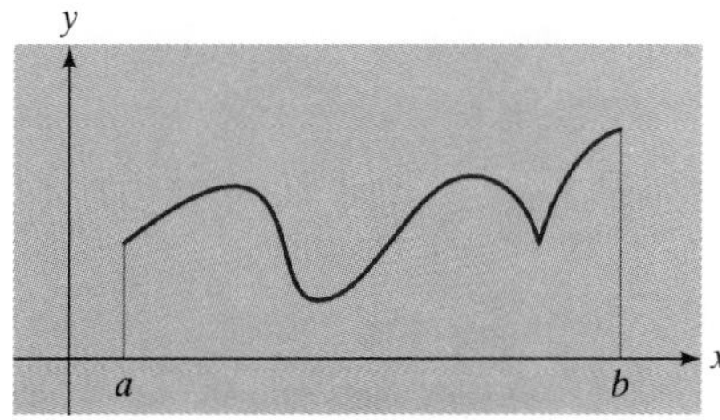

Figure 1

Figure 1 shows the graph of a function f whose domain is the closed interval $[a, b]$. A mathematician sees lots of interesting things going on here. There are hills and valleys and even a small chasm (called a "cusp") toward the right. For many purposes, the important features of this curve are the highs and lows.

Suppose, for example, you know that the price of the stock of a certain company will follow this graph during the course of a week. Although you would certainly make a handsome profit if you bought at time a and sold at time b, your best strategy would be to follow the old adage to "buy low and sell high," buying at all the lows and selling at all the highs.

Figure 2 shows the graph once again with the highs and lows marked.

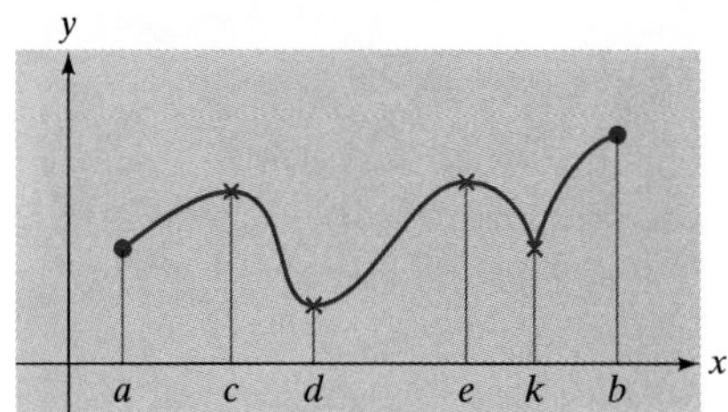

Figure 2

Mathematicians have names for these points: The highs (at the x values c, e, and b) are referred to as **relative maxima,** and the lows (at the x values a, d, and k) are referred to as **relative minima.** Collectively, these highs and lows are referred to as **relative extrema.** (A point of language: the singular forms of the plurals *minima, maxima,* and *extrema* are *minimum, maximum,* and *extremum.*)

Why do we refer to these points as "relative" extrema? Take a look at the point corresponding to $x = c$. It is the highest point of the graph *compared to other points nearby.* If you were an extremely nearsighted mountaineer standing at point c, you would *think* that you were at the highest point of the graph, not being able to see the distant peaks at $x = e$ and $x = b$.

Let's translate this idea into mathematical terms. We are talking about the heights of various points on the curve. The height of the curve at $x = c$ is $f(c)$, so we are saying that $f(c)$ is greater than $f(x)$ for every x near c. For instance, *$f(c)$ is the greatest value that $f(x)$ has for all choices of x between a and d* (see Figure 3).

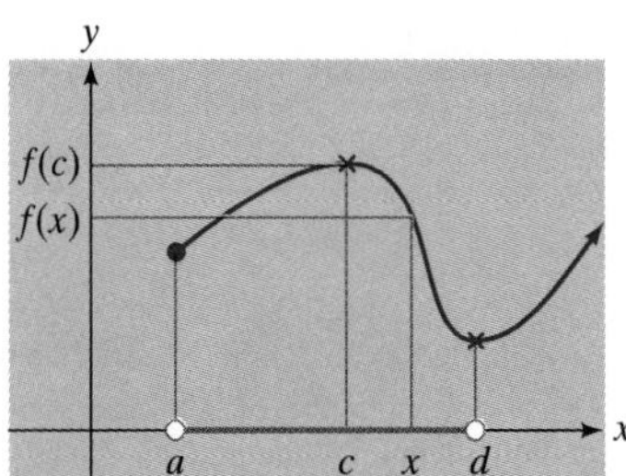

Figure 3

We can phrase the formal definition as follows.

Relative Extremum

f has a **relative maximum** at c if there is some interval (r, s) (even a very small one) containing c for which $f(c) \geq f(x)$ for all x between r and s for which $f(x)$ is defined.

f has a **relative minimum** at c if there is an interval (r, s) (even a very small one) containing c for which $f(c) \le f(x)$ for all x between r and s for which $f(x)$ is defined.

Relative maxima and minima are sometimes also referred to as **local** maxima and minima.

Quick Examples

In Figure 2, f has the following relative extrema:

1. A relative maximum at c, as shown by the interval (a, d).
2. A relative maximum at e, as shown by the interval (d, k).
3. A relative maximum at b, as shown by the interval $(k, b + 1)$. Note that $f(x)$ is not defined for $x > b$. However, $f(b) \ge f(x)$ for every x in the interval $(k, b + 1)$ *for which $f(x)$ is defined*—that is, for every x in $(k, b]$.
4. A relative minimum at d, as shown by the interval (c, e).
5. A relative minimum at k, as shown by the interval (e, b).
6. A relative minimum at a, as shown by the interval $(a - 1, c)$. See 3 above.

Note

Our definition of relative extremum allows f to have a relative extremum at an endpoint of its domain; the definitions used in some books do not. In view of examples like our stock market investing strategy, we see no good reason not to count endpoints as extrema.

Looking carefully at Figure 2, we can see that the lowest point on the whole graph is where $x = d$ and the highest point is where $x = b$. This means that $f(d)$ is the smallest value of f on the whole domain of f (the interval $[a, b]$), and $f(b)$ is the largest value. We call these the *absolute* minimum and maximum.

Absolute Extremum

f has an **absolute maximum** at c if $f(c) \ge f(x)$ for every x in the domain of f.
f has an **absolute minimum** at c if $f(c) \le f(x)$ for every x in the domain of f.

Absolute maxima and minima are sometimes also referred to as **global** maxima and minima.

Quick Examples

1. In Figure 2, f has an absolute maximum at b and an absolute minimum at d.
2. If $f(x) = x^2$, then $f(x) \ge f(0)$ for every real number x. Therefore, $f(x) = x^2$ has an absolute minimum at $x = 0$.

Question Is there only one absolute maximum of f?
Answer Although there is only one absolute maximum *value* of f, this value may occur at many different values of x. An extreme case is a constant function; since we use $\ge$ in the definition, a constant function has an absolute maximum (and minimum) at every point in its domain.

Now, how do we go about locating extrema? In many cases we can get a good idea by using graphing technology to zoom in on a maximum or minimum and approximate its coordinates. However, calculus gives us a way to find the exact locations of the extrema and at the same time to understand why the graph of a function behaves the way it does. In fact, it is often best to combine the powers of graphing technology with those of calculus, as we shall see.

In Figure 4 we see the graph from Figure 1 once more, but now we have labeled each extreme point as one of three types.

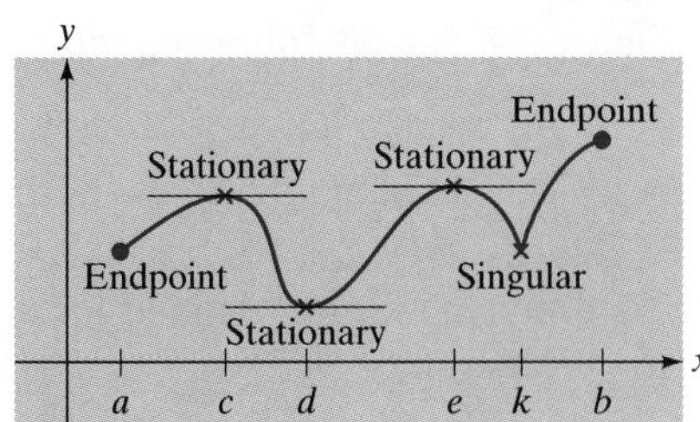

Figure 4

At the points labeled "stationary," the tangent lines to the graph are horizontal and so have slope zero. Remember that the slope of the tangent line is given by the derivative. Thus, the derivative of f is zero at each of the stationary points. In other words,

$$f'(c) = 0 \qquad f'(d) = 0, \qquad f'(e) = 0$$

To find the exact location of each of these extrema, we solve the equation $f'(x) = 0$. Any time $f'(x) = 0$ we say that f has a **stationary point** at x because the rate of change of f is zero there. We call an *extremum* that occurs at a stationary point a **stationary extremum.**

There is a relative minimum in Figure 4 at $x = k$, but there is no horizontal tangent there. In fact, there is no tangent line at all; $f'(k)$ is not defined. (Recall a similar situation with the graph of $f(x) = |x|$ at $x = 0$.) When $f'(x)$ does not exist, we say that f has a **singular point** at x. We call an extremum that occurs at a singular point a **singular extremum.** The points that are either stationary or singular we call collectively the **critical points** of f.

The remaining two extrema are at the **endpoints** of the domain.[1] As we see in the figure, they are almost always either relative maxima or relative minima.

Question Are there any other types of relative extrema?
Answer The answer is a *qualified no.* One can prove that if the function we are looking at is differentiable at every point of its domain except for a few isolated points, then the relative extrema all occur at critical points or endpoints (a rigorous proof is beyond the scope of this book).[2]

---

[1] Remember that we do allow local extrema at endpoints.

[2] Here is an outline of the argument. Suppose f has a local maximum, say, at $x = a$, at a point other than an endpoint of the domain. Then either f is differentiable there or it is not. If it is not, then we have a singular point. If f is differentiable at $x = a$, then consider the slope of the secant line through the points where $x = a$ and $x = a + h$ for small positive h. Since f has a local maximum at $x = a$, it is falling (or level) to the right of $x = a$, and so the slope of this secant line must be ≤ 0. Thus, we must have $f'(a) \leq 0$ in the limit as $h \to 0$. On the other hand, if h is small and *negative,* then the corresponding secant line must have slope ≥ 0 because f is also falling (or level) to the left of $x = a$, and so $f'(a) \geq 0$. Since $f'(a)$ is both ≥ 0 and ≤ 0, it must be zero, and so we have a stationary point at $x = a$.

Locating Candidates for Relative Extrema

If f is differentiable on its domain except at a few isolated points, then its relative extrema occur among the following types of points.

1. *Stationary points:* f has a stationary point at x if x is in the domain and $f'(x) = 0$. To locate stationary points, set $f'(x) = 0$ and solve for x.
2. *Singular points:* f has a singular point at x if x is in the domain and $f'(x)$ is not defined. To locate singular points, find values of x where $f'(x)$ is *not* defined but $f(x)$ *is* defined.
3. *Endpoints:* The x-coordinates of endpoints are the endpoints of the domain, if any. Recall that closed intervals contain endpoints, but open intervals do not.

Once we have the x-coordinates of a candidate for a relative extremum, we find the corresponding y-coordinate using $y = f(x)$.

Quick Examples

1. *Stationary points:* Let $f(x) = x^3 - 12x$. To locate the stationary points, set $f'(x) = 0$ and solve for x. This gives $3x^2 - 12 = 0$, so f has stationary points at $x = \pm 2$. Thus, the stationary points are $(-2, f(-2)) = (-2, 16)$ and $(2, f(2)) = (2, -16)$.
2. *Singular points:* Let $f(x) = 3(x - 1)^{1/3}$. Then $f'(x) = (x - 1)^{-2/3} = 1/(x - 1)^{2/3}$. $f'(1)$ is not defined, although $f(1)$ *is* defined. Thus, the (only) singular point occurs at $x = 1$. Its coordinates are $(1, f(1)) = (1, 0)$.
3. *Endpoints:* Let $f(x) = 1/x$, with domain $(-\infty, 0) \cup [1, +\infty)$. The only endpoint in the domain of f occurs when $x = 1$, and it has coordinates $(1, 1)$. The natural domain of $1/x$ has no endpoints.

Remember, though, that these points are only *candidates* for relative extrema. It is quite possible, as we shall see, to have a stationary point (or singular point) that is neither a relative maximum nor a relative minimum.

Now let us look at some examples of finding maxima and minima. In all of these examples we use the following procedure: First, we find the derivative, which we examine to find the stationary points and singular points. Next, we make a table that lists the x-coordinates of the critical points and endpoints, together with their y-coordinates. We use this table to make a rough sketch of the graph. From the table and rough sketch we usually have enough data to be able to say where the extreme points are and what kind they are.

example 1

Maxima and Minima

Find the relative and absolute maxima and minima of $f(x) = x^2 - 2x$ on the interval $[0, 4]$.

Solution

We first calculate $f'(x) = 2x - 2$. We use this derivative to locate the stationary and singular points.

- *Stationary points:* To locate the stationary points we solve the equation $f'(x) = 0$ or

  $$2x - 2 = 0$$

  and get $x = 1$. The domain of the function is $[0, 4]$, so $x = 1$ is in the domain. Thus, the only candidate for a stationary relative extremum occurs when $x = 1$.

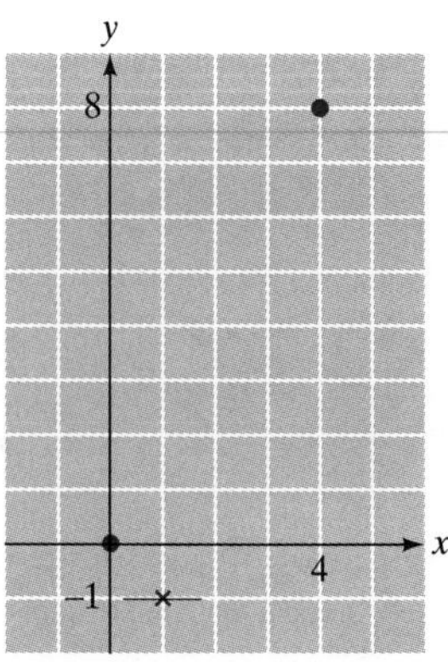

Figure 5

- *Singular Points:* We look for points where the derivative is not defined. However, the derivative is $2x - 2$, which is defined for every x. Thus, there are no singular points and hence no candidates for singular relative extrema.
- *Endpoints:* The domain is [0, 4], so the endpoints occur when $x = 0$ and $x = 4$.

We now record these values of x in a table, together with the corresponding y-coordinates (values of f):

x	0	1	4
$f(x) = x^2 - 2x$	0	−1	8

The table gives us three points on the graph, (0, 0), (1, −1), and (4, 8), which we plot in Figure 5. We remind ourselves that the point (1, −1) is a stationary point of the graph by drawing in a part of the horizontal tangent line. Connecting these points gives us a graph something like that in Figure 6. Notice that the graph has a horizontal tangent line at $x = 1$ but not at either of the endpoints because the endpoints are not stationary points.

From Figure 6 we can see that f has:

- A relative maximum at the point (0, 0)
- A relative minimum at the point (1, −1), which is also the absolute minimum on [0, 4]
- A relative maximum at the point (4, 8), which is also the absolute maximum on [0, 4]

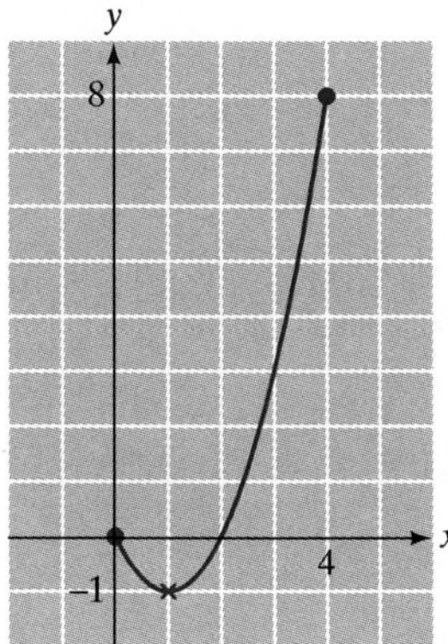

Figure 6

Before we go on . . . How can we be sure that the graph doesn't look like Figure 7? If it did, there would be another stationary point somewhere between $x = 1$ and $x = 4$. This would have shown up when we solved $f'(x) = 0$, and it didn't. The table we made listed all of the possible extrema; there can be no more.

It's also useful to consider the following: We found that $f'(1) = 0$; f has a stationary point at $x = 1$. What are the values of $f'(x)$ like to the left and right of 1? Let's make a table with some values:

x	0	1	2
$f'(x) = 2x - 2$	−2	0	2
	↘		↗

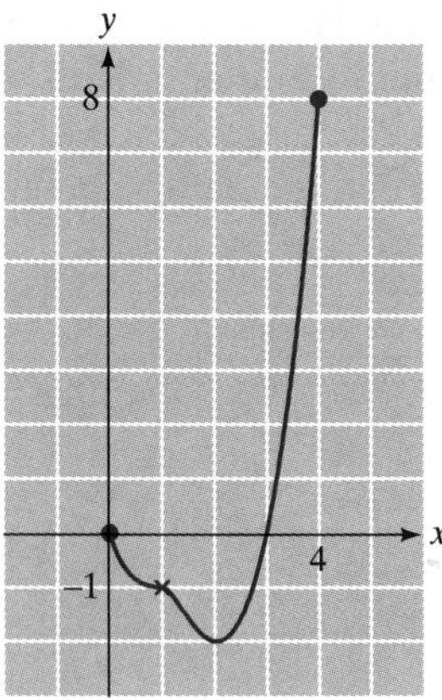

Figure 7

At $x = 0$, $f'(0) = -2 < 0$, so the graph has negative slope and f is **decreasing;** its values are going down as x increases. We note this with the downward pointing arrow in the chart. At $x = 2$, $f'(2) = 2 > 0$, so the graph has positive slope and f is **increasing;** its values are going up as x increases. In fact, since $f'(x) = 0$ only at $x = 1$, we know that $f'(x) < 0$ for all x in [0, 1), and we can say that f is decreasing on the interval [0, 1]. Similarly, f is increasing on [1, 4]. So, starting at $x = 0$, the graph of f goes down until we reach $x = 1$ and then it goes back up. This is another way of checking that the stationary point at $x = 1$ is a relative minimum. (Using the derivative in this way to check what happens at a critical point is called the **first derivative test.**)[3]

[3] The **second derivative** of a function (which we discuss later in this chapter) is the derivative of the ("first") derivative—hence the name "first derivative test."

Note Here is some terminology: If the point (a, b) is a maximum (or minumum) of f, we sometimes say that **f has a maximum (or minimum) value of b at $x = a$.** Thus, in the above example, we could have said the following:

- f has a relative maximum value of 0 at $x = 0$.
- f has a relative minimum value of -1 at $x = 1$.
- f has a relative maximum value of 8 at $x = 4$.

example 2

Unbounded Interval

Find all extrema of $f(x) = 3x^4 - 4x^3$ on $[-1, \infty)$.

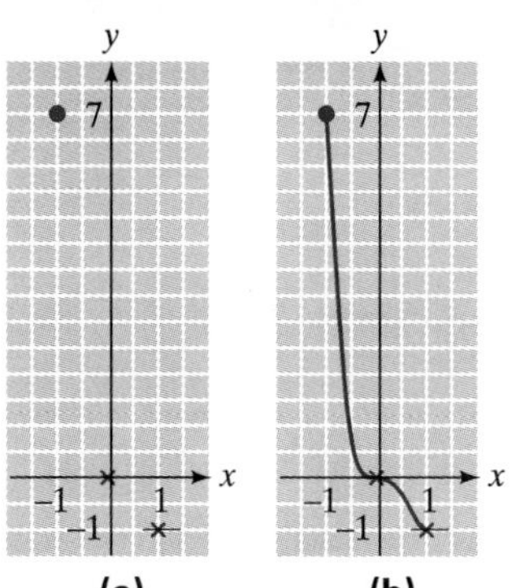

(a) (b)

Figure 8

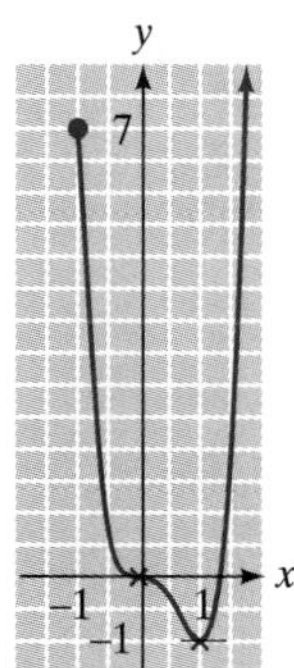

Figure 9

Solution

We first calculate $f'(x) = 12x^3 - 12x^2$.

- *Stationary points:* We solve the equation $f'(x) = 0$, which is

$$12x^3 - 12x^2 = 0$$

or

$$12x^2(x - 1) = 0$$

There are two solutions, $x = 0$ and $x = 1$, and both are in the domain. These are our candidates for the x-coordinates of stationary relative extrema.

- *Singular points:* There are no points where $f'(x)$ is not defined, so there are no singular points.
- *Endpoints:* The domain is $[-1, \infty)$, so there is one endpoint, at $x = -1$.

We record these points in a table with the corresponding y-coordinates:

x	-1	0	1
$f(x) = 3x^4 - 4x^3$	7	0	-1

We can either plot these points and sketch the graph by hand, or we can turn to technology to help us.

If we plot these points by hand, we obtain Figure 8(a), which suggests Figure 8(b). We can't be sure what happens to the right of $x = 1$. Does the curve go up, or does it go down? To find out, let us plot a "test point" to the right of $x = 1$. Choosing $x = 2$, we obtain $y = 3(2)^4 - 4(2)^3 = 16$, so $(2, 16)$ is another point on the graph. It must turn upward to the right of $x = 1$ as shown in Figure 9.

Technology

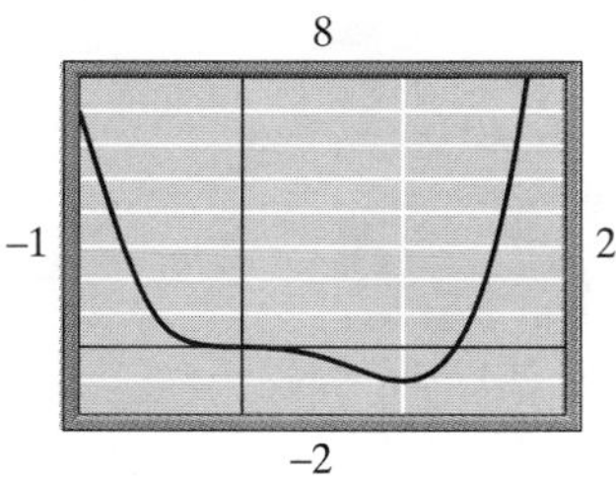

Figure 10

If we use a graphing calculator, we need to choose the viewing window ourselves, and we should choose it so that it contains the three interesting points we've found so far. Again, we can't be sure yet what happens to the right of $x = 1$; does the graph go up or down from that point? If we set the viewing window to an interval of $[-1, 2]$ for x and $[-2, 8]$ for y, we leave enough room to the right of $x = 1$ and below $y = -1$ to see what the graph will do. The result is something like Figure 10. Now we can tell what happens to the right of $x = 1$: The function increases. We know that it cannot later decrease again because if it did, there would have to be another critical point where it turns around, and we found that there are no other critical points.

From this graph we can see that f has:

- A relative maximum at the endpoint $(-1, 7)$
- A relative and absolute minimum at the point $(1, -1)$
- No absolute maximum

Before we go on . . . Notice that the stationary point at $x = 0$ is neither a relative maximum nor a relative minimum. It is simply a place where the graph of f flattens out for a moment before it continues to fall.

example 3

Singular Point
Find all extrema of $f(t) = t^{2/3}$ on $[-1, 1]$.

Solution
First, $f'(t) = \frac{2}{3}t^{-1/3}$.

- *Stationary points:* We need to solve

$$\frac{2}{3}t^{-1/3} = 0$$

We can rewrite this equation without the negative exponent:

$$\frac{2}{3t^{1/3}} = 0$$

Now, the only way that a fraction can equal 0 is if the numerator is 0. But the numerator here is 2, so the fraction can never equal 0. Thus, there are no stationary points.

- *Singular points:* Are there any points where

$$f'(t) = \frac{2}{3t^{1/3}}$$

is not defined? Yes, where $t = 0$. Notice that f itself is defined at $t = 0$, so 0 is in the domain. Thus, f has a singular point at $t = 0$.

- *Endpoints:* There are two endpoints, -1 and 1.

We now put these three points in a table with the corresponding y-coordinates:

t	−1	0	1
$f(t)$	1	0	1

Technology

1
−1
1
0

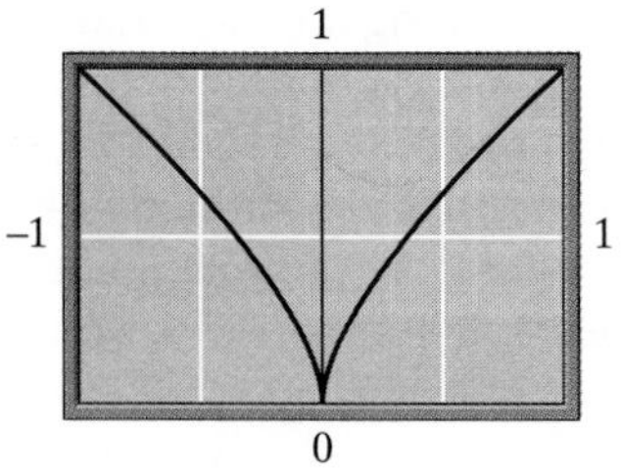

Figure 11

Since there is only one critical point, at $t = 0$, it is clear from this table that f must decrease from $t = -1$ to $t = 0$ and then increase to $t = 1$. To graph f using technology, we choose a viewing window with an interval of $[-1, 1]$ for t and $[0, 1]$ for y. The result is something like Figure 11.[4]

From this graph we can see that f has:

- A relative and absolute maximum at the point $(-1, 1)$
- A relative and absolute minimum at the point $(0, 0)$
- A relative and absolute maximum at the point $(1, 1)$

Notice that the absolute maximum value of f is achieved at two values of t: $t = -1$ and $t = 1$.

[4] Many graphing calculators give only the right-hand half of the graph shown in Figure 11 because fractional powers of negative numbers are not, in general, real numbers. To obtain the whole curve, enter the formula as `Y=(x^2)^(1/3)`, a fractional power of the nonnegative function x^2.

Before we go on . . . What exactly happens at the singular point at $x = 0$? The derivative is not defined at this point. Let us use limits to investigate what happens to the derivative as we approach zero from either side:

$$\lim_{t \to 0^-} f'(t) = \lim_{t \to 0^-} \frac{2}{3t^{1/3}} = -\infty$$

$$\lim_{t \to 0^+} f'(t) = \lim_{t \to 0^+} \frac{2}{3t^{1/3}} = +\infty$$

Thus, the graph decreases very steeply as it approaches $t = 0$ from the left and then rises very steeply as it leaves to the right.

In Examples 1 and 3 we could have found the absolute maxima and minima without doing any graphing. In Example 1, after finding the critical points and endpoints, we created the following table:

x	0	1	4
$f(x)$	0	−1	8

From this table we can see that f must decrease from its value of 0 at $x = 0$ to -1 at $x = 1$ and then increase to 8 at $x = 4$. The value of 8 must be the largest value it takes on, and the value of -1 must be the smallest, on the interval $[0, 4]$. Similarly, in Example 3 we created the following table:

t	−1	0	1
$f(t)$	1	0	1

From this table we can see that the largest value of f on the interval $[-1, 1]$ is 1, and the smallest value is 0. We are taking advantage of the following fact, whose proof uses some deep and beautiful mathematics (alas, beyond the scope of this book).

Absolute Extrema on a Closed Interval

If f is *continuous* on a closed interval $[a, b]$, then it will have an absolute maximum and an absolute minimum value on that interval. Each absolute extremum must occur at either an endpoint or a critical point. Therefore, the absolute maximum is the largest value in a table of the values of f at the endpoints and critical points, and the absolute minimum is the smallest value.

Quick Example

The function $f(x) = 3x - x^3$, on the interval $[0, 2]$, has one critical point, at $x = 1$. The values of f at the critical point and the endpoints of the interval are given in the following table:

x	0	1	2
$f(x)$	0	2	−2

From this table we can say that the absolute maximum value of f on $[0, 2]$ is 2, which occurs at $x = 1$, and the absolute minimum value of f is -2, which occurs at $x = 2$.

As we can see in Example 2 and the following examples, if the domain is not a closed interval, then f may not even have an absolute maximum and minimum, and a table of values as above is of little help in determining whether it does or not.

example 4

Domain Not a Closed Interval

Find all extrema of $f(x) = x + \frac{1}{x}$.

Solution

Since no domain is specified, we take the domain to be as large as possible. The function is not defined at $x = 0$, but it is at all other points, so we take its domain to be $(-\infty, 0) \cup (0, +\infty)$. We calculate

$$f'(x) = 1 - \frac{1}{x^2}$$

- *Stationary points:* Setting $f'(x) = 0$, we solve:

$$1 - \frac{1}{x^2} = 0$$

$$1 = \frac{1}{x^2}$$

$$x^2 = 1 \qquad \text{Multiply both sides by } x^2.$$

$$x = \pm 1$$

  Calculating the corresponding values of f, we get the two stationary points $(1, 2)$ and $(-1, -2)$.

- *Singular points:* The only value of x for which $f'(x)$ is not defined is $x = 0$, but then f is not defined there either, so there are no singular points in the domain.
- *Endpoints:* The domain, $(-\infty, 0) \cup (0, +\infty)$, has no endpoints.

From this scant information it is hard to tell what f does. If we are sketching the graph by hand, we will need to plot additional "test points" to the left and right of the stationary points $x = \pm 1$.

T Technology

Let us choose a viewing window with an interval of $[-3, 3]$ for x and $[-4, 4]$ for y, which should leave plenty of room to see how f behaves near the stationary points. The result is something like Figure 12.

From this graph we can see that f has:

- A relative maximum at the point $(-1, -2)$
- A relative minimum at the point $(1, 2)$

It is curious that the relative maximum is lower than the relative minimum! Notice also that, because of the break in the graph at $x = 0$, the graph did not need to rise to get from $(-1, -2)$ to $(1, 2)$.

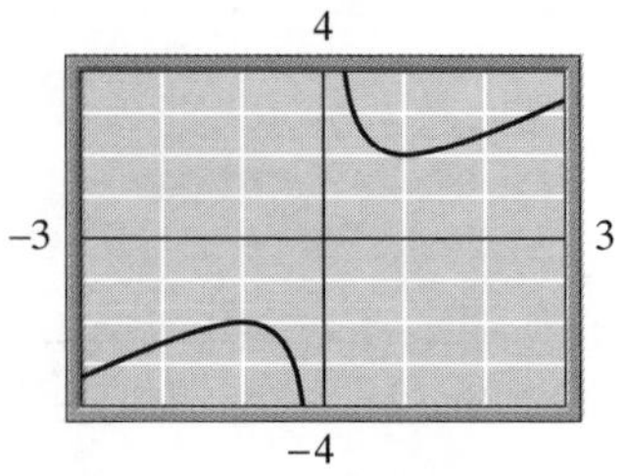

4
-3
3
-4

Figure 12

So far we have been solving the equation $f'(x) = 0$ to obtain the candidates for stationary extrema. However, it is often not easy—or even possible—to solve equations analytically. In the next example we show a way to get around this problem using graphing technology.

example 5

Using Technology to Locate Extrema

Find the approximate location (x-coordinates accurate to four decimal places) of all extrema of $f(x) = 3x^5 - 25x^3 - 15x^2 + 60x$.

Solution

Since the domain is not specified, we take the largest possible domain, which is $(-\infty, +\infty)$. Thus, there will be no endpoints. Next, we calculate the derivative as usual:

$$f'(x) = 15x^4 - 75x^2 - 30x + 60$$

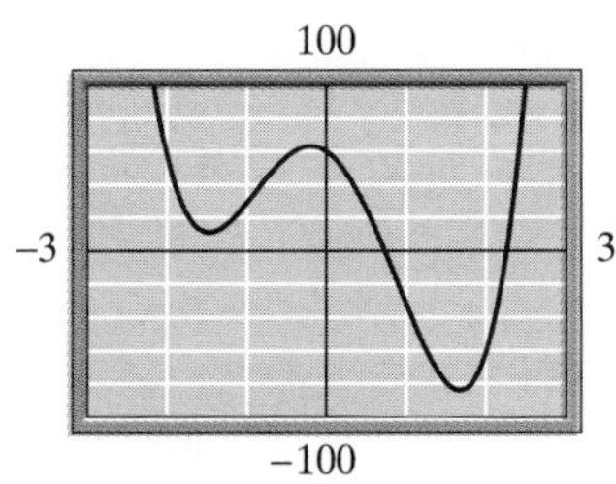

Figure 13

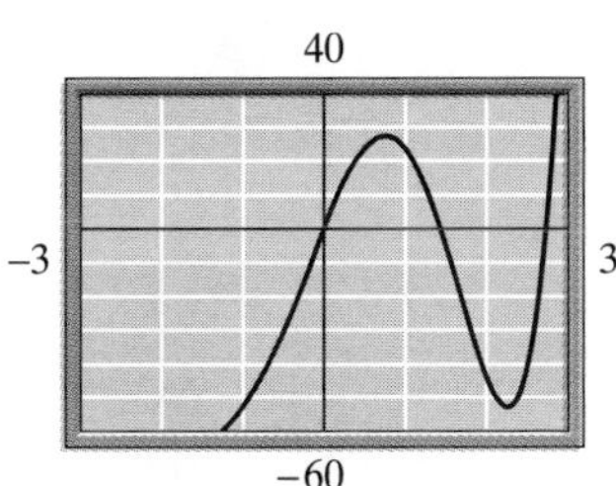

Figure 14

Now, the equation $f'(x) = 0$ is rather formidable, so we use technology to approximate the solutions by graphing $f'(x)$. After some experimentation, we settle on a viewing window with an interval of $[-3, 3]$ for x and $[-100, 100]$ for y, as shown in Figure 13. We can see that there are two places where $f'(x) = 0$ and hence two stationary points. By zooming in on each of these points,[5] we find that they occur at approximately $x = 0.7505$ and $x = 2.2586$. The corresponding stationary points on the graph of f are $(0.7505, 26.7276)$ and $(2.2586, -52.7199)$. This suggests that graphing the original function f in a viewing window with an interval of $[-3, 3]$ for x and $[-60, 40]$ for y will show us what happens at the two stationary points. The result is shown in Figure 14.

Thus, f has:

- A relative maximum of approximately 26.7276 at approximately $x = 0.7505$
- A relative minimum of approximately -52.7199 at approximately $x = 2.2586$

Neither of these is an absolute extremum.

Before we go on . . .

Question Why bother graphing the derivative in the first place? Why don't we just graph the original function and then zoom in to locate the relative extrema accurately?

Answer The answer depends on how intelligently your graphing software zooms in. For example, if you use a TI-83 and repeatedly zoom in on the relative maximum we just found (you will need to zoom in about six times), you will end up with a graph that looks something like Figure 15.

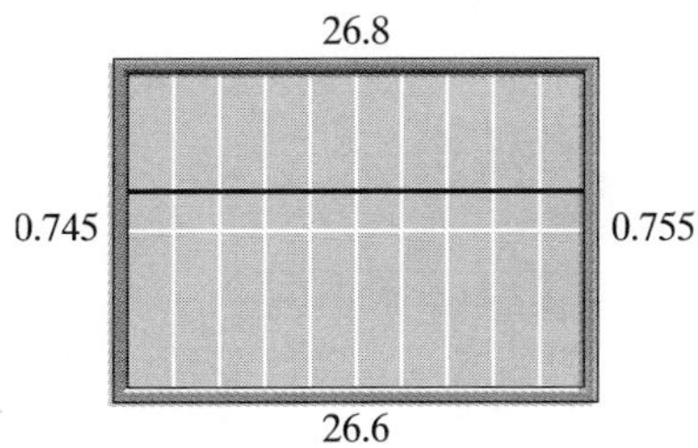

Figure 15

[5] We could also use the "Roots" feature under "Calc" in the TI-83.

When the graph starts looking horizontal, it's hard to see where the extremum lives.[6] It is usually much easier to zoom in on the graph of f' to see where it crosses the x-axis.[7]

5.1 exercises

Locate and classify all extrema in each of the graphs in Exercises 1–12. (By "classifying" the extrema, we mean listing whether each extremum is a relative or absolute maximum or minimum.) Also, locate any stationary points or singular points that are not relative extrema.

1.

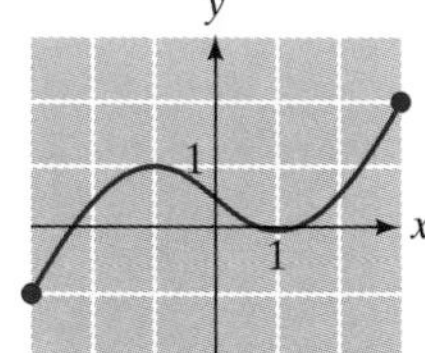

2.

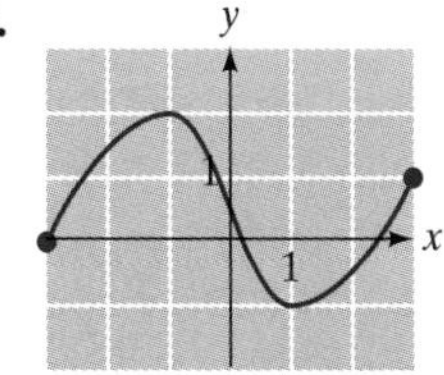
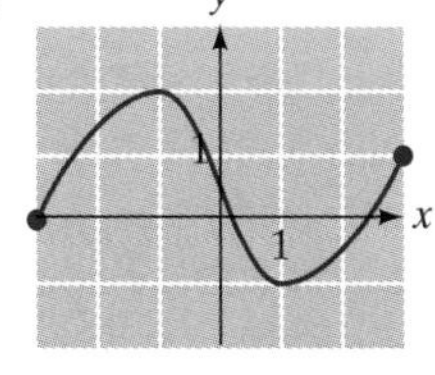

3.

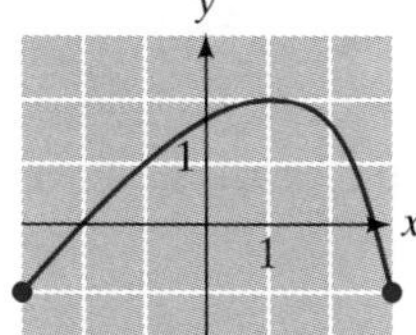

4.

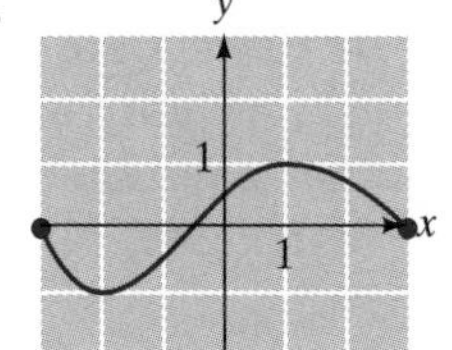

5.

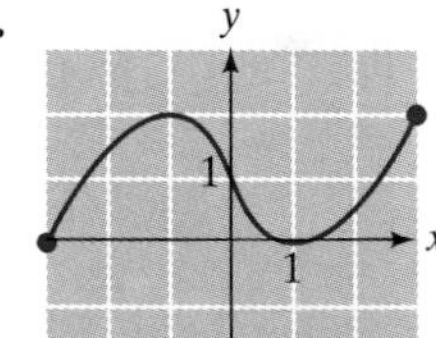

6.

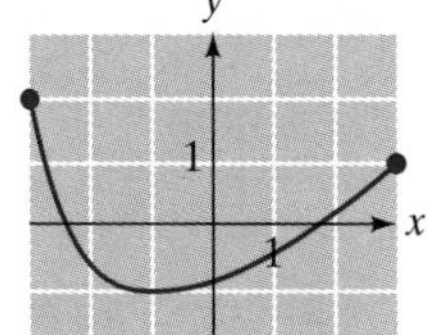

7.

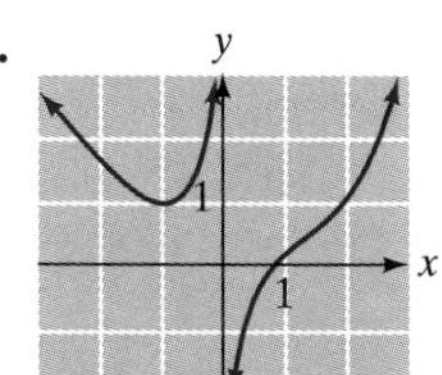

8.

9.

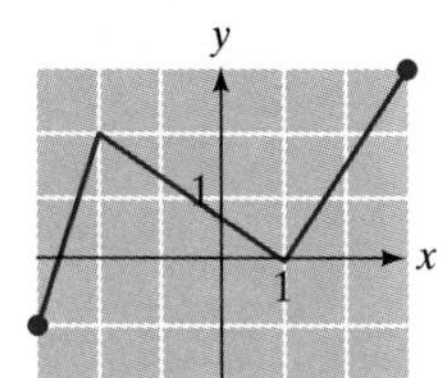

10.

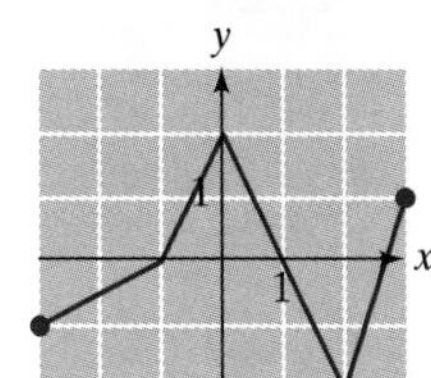
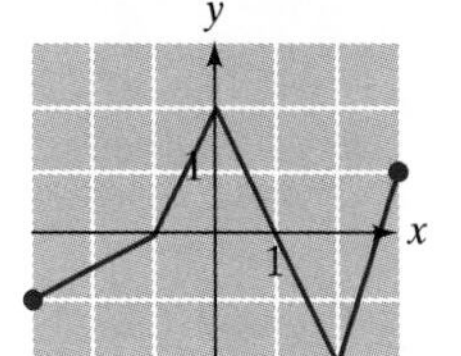

11.

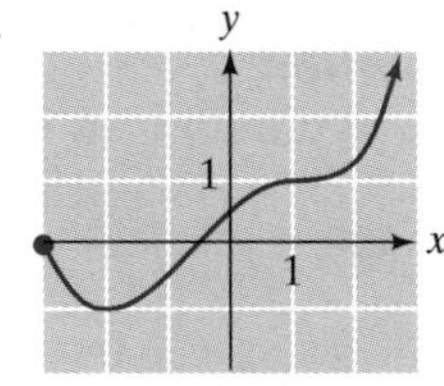

12.

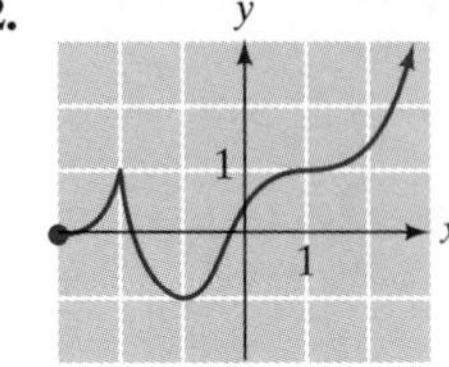

In each of Exercises 13–20, the graph of the derivative *of a function f is shown. Determine the x-coordinates of all stationary and singular points of f, and classify each as a relative maximum, relative minimum, or neither. (Assume that $f(x)$ is defined and continuous everywhere in $[-3, 3]$.)*

13.

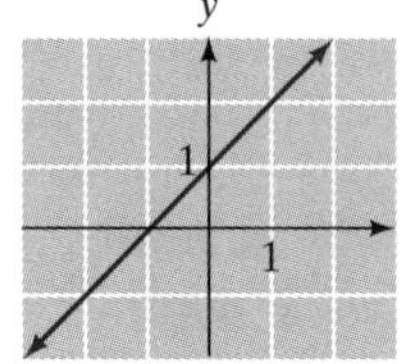

14.

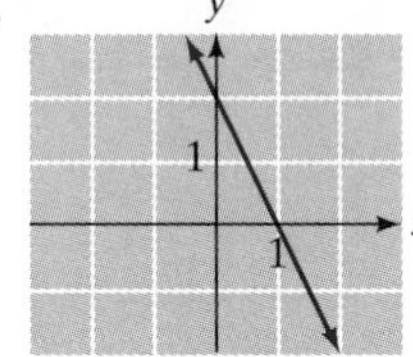

15.

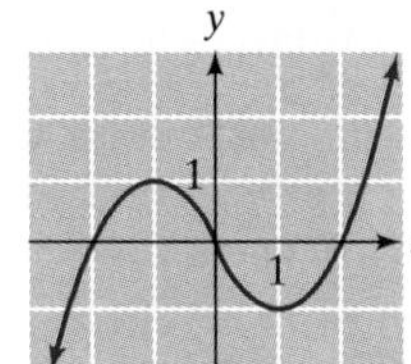

16.

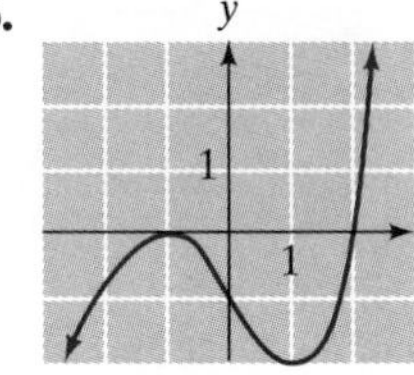

17.

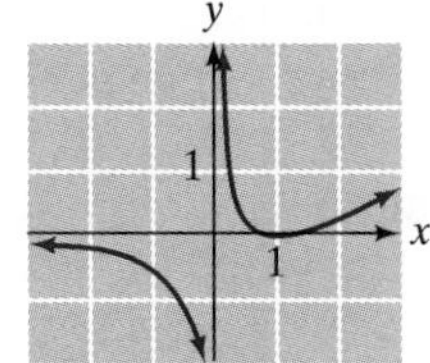

18.

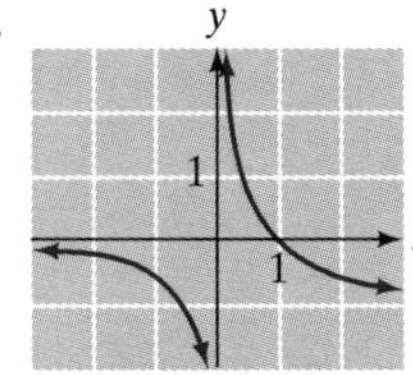

19.

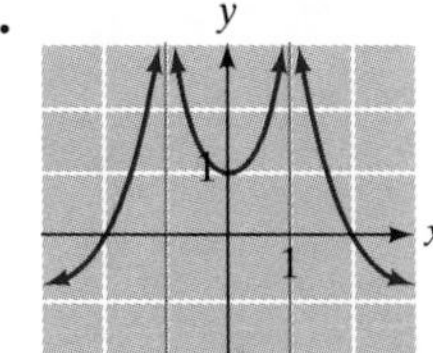

20.

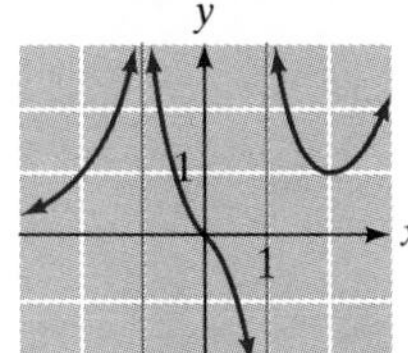

---

[6] There *are* ways of getting around this. On the TI-83 you can perform a "zoomfit" after every "zoom." But this will take about 12 zooming operations to locate the x-coordinate accurate to four decimal places.

[7] It is also more in the spirit of understanding what is actually going on!

Find the exact location of all the relative and absolute extrema of each function in Exercises 21–50.

21. $f(x) = x^2 - 4x + 1$ with domain $[0, 3]$

22. $f(x) = 2x^2 - 2x + 3$ with domain $[0, 3]$

23. $g(x) = x^3 - 12x$ with domain $[-4, 4]$

24. $g(x) = 2x^3 - 6x + 3$ with domain $[-2, 2]$

25. $f(t) = t^3 + t$ with domain $[-2, 2]$

26. $f(t) = -2t^3 - 3t$ with domain $[-1, 1]$

27. $h(t) = 2t^3 + 3t^2$ with domain $[-2, +\infty)$

28. $h(t) = t^3 - 3t^2$ with domain $[-1, +\infty)$

29. $f(x) = x^4 - 4x^3$ with domain $[-1, +\infty)$

30. $f(x) = 3x^4 - 2x^3$ with domain $[-1, +\infty)$

31. $g(t) = \frac{1}{4}t^4 - \frac{2}{3}t^3 + \frac{1}{2}t^2$ with domain $(-\infty, +\infty)$

32. $g(t) = 3t^4 - 16t^3 + 24t^2 + 1$ with domain $(-\infty, +\infty)$

33. $f(t) = \dfrac{t^2 + 1}{t^2 - 1}$; $-2 \le t \le 2, t \ne \pm 1$

34. $f(t) = \dfrac{t^2 - 1}{t^2 + 1}$ with domain $[-2, 2]$

35. $f(x) = \sqrt{x}\,(x - 1)\,; x \ge 0$

36. $f(x) = \sqrt{x}\,(x + 1)\,; x \ge 0$

37. $g(x) = x^2 - 4\sqrt{x}$

38. $g(x) = \dfrac{1}{x} - \dfrac{1}{x^2}$

39. $g(x) = \dfrac{x^3}{x^2 + 3}$

40. $g(x) = \dfrac{x^3}{x^2 - 3}$

41. $f(x) = x - \ln x$ with domain $(0, +\infty)$

42. $f(x) = x - \ln x^2$ with domain $(0, +\infty)$

43. $g(t) = e^t - t$ with domain $[-1, 1]$

44. $g(t) = e^{-t^2}$ with domain $(-\infty, +\infty)$

45. $f(x) = \dfrac{2x^2 - 24}{x + 4}$

46. $f(x) = \dfrac{x - 4}{x^2 + 20}$

47. $f(x) = xe^{1-x^2}$

48. $f(x) = x \ln x$ with domain $(0, +\infty)$

49. $g(x) = \sin x$ with domain $[0, 6\pi]$

50. $g(x) = \cos^2 x$ with domain $[0, 3\pi]$

In Exercises 51–54, use graphing technology and the method in Example 5 to find the x-coordinates of the critical points, accurate to two decimal places. Find all relative and absolute maxima and minima.

T **51.** $y = x^2 + \dfrac{1}{x - 2}$ with domain $(-3, 2) \cup (2, 6)$

T **52.** $y = x^2 - 10(x - 1)^{2/3}$ with domain $(-4, 4)$

T **53.** $f(x) = (x - 5)^2(x + 4)(x - 2)$ with domain $[-5, 6]$

T **54.** $f(x) = (x + 3)^2(x - 2)^2$ with domain $[-5, 5]$

Communication and Reasoning Exercises

55. Draw the graph of a function f with domain the set of all real numbers, such that f is not linear and has no relative extrema.

56. Draw the graph of a function g with domain the set of all real numbers, such that g has a relative maximum and minimum but no absolute extrema.

57. Draw the graph of a function that has stationary and singular points but no relative extrema.

58. Draw the graph of a function that has relative, not absolute, maxima and minima but has no stationary or singular points.

59. If a stationary point is not a relative maximum, then must it be a relative minimum? Explain your answer.

60. If one endpoint is a relative maximum, must the other be a relative minimum? Explain your answer.

5.2 Applications of Maxima and Minima

In many applications we would like to find the largest or smallest possible value of some quantity—for instance, the greatest possible profit or the lowest cost. We call this the *optimal* (best) value. In this section we consider several such examples and use calculus to find the optimal value in each.

In all applications the first step is to translate a written description into a mathematical problem. In the problems we look at in this section, there are some *unknowns* that we are asked to find, there is an expression involving those unknowns that must be made as large or as small as possible—the **objective**

function, and there may be **constraints**—equations or inequalities relating the variables.[1]

example 1

Minimizing Average Cost

Gymnast Clothing, Inc., manufactures expensive hockey jerseys for sale to college bookstores in runs of up to 500. Its cost (in dollars) for a run of x hockey jerseys is

$$C(x) = 2000 + 10x + 0.2x^2$$

How many jerseys should Gymnast Clothing, Inc., produce per run in order to minimize average cost?[2]

Solution

Here is the procedure we will follow to solve this problem.

1. *Identify the unknown(s).* There is one unknown: x, the number of hockey jerseys Gymnast should produce per run. (We know this because the question is How many jerseys . . . ?)
2. *Identify the objective function.* The objective function is the quantity that must be made as small (in this case) as possible. In this example it is the average cost, which is given by

$$\bar{C}(x) = \frac{C(x)}{x} = \frac{2000 + 10x + 0.2x^2}{x}$$

$$= \frac{2000}{x} + 10 + 0.2x \text{ dollars per jersey}$$

3. *Identify the constraints (if any).* At most 500 jerseys can be manufactured in a run. Also, $\bar{C}(0)$ is not defined. Thus, x is constrained by

$$0 < x \leq 500$$

Put another way, the domain of the objective function $\bar{C}(x)$ is $(0, 500]$.

4. *State and solve the resulting optimization problem.* Our optimization problem is to

$$\text{Minimize } \bar{C}(x) = \frac{2000}{x} + 10 + 0.2x \qquad \text{Objective function}$$

$$\text{subject to } 0 < x \leq 500 \qquad \text{Constraint}$$

We now solve this problem as in the previous section. We first calculate

$$\bar{C}'(x) = -\frac{2000}{x^2} + 0.2$$

[1] If you have studied linear programming, you will notice a similarity here, but unlike the situation in linear programming, neither the objective function nor the constraints need to be linear.

[2] Why don't we seek to minimize total cost? The answer would be uninteresting; to minimize total cost, we would make *no* jerseys at all. Minimizing average cost is a more practical objective.

- *Stationary points:* We set $\bar{C}'(x) = 0$ and solve for x:

$$-\frac{2000}{x^2} + 0.2 = 0$$

$$0.2 = \frac{2000}{x^2}$$

$$x^2 = \frac{2000}{0.2} = 10{,}000$$

$$x = \pm 100$$

We reject $x = -100$ because -100 is not in the domain of $\bar{C}$ (and makes no sense), so we have one stationary point, at $x = 100$. There, the average cost is $\bar{C}(100) = \$50$ per jersey.

- *Singular points:* The only point at which $\bar{C}'$ is not defined is $x = 0$, but that is not in the domain, so we have no singular points.

- *Endpoints:* We have one endpoint in the domain, at $x = 500$. There, the average cost is $\bar{C}(500) = \$114$.

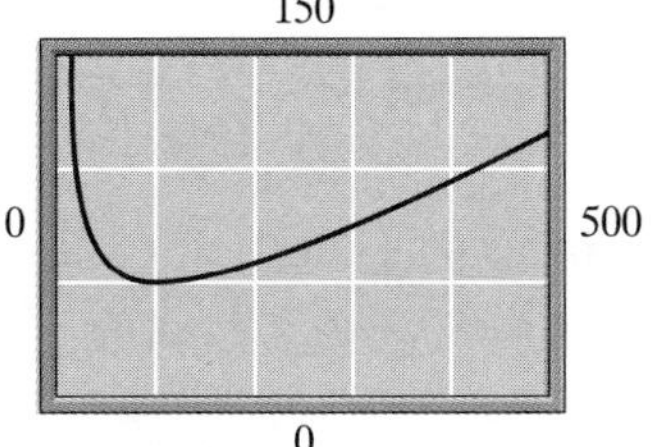

Figure 16

T Let us plot $\bar{C}$ in a viewing window with interval [0, 500] for x and [0, 150] for y, which will show the whole domain and the two interesting points we've found so far. The result is Figure 16. From the graph of $\bar{C}$ we can see that the stationary point at $x = 100$ gives the absolute minimum. We can say, therefore, that Gymnast Clothing, Inc., should produce 100 jerseys per run, for a lowest possible average cost of \$50 per jersey.

example 2

Maximizing Area

Slim wants to build a rectangular enclosure for his pet rabbit, Killer, against the side of his house, as shown in Figure 17. He has bought 100 feet of fencing. What are the dimensions of the largest area he can enclose?

Figure 17

Solution

1. *Identify the unknown(s).* To identify the unknown(s), we look at the question: What are the *dimensions* of the largest area he can enclose? Thus, the

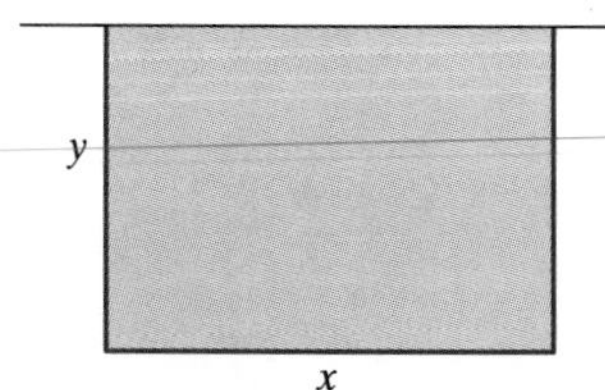

Figure 18

unknowns are the dimensions of the fence. We call these x and y, as shown in Figure 18.

2. *Identify the objective function.* We look for what it is that we are trying to maximize (or minimize). The phrase "largest area" tells us that our object is to *maximize the area,* which is the product of length and width, so the objective function is

$$A = xy$$

3. *Identify the constraints (if any).* What stops Slim from making the area as large as he wants? He has only 100 feet of fencing to work with. Looking again at Figure 18, we see that the sum of the lengths of the three sides must equal 100, so

$$x + 2y = 100$$

One more point: Because x and y represent the lengths of the sides of the enclosure, neither can be a negative number.

4. *State and solve the resulting optimization problem.* Our mathematical problem is:

Maximize $A = xy$ — Objective function

subject to $x + 2y = 100, x \geq 0$, and $y \geq 0$ — Constraints

We know how to find maxima and minima of a function of one variable, but A appears to depend on two variables. We can remedy this by using a constraint to express one variable in terms of the other. Let us take the constraint $x + 2y = 100$ and solve for x in terms of y:

$$x = 100 - 2y$$

Substituting into the objective function gives

$$A = xy = (100 - 2y)y = 100y - 2y^2$$

and we have eliminated x from the objective function. What about the inequalities? One says that $x \geq 0$, but we want to eliminate x from this as well. We substitute for x again, getting

$$100 - 2y \geq 0$$

Solving this inequality for y gives $y \leq 50$. The second inequality says that $y \geq 0$. Thus, we can restate our problem with x eliminated:

Maximize $A(y) = 100y - 2y^2$ subject to $0 \leq y \leq 50$.

We now proceed with our usual method of solving such problems. We calculate $A'(y) = 100 - 4y$.

- *Stationary points:* Solving $100 - 4y = 0$, we get one stationary point, at $y = 25$. There, $A(25) = 1250$.
- *Singular points:* There are no points at which $A'(y)$ is not defined.
- *Endpoints:* We have two endpoints, at $y = 0$ and $y = 50$. The corresponding areas are $A(0) = 0$ and $A(50) = 0$.

We record the three points we found in a table:

y	0	25	50
$A(y)$	0	1250	0

It's clear now how A must behave: It increases from 0 at $y = 0$ to 1250 at $y = 25$ and then decreases back to 0 at $y = 50$. Thus, the largest possible value of A is 1250 square feet, which occurs when $y = 25$. To completely answer the question that was asked, we need to know the corresponding value of x. We have $x = 100 - 2y$, so $x = 50$ when $y = 25$. Thus, Slim should build his enclosure 50 feet across and 25 feet deep (with the "missing" 50-foot side being formed by part of the house).

Before we go on . . . Notice that the problem here came down to finding the absolute maximum value of A on the closed and bounded interval [0, 50]. As we noted in the preceding section, the table of values of A at its critical points and the endpoints of the interval gives us enough information to find the absolute maximum.

Now we stop for a moment and summarize the steps we've taken in these two examples.

Solving an Optimization Problem

1. *Identify the unknown(s), possibly with the aid of a diagram.* These are usually the quantities asked for in the problem.
2. *Identify the objective function.* This is the quantity you are asked to maximize or minimize. You should name it explicitly, as in "Let S = surface area."
3. *Identify the constraint(s).* These can be equations relating variables or inequalities that express limitations on the values of variables.
4. *State the optimization problem.* This will have the form "Maximize [minimize] the objective function subject to the constraint(s)."
5. *Eliminate extra variables.* If the objective function depends on several variables, solve the constraint equations to express all variables in terms of one particular variable. Substitute these expressions into the objective function to rewrite it as a function of a single variable. Substitute the expressions into any inequality constraints to help determine the domain of the objective function.
6. *Find the absolute maximum (or minimum) of the objective function.* Use the techniques of the preceding section.

example 3

Maximizing Revenue

The Cozy Carriage Co. builds baby strollers. Market research estimates that if it sets the price of a stroller at p dollars, then the company can sell $q = 300{,}000 - 10p^2$ strollers per year. What price will bring in the greatest annual revenue?

Solution

The question we are asked identifies our main unknown, the price p. However, there is another quantity that we do not know—q, the number of strollers the company will sell per year. The question also identifies the objective function, revenue, which is

$$R = pq$$

Including the equality constraint given to us, $q = 300{,}000 - 10p^2$, and the "reality" inequality constraints $p \geq 0$ and $q \geq 0$, we can write our problem as

$$\text{Maximize } R = pq \text{ subject to } q = 300{,}000 - 10p^2,\ p \geq 0, \text{ and } q \geq 0.$$

We need to eliminate one of the unknowns so that R is expressed as a function of one variable alone. We are given q in terms of p, so we can substitute to eliminate q:

$$R = pq = p(300{,}000 - 10p^2) = 300{,}000p - 10p^3$$

Substituting in the inequality $q \geq 0$, we get

$$300{,}000 - 10p^2 \geq 0$$

Thus, $p^2 \leq 30{,}000$, which gives $-100\sqrt{3} \leq p \leq 100\sqrt{3}$. When we combine this with $p \geq 0$, we get the following restatement of our problem:

$$\text{Maximize } R(p) = 300{,}000p - 10p^3 \text{ such that } 0 \leq p \leq 100\sqrt{3}.$$

We solve this problem in much the same way we did the preceding one. We calculate $R'(p) = 300{,}000 - 30p^2$. Setting $300{,}000 - 30p^2 = 0$, we find one stationary point at $p = 100$. There are no singular points, and we have the endpoints $p = 0$ and $p = 100\sqrt{3}$. Putting these points in a table and computing the corresponding values of R, we get the following:

p	0	100	$100\sqrt{3}$
$R(p)$	0	20,000,000	0

Thus, the Cozy Carriage Co. should price its strollers at \$100 each, which will bring in the greatest possible revenue of \$20,000,000.

example 4

Minimizing Resources

The Metal Can Co. has an order to make cans with a volume of 250 cubic centimeters. What should the dimensions of the cans be in order to use the least metal in their production?

Figure 19

Solution

We are asked to find the dimensions of the cans. It is traditional to take as the dimensions of a cylinder the height h and the radius of the base r, as in Figure 19. We are also asked to minimize the total metal used in the can, which is the area of the surface of the cylinder. We can look up the formula or figure it out ourselves: Imagine removing the circular top and bottom and then cutting vertically and flattening out the hollow cylinder to get a rectangle, as shown in Figure 20.

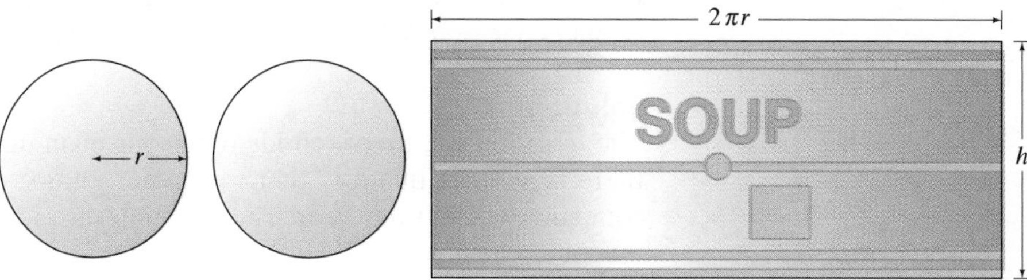

Figure 20

Our objective function is the (total) surface area S of the can. The area of each disc is πr^2, and the area of the rectangular piece is $2\pi rh$. Thus, our objective function is

$$S = 2\pi r^2 + 2\pi rh$$

As usual, there is a constraint; the volume must be exactly 250 cubic centimeters. The formula for the volume of a cylinder is $V = \pi r^2 h$, so we have

$$\pi r^2 h = 250$$

It is easiest to solve this constraint for h in terms of r:

$$h = \frac{250}{\pi r^2}$$

Substituting in the objective function, we get

$$S = 2\pi r^2 + 2\pi r \frac{250}{\pi r^2} = 2\pi r^2 + \frac{500}{r}$$

Now r cannot be negative or zero, but it can become very large (a very wide but very short can could have the right volume). We therefore take the domain of $S(r)$ to be $(0, +\infty)$, so our mathematical problem is as follows:

$$\text{Minimize } S(r) = 2\pi r^2 + \frac{500}{r} \text{ subject to } r > 0.$$

Now we calculate

$$S'(r) = 4\pi r - \frac{500}{r^2}$$

To find stationary points, we set this equal to zero and solve:

$$4\pi r - \frac{500}{r^2} = 0$$

$$4\pi r = \frac{500}{r^2}$$

$$4\pi r^3 = 500$$

$$r^3 = \frac{125}{\pi}$$

So $$r = \sqrt[3]{\frac{125}{\pi}} = \frac{5}{\sqrt[3]{\pi}} \approx 3.41$$

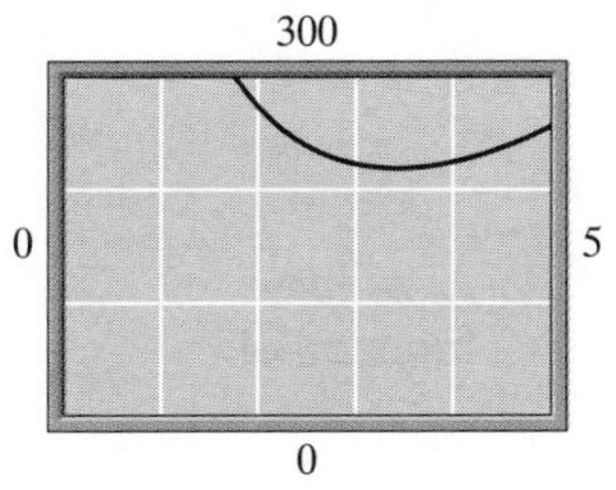

T The corresponding surface area is approximately $S(3.41) \approx 220$. There are no singular points or endpoints in the domain. To see how S behaves near the one stationary point, let us graph it in a viewing window with interval $[0, 5]$ for r and $[0, 300]$ for S. The result is Figure 21. From the graph we can clearly see that the smallest surface area occurs at the stationary point at $r \approx 3.41$. The height of the can will be

$$h = \frac{250}{\pi r^2} \approx 6.83$$

300

0 5

0

Figure 21

Thus, the can that uses the least metal has a height of approximately 6.83 centimeters and a radius of approximately 3.41 centimeters. Such a can will use approximately 220 square centimeters of metal.

Before we go on . . . If we substitute the exact expression

$$r = \sqrt[3]{\frac{125}{\pi}} = \frac{5}{\sqrt[3]{\pi}}$$

into the formula for h, we get

$$h = \frac{10}{\sqrt[3]{\pi}}$$

which is exactly twice r. Put another way, the height is exactly equal to the diameter, so that the cans look square when viewed from the side. Have you ever seen cans with that shape? Why do you think most cans do not have this shape?

---

example 5

Allocation of Labor

The Gym Sock Company manufactures cotton athletic socks. Production is partially automated through the use of robots. Daily operating costs amount to \$50 per laborer and \$30 per robot. The number of pairs of socks the company can manufacture in a day is given by a Cobb-Douglas production formula. (see Section 4.5):

$$q = 50n^{0.6}r^{0.4}$$

where q is the number of pairs of socks that can be manufactured by n laborers and r robots. Assuming that the company wishes to produce 1000 pairs of socks per day at a minimum cost, how many laborers and how many robots should it use?

Solution

The unknowns are n, the number of laborers, and r, the number of robots. The objective is to minimize the daily cost:

$$C = 50n + 30r$$

The constraints are given by the daily quota:

$$1000 = 50n^{0.6}r^{0.4}$$

and the fact that n and r are nonnegative. We solve the constraint equation for one of the variables; let us solve for n:

$$n^{0.6} = \frac{1000}{50r^{0.4}} = \frac{20}{r^{0.4}}$$

Taking the 1/0.6 power of both sides gives

$$n = \left(\frac{20}{r^{0.4}}\right)^{1/0.6} = \frac{20^{1/0.6}}{r^{0.4/0.6}} = \frac{20^{5/3}}{r^{2/3}} \approx \frac{147.36}{r^{2/3}}$$

Substituting in the objective equation gives us the cost as a function of r:

$$C(r) \approx 50\left(\frac{147.36}{r^{2/3}}\right) + 30r = 7368r^{-2/3} + 30r$$

The only remaining constraint on r is that $r > 0$. To find the minimum value of $C(r)$, we first take the derivative:

$$C'(r) \approx -4912r^{-5/3} + 30$$

Setting this equal to zero, we solve for r:

$$-4912r^{-5/3} + 30 = 0$$

$$-4912r^{-5/3} = -30$$

$$r^{-5/3} \approx 0.006107$$

$$r \approx (0.006107)^{-3/5} \approx 21.3$$

T The corresponding cost is $C(21.3) \approx \$1600$. There are no singular points or endpoints in the domain of C. To see how C behaves near its stationary point, let us draw its graph in a viewing window with an interval of [0, 40] for r and [0, 2000] for C. The result is Figure 22.

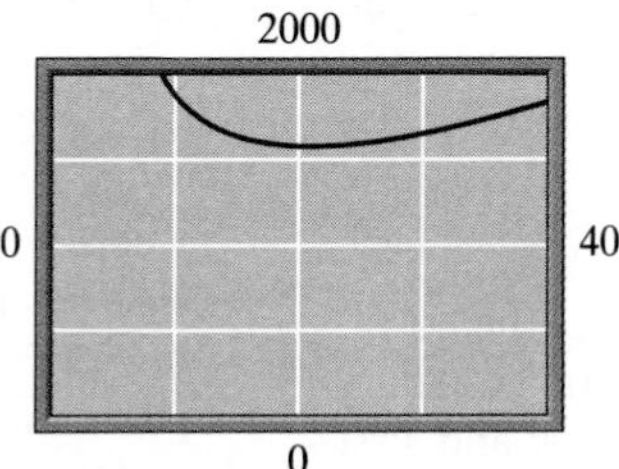

Figure 22

From the graph we can see that C does have its minimum at the stationary point. The corresponding value of n is

$$n \approx \frac{147.36}{r^{2/3}} \approx 19.2$$

At this point, our solution appears to be this: Use (approximately) 19.2 laborers and (approximately) 21.3 robots to meet the manufacturing quota at a minimum cost. However, we are not interested in fractions of robots or people, so we need to find integer solutions for n and r. If we round these numbers, we get the solution $(n, r) = (19, 21)$. However, a quick calculation shows that

$$q = 50(19)^{0.6}(21)^{0.4} \approx 989 \text{ pairs of socks}$$

which fails to meet the quota of 1000. Thus, we need to round at least one of the quantities n and r *upward* to meet the quota. The three possibilities, with corresponding values of q and C, are as follows:

$$(n, r) = (20, 21), \text{ with } q \approx 1020, \text{ and } C = \$1630$$

$$(n, r) = (19, 22), \text{ with } q \approx 1007, \text{ and } C = \$1610$$

$$(n, r) = (20, 22), \text{ with } q \approx 1039, \text{ and } C = \$1660$$

Of these, the solution that meets the quota at a minimum cost is $(n, r) = (19, 22)$. Thus, the Gym Sock Company should use 19 laborers and 22 robots, at a cost of $50 \times 19 + 30 \times 22 = \$1610$, to manufacture $50 \times 19^{0.6} \times 22^{0.4} \approx 1007$ pairs of socks.

5.2 exercises

Solve the optimization problems in Exercises 1–8.

1. Maximize $P = xy$ with $x + y = 10$.
2. Maximize $P = xy$ with $x + 2y = 40$.
3. Minimize $S = x + y$ with $xy = 9$ and both x and $y > 0$.
4. Minimize $S = x + 2y$ with $xy = 2$ and both x and $y > 0$.
5. Minimize $F = x^2 + y^2$ with $x + 2y = 10$.
6. Minimize $F = x^2 + y^2$ with $xy^2 = 16$.
7. Maximize $P = xyz$ with $x + y = 30$ and $y + z = 30$, and $x, y,$ and $z \geq 0$.
8. Maximize $P = xyz$ with $x + z = 12$ and $y + z = 12$, and $x, y,$ and $z \geq 0$.
9. For a rectangle with perimeter 20 to have the largest area, what dimensions should it have?
10. For a rectangle with area 100 to have the smallest perimeter, what dimensions should it have?

Applications

11. ***Fences*** I want to fence in a rectangular vegetable patch. The fencing for the east and west sides costs \$4 per foot, while the fencing for the north and south sides costs only \$2 per foot. I have a budget of \$80 for the project. What is the largest area I can enclose?

12. ***Fences*** Actually, my vegetable garden (see Exercise 11) abuts my house, so that the house itself forms the northern boundary. The fencing for the southern boundary costs \$4 per foot, while the fencing for the east and west sides costs \$2 per foot. If I have a budget of \$80 for the project, what is the largest area I can enclose this time?

13. ***Revenue*** Hercules Films is deciding on the price of the video release of its film *Son of Frankenstein*. Its marketing people estimate that at a price of p dollars, it can sell a total of $q = 200{,}000 - 10{,}000p$ copies. What price will bring in the greatest revenue?

14. ***Profit*** Hercules Films is also deciding on the price of the video release of its film *Bride of the Son of Frankenstein*. Again, marketing estimates that at a price of p dollars, it can sell $q = 200{,}000 - 10{,}000p$ copies, but each copy costs \$4 to make. What price will give the greatest profit?

15. ***Revenue*** The demand for rubies at Royal Ruby Retailers is given by the equation

$$q = -\frac{4}{3}p + 80$$

where p is the price RRR charges (in dollars) and q is the number of rubies RRR sells per week. At what price should RRR sell its rubies to maximize its weekly revenue?

16. ***Revenue*** The consumer demand curve for tissues is given by

$$q = (100 - p)^2 \qquad (0 \leq p \leq 100)$$

where p is the price per case of tissues and q is the demand in weekly sales. At what price should tissues be sold in order to maximize revenue?

17. ***Revenue*** Assume that the demand for tuna in a small coastal town is given by

$$p = \frac{500{,}000}{q^{1.5}}$$

where q is the number of pounds of tuna that can be sold in a month at p dollars per pound. Assume that the town's fishery wishes to sell at least 5000 pounds of tuna per month.
   a. How much should the town's fishery charge for tuna in order to maximize monthly revenue?
   b. How much tuna will it sell per month at that price?
   c. What will the fishery's resulting revenue be?

18. ***Revenue*** Economist Henry Schultz devised the following demand function for corn:

$$p = \frac{6{,}570{,}000}{q^{1.3}}$$

where q is the number of bushels of corn that could be sold at p dollars per bushel in 1 year.[3] Assume that at least 10,000 bushels of corn per year must be sold.
   a. How much should farmers charge per bushel of corn to maximize annual revenue?
   b. How much corn can farmers sell per year at that price?
   c. What will the farmers' resulting revenue be?

Source: Henry Schultz, *The Theory and Measurement of Demand,* as cited in *Introduction to Mathematical Economics* by A. L. Ostrosky, Jr., and J. V. Koch (Prospect Heights, IL: Waveland Press, 1979).

19. ***Revenue*** The wholesale price for chicken in the United States fell from 25¢ per pound to 14¢ per pound while, at the same time, per capita chicken consumption rose from 22 pounds per year to 27.5 pounds per year.[4] Assuming that the demand for chicken depends linearly on the price, what wholesale price for chicken maximizes revenues for poultry farmers, and what does that revenue amount to?

Source: Agricultural Statistics, U.S. Department of Agriculture.

---

[3] Based on data for the period 1915–1929.

[4] Data are provided for the years 1951–1958.

20. ***Revenue*** Your underground used book business is booming. Your policy is to sell all used versions of *Calculus and You* at the same price (regardless of condition). When you set the price at \$10, sales amounted to 120 volumes during the first week of classes. The following semester, you set the price at \$30 and sold not a single book. Assuming that the demand for books depends linearly on the price, what price gives you the maximum revenue, and what does that revenue amount to?

21. ***Profit*** The demand for rubies at Royal Ruby Retailers is given by the equation

$$q = -\frac{4}{3}p + 80$$

where p is the price RRR charges (in dollars) and q is the number of rubies RRR sells per week. Assuming that due to extraordinary market conditions, RRR can obtain rubies for \$25 each, how much should it charge per ruby to make the greatest possible weekly profit? What will that profit be?

22. ***Profit*** The consumer demand curve for tissues is given by

$$q = (100 - p)^2 \qquad (0 \le p \le 100)$$

where p is the price per case of tissues and q is the demand in weekly sales. If tissues cost \$30 per case to produce, at what price should tissues be sold for the greatest possible weekly profit? What will that profit be?

23. ***Profit*** The demand equation for your company's virtual reality video headsets is

$$p = \frac{1000}{q^{0.3}}$$

where q is the total number of headsets that your company can sell in a week at a price of p dollars. The total manufacturing and shipping cost amounts to \$100 per headset.

a. What is the greatest profit your company can make in a week, and how many headsets will your company sell at this level of profit? (Give answers to the nearest whole number.)

b. How much, to the nearest \$1, should your company charge per headset for the maximum profit?

24. ***Profit*** Due to sales by a competing company, your company's sales of virtual reality video headsets have dropped, and your financial consultant revises the demand equation to

$$p = \frac{800}{q^{0.35}}$$

where q is the total number of headsets that your company can sell in a week at a price of p dollars. The total manufacturing and shipping cost still amounts to \$100 per headset.

a. What is the greatest profit your company can make in a week, and how many headsets will your company sell at this level of profit? (Give answers to the nearest whole number.)

b. How much, to the nearest \$1, should your company charge per headset for the maximum profit?

25. ***Box Design*** The Chocolate Box Co. is going to make open-topped boxes out of 6″ × 16″ rectangles of cardboard by cutting squares out of the corners and folding up the sides. What is the largest volume box it can make this way?

26. ***Box Design*** A packaging company is going to make open-topped boxes, with square bases, that hold 108 cubic centimeters. What are the dimensions of the box that can be built with the least material?

27. ***Asset Appreciation*** As the financial consultant to a classic auto dealership, you estimate that the total value (in dollars) of its collection of 1959 Chevrolets and Fords is given by the formula

$$v = 300{,}000 + 1000t^2$$

where t is the number of years from now. You anticipate a continuous inflation rate of 5% per year, so that the discounted (present) value of an item that will be worth \$$v$ in t years' time is

$$p = ve^{-0.05t}$$

When would you advise the dealership to sell the vehicles to maximize their discounted value?

28. ***Plantation Management*** The value of a fir tree on your plantation increases with the age of the tree according to the formula

$$v = \frac{20t}{1 + 0.05t}$$

where t is the age of the tree in years. Given a continuous inflation rate of 5% per year, the discounted (present) value of a newly planted seedling is

$$p = ve^{-0.05t}$$

At what age (to the nearest year) should you harvest your trees to ensure the greatest possible discounted value?

29. ***Marketing Strategy*** The FeatureRich Software Co. has a dilemma. Its new program, Doors 2001, is almost ready to go on the market. However, the longer the company works on it, the better it can make the program and the more it can charge for it. The company's marketing analysts estimate that if it delays t days, it can set the price at $100 + 2t$ dollars. On the other hand, the longer it delays, the more market share they will lose to their main competitor (see the next exercise) so that if it delays t days, it

will be able to sell $400{,}000 - 2500t$ copies of the program. How many days should FeatureRich delay the release in order to get the greatest revenue?

30. ***Marketing Strategy*** FeatureRich Software's main competitor (see Exercise 29) is Moon Software, and Moon is in a similar predicament. Its product, Walls 2001, could be sold now for \$200, but for each day Moon delays, it could increase the price by \$4. On the other hand, it could sell 300,000 copies now, but each day it waits will cut sales by 1500. How many days should Moon delay the release in order to get the greatest revenue?

31. ***Agriculture*** The fruit yield per tree in an orchard that contains 50 trees is 100 pounds per tree each year. Due to crowding, the yield decreases by 1 pound per season for every additional tree planted. How may additional trees should be planted for a maximum total annual yield?

32. ***Agriculture*** Two years ago your orange orchard contained 50 trees, and the total yield was 75 bags of oranges. Last year you removed ten of the trees and noticed that the total yield increased to 80 bags. Assuming that the yield per tree depends linearly on the number of trees in the orchard, what should you do this year to maximize your total yield?

33. ***Average Cost*** The cost function for the manufacture of portable CD players is given by

$$C(x) = 150{,}000 + 20x + \frac{x^2}{10{,}000} \text{ dollars}$$

where x is the number of CD players manufactured. How many CD players should be manufactured to minimize average cost? What is the resulting average cost of a CD player? (Give your answer to the nearest dollar.)

34. ***Average Cost*** Repeat Exercise 33 using the revised cost function

$$C(x) = 150{,}000 + 20x + \frac{x^2}{100} \text{ dollars}$$

35. ***Pollution Control*** The cost of controlling emissions at a firm goes up rapidly as the amount of emissions reduced goes up. Here is a possible model:

$$C(q) = 4000 + 100q^2$$

where q is the reduction in emissions (in pounds of pollutant per day) and C is the daily cost to the firm (in dollars) of this reduction. Government clean air subsidies amount to \$500 per pound of pollutant removed. How many pounds of pollutant should the firm remove each day to minimize *net* cost (cost minus subsidy)?

36. ***Pollution Control*** Repeat Exercise 35 using the following cost function:

$$C(q) = 2000 + 200q^2$$

with government subsidies amounting to \$100 per pound of pollutant removed per day.

37. ***Luggage Dimensions*** TransWorld Airlines (TWA) has a rule for checked baggage: The total dimensions (length + width + height) may not exceed 62 inches for each bag.[5] Suppose you wish to check a bag whose height equals its width. What is the largest volume bag of this shape that you can check on a TWA flight?

38. ***Luggage Dimensions*** American Airlines has the same rule as TWA for the first bag you check on a flight, but the second bag you check may have total outside dimensions (length + width + height) of no more than 55 inches.[6] Suppose you want to check a bag whose length is twice its height. What is the largest volume bag of this shape that you can check as the second bag on an American flight?

39. ***Luggage Dimensions*** Fly-by-Night Airlines has a peculiar rule about luggage: The length and width of a bag must add to 45 inches, while the width and height must also add to 45 inches. What are the dimensions of the bag with largest volume that it will accept?

40. ***Luggage Dimensions*** Fair Weather Airlines will accept only bags for which the sum of the length and width is 36 inches, while the sum of length, height, and twice the width is 72 inches. What are the dimensions of the bag with the largest volume that Fair Weather will accept?

41. ***Package Dimensions*** The U.S. Postal Service (USPS) will accept packages only if length plus girth is no more than 108 inches.[7] (See the figure.)

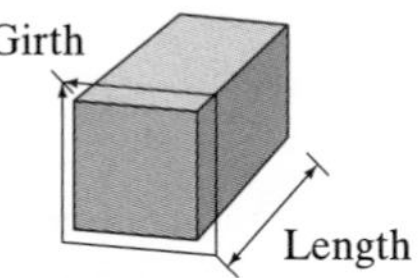

Assuming that the front face of the package (as shown in the figure) is square, what is the largest volume package that the USPS will accept?

42. ***Package Dimensions*** United Parcel Service (UPS) will only accept packages with a length no more than 108 inches and length plus girth no more than 130 inches.[8] (See the figure for Exercise 41.) Assuming that the front face of the package (as shown in the figure) is square, what is the largest volume package that UPS will accept?

---

[5] According to information on its web site (http://www.twa.com/) as of March 1999.

[6] According to information on its web site (http://www.aa.com/) as of March 1999.

[7] The requirement at the time this book was written.

[8] The requirement at the time this book was written.

43. ***Average Profit*** The FeatureRich Software Company sells its graphing program, Dogwood, with a volume discount. If a customer buys x copies, then that customer pays[9] $\$500\sqrt{x}$. It costs the company \$10,000 to develop the program and \$2 to manufacture each copy. If just one customer were to buy all the copies of Dogwood, how many copies would the customer have to buy for FeatureRich Software's average profit per copy to be maximized? How are average profit and marginal profit related at this number of copies?

44. ***Average Profit*** Repeat Exercise 43 with the charge to the customer $\$600\sqrt{x}$ and the cost to develop the program \$9000.

45. ***Prison Population*** The size of the U.S. prison population followed the curve

$$N(t) = 0.028234t^3 - 1.0922t^2 + 13.029t + 146.88 \quad (0 \le t \le 39)$$

in the years 1950–1989. Here t is the number of years since 1950, and N is the number of prisoners in thousands. When, to the nearest year, was the prison population decreasing most rapidly? When was it increasing most rapidly?

Source: The authors' model from data in *Sourcebook of Criminal Justice Statistics*, 1990, U.S. Dept. of Justice, Bureau of Justice Statistics (Washington D.C.: Government Printing Office), p. 604.

46. ***Test Scores*** Combined SAT scores in the United States in the years 1967–1991 could be approximated by

$$T(t) = -0.01085t^3 + 0.5804t^2 - 10.12t + 962.4 \quad (0 \le t \le 22)$$

where t is the number of years since 1967, and T is the combined SAT score average. Based on this model, when (to the nearest year) was the average SAT score decreasing most rapidly? When was it increasing most rapidly?

Source: The authors' model based on data from Educational Testing Service.

47. ***Embryo Development*** The oxygen consumption of a bird embryo increases from the time the egg is laid through the time the chick hatches. In a typical galliform bird, the oxygen consumption (in milliliters per hour) can be approximated by

$$c(t) = -0.00271t^3 + 0.137t^2 - 0.892t + 0.149 \quad (8 \le t \le 30)$$

where t is the time (in days) since the egg was laid.[10] (An egg will typically hatch at around $t = 28$.) When, to the nearest day, is $c'(t)$ a maximum? What does the answer tell you?

48. ***Embryo Development*** The oxygen consumption of a turkey embryo increases from the time the egg is laid through the time the chick hatches. In a brush turkey, the oxygen consumption (in milliliters per hour) can be approximated by

$$c(t) = -0.00118t^3 + 0.119t^2 - 1.83t + 3.972 \quad (20 \le t \le 50)$$

where t is the time (in days) since the egg was laid. (An egg will typically hatch at around $t = 50$.) When, to the nearest day, is $c'(t)$ a maximum? What does the answer tell you?

49. ***Minimizing Resources*** Basic Buckets, Inc., has an order for plastic buckets that hold 5000 cubic centimeters. The buckets are open-topped cylinders, and the company wants to know what dimensions will use the least plastic per bucket. (The volume of an open-topped cylinder with height h and radius r is $\pi r^2 h$; the surface area is $\pi r^2 + 2\pi rh$.)

50. ***Optimizing Capacity*** Basic Buckets would like to build a bucket with a surface area of 1000 square centimeters. What is the volume of the largest bucket they can build? (See Exercise 49.)

The use of either a graphing calculator or graphing computer software is required for Exercises 51–58.

T **51.** ***Book Sales*** The number of books sold in the United States from 1990 to 1997 can be approximated by

$$n(t) = 2 + \frac{1}{2(1 + 3e^{-0.5t})} \quad (0 \le t \le 7)$$

where $n(t)$ is the number of books sold, in billions per year, t years after 1990. When, to the nearest year, were book sales increasing most rapidly?

Source: The authors' model from data by Book Industry Study Group/*New York Times*, August 13, 1995, p. F2.

T **52.** ***Employment*** The number of employees at Amerada Hess Corp. for the period 1984 ($t = 0$) through 1993 ($t = 9$) could be approximated by the formula[11]

$$n(t) = 7500 + \frac{23{,}000}{8 + 100e^{-0.5t}}$$

When, to the nearest year, was employment at Amerada Hess changing most rapidly? When was it changing most slowly?

T **53.** ***Education*** In 1991 the expected income of an individual depended on his or her educational level, according to the following formula:

$$I(n) = 2928.8n^3 - 115{,}860n^2 + 1{,}532{,}900n - 6{,}760{,}800 \quad (12 \le n \le 15)$$

---

[9]This is similar to the way site licences have been structured for the program Maple®.

[10]The models in Exercises 47 and 48 approximate graphical data published in the article "The Brush Turkey" by Roger S. Seymour in *Scientific American*, December 1991, pp. 108–114.

[11]The model is based on a best-fit logistic curve. *Source for data:* Hoover's Handbook Database (World Wide Web site), The Reference Press, Inc., Austin, Texas, 1995.

where n is the number of school years completed and $I(n)$ is the individual's expected annual income in dollars.[12] Using [12, 15] as the domain, locate and classify the absolute extrema of $I'(n)$, and interpret the results.

T 54. ***Marriage*** Based on statistics published by the U.S. Bureau of the Census, the median age of an individual at his or her first marriage can be modeled by the following functions:

Females:

$$F(n) = 0.000023453n^3 - 0.0026363n^2 + 0.050582n + 21.766$$

Males:

$$M(n) = 0.000023807n^3 - 0.0025184n^2 + 0.015754n + 25.966$$

where n = number of years since 1890; $0 \le n \le 102$.

a. Using [1, 102] as the domain, locate the relative extrema of both functions. Round your answers to the nearest integer and interpret the results.

b. What do the relative extrema of $F'(n)$ and $M'(n)$ tell?

c. What do the relative extrema of $M(n) - F(n)$ tell?

Source: U.S. Bureau of the Census, "Marital Status and Living Arrangements: March 1992," Current Population Reports, *Population Characteristics,* Series P-20, No. 468, March 1992, p. vii.

T 55. ***Asset Appreciation*** You manage a small antique company that owns a collection of Louis XVI jewelry boxes. Their value v is increasing according to the formula

$$v = \frac{10{,}000}{1 + 500e^{-0.5t}}$$

where t is the number of years from now. You anticipate an inflation rate of 5% per year, so that the present value of an item that will be worth $\$v$ in t years' time is given by

$$p = v(1.05)^{-t}$$

When (to the nearest year) should you sell the jewelry boxes to maximize their present value? How much (to the nearest constant dollar) will they be worth at that time?

T 56. ***Harvesting Forests*** The following equation models the approximate volume in cubic feet of a typical Douglas fir tree of age t years:

$$V = \frac{22{,}514}{1 + 22{,}514t^{-2.55}}$$

The lumber will be sold at \$10 per cubic foot, and you do not expect the price of lumber to appreciate in the foreseeable future. On the other hand, you anticipate a general inflation rate of 5% per year, so that the present value of an item that will be worth $\$v$ in t years' time is given by

$$p = v(1.05)^{-t}$$

At what age (to the nearest year) should you harvest a Douglas fir tree to maximize its present value? How much (to the nearest constant dollar) will a Douglas fir tree be worth at that time?

Source: The authors' model based on data in *Environmental and Natural Resource Economics,* 3d ed., by Tom Tietenberg (New York: HarperCollins, 1992), p. 282.

T 57. ***Resource Allocation*** Your automobile assembly plant has a Cobb-Douglas production function given by

$$q = x^{0.4}y^{0.6}$$

where q is the number of automobiles it produces per year, x is the number of employees, and y is the daily operating budget (in dollars). Annual operating costs amount to an average of \$20,000 per employee plus the operating budget of $\$365y$. Assume you wish to produce 1000 automobiles per year at a minimum cost. How many employees should you hire?

T 58. ***Resource Allocation*** Repeat Exercise 57 using the production formula

$$q = x^{0.5}y^{0.5}$$

59. ***Revenue*** *(based on a question on the GRE economics test)* If total revenue (TR) is specified by $TR = a + bQ - cQ^2$, where Q is quantity of output and a, b, and c are positive parameters, then TR is maximized for this firm when it produces Q equal to:

a. $b/2ac$ **b.** $b/4c$ **c.** $(a+b)/c$ **d.** $b/2c$ **e.** $c/2b$

Source: GRE Economics Test, by G. Gallagher, G. E. Pollock, W. J. Simeone, and G. Yohe (Piscataway, NJ: Research and Education Association, 1989).

60. ***Revenue*** *(based on a question on the GRE economics test)* If total demand (Q) is specified by $Q = -aP + b$, where P is unit price and a and b are positive parameters, then total revenue is maximized for this firm when it charges P equal to:

a. $b/2a$ **b.** $b/4a$ **c.** a/b **d.** $a/2b$ **e.** $-b/2a$

Source: See the source in Exercise 59.

[12] The model is a best-fit cubic based on Table 358, U.S. Department of Education, *Digest of Education Statistics, 1991* (Washington, DC: Government Printing Office, 1991).

Communication and Reasoning Exercises

61. Explain why the following problem is uninteresting: A packaging company wishes to make cardboard boxes with open tops by cutting square pieces from the corners of a square sheet of cardboard and folding up the sides. What is the box with the smallest surface area it can make this way?

62. Explain why finding the production level that minimizes a cost function is frequently uninteresting. What would a more interesting objective be?

63. If demand q decreases as price p increases, what does the minimum value of dq/dp measure?

64. Explain how you would solve an optimization problem of the following form: Maximize $P = f(x, y, z)$ subject to $z = g(x, y)$ and $y = h(x)$.

5.3 The Second Derivative and Analyzing Graphs

Acceleration

Recall that if $s(t)$ represents the position of a car at time t, then its velocity is given by the derivative: $v(t) = s'(t)$. But one rarely drives a car at a constant speed; the velocity itself is changing. The rate at which the velocity is changing is the **acceleration.** Since the derivative measures the rate of change, acceleration is the derivative of velocity: $a(t) = v'(t)$. Since v is the derivative of s, we can express the acceleration in terms of s:

$$a(t) = v'(t) = (s')'(t) = s''(t)$$

That is, a is the derivative of the derivative of s, which we call the **second derivative** of s and write as s''. (In this context you will often hear the derivative s' referred to as the **first derivative.**)

Acceleration and the Second Derivative

The **acceleration** of a moving object is the derivative of its velocity—that is, the second derivative of the position function.

Quick Example
If t is time in hours and the position of a car at time t is $s(t) = t^3 + 2t^2$ miles, then the car's velocity is $v(t) = s'(t) = 3t^2 + 4t$ miles per hour and its acceleration is $a(t) = s''(t) = 6t + 4$ miles per hour per hour.

example 1

Acceleration Due to Gravity
According to the laws of physics, an object near the surface of the earth falling in a vacuum from an initial rest position under the influence of gravity will travel a distance of approximately

$$s(t) = 16t^2 \text{ feet}$$

in t seconds. Find its acceleration.

Solution
The velocity of the object is

$$v(t) = s'(t) = 32t \text{ ft/s}$$

Hence, the acceleration is

$$a(t) = s''(t) = 32 \text{ ft/s}^2$$

(We write ft/s^2 as an abbreviation for feet/second/second—that is, feet per second per second. It is often read "feet per second squared.") Thus, the velocity is increasing at a rate of 32 ft/s every second, and we call 32 ft/s^2 the **acceleration**

due to gravity. If we ignore air resistance, all falling bodies, no matter what their weight, will fall with this acceleration.[1]

Before we go on... This was one of Galileo's most important discoveries: In very careful experiments using balls rolling down inclined planes, he discovered that the acceleration due to gravity is constant and that it does not depend on the weight or composition of the object falling.[2] A famous, though probably apocryphal, story has him dropping cannonballs of different weights off the Leaning Tower of Pisa to prove his point.[3]

We have written the second derivative of $f(x)$ as $f''(x)$. We could also use differential notation:

$$f''(x) = \frac{d^2f}{dx^2}$$

This notation comes from writing the second derivative as the derivative of the derivative in differential notation:

$$f''(x) = \frac{d}{dx}\left(\frac{df}{dx}\right) = \frac{d^2f}{dx^2}$$

example 2

Acceleration of Sales

For the first 15 months after the introduction of a new video game, the total number of units sold (total sales) can be modeled by the curve

$$S(t) = 20e^{0.4t}$$

where t is the time in months since the game was introduced. After about 25 months, total sales follow more closely the curve

$$S(t) = 100{,}000 - 20e^{17-0.4t}$$

How fast are total sales accelerating after 10 months? How fast are they accelerating after 30 months? What do these numbers mean?

Solution

By acceleration we mean the rate of change of the rate of change, which is the second derivative. During the first 15 months, the first derivative of sales is

$$\frac{dS}{dt} = 8e^{0.4t}$$

[1] There is nothing unique about the number 32. On other planets the number is different. For example, on Jupiter the acceleration due to gravity is about three times as large.

[2] An interesting aside: Galileo's experiments depended on getting extremely accurate timings. Since the timepieces of his day were very inaccurate, he used the most accurate time measurement he could: He sang and used the beat as his stopwatch.

[3] A true story: The point was made again during the Apollo 15 mission to the moon (July 1971) when astronaut David R. Scott dropped a feather and a hammer from the same height. The moon has no atmosphere, so the two hit the surface of the moon simultaneously.

and so the second derivative is

$$\frac{d^2S}{dt^2} = 3.2e^{0.4t}$$

Thus, after 10 months the acceleration of sales is

$$\left.\frac{d^2S}{dt^2}\right|_{t=10} = 3.2e^4 \approx 175 \text{ units/month per month, or units/month}^2$$

We can also compute total sales:

$$S(10) = 20e^4 \approx 1092 \text{ units}$$

and the rate of change of sales:

$$\left.\frac{dS}{dt}\right|_{t=10} = 8e^4 \approx 437 \text{ units/month}$$

What do these numbers mean? By the end of the tenth month, a total of 1092 video games have been sold. At that time, the game is selling at the rate of 437 units per month. This rate of sales is increasing by 175 units per month per month. More games will be sold each month than the month before.

To analyze the sales after 30 months is similar, using the formula

$$S(t) = 100{,}000 - 20e^{17-0.4t}$$

The derivative is

$$\frac{dS}{dt} = 8e^{17-0.4t}$$

and the second derivative is

$$\frac{d^2S}{dt^2} = -3.2e^{17-0.4t}$$

After 30 months,

$$S(30) = 100{,}000 - 20e^{17-12} \approx 97{,}032 \text{ units}$$

$$\left.\frac{dS}{dt}\right|_{t=30} = 8e^{17-12} \approx 1187 \text{ units/month}$$

$$\left.\frac{d^2S}{dt^2}\right|_{t=30} = -3.2e^{17-12} \approx -475 \text{ units/month}^2$$

By the end of the 30th month, 97,032 video games have been sold, the game is selling at a rate of 1187 units per month, and the rate of sales is *decreasing* by 475 units per month per month. Fewer games will be sold each month than the month before.

Concavity

Question I know that the first derivative of f tells where the graph of f is rising [where $f'(x) > 0$] and where it is falling [where $f'(x) < 0$]. What, if anything, does the second derivative tell about the graph of f?

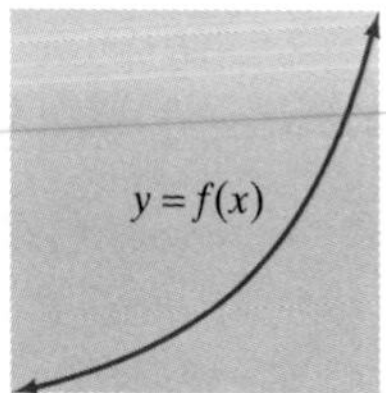

Figure 23

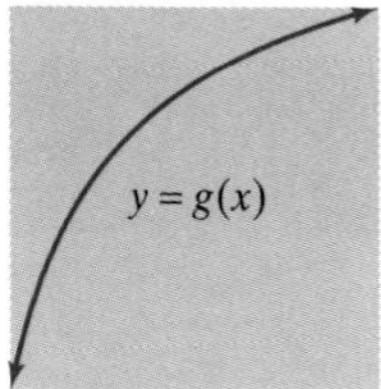

Figure 24

Answer The second derivative tells in what direction the graph of f *curves* or *bends.* Consider the graphs in Figures 23 and 24. Now think of a car driving from left to right along each of the roads shown in the two figures. A car driving along the graph of f in Figure 23 will turn to the left (upward); a car driving along the graph of g in Figure 24 will turn to the right (downward). We say that the graph of f is **concave up,** while the graph of g is **concave down.** Now think about the derivatives of f and g. The derivative $f'(x)$ starts small but *increases* as the graph gets steeper. Since $f'(x)$ is increasing, its derivative, $f''(x)$, must be positive. On the other hand, $g'(x)$ *decreases* as we go to the right. Since $g'(x)$ is decreasing, its derivative, $g''(x)$, must be negative. Summarizing, we have the following rules.

Concavity and the Second Derivative

A curve is **concave up** if its slope is increasing, in which case the second derivative is positive. A curve is **concave down** if its slope is decreasing, in which case the second derivative is negative. A point where the graph of f changes concavity, from concave up to concave down or vice versa, is called a **point of inflection.** At a point of inflection the second derivative is either zero or undefined.

Quick Example
Consider $f(x) = x^3 - 3x$, whose graph is shown in the figure.

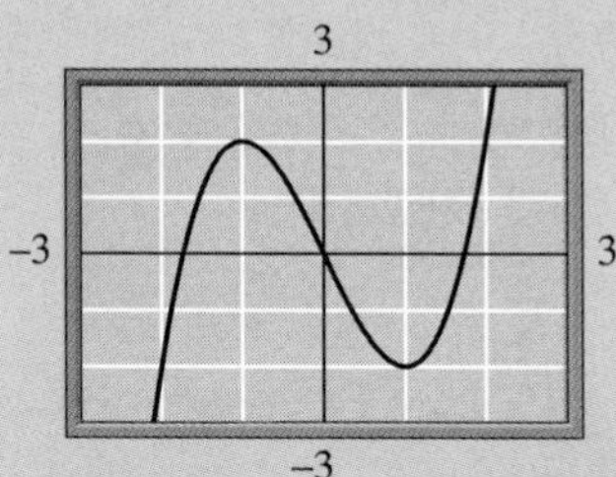

Here, $f''(x) = 6x$ is negative when $x < 0$ and positive when $x > 0$. The graph of f is concave down when $x < 0$ and concave up when $x > 0$. Also, f has a point of inflection at $x = 0$, where the second derivative is zero.

example 3

The Point of Diminishing Returns
After the introduction of a new video game, the total number of units sold (total sales) is modeled by the curve

$$S(t) = \frac{100{,}000}{1 + 5000e^{-0.4t}}$$

where t is the time in months since the game was introduced (compare Example 2). Where is the graph of S concave up and where is it concave down? Where are any points of inflection? What does this all mean?

Figure 25

Solution
The graph of S is shown in Figure 25. We see that the graph of S is concave up in the early months and then becomes concave down later. The point of inflection,

where the concavity changes, is somewhere around 20 months. With some calculus, we can determine exactly where the changeover occurs. Using the quotient rule, we compute

$$\frac{dS}{dt} = \frac{200{,}000{,}000e^{-0.4t}}{(1 + 5000e^{-0.4t})^2}$$

Using the quotient rule again (and simplifying), we get

$$\frac{d^2S}{dt^2} = \frac{80{,}000{,}000e^{-0.4t}(5000e^{-0.4t} - 1)}{(1 + 5000e^{-0.4t})^3}$$

Now, at the point of inflection this second derivative is either undefined or zero. The denominator is never zero, so there is no point at which the second derivative is not defined. For the second derivative to be zero, the numerator has to be zero, so we set

$$80{,}000{,}000e^{-0.4t}(5000e^{-0.4t} - 1) = 0$$

The factor $80{,}000{,}000e^{-0.4t}$ can never be zero, so we must have

$$5000e^{-0.4t} - 1 = 0$$

We can now solve for t:

$$5000e^{-0.4t} = 1$$

$$e^{-0.4t} = \frac{1}{5000} = 0.0002 \qquad \text{Exponential form}$$

$$-0.4t = \ln(0.0002) \qquad \text{Logarithmic form}$$

$$t = \frac{\ln(0.0002)}{-0.4} \approx 21.3$$

Thus, the graph of S is concave up for the first 21 months and is concave down after the 22nd month, with the point of inflection being at approximately $t = 21.3$ months.

What does this mean? In Example 2 we noted that where the graph of S is concave up, monthly sales are increasing; more units are being sold each month than the month before. Where the graph of S is concave down, monthly sales are decreasing; fewer units are being sold each month than the month before. The point of inflection is the time when sales stop increasing and start to fall off. This is sometimes called the **point of diminishing returns.** Although the total sales figure continues to rise (game units continue to be sold), the rate at which units are sold starts to drop.

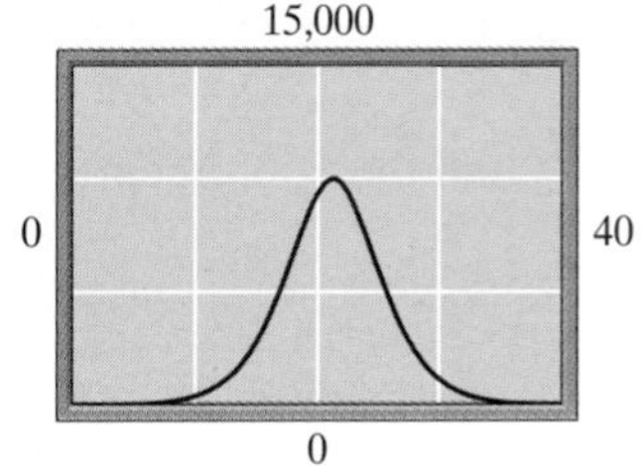

Figure 26

Before we go on . . . Since a point of inflection is the point at which the derivative changes from increasing to decreasing or vice versa, a point of inflection is always a relative maximum or minimum of the derivative. The graph of the derivative in this example is shown in Figure 26. We can see that the derivative has a relative (and absolute) maximum of approximately 10,000 at $t = 21.3$ months. That is, the monthly sales are highest in the 22nd month, when the game is selling at a rate of about 10,000 units per month.

Analyzing Graphs

We now have the tools we need to find the most interesting points on the graph of a function. It is easy to use graphing technology to draw the graph, but we need to use calculus to understand what we are seeing. The most interesting features of a graph are the following.

Features of a Graph

1. *The x- and y-intercepts:* If $y = f(x)$, find the x-intercept(s) by setting $y = 0$ and solving for x; find the y-intercept by setting $x = 0$.
2. *Relative extrema:* Use the technique of Section 5.1 to locate the relative extrema.
3. *Points of inflection:* Use the technique of this section to find the points of inflection.
4. *Behavior near points where the function is not defined:* If $f(x)$ is not defined at $x = a$, consider $\lim_{x\to a^-} f(x)$ and $\lim_{x\to a^+} f(x)$ to see how the graph of f approaches this point.
5. *Behavior at infinity:* Consider $\lim_{x\to -\infty} f(x)$ and $\lim_{x\to +\infty} f(x)$, if appropriate, to see how the graph of f behaves far to the left and right.

example 4

Analyzing a Graph

Analyze the graph of $f(x) = \dfrac{1}{x} - \dfrac{1}{x^2}$.

T Solution

The graph, as drawn using graphing technology, is shown in Figure 27. (Note that $x = 0$ is not in the domain of f.)

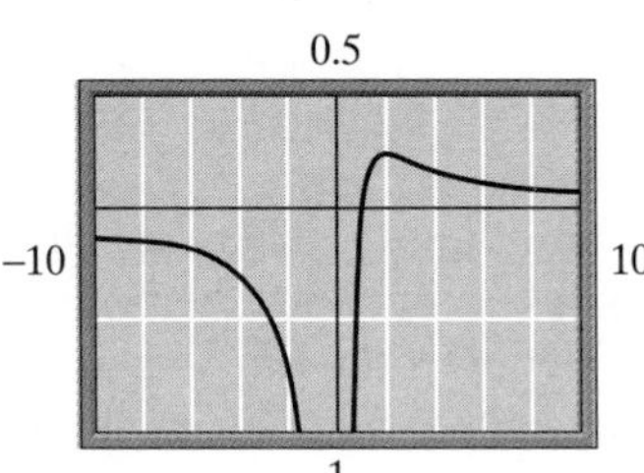

Figure 27

In the process of locating the interesting features of the graph, we can check that we've included them all in the viewing window.

1. *The x- and y-intercepts:* We consider $y = 1/x - 1/x^2$. To find the x-intercept(s), we set $y = 0$ and solve for x:

$$0 = \frac{1}{x} - \frac{1}{x^2}$$

$$\frac{1}{x} = \frac{1}{x^2}$$

Multiplying both sides by x^2 (we know that x cannot be zero, so we are not multiplying both sides by zero) gives

$$x = 1$$

Thus, there is one x-intercept (which we can see in Figure 27), at $x = 1$. We cannot substitute $x = 0$; since $f(0)$ is not defined, the graph does not meet the y-axis.

2. *Relative extrema:* We calculate $f'(x) = -1/x^2 + 2/x^3$. To find any stationary points, we set the derivative equal to zero and solve for x:

$$-\frac{1}{x^2} + \frac{2}{x^3} = 0$$

$$\frac{1}{x^2} = \frac{2}{x^3}$$

$$x = 2$$

Thus, there is one stationary point, at $x = 2$. The graph in Figure 27 shows that f has a relative maximum at this point. The only possible singular point is at $x = 0$ because $f'(0)$ is not defined. However, $f(0)$ is not defined either, so there are no singular points.

3. *Points of inflection:* We calculate $f''(x) = 2/x^3 - 6/x^4$. To find points of inflection, we set the second derivative equal to zero and solve for x:

$$\frac{2}{x^3} - \frac{6}{x^4} = 0$$

$$\frac{2}{x^3} = \frac{6}{x^4}$$

$$2x = 6$$

$$x = 3$$

Figure 27 confirms that the graph of f changes from being concave down to being concave up at $x = 3$, so this is a point of inflection. $f''(x)$ is not defined at $x = 0$, but that is not in the domain, so there are no other points of inflection. Thus, for example, the graph must be concave down in the whole region $(-\infty, 0)$.

4. *Behavior near points where f is not defined:* The only point where $f(x)$ is not defined is $x = 0$. From the graph, $f(x)$ appears to go to $-\infty$ as x approaches zero from either side. To calculate these limits, we rewrite $f(x)$:

$$f(x) = \frac{1}{x} - \frac{1}{x^2} = \frac{x-1}{x^2}$$

Now, if x is close to zero (on either side), the numerator $x - 1$ is close to -1 and the denominator is a very small but positive number. The quotient, therefore, is a negative number of very large magnitude. We have

$$\lim_{x \to 0^-} f(x) = -\infty$$

and $$\lim_{x \to 0^+} f(x) = -\infty$$

as appears to happen in Figure 27. We say that f has a **vertical asymptote** at $x = 0$, meaning that the graph approaches the line $x = 0$ without touching it.

5. *Behavior at infinity:* Both $1/x$ and $1/x^2$ go to zero as x goes to $-\infty$ or $+\infty$; that is,

$$\lim_{x \to -\infty} f(x) = 0$$

and
$$\lim_{x \to +\infty} f(x) = 0$$

as we can see happening in Figure 27. We say that f has a **horizontal asymptote** at $y = 0$.

In summary, there is one x-intercept, at $x = 1$; there is one relative maximum (which, we can now see, is also an absolute maximum) at $x = 2$; and there is one point of inflection at $x = 3$, where the graph changes from being concave down to being concave up. The graph approaches zero to the left and to the right, and it goes to $-\infty$ approaching $x = 0$. All of these features can be seen in Figure 27.

5.3 exercises

Calculate d^2y/dx^2 in Exercises 1–10.

1. $y = 3x^2 - 6$
2. $y = -x^2 + x$
3. $y = 2/x$
4. $y = -2/x^2$
5. $y = 4x^{0.4} - x$
6. $y = 0.2x^{-0.1}$
7. $y = e^{-(x-1)} - x$
8. $y = e^{-x} + e^x$
9. $y = 1/x - \ln x$
10. $y = x^{-2} + \ln x$

In Exercises 11–16, the position s of a point (in feet) is given as a function of time t (in seconds). Find **a.** *its acceleration as a function of t;* **b.** *its acceleration at the specified time.*

11. $s = 12 + 3t - 16t^2; t = 2$
12. $s = -12 + t - 16t^2; t = 2$
13. $s = \dfrac{1}{t} + \dfrac{1}{t^2}; t = 1$
14. $s = \dfrac{1}{t} - \dfrac{1}{t^2}; t = 2$
15. $s = \sqrt{t} + t^2; t = 4$
16. $s = 2\sqrt{t} + t^3; t = 1$

In each of Exercises 17–24, the graph of a function is given. Find the approximate coordinates of all points of inflection of each function (if any).

17.

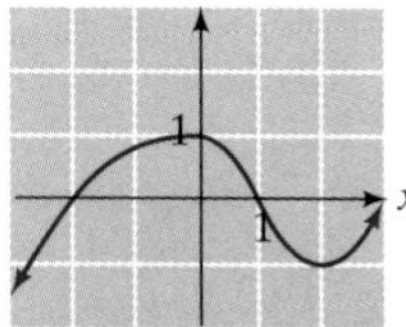

18.

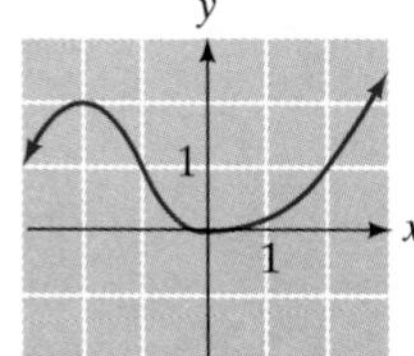

19.

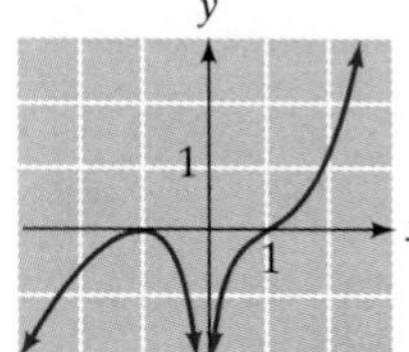

20.

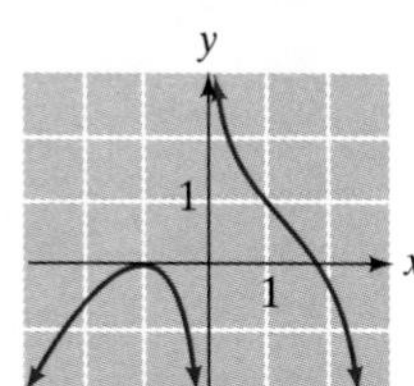

21. y, 1, 1, x

22. y, 1, 1, x

23. y, 1, 1, x

24. y, 1, 1, x

In each of Exercises 25–28, the graph of the derivative, $f'(x)$, *is given. Determine the x-coordinates of all points of inflection of $f(x)$, if any. (Assume that $f(x)$ is defined and continuous everywhere in $[-3, 3]$.)*

25.

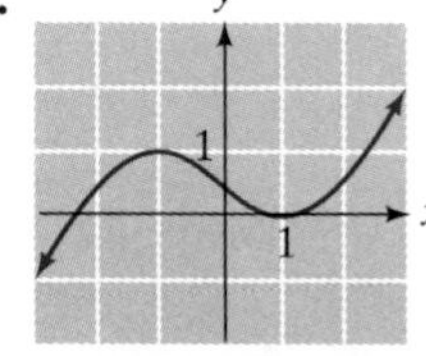

26.

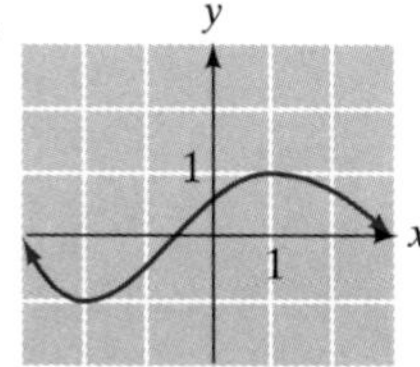

27.

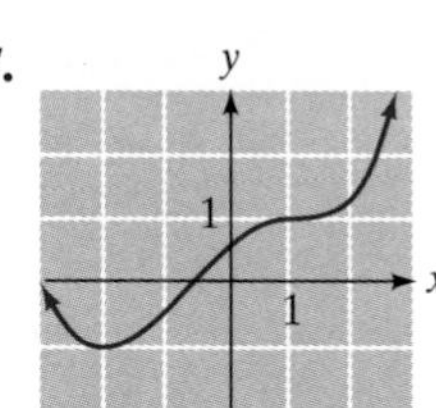

28.

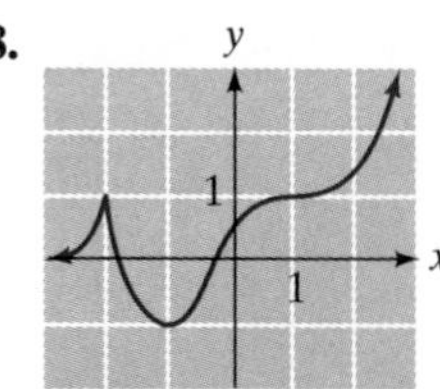

In each of Exercises 29–34, the graph of the second derivative, $f''(x)$, *is given. Determine the x-coordinates of all points of inflection of $f(x)$, if any. (Assume that $f(x)$ is defined and continuous everywhere in $[-3, 3]$.)*

29. 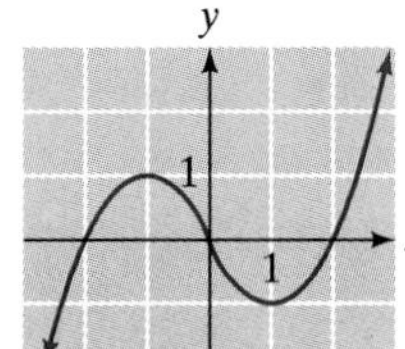

30. 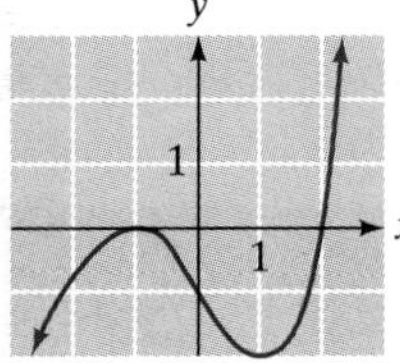

31. 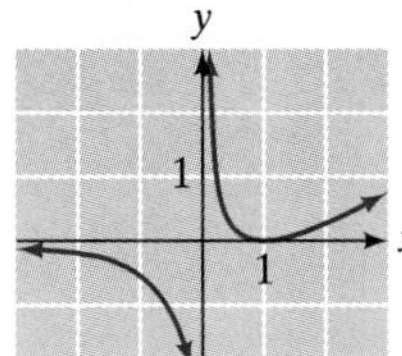

32. 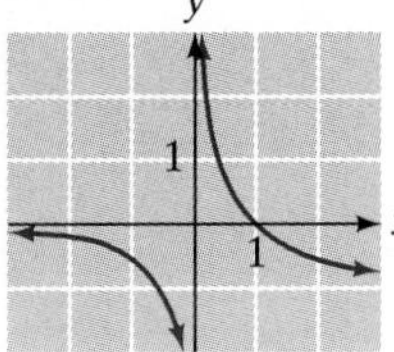

33. 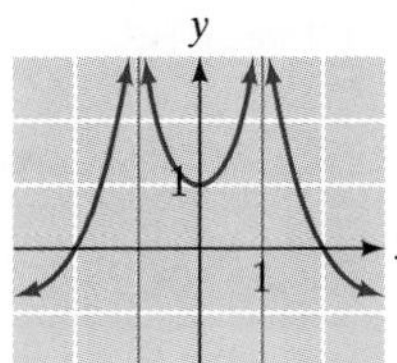

34. 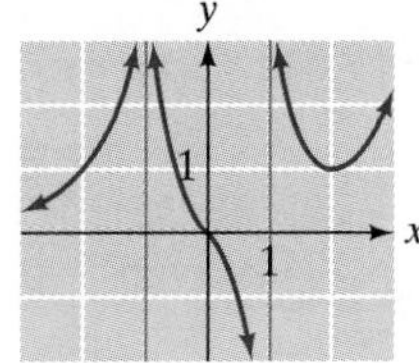

In Exercises 35–64, use technology to draw the graph of the given function, labeling all relative and absolute extrema and points of inflection. Find the coordinates of these points exactly, where possible. Also, find any vertical or horizontal asymptotes. In the marked exercises, use graphing technology to approximate the coordinates of the extrema and points of inflection to two decimal places.

35. $f(x) = x^2 + 2x + 1$ **36.** $f(x) = -x^2 - 2x - 1$

37. $f(x) = 2x^3 + 3x^2 - 12x + 1$

38. $f(x) = 4x^3 + 3x^2 + 2$

39. $g(x) = x^3 - 12x$; domain $[-4, 4]$

40. $g(x) = 2x^3 - 6x$; domain $[-4, 4]$

41. $g(t) = \frac{1}{4}t^4 - \frac{2}{3}t^3 + \frac{1}{2}t^2$

42. $g(t) = 3t^4 - 16t^3 + 24t^2 + 1$

43. $f(t) = \dfrac{t^2 + 1}{t^2 - 1}$; domain $[-2, 2]$

44. $f(t) = \dfrac{t^2 - 1}{t^2 + 1}$; domain $[-2, 2]$

45. $g(x) = x^3/(x^2 + 3)$ **46.** $g(x) = x^3/(x^2 - 3)$

47. $f(x) = x + \dfrac{1}{x}$ **48.** $f(x) = x^2 + \dfrac{1}{x^2}$

49. $g(x) = (x - 3)\sqrt{x}$ **50.** $g(x) = (x + 3)\sqrt{x}$

51. $f(x) = x - \ln x$; domain $(0, +\infty)$

52. $f(x) = x - \ln x^2$; domain $(0, +\infty)$

53. $f(x) = x^2 + \ln x^2$ **54.** $f(x) = 2x^2 + \ln x$

55. $g(t) = e^t - t$; domain $[-1, 1]$

56. $g(t) = e^{-t^2}$

57. $f(x) = \sin x$; domain $[-\pi, \pi]$

58. $f(x) = \cos x$; domain $[-\pi, \pi]$

T **59.** $f(x) = x^4 - 2x^3 + x^2 - 2x + 1$

T **60.** $f(x) = x^4 + x^3 + x^2 + x + 1$

T **61.** $f(x) = e^x - x^3$ T **62.** $f(x) = e^x - \dfrac{x^4}{4}$

T **63.** $f(x) = \cos x + 0.1x^2$; domain $[-6, 6]$

T **64.** $f(x) = \sin x + 0.01x^4$; domain $[-4, 4]$

Applications

65. ***Epidemics*** The following graph shows the total number, n, of people (in millions) infected in an epidemic as a function of time t (in years):

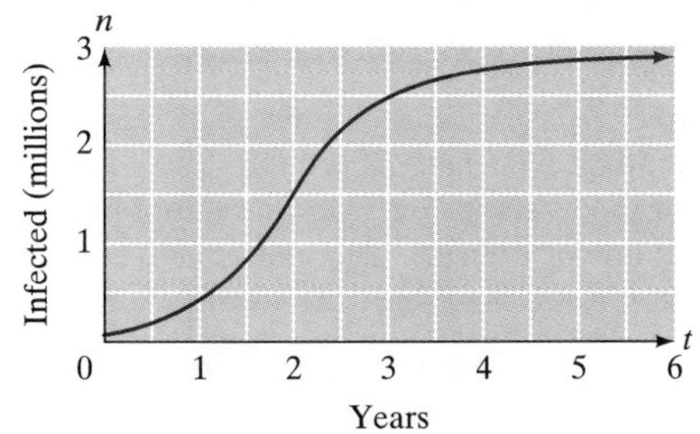

a. When, to the nearest year, was the highest rate of new infection?

b. When could the Centers for Disease Control announce that the rate of new infection was beginning to drop?

66. ***Sales*** The following graph shows the total number of Pomegranate II computers sold since their release:

a. When were the computers selling fastest?
b. Explain why this graph might look as it does.

67. ***Industrial Output*** The following graph shows the yearly industrial output of a developing country (measured in billions of dollars) over a 7-year period:

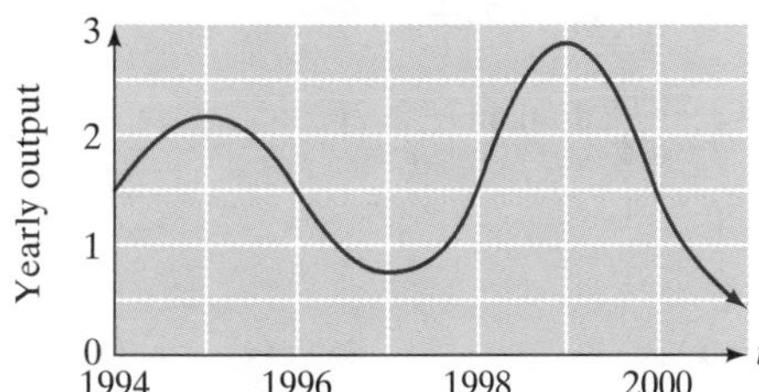

a. When (to the nearest year) did the rate of change of yearly industrial output reach a maximum?
b. When (to the nearest year) did the rate of change of yearly industrial output reach a minimum?
c. When (to the nearest year) did the rate of change of yearly industrial output first start to increase?

68. ***Profits*** The following graph shows the yearly profits (in billions of dollars) of Gigantic Conglomerate, Inc. (GCI), from 1990 to 2005:

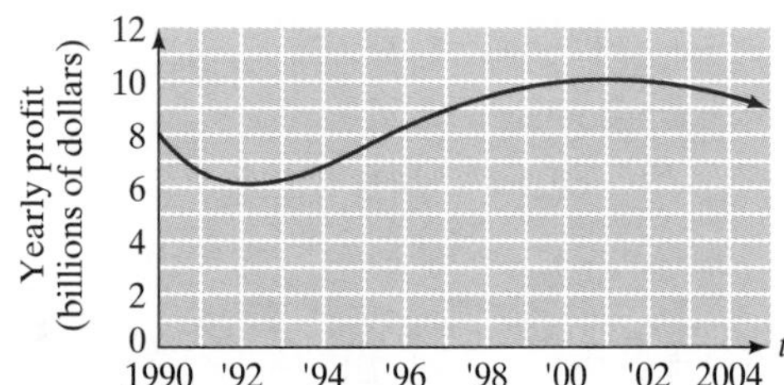

a. When were the profits rising most rapidly?
b. When were the profits falling most rapidly?
c. When could GCI's board of directors legitimately tell stockholders that they had "turned the company around"?

69. ***Prison Population*** The size of the U.S. prison population followed the curve

$$N(t) = 0.028234t^3 - 1.0922t^2 + 13.029t + 146.88 \quad (0 \le t \le 39)$$

in the years 1950–1989. Here t is the number of years since 1950, and N is the number of prisoners in thousands. Locate all points of inflection on the graph of N, and interpret the result.

Source: The authors' model from data in *Sourcebook of Criminal Justice Statistics,* 1990, U.S. Dept. of Justice, Bureau of Justice Statistics (Washington, D.C.: Government Printing Office), p. 604.

70. ***Test Scores*** Combined SAT scores in the United States could be approximated by

$$T(t) = -0.01085t^3 + 0.5804t^2 - 10.12t + 962.4 \quad (0 \le t \le 22)$$

in the years 1967–1991. Here t is the number of years since 1967, and T is the combined SAT score average for the United States. Locate all points of inflection on the graph of T, and interpret the result.

Source: The authors' model based on data from Educational Testing Service.

71. ***Education and Crime*** The following graph shows a striking relationship between the total prison population and the average combined SAT score in the United States:

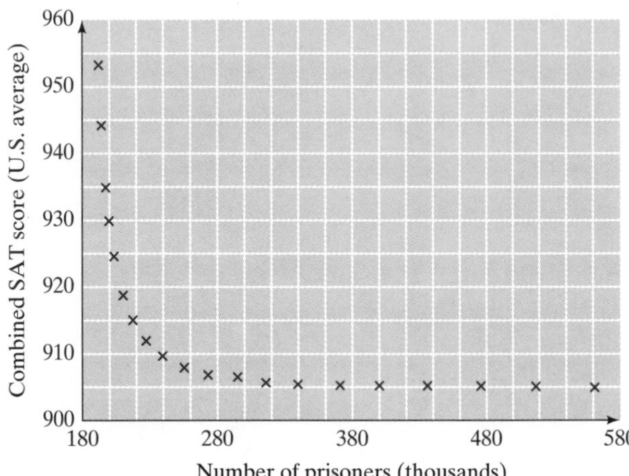

These data for the years 1967–1989 can be accurately modeled by

$$S(n) = 904 + \frac{1326}{(n - 180)^{1.325}} \quad (192 \le n \le 563)$$

here, $S(n)$ is the combined U.S. average SAT score at a time when the total U.S. prison population was n thousand.

a. Are there any points of inflection on the graph of S?
b. What does the concavity of the graph of S tell you about prisons and SAT scores?

Source: The authors' model from data in *Sourcebook of Criminal Justice Statistics,* 1990, p. 604/Educational Testing Service.

72. ***Education and Crime*** Refer back to the model in Exercise 71.

a. Are there any points of inflection on the graph of S'?
b. When is S'' a maximum? Interpret your answer in terms of prisoners and SAT scores.

73. ***Patents*** In 1965 the economist F. M. Scherer modeled the number, n, of patents produced by a firm as a function of the size, s, of the firm (measured in annual sales in millions of dollars). He came up with the following equation based on a study of 448 large firms:

$$n = -3.79 + 144.42s - 23.86s^2 + 1.457s^3$$

a. Find $\left.\dfrac{d^2n}{ds^2}\right|_{s=3}$. Is the rate at which patents are produced as the size of a firm goes up increasing or decreasing with size when $s = 3$? Comment on Scherer's words, "... we find diminishing returns dominating."

b. Find $\left.\dfrac{d^2n}{ds^2}\right|_{s=7}$ and interpret the answer.

c. Find the s-coordinate of any points of inflection and interpret the result.

Source: F. M. Scherer, "Firm Size, Market Structure, Opportunity, and the Output of Patented Inventions," *American Economic Review,* Vol. 55, December 1965, pp. 1097–1125.

74. ***Return on Investments*** A company finds that the number of new products it develops per year depends on the size of its annual R&D budget, x (in thousands of dollars), according to the formula

$$n(x) = -1 + 8x + 2x^2 - 0.4x^3$$

a. Find $n''(1)$ and $n''(3)$, and interpret the results.

b. Find the size of the budget that gives the highest rate of return as measured in new products per dollar (again, called the point of diminishing returns).

T **75.** ***Asset Appreciation*** You manage a small antique store that owns a collection of Louis XVI jewelry boxes. Their value v is increasing according to the formula

$$v = \frac{10{,}000}{1 + 500e^{-0.5t}}$$

where t is the number of years from now. You anticipate an inflation rate of 5% per year, so that the present value of an item that will be worth \$$v$ in t years' time is given by

$$p = v(1.05)^{-t}$$

What is the greatest rate of increase of the present value of your antiques, and when is this rate attained?

T **76.** ***Harvesting Forests*** The following equation models the approximate volume in cubic feet of a typical Douglas fir tree of age t years.

$$V = \frac{22{,}514}{1 + 22{,}514t^{-2.55}}$$

The lumber will be sold at \$10 per cubic foot, and you do not expect the price of lumber to appreciate in the foreseeable future. On the other hand, you anticipate a general inflation rate of 5% per year, so that the present value of an item that will be worth \$$v$ in t years' time is given by

$$p = v(1.05)^{-t}$$

What is the greatest rate of increase of the present value of a fir tree, and when is this rate attained?

Source: The authors' model based on data in *Environmental and Natural Resource Economics,* 3d ed., by Tom Tietenberg (New York: HarperCollins, 1992), p. 282.

T **77.** ***Asset Appreciation*** As the financial consultant to a classic auto dealership, you estimate that the total value of its collection of 1959 Chevrolets and Fords is given by the formula

$$v = 300{,}000 + 1000t^2$$

where t is the number of years from now. You anticipate a continuous inflation rate of 5% per year, so that the discounted (present) value of an item that will be worth \$$v$ in t years' time is given by

$$p = ve^{-0.05t}$$

When is the discounted value of the collection of classic cars increasing most rapidly? When is it decreasing most rapidly?

T **78.** ***Plantation Management*** The value of a fir tree on your plantation increases with the age of the tree according to the formula

$$v = \frac{20t}{1 + 0.05t}$$

where t is the age of the tree in years. With a continuous inflation rate of 5% per year, the discounted (present) value of a newly planted seedling is

$$p = ve^{-0.05t}$$

When is the discounted value of a tree increasing most rapidly? When is it decreasing most rapidly?

T **79.** ***Venture Capital*** Venture capital firms raise money to assist promising entepreneurs until they can sell shares to the public. The following graphs show the total amount of money controlled by venture capital firms as well as the number of venture capital firms:

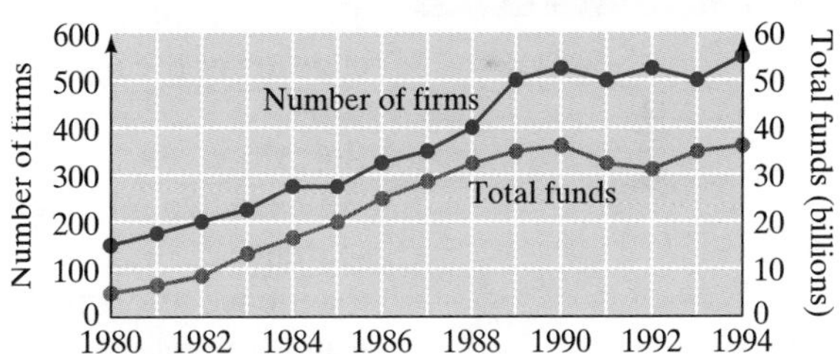

Source: Venture Economics Information Services, Boston/*New York Times,* May 25, 1995, p. D5.

Let $N(t)$ be the number of firms at time t years since 1980, and let $R(t)$ be the total amount of money these firms control. These functions can be approximated by the following linear functions (values are rounded):

$$N(t) = 31t + 145$$

$$R(t) = 2.4t + 7.3$$

a. Graph the function $R(t)/N(t)$, and use the graph to estimate $\lim_{t\to+\infty} R(t)/N(t)$.

b. What does $R(t)/N(t)$ represent? What does $\lim_{t\to+\infty} R(t)/N(t)$ represent?

T **80.** ***Book Sales*** Revenue from book sales in the United States increased at a more or less constant rate from $18 billion in 1990 to a projected $27 billion in 1997. The number of books sold can be approximated by

$$n(t) = 2 + \frac{1}{2(1 + 3e^{-0.5t})} \qquad (0 \le t \le 7)$$

where $n(t)$ is the number of books sold, in billions, in the tth year since 1990.

a. Construct a linear model for the annual revenue, $R(t)$, from book sales in the tth year since 1990.

b. Use technology to graph the curve $y = R(t)/n(t)$ for $0 \le t \le 7$.

c. Locate the approximate t-coordinate (to within ± 1) of any point(s) of inflection, and interpret the result.

Source: The authors' models based on data in Book Industry Study Group/*New York Times,* August 13, 1995, p. F2. Data are rounded.

Communication and Reasoning Exercises

81. Complete the following: If the graph of a function is concave up on its entire domain, then its second derivative is ______ on the domain.

82. Complete the following: If the graph of a function is concave up on its entire domain, then its first derivative is ______ on the domain.

83. Explain geometrically why the derivative of a function has a relative extremum at a point of inflection, if it is defined there. Which points of inflection give rise to relative maxima in the derivative?

84. If we regard position, s, as a function of time, t, what is the significance of the *third* derivative, $s'''(t)$? Describe an everyday scenario in which it arises.

5.4 Related Rates

We know that if Q is a quantity that is changing over time, then the derivative dQ/dt is the rate at which Q changes over time. In this section we are concerned with what are called **related rates** problems. In such a problem we have two (sometimes more) related quantities, we know the rate at which one is changing, and we wish to find the rate at which another is changing. A typical example follows.

example 1

The Expanding Circle

The radius of a circle is increasing at a rate of 10 cm/s. How fast is the area increasing at the instant when the radius has reached 5 cm?

Solution

We have two related quantities: the radius of the circle, r, and its area, A. The first sentence of the problem tells us that r is increasing at a certain rate. When we see a sentence referring to speed or change, it is very helpful to rephrase the sentence using "the rate of change of." Here, we can say

The rate of change of r is 10 cm/s.

Since the rate of change is the derivative, we can rewrite this sentence as the equation

$$\frac{dr}{dt} = 10$$

Similarly, the second sentence of the problem asks how A is changing. We can also rewrite that question:

What is the rate of change of A when the radius is 5 cm?

Using mathematical notation, we state the question as:

$$\textit{What is } \frac{dA}{dt} \textit{ when } r = 5?$$

Thus, knowing one rate of change, dr/dt, we wish to find a related rate of change, dA/dt. To find exactly how these derivatives are related, we need the equation relating the variables, which is

$$A = \pi r^2$$

To find the relationship between the derivatives, we take the derivative of both sides of this equation *with respect to t*. On the left we get dA/dt. On the right we need to be careful. We do not get $2\pi r$, because that would be the derivative with respect to r, not t. We need to remember that r is a function of t and use the chain rule. We get

$$\frac{dA}{dt} = 2\pi r \frac{dr}{dt}$$

Now we substitute the given values $r = 5$ and $dr/dt = 10$ to get

$$\left.\frac{dA}{dt}\right|_{r=5} = 2\pi(5)(10) = 100\pi \approx 314 \text{ cm}^2/\text{s}$$

We can organize our work as follows.

Solving a Related Rates Problem

A. The problem
  1. List the related, changing quantities.
  2. Restate the problem in terms of rates of change. Rewrite the problem using mathematical notation for the changing quantities and their derivatives.

B. The relationship
  1. Draw a diagram, if appropriate, showing the changing quantities.
  2. Find an equation or equations relating the changing quantities.
  3. Take the derivative with respect to time of the equation(s) relating the quantities to get the **derived equation(s),** relating the rates of change of the quantities.

C. The solution
  1. Substitute into the derived equation(s) the given values of the quantities and their derivatives.
  2. Solve for the derivative required.

We can illustrate the procedure with the "ladder problem" found in almost every calculus textbook.

example 2

The Falling Ladder

Jane is at the top of a 5-foot ladder when it starts to slide down the wall at a rate of 3 feet per minute. Jack is standing on the ground behind her. How fast is the base of the ladder moving when it hits him if Jane is 4 feet from the ground at that instant?

Solution

The first sentence talks about the (top of the) ladder sliding down the wall. Thus, one of the changing quantities is the height of the top of the ladder. The question asked refers to the motion of the base of the ladder, so another changing quantity is the distance of the base of the ladder from the wall. Let us record these variables and follow the outline above to obtain the solution.

A. The Problem

1. The changing quantities are h = the height of the top of the ladder and b = the distance of the base of the ladder from the wall.
2. We rephrase the problem in words, using the phrase "rate of change": The rate of change of the height of the top of the ladder is -3 feet per minute. What is the rate of change of the distance of the base from the wall when the top of the ladder is 4 feet from the ground? We can now rewrite the problem mathematically:

$$\frac{dh}{dt} = -3. \text{ Find } \frac{db}{dt} \text{ when } h = 4.$$

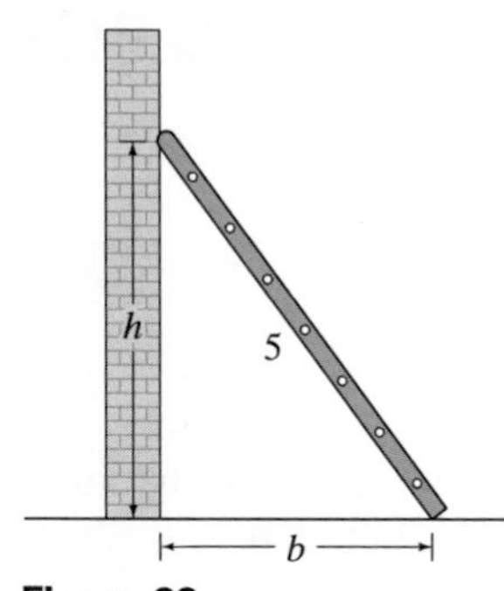

Figure 28

B. The Relationship

1. Figure 28 shows the ladder and the variables h and b. Notice that we put in the figure the fixed length, 5, of the ladder, but any changing quantities, like h and b, we leave as variables. We do not use any specific values for h or b until the very end.
2. From the figure we can see that h and b are related by the Pythagorean theorem:

$$h^2 + b^2 = 25$$

3. Taking the derivative with respect to time of the equation above gives us the derived equation:

$$2h\frac{dh}{dt} + 2b\frac{db}{dt} = 0$$

C. The Solution

1. We substitute the known values $dh/dt = -3$ and $h = 4$ into the derived equation:

$$2(4)(-3) + 2b\frac{db}{dt} = 0$$

We would like to solve for db/dt, but what do we do with the b in the equation? We can determine b from the equation $h^2 + b^2 = 25$, using the value $h = 4$:

$$16 + b^2 = 25$$
$$b^2 = 9$$
$$b = 3$$

Substituting into the derived equation, we get

$$-24 + 2(3)\frac{db}{dt} = 0$$

2. Solving for db/dt gives

$$\frac{db}{dt} = \frac{24}{6} = 4$$

Thus, the base of the ladder is sliding away from the wall at a speed of 4 ft/min when it hits Jack.

---

example 3

Average Cost

The cost to manufacture x portable pagers is

$$C(x) = \$10{,}000 + 3x + \frac{x^2}{10{,}000}$$

The production level is currently $x = 5000$ and it is increasing by 100 units per day. How is the average cost changing?

Solution

A. The Problem

1. The changing quantities are the production level, x, and the average cost, $\overline{C}$.
2. We rephrase the problem as follows: The production level x is 5000 units, and the rate of change of x is 100 units/day. What is the rate of change of the average cost, $\overline{C}$? In mathematical notation,

$$x = 5000 \text{ and } \frac{dx}{dt} = 100. \text{ Find } \frac{d\overline{C}}{dt}.$$

B. The Relationship

1. The changing quantities cannot easily be depicted geometrically.
2. We are given a formula for the *total* cost. We get the *average* cost by dividing the total cost by x:

$$\overline{C} = \frac{C}{x}$$

So, $$\overline{C} = \frac{10{,}000}{x} + 3 + \frac{x}{10{,}000}$$

3. Taking derivatives with respect to t of both sides, we get the derived equation:

$$\frac{d\overline{C}}{dt} = \left(-\frac{10{,}000}{x^2} + \frac{1}{10{,}000}\right)\frac{dx}{dt}$$

C. The Solution

Substituting the values from part A into the derived equation, we get

$$\frac{d\overline{C}}{dt} = \left(-\frac{10{,}000}{5000^2} + \frac{1}{10{,}000}\right)100 = -\,0.03 \text{ dollar/day}$$

Thus, the average cost is decreasing by 3¢ per day.

The scenario in the following example is similar to Example 5 in Section 5.2.

example 4

Automation

The Gym Sock Company manufactures cotton athletic socks. Production is partially automated through the use of robots. The number of pairs of socks the company can manufacture in a day is given by a Cobb-Douglas production formula:

$$q = 50n^{0.6}r^{0.4}$$

where q is the number of pairs of socks that can be manufactured by n laborers and r robots. The company currently produces 1000 pairs of socks per day and employs 20 laborers. It is bringing one new robot on line every month. At what rate are laborers being laid off, assuming that daily productivity remains constant?

Solution

A. The Problem

1. The changing quantities are the number of laborers, n, and the number of robots, r.
2. $\dfrac{dr}{dt} = 1$. Find $\dfrac{dn}{dt}$ when $n = 20$.

B. The Relationship

1. No diagram is appropriate here.
2. The equation relating the changing quantities is

$$1000 = 50n^{0.6}r^{0.4}$$

or

$$20 = n^{0.6}r^{0.4}$$

(Productivity is constant at 1000 pairs of socks per day.)

3. For the derived equation we have

$$\begin{aligned} 0 &= 0.6n^{-0.4}\left(\frac{dn}{dt}\right)r^{0.4} + 0.4n^{0.6}r^{-0.6}\left(\frac{dr}{dt}\right) \\ &= 0.6\left(\frac{r}{n}\right)^{0.4}\left(\frac{dn}{dt}\right) + 0.4\left(\frac{n}{r}\right)^{0.6}\left(\frac{dr}{dt}\right) \end{aligned}$$

We solve this equation for dn/dt because we shall want to find dn/dt below and because the equation becomes simpler when we do this:

$$0.6\left(\frac{r}{n}\right)^{0.4}\left(\frac{dn}{dt}\right) = -0.4\left(\frac{n}{r}\right)^{0.6}\left(\frac{dr}{dt}\right)$$

$$\frac{dn}{dt} = -\frac{0.4}{0.6}\left(\frac{n}{r}\right)^{0.6}\left(\frac{n}{r}\right)^{0.4}\left(\frac{dr}{dt}\right) = -\frac{2}{3}\left(\frac{n}{r}\right)\left(\frac{dr}{dt}\right)$$

C. The Solution

1. Substituting the known numbers into the preceding equation, we get

$$\frac{dn}{dt} = -\frac{2}{3}\left(\frac{20}{r}\right)(1)$$

We need to compute r by substituting the known value of n in the original formula:

$$20 = n^{0.6}r^{0.4}$$

$$20 = 20^{0.6}r^{0.4}$$

$$r^{0.4} = \frac{20}{20^{0.6}} = 20^{0.4}$$

giving $r = 20$.

2. Thus,

$$\frac{dn}{dt} = -\frac{2}{3}\left(\frac{20}{20}\right)(1) = -\frac{2}{3} \text{ laborer per month}$$

The company is laying off laborers at a rate of $\frac{2}{3}$ per month, or two every three months.

Before we go on . . . We can interpret this result as saying that, at the current level of production and number of laborers, one robot is as productive as $\frac{2}{3}$ of a laborer, or three robots are as productive as two laborers.

5.4 exercises

Rewrite the statements in Exercises 1–8 in mathematical notation.

1. The population P is currently 10,000 and growing at a rate of 1000 per year.

2. There are presently 400 cases of Bangkok flu, and the number is growing by 30 new cases every month.

3. The annual revenue of your tie-dye T-shirt operation is currently \$7000 and growing by 10% each year. How fast are annual sales increasing?

4. A ladder is sliding down a wall so that the distance between the top of the ladder and the floor is decreasing at a rate of 3 feet per second. How fast is the base of the ladder receding from the wall?

5. The price of shoes is rising \$5 per year. How fast is the demand changing?

6. Stock prices are rising \$1000 per year. How fast is the value of your portfolio increasing?

7. The average global temperature is 60°F and rising by 0.1°F per decade. How fast are annual sales of Bermuda shorts increasing?

8. The country's population is now 260,000,000 and is increasing by 1,000,000 people per year. How fast is the annual demand for diapers increasing?

Applications

9. ***Sun Spots*** The area of a circular sun spot is growing at a rate of 1200 km²/s.

a. How fast is the radius growing at the instant when it equals 10,000 km?

b. How fast is the radius growing at the instant when the sun spot has an area of 640,000 km²?

10. ***Puddles*** The radius of a circular puddle is growing at a rate of 5 cm/s.

a. How fast is its area growing at the instant when the radius is 10 cm?

b. How fast is the area growing at the instant when it equals 36 cm²?

11. ***Sliding Ladders*** The base of a 50-foot ladder is being pulled away from a wall at a rate of 10 feet per second. How fast is the top of the ladder sliding down the wall at the instant when the base of the ladder is 30 feet from the wall?

12. ***Sliding Ladders*** The top of a 5-foot ladder is sliding down a wall at a rate of 10 feet per second. How fast is the base of the ladder sliding away from the wall at the instant when the top of the ladder is 3 feet from the ground?

13. ***Demand*** Assume that the demand function for tuna in a small coastal town is given by

$$p = \frac{50,000}{q^{1.5}}$$

where q is the number of pounds of tuna that can be sold in 1 month at the price of p dollars per pound. The town's fishery finds that the demand for tuna is currently 900 pounds per month and is increasing at a rate of 100 pounds per month. How fast is the price changing?

14. ***Demand*** The demand equation for rubies at Royal Ruby Retailers is given by

$$q = -\frac{4}{3}p + 80$$

where q is the number of rubies RRR can sell per week at p dollars per ruby. RRR finds that the demand for its rubies is currently 20 rubies per week and is dropping at a rate of one per week. How fast is the price changing?

15. ***Demand*** Demand for your tie-dyed T-shirts is given by the formula

$$p = 5 + \frac{100}{\sqrt{q}}$$

where p is the price (in dollars) you can charge to sell q T-shirts per month. If you currently sell T-shirts for \$15 each and you raise your price by \$2 per month, how fast will the demand drop?

16. ***Supply*** The number of portable CD players you are prepared to supply to a retail outlet every week is given by the formula

$$p = 0.1q^2 + 3q$$

where p is the price the outlet offers. The retail outlet is currently offering you \$40 per CD player. If the price it offers decreases at a rate of \$10 per week, how will this affect the supply?

17. ***Revenue*** You can now sell 50 cups of lemonade per week at 30¢ per cup, but demand is dropping at a rate of 5 cups per week each week. Assuming that raising the price does not affect demand, how fast do you have to raise your price if you want to keep your weekly revenue constant?

18. ***Revenue*** You can now sell 40 cars per month at \$20,000 per car, and demand is increasing at a rate of 3 cars per month each month. What is the fastest you could drop your price before your monthly revenue starts to drop?

19. ***Production*** The automobile assembly plant you manage has a Cobb-Douglas production function given by

$$P = 10x^{0.3}y^{0.7}$$

where P is the number of automobiles the plant produces per year, x is the number of employees, and y is the daily operating budget (in dollars). You maintain a production level of 1000 automobiles per year. If you currently employ 150 workers and are hiring new workers at a rate of 10 per year, how fast is your daily operating budget changing?

20. ***Production*** Refer back to the Cobb-Douglas production formula in Exercise 19. Assume that you maintain a work force of 200 workers and wish to increase production to meet a demand that is increasing by 100 automobiles per year. The current demand is 1000 automobiles per year. How fast should your daily operating budget be increasing?

21. ***Balloons*** A spherical party balloon is being inflated by helium pumped in at a rate of 3 cubic feet per minute. How fast is the radius growing at the instant when the radius has reached 1 foot? (The volume of a sphere of radius r is $V = \frac{4}{3}\pi r^3$.)

22. ***More Balloons*** A rather flimsy spherical balloon is designed to pop at the instant its radius has reached 10 cm. Assuming the balloon is filled with helium at a rate of 10 cm^3/s, calculate how fast the diameter is growing at the instant it pops. (The volume of a sphere of radius r is $V = \frac{4}{3}\pi r^3$.)

23. ***Movement Along a Graph*** A point on the graph of $y = 1/x$ is moving along the curve in such a way that its x-coordinate is increasing at a rate of 4 units per second. What is happening to the y-coordinate at the instant the y-coordinate is equal to 2?

24. ***Motion Around a Circle*** A point is moving along the circle $x^2 + (y - 1)^2 = 8$ in such a way that its x-coordinate is decreasing at a rate of 1 unit per second. What is happening to the y-coordinate at the instant when the point has reached $(-2, 3)$?

25. ***Ships Sailing Apart*** The HMS *Dreadnaught* is 40 miles north of Montauk and steaming due north at 20 mph, while the USS *Mona Lisa* is 50 miles east of Montauk and steaming due east at an even 30 mph. How fast is their distance apart increasing?

26. ***Near Miss*** My aunt and I were approaching the same intersection, she from the south and I from the west. She was traveling at a steady speed of 10 mph, while I was approaching the intersection at 60 mph. At a certain instant in time, I was one-tenth of a mile from the intersection, while she was one-twentieth of a mile from it. How fast were we approaching each other at that instant?

27. ***Education*** In 1991 the expected income of an individual depended on his or her educational level, according to the following formula:

$$I(n) = 2928.8n^3 - 115{,}860n^2 + 1{,}532{,}900n - 6{,}760{,}800 \qquad (12 \le n \le 15)$$

where n is the number of school years completed and $I(n)$ is the individual's expected income.[1] You have completed 13 years of school and are currently a part-time student. Your schedule is such that you will complete the equivalent of 1 year of college every 3 years. Assuming that your salary is linked to the above model, how fast is your income going up? (Round your answer to the nearest $1.)

28. ***Education*** Refer back to the model in Exercise 27. Assume that someone has completed 14 years of school and that her income is increasing by $10,000 per year. How much schooling per year is this rate of increase equivalent to?

29. ***Employment*** An employment research company estimates that the value of a recent MBA graduate to an accounting company is

$$V = 3e^2 + 5g^3$$

where V is the value of the graduate, e is the number of years of prior business experience, and g is the graduate school grade point average. A company that currently employs graduates with a 3.0 average wishes to maintain a constant employee value of $V = 200$ but finds that the grade point average of its new employees is dropping at a rate of 0.2 per year. How fast must the experience of its new employees be growing to compensate for the decline in grade point average?

30. ***Grades*** A production formula for a student's performance on a difficult English examination is given by

$$g = 4hx - 0.2h^2 - 10x^2$$

where g is the grade the student can expect to obtain, h is the number of hours of study for the examination, and x is the student's grade point average. The instructor finds that students' grade point averages have remained constant at 3.0 over the years and that students currently spend an average of 15 hours studying for the examination. However, scores on the examination are dropping at a rate of 10 points per year. At what rate is the average study time decreasing?

Source: Based on an exercise in *Introduction to Mathematical Economics* by A. L. Ostrosky, Jr., and J. V. Koch (Prospect Heights, IL: Waveland Press, 1979).

31. ***Cones*** A right circular conical vessel is being filled with green industrial waste at a rate of 100 cubic meters per second. How fast is the level rising after 200π cubic meters have been poured in? (The cone has height 50 m and radius 30 m at its brim. The volume of a cone of height h and cross-sectional radius r at its brim is given by $V = \frac{1}{3}\pi r^2 h$.)

32. ***More Cones*** A circular conical vessel is being filled with ink at a rate of 10 cm³/s. How fast is the level rising after 20 cm³ have been poured in? (The cone has height 50 cm and radius 20 cm at its brim. The volume of a cone of height h and cross-sectional radius r at its brim is given by $V = \frac{1}{3}\pi r^2 h$.)

33. ***Cylinders*** The volume of paint in a right cylindrical can is given by $V = 4t^2 - t$, where t is time in seconds and V is the volume in cm³. How fast is the level rising when the height is 2 cm? The can has a height of 4 cm and a radius of 2 cm. [*Hint:* To get h as a function of t, first solve the volume $V = \pi r^2 h$ for h.]

34. ***Cylinders*** A cylindrical bucket is being filled with paint at a rate of 6 cm³/min. How fast is the level rising when the bucket starts to overflow? The bucket has radius 30 cm and height 60 cm.

Education and Crime *The following graph shows a striking relationship between the total prison population and the average combined SAT score in the United States:*

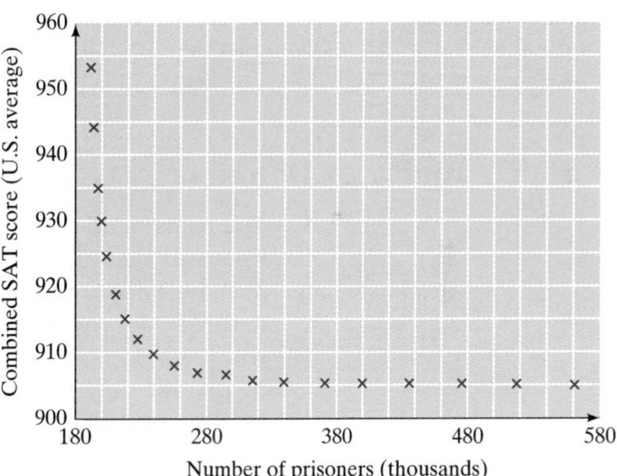

Exercises 35 and 36 are based on the following model for these data for the years 1967–1989:

$$S(n) = 904 + \frac{1326}{(n - 180)^{1.325}} \qquad (192 \le n \le 563)$$

Here, $S(n)$ is the combined average SAT score at a time when the total prison population is n thousand.

Source: The authors' model from data in *Sourcebook of Criminal Justice Statistics,* 1990, p. 604/Educational Testing Service.

[1] The model is a best-fit cubic based on Table 358, U.S. Department of Education, *Digest of Education Statistics, 1991* (Washington, DC: Government Printing Office, 1991).

35. In 1985 the U.S. prison population was 475,000 and increasing at a rate of 35,000 per year. What was the average SAT score, and how fast, and in what direction, was it changing? (Round your answers to two decimal places.)

36. In 1970 the U.S. combined SAT average was 940 and dropping by 10 points per year. What was the U.S. prison population, and how fast, and in what direction, was it changing? (Round your answers to the nearest 100.)

Divorce Rates *A study found that the divorce rate d (given as a percentage) appears to depend on the ratio r of available men to available women.*[2] *This function can be approximated by*

$$d(r) = \begin{cases} -40r + 74 & \text{if } r \le 1.3 \\ \dfrac{130r}{3} - \dfrac{103}{3} & \text{if } r > 1.3 \end{cases}$$

Exercises 37 and 38 are based on this model.

37. There are currently 1.1 available men per available woman in Littleville, and this ratio is increasing by 0.05 per year. What is happening to the divorce rate?

38. There are currently 1.5 available men per available woman in Largeville, and this ratio is decreasing by 0.03 per year. What is happening to the divorce rate?

[2] The cited study, by Scott J. South and associates, appeared in the *American Sociological Review,* February 1995. Figures are rounded. *Source: New York Times,* February 19, 1995, p. 40.

Communication and Reasoning Exercises

39. If you know how fast one quantity is changing and you need to compute how fast a second quantity is changing, what kind of information do you need?

40. If three quantities are related by a single equation, how do you go about computing how fast one of them is changing based on a knowledge of the other two?

41. Why is this section titled "related rates"?

42. In a recent exam you were given a related rates problem based on an algebraic equation relating two variables x and y. Your friend told you that the correct relationship between dx/dt and dy/dt is given by

$$\frac{dx}{dt} = \left(\frac{dy}{dt}\right)^2$$

Could he be correct?

43. Transform the following into a mathematical statement about derivatives: If my grades are improving at twice the speed of yours, then your grades are improving at half the speed of mine.

44. If two quantities x and y are related by a linear equation, how are their rates of change related?

5.5 Elasticity of Demand

You manufacture an extremely popular brand of sneakers, and you want to know what will happen if you increase the selling price. Common sense tells you that demand will drop as you raise the price. But, will the drop in demand be enough to cause your revenue to go down? Or will it be small enough that your revenue will go up because of the higher selling price? For example, if you raise the price by 1%, you might suffer only a 0.5% loss in sales. In this case, the loss in sales will be more than offset by the increase in price and your revenue will rise. In such a case, we say that the demand is **inelastic** because it is not very sensitive to the increase in price. On the other hand, if your 1% price increase results in a 2% drop in demand, then raising the price will cause a drop in revenues. We then say that the demand is **elastic** because it reacts strongly to a price change.

We can use calculus to measure the response of demand to price changes if we have a demand equation for the item we are selling.[1] We need to know by

[1] Coming up with a good demand equation is not always easy. We saw in Chapter 1 that it is possible to find a linear demand equation if we know the sales figures at two different prices. However, such an equation is only a first approximation. To come up with a more accurate demand equation, we might need to gather data corresponding to sales at several different prices and then use "curve-fitting" techniques. Another approach would be an analytic one, based on mathematical modeling techniques an economist might use.

what percent demand will drop for each percent increase in price. In other words, we want the *percentage drop in demand per percentage increase in price.* This ratio is called the **elasticity of demand,** or **price elasticity of demand,** and is usually denoted by E. Let us derive a formula for E in terms of the demand equation.

Assume that we have a demand equation

$$q = f(p)$$

where q stands for the number of items we would sell (per week, per month, or what have you) if we set the price per item at p. Now suppose we increase the price p by a very small amount, Δp. Then our percentage increase in price is $(\Delta p/p) \times 100\%$ (we multiply by 100 to get a percentage). This increase in p will presumably result in a decrease in the demand q. Let us denote this corresponding decrease in q by $-\Delta q$ (we use the minus sign because, by convention, Δq stands for the *increase* in demand). Thus, the percentage decrease in demand is $(-\Delta q/q) \times 100\%$.

Now E is the ratio

$$E = \frac{\text{Percentage decrease in demand}}{\text{Percentage increase in price}}$$

so

$$E = \frac{-\dfrac{\Delta q}{q} \times 100\%}{\dfrac{\Delta p}{p} \times 100\%}$$

Canceling the 100%s and reorganizing, we get

$$E = -\frac{\Delta q}{\Delta p} \cdot \frac{p}{q}$$

Question Fine, but what small change in price will we use for Δp?
Answer It should probably be pretty small. If, say, we increased the price of sneakers to \$1,000,000 per pair, the sales would likely drop to zero. But knowing this tells us nothing about how the market would respond to a modest increase in price. In fact, we'll do the usual thing we do in calculus and let Δp approach zero.

In the expression for E, if we let Δp go to zero, then the ratio $\Delta q/\Delta p$ goes to the derivative dq/dp. This gives us our final, and most useful, definition of the elasticity.

Elasticity of Demand

The **elasticity of demand, E,** is the percentage rate of decrease of demand per percentage increase in price. E is given by the formula

$$E = -\frac{dq}{dp} \cdot \frac{p}{q}$$

We say that the demand is **elastic** if $E > 1$, is **inelastic** if $E < 1$, and has **unit elasticity** if $E = 1$.

Quick Example

Suppose that the demand equation is $q = 20{,}000 - 2p$. Then

$$E = -(-2)\frac{p}{20{,}000 - 2p} = \frac{p}{10{,}000 - p}$$

If $p = 2000$, then $E = \frac{1}{4}$, and demand is inelastic at this price.
If $p = 8000$, then $E = 4$, and demand is elastic at this price.
If $p = 5000$, then $E = 1$, and the demand has unit elasticity at this price.

example 1

Rubies

The demand for rubies at Royal Ruby Retailers is given by the equation

$$q = -\frac{4p}{3} + 80$$

where p is the price RRR charges (in dollars) and q is the number of rubies it sells per week. Find the elasticity of demand at the price level of \$40 per ruby and determine whether RRR should increase or decrease the price in order to increase revenue.

Solution

We begin by finding the elasticity using the formula

$$E = -\frac{dq}{dp} \cdot \frac{p}{q}$$

Taking the derivative and substituting for q give

$$E = -\left(-\frac{4}{3}\right)\frac{p}{-4p/3 + 80} = \frac{p}{-p + 60}$$

When $p = \$40$, the elasticity is

$$E = \frac{40}{-40 + 60} = 2$$

Thus, each percent increase in the price of rubies will cause a decrease in sales of twice the percentage. RRR will lose more in lost sales than it will gain due to the higher price. Therefore, it should not increase the price. In fact, RRR should *decrease* the price because then it will more than make up in increased sales what it will lose due to the lower price.

Question What price would bring in the greatest revenue?

Answer If demand is elastic, as it was here, then the price is too high. If demand is inelastic, then the price is too low because we can safely raise the price without losing revenue. The optimal price must be at unit elasticity.

In this example, unit elasticity occurs when

$$\frac{p}{-p + 60} = 1$$

or

$$p = -p + 60$$

$$2p = 60$$

$$p = \$30$$

Before we go on . . . We can check the answer by finding the maximum revenue directly. Revenue is

$$R = pq = p\left(-\frac{4p}{3} + 80\right) = -\frac{4p^2}{3} + 80p$$

To find the maximum we look for critical points. We first calculate

$$\frac{dR}{dp} = -\frac{8p}{3} + 80$$

There are no singular points, but there is a stationary point at $p = 30$. The domain of R is $[0, 60]$, and R is 0 at both endpoints. Therefore, $p = 30$ does give the maximum revenue.

Question Is it really true that the revenue is greatest where the elasticity is 1?
Answer Usually, and here's why. As we did in *Before we go on . . .* in the example above, we can consider the revenue directly and see where it has a maximum:

$$R = pq \qquad \text{Remember that } q \text{ is a function of } p.$$

$$\frac{dR}{dp} = q + p\frac{dq}{dp} \qquad \text{Product rule}$$

If we assume that there are no singular points, then we are interested in the stationary points, which occur when

$$q + p\frac{dq}{dp} = 0$$

Now, we can relate this equation to the elasticity by rearranging a bit:

$$p\frac{dq}{dp} = -q$$

$$-\frac{dq}{dp} \cdot \frac{p}{q} = 1$$

which says $E = 1$. Thus, the revenue has a stationary point where the elasticity is 1. In ordinary cases the revenue will have a maximum at such a stationary point. The usual case is that there is a single point of unit elasticity. At the extremes, a price of zero or a price so high that demand is zero will produce zero revenue. Thus, the single stationary point in between must give the maximum revenue.

example 2

Dolls

Suppose that the demand equation for Bobby Dolls is given by $q = 216 - p^2$, where p is the price per doll in dollars and q is the number of dolls sold per week. Find the range of prices for which (a) the demand is elastic, (b) the demand is inelastic, and (c) the weekly revenue is maximized. Also, calculate the maximum weekly revenue.

Solution

Let us calculate the elasticity:

$$E = -\frac{dq}{dp} \cdot \frac{p}{q}$$

$$= 2p \cdot \frac{p}{216 - p^2} = \frac{2p^2}{216 - p^2}$$

We answer part (c) first. Setting $E = 1$, we get

$$\frac{2p^2}{216 - p^2} = 1$$

or

$$2p^2 = 216 - p^2$$

$$3p^2 = 216$$

$$p^2 = 72$$

Thus, we conclude that the maximum revenue occurs when $p = \sqrt{72} \approx \$8.49$. We can now answer parts (a) and (b) without further calculation: The demand is elastic when $p > \$8.49$ (the price is too high), and the demand is inelastic when $p < \$8.49$ (the price is too low). Finally, we calculate the maximum weekly revenue, which equals the revenue corresponding to the price of \$8.49:

$$R = qp = (216 - p^2)p = (216 - 72)\sqrt{72} = 144\sqrt{72} \approx \$1222$$

5.5 exercises

Applications

1. ***Demand for Oranges*** The weekly sales of Honolulu Red Oranges is given by $q = 1000 - 20p$. Calculate the elasticity of demand for a price of \$30 per orange (yes, \$30 per orange[2]). Interpret your answer. Also, calculate the price that gives a maximum weekly revenue, and find this maximum revenue.

2. ***Demand for Oranges*** Repeat Exercise 1 for weekly sales of $1000 - 10p$.

3. ***Tissues*** The consumer demand curve for tissues is given by $q = (100 - p)^2$, where p is the price per case of tissues and q is the demand in weekly sales.
   a. Determine the elasticity of demand E when the price is set at \$30, and interpret your answer.
   b. At what price should tissues be sold to maximize the revenue?
   c. Approximately how many cases of tissues are demanded at that price?

4. ***Bodybuilding*** The consumer demand curve for Professor Stefan Schwartzenegger dumbbells is given by $q = (100 - 2p)^2$, where p is the price per dumbbell and q is the demand in weekly sales. Find the price Professor Schwartzenegger should charge for his dumbbells in order to maximize revenue.

5. ***College Tuition*** A study of about 1800 U.S. colleges and universities resulted in the demand equation $q = 9859.39 - 2.17p$, where q is the enrollment at a college or university and p is the average annual tuition (plus fees) it charges.
   a. The study also found that the average tuition charged by universities and colleges was \$2867. What is the corresponding elasticity of demand? Interpret your answer.
   b. Based on the study, what would you advise a college to charge its students in order to maximize total revenue? What should the revenue be?

Source: Based on a study by A. L. Ostrosky, Jr., and J. V. Koch, as cited in their book, *Introduction to Mathematical Economics* (Prospect Heights, IL: Waveland Press, 1979), p. 133.

[2] They are very hard to find, and their possession confers considerable social status.

6. ***Demand for Fried Chicken*** A fried chicken franchise finds that the demand equation for its new roast chicken product, "Roasted Rooster," is given by

$$p = \frac{40}{q^{1.5}}$$

where p is the price (in dollars) per quarter-chicken serving and q is the number of quarter-chicken servings that can be sold per hour at this price. Express q as a function of p, and find the elasticity of demand when the price is set at $4 per serving. Interpret the result.

7. ***Linear Demand Functions*** A general linear demand function has the form $q = mp + b$ (m and b constants, $m \neq 0$).
   a. Obtain a formula for the elasticity of demand at a unit price of p.
   b. Obtain a formula for the price that maximizes revenue.

8. ***Exponential Demand Functions*** A general exponential demand function has the form $q = Ae^{-bp}$ (A, b nonzero constants).
   a. Obtain a formula for the elasticity of demand at a unit price of p.
   b. Obtain a formula for the price that maximizes revenue.

9. ***Hyperbolic Demand Functions*** A general hyperbolic demand function has the form $q = k/p^r$ (r, k nonzero constants).
   a. Obtain a formula for the elasticity of demand at unit price p.
   b. How does E vary with p?
   c. What does the answer to part (b) say about the model?

10. ***Quadratic Demand Functions*** A general quadratic demand function has the form $q = ap^2 + bp + c$ (a, b, c constants with $a \neq 0$).
   a. Obtain a formula for the elasticity of demand at a unit price p.
   b. Obtain a formula for the price or prices that could maximize revenue.

11. ***Exponential Demand Functions*** The estimated monthly sales of Mona Lisa paint-by-number sets is given by the formula $q = 100e^{-3p^2+p}$, where q is the demand in monthly sales and p is the retail price in yen.
   a. Determine the elasticity of demand E when the retail price is set at 3 yen, and interpret your answer.
   b. At what price will revenue be a maximum?
   c. Approximately how many paint-by-number sets will be sold per week at the price in part (b)?

12. ***Exponential Demand Functions*** Repeat Exercise 11 using the demand equation $q = 100e^{p - 3p^2/2}$.

13. ***Modeling Linear Demand*** You have been hired as a marketing consultant to Johannesburg Burger Supply, Inc., and you wish to come up with a unit price for its hamburgers in order to maximize its weekly revenue. To make life as simple as possible, you assume that the demand equation for Johannesburg hamburgers has the linear form $q = mp + b$, where p is the price per hamburger, q is the demand in weekly sales, and m and b are certain constants you must determine.
   a. Your market studies reveal the following sales figures: When the price is set at $2.00 per hamburger, the sales amount to 3000 per week, but when the price is set at $4.00 per hamburger, the sales drop to zero. Use these data to calculate the demand equation.
   b. Now estimate the unit price in order to maximize weekly revenue and predict what the weekly revenue will be at that price.

14. ***Modeling Linear Demand*** You have been hired as a marketing consultant to Big Book Publishing, Inc., and you have been approached to determine the best selling price for the hit calculus text by Whiner and Istanbul entitled *Fun with Derivatives.* You decide to make life easy and assume that the demand equation for *Fun with Derivatives* has the linear form $q = mp + b$, where p is the price per book, q is the demand in annual sales, and m and b are certain constants you'll have to figure out.
   a. Your market studies reveal the following sales figures: When the price is set at $50.00 per book, the sales amount to 10,000 per year; when the price is set at $80.00 per book, the sales drop to 1000 per year. Use these data to calculate the demand equation.
   b. Now estimate the unit price in order to maximize annual revenue and predict what Big Book Publishing's annual revenue will be at that price.

15. ***Modeling Exponential Demand*** As the new owner of a supermarket, you have inherited a large inventory of unsold imported Limburger cheese, and you would like to set the price so that your revenue from selling it is as great as possible. Previous sales figures of the cheese are shown in the following table:

Price per pound, p	$3.00	$4.00	$5.00
Monthly sales (pounds), q	407	287	223

   a. Use the sales figures for the prices $3 and $5 per pound to construct a demand function of the form $q = Ae^{-bp}$, where A and b are constants you must determine. (Round A and b to two significant digits.)
   b. Use your demand function to find the elasticity of demand at each of the prices listed.
   c. At what price should you sell the cheese in order to maximize monthly revenue?
   d. If your total inventory of cheese amounts to only 200 pounds, and it will spoil 1 month from now, how should you price it to make the greatest revenue? Is this the same answer you got in part (c)? If not, give a brief explanation.

16. ***Modeling Exponential Demand*** Repeat Exercise 15, but this time use the sales figures for \$4 and \$5 per pound to construct the demand function.

17. ***Income Elasticity of Demand*** *(based on a question on the GRE economics test)* If $Q = aP^{\alpha}Y^{\beta}$ is the individual's demand function for a commodity, where P is the (fixed) price of the commodity, Y is the individual's income, and a, α, and β are parameters, explain why β can be interpreted as the income elasticity of demand.

18. ***College Tuition*** *(from the GRE economics test)* A time-series study of the demand for higher education, using tuition charges as a price variable, yields the following result:

$$\frac{dq}{dp} \cdot \frac{p}{q} = -0.4$$

where p is tuition and q is the quantity of higher education. Which of the following is suggested by the result?

a. As tuition rises, students want to buy a greater quantity of education.
b. As a determinant of the demand for higher education, income is more important than price.
c. If colleges lowered tuition slightly, their total tuition receipts would increase.
d. If colleges raised tuition slightly, their total tuition receipts would increase.
e. Colleges cannot increase enrollments by offering larger scholarships.

Communication and Reasoning Exercises

19. Complete the following: When demand is inelastic, revenue will decrease if _____.

20. Complete the following: When demand has unit elasticity, revenue will decrease if _____.

21. Your calculus study group is discussing elasticity of demand, and a member of the group asks the following question: "Since elasticity of demand measures the response of demand to a change in unit price, what is the difference between elasticity of demand and the quantity $-dq/dp$?" How would you respond?

22. Another member of your study group claims that unit elasticity of demand need not always correspond to maximum revenue. Is he correct? Explain your answer.

You're the Expert

Production Lot Size Management

Your publishing company, Knockem Dead Paperbacks, Inc., is about to release its next bestseller, *Henrietta's Heaving Heart* by Celestine A. Lafleur. The company expects to sell 100,000 books per month in the next year. You have been given the job of scheduling print runs to meet the anticipated demand and minimize total costs to the company. Each print run has a setup cost of \$5000, each book costs \$1 to produce, and monthly storage costs for books awaiting shipment average 1¢ per book. What will you do?

If you decide to print all 1,200,000 books (the total demand for the year, or 100,000 books per month for 12 months) in a single run at the start of the year and sales run as predicted, then the number of books in stock would begin at 1,200,000 and decrease to zero by the end of the year, as shown in Figure 29.

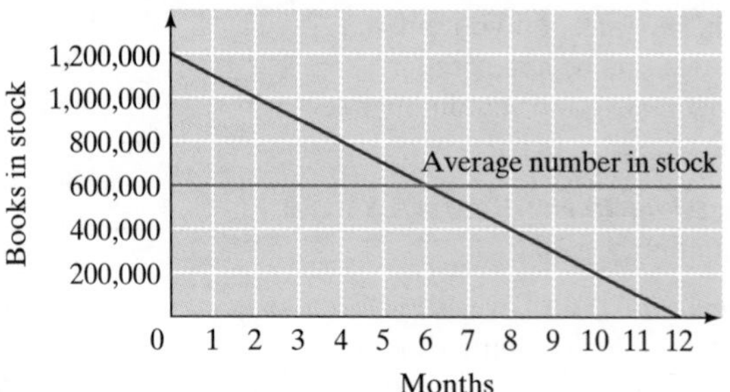

Figure 29

On average, you would be storing 600,000 books for 12 months at 1¢ per book, which gives a total storage cost of $600{,}000 \times 12 \times .01 = \$72{,}000$. The setup cost for the single print run would be $5000. When you add to these costs the total cost of producing 1,200,000 books at $1 per book, your total cost would be $1,277,000.

If, on the other hand, you decide to cut down on storage costs by printing the book in two runs of 600,000 each, you would get the picture shown in Figure 30.

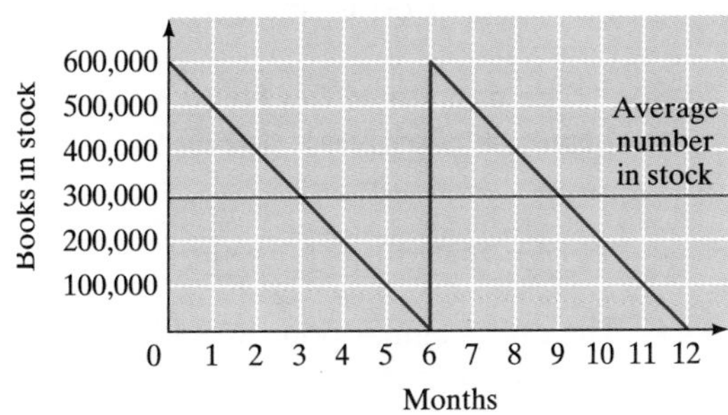

Figure 30

Now the storage cost would be cut in half because on average there would be only 300,000 books in stock. Thus, the total storage cost would be $36,000, and the setup cost would double to $10,000 (there would now be two runs). The production costs would be the same: 1,200,000 books at $1 per book. The total cost would therefore be reduced to $1,246,000, a savings of $31,000 compared to the first scenario.

"Aha!" you say to yourself, after doing these calculations. "Why not drastically cut costs by setting up a run every month?" You calculate that the setup costs alone would be $12 \times \$5000 = \$60{,}000$, which is already more than the setup plus storage costs for two runs, so a run every month would cost too much. Perhaps, then, you should investigate three runs, four runs, and so on, until you find the lowest cost. This strikes you as too laborious a process, especially considering that you will have to do it all over again when planning for Lafleur's sequel, *Lorenzo's Lost Love,* due to be released next year. Realizing that this is an optimization problem, you decide to use some calculus to help you come up with a *formula* that you can use for all future plans. So you get to work.

Instead of working with the number 1,200,000, you use the letter N so that you can be as flexible as possible. (What if *Lorenzo's Lost Love* sells more copies?) Thus, you have a total of N books to be produced for the year. You now calculate the total cost of using x print runs per year. Because you are to produce a total of N books in x print runs, you will produce N/x books in each print run. N/x is called the **lot size.** As you can see from the diagrams above, the average number of books in storage will be half that amount, $N/(2x)$.

Now you can calculate the total cost for a year. Write P for the setup cost of a single print run ($P = \$5000$ in your case) and c for the *annual* cost of storing a book (to convert all of the time measurements to years; $c = \$0.12$ here). Finally, write b for the cost of producing a single book ($b = \$1$ here). The costs break down as follows:

Setup costs: x print runs @ P dollars per run	Px
Storage costs: $N/(2x)$ books stored @ c dollars per year	$cN/(2x)$
Production costs: N books @ b dollars per book	Nb
Total cost:	$Px + \dfrac{cN}{2x} + Nb$

Remember that P, N, c, and b are all constants, and x is the only variable. Thus, your cost function is

$$C(x) = Px + \frac{cN}{2x} + Nb$$

and you need to find the value of x that will minimize $C(x)$. But that's easy! All you need to do is find the relative extrema and select the absolute minimum (if any).

The domain of $C(x)$ is $(0, +\infty)$ because there is an x in the denominator and x can't be negative. To locate the extrema you start by locating the critical points:

$$C'(x) = P - \frac{cN}{2x^2}$$

The only singular point would be at $x = 0$, but 0 is not in the domain. To find stationary points, you set $C'(x) = 0$ and solve for x:

$$P - \frac{cN}{2x^2} = 0$$

$$2x^2 = \frac{cN}{P}$$

so

$$x = \sqrt{\frac{cN}{2P}}$$

There is only one stationary point, and there are no singular points or endpoints. To graph the function you need to put in numbers for the various constants. Substituting $N = 1{,}200{,}000$, $P = 5000$, $c = 0.12$, and $b = 1$, you get

$$C(x) = 5000x + \frac{72{,}000}{x} + 1{,}200{,}000$$

with the stationary point at

$$x = \sqrt{\frac{(0.12)(1{,}200{,}000)}{2(5000)}}$$

$$\approx 3.79$$

The total cost at the stationary point is

$$C(3.79) \approx 1{,}240{,}000$$

You now graph $C(x)$ in a window that includes the stationary point, getting Figure 31.

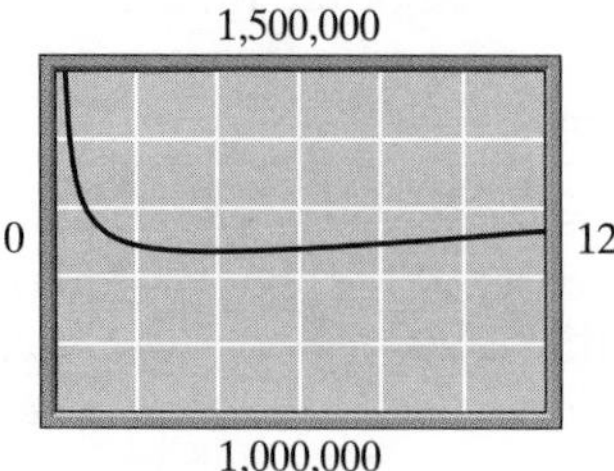

Figure 31

From the graph you can see that the stationary point is an absolute minimum. It appears that the graph is always concave up, which also tells you that your stationary point is a minimum. You can check the concavity by computing the second derivative:

$$C''(x) = \frac{cN}{x^3} > 0$$

The second derivative is always positive because c, N, and x are all positive numbers, so indeed the graph is always concave up. Now you also know that the formula works regardless of the particular values of the constants.

So now you are practically done! You know that the absolute minimum cost occurs when you have $x \approx 3.79$ print runs per year. Don't be disappointed that the answer is not a whole number; whole number solutions are rarely found in real scenarios. What the answer and the graph do indicate is that either three or four print runs per year will cost the least money. If you take $x = 3$, you get a total cost of

$$C(3) = \$1{,}239{,}000$$

If you take $x = 4$, you get a total cost of

$$C(4) = \$1{,}238{,}000$$

So, four print runs per year will allow you to minimize your total costs.

exercises

1. *Lorenzo's Lost Love* will sell 2,000,000 copies in a year. The remaining costs are the same. How many print runs should you use now?
2. In general, what happens to the number of runs that minimizes cost if both the setup cost and the total number of books are doubled?
3. In general, what happens to the number of runs that minimizes cost if the setup cost increases by a factor of 4?
4. Assuming that the total number of copies and storage costs are as originally stated, find the setup cost that will result in a single print run.
5. Assuming that the total number of copies and setup cost are as originally stated, find the storage cost that will result in a print run each month.
6. In Figure 30 we assumed that all the books in each run are manufactured in a very short time; otherwise, the figure might look more like the following one, which shows the inventory assuming a slower rate of production.

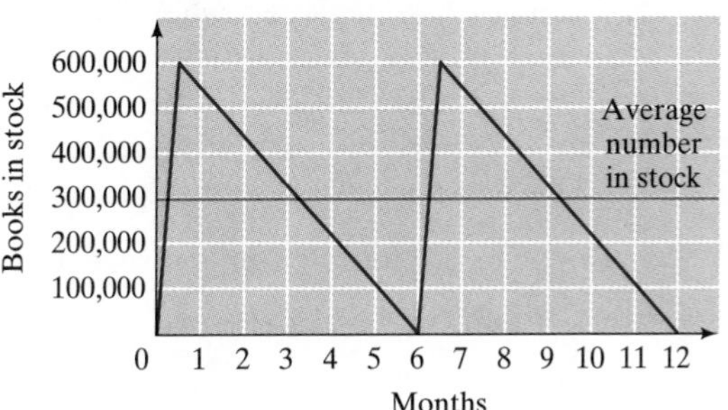

How would this affect the answer?

7. Referring to the general situation discussed in the text, find the cost as a function of the total number of books produced, assuming that the number of runs is chosen to minimize total cost. Also find the average cost per book.
8. Let $\overline{C}$ be the average cost function found in Exercise 7. Calculate $\lim_{N \to +\infty} \overline{C}(N)$, and interpret the result.

chapter 5 review test

1. In each of the following, find all the relative and absolute extrema of the given functions on the given domain (if supplied) or on the largest possible domain (if no domain is supplied).
   - **a.** $f(x) = 2x^3 - 6x + 1$ on $[-2, +\infty)$
   - **b.** $f(x) = \dfrac{x+1}{(x-1)^2}$ on $[-2, 2]$
   - **c.** $g(x) = (x-1)^{2/3}$
   - **d.** $g(x) = x^2 + \ln x$ on $(0, +\infty)$
   - **e.** $h(x) = \dfrac{1}{x} + \dfrac{1}{x^2}$
   - **f.** $h(x) = e^{x^2} + 1$

2. In each of the following, the graph of the function f or its derivative is given. Find the approximate x-coordinates of all relative extrema and points of inflection of the original function f (if any).

a. Graph of f: **b.** Graph of f:

c. Graph of f':

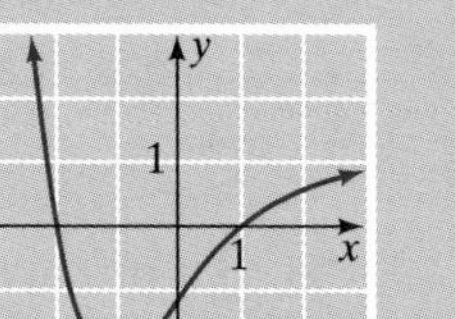

d. Graph of f':

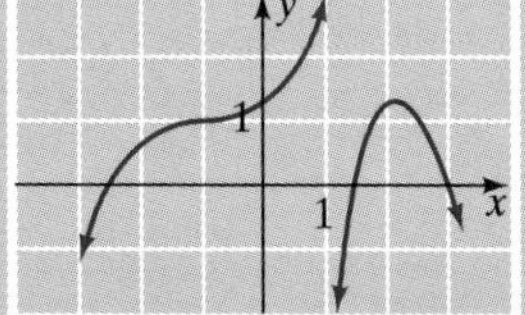

T 3. In each of the following, use technology to draw the graph of the given function, indicating all relative and absolute extrema and points of inflection. Find the coordinates of these points exactly, where possible. Also indicate any horizontal and vertical asymptotes.
   - **a.** $f(x) = x^3 - 12x$ on $[-2, +\infty)$
   - **b.** $f(x) = \dfrac{x^2 - 3}{x^3}$
   - **c.** $f(x) = (x-1)^{2/3} + \dfrac{2x}{3}$

Applications

4. Demand for the latest bestseller at OHaganBooks.com, *A River Burns Through It,* is given by

$$q = -p^2 + 33p + 9 \qquad (18 \le p \le 28)$$

copies sold per week, when the price is p dollars.
   - **a.** Find the price elasticity of demand as a function of p.
   - **b.** Find the elasticity of demand for this book at a price of \$20 and at a price of \$25. (Round your answers to two decimal places.) Interpret the answers.
   - **c.** What price should the company charge to obtain the greatest revenue?

5. If we take into account storage and shipping, it costs OHaganBooks.com

$$C = 9q + 100$$

dollars to sell q copies of *A River Burns Through It* in a week.
   - **a.** If demand is as in Question 4, express the weekly profit earned by OHaganBooks.com from the sale of *A River Burns Through It* as a function of unit price p.
   - **b.** What price should the company charge to get the greatest weekly profit?
   - **c.** What is the maximum possible weekly profit?
   - **d.** Compare your answer to part (b) with the price the company should charge to obtain the greatest revenue. Explain any difference.

T 6. OHaganBooks.com modeled its weekly sales over a period of time with the function

$$s(t) = 6053 + \frac{4474}{1 + e^{-0.55(t-4.8)}}$$

as shown in the following graph:

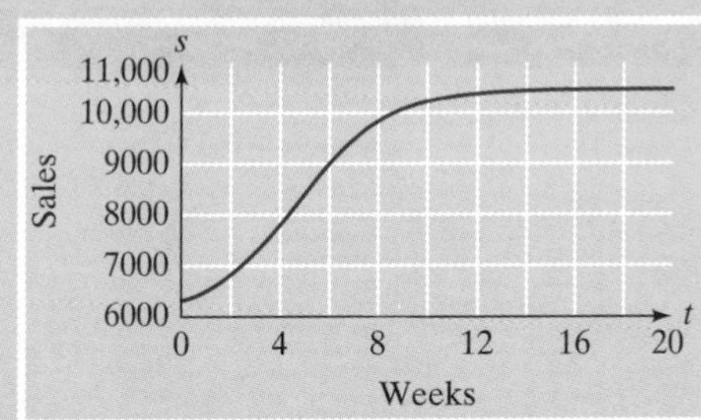

a. Compute and graph $s'(t)$, and use your graph (or some other method) to estimate when, to the nearest week, the weekly sales were growing fastest.

b. To what feature on the graph of s does your answer to part (a) correspond?

c. The graph of s has a horizontal asymptote. What is the value (s-coordinate) of this asymptote, and what is its significance in terms of weekly sales at OHaganBooks.com?

d. The graph of s' has a horizontal asymptote. What is the value (s-coordinate) of this asymptote, and what is its significance in terms of weekly sales at OHaganBooks.com?

7. OHaganBooks.com's web site has an animated graphic with its name in a rectangle whose height and width change. On either side of the rectangle are semicircles, as in the figure, whose diameters are the same as the height of the rectangle.

For reasons too complicated to explain, the designer wanted the combined area of the rectangle and semicircles to remain constant. At one point during the animation, the width of the rectangle is 1 inch, growing at a rate of 0.5 inch per second, and the height is 3 inches. How fast is the height changing?

Additional On-Line Review

If you follow the path

Web site → Everything for Calculus → Chapter 5

you will find the following additional resources to help you review:

- A comprehensive chapter summary (including examples and interactive features)
- Additional review exercises (including interactive exercises and many with help)
- A true–false chapter quiz
- An on-line tutorial for related rates

You're the Expert

Wage Inflation

As assistant personnel manager for a large corporation, you have been asked to estimate the average annual wage earned by a worker in your company, from the time the worker is hired to the time the worker retires. You have data about wage increases. How can you estimate this average?

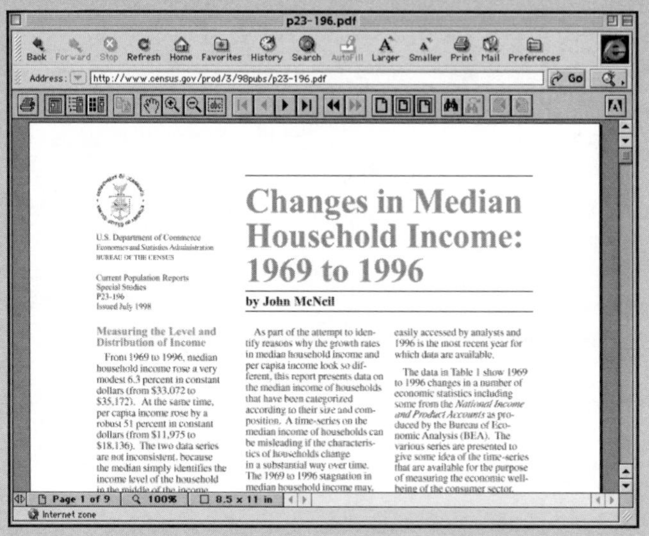

Internet Resources for This Chapter

At the web site, follow the path

Web site → Everything for Calculus → Chapter 6

where you will find links to step-by-step tutorials for the main topics in this chapter, a detailed chapter summary you can print out, a true–false quiz, and a collection of sample test questions. You will also find downloadable Excel tutorials for each section, an on-line numerical integration utility, and other resources. Complete text and interactive exercises have been placed on the web site covering the following optional topic:

- Numerical Integration

The Integral

chapter 6

6.1 The Indefinite Integral
6.2 Substitution
6.3 The Definite Integral As a Sum: A Numerical Approach
6.4 The Definite Integral As Area: A Geometric Approach
6.5 The Definite Integral: An Algebraic Approach and the Fundamental Theorem of Calculus

Introduction

Roughly speaking, calculus is divided into two parts: **differential calculus** (the calculus of derivatives) and **integral calculus,** which is the subject of this chapter and the next. Integral calculus is concerned with problems that are in some sense the reverse of the problems seen in differential calculus. For example, whereas differential calculus shows how to compute the rate of change of a quantity, integral calculus shows how to find the quantity if we know its rate of change. This idea is made precise in the **Fundamental Theorem of Calculus.** Integral calculus and the Fundamental Theorem of Calculus allow us to solve many problems in economics, physics, and geometry, including one of the oldest problems in mathematics—computing areas of regions with curved boundaries.

6.1 The Indefinite Integral

Having studied differentiation in the preceding chapters, we now discuss how to *reverse* the process.

Question If the derivative of $F(x)$ is $4x^3$, what was $F(x)$?
Answer After a moment's thought, we recognize $4x^3$ as the derivative of x^4. So, we might have $F(x) = x^4$. However, on thinking further, we realize that, for example $F(x) = x^4 + 7$ works just as well. In fact, $F(x) = x^4 + C$ works for any number C. Thus, there are *infinitely many* possible answers to this question.

In fact, we will see shortly that the formula $F(x) = x^4 + C$ covers *all* possible answers to the question. Let us give a name to what we are doing.

Antiderivative

An **antiderivative** of a function f is a function F such that $F' = f$.

Quick Examples

1. An antiderivative of $4x^3$ is x^4. — Because the derivative of x^4 is $4x^3$

2. Another antiderivative of $4x^3$ is $x^4 + 7$. — Because the derivative of $x^4 + 7$ is $4x^3$

3. An antiderivative of $2x$ is $x^2 + 12$. — Because the derivative of $x^2 + 12$ is $2x$

Thus,

If the derivative of $A(x)$ is $B(x)$, then an antiderivative of $B(x)$ is $A(x)$.

We call the set of *all* antiderivatives of a function the **indefinite integral** of the function.

Indefinite Integral

The expression

$$\int f(x)\,dx$$

is read "the **indefinite integral** of $f(x)$ with respect to x" and stands for the set of all antiderivatives of f. Thus, $\int f(x)\,dx$ is a *collection of functions;* it is not a single function nor a number. The function f that is being **integrated** is called the **integrand,** and the variable x is called the **variable of integration.**

Quick Examples

1. $\int 4x^3\,dx = x^4 + C$ Every possible antiderivative of $4x^3$ has the form $x^4 + C$.
2. $\int 2x\,dx = x^2 + C$ Every possible antiderivative of $2x$ has the form $x^2 + C$.

The **constant of integration,** C, reminds us that we can add any constant and get a different antiderivative.

example 1

Indefinite Integral

Check that $\int x\,dx = \frac{x^2}{2} + C$.

Solution

We check by taking the derivative of the right-hand side:

$$\frac{d}{dx}\left(\frac{x^2}{2} + C\right) = \frac{2x}{2} + 0 = x \ ✔$$

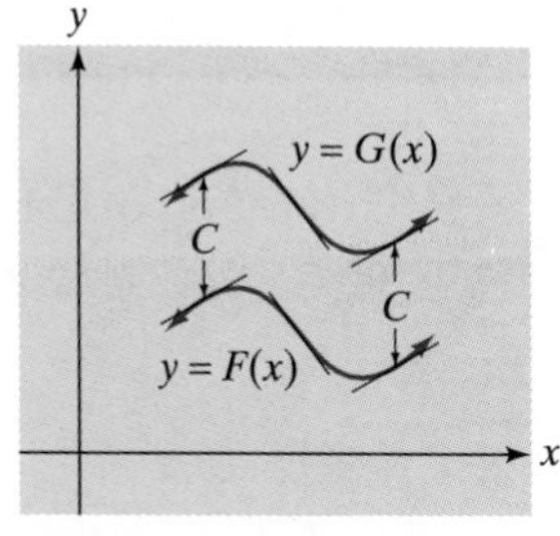

Figure 1

Question If $F(x)$ is one antiderivative of $f(x)$, then why must all other antiderivatives have the form $F(x) + C$?

Answer Suppose $F(x)$ and $G(x)$ are both antiderivatives of $f(x)$, so that $F'(x) = G'(x)$. Consider what this means by looking at Figure 1. If $F'(x) = G'(x)$ for all x, then F and G have the *same slope* at each value of x. This means that their graphs must be *parallel* and hence remain exactly the same vertical distance apart. But that is the same as saying that the functions differ by a constant—that is, that $G(x) = F(x) + C$ for some constant C.[1]

[1]This argument can be turned into a more rigorous proof—that is, a proof that does not rely on geometric concepts such as "parallel graphs." We should also say that the result requires that F and G have the same derivative *on an interval* $[a, b]$.

Now, we would like to make the process of finding indefinite integrals (antiderivatives) more mechanical. For example, it would be nice to have a power rule for indefinite integrals similar to the one we already have for derivatives. Two cases suggested by the examples above are

$$\int x\,dx = \frac{x^2}{2} + C \qquad \text{and} \qquad \int x^3\,dx = \frac{x^4}{4} + C$$

We can check the last equation by taking the derivative of its right-hand side. These cases suggest the following general statement.

Power Rule for the Indefinite Integral, Part I

$$\int x^n\,dx = \frac{x^{n+1}}{n+1} + C \qquad \text{if } n \neq -1$$

In Words

To find the integral of x^n, add 1 to the exponent and then divide by the new exponent. This rule works provided n is not -1.

Quick Examples

1. $\int x^{55}\,dx = \frac{x^{56}}{56} + C$

2. $\int \frac{1}{x^{55}}\,dx = \int x^{-55}\,dx$ Exponent form

$= \frac{x^{-54}}{-54} + C$ When we add 1 to -55, we get -54, *not* -56.

$= -\frac{1}{54x^{54}} + C$

3. $\int 1\,dx = x + C$ Since $1 = x^0$. This is an important special case.

Notes

The integral $\int 1\,dx$ is commonly written as $\int dx$.

Similarly, the integral $\int \frac{1}{x^{55}}\,dx$ may be written as $\int \frac{dx}{x^{55}}$.

We can easily check the power rule formula by taking the derivative of the right-hand side:

$$\frac{d}{dx}\left(\frac{x^{n+1}}{n+1} + C\right) = \frac{(n+1)x^n}{n+1} = x^n \checkmark$$

Question What is the reason for the restriction $n \neq -1$?

Answer Let us answer the question with a question: Does the right-hand side of the power rule formula make sense if $n = -1$?

Question Well, no, so what is

$$\int x^{-1}\,dx = \int \frac{1}{x}\,dx$$

Answer Think before reading on: Have you ever seen a function whose derivative is $1/x$? Prodding our memories a little, we recall that $\ln x$ has derivative $1/x$. In fact, as we pointed out when we first discussed it, $\ln|x|$ also has derivative $1/x$, but it has the advantage that its domain is the same as that of $1/x$. Thus, we can fill in the missing case as follows.

Power Rule for the Indefinite Integral, Part II

$$\int x^{-1}\,dx = \ln|x| + C \qquad \text{Equivalently, } \int \frac{1}{x}\,dx = \ln|x| + C.$$

Note Consider the function

$$F(x) = \begin{cases} \ln|x| + C_1 & \text{if } x > 0 \\ \ln|x| + C_2 & \text{if } x < 0 \end{cases}$$

where C_1 and C_2 are possibly different constants. This function also has derivative $1/x$, and so we should really write

$$\int x^{-1}\,dx = \begin{cases} \ln|x| + C_1 & \text{if } x > 0 \\ \ln|x| + C_2 & \text{if } x < 0 \end{cases}$$

However, most books ignore this subtle point, as we shall also, and implicitly assume that $C_1 = C_2$. Thus, we will continue to write

$$\int x^{-1}\,dx = \ln|x| + C$$

Here are some other indefinite integrals that come from the corresponding formulas for differentiation.

Indefinite Integrals of Some Exponential and Trig Functions

$$\int e^x\,dx = e^x + C \qquad \text{Because } \frac{d}{dx}(e^x) = e^x$$

$$\int \cos x\,dx = \sin x + C \qquad \text{Because } \frac{d}{dx}(\sin x) = \cos x$$

$$\int \sin x\,dx = -\cos x + C \qquad \text{Because } \frac{d}{dx}(-\cos x) = \sin x$$

$$\int \sec^2 x\,dx = \tan x + C \qquad \text{Because } \frac{d}{dx}(\tan x) = \sec^2 x$$

Question What about more complicated functions, such as $2x^3 + 6x^5 - 1$?
Answer We need the following rules for integrating sums, differences, and constant multiples.

Rules for the Indefinite Integral

Sum and Difference Rules

$$\int [f(x) \pm g(x)]\,dx = \int f(x)\,dx \pm \int g(x)\,dx$$

In Words
The integral of a sum is the sum of the integrals, and the integral of a difference is the difference of the integrals.

Constant Multiple Rule

$$\int kf(x)\,dx = k\int f(x)\,dx \qquad (k \text{ constant})$$

In Words

The integral of a constant times a function is the constant times the integral of the function. (In other words, *the constant "goes along for the ride."*)

Quick Examples

1. Sum rule: $\int (x^3 + 1)\,dx = \int x^3\,dx + \int 1\,dx = \frac{x^4}{4} + x + C$ $\quad f(x) = x^3;\ g(x) = 1$
2. Constant multiple rule: $\int 5x^3\,dx = 5\int x^3\,dx = 5\frac{x^4}{4} + C$ $\quad k = 5;\ f(x) = x^3$
3. Constant multiple rule: $\int 4\,dx = 4\int 1\,dx = 4x + C$ $\quad k = 4;\ f(x) = 1$

Why are these rules true? Because the derivative of a sum is the sum of the derivatives, and similarly for differences and constant multiples.

example 2

Using the Sum and Difference Rules

Find the integrals.

a. $\int (x^3 + x^5 - 1)\,dx$ **b.** $\int \left(x^{2.1} + \frac{1}{x^{1.1}} + \frac{1}{x}\right)dx$

c. $\int (e^x - \sin x + \cos x)\,dx$

Solution

a. $\int (x^3 + x^5 - 1)\,dx = \int x^3\,dx + \int x^5\,dx - \int 1\,dx$ Sum/difference rule

$= \frac{x^4}{4} + \frac{x^6}{6} - x + C$ Power rule

b. $\int \left(x^{2.1} + \frac{1}{x^{1.1}} + \frac{1}{x}\right)dx = \int (x^{2.1} + x^{-1.1} + x^{-1})\,dx$ Exponent form

$= \int x^{2.1}\,dx + \int x^{-1.1}\,dx + \int x^{-1}\,dx$ Sum rule

$= \frac{x^{3.1}}{3.1} + \frac{x^{-0.1}}{-0.1} + \ln|x| + C$ Power rule

$= \frac{x^{3.1}}{3.1} - \frac{10}{x^{0.1}} + \ln|x| + C$ Back to fraction form

c. $\int (e^x - \sin x + \cos x)\,dx = e^x + \cos x + \sin x + C$ Two steps in one

Before we go on . . . As usual, you should check each of the answers by differentiating.

Question Why is there only a single arbitrary constant C in each of the answers?
Answer We could have written the answer to part (a) as

$$\frac{x^4}{4} + D + \frac{x^6}{6} + E - x + F$$

where D, E, and F are all arbitrary constants. Now suppose that, for example, we set $D = 1$, $E = -2$, and $F = 6$. Then the particular antiderivative we get is $x^4/4 + x^6/6 - x + 5$, which has the form $x^4/4 + x^6/6 - x + C$. Thus, we could have chosen the single constant C to be 5 and obtained the same answer. In other words, the answer $x^4/4 + x^6/6 - x + C$ is just as general as the answer $x^4/4 + D + x^6/6 + E - x + F$, but simpler.

In practice we do not explicitly write the integral of a sum as a sum of integrals. We just "integrate term by term" [see part (c) in Example 2] much as we learned to differentiate term by term.

example 3

Combining the Rules

Find the integrals.

a. $\int (10x^4 + 2x^2 - 3e^x)\, dx$ **b.** $\int \left(\frac{2}{x^{0.1}} + \frac{x^{0.1}}{2} - 3 \sin x\right) dx$

Solution

a. We need to integrate separately each of the terms $10x^4$, $2x^2$, and $3e^x$. To integrate $10x^4$ we use the rules for constant multiples and powers:

$$\int 10x^4\, dx = 10\int x^4\, dx = 10\frac{x^5}{5} + C = 2x^5 + C$$

The other two terms are similar. We get

$$\int (10x^4 + 2x^2 - 3e^x)\, dx = 10\frac{x^5}{5} + 2\frac{x^3}{3} - 3e^x + C = 2x^5 + \frac{2}{3}x^3 - 3e^x + C$$

b. We first convert to exponent form and then integrate term by term:

$$\int \left(\frac{2}{x^{0.1}} + \frac{x^{0.1}}{2} - 3 \sin x\right) dx = \int \left(2x^{-0.1} + \frac{1}{2}x^{0.1} - 3 \sin x\right) dx \quad \text{Exponent form}$$

$$= 2\frac{x^{0.9}}{0.9} + \frac{1}{2}\frac{x^{1.1}}{1.1} + 3 \cos x + C \quad \text{Integrate term by term.}$$

$$= \frac{20x^{0.9}}{9} + \frac{x^{1.1}}{2.2} + 3 \cos x + C$$

example 4

Different Variable Name

Find $\int \left(\frac{1}{u} + \frac{1}{u^2}\right) du.$

Solution

This integral may look a little strange because we are using the letter u instead of x, but there is really nothing special about x. We get

$$\int \left(\frac{1}{u} + \frac{1}{u^2}\right) du = \int (u^{-1} + u^{-2})\, du \quad \text{Exponent form}$$

$$= \ln|u| + \frac{u^{-1}}{-1} + C \quad \text{Integrate term by term.}$$

$$= \ln|u| - \frac{1}{u} + C \quad \text{Simplify the result.}$$

Before we go on . . . When we compute an indefinite integral, we want the independent variable in the answer to be the same as the variable of integration. Thus, if the integral had been written in terms of x rather than u, we would have written

$$\int \left(\frac{1}{x} + \frac{1}{x^2}\right) dx = \ln|x| - \frac{1}{x} + C$$

Application: Cost and Marginal Cost

example 5

Finding Cost from Marginal Cost

The marginal cost to produce baseball caps at a production level of x caps is $3.20 - 0.001x$ dollars per cap, and the cost of producing 50 caps is \$200. Find the cost function.

Solution

We are asked to find the cost function $C(x)$, given that the *marginal* cost function is $3.20 - 0.001x$. Recalling that the marginal cost function is the derivative of the cost function, we have

$$C'(x) = 3.20 - 0.001x$$

and we must find $C(x)$. Now $C(x)$ must be an antiderivative of $C'(x)$, so we write

$$\begin{aligned} C(x) &= \int (3.20 - 0.001x)\, dx \\ &= 3.20x - 0.001\frac{x^2}{2} + K \qquad K \text{ is the constant of integration.} \\ &= 3.20x - 0.0005x^2 + K \end{aligned}$$

(Why did we use K and not C for the constant of integration?) Now, unless we know a value for K, we don't really know what the cost function is. However, there is another piece of information we have ignored: The cost of producing 50 baseball caps is \$200. In symbols,

$$C(50) = 200$$

Substituting in our formula for $C(x)$, we have

$$\begin{aligned} C(50) &= 3.20(50) - 0.0005(50)^2 + K \\ 200 &= 158.75 + K \\ K &= 41.25 \end{aligned}$$

Now that we know what K is, we can write the cost function.

$$C(x) = 3.20x - 0.0005x^2 + 41.25$$

Before we go on . . .

Question What is the significance of the term 41.25?

Answer If we substitute $x = 0$, we get

$$C(0) = 3.20(0) - 0.0005(0)^2 + 41.25$$

or $$C(0) = 41.25.$$

Thus, \$41.25 is the cost of producing zero items; in other words, it is the **fixed cost.**

Application: Motion in a Straight Line

An important application of the indefinite integral is in the study of motion. The application of calculus to problems about motion is an example of the intertwining of mathematics and physics that is an important part of both. We begin by bringing together some facts, scattered throughout the last several chapters, that have to do with an object moving in a straight line.

Position, Velocity, and Acceleration: Derivative Form

If $s = s(t)$ is the **position** of an object at time t, then its **velocity** is given by the derivative

$$v = \frac{ds}{dt}$$

In other words, *velocity is the derivative of position.*

The **acceleration** of an object is given by the derivative

$$a = \frac{dv}{dt}$$

In other words, *acceleration is the derivative of velocity.* On the planet Earth, a freely falling body experiencing no air resistance accelerates at approximately 32 feet per second per second, or 32 ft/s^2 (or 9.8 m/s^2).

We may rewrite the derivative formulas above as integral formulas.

Position, Velocity, and Acceleration: Integral Form

$$s(t) = \int v(t)\, dt \qquad \text{because } v = \frac{ds}{dt}$$

$$v(t) = \int a(t)\, dt \qquad \text{because } a = \frac{dv}{dt}$$

example 6

Motion in a Straight Line

You toss a stone upward at a speed of 30 feet per second.

a Find the stone's velocity as a function of time. How fast and in what direction is it going after 5 seconds?

b. Find the position of the stone as a function of time. Where will it be after 5 seconds?

c. When and where will the stone reach its **zenith,** its highest point?

Solution

a. Let us measure heights above the ground as positive, so that a rising object has positive velocity and the acceleration downward due to gravity is negative. Thus, the acceleration of the stone is given by

$$a(t) = -32 \text{ ft/s}^2$$

We wish to know the velocity, which is an antiderivative of acceleration, so we compute

$$v(t) = \int (-32)\, dt = -32t + C$$

This is the velocity as a function of time t. But what is the value of C? Now, we are told that you tossed the stone upward at 30 ft/s, so when $t = 0$, $v = 30$; that is, $v(0) = 30$. Thus,

$$30 = v(0) = -32(0) + C$$

so $C = 30$ and the formula for velocity is $v(t) = -32t + 30$. In particular, after 5 seconds the velocity is

$$v(5) = -32(5) + 30 = -130 \text{ ft/s}$$

Thus, after 5 seconds the stone is *falling* with a speed of 130 feet per second.

b. We wish to know the position, but position is an antiderivative of velocity. Thus,

$$s(t) = \int v(t)\,dt = \int (-32t + 30)\,dt = -16t^2 + 30t + C$$

Now, to find C we need to know the initial position $s(0)$. We are not told this, so let us measure heights so that the initial position is zero. Then

$$0 = s(0) = C$$

and $s(t) = -16t^2 + 30t$. In particular, after 5 seconds the stone has a height of

$$s(5) = -16(5)^2 + 30(5) = -250 \text{ ft}$$

In other words, the stone is now 250 ft *below* where it was when you first threw it, as shown in Figure 2.

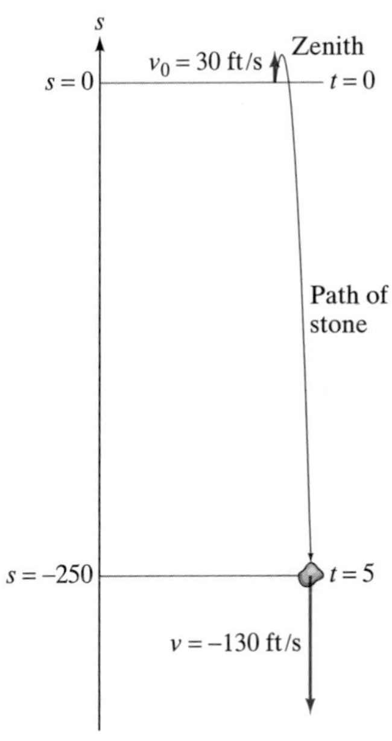

Figure 2

c. The stone reaches its zenith when its height $s(t)$ is at its maximum value, which occurs when $v(t) = s'(t)$ is zero. So we solve

$$v(t) = -32t + 30 = 0$$

to get $t = \frac{30}{32} = \frac{15}{16} = 0.9375$ s. This is the time when the stone reaches its zenith. The height of the stone at that time is

$$s\left(\frac{15}{16}\right) = -16\left(\frac{15}{16}\right)^2 + 30\left(\frac{15}{16}\right) = 14.0625 \text{ feet}$$

6.1 exercises

Evaluate the integrals in Exercises 1–10 mentally.

1. $\int x^5\,dx$

2. $\int x^7\,dx$

3. $\int 6\,dx$

4. $\int (-5)\,dx$

5. $\int x\,dx$

6. $\int (-x)\,dx$

7. $\int (x^2 - x)\,dx$

8. $\int (x + x^3)\,dx$

9. $\int (1 + x)\,dx$

10. $\int (4 - x)\,dx$

Evaluate the integrals in Exercises 11–38.

11. $\int x^{-5}\,dx$

12. $\int x^{-7}\,dx$

13. $\int (x^{2.3} + x^{-1.3})\,dx$

14. $\int (x^{-0.2} - x^{0.2})\,dx$

15. $\int (u^2 - 1/u)\,du$

16. $\int (1/v^2 + 2/v)\,dv$

17. $\int \sqrt{x}\,dx$

18. $\int \sqrt[3]{x}\,dx$

19. $\int (3x^4 - 2x^{-2} + x^{-5} + 4)\,dx$

20. $\int (4x^7 - x^{-3} + 1)\,dx$

21. $\int \left(\frac{1}{x} + \frac{2}{x^2} - \frac{1}{x^3}\right) dx$

22. $\int \left(\frac{3}{x} - \frac{1}{x^5} + \frac{1}{x^7}\right) dx$

23. $\int (3x^{0.1} - x^{4.3} - 4.1)\,dx$

24. $\int \left(\frac{x^{2.1}}{2} - 2.3\right) dx$

25. $\int \left(\frac{3}{x^{0.1}} - \frac{4}{x^{1.1}}\right) dx$

26. $\int \left(\frac{1}{x^{1.1}} - \frac{1}{x}\right) dx$

27. $\int (2e^x + 5/x + 1/4)\,dx$

28. $\int (-e^x + x^{-2} - 1/8)\,dx$

29. $\int \left(\frac{6.1}{x^{0.5}} + \frac{x^{0.5}}{6} - e^x\right) dx$

30. $\int \left(\frac{4.2}{x^{0.4}} + \frac{x^{0.4}}{3} - 2e^x\right) dx$

Δ 31. $\int (\sin x + \cos x)\, dx$

Δ 32. $\int (\cos x - \sin x)\, dx$

Δ 33. $\int (2 \cos x - 4.3 \sin x - 9.33)\, dx$

Δ 34. $\int (4.1 \sin x + \cos x - 9.33/x)\, dx$

Δ 35. $\int \left(3.4 \sec^2 x + \frac{\cos x}{1.3} - 3.2e^x\right) dx$

Δ 36. $\int \left(\frac{3 \sec^2 x}{2} + 1.3 \sin x - \frac{e^x}{3.2}\right) dx$

37. $\int \frac{x+2}{x^3}\, dx$

38. $\int \frac{x^2-2}{x}\, dx$

39. Find $f(x)$ if $f(0) = 1$ and the tangent line at $(x, f(x))$ has slope x.

40. Find $f(x)$ if $f(1) = 1$ and the tangent line at $(x, f(x))$ has slope $1/x$.

41. Find $f(x)$ if $f(0) = 0$ and the tangent line at $(x, f(x))$ has slope $e^x - 1$.

42. Find $f(x)$ if $f(1) = -1$ and the tangent line at $(x, f(x))$ has slope $2e^x + 1$.

Applications

43. ***Marginal Cost*** The marginal cost of producing the xth box of lightbulbs is $5 - (x/10{,}000)$ and the fixed cost is \$20,000. Find the cost function $C(x)$.

44. ***Marginal Cost*** The marginal cost of producing the xth box of computer disks is $10 + (x^2/100{,}000)$ and the fixed cost is \$100,000. Find the cost function $C(x)$.

45. ***Marginal Cost*** The marginal cost of producing the xth roll of film is $5 + 2x + 1/x$. The total cost to produce one roll is \$1000. Find the cost function $C(x)$.

46. ***Marginal Cost*** The marginal cost of producing the xth box of videotape is $10 + x + 1/x^2$. The total cost to produce 100 boxes is \$10,000. Find the cost function $C(x)$.

47. ***Interest Rates*** Between 1990 and 1998 the discount interest rate in Japan declined at a rate of 0.7 percentage point per year.[2] Given that the discount interest rate was 6% in 1992, use an indefinite integral to find a formula for the interest rate I as a function of time t since 1990 ($t = 0$ represents 1990), and use your formula to calculate the interest rate in 1998.
Source: Bloomberg Financial Markets/Japan External Trade Organization/*New York Times,* September 20, 1998, p. WK5.

48. ***Real Estate*** Between 1990 and 1997 the price of a square foot of land in Tokyo declined at an average rate of \$25 per year. Given that the price in 1990 was \$375, use an indefinite integral to find a formula for the price p as a function of time t since 1990 ($t = 0$ represents 1990), and use your formula to calculate the price in 1997.
Source: See the source in Exercise 47.

49. ***Employment*** In 1988 statewide employment in Massachusetts was approximately 3,100,000. The following quadratic model approximates the rate of increase in employment, in thousands of people per year, in Massachusetts from 1988 through 1994:[3]

$$C(t) = 25t^2 - 137t + 68$$

where t is the number of years since 1988. Use the model and the 1988 employment figure for Massachusetts to obtain a model for the total number of people $N(t)$ employed in Massachusetts as a function of t.

50. ***Hawaiian Tourism*** The rate of visitor spending, in billions of dollars per year, in Hawaii during the years 1985 to 1993 can be approximated by[4]

$$r(t) = -0.164t^2 + 1.60t + 6.71$$

where $t = 0$ represents June 30, 1985. According to the model, how much revenue did Hawaii earn from visitor spending between June 30, 1985, and June 30, 1990? (Give your answer to the nearest billion dollars.)

51. ***Medicare Spending*** The rate of federal spending on Medicare (in constant 1994 dollars) increased more or less linearly from \$50 billion per year in 1972 to \$160 billion per year in 1994.[5] Obtain a linear model for the rate of increase of federal spending as a function of the time t in years since 1972, and use your model to obtain the total amount $C(t)$ spent on Medicare since 1972.
Source: Health Care Financing Administration, Economic Report of the President/*New York Times,* July 23, 1995, p. 1.

52. ***Certified Financial Planners*** The number of people who passed the exam to become a certified financial plan-

[2] This was the average rate of change over the given period.

[3] The model is a quadratic regression based on actual data from the Massachusetts Department of Employment, DRI/McGraw Hill/*New York Times,* January 27, 1995, p. D4.

[4] The model is based on a best-fit quadratic for (approximate) tourism data, as measured in constant 1993 dollars. *Source for data:* Hawaii Visitors Bureau/*New York Times,* September 5, 1995, p. A12.

[5] Figures are rounded to the nearest \$10 billion.

ner (CFP) rose from 400 in 1992 to 2000 in 1995. In 1995 a total of 30,000 people had CFP designation. Obtain a linear model for the number of people who passed the exam (in year t, with $t = 0$ corresponding to 1992). Hence, estimate the total number of people with CFP designation in 1992 ($t = 0$).

Source: See the source in Exercise 51. Figures are approximate.

53. ***Motion in a Straight Line*** The velocity of a particle moving in a straight line is given by $v(t) = t^2 + 1$.

a. Find an expression for the position s after a time t.

b. Given that $s = 1$ at time $t = 0$, find the constant of integration C, and hence find an expression for s in terms of t without any unknown constants.

54. ***Motion in a Straight Line*** The velocity of a particle moving in a straight line is given by $v = 3e^t + t$.

a. Find an expression for the position s after a time t.

b. Given that $s = 3$ at time $t = 0$, find the constant of integration C, and hence find an expression for s in terms of t without any unknown constants.

Vertical Motion *In Exercises 55–64, neglect the effects of air resistance.*

55. If a stone is dropped from a rest position above the ground, how fast and in what direction will it be traveling after 10 seconds?

56. If a stone is thrown upward at 10 feet per second, how fast and in what direction will it be traveling after 10 seconds?

57. Show that if a projectile is thrown upward with a velocity of v_0 ft/s, it will reach its highest point after $v_0/32$ seconds.

58. Use your solution to Exercise 57 to show that if a projectile is thrown upward with a velocity of v_0 ft/s, its highest point will be $v_0^2/64$ feet above the starting point.

Exercises 59–64 use the results of Exercises 57 and 58.

59. I threw a ball up in the air to a height of 20 feet. How fast was the ball traveling when it left my hand?

60. I threw a ball up in the air to a height of 40 feet. How fast was the ball traveling when it left my hand?

61. A piece of chalk is tossed vertically upward by Professor Schwartzenegger and hits the ceiling 100 feet above with a *BANG*.

a. What is the minimum speed the piece of chalk must have been traveling to enable it to hit the ceiling?

b. Assuming that Prof. Schwartzenegger in fact tossed the piece of chalk up at 100 ft/s, how fast was it moving when it struck the ceiling?

c. Assuming that Prof. Schwartzenegger tossed the chalk up at 100 ft/s, and that it recoils from the ceiling with the same speed it had at the instant it hit, how long will it take the chalk to make the return journey and hit the ground?

62. A projectile is fired vertically upward from ground level at 16,000 feet per second.

a. How high does the projectile go?

b. How long does it take to reach its zenith (highest point)?

c. How fast is it traveling when it hits the ground?

63. ***Strength*** Professor Strong can throw a 10-pound dumbbell twice as high as Professor Weak can. How much faster can Prof. Strong throw it?

64. ***Weakness*** Professor Weak can throw a computer disc three times as high as Professor Strong can. How much faster can Prof. Weak throw it?

Communication and Reasoning Exercises

65. Give an argument for the rule that the integral of a sum is the sum of the integrals.

66. Is it true that $\int(1/x^3)\,dx = \ln(x^3) + C$? Give a reason for your answer.

67. Give an example to show that the integral of a product is not the product of the integrals.

68. Give an example to show that the integral of a quotient is not the quotient of the integrals.

69. If x represents the number of items manufactured and $f(x)$ represents dollars per item, what does $\int f(x)\,dx$ represent? In general, how are the units of $f(x)$ and the units of $\int f(x)\,dx$ related?

70. Complete the following. $-(1/x)$ is an ______ of $(1/x^2)$, whereas $\ln x^2$ is not. Also, $-(1/x) + C$ is the ______ of $(1/x^2)$ because the ______ of $-(1/x) + C$ is ______.

71. Complete the following: If you take the ______ of the ______ of $f(x)$, you obtain $f(x)$ back. On the other hand, if you take the ______ of the ______ of $f(x)$, you obtain $f(x) + C$.

72. If a Martian told you that the Institute of Alien Mathematics, after a long and difficult search, has announced the discovery of a new antiderivative of $x - 1$ called $M(x)$ [the formula for $M(x)$ is classified information and cannot be revealed here], how would you respond?

6.2 Substitution

The chain rule for derivatives gives us an extremely useful technique for finding antiderivatives. This technique is called **substitution** or **change of variables.**

Consider the integral $\int 4x(x^2+1)^{4.4}\,dx$. If we were *differentiating,* rather than integrating, the function $4x(x^2+1)^{4.4}$, then, in using the product rule, we would need to differentiate $(x^2+1)^{4.4}$, and would use the chain rule with $u = x^2+1$ to do this. The integration technique of substitution gives us a procedure, analogous to the chain rule, for integrating expressions such as $4x(x^2+1)^{4.4}$ that involve some function u of x. Think of the integral $\int 4x(x^2+1)^{4.4}\,dx$ as having the form $\int f\,dx$, where f is a complicated function involving the function u of x. We write

$$I = \int f\,dx$$

for the integral we are trying to compute. By the chain rule,

$$\frac{dI}{dx} = \frac{dI}{du}\frac{du}{dx}$$

Solving for dI/du gives

$$\frac{dI}{du} = \frac{dI/dx}{du/dx}$$

$$= \frac{f}{du/dx} \qquad \text{Since } \frac{dI}{dx} = f$$

Writing this in integral form gives a new expression for the integral I we are trying to compute:

$$I = \int\left(\frac{f}{du/dx}\right)du$$

Strange as it may seem, this integral will often turn out to be easier to compute!

Notice that we could have obtained the new expression by "cheating" as follows: Write the differential dx as

$$dx = \frac{dx}{du}\,du \qquad \text{As if we could cancel } du \text{ on the right}$$

$$= \frac{1}{du/dx}\,du \qquad \text{As if derivatives behaved like fractions}$$

Now use this formula to substitute for dx in the expression $I = \int f\,dx$:

$$I = \int f\,dx = \int (f)\cdot\left(\frac{1}{du/dx}\right)du = \int\left(\frac{f}{du/dx}\right)du$$

This interpretation of the formula gives us a procedure for evaluating complicated integrals.

Substitution Rule

If u is a function of x, then we can use the following formula to evaluate an integral:

$$\int f\,dx = \int\left(\frac{f}{du/dx}\right)du$$

Using the formula is equivalent to the following procedure:

1. Write u as a function of x.
2. Take the derivative du/dx and solve for the quantity dx in terms of du.
3. Use the expression you obtain in step 2 to substitute for dx in the given integral, and substitute u for its defining expression.

example 1

Find $\int 4x(x^2 + 1)^{4.4}\,dx$.

Solution

To use substitution we need to choose an expression to be u. There is no hard and fast rule, but here is one hint that often works:

Take u to be an expression that is being raised to a power.

In this case, let us set $u = x^2 + 1$. Continuing the procedure above, we place the calculations for step 2 in a box.

$u = x^2 + 1$	Write u as a function of x.
$\dfrac{du}{dx} = 2x$	Take the derivative of u with respect to x.
$dx = \dfrac{1}{2x}\,du$	Solve for dx: $dx = \dfrac{1}{du/dx}\,du$.

Now we *substitute u for its defining expression and substitute for dx* in the original integral:

$$\int 4x(x^2 + 1)^{4.4}\,dx = \int 4x\,u^{4.4}\frac{1}{2x}\,du \qquad \text{Substitute[1] for } u \text{ and[2] } dx.$$

$$= \int 2u^{4.4}\,du \qquad \text{Cancel the } x\text{s and simplify.}$$

We have boiled the given integral down to the much simpler integral $\int 2u^{4.4}\,du$, and we can now write the solution:

$$2\frac{u^{5.4}}{5.4} + C = \frac{2(x^2 + 1)^{5.4}}{5.4} + C \qquad \text{Substitute } (x^2 + 1) \text{ for } u \text{ in the answer.}$$

Before we go on . . . There are two points to notice here. First, before we can actually integrate with respect to u, *we must eliminate all xs from the integrand.* If we cannot, we may have chosen the wrong expression for u. Second, after integrating, we must substitute back to obtain an expression involving x.

It is easy to check our answer. We differentiate:

$$\frac{d}{dx}\left(\frac{2(x^2 + 1)^{5.4}}{5.4}\right) = \frac{2(5.4)(x^2 + 1)^{4.4}(2x)}{5.4} = 4x(x^2 + 1)^{4.4}$$

Notice how we used the chain rule to check the result obtained by substitution.

[1] Can you see how this step is equivalent to using the formula $\int f\,dx = \int\left(\frac{f}{du/dx}\right)du$?

[2] If it bothers you that the integral contains both x and u, note that x is now a function of u.

When we use substitution, the first step is always to decide what to take as u. Again, there are no set rules, but we see some common cases in the examples.

example 2

Calculate $\int x^2(x^3+1)^2\,dx$.

Solution

As in the preceding example, it often works to take u to be an expression that is being raised to a power. We usually also want to see the derivative of u as a factor in the integrand. In this case, x^3+1 is being raised to a power, so let us set $u = x^3+1$. Its derivative is $3x^2$; in the integrand is x^2, which is missing the factor 3, but missing or incorrect constant factors are not a problem.

$u = x^3 + 1$	Write u as a function of x.
$\dfrac{du}{dx} = 3x^2$	Take the derivative of u with respect to x.
$dx = \dfrac{1}{3x^2}\,du$	Solve for dx: $dx = \dfrac{1}{du/dx}\,du$.

$$\int x^2(x^3+1)^2\,dx = \int x^2 u^2 \frac{1}{3x^2}\,du \qquad \text{Substitute for } u \text{ and } dx.$$

$$= \int \frac{1}{3}u^2\,du \qquad \text{Cancel the terms with } x.$$

$$= \frac{1}{9}u^3 + C \qquad \text{Take the antiderivative.}$$

$$= \frac{1}{9}(x^3+1)^3 + C \qquad \text{Substitute for } u \text{ in the answer.}$$

example 3

Evaluate $\int 3xe^{x^2}\,dx$.

Solution

When we have an exponential with an expression in the exponent, it often works to substitute u for that expression. In this case, let us take $u = x^2$.

$$u = x^2$$

$$\frac{du}{dx} = 2x$$

$$dx = \frac{1}{2x}\,du$$

Substituting into the integral, we get

$$\int 3xe^{x^2}\,dx = \int 3xe^u \frac{1}{2x}\,du = \int \frac{3}{2}e^u\,du = \frac{3}{2}e^u + C = \frac{3}{2}e^{x^2} + C$$

example 4

Evaluate $\int \frac{1}{2x+5}\,dx$.

Solution

We begin by rewriting the integrand as a power:

$$\int \frac{1}{2x+5}\,dx = \int (2x+5)^{-1}\,dx$$

Now we take our earlier advice and set u equal to the expression that is being raised to a power:

$$u = 2x+5$$
$$\frac{du}{dx} = 2$$
$$dx = \frac{1}{2}\,du$$

Substituting, we have

$$\int \frac{1}{2x+5}\,dx = \int \frac{1}{2}u^{-1}\,du = \frac{1}{2}\ln|u| + C = \frac{1}{2}\ln|2x+5| + C$$

example 5

Evaluate $\int (x+3)\sin(x^2+6x)\,dx$.

Solution

There are two parenthetical expressions. Notice, however, that the derivative of the expression (x^2+6x) is $2x+6$, which is twice the term $(x+3)$ in front of the sine. Recall that we would like the derivative of u to appear as a factor. Thus, let us take $u = x^2+6x$.

$$u = x^2+6x$$
$$\frac{du}{dx} = 2x+6 = 2(x+3)$$
$$dx = \frac{1}{2(x+3)}\,du$$

Substituting into the integral, we get

$$\int (x+3)\sin(x^2+6x)\,dx = \int (x+3)\sin u\left(\frac{1}{2(x+3)}\right)du$$
$$= \int \frac{1}{2}\sin u\,du$$
$$= -\frac{1}{2}\cos u + C = -\frac{1}{2}\cos(x^2+6x) + C$$

There are some cases where a little more work is required.

example 6

When the x Terms Do Not Cancel

Evaluate $\int \frac{2x}{(x-5)^2}\,dx$.

Solution

We first rewrite:

$$\int \frac{2x}{(x-5)^2}\,dx = \int 2x(x-5)^{-2}\,dx$$

This suggests that we should set $u = x - 5$.

$$u = x - 5$$
$$\frac{du}{dx} = 1$$
$$dx = du$$

Substituting, we have

$$\int \frac{2x}{(x-5)^2}\,dx = \int 2xu^{-2}\,du$$

Now, there is nothing in the integrand to cancel the x that appears. We can do the following: If, after we substitute, there is still an x in the integrand, we go back to the expression for u, solve for x, and substitute the expression we obtain for x in the integrand. So, we take $u = x - 5$ and solve for $x = u + 5$. Substituting, we get

$$\begin{aligned}\int 2xu^{-2}\,du &= \int 2(u+5)u^{-2}\,du \\ &= 2\int (u^{-1} + 5u^{-2})\,du \\ &= 2\ln|u| - \frac{10}{u} + C \\ &= 2\ln|x-5| - \frac{10}{x-5} + C\end{aligned}$$

example 7

Medicare Costs

The rate of increase of accumulated Medicare costs for hospice care in the United States, in billions of dollars per year, was given by[3]

$$R(t) = \frac{2.2e^{0.67t}}{13 + e^{0.67t}} \qquad (0 \le t \le 8;\ t = 0 \text{ represents } 1989)$$

[3]The logistic model is based on actual data from Health Care Financing Administration/*New York Times,* May 10, 1998, p. 18.

a. Find an expression for the accumulated costs from 1989.
b. How much had Medicare spent on hospice care from 1989 to 1997?

Solution

a. If we write the accumulated costs to year t as $A(t)$, then we are told that

$$A'(t) = \frac{2.2e^{0.67t}}{13 + e^{0.67t}}$$

Thus, $$A(t) = \int \frac{2.2e^{0.67t}}{13 + e^{0.67t}}\, dt$$

is the function we are after. To integrate the expression, we take u to be the denominator of the integrand:

$$u = 13 + e^{0.67t}$$

$$\frac{du}{dt} = 0.67e^{0.67t}$$

$$dt = \frac{1}{0.67e^{0.67t}}\, du$$

Substituting

$$A(t) = \int \frac{2.2e^{0.67t}}{13 + e^{0.67t}\, dt} = \int \frac{2.2e^{0.67t}}{u} \cdot \frac{1}{0.67e^{0.67t}}\, du$$

$$= \frac{2.2}{0.67} \int \frac{1}{u}\, du$$

$$\approx 3.28 \ln|u| + C$$

$$= 3.28 \ln(13 + e^{0.67t}) + C$$

Why could we drop the absolute value?

Now what is C? Since $A(t)$ represents accumulated costs *since time* $t = 0$, we have $A(0) = 0$ (because that is when we started counting). Thus,

$$0 = 3.28 \ln(13 + e^0) + C$$

$$= 3.28 \ln(14) + C$$

Hence, $C = -3.28 \ln(14) \approx -8.7$

Thus, the accumulated cost is

$$A(t) = 3.28 \ln(13 + e^{0.67t}) - 8.7$$

b. Since 1997 is represented by $t = 8$, the accumulated costs from 1989 to 1997 are given by

$$A(8) = 3.28 \ln(13 + e^{0.67(8)}) - 8.7 \approx \$9.1 \text{ billion}$$

Antiderivatives of the Six Trigonometric Functions

The following table gives indefinite integrals of the six trigonometric functions.

Integrals of the Trigonometric Functions

$\int \sin x \, dx = -\cos x + C$	We saw this in the preceding section.
$\int \cos x \, dx = \sin x + C$	We saw this in the preceding section.
$\int \tan x \, dx = -\ln\lvert \cos x \rvert + C$	Shown below
$\int \cot x \, dx = \ln\lvert \sin x \rvert + C$	See the exercise set.
$\int \sec x \, dx = \ln\lvert \sec x + \tan x \rvert + C$	Shown below
$\int \csc x \, dx = -\ln\lvert \csc x + \cot x \rvert + C$	See the exercise set.

Question Why is $\int \tan x \, dx = -\ln\lvert \cos x \rvert + C$?
Answer We can write $\tan x$ as $\sin x/\cos x$ and put $u = \cos x$ in the integral:

$$\begin{aligned} \int \tan x \, dx &= \int \frac{\sin x}{\cos x} dx \\ &= -\int \frac{\sin x}{u} \frac{du}{\sin x} \\ &= -\int \frac{du}{u} \\ &= -\ln\lvert u \rvert + C \\ &= -\ln\lvert \cos x \rvert + C \end{aligned}$$

$$u = \cos x \qquad \frac{du}{dx} = -\sin x \qquad dx = -\frac{du}{\sin x}$$

Question Why is $\int \sec x \, dx = \ln\lvert \sec x + \tan x \rvert + C$?
Answer To get started, we use a little "trick": We write $\sec x$ as $\sec x \left(\frac{\sec x + \tan x}{\sec x + \tan x}\right)$ and put u equal to the denominator:

$$\begin{aligned} \int \sec x \, dx &= \int \sec x \left(\frac{\sec x + \tan x}{\sec x + \tan x}\right) dx \\ &= \int \sec x \frac{\sec x + \tan x}{u} \frac{du}{\sec x(\tan x + \sec x)} \\ &= \int \frac{du}{u} \\ &= \ln\lvert u \rvert + C \\ &= \ln\lvert \sec x + \tan x \rvert + C \end{aligned}$$

$$\begin{aligned} u &= \sec x + \tan x \\ \frac{du}{dx} &= \sec x \tan x + \sec^2 x \\ &= \sec x\,(\tan x + \sec x) \\ dx &= \frac{du}{\sec x\,(\tan x + \sec x)} \end{aligned}$$

Shortcuts

If a and b are constants with $a \neq 0$, then we have the following formulas. (We have already seen one of them in Example 4. All of them can be obtained using the substitution $u = ax + b$. They will appear in the exercises.)

Shortcuts: Integrals of Expressions Involving $(ax + b)$

Rule	Quick Example
$\int (ax+b)^n\,dx = \frac{(ax+b)^{n+1}}{a(n+1)} + C$ (if $n \neq -1$)	$\int (3x-1)^2\,dx = \frac{(3x-1)^3}{3(3)} + C = \frac{(3x-1)^3}{9} + C$
$\int (ax+b)^{-1}\,dx = \frac{1}{a}\ln\lvert ax+b\rvert + C$	$\int (3-2x)^{-1}\,dx = \frac{1}{(-2)}\ln\lvert 3-2x\rvert + C = -\frac{1}{2}\ln\lvert 3-2x\rvert + C$
$\int e^{ax+b}\,dx = \frac{1}{a}e^{ax+b} + C$	$\int e^{-x+4}\,dx = \frac{1}{(-1)}e^{-x+4} + C = -e^{-x+4} + C$
$\int \sin(ax+b)\,dx = -\frac{1}{a}\cos(ax+b) + C$	$\int \sin(-4x)\,dx = \frac{1}{4}\cos(-4x) + C$
$\int \cos(ax+b)\,dx = \frac{1}{a}\sin(ax+b) + C$	$\int \cos(x+1)\,dx = \sin(x+1) + C$

6.2 exercises

In Exercises 1–44, evaluate the given integral.

1. $\int (3x+1)^5\,dx$
2. $\int (-x-1)^7\,dx$
3. $\int (-2x+2)^{-2}\,dx$
4. $\int (2x)^{-1}\,dx$
5. $\int x(3x^2+3)^3\,dx$
6. $\int x(-x^2-1)^3\,dx$
7. $\int x(x^2+1)^{1.3}\,dx$
8. $\int \frac{x}{(3x^2-1)^{0.4}}\,dx$
9. $\int (1+9.3e^{3.1x-2})\,dx$
10. $\int (3.2-4e^{1.2x-3})\,dx$
11. ∆ $\int 7.6\cos(3x-4)\,dx$
12. ∆ $\int 4.4\sin(-3x+4)\,dx$
13. ∆ $\int x\sin(3x^2-4)\,dx$
14. ∆ $\int x\cos(-3x^2+4)\,dx$
15. ∆ $\int (4x+2)\sin(x^2+x)\,dx$
16. ∆ $\int (x+1)[\cos(x^2+2x)+(x^2+2x)]\,dx$
17. ∆ $\int (x+x^2)\sec^2(3x^2+2x^3)\,dx$
18. ∆ $\int (4x+2)\sec^2(x^2+x)\,dx$
19. ∆ $\int (x^2)\tan(2x^3)\,dx$
20. ∆ $\int (4x)\tan(x^2)\,dx$
21. ∆ $\int 6\sec(2x-4)\,dx$
22. ∆ $\int 3\csc(3x)\,dx$
23. $\int 2x\sqrt{3x^2-1}\,dx$
24. $\int 3x\sqrt{-x^2+1}\,dx$
25. $\int xe^{-x^2+1}\,dx$
26. $\int xe^{2x^2-1}\,dx$
27. $\int (x+1)e^{-(x^2+2x)}\,dx$
28. $\int (2x-1)e^{2x^2-2x}\,dx$
29. ∆ $\int e^{2x}\cos(e^{2x}+1)\,dx$
30. ∆ $\int e^{-x}\sin(e^{-x})\,dx$
31. $\int \frac{-2x-1}{(x^2+x+1)^3}\,dx$
32. $\int \frac{x^3-x^2}{3x^4-4x^3}\,dx$
33. $\int \frac{x^2+x^5}{\sqrt{2x^3+x^6-5}}\,dx$
34. $\int \frac{2(x^3-x^4)}{(5x^4-4x^5)^5}\,dx$
35. $\int x(x-2)^5\,dx$
36. $\int x(x-2)^{1/3}\,dx$

37. $\int 2x\sqrt{x+1}\,dx$

38. $\int \frac{x}{\sqrt{x+1}}\,dx$

39. $\int \frac{3e^{-1/x}}{x^2}\,dx$

40. $\int \frac{2e^{2/x}}{x^2}\,dx$

41. $\int \frac{e^{-0.05x}}{1-e^{-0.05x}}\,dx$

42. $\int \frac{3e^{1.2x}}{2+e^{1.2x}}\,dx$

43. $\int [(2x-1)e^{2x^2-2x} + xe^{x^2}]\,dx$

44. $\int (xe^{-x^2+1} + e^{2x})\,dx$

In Exercises 45–50, derive each equation, where a and b are constants with $a \neq 0$.

45. $\int (ax+b)^n\,dx = \frac{(ax+b)^{n+1}}{a(n+1)} + C \quad (\text{if } n \neq -1)$

46. $\int (ax+b)^{-1}\,dx = \frac{1}{a}\ln|ax+b| + C$

47. $\int e^{ax+b}\,dx = \frac{1}{a}e^{ax+b} + C$

△ 48. $\int \cos(ax+b)\,dx = \frac{1}{a}\sin(ax+b) + C$

△ 49. $\int \cot x\,dx = \ln|\sin x| + C$

△ 50. $\int \csc x\,dx = -\ln|\csc x + \cot x| + C$

Use the shortcut formulas in Exercises 45–50 to calculate the integrals in Exercises 51–62 mentally.

51. $\int e^{-x}\,dx$

52. $\int e^{x-1}\,dx$

53. $\int e^{2x-1}\,dx$

54. $\int e^{-3x}\,dx$

55. $\int (2x+4)^2\,dx$

56. $\int (3x-2)^4\,dx$

57. $\int \frac{1}{5x-1}\,dx$

58. $\int (x-1)^{-1}\,dx$

59. $\int (1.5x)^3\,dx$

60. $\int e^{2.1x}\,dx$

△ 61. $\int \sin(-1.1x-1)\,dx$

△ 62. $\int \cos(4.2x-1)\,dx$

63. Find $f(x)$ if $f(0) = 0$ and the tangent line at $(x, f(x))$ has slope $x(x^2+1)^3$.

64. Find $f(x)$ if $f(1) = 0$ and the tangent line at $(x, f(x))$ has slope $\frac{x}{x^2+1}$.

65. Find $f(x)$ if $f(1) = \frac{1}{2}$ and the tangent line at $(x, f(x))$ has slope xe^{x^2-1}.

66. Find $f(x)$ if $f(2) = 1$ and the tangent line at x has slope $(x-1)e^{x^2-2x}$.

Applications

67. ***Cost*** The marginal cost of producing the xth roll of film is given by $5 + 1/(x+1)^2$. The total cost to produce one roll is \$1000. Find the total cost function $C(x)$.

68. ***Cost*** The marginal cost of producing the xth box of videotape is given by $10 - x/(x^2+1)^2$. The total cost to produce two boxes is \$1000. Find the total cost function $C(x)$.

69. ***Employment*** In 1988 statewide employment in Massachusetts was approximately 3,100,000. The following quadratic model approximates the rate of increase in employment, in people per year, in Massachusetts from 1988 through 1994:[4]

$C(t) = 25{,}000(t-1988)^2 - 137{,}000(t-1988) + 68{,}000$
$(1988 \le t \le 1994)$

where t is the year. Use the model and the 1988 employment figure for Massachusetts to obtain a model that gives the total number of people $N(t)$ employed in Massachusetts as a function of t.

70. ***Hawaiian Tourism*** The rate of visitor spending, in billions of dollars per year, in Hawaii during the years 1985 to 1993 can be approximated by[5]

$r(t) = -0.164(t-1985)^2 + 1.60(t-1985) + 6.71$
$(1985 \le t \le 1993)$

where t is the year. According to the model, how much revenue did Hawaii earn from visitor spending from 1985 to 1990? (Give your answer to the nearest billion dollars.)

71. ***Motion in a Straight Line*** The velocity of a particle moving in a straight line is given by $v = t(t^2+1)^4 + t$.
   a. Find an expression for the position s after a time t.
   b Given that $s = 1$ at time $t = 0$, find the constant of integration C, and hence find an expression for s in terms of t without any unknown constants.

72. ***Motion in a Straight Line*** The velocity of a particle moving in a straight line is given by $v = 3te^{t^2} + t$.
   a. Find an expression for the position s after a time t.
   b. Given that $s = 3$ at time $t = 0$, find the constant of integration C, and hence find an expression for s in terms of t without any unknown constants.

[4]The model is a quadratic regression based on actual data from the Massachusetts Department of Employment, DRI/McGraw Hill/*New York Times,* January 27, 1995, p. D4.

[5]The model is based on a best-fit quadratic for (approximate) tourism data, as measured in constant 1993 dollars, from the Hawaii Visitors Bureau/*New York Times,* September 5, 1995, p. A12.

Communication and Reasoning Exercises

73. Are there any circumstances in which one should use the substitution $u = x$? Illustrate your answer by giving an example that shows the effect of this substitution.

74. Give an example of an integral that can be calculated by using the substitution $u = x^2 + 1$, and then carry out the calculation.

75. Give an example of an integral that can be calculated by using the power rule for antiderivatives and also by using the substitution $u = x^2 + x$, and then carry out the calculations.

76. At what stage of a calculation using a u substitution should one substitute back for u in terms of x: before or after taking the antiderivative?

77. You are asked to calculate $\int \frac{u}{u^2+1}\,du$. What is wrong with the substitution $u = u^2 + 1$?

78. What is wrong with the following "calculation" of $\int \frac{1}{x^2-1}\,dx$?

$$\int \frac{1}{x^2-1} = \int \frac{1}{u} \qquad \text{Using the substitution } u = x^2 - 1$$
$$= \ln|u| + C$$
$$= \ln|x^2-1| + C$$

6.3 The Definite Integral As a Sum: A Numerical Approach

In the first two sections of this chapter we discussed the indefinite integral. There is a related concept called the **definite integral.** Let us introduce this new idea with an example. (We shall drop hints now and then about how the two types of integral are related. In Section 6.5 we discuss the exact relationship, which is one of the most important results in calculus.)

In the first section we used antiderivatives to answer questions of the form "Given the marginal cost, compute the total cost" (see Example 5 in Section 6.1). In this section we approach such questions more directly, using a numerical approach, as though we had never even heard of antiderivatives.

example 1

Total Cost

A cellular phone company offers you an innovative pricing scheme. When you make a call, the *marginal* cost of the tth minute of the call is

$$c(t) = \frac{20}{t+100} \text{ dollars per minute}$$

Use a numerical calculation to estimate the cost of a 60-minute phone call.

Solution

As we pointed out above, we *could* solve this problem using an antiderivative, similar to Example 5 in Section 6.1. However, let us forget about antiderivatives for the moment and do the computation numerically. As we did when first computing derivatives, we approach the answer by making better and better approximations.

Let us start with a very crude estimate. The marginal cost at the beginning of your call is $c(0) = \$0.20$/min. If this cost were to remain constant for the length of your call, the total cost of the call would be

$$\text{Cost of call} = \text{Cost per minute} \times \text{Number of minutes} = 0.20 \times 60 = \$12$$

But the marginal cost does not remain constant. It goes down over the course of the call. (Why?) We can obtain a much more accurate estimate of the cost by looking at the call minute by minute. We estimate the cost of each minute using the marginal cost at the beginning of that minute. The first minute costs (approximately) $c(0) = \$0.20$, but the second minute costs (approximately) $c(1) = \$0.198$. Adding together the costs of all 60 minutes gives us the estimate

$$c(0) \times 1 + c(1) \times 1 + \cdots + c(59) \times 1 = \$9.44 \qquad \text{Minute-by-minute calculation}$$

If we assume that the phone company is honest about $c(t)$ being the marginal cost and is actually calculating your cost continuously, we get an even better estimate of the cost by looking at the call second by second. For example, the marginal cost at the beginning of the first second is \$0.20/min, and the first second is $\frac{1}{60}$ of a minute, so the first second costs you approximately $\$0.20 \times \frac{1}{60}$. The marginal cost at the beginning of the next second is $c\left(\frac{1}{60}\right) = \$0.199967$, so the cost is approximately $\$0.199967 \times \frac{1}{60}$. Adding together the costs of all 3600 seconds in the hour gives us the estimate

$$c(0) \times 1/60 + c(1/60) \times 1/60 + c(2/60) \times 2/60 + \cdots + c(3599/60) \times 1/60 = \$9.94 \qquad \text{Second-by-second calculation}$$

Before we go on . . . The minute-by-minute calculation above is tedious to do by hand, and no one in their right mind would even *attempt* to do the second-by-second calculation by hand. Below we discuss ways of doing these calculations with the aid of technology.

The type of calculation done in this example is useful in many applications. Let us look at the general case and give the result a name.

In general, we have a function f (such as the function c in the example) and we consider an interval $[a, b]$ of possible values of the independent variable x (the time interval $[0, 60]$ in the example). We break the interval $[a, b]$ into some number of segments of equal length. In the example above we considered $[0, 60]$ first as a single segment, then we broke it into 60 segments, and then into 3600 segments. We write n for the number of segments.

Next we label the endpoints of these segments x_0 for a, x_1 for the end of the first segment, x_2 for the end of the second segment, and so on until we get to x_n, the end of the nth segment, so that $x_n = b$. Thus,

$$a = x_0 < x_1 < \cdots < x_n = b$$

The first segment is the interval $[x_0, x_1]$, the second segment is $[x_1, x_2]$, and so on until we get to the last segment, which is $[x_{n-1}, x_n]$. In the example, when n was 60, the segments were $[0, 1]$, $[1, 2]$, and so on, with the last segment being $[59, 60]$. In the general case, we are dividing the interval $[a, b]$ into n segments of equal length, so each segment has length $(b - a)/n$. We write Δx for $(b - a)/n$ (Figure 3). The widths of the intervals in the example when $n = 60$ were $\Delta x = 1$.

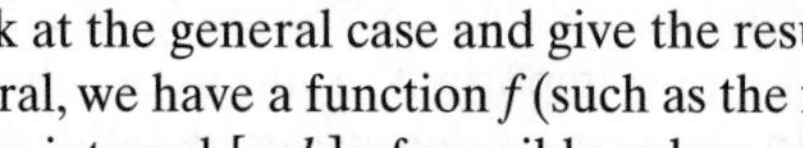

$x_0 \quad x_1 \quad x_2 \quad x_3 \quad \ldots \quad x_{n-1} \quad x_n$

$a \quad \leftarrow \Delta x \rightarrow \quad b$

Figure 3

Having established this notation, we can write the calculation we want to do as follows: For each segment $[x_{k-1}, x_k]$, we compute $f(x_{k-1})$, the value of the function f at the left endpoint. We multiply this value by the length of the interval, which is Δx. Then we add together all n of these products to get the number

$$f(x_0)\,\Delta x + f(x_1)\,\Delta x + \cdots + f(x_{n-1})\,\Delta x$$

This sum is called a (left) **Riemann sum**[1] for f. (In Example 1 we computed three different Riemann sums.)

Since sums are used often in mathematics, mathematicians have developed a shorthand notation for them. We write

$$\sum_{k=0}^{n-1} f(x_k)\,\Delta x = f(x_0)\,\Delta x + f(x_1)\,\Delta x + \cdots + f(x_{n-1})\,\Delta x$$

The symbol Σ is the Greek letter *sigma* and stands for **summation.** The letter k here is called the index of summation, and we can think of it as counting off the segments. We read the notation as "the sum from $k = 0$ to $n - 1$ of the quantities $f(x_k)\,\Delta x$." Think of it as a set of instructions:

	Telephone Example ($n = 60$)
Set $k = 0$ and calculate $f(x_0)\,\Delta x$.	Cost of the 1st minute: $c(0) \times 1$
Set $k = 1$ and calculate $f(x_1)\,\Delta x$.	Cost of the 2nd minute: $c(1) \times 1$
$\vdots$	
Set $k = n - 1$ and calculate $f(x_{n-1})\,\Delta x$.	Cost of the 60th minute: $c(59) \times 1$
Now sum all the quantities so calculated.	

Riemann Sum

If f is a continuous function, then the **left Riemann sum** with n equal subdivisions for f over the interval $[a, b]$ is defined to be

$$\begin{aligned}\text{Riemann sum} &= \sum_{k=0}^{n-1} f(x_k)\,\Delta x \\ &= f(x_0)\,\Delta x + f(x_1)\,\Delta x + \cdots + f(x_{n-1})\,\Delta x \\ &= \left[f(x_0) + f(x_1) + \cdots + f(x_{n-1})\right]\Delta x\end{aligned}$$

where $a = x_0 < x_1 < \cdots < x_n = b$ are the subdivisions, and $\Delta x = (b - a)/n$.

Quick Example

In Example 1 we computed three Riemann sums:

$n = 1$: Riemann sum $= c(0)\,\Delta t = 0.20 \times 60 = \$12$

$n = 60$: Riemann sum $= \left[c(t_0) + c(t_1) + \cdots + c(t_{n-1})\right]\Delta t$
$= \left[c(0) + c(1) + \cdots + c(59)\right](1) = \$9.44$

$n = 3600$: Riemann sum $= \left[c(t_0) + c(t_1) + \cdots + c(t_{n-1})\right]\Delta t$
$= \left[c(0) + c(1/60) + \cdots + c(3599/60)\right](1/60) = \$9.40$

As in Example 1, we are most interested in what happens to the Riemann sum when we let n get very large. When f is continuous,[2] the Riemann sums always approach a limit as n goes to infinity. (This is not meant to be an obvious fact. Proofs may be found in advanced calculus texts.) We give the limit a name.

[1] After Georg Friedrich Bernhard Riemann (1826–1866).

[2] And for some other functions as well.

The Definite Integral

If f is a continuous function, the **definite integral of f from a to b** is defined to be

$$\int_a^b f(x)\,dx = \lim_{n\to\infty} \sum_{k=0}^{n-1} f(x_k)\,\Delta x$$

The function f is called the **integrand,** the numbers a and b are the **limits of integration,** and the variable x is the **variable of integration.** A Riemann sum with a large number of subdivisions may be used to approximate the definite integral.

Quick Example

We approximated the definite integral in Example 1 using the Riemann sum with $n = 3600$:

$$\int_0^{60} c(t)\,dt = \int_0^{60} \frac{20}{t+100}\,dt \approx 9.40$$

Notes

1. Remember that $\int_a^b f(x)\,dx$ stands for a number that depends on f, a, and b. The variable x that appears is called a **dummy variable** because it has no effect on the answer. We could just as well use any other variable, like t, as we did in the Quick Example above. In other words,

$$\int_a^b f(x)\,dx = \int_a^b f(t)\,dt \qquad x \text{ or } t \text{ is just a name we give the variable.}$$

2. The notation for the definite integral (due to Leibniz) comes from the notation for the Riemann sum. The integral sign $\int$ is an elongated S, the Roman equivalent of the Greek Σ. The d in dx is the lowercase Roman equivalent of the Greek Δ.
3. The definition above is adequate for continuous functions, but more complicated definitions are needed to handle other functions. For example, we broke the interval $[a, b]$ into n segments of equal length, but other definitions allow a **partition** of the interval into segments of possibly unequal lengths. We evaluated f at the left endpoint of each segment, but we could equally well have used the right endpoint or any other point in the segment. All of these variations lead to the same answer when f is continuous.
4. $\int_a^b f(x)\,dx$ is usually read "the integral, from a to b, of $f(x)\,dx$."
5. The similarity between the notations for the definite integral and the indefinite integral is no mistake. However, we wait to discuss the exact connection until the last section of this chapter.

example 2

Computing a Riemann Sum

Compute the Riemann sum for the integral $\int_{-1}^{1} (x^2 + 1)\,dx$ using $n = 5$ subdivisions.

Solution

Since the interval is $[a, b] = [-1, 1]$ and $n = 5$, we have

$$\Delta x = \frac{b-a}{n} = \frac{1-(-1)}{5} = 0.4 \qquad \text{Width of segments}$$

Thus, the subdivisions of $[-1, 1]$ are given by

$$-1 < -0.6 < -0.2 < 0.2 < 0.6 < 1$$ Start with -1 and keep adding $\Delta x = 0.4$.

The Riemann sum we want is

$$[f(x_0) + f(x_1) + \cdots + f(x_{n-1})]\Delta x = [f(-1) + f(-0.6) + f(-0.2) + f(0.2) + f(0.6)]0.4$$

We can conveniently organize this calculation in a table as follows:

x	−1	−0.6	−0.2	0.2	0.6	Total
$f(x) = x^2 + 1$	2	1.36	1.04	1.04	1.36	6.8

The Riemann sum is therefore

$$6.8\,\Delta x = 6.8 \times 0.4 = 2.72$$

Before we go on . . . We suggest you try the computation above on a spreadsheet, in preparation for the next example, where we use a large number of subdivisions.

example 3

Using Technology to Approximate the Definite Integral

Use technology to approximate the integral $\int_0^{60} \frac{20}{t + 100}\,dt$ using a Riemann sum with $n = 100$.

Solution

As in Example 1, we write

$$c(t) = \frac{20}{t + 100}$$

The Riemann sum with $n = 100$ has $\Delta t = (b - a)/n = (60 - 0)/100 = 0.6$ and is given by

$$\sum_{k=0}^{99} c(t_k)\,\Delta t = [c(0) + c(0.6) + \cdots + c(59.4)](0.6)$$

Spreadsheet

We first need to calculate the numbers $c(0)$, $c(0.6)$, and so on. To enter the numbers 0, 0.6, 1.2, . . . in the first column, we enter 0 in cell A1 and select this cell. From the Edit menu, we select Fill→Series . . . . In the dialog box, we choose Series in Columns, Type Linear, and enter 0.6 as the step value and 59.4 as the stop value. Excel then fills in the cells through A100 with the correct values. (Scroll down to cell A100 to check.) Now we enter the formula for $c(t)$ in cell B1, as

```
=20/(A1+100)
```

We copy this formula to cells B2 through B100.

	A	B	C
1	0	=20/(A1+100)	
2	0.6		
3	1.2		
100	59.4		
	values of t	values of $c(t)$	

Now to complete our calculation, we enter in C1 the formula

```
=SUM(B1:B100)*0.6
```

for the Riemann sum, giving us the result.

	A	B	C
1	0	0.2	9.422609147
2	0.6	0.198807157	
3	1.2	0.197628458	
100	59.4	0.125470514	

Thus, we get the estimate of 9.4226 for the Riemann sum.

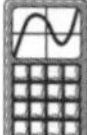

Graphing Calculator

There are several ways to compute Riemann sums with a graphing calculator. We illustrate two methods.

Method 1: *Using the LIST functions* We first need to calculate the numbers $c(0)$, $c(0.6)$, and so on, and then add them up. The TI-83 has a built-in `sum` function (available in the LIST MATH menu), which, like the SUM function in a spreadsheet, sums the entries in a list. To generate a list that contains the numbers we want to add together, we use the `seq` function (available in the LIST OPS menu). If we enter

```
seq(20/(X+100),X,0,59.4,0.6)
```

the calculator will calculate a list by evaluating `20/(X+100)` for values of X from 0 to 59.4 in steps of 0.6.

If we want the sum of all these numbers, we wrap the `seq` function in a call to `sum` (available in the LIST MATH menu):

```
sum(seq(20/(X+100),X,0,59.4,0.6))
```

This gives the sum

$$c(0) + c(0.6) + \cdots + c(59.4) = 15.70434858$$

To obtain the Riemann sum, we need to multiply this sum by $\Delta t = 0.6$, and we obtain the estimate of $15.70434858 \times 0.6 \approx 9.4226$ for the Riemann sum. One disadvantage of this method is that the TI-83 can generate and sum a list of at most 999 entries.

Method 2: *Using the LEFTSUM Program* An alternative is to use the LEFTSUM program given at the end of this section, which can calculate Riemann sums for any n. Assuming that you have entered this program in your calculator and called it LEFTSUM, here is how you use it. First, in the "`Y=`" window enter the function as Y_1:

```
Y1 = 20/(X+100)
```

Next, press PRGM, select the program LEFTSUM, and press ENTER. The program will ask for the left endpoint, the right endpoint, and n. To approximate the integral above using $n = 100$, you should enter 0 for the left endpoint, 60 for the right endpoint, and 100 for n. After a while, the program will return and tell you that the left-hand sum is 9.422609147. If you run the LEFTSUM program repeatedly with larger and larger values of n, you will notice that the Riemann sum appears to get closer and closer to a fixed value, 9.40007 . . . , so we can say with some certainty that the given integral is approximately 9.40007. The TI-83 also has a built-in function `fnInt`, which finds a very accurate approximation of a definite integral, but it uses a more sophisticated technique than we are discussing here.

Web Site

The path

Web site → On-Line Utilities → Numerical Integration Utility

will take you to a page that evaluates Riemann sums. Enter the function, the limits of integration, and n into the appropriate places, and click Left Sum to calculate the Riemann sum we are using here.

Applications: The Definite Integral As a Total

In Example 1 we saw that if $c(t)$ is the *marginal* cost of a phone call, in dollars per minute, then the *total* cost of a 60-minute call is $\int_0^{60} c(t)\,dt$ dollars. If, instead of computing the total cost of the call over the interval [0, 60] minutes, we computed the cost over the interval $[a, b]$ minutes, we would have obtained a total cost of $\int_a^b c(t)\,dt$ dollars. Thus, for example, the total cost of the call from $t = 5$ minutes to $t = 10$ minutes is $\int_5^{10} c(t)\,dt$ dollars.

The Definite Integral As a Total

If $r(x)$ is the rate of change of a quantity Q, then the total or accumulated change of the quantity as x changes from a to b is given by

$$\text{Total change in quantity } Q = \int_a^b r(x)\,dx \qquad \text{Total change in } Q \text{ over } [a, b]$$

Units

Note that we measure the rate of change r in units of Q per unit of x, and we measure the total $\int_a^b r(x)\,dx$ in units of Q.

Quick Examples

1. If, at time t hours, you are selling wall posters at a rate of $s(t)$ posters per hour, then

$$\text{Total number of posters sold from hour 3 to hour 5} = \int_3^5 s(t)\,dt$$

2. If, t days into the surf season, the water temperature at your favorite surf spot is changing at a rate of $0.05t^2 + 4$ degrees per day, then the total change in the water temperature from the start of the season ($t = 0$) to the start of day 10 is given by

$$\text{Total change in temperature} = \int_0^{10} (0.05t^2 + 4)\,dt \text{ degrees}$$

3. If a skateboard is moving at $v(t)$ feet per second at time t seconds, then the total distance the skateboard has moved during the interval [2, 5] seconds is

$$\text{Total distance moved} = \int_2^5 v(t)\,dt \text{ feet.}$$

example 4

Prozac® Prescriptions

Over the period 1988–1998, doctors were writing prescriptions for Prozac at the rate of approximately $-0.015t^2 + t + 1.9$ million new prescriptions per year (t is measured in years; $t = 0$ represents June 1988).[3] Use a Riemann sum with $n = 5$ subdivisions to obtain the approximate number of new prescriptions for Prozac written from June 1990 ($t = 2$) to June 1995 ($t = 7$).

Solution

We see that

$$\text{Total number of new prescriptions} = \int_2^7 (-0.015t^2 + t + 1.9)\, dt$$

Since we are approximating the integral with a Riemann sum with $n = 5$ subdivisions of the interval $[2, 7]$, we obtain

$$\Delta t = \frac{b - a}{n} = \frac{7 - 2}{5} = 1$$

and the subdivisions of $[2, 7]$ are

$$a = 2 < 3 < 4 < 5 < 6 < 7 = b$$

The rest of the calculation is done in the following table:

t	2	3	4	5	6	Total
$-0.015t^2 + t + 1.9$	3.84	4.765	5.66	6.525	7.36	28.15

The Riemann sum is therefore

$$28.15\, \Delta t = 28.15 \times 1 = 28.15 \text{ million prescriptions}$$

Before we go on . . . Try to do this example using antiderivatives instead of Riemann sums. Then go back and do Example 1 using antiderivatives. Do you notice a relationship between antiderivatives and definite integrals?

example 5

Motion

A ball thrown in the air has velocity $v(t) = -32t + 100$ ft/s after t seconds (as measured by a radar gun). Use several values of n to find the ball's change in height from time $t = 3$ seconds to time $t = 4$ seconds.

Solution

Since the velocity $v(t)$ is the rate of change of height, the total change in height over the interval $[3, 4]$ is

$$\text{Total change in height} = \int_3^4 v(t)\, dt = \int_3^4 (-32t + 100)\, dt$$

As in Example 1, we can subdivide the 1-second interval $[3, 4]$ into smaller and smaller parts to get more and more accurate approximations of the integral. By computing Riemann sums for various values of n, we get the following results:

[3]The quadratic approximation is based on published data from IMS Health/*New York Times,* October 11, 1998, p. BU8.

$$n = 10: \sum_{k=0}^{9} v(t_k)\,\Delta t = -10.4 \qquad n = 100: \sum_{k=0}^{99} v(t_k)\,\Delta t = -11.84$$

$$n = 1000: \sum_{k=0}^{999} v(t_k)\,\Delta t = -11.984 \qquad n = 10{,}000: \sum_{k=0}^{9999} v(t_k)\,\Delta t = -11.9984$$

These calculations suggest that the total change in height, the value of the definite integral, is -12 feet, which means that the ball is 12 feet lower at time $t = 4$ seconds than it was at time $t = 3$ seconds.

Optional Internet Topic: Numerical Integration

At the web site you will find on-line text, examples, and exercises on other techniques for approximating integrals. The particular Riemann sums we are using here are fine for defining the integral, but there are sums that get closer to the correct value of the integral using fewer terms.

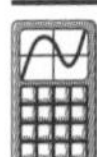

The LEFTSUM program for TI-83s

The following program calculates (left-hand) Riemann sums for any n. See Example 3 for instructions on how to use it. The latest version of this program (and others) is available at the web site.

```
PROGRAM: LEFTSUM
:Input "LEFT ENDPOINT? ",A      Prompts for the left endpoint a
:Input "RIGHT ENDPOINT? ",B     Prompts for the right endpoint b
:Input "N? ",N                  Prompts for the number of subdivisions
:(B-A)/N→D                      D is Δx = (b − a)/n
:Ø→L                            L will eventually be the left-hand sum
:A→X                            X is the current x-coordinate
:For(I,1,N)                     Start of a loop—recall the sigma notation
:L+Y₁→L                         Add f(x_{i−1}) to L
:A+I*D→X                        Corresponds to the formula x_i = a + i Δx
:End                            End of loop
:L*D→L                          Multiply by Δx
:Disp "LEFT SUM IS ",L
:Stop
```

6.3 exercises

Calculate the Riemann sums for the integrals in Exercises 1–12 using the given values of n.

1. $\int_0^2 (4x - 1)\,dx,\ n = 4$

2. $\int_{-1}^1 (1 - 3x)\,dx,\ n = 4$

3. $\int_{-2}^2 x^2\,dx,\ n = 4$

4. $\int_1^5 x^2\,dx,\ n = 4$

5. $\int_0^1 \frac{1}{1 + x}\,dx,\ n = 5$

6. $\int_0^1 \frac{x}{1 + x^2}\,dx,\ n = 5$

7. $\int_0^{10} e^{-x}\,dx,\ n = 5$

8. $\int_{-5}^5 e^{-x}\,dx,\ n = 5$

9. $\int_0^{10} e^{-x^2}\,dx,\ n = 4$

10. $\int_0^{100} e^{-x^2}\,dx,\ n = 4$

Δ **11.** $\int_0^3 \sin x\,dx,\ n = 4$

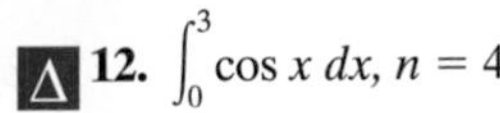

Δ **12.** $\int_0^3 \cos x\,dx,\ n = 4$

Calculate the Riemann sums for the integrals in Exercises 13–18 using $n = 10$, $n = 100$, and $n = 1000$. Round all answers to four decimal places.

T **13.** $\int_0^1 4\sqrt{1 - x^2}\,dx$

T **14.** $\int_0^1 \frac{4}{1 + x^2}\,dx$

T **15.** $\int_2^3 \frac{2x^{1.2}}{1 + 3.5x^{4.7}}\,dx$

T **16.** $\int_3^4 3xe^{1.3x}\,dx$

T **17.** $\int_0^{\pi} \sin x\,dx$

T **18.** $\int_0^{\pi} 2(\sin x)^2\,dx$

Applications

19. ***Cost*** The marginal cost function for the manufacture of portable CD players is given by

$$C'(x) = 20 - \frac{x}{200}$$

where x is the number of CD players manufactured. Use a Riemann sum with $n = 5$ to estimate the cost of producing the first five CD players.

20. ***Cost*** Repeat Exercise 19 using the marginal cost function

$$C'(x) = 25 - \frac{x}{50}$$

21. ***Employment*** The following quadratic model approximates the rate of increase in employment, in thousands of people per year, in Massachusetts from 1988 through 1995:[4]

$$C(t) = 25t^2 - 137t + 68 \text{ thousand people per year}$$

where t is the number of years since 1988. Use a Riemann sum with $n = 5$ to estimate the change in the number of people employed in Massachusetts from 1990 ($t = 2$) to 1995 ($t = 7$).

22. ***Hawaiian Tourism*** The rate of visitor spending, in billions of dollars per year, in Hawaii during the years 1985 to 1993 can be approximated by[5]

$$r(t) = -0.164t^2 + 1.60t + 6.71$$

where $t = 0$ represents June 30, 1985. Use a Riemann sum with $n = 5$ to estimate how much revenue Hawaii earned from visitor spending between June 30, 1985, and June 30, 1990. (Give your answer to the nearest billion dollars.)

23. ***Sales*** Estimated sales of IBM's mainframe computers and related peripherals showed the following pattern of decline over the years 1990 through 1994:

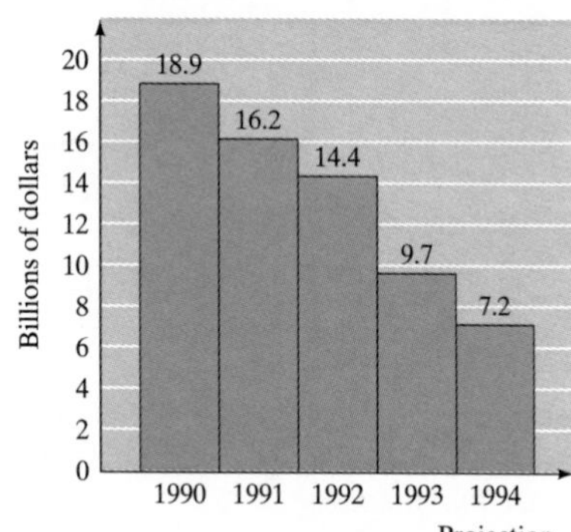

Source: Salomon Brothers/*New York Times,* October 26, 1993, p. D1.

Let $S(t)$ be the sales per year in year t, with $t = 0$ representing January 1990. Estimate $\int_0^5 S(t)\,dt$ using a Riemann sum with $n = 5$. What does this number represent?

24. ***Total Profit*** The profitability of United Airlines for the years 1988 through 1992 showed the following decline:

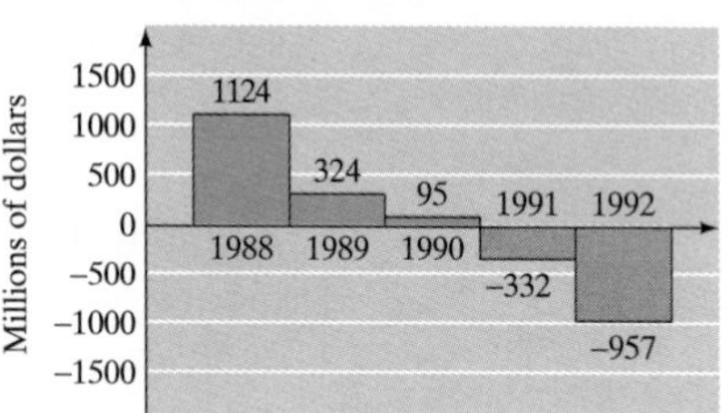

Source: Company reports/*New York Times,* December 24, 1993, p. D1.

Let $p(t)$ be United Airlines' annual profit in year t, where $t = 0$ represents January 1988. Estimate $\int_0^5 p(t)\,dt$ using a Riemann sum with $n = 5$. What does this number represent?

25. ***Motion*** A model rocket has upward velocity $v(t) = 40t^2$ ft/s, t seconds after launch. Use a Riemann sum with $n = 10$ to estimate how high the rocket is 2 seconds after launch.

26. ***Motion*** A race car has a velocity of $v(t) = 600(1 - e^{-0.5t})$ ft/s, t seconds after starting. Use a Riemann sum with $n = 10$ to estimate how far the car travels in the first 4 seconds. (Round your answer to the nearest whole number.)

T **27.** ***Medicare Costs*** (Compare Example 7 in Section 6.2.) The rate of spending by Medicare on hospice care in the United States, in billions of dollars per year, was given by[6]

$$R(t) = \frac{2.2e^{0.67t}}{13 + e^{0.67t}} \qquad (0 \le t \le 8;\ t = 0 \text{ represents } 1989)$$

Estimate $\int_0^8 R(t)\,dt$ using a Riemann sum with $n = 100$. (Round your answer to four decimal places.) Interpret the answer.

T **28.** ***Sales of Sport Utility Vehicles*** The rate at which sport utility vehicles (SUVs) were sold in the United States each year from 1980 through 1994 can be approximated by

$$r(t) = 0.4t^4 - 10t^3 + 73t^2 - 60t + 200$$

measured in thousands of SUVs per year, where t is the number of years since June 1980. Estimate $\int_0^{14} r(t)\,dt$

[4]The model is a quadratic regression based on the actual data through 1994. Hence, the 1995 figure is projected. *Source for data:* Massachusetts Department of Employment, DRI/McGraw Hill/*New York Times,* January 27, 1995, p. D4.

[5]The model is based on a best-fit quadratic for (approximate) tourism data, as measured in constant 1993 dollars, from the Hawaii Visitors Bureau/*New York Times,* September 5, 1995, p. A12.

[6]This logistic model is based on actual data from the Health Care Financing Administration/*New York Times,* May 10, 1998, p. 18.

using a Riemann sum with $n = 100$. (Round the Riemann sum to the nearest whole number.) Interpret the answer.
Source: Ford Motor Company/*New York Times*, February 9, 1995, p. D17.

The Normal Curve *The* normal distribution curve, *which models distributions of data in a wide range of applications, is given by the function*

$$p(x) = \frac{1}{\sqrt{2\pi}\,\sigma} e^{-(x-\mu)^2/2\sigma^2}$$

where $\pi = 3.14159265\ldots$ and σ and μ are constants called the standard deviation *and* the mean, *respectively. Its graph (when $\sigma = 1$ and $\mu = 2$) is shown in the figure. Exercises 29 and 30 illustrate its use.*

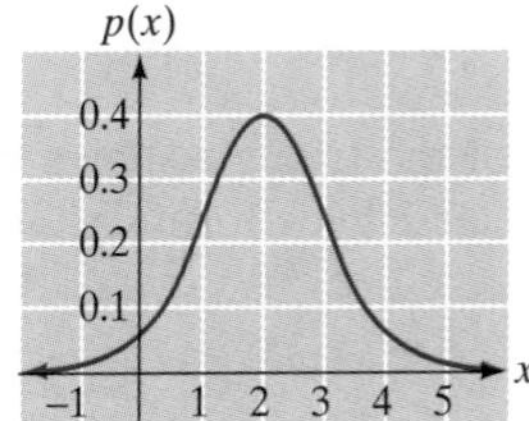

T **29.** ***Test Scores*** Enormous State University's Calculus I test scores are modeled by a normal distribution curve with $\mu = 72.6$ and $\sigma = 5.2$. The percentage of students who obtained scores between a and b on the test is given by

$$\int_a^b p(x)\,dx$$

a. Use a Riemann sum with $n = 40$ to estimate the percentage of students who obtained between 60 and 100 on the test.

b. What percentage of students scored less than 30?

T **30.** ***Consumer Satisfaction*** In a survey, consumers were asked to rate a new toothpaste on a scale of 1–10. The resulting data are modeled by a normal distribution with $\mu = 4.5$ and $\sigma = 1.0$. The percentage of consumers who gave the toothpaste a score between a and b on the test is given by

$$\int_a^b p(x)\,dx$$

a. Use a Riemann sum with $n = 10$ to estimate the percentage of customers who rated the toothpaste 5 or higher. (Use the range 4.5 to 10.5.)

b. What percentage of customers rated the toothpaste 0 or 1? (Use the range -0.5 to 1.5.)

Communication and Reasoning Exercises

31. When approximating a definite integral by computing Riemann sums, how might you judge whether you have chosen n large enough to get your answer accurate to, say, three decimal places?

32. Compare the discussions of marginal cost in the first section and this section. What do these discussions suggest about the relationship between the indefinite and definite integrals?

6.4 The Definite Integral As Area: A Geometric Approach

One of the oldest problems in mathematics is the problem of calculating areas of shapes that have curved boundaries. In this section we consider the problem of finding areas of a special sort. Suppose we have a function f with positive values when evaluated on an interval $[a, b]$. We wish to find the area between the graph of f and the x-axis, over the interval $[a, b]$, as shown in Figure 4.

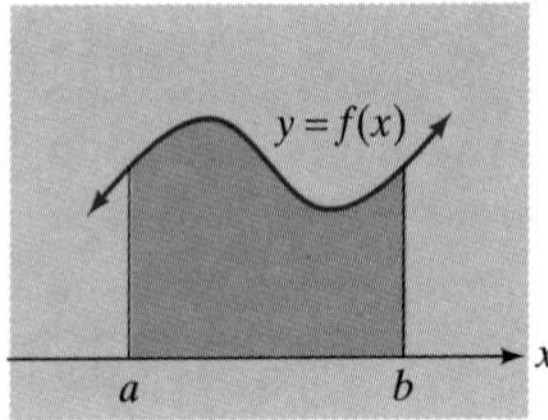

Figure 4

example 1

Area under a Graph
Find the area under the graph of $f(x) = 2$ over the interval $[1, 4]$.

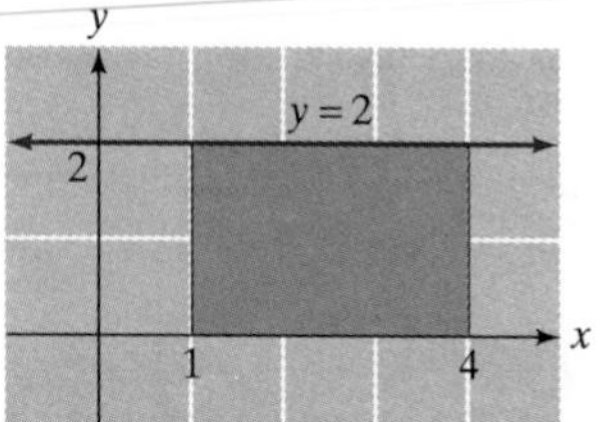

Figure 5

Solution
We are being asked to find the area in the shaded rectangle shown in Figure 5. This rectangle has a height of 2 units and a width of $4 - 1 = 3$ units, so it has an area of $2 \times 3 = 6$ square units.

Before we go on... If the units of measurement on the axes are not given, we sometimes do not mention units at all and simply say, for example, that "the area is 6."

example 2

Area under a Graph
Find the area under the graph of $f(x) = x$ over the interval $[0, 2]$.

Solution
This time we are being asked to find the area of the triangle shown in Figure 6. This triangle has a height of 2 and a base of 2, so, by a formula familiar from geometry, the area is $\frac{1}{2} \times 2 \times 2 = 2$.

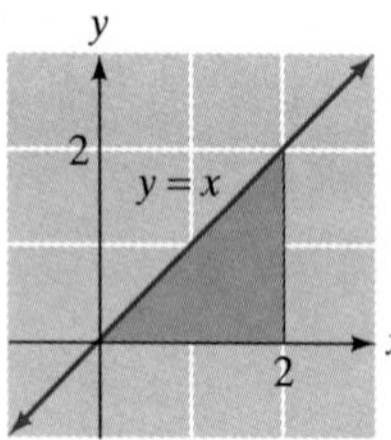

Figure 6

example 3

Area under a Graph
Find the area under the graph of $f(x) = 1 - x^2$ over the interval $[0, 1]$.

Solution
This time we need to find the area under the parabola shown in Figure 7. This is a considerably harder problem than finding the area of a rectangle or a triangle. Archimedes's solution of this problem was one of the high points of ancient Greek mathematics.

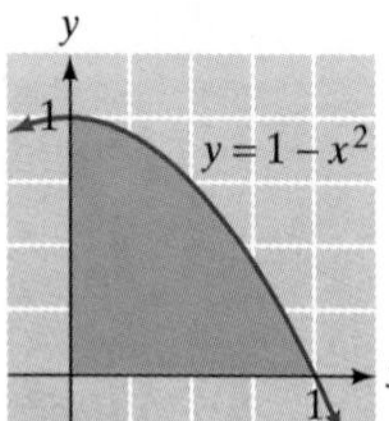

Figure 7

Archimedes used what was known as the "method of exhaustion," which is similar in spirit to what we are about to do. We *approximate* the area under the parabola with areas that we can easily calculate—namely, areas of rectangles.

Figure 8 shows how we might approximate the area under the parabola using four rectangles. Each of the rectangles in Figure 8 has width 0.25. The heights of the rectangles are $f(0) = 1$, $f(0.25) = 0.9375$, $f(0.5) = 0.75$, and $f(0.75) = 0.4375$. Thus, the total area enclosed by the four rectangles is

$$1 \times 0.25 + 0.9375 \times 0.25 + 0.75 \times 0.25 + 0.4375 \times 0.25 = 0.78125$$

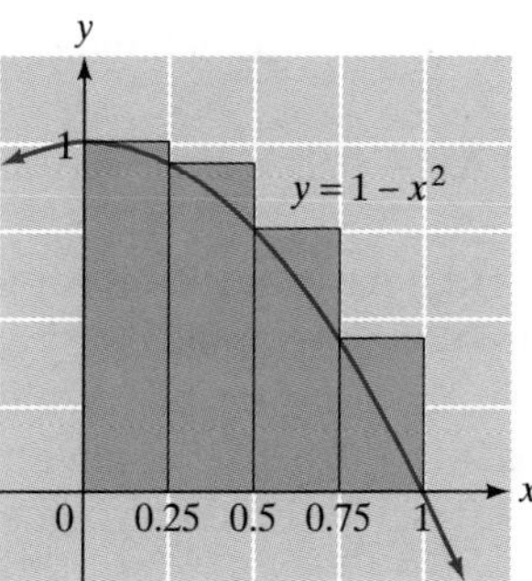

Figure 8

Now consider what happens when we use more than four rectangles. Figure 9 shows an approximation by ten rectangles.

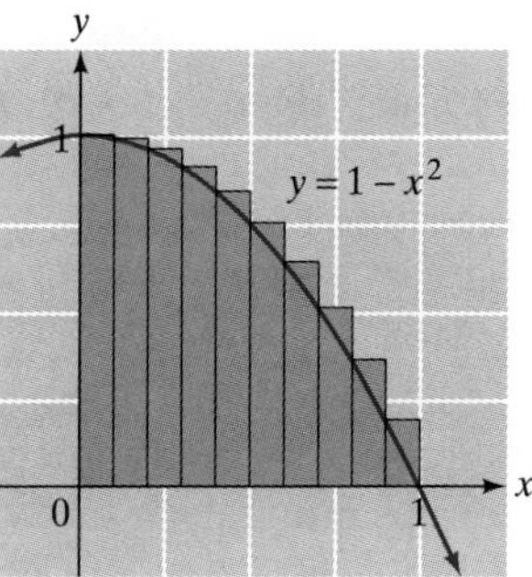

Figure 9

We can see that this is a better approximation; there is less extraneous area. If you're not yet convinced, consider Figures 10 and 11.

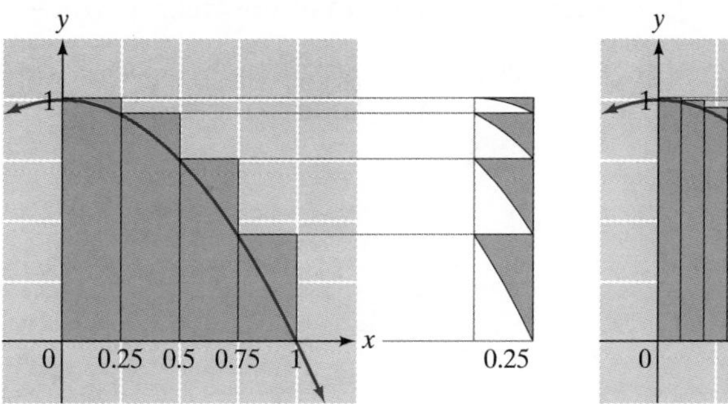

Figure 10

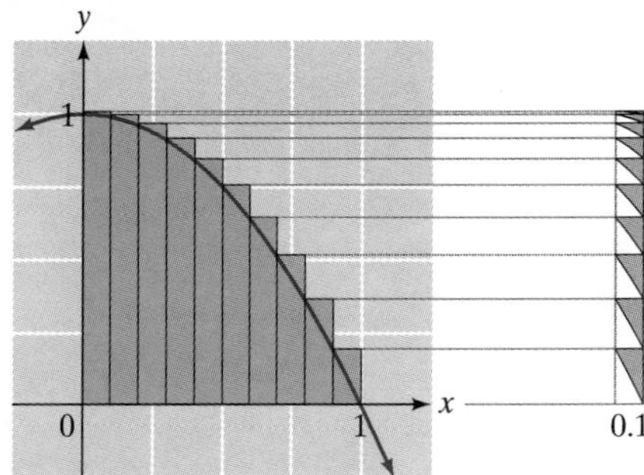

Figure 11

In Figure 10 we've taken the extraneous areas from Figure 8 and stacked them up. They fit inside a rectangle of area 0.25, so the error is less than 0.25. In Figure 11 we've done the same for the extraneous areas from Figure 9, but now they fit inside a rectangle of area 0.1. So, we've reduced the possible error to less than 0.1.

Each of the rectangles in Figure 9 has width 0.1, and the heights are given by the numbers $f(0)$, $f(0.1)$, and so on through $f(0.9)$. Using technology, we can calculate the total area enclosed by the rectangles as

$$f(0) \times 0.1 + f(0.1) \times 0.1 + \cdots + f(0.9) \times 0.1 = 0.715$$

By now you may have noticed that the sum of the areas of these rectangles is a Riemann sum of the sort discussed in the preceding section. For example, the sum of the areas of the ten rectangles in Figure 9 can be written as

$$\sum_{k=0}^{9} f(x_k)\,\Delta x = f(0) \times 0.1 + f(0.1) \times 0.1 + \cdots + f(0.9) \times 0.1 = 0.715$$

Each product $f(x_k)\,\Delta x$ in this sum is the height of a rectangle times its width and hence is the area of a rectangle.

As we let n, the number of rectangles, get larger and larger, we should get closer and closer to the actual area under the parabola. But, since we are computing Riemann sums, we should also get closer and closer to the value of the definite integral:

$$\text{Area} = \lim_{n\to\infty} \sum_{k=0}^{n-1} f(x_k)\,\Delta x$$

$$= \int_0^1 (1 - x^2)\,dx \qquad \text{The limit of the Riemann sums is the definite integral.}$$

That is, *the area under the parabola is the value of this definite integral.* As in the preceding section, we can use some form of technology to approximate the integral by Riemann sums. For example,

$$\sum_{k=0}^{99} f(x_k)\,\Delta x = 0.67165 \qquad \sum_{k=0}^{999} f(x_k)\,\Delta x = 0.6671665 \qquad \sum_{k=0}^{9999} f(x_k)\,\Delta x = 0.666716665$$

These numbers appear to be approaching $\frac{2}{3}$. Indeed, we see in the next section that the value of this integral is exactly $\frac{2}{3}$.

There is nothing particularly special about $1 - x^2$ in the preceding example. We can make the following generalization.

Geometric Interpretation of the Definite Integral (Nonnegative Functions)

If $f(x) \geq 0$ for all x in $[a, b]$, then $\int_a^b f(x)\,dx$ is the area under the graph of f over the interval $[a, b]$, as shaded in the figure.

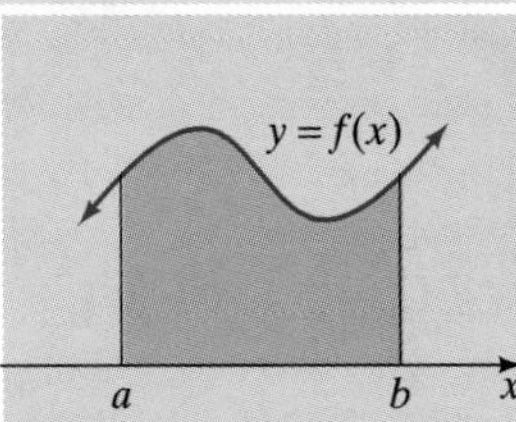

Quick Examples

1.

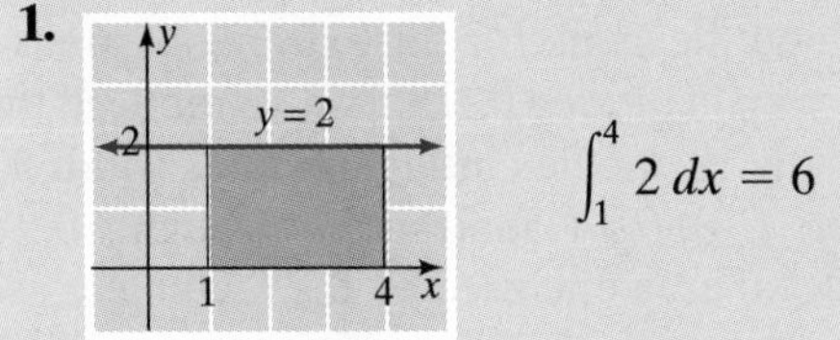

$\int_1^4 2\,dx = 6$ See Example 1.

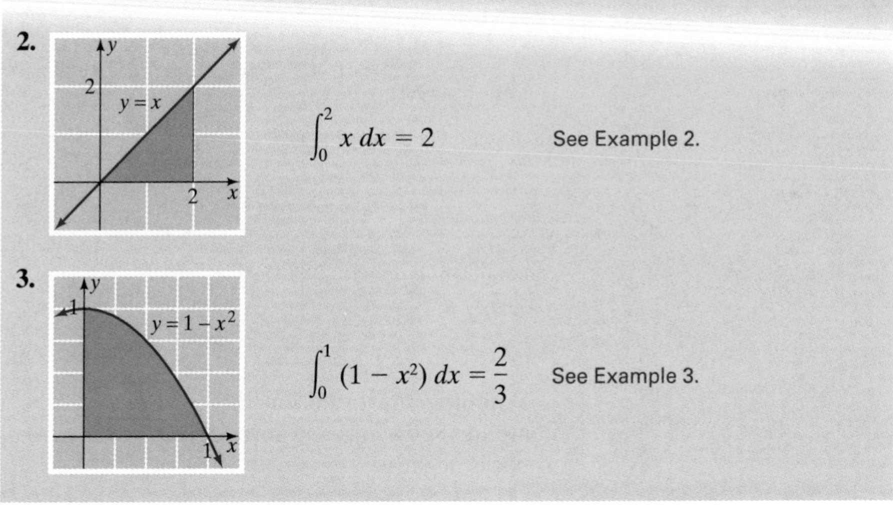

In fact, to make the idea of area precise, we *define* the area under the graph to be the value of the integral.

Question Why did we consider only nonnegative functions?
Answer If $f(x_k) < 0$, then $f(x_k)\,\Delta x < 0$ and is the *negative* of the area of the rectangle shown in Figure 12.

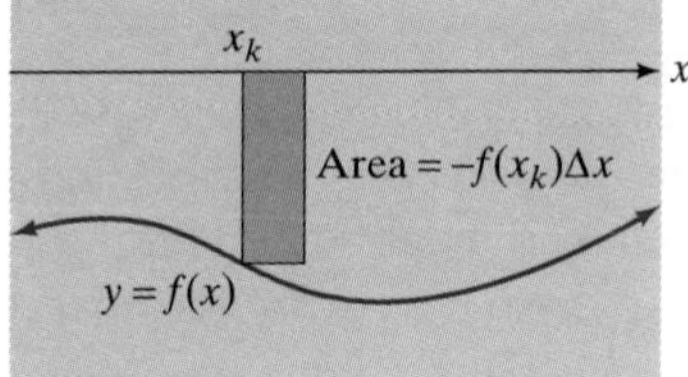

Figure 12

The result is that, in regions where $f(x) < 0$, the area below the x-axis and above the graph of f is *subtracted* from the value of the definite integral. This leads us to the following statement.

Geometric Interpretation of the Definite Integral (All Functions)

$\int_a^b f(x)\,dx$ is the area between $x = a$ and $x = b$ that is above the x-axis and below the graph of f, minus the area that is below the x-axis and above the graph of f:

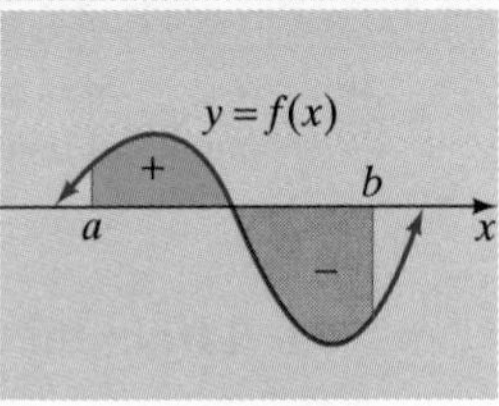

$$\int_a^b f(x)\,dx = \text{Area above } x\text{-axis} - \text{Area below } x\text{-axis}$$

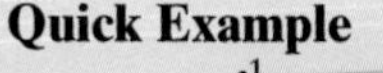

Quick Example

$$\int_{-1}^{1} x\,dx = 0 \qquad \text{Since the areas above and below the } x\text{-axis are equal}$$

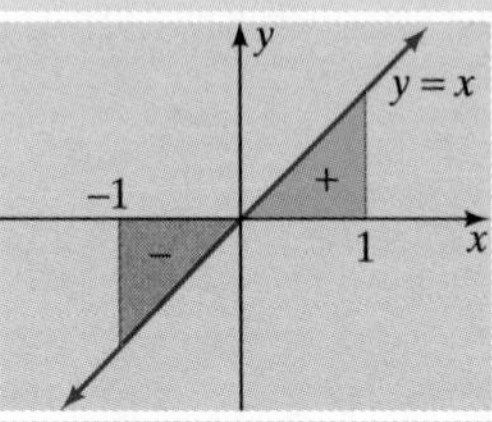

Another interpretation that you sometimes see is that the definite integral computes *signed* area, where area above the x-axis is positive and area below the x-axis is negative.

example 4

Total Cost

Recall the following example from the preceding section. A cellular phone company offers you an innovative pricing scheme. When you make a call, the *marginal* cost of the tth minute of the call is

$$c(t) = \frac{20}{t + 100} \text{ dollars per minute}$$

How is the total cost to make a 60-minute phone call related to the graph of c?

Solution

We saw in the preceding section that the total cost to make a 60-minute phone call is given by the definite integral

$$\int_0^{60} c(t)\,dt$$

Now we know that this integral also represents the area under the graph of c (which is always positive) over the interval $[0, 60]$; that is, the total cost is the area of the region shown in Figure 13.

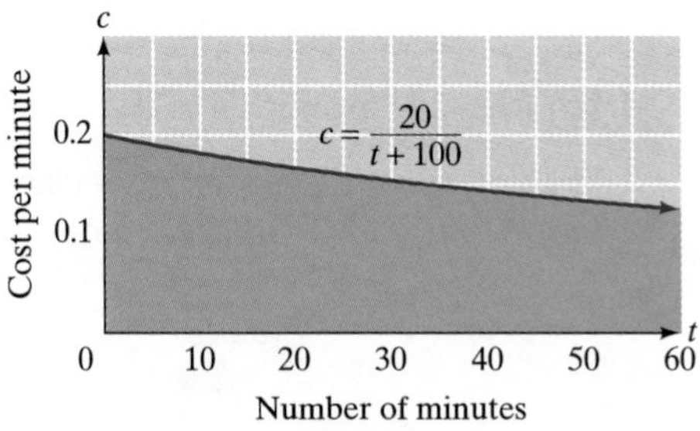

Figure 13

Question How can we tell from the graph that the total cost is represented by the shaded area?

Answer Take a look at how some Riemann sum approximations relate to the graph. When $n = 1$ partition, we have

$$\text{Cost} = C(0) \times 60 = 0.20 \times 60 = \$12 = \text{Shaded area in Figure 14}$$

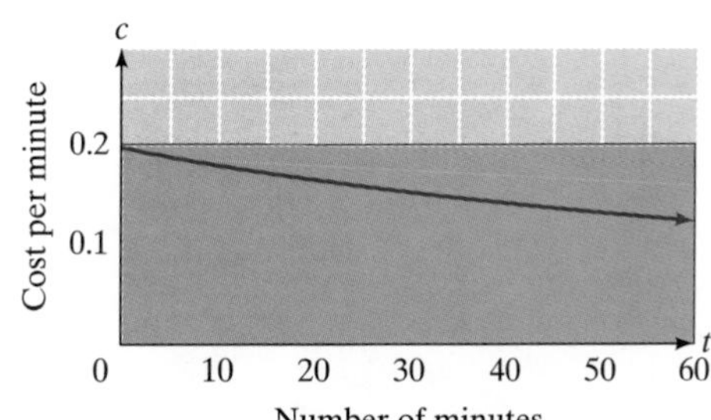

Figure 14

When $n = 6$ partitions, we have

$$\text{Cost} = C(0) \times 10 + \cdots + C(50) \times 10 = \text{Shaded area in Figure 15}$$

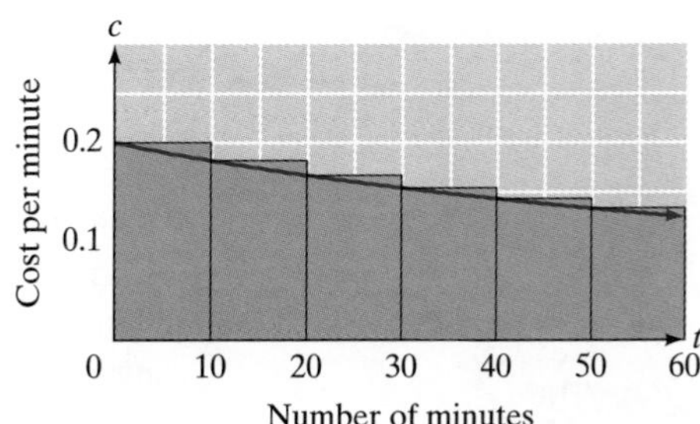

Figure 15

The areas of the rectangles are the approximate costs for successive 10-minute periods. Figures 14 and 15 illustrate how adding the costs for successive periods amounts to adding the areas of the rectangles that represent the Riemann sum.

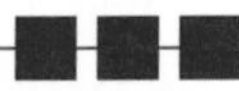

6.4 exercises

Use geometry (not Riemann sums) to compute the integrals in Exercises 1–12.

1. $\int_0^1 1\,dx$

2. $\int_0^2 5\,dx$

3. $\int_0^1 x\,dx$

4. $\int_1^2 x\,dx$

5. $\int_0^1 \frac{1}{2}x\,dx$

6. $\int_{-2}^2 \frac{1}{2}x\,dx$

7. $\int_2^4 (x-2)\,dx$

8. $\int_3^6 (x-3)\,dx$

9. $\int_{-1}^1 x^3\,dx$

10. $\int_1^2 \frac{1}{2}x\,dx$

Δ 11. $\int_{-\pi/2}^{\pi/2} 2\sin x\,dx$

Δ 12. $\int_0^{\pi} \cos x\,dx$

Use technology to estimate the areas in Exercises 13–20 using Riemann sums with $n = 10$ and $n = 100$. Round all answers to five significant digits.

T 13. Bounded by the line $y = x$, the x-axis, and the lines $x = 0$ and $x = 1$

T 14. Bounded by the line $y = 2x$, the x-axis, and the lines $x = 1$ and $x = 2$

Δ T 15. Bounded by the curve $y = \sin x$, the x-axis, and the lines $x = 0$ and $x = \pi$

Δ T 16. Bounded by the curve $y = \cos x$, the x-axis, and the lines $x = 0$ and $x = \pi/2$

T 17. Bounded by the curve $y = x^2 - 1$, the x-axis, and the lines $x = 0$ and $x = 4$

T 18. Bounded by the curve $y = 1 - x^2$, the x-axis, and the lines $x = -1$ and $x = 2$

19. Bounded by the x-axis, the curve $y = e^{x^2}$, and the lines $x = 0$ and $x = 1$

20. Bounded by the x-axis, the curve $y = e^{x^2-1}$, and the lines $x = 0$ and $x = 2$

Applications

21. ***Sales*** Estimated sales of IBM's mainframe computers and related peripherals showed the following pattern of decline over the years 1990 through 1994:

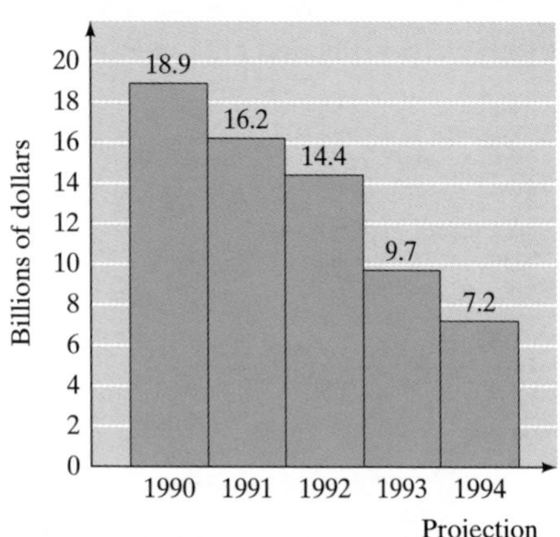

Source: Salomon Brothers/*New York Times,* October 26, 1993, p. D1.

What is the area under this graph? What does this number represent?

22. ***Total Profit*** The profitability of United Airlines for the years 1988 through 1992 showed the following decline:

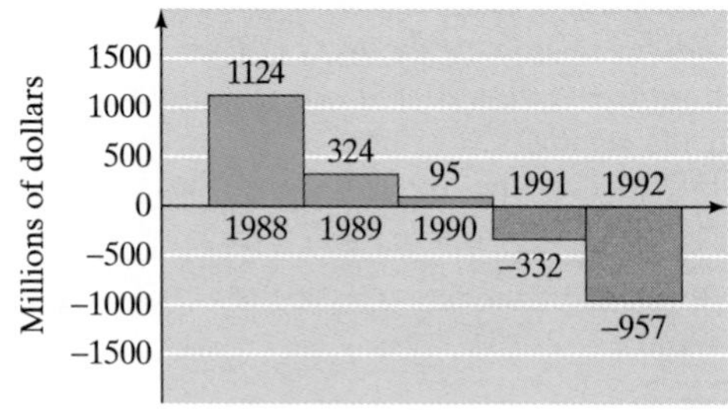

Source: Company reports/*New York Times,* December 24, 1993, p. D1.

What is the net area between this graph and the time axis (subtract the area under the axis as usual)? What does this number represent?

23. ***Foreign Investments*** The following chart shows the annual flow of private investment to developing countries from more industrialized countries for the years 1986 to 1993:

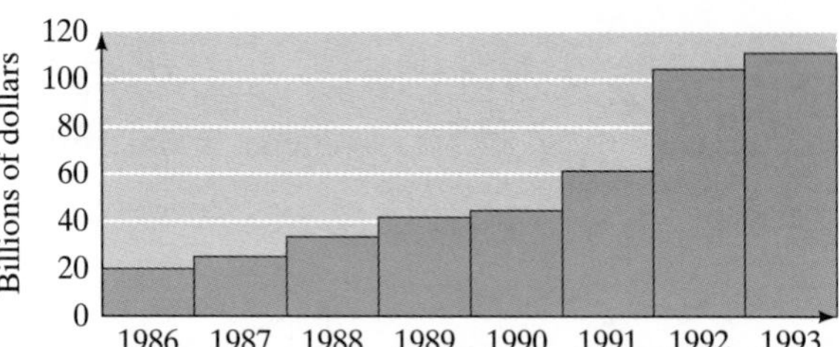

Source: World Bank, *New York Times,* December 17, 1993, p. D1.

What does the area under this graph represent?

24. ***Foreign Investments*** The following chart shows the annual flow of government loans and grants to developing countries from more industrialized countries:

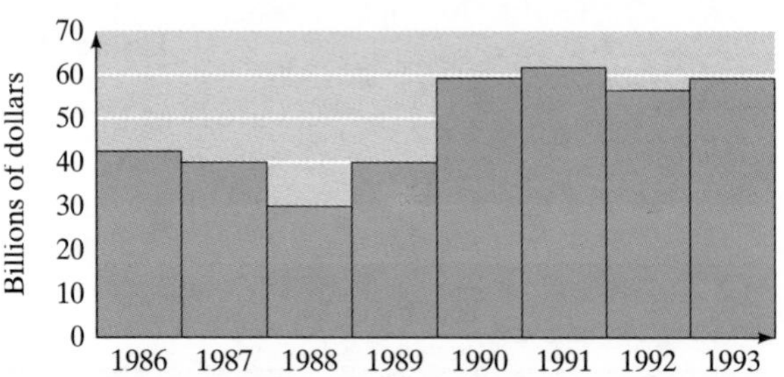

Source: World Bank, *New York Times,* December 17, 1993, p. D1.

What does the area under this graph represent?

25. ***Medicare Costs*** (Compare Example 7 in Section 6.2.) The rate of spending by Medicare on hospice care in the United States, in billions of dollars per year, was given by[1]

$$R(t) = \frac{2.2e^{0.67t}}{13 + e^{0.67t}} \qquad (0 \le t \le 8;\ t = 0 \text{ represents } 1989)$$

a. Graph the function R, indicating the area that represents $\int_1^7 R(t)\,dt$. What does this area represent?

b. Estimate the area in part (a) using a Riemann sum with $n = 100$. Interpret the answer.

26. ***Sales of Sport Utility Vehicles*** The following graph shows the number of sport utility vehicles sold in the United States each year from 1980 through 1994:

[1]This is a logistic model based on actual data from the Health Care Financing Administration/*New York Times,* May 10, 1998, p. 18.

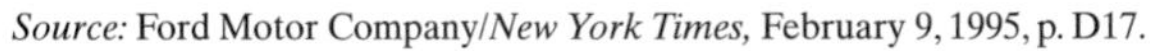

Source: Ford Motor Company/*New York Times,* February 9, 1995, p. D17.

A quartic model that approximates these data is

$$r(t) = 0.4t^4 - 10t^3 + 73t^2 - 60t + 200$$

where t is the number of years since June 1980 and $r(t)$ represents annual sales at time t.

a. Use the model and a Riemann sum with $n = 100$ to estimate the total sales of sport utility vehicles in the United States from June 1980 to June 1994. (Round your answer to the nearest 1000 vehicles.)

b. The answer you obtained in part (a) approximates part of the shaded area shown in the graph. Which part?

27. ***Surveying*** My uncle intends to build a kidney-shaped swimming pool in his small yard, and the town zoning board will approve the project only if the total area of the pool does not exceed 500 square feet. The accompanying figure shows a diagram of the planned swimming pool, with measurements of its width at the indicated points. Will my uncle's plans be approved? (Use a Riemann sum to approximate the area.)

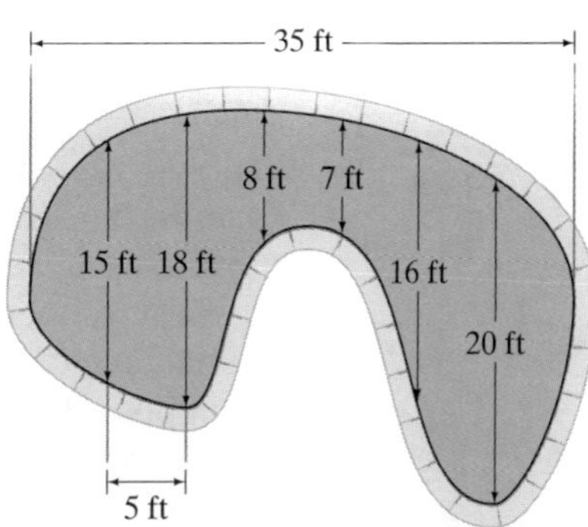

28. ***Pollution*** An aerial photograph of an ocean oil spill shows the pattern in the accompanying diagram. Assuming that the oil slick has a uniform depth of 0.01 meter, how many cubic meters of oil do you estimate to be in the spill? (Volume = Area × Thickness)

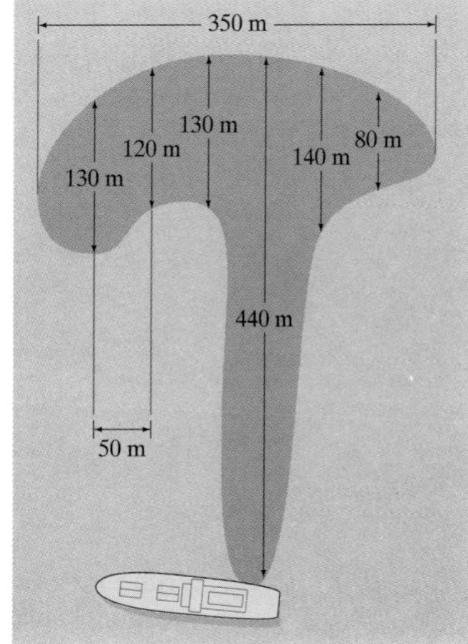

Communication and Reasoning Exercises

29. We have always used the *left-hand* Riemann sum. The **right-hand** Riemann sum is similar, except that we evaluate the function f on the right endpoint of each interval. Thus, the right-hand Riemann sum is

$$\sum_{k=1}^{n} f(x_k)\,\Delta x$$

Let f be a decreasing function on the interval $[a, b]$. Demonstrate by means of a sketch that the left-hand sum is always greater than or equal to the right-hand sum. What can you say about *increasing* functions?

30. Let f be an increasing function. Draw a picture and demonstrate that the difference between the right- and left-hand sums for $\int_a^b f(x)\,dx$ is $[f(b) - f(a)]\,\Delta x$. Conclude that the difference approaches zero as $n \to +\infty$.

31. Another approximation of the integral is the **midpoint** approximation, in which we compute the sum

$$\sum_{k=1}^{n} f(\bar{x}_k)\,\Delta x$$

where $\bar{x}_k = (x_{k-1} + x_k)/2$ is the point midway between the left and right endpoints of the interval $[x_{k-1}, x_k]$. Why is it true that the midpoint approximation is exact if f is linear? (Draw a picture.)

32. If $\int_a^b f(x)\,dx = 0$, what can you say about the graph of f?

6.5 The Definite Integral: An Algebraic Approach and the Fundamental Theorem of Calculus

So far we have calculated definite integrals by approximating them by Riemann sums. In this section we see an algebraic way of calculating many definite integrals, using antiderivatives.

The connection between antiderivatives and definite integrals is given by the Fundamental Theorem of Calculus. Here is the central idea behind the theorem: Suppose that f is a continuous function defined on $[a, b]$. We may define a new function with domain $[a, b]$ by letting

$$F(x) = \int_a^x f(t)\, dt$$

Consider the following examples.

example 1

An Area Function
Let $A(x) = \int_0^x t\, dt$ for $x > 0$. Find an explicit formula for A.

Solution
$A(x)$ is the area of the triangle shaded in Figure 16.

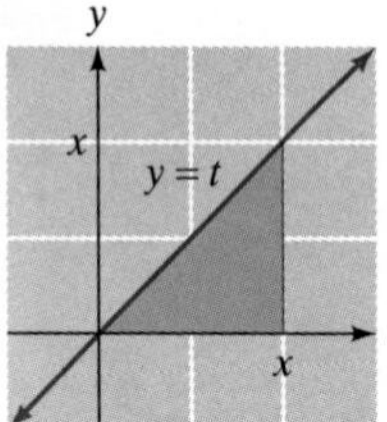

Figure 16

This triangle has base x and height x, so its area is $\frac{1}{2}x^2$. Therefore,

$$A(x) = \frac{1}{2}x^2$$

Before we go on . . . One other important fact to notice: $A'(x) = x$, which is the function being integrated (except that t has been replaced with x). In other words, $A(x)$ is an antiderivative of the integrand.

example 2

Graphing the Integral
Graph the function $F(x) = \int_0^x (1 - t^2)\, dt$ for $0 \le x \le 2$.

Solution
We can use technology to get a table of approximate values of F and draw the graph.

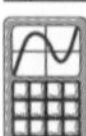

Graphing Calculator

The simplest way to graph a function and its integral on the TI-83 is to use the calculator's built-in numeric integration function, `fnInt`. In the `Y=` window, enter

```
Y1= 1 - X²
Y2=fnInt(Y1(T),T,0,X)
```
Format for $\int_0^x Y_1(t)\, dt$

This defines Y_1 to be f and Y_2 to be F. Graphing these two functions in the window defined by Xmin $= 0$, Xmax $= 2$, Ymin $= -3$, and Ymax $= 3$ gives Figure 17.

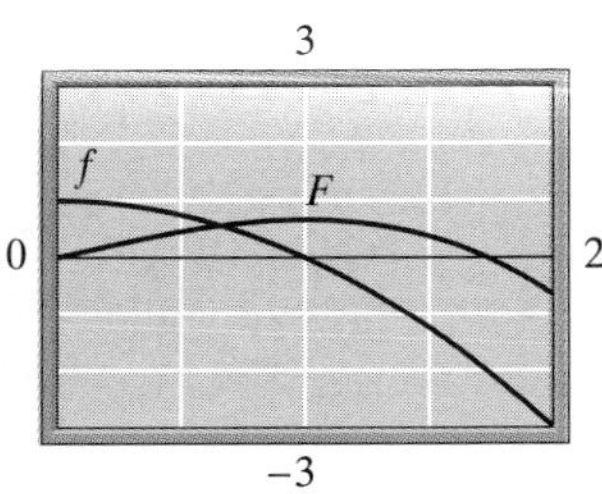

Figure 17

One of the graphs is the derivative of the other. Can you see which?

Warning: It will take a while to graph F. You can speed things up considerably, at the price of getting a slightly cruder graph, by increasing the value of the window variable `Xres` (the largest possible value is 8, and the smallest value is 1).

Spreadsheet

To see these graphs in Excel, first set up the spreadsheet to generate a table of 200 values of $f(t) = 1 - t^2$ as follows. (As in the first section, you can use the Fill→Series dialog box from the Edit menu to enter a series in the first column, starting at 0, with a step value of 0.01, stopping at 2.)

	A	B	C
1	0	=1-A1^2	
2	0.01		
3	0.02		
201	2		

Now use column C to compute values of $F(x)$ as shown.

	A	B	C
1	0	1	0
2	0.01	0.9999	=0.01*SUM(B$1:B1)
3	0.02	0.9996	
201	2	-3	

The formula in cell C2 gives a (crude) estimate of the value of $F(0.01)$. After copying as shown, the formula in cell C201 should be `=0.01*SUM(B$1:B200)`. The value in cell Cn is an approximation of the value of F at An. For example, the number in cell C201 is an approximation of $F(2)$.

	A	B	C
1	0	1	0
2	0.01	0.9999	0.01
3	0.02	0.9996	0.019999
201	2	-3	-0.6467

To graph both f and F, select cells A1 through C201 and ask Excel for an XY scatter graph. The result should look similar to Figure 17.

Before we go on . . . Notice that in Figure 17 the area function F begins at 0 when $x = 0$, increases as long as f is positive, and begins to decrease when f becomes negative. Remember that the area below the x-axis is subtracted from the value of the integral.

Notice also that, in the Excel worksheet above, once we have the values of F, we can find an approximation of F'. Since $\Delta x = 0.01$, we could approximate $F'(0.01)$ using the formula

```
=(C2-C1)/0.01
```

in cell D2. We could then copy this formula and paste it into cells D3 through D200. However, the result would be a copy of column B, the values of f. Why does this happen?

These examples suggest the first part of the following result. The second part allows us to calculate integrals algebraically.

The Fundamental Theorem of Calculus (FTC)

Let f be a continuous function defined on the interval $[a, b]$.

1. If $A(x) = \int_a^x f(t)\,dt$, then $A'(x) = f(x)$; that is, A is an antiderivative of f.
2. If F is any continuous antiderivative of f and is defined on $[a, b]$, then

$$\int_a^b f(x)\,dx = F(b) - F(a)$$

Quick Examples

1. If $A(x) = \int_0^x e^{t^2}\,dt$, then $A'(x) = e^{x^2}$.
2. Since $F(x) = x^2$ is an antiderivative of $f(x) = 2x$,

$$\int_0^1 2x\,dx = F(1) - F(0) = 1^2 - 0^2 = 1$$

We sketch out a proof of this result at the end of the section. Part 2 tells us that to calculate a definite integral, we should first look for an antiderivative.

example 3

Using the FTC to Calculate a Definite Integral

Calculate $\int_0^1 (1 - x^2)\,dx$

Solution

To use part 2 of the FTC, we need to find an antiderivative of $1 - x^2$. But we know that

$$\int (1 - x^2)\,dx = x - \frac{x^3}{3} + C$$

We need only one antiderivative, so let us take $F(x) = x - x^3/3$. The FTC tells us that

$$\int_0^1 (1 - x^2)\,dx = F(1) - F(0) = \left(1 - \frac{1}{3}\right) - (0) = \frac{2}{3}$$

which is the value we estimated in the preceding section.

Before we go on . . . There is a useful piece of notation that is often used here. We write[1]

$$\left[F(x)\right]_a^b = F(b) - F(a)$$

Thus, we can rewrite the computation above as

$$\int_0^1 (1 - x^2)\,dx = \left[x - \frac{x^3}{3}\right]_0^1 = \left(1 - \frac{1}{3}\right) - (0) = \frac{2}{3}$$

---

example 4

Compute the following definite integrals.

a. $\int_0^1 (2x^3 + 10x + 1)\,dx$ **b.** $\int_1^5 \left(\frac{1}{x^2} + \frac{1}{x}\right) dx$ **c.** $\int_0^\pi \sin x\,dx$

Solution

a. $$\int_0^1 (2x^3 + 10x + 1)\,dx = \left[\frac{1}{2}x^4 + 5x^2 + x\right]_0^1 = \left(\frac{1}{2} + 5 + 1\right) - (0) = \frac{13}{2}$$

b. $$\int_1^5 \left(\frac{1}{x^2} + \frac{1}{x}\right) dx = \int_1^5 (x^{-2} + x^{-1})\,dx = \left[-x^{-1} + \ln|x|\right]_1^5 = \left(-\frac{1}{5} + \ln 5\right) - (-1 + \ln 1) = \frac{4}{5} + \ln 5$$

c. $$\int_0^\pi \sin x\,dx = \left[-\cos x\right]_0^\pi = (-\cos \pi) - (-\cos 0) = -(-1) - (-1) = 2$$

Thus, the area under one "arch" of the sine curve is exactly 2 square units!

When calculating a definite integral, we may have to use substitution to find the necessary antiderivative. We could substitute, evaluate the indefinite integral with respect to u, express the answer in terms of x, and then evaluate at the limits of integration. However, there is a shortcut, as we shall see in the next example.

[1]There seem to be several notations in use actually. Another common notation is $F(x)\big|_a^b$.

example 5

Using the FTC with Substitution
Evaluate $\int_1^2 (2x - 1)e^{2x^2-2x}\,dx$.

Solution

The shortcut we mentioned is to put *everything* in terms of u, including the limits of integration.

$$u = 2x^2 - 2x$$

$$\frac{du}{dx} = 4x - 2$$

$$dx = \frac{1}{4x-2}\,du$$

When $x = 1, u = 0$. — Substitute $x = 1$ in the formula for u.

When $x = 2, u = 4$. — Substitute $x = 2$ in the formula for u.

We get the value $u = 0$, for example, by substituting $x = 1$ in the equation $u = 2x^2 - 2x$. We can now rewrite the integral.

$$\int_1^2 (2x - 1)e^{2x^2-2x}\,dx = \int_0^4 (2x - 1)e^u \frac{1}{4x-2}\,du$$

$$= \int_0^4 \frac{1}{2} e^u\,du = \left[\frac{1}{2}e^u\right]_0^4 = \frac{1}{2}e^4 - \frac{1}{2}$$

Before we go on . . . The alternative, longer calculation is first to calculate the indefinite integral:

$$\int (2x - 1)e^{2x^2-2x}\,dx = \int \frac{1}{2}e^u\,du = \frac{1}{2}e^u + C = \frac{1}{2}e^{2x^2-2x} + C$$

Then we can say that

$$\int_1^2 (2x - 1)e^{2x^2-2x}\,dx = \left[\frac{1}{2}e^{2x^2-2x}\right]_1^2 = \frac{1}{2}e^4 - \frac{1}{2}$$

Applications

example 6

Total Cost
We considered this example earlier. A cellular phone company offers you an innovative pricing scheme. When you make a call, the marginal cost of the tth minute of the call is

$$c(t) = \frac{20}{t + 100} \text{ dollars per minute}$$

If you make a cellular call that lasts 60 minutes, how much will it cost you?

Solution

We need to calculate

$$\int_0^{60} \frac{20}{t + 100}\,dt$$

We can evaluate this integral either by using the substitution $u = t + 100$ or by using the shortcut formula:

$$\int \frac{1}{ax+b}\,dx = \frac{1}{a}\ln|ax+b| + C$$

from Section 6.2. Using the shortcut formula gives

$$\begin{aligned}\int_0^{60} \frac{20}{t+100}\,dt &= 20\int_0^{60} \frac{1}{t+100}\,dt \\ &= 20\Big[\ln|t+100|\Big]_0^{60} \\ &= 20(\ln 160 - \ln 100) = \$9.40\end{aligned}$$

Compare this with Example 1 of Section 6.3, where we found the same answer by approximating with Riemann sums.

---

example 7

Computing Area

Find the total area of the region enclosed by the graph of $y = xe^{x^2}$, the x-axis, and the vertical lines $x = -1$ and $x = 1$.

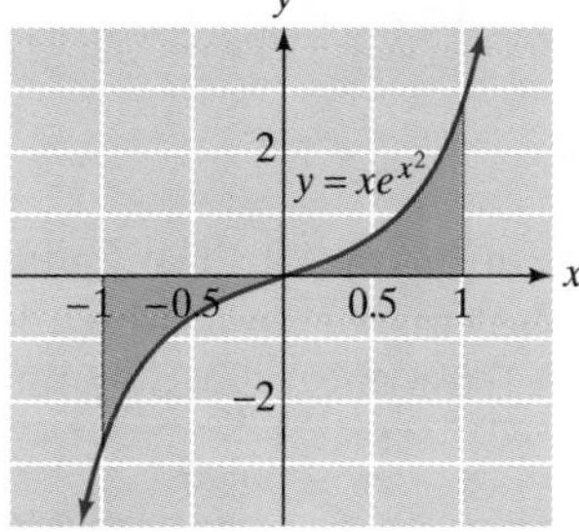

Figure 18

Solution

The region with the area we want is shown in Figure 18. Notice the symmetry of the graph. Also, half the region we are interested in is above the x-axis and the other half is below. If we calculated the integral $\int_{-1}^{1} xe^{x^2}\,dx$, the result would be

$$\text{Area above } x\text{-axis} - \text{Area below } x\text{-axis} = 0$$

which does not give us the total area. To prevent the area below the x-axis from being combined with the area above the axis, we do the calculation in two parts, as illustrated in Figure 19.

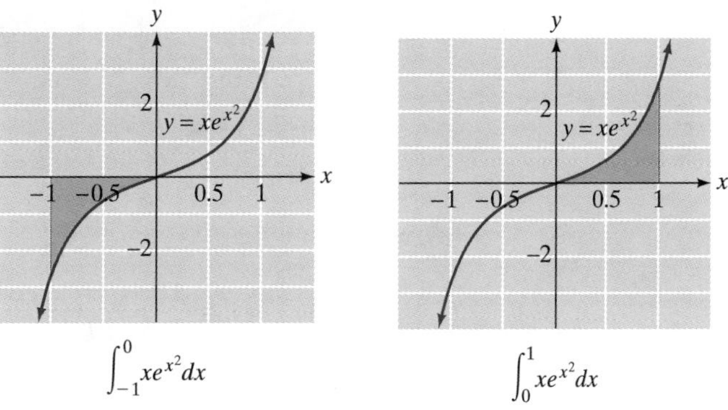

Figure 19

(We broke the integral at $x = 0$ because that is where the graph crosses the x-axis.) These integrals can be calculated using the substitution $u = x^2$:

$$\int_{-1}^{0} xe^{x^2}\,dx = \frac{1}{2}\Big[e^{x^2}\Big]_{-1}^{0} = \frac{1}{2}(1-e) \approx -0.85914$$

Why is it negative?

$$\int_{0}^{1} xe^{x^2}\,dx = \frac{1}{2}\Big[e^{x^2}\Big]_{0}^{1} = \frac{1}{2}(e-1) \approx 0.85914$$

To obtain the total area, we should add the *absolute values* of these answers because we don't wish to count any area as negative. Thus,

$$\text{Total area} \approx 0.85914 + 0.85914 = 1.71828$$

Proof of the FTC

We end this section with a sketch of the proof of the Fundamental Theorem of Calculus. The FTC was first shown by the English mathematician Isaac Barrow (1630–1677), who was Newton's teacher. It was Newton, however, who recognized its real importance.

We begin with a continuous function f defined on an interval $[a, b]$, and we define the function

$$A(x) = \int_a^x f(t)\,dt$$

Remember that the definite integral is defined by a limit. Technically, one of the hardest parts of the proof is to show that if f is continuous, then the limit always exists, so that $A(x)$ is defined. This part of the proof is beyond the scope of this book, so we assume it shown and proceed to calculate $A'(x)$ in order to verify part 1 of the FTC.

Recall that $A'(x)$ is also defined by a limit:

$$A'(x) = \lim_{h \to 0} \frac{A(x+h) - A(x)}{h}$$

We can think of $A(x+h)$ and $A(x)$ as the areas shown in Figure 20.[2] The figure shows that $A(x+h) - A(x)$ is the larger area minus the smaller area, which is the area of the "sliver" shown in Figure 21. (We take $h > 0$ for this discussion, but the case $h < 0$ is dealt with by a similar argument.)

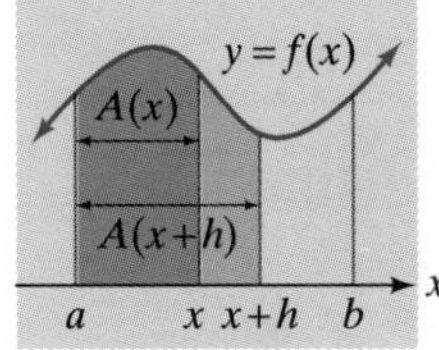

Figure 20

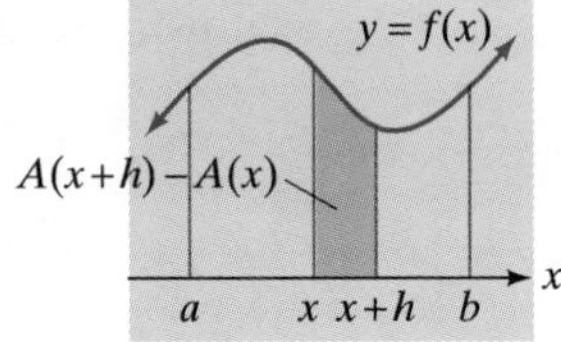

Figure 21

Thus, we can say that

$$A'(x) = \lim_{h \to 0} \frac{A(x+h) - A(x)}{h} = \lim_{h \to 0} \frac{\text{Area of the sliver}}{h}$$

Here is an intuitive argument that the value of this limit is $f(x)$: The area of the sliver is approximately the same as the area of a rectangle of height $f(x)$ and width h, so the area of the sliver is approximately $f(x)h$. If we divide this area by h, we get

$$\frac{\text{Area of the sliver}}{h} \approx f(x)$$

[2]To simplify the argument, we assume that $f(x) \geq 0$ as shown in the graph.

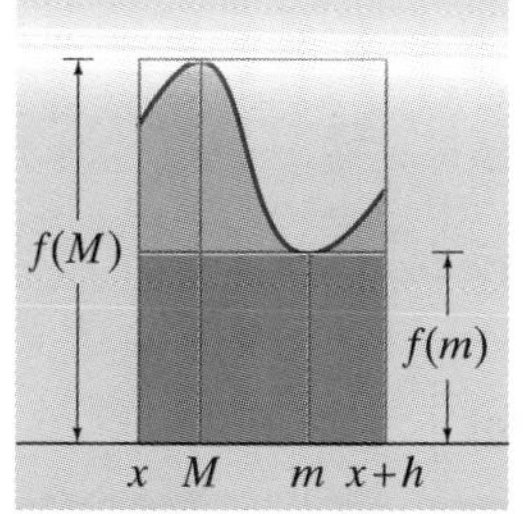

Figure 22

Here is a more careful argument. In Figure 22 we see an enlarged view of the sliver, showing also that there might be more "wobbles" in the function than in the picture we were looking at originally. In this figure we have also marked points M and m, where $f(M)$ is the maximum value of f on $[x, x + h]$ and $f(m)$ is the minimum value of f on $[x, x + h]$. We can see from the figure that the area of the sliver must be between the areas of two rectangles, the rectangles with width h and heights $f(m)$ and $f(M)$, respectively; that is,

$$f(m)h \leq \text{Area of the sliver} \leq f(M)h$$

Dividing by h, we get

$$f(m) \leq \frac{\text{Area of the sliver}}{h} \leq f(M)$$

Now consider what happens as h approaches zero. The two points m and M, caught between x and $x + h$, must both approach x. Since f is continuous, $f(m)$ and $f(M)$ must therefore both approach $f(x)$. The term in the middle, Area of the sliver/h, must therefore also approach $f(x)$. Thus,

$$A'(x) = \lim_{h\to 0} \frac{A(x + h) - A(x)}{h} = \lim_{h\to 0} \frac{\text{Area of the sliver}}{h} = f(x)$$

This completes the proof of the first part of the FTC.

For the second part, let F be any antiderivative of f on $[a, b]$. Since F and A are both antiderivatives of f, they must differ by a constant (see Section 6.1): $F(x) = A(x) + C$ for some constant C. Therefore,

$$\begin{aligned} F(b) - F(a) &= [A(b) + C] - [A(a) + C] \\ &= A(b) - A(a) \\ &= \int_a^b f(x)\,dx - 0 = \int_a^b f(x)\,dx \end{aligned}$$

6.5 exercises

In Exercises 1–6, use geometry (not antiderivatives) to compute the given area function $A(x) = \int_a^x f(t)\,dt$.

1. $A(x) = \int_0^x 1\,dt$; verify that $A'(x) = 1$.

2. $A(x) = \int_0^x t\,dt$; verify that $A'(x) = x$.

3. $A(x) = \int_2^x t\,dt$; verify that $A'(x) = x$.

4. $A(x) = \int_2^x 1\,dt$; verify that $A'(x) = 1$.

5. $A(x) = \int_a^x 4\,dt$; verify that $A'(x) = 4$.

6. $A(x) = \int_{-a}^x 4\,dt$; verify that $A'(x) = 4$.

In Exercises 7–16, use antiderivatives to find the area function $A(x) = \int_a^x f(t)\,dt$ for the given $f(x)$ and a. Sketch both $f(x)$ and $A(x)$.

7. $f(x) = x$, $a = 0$

8. $f(x) = 2x$, $a = 1$

9. $f(x) = x^2$, $a = 0$

10. $f(x) = x^2 + 1$, $a = 0$

11. $f(x) = e^x$, $a = 0$

12. $f(x) = e^{-x}$, $a = 0$

13. $f(x) = 1/x$, $a = 1$

14. $f(x) = 1/(x + 1)$, $a = 0$

15. $f(x) = \begin{cases} x & \text{if } 0 \leq x \leq 1 \\ 2 & \text{if } x > 1 \end{cases}$, $a = 0$

16. $f(x) = \begin{cases} 2x & \text{if } 0 \leq x \leq 2 \\ 1 & \text{if } x > 2 \end{cases}$, $a = 0$

Evaluate the integrals in Exercises 17–50.

17. $\int_{-1}^{1} (x^2 + 2)\, dx$

18. $\int_{-2}^{1} (x - 2)\, dx$

19. $\int_{-2}^{2} (x^3 - 2x)\, dx$

20. $\int_{-1}^{1} (2x^3 + x)\, dx$

21. $\int_{1}^{3} \left(\frac{2}{x^2} + 3x\right) dx$

22. $\int_{2}^{3} \left(x + \frac{1}{x}\right) dx$

23. $\int_{0}^{1} (2.1x - 4.3x^{1.2})\, dx$

24. $\int_{-1}^{0} (4.3x^2 - 1)\, dx$

△ 25. $\int_{-\pi}^{0} \sin x\, dx$

△ 26. $\int_{\pi/2}^{\pi} \cos x\, dx$

△ 27. $\int_{0}^{\pi/3} \tan x\, dx$

△ 28. $\int_{\pi/6}^{\pi/2} \cot x\, dx$

29. $\int_{0}^{1} 2e^x\, dx$

30. $\int_{-1}^{0} 3e^x\, dx$

31. $\int_{0}^{1} \sqrt{x}\, dx$

32. $\int_{-1}^{1} \sqrt[3]{x}\, dx$

33. $\int_{-1}^{1} e^{2x-1}\, dx$

34. $\int_{0}^{2} e^{-x+1}\, dx$

35. $\int_{0}^{50} e^{-0.02x-1}\, dx$

36. $\int_{-20}^{0} 3e^{2.2x}\, dx$

37. $\int_{-1.1}^{1.1} e^{x+1}\, dx$

38. $\int_{0}^{\sqrt{2}} x\sqrt{2x^2 + 1}\, dx$

39. $\int_{-\sqrt{2}}^{\sqrt{2}} 3x\sqrt{2x^2 + 1}\, dx$

40. $\int_{-1.2}^{1.2} e^{-x-1}\, dx$

41. $\int_{0}^{1} 5xe^{x^2+2}\, dx$

42. $\int_{0}^{2} \frac{3x}{x^2 + 2}\, dx$

43. $\int_{0}^{1} x\sqrt{2x + 1}\, dx$

44. $\int_{-1}^{0} 2x\sqrt{x + 1}\, dx$

△ 45. $\int_{1}^{\sqrt{\pi+1}} x\cos(x^2 - 1)\, dx$

△ 46. $\int_{0.5}^{(\pi+1)/2} \sin(2x - 1)\, dx$

△ 47. $\int_{1/\pi}^{2/\pi} \frac{\sin(1/x)}{x^2}\, dx$

△ 48. $\int_{0}^{\pi/3} \frac{\sin x}{\cos^2 x}\, dx$

49. $\int_{2}^{3} \frac{x^2}{x^3 - 1}\, dx$

50. $\int_{2}^{3} \frac{x}{2x^2 - 5}\, dx$

Calculate the total area of the regions described in Exercises 51–58. Do not count the area below the x-axis as negative.

51. Bounded by the line $y = x$, the x-axis, and the lines $x = 0$ and $x = 1$

52. Bounded by the line $y = 2x$, the x-axis, and the lines $x = 1$ and $x = 2$

53. Bounded by the curve $y = \sqrt{x}$, the x-axis, and the lines $x = 0$ and $x = 4$

54. Bounded by the curve $y = 2\sqrt{x}$, the x-axis, and the lines $x = 0$ and $x = 16$

55. Bounded by the curve $y = x^2 - 1$, the x-axis, and the lines $x = 0$ and $x = 4$

56. Bounded by the curve $y = 1 - x^2$, the x-axis, and the lines $x = -1$ and $x = 2$

57. Bounded by the x-axis, the curve $y = xe^{x^2}$, and the lines $x = 0$ and $x = (\ln 2)^{1/2}$

58. Bounded by the x-axis, the curve $y = xe^{x^2-1}$, and the lines $x = 0$ and $x = 1$

Applications

59. ***Cost*** The marginal cost of producing the xth box of lightbulbs is $5 + x^2/1000$. Determine how much is added to the total cost by a change in production from $x = 10$ to $x = 100$ boxes.

60. ***Revenue*** The marginal revenue of the xth box of computer disks sold is $100e^{-0.001x}$. Find the revenue generated by selling items 101 through 1000.

61. ***Employment*** The following quadratic model approximates the rate of increase in employment in Massachusetts from 1988 through 1995:[3]

$$C(t) = 25t^2 - 137t + 68 \text{ thousand people per year}$$

where t is the number of years since 1988. Use the Fundamental Theorem of Calculus to estimate the change in the number of people employed in Massachusetts from 1990 ($t = 2$) to 1995 ($t = 7$). Round your answer to the nearest 1000 people.

62. ***Hawaiian Tourism*** The rate of visitor spending in Hawaii during the years 1985 to 1993 can be approximated by[4]

$$r(t) = -0.164t^2 + 1.60t + 6.71 \text{ billion dollars per year}$$

where $t = 0$ represents June 30, 1985. Use the Fundamental Theorem of Calculus to estimate how much revenue Hawaii earned from visitor spending between June 30, 1985, and June 30, 1990. (Give your answer to the nearest billion dollars.)

63. ***Revenue*** The EMC Corporation, which specializes in the manufacture of computer data storage systems, under-

[3]The model is a quadratic regression based on the actual data through 1994. Hence, the 1995 figure is projected. *Source for data:* Massachusetts Department of Employment, DRI/McGraw Hill/*New York Times,* January 27, 1995, p. D4.

[4]The model is based on a best-fit quadratic for (approximate) tourism data, as measured in constant 1993 dollars, from the Hawaii Visitors Bureau/*New York Times,* September 5, 1995, p. A12.

went rapid growth during the period 1989–1994, and its annual revenue (in billions of dollars) was approximately

$$r(t) = 0.13e^{0.44t} \qquad (0 \le t \le 5)$$

where t is the number of years since June 1989. Use this model to estimate the total revenue (to the nearest \$0.1 billion) earned by EMC from June 1989 to June 1994.

Source: The authors' model from data in company reports/IDC/*New York Times,* November 9, 1994, p. D8.

64. ***Revenue*** Carnival Corporation's airline services grew rapidly during the period 1989–1994, and its resulting annual revenues (in millions of dollars) can be approximated by

$$r(t) = 0.33t^3 + 0.929t^2 + 15.02t + 6.43 \qquad (0 \le t \le 5)$$

where t is the number of years since June 1989. Use this model to estimate the total revenue (to the nearest \$1 million) earned by Carnival from its airline service from June 1990 to June 1994.

Source: The authors' model from data in company reports/*New York Times,* October 23, 1994, p. F5.

65. ***Displacement*** A car traveling down a road has a velocity of $v(t) = 60 - e^{-t/10}$ mph at time t hours. Find the total distance the car travels from time $t = 1$ to time $t = 6$. (Round your answer to the nearest mile.)

66. ***Displacement*** A ball thrown in the air has a velocity of $v(t) = 100 - 32t$ ft/s at time t seconds. Find the total displacement of the ball between times $t = 1$ second and $t = 7$ seconds, and interpret your answer.

67. ***Oil Exploration Costs*** Annual spending by U.S. oil companies on domestic oil exploration dropped from \$9.2 billion in 1981 to \$0.8 billion in 1992.

a. Assuming a linear decrease in annual spending over the given period, give the annual amount $c(t)$ (in billions of dollars) spent on domestic oil exploration as a function of time t in years since 1981.

b. Use your linear model to estimate, to the nearest \$1 billion, the total amount spent on domestic oil exploration over the given period.

Source: American Petroleum Institute/*New York Times,* November 8, 1993, p. D3.

68. ***Oil Exploration Costs*** Annual spending by U.S. oil companies on oil exploration in foreign countries dropped from \$4.5 billion in 1981 to \$4.0 billion in 1992.

a. Assuming a linear decrease in annual spending over the given period, give the annual amount $c(t)$ (in billions of dollars) spent on foreign oil exploration as a function of time t in years since 1981.

b. Use your linear model to estimate, to the nearest \$1 billion, the total amount spent on foreign oil exploration over the given period.

Source: American Petroleum Institute/*New York Times,* November 8, 1993, p. D3.

69. ***Embryo Development*** The oxygen consumption of a bird embryo increases from the time the egg is laid through the time the chick hatches. In a typical galliform bird, the oxygen consumption can be approximated by[5]

$c(t) = -0.00271t^3 + 0.137t^2 - 0.892t + 0.149$ milliliters per hour $\quad (8 \le t \le 30)$

where t is the time (in days) since the egg was laid. (An egg will typically hatch at around $t = 28$.) Find the total amount of oxygen consumed during the ninth and tenth days ($t = 8$ to $t = 10$). Round your answer to the nearest milliliter.

70. ***Embryo Development*** The oxygen consumption of a turkey embryo increases from the time the egg is laid through the time the chick hatches. In a brush turkey, the oxygen consumption can be approximated by

$c(t) = -0.00118t^3 + 0.119t^2 - 1.83t + 3.972$ milliliters per hour $\quad (20 \le t \le 50)$

where t is the time (in days) since the egg was laid. (An egg will typically hatch at around $t = 50$.) Find the total amount of oxygen consumed during the 21st and 22nd days ($t = 20$ to $t = 22$). Round your answer to the nearest milliliter.

71. ***Total Cost*** Use the Fundamental Theorem of Calculus to show that if $m(x)$ is the marginal cost at a production level of x items, then the cost function $C(x)$ is given by

$$C(x) = C(0) + \int_0^x m(t)\,dt$$

What term do we use for $C(0)$?

72. ***Total Sales*** The total cost of producing x items is given by

$$C(x) = 246.76 + \int_0^x 5t\,dt$$

Find the fixed cost and the marginal cost of producing the tenth item.

T **73.** ***Foreign Investments*** The following chart shows the annual flow of private investment to developing countries from more industrialized countries for the years 1986 to 1993:

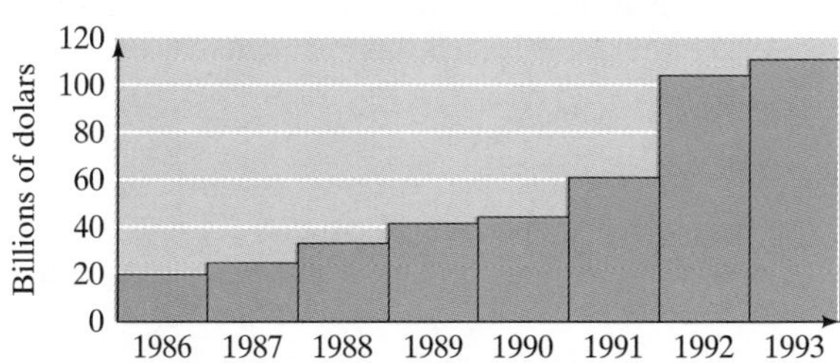

Source: World Bank/*New York Times,* December 17, 1993, p. D1.

[5]The models in Exercises 69 and 70 approximate graphical data published in the article "The Brush Turkey" by Roger S. Seymour in *Scientific American,* December 1991, pp. 108–114.

a. Let $q(t)$ represent the annual quantity of private investments (in billions of dollars), where t is the number of years since the start of 1986. Which of the following models best fits the data shown? (Feel free to use graphing technology.)

(A) $q(t) = 10 + 12.5t$

(B) $q(t) = 10 + 35.5\sqrt{t}$

(C) $q(t) = 10 + 1.56t^2$

b. Use the model from part (a) to predict how much money will have been sent to developing countries from January 1, 1986, through January 1, 1996.

T **74. *Foreign Investments*** The following chart shows the annual flow of government loans and grants to developing countries from more industrialized countries:

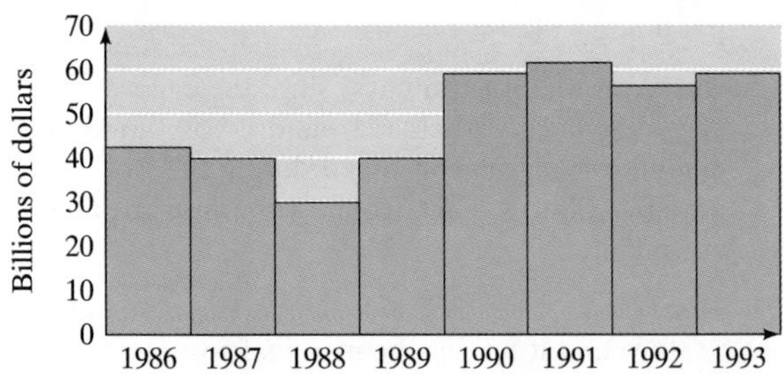

Source: World Bank/*New York Times,* December 17, 1993, p. D1.

a. Let $q(t)$ represent the annual quantity of loans (in billions of dollars), where t is the number of years since the start of 1986. Which of the following models best fits the data shown? (Feel free to use a graphing calculator.)

(A) $q(t) = 38 - 0.6(t - 3)^2$

(B) $q(t) = 38 + 0.6(t - 3)^2$

(C) $q(t) = 20 + e^{0.54t}$

b. Use the model from part (a) to predict how much money will have been sent to developing countries from January 1, 1986, through January 1, 1996.

75. *Exports* Based on figures released by the U.S. Agriculture Department, the value of U.S. pork exports from 1985 through 1993 can be approximated by the equation[6]

$$q = \frac{460e^{(t-4)}}{1 + e^{(t-4)}}$$

where q represents annual exports (in millions of dollars) and t the time in years, with $t = 0$ corresponding to January 1, 1985. The figure shows the actual data with the graph of the equation superimposed.

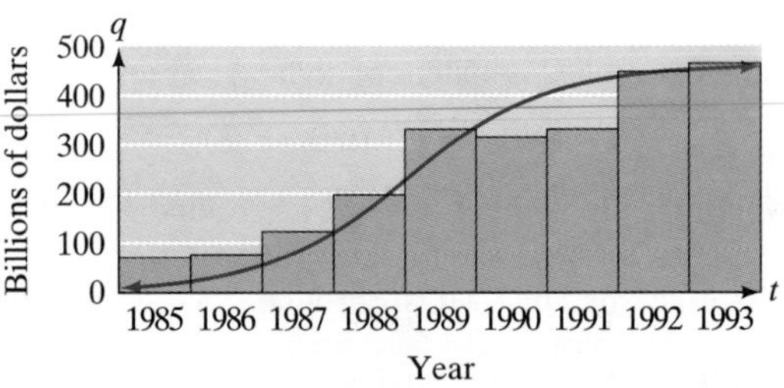

Source: New York Times, December 12, 1998, p. D1.

Use the model to predict the total value of U.S. pork exports, to the nearest million dollars, from January 1, 1985, to January 1, 2000.

76. *Sales* The weekly demand for your company's Lo-Cal Chocolate Mousse can be modeled by the equation

$$q = \frac{50e^{2t-1}}{1 + e^{2t-1}}$$

where q is the number of gallons sold per week and t is the time in weeks. Estimate the total sales of Lo-Cal mousse for the coming year.

77. *Health Care Spending* The following chart shows approximate figures based on a graph for annual spending on health care in the United States from 1981 to 1994, in billions of dollars:

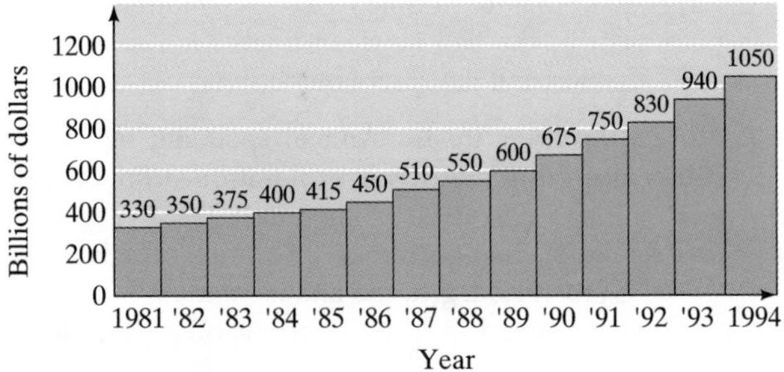

Source: Department of Health and Human Services, Department of Commerce/ *New York Times,* December 29, 1993, p. A12.

a. Take t to be time in years since January 1981, and use the figures for 1981 ($t = 0$) and 1993 ($t = 12$) to construct an exponential model of the form $s(t) = Pe^{kt}$ for annual spending, where $s(t)$ is the annual spending (in billions of dollars) at time t.

b. Use your model to calculate the total amount spent on health care in the United States for the period shown, and compare it with the result given by the data in the chart. [*Hint:* The period shown extends to January 1995.]

c. Use your model to predict total spending on health care in the United States from January 1995 to January 2000.

78. *Health Care Spending* Repeat Exercise 77, but this time base your model on the figures for 1982 ($t = 1$) and 1994 ($t = 13$).

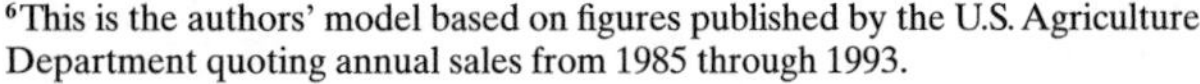

[6]This is the authors' model based on figures published by the U.S. Agriculture Department quoting annual sales from 1985 through 1993.

79. ***Sales*** Weekly sales of your Jurassic Park T-shirts have been falling by 5% per week. Assuming you are now selling 50 T-shirts per week, how many shirts will you sell during the coming year? (Round your answer to the nearest shirt.)

80. ***Sales*** Annual sales of fountain pens in Littleville are presently 4000 per year and are increasing by 10% per year. How many fountain pens will be sold over the next 5 years?

81. ***Fuel Consumption*** The way Professor Waner drives, he burns gas at the rate of $1 - e^{-t}$ gallons each hour, t hours after a fill-up. Find the number of gallons of gas he burns in the first 10 hours after a fill-up.

82. ***Fuel Consumption*** The way Professor Costenoble drives, he burns gas at the rate of $1/(t + 1)$ gallons each hour, t hours after a fill-up. Find the number of gallons of gas he burns in the first 10 hours after a fill-up.

Using the Logistic Equation to Predict Sales[7] *The* ***logistic equation*** *has the form*

$$y(x) = \frac{NP_0}{P_0 + (N - P_0)e^{-kx}}$$

where N, P_0, and k are positive constants. (See the accompanying graph.)

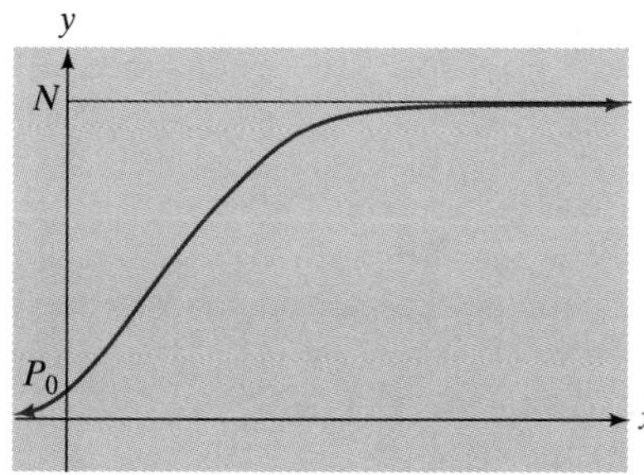

This equation is useful in modeling and predicting the demand for a new product, where:

y = demand in annual sales

x = number of years since the introduction of the product to the market

P_0 = initial demand (demand at time $x = 0$)

N = annual total potential demand

k = approximate initial rate of growth of demand (a good approximation if N is large compared with P_0)

[7]See the Expert Opinion Section in Chapter 2 on logarithmic and exponential functions for a discussion of the use of the logistic equation in modeling epidemics.

In Exercises 83–88, we use the logistic curve to model the growing demand for several therapeutic drugs.

83. Show that the logistic equation can be rewritten in the form

$$y(x) = \frac{NP_0e^{kx}}{P_0e^{kx} + (N - P_0)}$$

84. Using the result in Exercise 83 and a suitable substitution, show that

$$\int y(x)\,dx = \frac{N}{k}\ln\left[P_0e^{kx} + (N - P_0)\right] + C$$

85. Eli Lilly Corp.'s human growth hormone Humatrope has a potential market estimated at roughly N = \$160 million per year, and its total Humatrope sales can be approximated by the logistic function with $k = 2.7$ and $P_0 = 0.0025$. We let $y(x)$ represent annual sales in millions of dollars and x represent the number of years since the drug's approval by the FDA in 1987.[8] Use the model to give an estimate (to the nearest \$10 million) of the value of total sales of Humatrope over the 10-year period beginning in 1987.

86. Genentech's human growth hormone Protropin has a potential market estimated at roughly N = \$300 million per year, and its total Protropin sales can be approximated by the logistic function with $k = 1.7$ and $P_0 = 0.20$. We let $y(x)$ represent annual sales in millions of dollars and x represent the number of years since the drug's approval by the FDA in 1986. Use the model to give an estimate (to the nearest \$10 million) of the value of total sales of Protropin over the 10-year period beginning in 1986.

T 87. Genzyme's drug against Gaucher's disease, Ceredase, cost the company \$30 million to develop and has a potential market of \$900 million per year. At the time of FDA approval (1991), annual sales were estimated at \$50 million. Assuming an initial rate of growth specified by $k = 0.5$ (that is, approximately a 50% initial growth rate), use a graphing calculator to estimate how long (to the nearest year) it will take Genzyme to earn 10 times the development costs from sales of Ceredase.[9]

T 88. Repeat Exercise 87 using an initial growth rate specified by $k = 0.3$ (that is, approximately a 30% initial growth rate).

[8]The models in Exercises 85 and 86 are very crude, based on 1991 sales data, total sales data through May 1992, and very rough estimates of the potential market and selling price. *Source:* Senate Judiciary Committee; Subcommittee on Antitrust and Monopoly/*New York Times,* May 14, 1992, p. D1.

[9]Initial sales figures are based on sales for 9 months of 1991 immediately after FDA approval. See the source from Exercises 85 and 86.

89. ***Kinetic Energy*** The work done in accelerating an object from velocity v_0 to velocity v_1 is given by

$$W = \int_{v_0}^{v_1} v \frac{d}{dv}(p)\, dv$$

where p is its momentum, given by $p = mv$ (m = mass). Assuming that m is a constant, show that

$$W = \frac{1}{2}mv_1^2 - \frac{1}{2}mv_0^2$$

The quantity $\frac{1}{2}mv^2$ is referred to as the **kinetic energy** of the object, so the work required to accelerate an object is equal to its change in kinetic energy.

90. ***Einstein's Energy Equation*** According to the special theory of relativity, the apparent mass of an object depends on its velocity according to the formula

$$m = \frac{m_0}{\left(1 - \frac{v^2}{c^2}\right)^{1/2}}$$

where v is its velocity, m_0 is the "rest mass" of the object (that is, its mass when $v = 0$), and c is the velocity of light: approximately 3×10^8 meters per second.

a. Show that if $p = mv$ is the momentum, then

$$\frac{d}{dv}(p) = \frac{m}{\left(1 - \frac{v^2}{c^2}\right)^{3/2}}$$

b. Use the integral formula for W in Exercise 89 together with the result in part (a) to show that the work required to accelerate an object from a velocity of v_0 to v_1 is given by

$$W = \frac{m_0c^2}{\sqrt{1 - \frac{v_1^2}{c^2}}} - \frac{m_0c^2}{\sqrt{1 - \frac{v_0^2}{c^2}}}$$

We call the quantity $(m_0c^2)/\sqrt{1 - (v^2/c^2)}$ the **total relativistic energy** of an object moving at velocity v. Thus, the work to accelerate an object from one velocity to another is equal to the change in its total relativistic energy.

c. Deduce (as Albert Einstein did) that the total relativistic energy E of a body at rest with rest mass m is given by the famous equation $E = mc^2$.

91. ***The Natural Logarithm Returns*** Suppose that you had never heard of the natural logarithm function and were therefore stuck when you tried to find an antiderivative of $1/x$. Here is how you might proceed.

a. What does part 1 of the Fundamental Theorem of Calculus give as an antiderivative of $1/x$?

b. Choosing $a = 1$, give this "new" function the name $M(x)$ and derive the following properties: $M(1) = 0$; $M'(x) = 1/x$.

c. Use the chain rule for derivatives to show that if a is any positive constant, then

$$\frac{d}{dx}\left[M(ax)\right] = \frac{1}{x}$$

d. Let $F(x) = M(ax) - M(x)$. Deduce that $F'(x) = 0$ and hence that $F(x)$ = constant.

e. By setting $x = 1$, find the value of this constant and hence deduce that $M(ax) = M(a) + M(x)$.

92. ***The Error Function*** The **error function,** erf(x), is defined by

$$\text{erf}(x) = \frac{2}{\sqrt{\pi}}\int_0^x e^{-t^2}\, dt$$

This function is very important in statistics.

a. Find erf$'(x)$.

b. Use Reimann sums with $n = 100$ to approximate erf(1) and erf(2).

c. Find an antiderivative of e^{-x^2} in terms of erf(x).

d. Use the answers to parts (b) and (c) to approximate $\int_1^2 e^{-x^2}\, dx$.

Communication and Reasoning Exercises

93. What does the Fundamental Theorem of Calculus permit one to do?

94. Your friend has just told you that the function $f(x) = e^{-x^2}$ can't be integrated and hence has no antiderivative. Is your friend correct? Explain your answer.

95. Use the FTC to find an antiderivative F of

$$f(x) = \begin{cases} 0 & \text{if } x < 0 \\ 1 & \text{if } x \geq 0 \end{cases}$$

Is it true that $F'(x) = f(x)$ for *all* x?

96. According to the Fundamental Theorem of Calculus as stated in this text, which functions are guaranteed to have antiderivatives? What does the answer to Exercise 95 tell about the theorem as stated in this text?

97. Explain how the indefinite integral and the definite integral are related.

98. Give an example of a nonzero function whose definite integral is zero.

99. Give a nontrivial example of a velocity function that will produce a displacement of zero from time $t = 0$ to time $t = 10$.

100. Complete the following: The total sales from time a to time b are obtained from the marginal sales by taking its _____ _____ from _____ to _____.

101. Give an example of a decreasing function $f(x)$ with the property that $\int_a^b f(x)\,dx$ is positive for every choice of a and $b > a$.

102. Explain why, in computing the total of a quantity from its rate of change, it is useful to have the definite integral subtract the area below the x-axis.

You're the Expert

Wage Inflation

You are the assistant personnel manager at ABC Development Enterprises, a large corporation, and yesterday you received the following memo from the personnel manager.

> **To**: SW
>
> **From**: SC
>
> **Subject**: Cost of labor
>
> Yesterday, the CEO asked me to find some mathematical formulas to estimate (1) the trend in annual wage increases and (2) the average annual wage of assembly line workers from the time they join the company to the time they retire. (She needs the information for next week's stockholder meeting.) So far, I have had very little luck. All I have been able to find is a table giving annual percentage wage increases (attached).[1] Also, I know that the average wage for an assembly line worker in 1981 was $25,000 per year. Do you have any ideas?

Year	1981	1982	1983	1984	1985	1986	1987	1988	1989	1990	1991	1992	1993	1994	1995
Annual change (%)	9.3	8.0	5.3	5.0	4.2	4.0	3.2	3.4	4.2	4.2	4.0	3.4	2.8	3.0	2.8

Getting to work, you decide that the first thing to do is fit these data to a mathematical curve (a "trendline") that you can use to project future changes in wages. You graph the data to get a sense of what mathematical models might be appropriate (Figure 23).

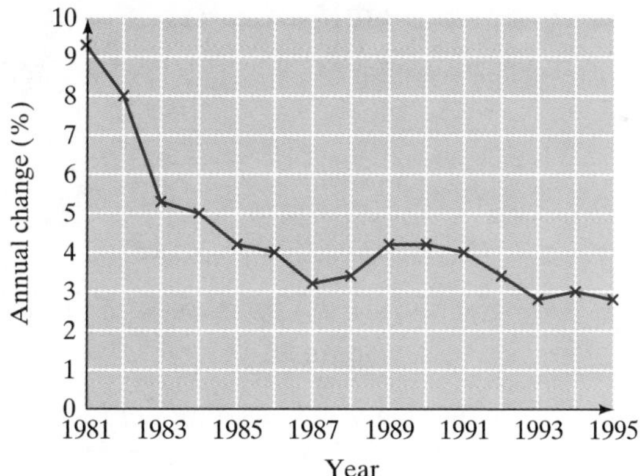

Figure 23

The graph suggests a decreasing trend, leveling off at about 3%. You recall that there are a variety of curves that behave this way. One of the simplest is the curve

$$y = \frac{a}{t} + b \qquad (t \ge 1)$$

[1]The data show approximate year-to-year percentage changes in U.S. wages. For example, the wages increased 9.3% from 1981 to 1982. *Source:* Datastream/*New York Times,* August 13, 1995, p. 26.

where a and b are constants (Figure 24).[2]

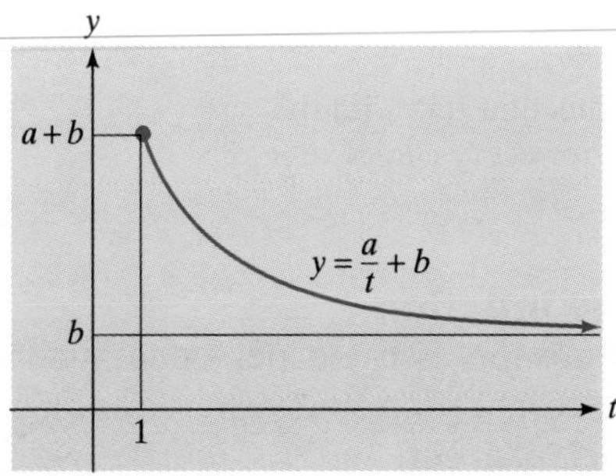

Figure 24

You take $t = 1$ to correspond to the time of the first data point, 1981, and convert all the percentages to decimals, giving the following table of data:

t	1	2	3	4	5	6	7	8	9	10	11	12	13	14	15
y	0.093	0.080	0.053	0.050	0.042	0.040	0.032	0.034	0.042	0.042	0.040	0.034	0.028	0.030	0.028

You then find the values of a and b that best fit the given data.[3] This gives you the following model for wage inflation (with figures rounded to five significant digits):

$$y = \frac{0.071813}{t} + 0.028647$$

(It is interesting that the model predicts wage inflation leveling off to about 2.86%.) Figure 25 shows the graph of y superimposed on the data.

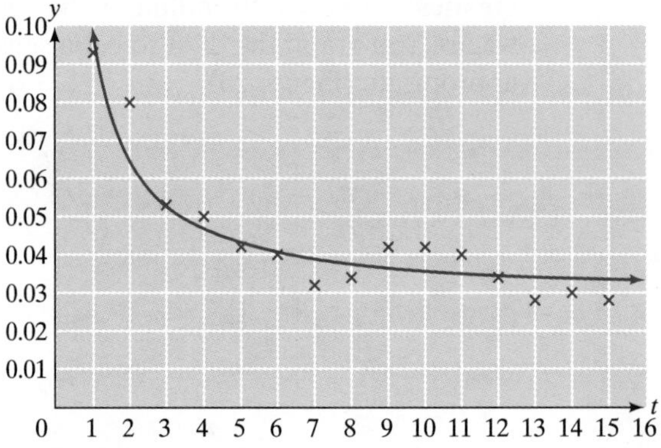

Figure 25

Now that you have a model for wage inflation, you must use it to find the annual wage. First, you realize that the model gives the *fractional rate of increase* of wages (since it is specified as a percentage, or fraction, of the total wage). In other words, if $w(t)$ represents a worker's annual wage at time t, then

$$y = \frac{dw/dt}{w} = \frac{d}{dt}(\ln w) \qquad \text{(by the chain rule for derivatives)}$$

[2]There is a good mathematical reason for choosing a curve of this form: It is a first approximation (for $t \geq 1$) to a general rational function that approaches a constant as $t \to +\infty$.

[3]To do this, you note that y is a linear function of $1/t$—namely, $y = a(1/t) + b$. Then you run a linear regression on the data $(1/t, y)$. (For more information on linear regression, see Chapter 1.)

You find an equation for a worker's annual wage at time t by solving for w:

$$\ln w = \int y\,dt = \int\left(\frac{a}{t} + b\right)dt = a\ln t + bt + C$$

where a and b are as above and C is the constant of integration, so

$$w = e^{a\ln t + bt + C}$$

To compute C, you substitute the initial data from the memo: $w(1) = 25{,}000$. Thus,

$$25{,}000 = e^{a\ln 1 + b + C} = e^{b+C} = e^{0.028647 + C}$$

Thus,

$$\ln(25{,}000) = 0.028647 + C$$

which gives

$$C = \ln(25{,}000) - 0.028647 \approx 10.098 \quad \text{(to five significant digits)}$$

Now you can write down the following formula for the annual wage of an assembly line worker as a function of t, the number of years since 1980:

$$w(t) = e^{a\ln t + bt + C} = e^{0.071813\ln t + 0.028647t + 10.098} = e^{0.071813\ln t}\,e^{0.028647t}\,e^{10.098} = t^{0.071813}\,e^{0.028647t}\,e^{10.098}$$

What remains is the calculation of the average annual wage. The average is the total wage earned over the worker's career divided by the number of years worked:

$$\overline{w} = \frac{1}{s-r}\int_r^s w(t)\,dt$$

where r is the time an employee begins working at the company and s is the time he or she retires. Substituting the formula for $w(t)$ gives

$$\overline{w} = \frac{1}{s-r}\int_r^s t^{0.071813}\,e^{0.028647t}\,e^{10.098}\,dt = \frac{e^{10.098}}{s-r}\int_r^s t^{0.071813}\,e^{0.028647t}\,dt$$

You cannot find an explicit antiderivative for the integrand, so you decide that the only way to compute it is numerically. You send the following memo to SC.

TO: SC

FROM: SW

SUBJECT: The formula you wanted

The average annual salary of an assembly line worker here at ABC Development Enterprises is given by the formula

$$\overline{w} = \frac{e^{10.098}}{s-r}\int_r^s t^{0.071813}\,e^{0.028647t}\,dt$$

where r is the time (in years after 1980) that a worker joins ABC and s is the time (in years after 1980) the worker retires. (The formula is valid only from 1981 on.) To calculate it easily (and impress the board members) I suggest you enter the following on your graphing calculator:

```
Y1=(e^(10.098)/(S-R))fnInt(T^0.071813e^(0.028647T),T,R,S)
```

Then suppose, for example, a worker joined the company in 1983 ($r = 3$) and retired in 2001 ($s = 21$). All you do is enter

```
3→R
21→S
Y1
```

and your calculator will give you the result: The average salary of the worker is \$41,307.16.

Have a nice day.

SW

exercises

1. Use the model developed above to compute the average annual income of a worker who joined the company in 1998 and left 3 years later.
2. What was the total amount paid by ABC Development Enterprises to the worker in Exercise 1?
3. What (if any) advantages are there to using a model for the annual wage inflation rate when the actual annual wage inflation rates are available?
4. The formula in the model was based on a 1981 salary of \$25,000. Change the model to allow for an arbitrary 1981 salary of $\$w_0$.
5. T If we had used exponential regression to model the wage inflation data, we would have obtained

$$y = 0.07142e^{-0.067135t}$$

Graph this equation along with the actual wage data and the earlier model for y. Is this a better or worse model?

6. Use the actual data in the table to calculate the average salary of an assembly line worker for the 6-year period from 1981 through the end of 1986, and compare it with the figure predicted by the model in the text.

chapter 6 review test

1. Evaluate the following indefinite integrals.

a. $\int (x^2 - 10x + 2)\,dx$ **b.** $\int (e^x + \sqrt{x})\,dx$

c. $\int x(x^2 + 4)^{10}\,dx$ **d.** $\int \frac{x^2 + 1}{(x^3 + 3x + 2)^2}\,dx$

e. $\int 5e^{-2x}\,dx$ **f.** $\int xe^{-x^2/2}\,dx$

2. Evaluate the following definite integrals using the Fundamental Theorem of Calculus.

a. $\int_0^1 (x - x^3)\,dx$ **b.** $\int_0^9 \frac{1}{x + 1}\,dx$

c. $\int_0^2 x^2\sqrt{x^3 + 1}\,dx$ **d.** $\int_0^{\pi} \cos(x + \pi/2)\,dx$

3. Find the areas of the following regions. Do not count the area below the x-axis as negative.
   a. The area bounded by $y = 4 - x^2$, the x-axis, and the lines $x = -2$ and $x = 2$
   b. The area bounded by $y = 4 - x^2$, the x-axis, and the lines $x = 0$ and $x = 5$
   c. The area bounded by $y = xe^{-x^2}$, the x-axis, and the lines $x = 0$ and $x = 5$
   d. The area bounded by $y = \sin x$, the x-axis, and the lines $x = 0$ and $x = 2\pi$

T **4.** Approximate the following definite integrals using Riemann sums with $n = 10, n = 100$, and $n = 1000$.

a. $\int_0^1 e^{-x^2}\,dx$

b. $\int_1^3 x^{-x}\,dx$

Applications

5. If OHaganBooks.com were to give away its latest bestseller, *A River Burns through It,* the demand would be 100,000 books. The marginal demand for the book is $-20p$ at a price of p dollars.

a. What is the demand function for this book?

b. At what price does demand drop to zero?

6. When OHaganBooks.com was about to go on-line, it estimated that its weekly sales would begin at about 6400 books per week, with sales increasing at such a rate that weekly sales would double about every 2 weeks.

a. If these estimates had been correct, how many books would the company have sold in the first 5 weeks?

b. In fact, OHaganBooks.com modeled its weekly sales over a period of time after it went on-line with the function

$$s(t) = 6053 + \frac{4474e^{0.55t}}{e^{0.55t} + 14.01}$$

where t is the time in weeks after it went on-line. According to this model, how many books did it sell in the first 5 weeks?

7. An overworked employee at OHaganBooks.com goes to the top of the company's 100-foot-tall headquarters building and flings a book up into the air at a speed of 60 feet per second.

a. When will the book hit the ground 100 feet below? (Neglect air resistance.)

b. How fast will it be traveling when it hits the ground?

c. How high will the book go?

Additional On-Line Review

If you follow the path

Web site → Everything for Calculus → Chapter 6

you will find the following additional resources to help you review:

- A comprehensive chapter summary (including examples and interactive features)
- Additional review exercises (including interactive exercises and many with help)
- A true–false chapter quiz
- A numerical integration utility

You're the Expert

Estimating Tax Revenues

You have just been hired by the incoming administration to coordinate national tax policy, and the so-called experts on your staff can't seem to agree on which of two tax proposals will result in more revenue for the government. The data you have are the two income tax proposals (graphs of tax vs. income) and the distribution of incomes in the country. How do you use this information to decide which tax policy will result in more revenue?

Internet Resources for This Chapter

At the web site, follow the path

Web site → Everything for Calculus → Chapter 7

where you will find a detailed chapter summary you can print out, a true–false quiz, and a collection of sample test questions. You will also find an on-line grapher, a numerical integration utility and other useful resources.

Further Integration Techniques and Applications of the Integral

chapter 7

7.1 Integration by Parts
7.2 Area Between Two Curves and Applications
7.3 Averages and Moving Averages
7.4 Continuous Income Streams
7.5 Improper Integrals and Applications
7.6 Differential Equations and Applications

Introduction

In Chapter 6 we learned how to compute many integrals and saw some of the applications of the integral. In this chapter we look at some more techniques for computing integrals and then at more applications of the integral. We also see how to extend the definition of the definite integral to include integrals over infinite intervals, and we show how such integrals can be used for long-term forecasting. Finally, we introduce the beautiful theory of differential equations and some of its numerous applications.

7.1 Integration by Parts

Integration by parts is an integration technique that comes from the product rule for derivatives. The tabular method we present here has been around for some time and makes integration by parts quite simple, particularly in problems where it has to be used several times. This particular version of the tabular method was developed and taught to us by Dan Rosen at Hofstra University.

We start with a little notation to simplify things while we introduce integration by parts (we use this notation only in the next few pages). If $f(x)$ is a function, denote its derivative by $D(f(x))$ and an antiderivative by $I(f(x))$. Thus, for example, if $f(x) = 2x^2$, then

$$D(f(x)) = 4x \qquad \text{and} \qquad I(f(x)) = \frac{2x^3}{3}$$

[If we wished, we could instead take $I(f(x)) = 2x^3/3 + 46$, but we opt to take the simplest antiderivative we can.]

Integration by Parts

$$\int f(x)g(x)\,dx = f(x)\,I(g(x)) - \int D(f(x))\,I(g(x))\,dx$$

Quick Example
(discussed more fully in Example 1 below)

$$\int x \cdot e^x\,dx = x\,I(e^x) - \int D(x)\,I(e^x)\,dx$$

$$= xe^x - \int 1 \cdot e^x\,dx \qquad I(e^x) = e^x;\ D(x) = 1$$

$$= xe^x - e^x + C \qquad \int e^x\,dx = e^x + C$$

As the Quick Example shows, although we could not immediately integrate $f(x)g(x) = x \cdot e^x$, we could more easily integrate $D(f(x))I(g(x)) = 1 \cdot e^x = e^x$.

Question Where does the integration-by-parts formula come from?
Answer As we mentioned, it comes from the product rule for derivatives. We apply the product rule to the function $f(x)I(g(x))$:

$$D[f(x)I(g(x))] = D(f(x))\,I(g(x)) + f(x)\,D(I(g(x))) = D(f(x))\,I(g(x)) + f(x)g(x)$$

because $D(I(g(x)))$ is the derivative of the antiderivative of $g(x)$, which is $g(x)$. Integrating both sides gives

$$f(x)\,I(g(x)) = \int D(f(x))\,I(g(x))\,dx + \int f(x)g(x)\,dx$$

A simple rearrangement of the terms gives us the integration-by-parts formula.

The integration-by-parts formula is easiest to use with the tabular method illustrated in the next example, where we repeat the calculation we did above.

example 1

Calculate $\int xe^x\,dx$.

Solution

First, the reason we *need* to use integration by parts to evaluate this integral is that none of the other techniques of integration that we've talked about up to now will help us. In particular, we cannot simply find antiderivatives of x and e^x and multiply them together. [You should check that $(x^2/2)\,e^x$ is *not* an antiderivative for xe^x.] However, as we saw above, this integral can be found by integration by parts. We want to find the integral of the *product* of x and e^x. We must make a decision: Which function will play the role of f and which will play the role of g in the integration-by-parts formula? Since the derivative of x is just 1, differentiating makes it simpler, so we try letting x be $f(x)$ and letting e^x be $g(x)$. We need to calculate $D(f(x))$ and $I(g(x))$, which we record in the following table:

	D		I
$+$	x		e^x
		↘	
$-\int$	1	→	e^x

$$+x \cdot e^x - \int 1 \cdot e^x\,dx$$

Below x in the D column we put $D(x) = 1$; below e^x in the I column we put $I(e^x) = e^x$. The arrow at an angle connecting x and $I(e^x)$ reminds us that the product $xI(e^x)$ will appear in the answer; the plus sign on the left of the table reminds us that it is $+xI(e^x)$ that appears. The integral sign and horizontal arrow connecting $D(x)$ and $I(e^x)$ remind us that the *integral* of the product $D(x)\,I(e^x)$

also appears in the answer; the minus sign on the left reminds us that we need to subtract this integral. Combining these two contributions, we get

$$\int xe^x\,dx = xe^x - \int e^x\,dx$$

The integral that appears on the right is much easier than the one we began with, so we can complete the problem:

$$\int xe^x\,dx = xe^x - \int e^x\,dx = xe^x - e^x + C$$

Before we go on . . . What if we had made the opposite decision and put e^x in the D column and x in the I column? Then we would have had the following table:

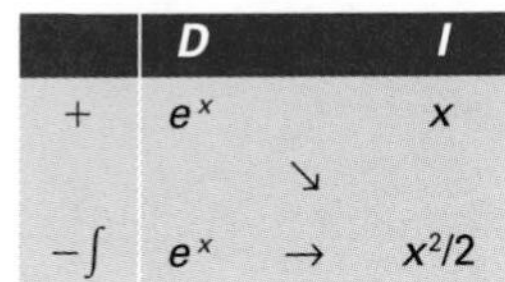

	D		I
$+$	e^x		x
		$\searrow$	
$-\int$	e^x	$\rightarrow$	$x^2/2$

This gives

$$\int xe^x\,dx = \frac{x^2}{2}e^x - \int \frac{x^2}{2}e^x\,dx$$

The integral on the right is harder than the one we started with, not easier! How do we know beforehand which way to go? We don't. We have to be willing to do a little trial and error. We try it one way, and if it doesn't make things simpler, we try it another way. *Remember, though, that the function we put in the I column must be one that we can integrate.*

example 2

Calculate $\int x^2e^{-x}\,dx$.

Solution

Again, we have a product—the integrand is the product of x^2 and e^{-x}. Since differentiating x^2 makes it simpler, we put it in the D column and get the following table:

	D		I
$+$	x^2		e^{-x}
		$\searrow$	
$-\int$	$2x$	$\rightarrow$	$-e^{-x}$

This table gives us

$$\int x^2e^{-x}\,dx = x^2(-e^{-x}) - \int 2x(-e^{-x})\,dx$$

The last integral is simpler than the one we started with, but it still involves a product. It's a good candidate for another integration by parts. The table we would use would start with $2x$ in the D column and $-e^{-x}$ in the I column, which is exactly what we see in the last row of the table we've already made. Therefore, we *continue the process,* elongating the table above:

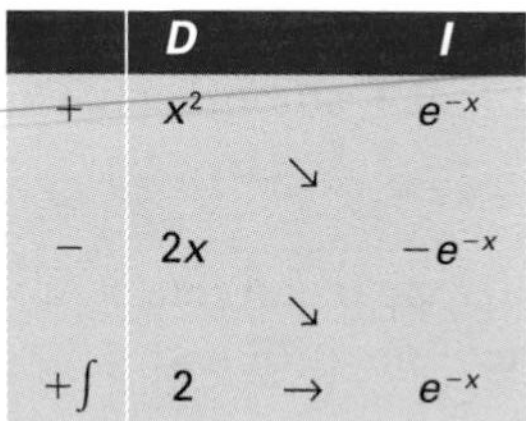

	D		I
$+$	x^2		e^{-x}
		↘	
$-$	$2x$		$-e^{-x}$
		↘	
$+\int$	2	→	e^{-x}

Notice how the signs on the left alternate. To compute $\int 2x(-e^{-x})\,dx$ we would use the following table:

	D		I
$+$	$2x$		$-e^{-x}$
		↘	
$-\int$	2	→	e^{-x}

However, we really need to compute its negative, $-\int 2x(-e^{-x})\,dx$, so we reverse all the signs. Now, we still have to compute an integral (the integral of the product of the functions in the bottom row) to complete the computation. But why stop here? Let us continue the process one more step:

	D		I
$+$	x^2		e^{-x}
		↘	
$-$	$2x$		$-e^{-x}$
		↘	
$+$	2		e^{-x}
		↘	
$-\int$	0	→	$-e^{-x}$

In the bottom line we see that all that is left to integrate is $0(-e^{-x}) = 0$. Since the indefinite integral of 0 is C, we can read the answer from the table as

$$\int x^2e^{-x}\,dx = x^2(-e^{-x}) - 2x(e^{-x}) + 2(-e^{-x}) + C$$
$$= -x^2e^{-x} - 2xe^{-x} - 2e^{-x} + C = -e^{-x}(x^2 + 2x + 2) + C$$

Before we go on . . . Since that took several steps, let's check our work:

$$\frac{d}{dx}\left[-e^{-x}(x^2 + 2x + 2) + C\right] = e^{-x}(x^2 + 2x + 2) - e^{-x}(2x + 2) = x^2e^{-x} \quad ✔$$

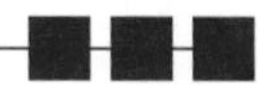

In Example 2 we saw a technique that we can summarize as follows.

Integrating a Polynomial Times a Function

If one of the factors in the integrand is a polynomial and the other factor is a function that can be integrated repeatedly, put the polynomial in the D column and keep differentiating until you get zero. Then complete the I column to the same depth and read off the answer.

For practice, redo Example 1 using this technique.

It is not always the case that the integrand is a polynomial times something easy to integrate, so we can't always expect to end up with a zero in the D column. In that case we hope that at some point we will be able to integrate the product of the functions in the last row. Here is an example.

example 3

Calculate $\int x \ln x \, dx$.

Solution

This is a product and therefore a good candidate for integration by parts. Our first impulse is to differentiate x, but that would mean integrating $\ln x$, and we do not (yet) know how to do that. So we try it the other way around and hope for the best.

	D		I
$+$	$\ln x$		x
		$\searrow$	
$-\int$	$\frac{1}{x}$	$\rightarrow$	$\frac{x^2}{2}$

Why did we stop? If we continued the table, both columns would get more complicated. However, if we stop here, we get

$$\int x \ln x \, dx = (\ln x)\left(\frac{x^2}{2}\right) - \int \left(\frac{1}{x}\right)\left(\frac{x^2}{2}\right) dx = \frac{x^2}{2} \ln x - \frac{1}{2} \int x \, dx = \frac{x^2}{2} \ln x - \frac{x^2}{4} + C$$

example 4

Calculate $\int \ln x \, dx$.

Solution

The integrand $\ln x$ does not look like a product. (It is the natural logarithm *of* x, *not* "the natural logarithm times x.") We can, however, *make* it into a product by thinking of it as $1 \cdot \ln x$. Now we are very tempted to put the 1 in the D column because its derivative is zero. But this would force us to put the $\ln x$ in the I column, and we don't yet know what the integral of $\ln x$ is—this is the problem we were given! So we do what we did in the preceding example and put $\ln x$ in the D column:

	D		I
$+$	$\ln x$		1
		$\searrow$	
$-\int$	$1/x$	$\rightarrow$	x

The product of $1/x$ and x is just 1, and we certainly know how to integrate that. Thus, we have

$$\int \ln x \, dx = x \ln x - \int \left(\frac{1}{x}\right) x \, dx = x \ln x - \int 1 \, dx = x \ln x - x + C$$

Now we know what the integral of $\ln x$ is!

Here is another trick that is sometimes useful.

example 5

Calculate $\int e^x \sin x\, dx$.

Solution

The integrand is the product of e^x and $\sin x$, so we put one in the D column and the other in the I column. For this example, it doesn't matter much which we put where.

	D		I
$+$	$\sin x$		e^x
		$\searrow$	
$-$	$\cos x$		e^x
		$\searrow$	
$+\int$	$-\sin x$	$\rightarrow$	e^x

It looks like we're just spinning our wheels. Let's stop and see what we have:

$$\int e^x \sin x\, dx = e^x \sin x - e^x \cos x - \int e^x \sin x\, dx$$

At first glance, it appears that we are back where we started, still having to evaluate $\int e^x \sin x\, dx$. However, if we add this integral to both sides of the equation above, we can solve for it:

$$2\int e^x \sin x\, dx = e^x \sin x - e^x \cos x + C$$

(Why $+C$?) So,

$$\int e^x \sin x\, dx = \frac{1}{2} e^x \sin x - \frac{1}{2} e^x \cos x + \frac{C}{2}$$

Since $C/2$ is just as arbitrary as C, we write C instead of $C/2$ and obtain

$$\int e^x \sin x\, dx = \frac{1}{2} e^x \sin x - \frac{1}{2} e^x \cos x + C$$

Hints and General Guidelines for Integration by Parts

1. To integrate a product in which one factor is a polynomial and the other can be integrated several times, put the polynomial in the D column and the other factor in the I column. Then differentiate the polynomial until you get zero.
2. If one of the factors is a polynomial but the other factor cannot be integrated easily, put the polynomial in the I column and the other factor in the D column. Stop when the product of the functions in the bottom row can be integrated.
3. If neither factor is a polynomial, put the factor that seems easiest to integrate in the I column and the other factor in the D column. Again, stop the table as soon as the product of the functions in the bottom row can be

integrated or when the integral you started with appears again (as in Example 5 above).

4. If your method doesn't work, try switching the functions in the D and I columns or try breaking the integrand into a product in a different way. If none of this works, maybe integration by parts isn't the technique to use on this problem.

7.1 exercises

Evaluate the integrals in Exercises 1–34.

1. $\int 2xe^x\,dx$
2. $\int 3xe^{-x}\,dx$
3. $\int (3x-1)e^{-x}\,dx$
4. $\int (1-x)e^x\,dx$
5. $\int (x^2-1)e^{2x}\,dx$
6. $\int (x^2+1)e^{-2x}\,dx$
7. $\int (x^2+1)e^{-2x+4}\,dx$
8. $\int (x^2+1)e^{3x+1}\,dx$
9. $\int \frac{x^2-x}{e^x}\,dx$
10. $\int \frac{2x+1}{e^{3x}}\,dx$
11. $\int x(x+2)^6\,dx$
12. $\int x^2(x-1)^6\,dx$
13. $\int \frac{x}{(x-2)^3}\,dx$
14. $\int \frac{x}{(x-1)^2}\,dx$
15. $\int x^3 \ln x\,dx$
16. $\int x^2 \ln x\,dx$
17. $\int (t^2+1)\ln(2t)\,dt$
18. $\int (t^2-t)\ln(-t)\,dt$
19. $\int t^{1/3}\ln t\,dt$
20. $\int t^{-1/2}\ln t\,dt$
21. ▲ $\int x\sin x\,dx$
22. ▲ $\int x^2\cos x\,dx$
23. ▲ $\int e^{-x}\sin x\,dx$
24. ▲ $\int e^{2x}\cos x\,dx$
25. $\int_0^1 (x+1)e^x\,dx$
26. $\int_{-1}^1 (x^2+x)e^{-x}\,dx$
27. $\int_0^1 x^2(x+1)^{10}\,dx$
28. $\int_0^1 x^3(x+1)^{10}\,dx$
29. $\int_1^2 x\ln(2x)\,dx$
30. $\int_1^2 x^2\ln(3x)\,dx$
31. $\int_0^1 x\ln(x+1)\,dx$
32. $\int_0^1 x^2\ln(x+1)\,dx$
33. ▲ $\int_0^{\pi} x^2\sin x\,dx$
34. ▲ $\int_0^{\pi/2} x\cos x\,dx$
35. Find the area bounded by the curve $y = xe^{-x}$, the x-axis, and the lines $x = 0$ and $x = 10$.
36. Find the area bounded by the curve $y = x\ln x$, the x-axis, and the lines $x = 1$ and $x = e$.
37. Find the area bounded by the curve $y = (x+1)\ln x$, the x-axis, and the lines $x = 1$ and $x = 2$.
38. Find the area bounded by the curve $y = (x-1)e^x$, the x-axis, and the lines $x = 0$ and $x = 2$.

Applications

39. ***Displacement*** A rocket rising from the ground has a velocity of $2000te^{-t/120}$ ft/s, after t seconds. How far does it rise in the first 2 minutes?

40. ***Sales*** Weekly sales of graphing calculators can be modeled by the equation

$$s(t) = 10 - te^{-t/20}$$

where s is the number of calculators sold per week after t weeks. How many graphing calculators (to the nearest unit) will be sold in the first 20 weeks?

41. ***Total Cost*** The marginal cost of the xth box of lightbulbs is $10 + [\ln(x+1)]/(x+1)^2$, and the fixed cost is \$5000. Find the total cost to make x boxes of bulbs.

42. ***Total Revenue*** The marginal revenue for selling the xth box of light bulbs is $10 + 0.001x^2e^{-x/100}$. Find the total revenue generated by selling 200 boxes of bulbs.

43. ***Revenue*** You have been raising the price of your *Titanic* T-shirts by 50¢ per week, and sales have been falling continuously at a rate of 2% per week. Assuming you are now selling 50 T-shirts per week and charging \$10 per T-shirt, how much revenue will you generate during the coming year? (Round your answer to the nearest dollar.)

44. ***Revenue*** Luckily, sales of your *Star Wars* T-shirts are now 50 T-shirts per week and increasing continuously at a rate of 5% per week. You are now charging \$10 per T-shirt and decreasing the price by 50¢ per week. How

much revenue will you generate during the next 6 weeks?

45. ***Revenue*** About 70.5 million magazines were sold in the United States in 1982, and sales have been falling by about 2.5 million per year since then.[1] Assume that the average price of magazines was \$2.00 in 1982 and that this price has been increasing continuously at 5% per year. How much revenue was generated by sales of magazines in the period 1982–1992? (Give your answer to the nearest \$1 million.) [*Hint:* Annual revenue after t years = Annual sales × Price per magazine.]

46. ***Lost Revenue*** In 1982 about 35 million magazines placed on newsstands in the United States were not sold. This number has increased by about 0.6 million per year. Assume that the average price of magazines was \$2.00 in 1982 and that this price has been increasing continuously at 5% per year. How much revenue was lost through unsold magazines in the period 1982–1992? (Give your answer to the nearest \$1 million.) [*Hint:* Annual lost revenue = Annual lost sales × Price per magazine.]

[1]The figures in Exercises 45 and 46 were for sales of 123 leading magazines at newstands and appeared in the *New York Times,* December 6, 1993, p. D6.

Communication and Reasoning Exercises

47. ***Hermite's Identity*** If $f(x)$ is a polynomial of degree n, show that

$$\int_0^b f(x)e^{-x}\,dx = F(0) - F(b)e^{-b}$$

where $F(x) = f(x) + f'(x) + f''(x) + \cdots + f^{(n)}(x)$ (this is the sum of f and all of its derivatives).

48. Write down a formula similar to Hermite's identity for $\int_0^b f(x)e^x\,dx$ when $f(x)$ is a polynomial of degree n.

7.2 Area Between Two Curves and Applications

As we saw in Chapter 6, we can use the definite integral to calculate the area between the graph of a function and the x-axis. With only a little more work, we can use it to calculate the area between two curves. Figure 1 shows the graphs of two functions, $f(x)$ and $g(x)$, with $f(x) \geq g(x)$ for every x in the interval $[a, b]$.

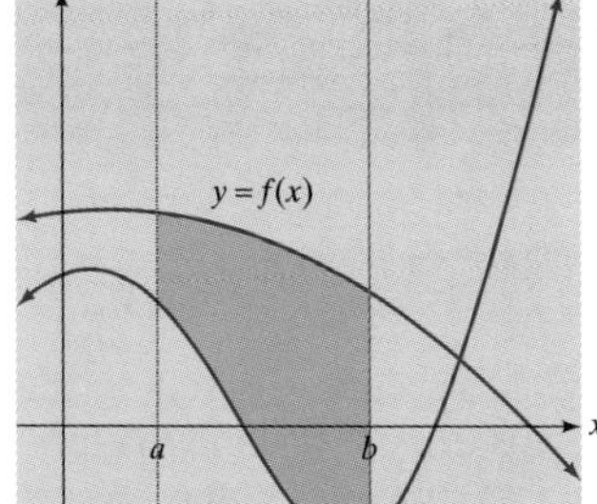

Figure 1

Question How can we find the shaded area between the graphs of the two functions?

Answer Pretty easily, actually, using the following formula.

> **Area Between Two Graphs**
>
> If $f(x) \geq g(x)$ for all x in $[a, b]$, then the area of the region between the graphs of f and g and between $x = a$ and $x = b$ is given by
>
> $$A = \int_a^b \left[f(x) - g(x)\right] dx$$

Let's look at an example and then discuss why the formula works.

example 1

The Area Between Two Curves

Find the area of the region between $f(x) = -x^2 - 3x + 4$ and $g(x) = x^2 - 3x - 4$ and between $x = -1$ and $x = 1$.

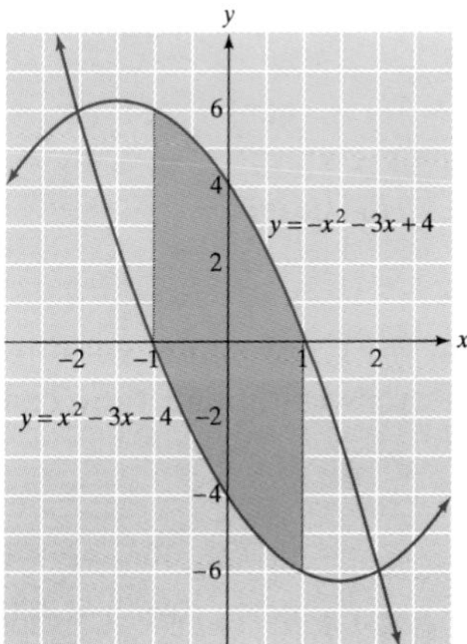

Figure 2

Solution

The area in question is shown in Figure 2.

Since the graph of f lies above the graph of g in the interval $[-1, 1]$, we have $f(x) \geq g(x)$ for all x in $[-1, 1]$. Therefore, we can use the formula given above and calculate the area as follows:

$$A = \int_{-1}^{1} [f(x) - g(x)]\, dx$$

$$= \int_{-1}^{1} (8 - 2x^2)\, dx = \left[8x - \frac{2}{3}x^3\right]_{-1}^{1} = \frac{44}{3}$$

Question Why does the formula for the area between two curves work?

Answer Let us go back once again to the general case illustrated in Figure 1, where we are given two functions f and g with $f(x) \geq g(x)$ for every x in the interval $[a, b]$. We would like to avoid complicating the argument by the fact that the graph of g, or f, or both may dip below the x-axis in the interval $[a, b]$ (as occurs in Figure 1 and also in Example 1). We can get around this issue as follows: We shift both graphs vertically upward by adding a big enough constant M to lift them both above the x-axis in the interval $[a, b]$, as shown in Figure 3.

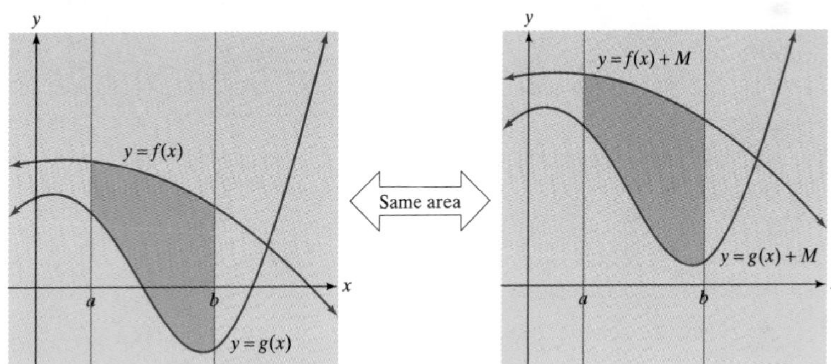

Figure 3

As the figure illustrates, the area of the region between the graphs is not affected, so we calculate the area of the region shown on the right of Figure 3. That calculation is shown in Figure 4.

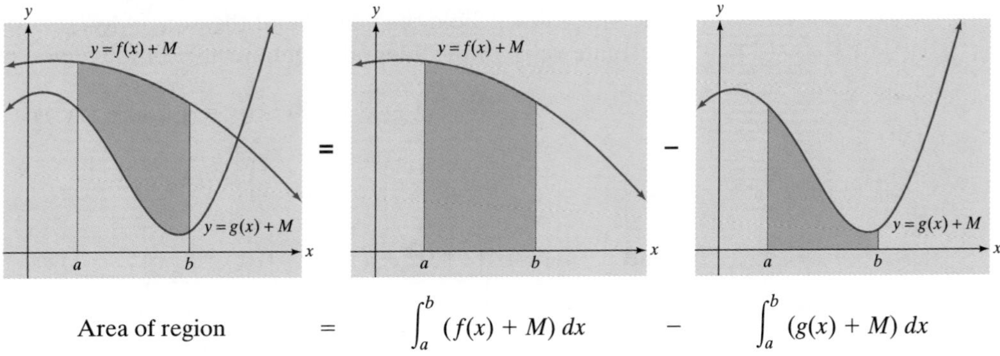

Area of region $= \int_a^b (f(x) + M)\, dx - \int_a^b (g(x) + M)\, dx$

Figure 4

From the figure, the area we want is

$$\int_a^b (f(x) + M)\,dx - \int_a^b (g(x) + M)\,dx = \int_a^b \left[(f(x) + M) - (g(x) + M)\right] dx$$
$$= \int_a^b \left[f(x) - g(x)\right] dx$$

which is the formula we gave originally.

So far, we've been assuming that $f(x) \geq g(x)$, so that the graph of f never dips below the graph of g and so that the graphs cannot cross (although they can touch). What if the two graphs *do* cross?

example 2

Regions Enclosed by Crossing Curves

Find the area of the region between $y = 3x^2$ and $y = 1 - x^2$ and between $x = 0$ and $x = 1$.

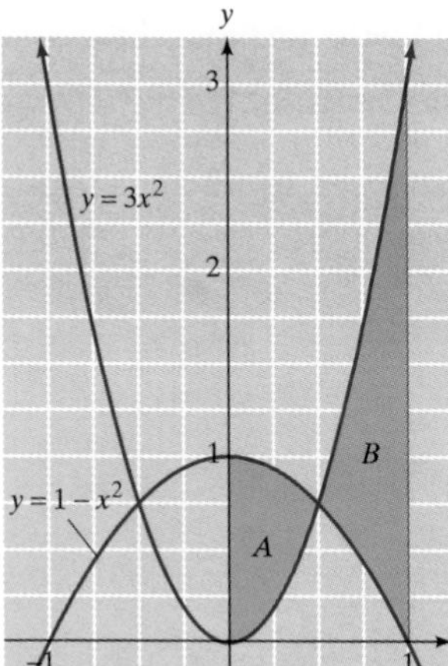

Figure 5

Solution

The area we wish to calculate is shown in Figure 5. From the figure we can see that neither graph lies above the other over the whole interval. To get around this, we break the area into the two pieces on either side of the point at which the graphs cross and then compute each area separately. To do this, we need to know exactly where that crossing point is. The crossing point is where $3x^2 = 1 - x^2$, so we solve for x:

$$3x^2 = 1 - x^2$$
$$4x^2 = 1$$
$$x^2 = \frac{1}{4}$$
$$x = \pm\frac{1}{2}$$

Since we are interested only in the interval $[0, 1]$, the crossing point we're interested in is at $x = \frac{1}{2}$.

Now, to compute the areas A and B we need to know which graph is on top in each of these areas. We can see that from the figure, but what if the functions were more complicated and we could not easily draw the graphs? If we want to be sure, we can test the values of the two functions at some point in each region. But we really need not worry. If we make the wrong choice for the top function, the integral will yield the negative of the area (why?), so we can simply take the absolute value of the integral to get the area of the region in question. We have

$$A = \int_0^{1/2} \left[(1 - x^2) - 3x^2\right] dx = \int_0^{1/2} (1 - 4x^2)\,dx$$
$$= \left[x - \frac{4x^3}{3}\right]_0^{1/2} = \left(\frac{1}{2} - \frac{1}{6}\right) - (0 - 0) = \frac{1}{3}$$

and
$$B = \int_{1/2}^1 \left[3x^2 - (1 - x^2)\right] dx = \int_{1/2}^1 (4x^2 - 1)\,dx$$
$$= \left[\frac{4x^3}{3} - x\right]_{1/2}^1 = \left(\frac{4}{3} - 1\right) - \left(\frac{1}{6} - \frac{1}{2}\right) = \frac{2}{3}$$

This gives a total area of $A + B = \frac{1}{3} + \frac{2}{3} = 1$.

Before we go on . . . What would have happened if we had not broken the area into two pieces but just calculated the integral of the difference of the two functions? We would have calculated

$$\int_0^1 \left[(1 - x^2) - 3x^2\right] dx = \int_0^1 (1 - 4x^2)\, dx = \left[x - \frac{4x^3}{3}\right]_0^1 = -\frac{1}{3}$$

which is not even close to the right answer. What this integral calculated was actually $A - B$ rather than $A + B$. Why?

example 3

The Area Enclosed by Two Curves
Find the area enclosed by $y = x^2$ and $y = x^3$.

Solution

This example has a new wrinkle: We are not told what interval to use for x. However, if we look at the graph in Figure 6, we see that the question can have only one meaning.

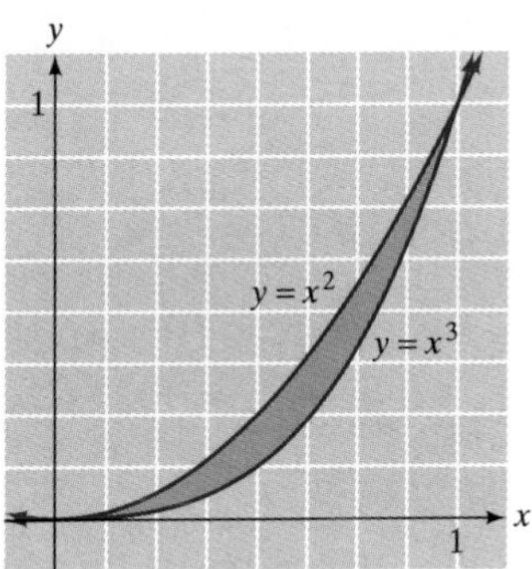

Figure 6

We are being asked to find the area of the shaded sliver, which is the only part of the picture that is actually *enclosed* by the two graphs. This sliver is bounded on either side by the two points where the graphs cross, so our first task is to find those points. They are the points where $x^2 = x^3$, so we solve for x:

$$x^2 = x^3$$

$$x^3 - x^2 = 0$$

$$x^2(x - 1) = 0$$

$$x = 0 \quad \text{or} \quad x = 1$$

Thus, we must integrate over the interval [0, 1]. Although we see from the diagram (or by substituting $x = \frac{1}{2}$) that the graph of $y = x^2$ is above that of $y = x^3$, let us pretend that we didn't notice that and calculate

$$\int_0^1 (x^3 - x^2)\, dx = \left[\frac{x^4}{4} - \frac{x^3}{3}\right]_0^1 = -\frac{1}{12}$$

This tells us that the required area is $\frac{1}{12}$ square unit and also that we had our integral reversed. Had we calculated $\int_0^1 (x^2 - x^3)\, dx$ instead, we would have found the correct answer, $\frac{1}{12}$.

We can summarize the procedure we used in the preceding two examples.

Finding the Area Between the Graphs of $f(x)$ and $g(x)$

1. Find all points of intersection by solving $f(x) = g(x)$ for x. This either determines the interval over which you will integrate or breaks up a given interval into regions between the intersection points.
2. Find the area of each region you found by integrating the difference of the larger and the smaller functions. (If you accidentally take the smaller minus the larger, the integral will give the negative of the area, so just take the absolute value.)
3. Add together the areas you found in Step 2 to get the total area.

Consumers' Surplus

Consider a general demand curve presented, as is traditional in economics, as $p = D(q)$, where p is unit price and q is demand measured, say, in annual sales (Figure 7).

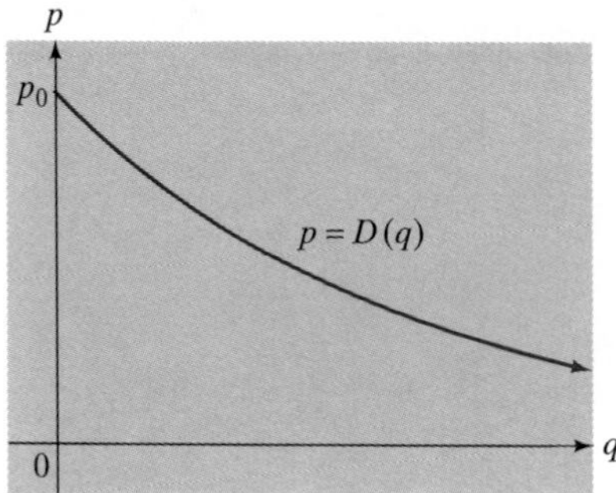

Figure 7

Thus, $D(q)$ is the price at which the demand will be q units per year. The price p_0 shown on the graph is the highest price that customers are willing to pay.

For a more concrete example, suppose that Figure 7 is the demand curve for a particular new model of computer. When the computer first comes out and supplies are low (q is small), "early adopters" are willing to pay a high price. This is the part of the graph on the left, near the p-axis. As supplies increase and the price drops, more consumers are willing to pay and more computers are sold. We can ask the following question.

Question How much are consumers willing to spend for the first $\bar{q}$ units?
Answer We can approximate consumers' willingness to spend as follows: We partition the interval $[0, \bar{q}]$ into n subintervals of equal length, as we did when discussing Riemann sums. Figure 8 shows a typical subinterval, $[q_{k-1}, q_k]$.

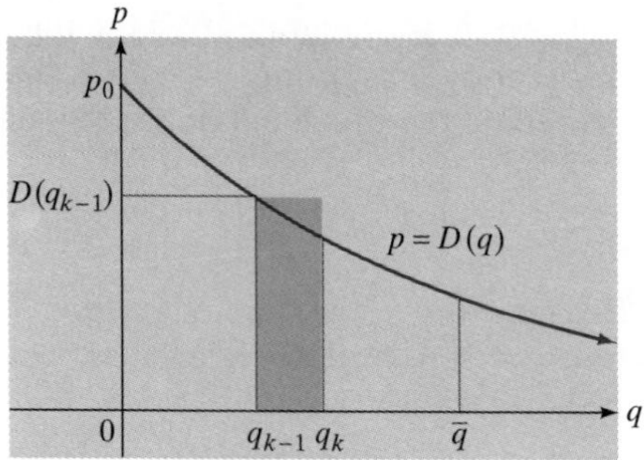

Figure 8

The price consumers are willing to pay for each of units q_{k-1} through q_k is approximately $D(q_{k-1})$, so the total that consumers are willing to spend for these units is approximately $D(q_{k-1})(q_k - q_{k-1}) = D(q_{k-1})\,\Delta q$, the area of the shaded region in Figure 8. Thus, the total amount that consumers are willing to spend for items 0 through $\overline{q}$ is

$$W \approx \sum_{k=1}^{n} D(q_{k-1})\,\Delta q$$

which is a Riemann sum. The approximation becomes better the larger n becomes, and in the limit the Riemann sums converge to the integral

$$W = \int_0^{\overline{q}} D(q)\,dq$$

This is the total **consumers' willingness to spend** to buy the first $\overline{q}$ units. It is the area shaded in Figure 9.

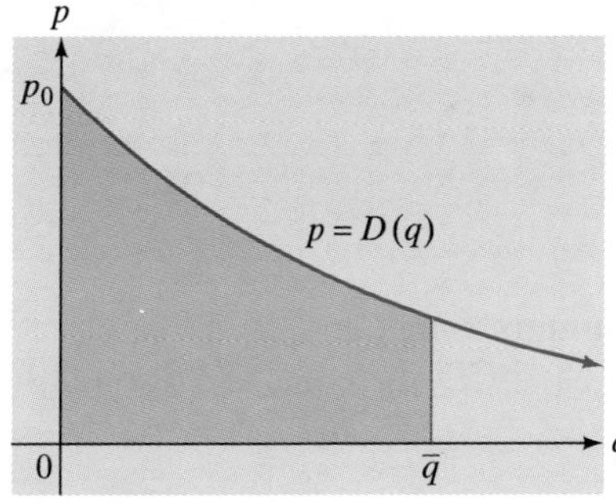

Figure 9

Now suppose that the manufacturer simply sets the price at $\overline{p}$, with a corresponding demand of $\overline{q}$, so $D(\overline{q}) = \overline{p}$.

Question How much will consumers actually spend to buy these $\overline{q}$ items?
Answer That's easy: $\overline{p}\,\overline{q}$, the product of the unit price and the quantity sold. This is the **consumers' expenditure,** the area of the rectangle shown in Figure 10.

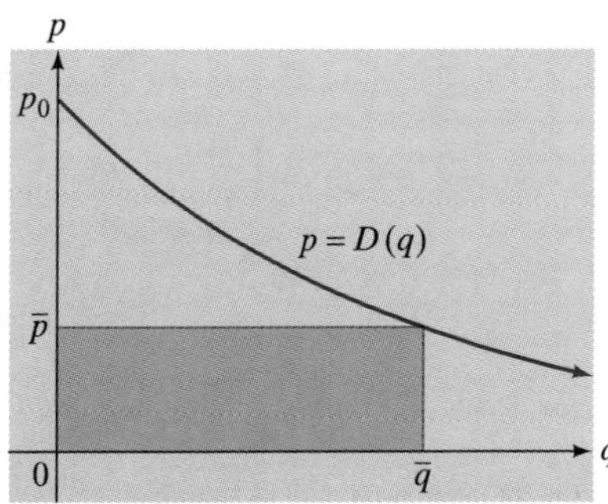

Figure 10

Notice that we can write $\overline{p}\,\overline{q} = \int_0^{\overline{q}} \overline{p}\,dq$, as suggested by the figure.

The difference between what consumers are willing to pay and what they actually pay is money in their pockets and is called the **consumers' surplus.**

Consumers' Surplus

If the demand for an item is given by $p = D(q)$, the selling price is $\overline{p}$, and $\overline{q}$ is the corresponding demand [so that $D(\overline{q}) = \overline{p}$], then the **consumers' surplus** is the difference between their willingness to spend and their actual expenditure:

$$CS = \int_0^{\overline{q}} D(q)\, dq - \overline{p}\,\overline{q} = \int_0^{\overline{q}} \left[D(q) - \overline{p}\right] dq$$

Graphically, it is the area between the graphs of $p = D(q)$ and $p = \overline{p}$, as shown in the figure.

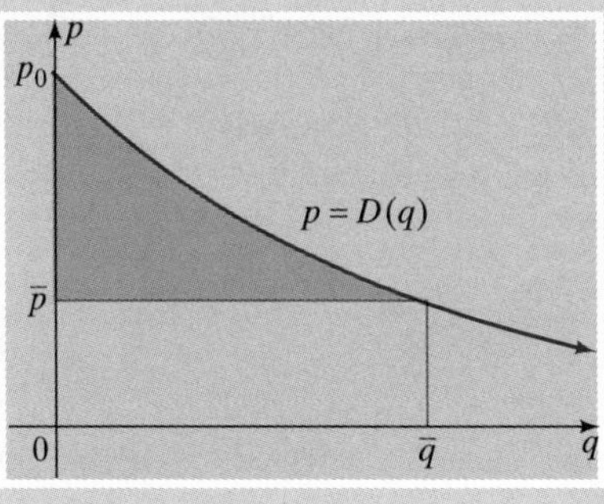

example 4

Consumers' Surplus

Your used-CD store has an exponential demand equation of the form

$$p = 15e^{-0.01q}$$

where q represents daily sales of CDs and p is the price you charge per used CD. Calculate the daily consumers' surplus if you sell used CDs at \$5 each.

Solution

We are given $D(q) = 15e^{-0.01q}$ and $\overline{p} = 5$. We also need $\overline{q}$. By definition,

$$D(\overline{q}) = \overline{p}$$

or

$$15e^{-0.01\overline{q}} = 5$$

which we must solve for $\overline{q}$:

$$e^{-0.01\overline{q}} = \frac{1}{3}$$

$$-0.01\overline{q} = \ln\left(\frac{1}{3}\right) = -\ln 3$$

$$\overline{q} = \frac{\ln 3}{0.01} \approx 109.8612$$

We now have

$$CS = \int_0^{\overline{q}} \left[D(q) - \overline{p}\right] dq$$

$$= \int_0^{109.8612} (15e^{-0.01q} - 5)\, dq$$

$$= \left[\frac{15}{-0.01} e^{-0.01q} - 5q\right]_0^{109.8612} \approx (-500 - 549.31) - (-1500 - 0) = \$450.69 \text{ per day}$$

Before we go on . . . We could have also calculated the consumers' willingness to spend and their actual expenditures. Their willingness to spend is

$$
\begin{aligned}
W &= \int_0^{\bar{q}} D(q)\, dq \\
&= \int_0^{109.8612} 15e^{-0.01q}\, dq \\
&= \left[\frac{15}{-0.01} e^{-0.01q}\right]_0^{109.8612} \approx -500 - (-1500) = \$1000 \text{ per day}
\end{aligned}
$$

Their actual expenditure is $\bar{p}\,\bar{q} = 5 \times 109.8612 = \$549.31$ per day. The difference is their surplus: $1000 - 549.31 = \$450.69$ per day.

Producers' Surplus

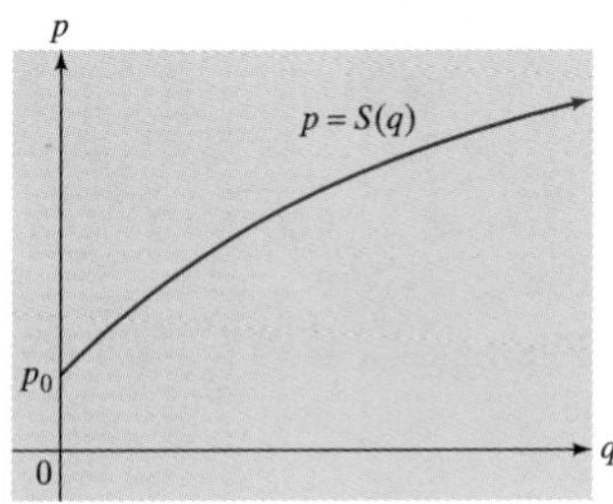

Figure 11

We can also calculate extra income earned by producers. Consider a supply equation of the form $p = S(q)$, where $S(q)$ is the price at which a supplier is willing to supply q items (per time period). Because a producer is generally willing to supply more units at a higher price per unit, a supply curve usually has a positive slope, as shown in Figure 11. The price p_0 is the lowest price a producer is willing to charge.

Question What is the minimum amount of money producers are willing to receive in exchange for $\bar{q}$ items?
Answer Arguing as before, we say it is $\int_0^{\bar{q}} S(q)\, dq$.

Question If the producers charge $\bar{p}$ per item for $\bar{q}$ items, what is their total revenue?
Answer As before, the total revenue is $\bar{p}\,\bar{q} = \int_0^{\bar{q}} \bar{p}\, dq$.

The difference between the producers' total revenue and the minimum they would have been willing to receive is the **producers' surplus.**

Producers' Surplus

The **producers' surplus** is the extra amount earned by producers who were willing to charge less than the selling price of $\bar{p}$ per unit and is given by

$$PS = \int_0^{\bar{q}} \left[\bar{p} - S(q)\right] dq$$

where $S(\bar{q}) = \bar{p}$. Graphically, it is the area of the region between the graphs of $p = \bar{p}$ and $p = S(q)$ for $0 \le q \le \bar{q}$, as in the figure.

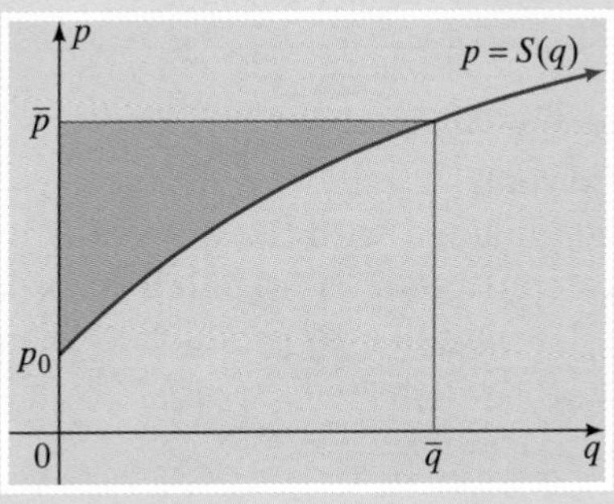

example 5

Producers' Surplus

My tie-dye T-shirt enterprise has grown to the extent that I am now able to produce T-shirts in bulk, and several campus groups have begun placing orders. I have informed one group that I am prepared to supply $20\sqrt{p-4}$ T-shirts at a price of p dollars per shirt. What is my total surplus if I sell T-shirts to the group at \$8 each?

Solution

We need to calculate the producers' surplus when $\bar{p} = 8$. The supply equation is

$$q = 20\sqrt{p-4}$$

but in order to use the formula for producers' surplus, we need to express p as a function of q. First, we square both sides to remove the radical sign:

$$q^2 = 400(p-4)$$

so
$$p - 4 = \frac{q^2}{400}$$

giving
$$p = S(q) = \frac{q^2}{400} + 4$$

We now need the value of $\bar{q}$ corresponding to $\bar{p} = 8$. The easiest way to find it is to substitute $p = 8$ in the original equation, which gives

$$\bar{q} = 20\sqrt{8-4} = 20\sqrt{4} = 40$$

Thus,

$$\begin{aligned} PS &= \int_0^{\bar{q}} [\bar{p} - S(q)]\, dq \\ &= \int_0^{40} \left[8 - \left(\frac{q^2}{400} + 4\right)\right] dq \\ &= \int_0^{40} \left(4 - \frac{q^2}{400}\right) dq = \left[4q - \frac{q^3}{1200}\right]_0^{40} \approx \$106.67 \end{aligned}$$

Before we go on . . . The amount I am willing to receive for these T-shirts is

$$\int_0^{40} S(q)\, dq = \int_0^{40} \left(\frac{q^2}{400} + 4\right) dq = \left[\frac{q^3}{1200} + 4q\right]_0^{40} \approx \$213.33$$

By selling T-shirts for \$8 a piece, my total revenue will be $8 \times 40 = \$320$. The difference, $320 - 213.33 = \$106.67$, is my surplus.

---

example 6

Equilibrium

We continue on with the preceding example. A representative informs me that the campus group is prepared to order only $\sqrt{200(16-p)}$ T-shirts at p dollars each. I would like to produce as many T-shirts for them as possible but avoid being left with unsold T-shirts. Given the supply curve from the preceding example, what price should I charge per T-shirt, and what are the consumers' and producers' surpluses at that price?

Solution

The price that guarantees neither a shortage nor a surplus of T-shirts is the **equilibrium price,** the price where supply equals demand. We have

$$\text{Supply:} \quad q = 20\sqrt{p-4}$$
$$\text{Demand:} \quad q = \sqrt{200(16-p)}$$

Equating these gives

$$\begin{aligned} 20\sqrt{p-4} &= \sqrt{200(16-p)} \\ 400(p-4) &= 200(16-p) \\ 400p - 1600 &= 3200 - 200p \\ 600p &= 4800 \\ p &= \$8 \text{ per T-shirt} \end{aligned}$$

We therefore take $\bar{p} = 8$ (which happens to be the price we used in the preceding example). We get the corresponding value for q by substituting $p = 8$ into either the demand or supply equation:

$$\bar{q} = 20\sqrt{8-4} = 40$$

Thus, $\bar{p} = 8$ and $\bar{q} = 40$.

We must now calculate the consumers' surplus and the producers' surplus. We calculated the producers' surplus for $\bar{p} = 8$ in the preceding example:

$$PS = \$106.67$$

For the consumers' surplus, we must first express p as a function of q for the demand equation. Thus, we solve the demand equation for p as we did for the supply equation, and we obtain

$$\text{Demand:} \quad D(q) = 16 - \frac{q^2}{200}$$

Therefore,

$$\begin{aligned} CS &= \int_0^{\bar{q}} \left[D(q) - \bar{p}\right] dq \\ &= \int_0^{40} \left[\left(16 - \frac{q^2}{200}\right) - 8\right] dq \\ &= \int_0^{40} \left(8 - \frac{q^2}{200}\right) dq = \left[8q - \frac{q^3}{600}\right]_0^{40} \approx \$213.33 \end{aligned}$$

Before we go on . . . Figure 12 shows both the consumers' surplus (top portion) and the producers' surplus (bottom portion).

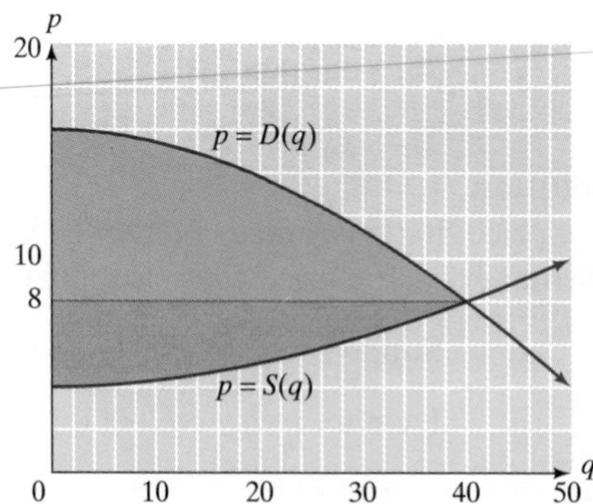

Figure 12

Since extra money in people's pockets is a good thing, the total of the consumers' and the producers' surpluses is called the **total social gain.** In this case it is

$$\text{Social gain} = CS + PS = 213.33 + 106.67 = \$320.00$$

As you can see from Figure 12, the total social gain is also the area between two curves and equals

$$\int_0^{40} [D(q) - S(q)]\, dq$$

7.2 exercises

Find the areas of the indicated regions in Exercises 1–24. (Technology may be useful for Exercises 21–24.)

1. Between $y = x^2$ and $y = -1$ for x in $[-1, 1]$
2. Between $y = x^3$ and $y = -1$ for x in $[-1, 1]$
3. Between $y = -x$ and $y = x$ for x in $[0, 2]$
4. Between $y = -x$ and $y = x/2$ for x in $[0, 2]$
5. Between $y = x$ and $y = x^2$ for x in $[-1, 1]$
6. Between $y = x$ and $y = x^3$ for x in $[-1, 1]$
7. Between $y = e^x$ and $y = x$ for x in $[0, 1]$
8. Between $y = e^{-x}$ and $y = -x$ for x in $[0, 1]$
9. Between $y = (x - 1)^2$ and $y = -(x - 1)^2$ for x in $[0, 1]$
10. Between $y = x^2(x^3 + 1)^{10}$ and $y = -x(x^2 + 1)^{10}$ for x in $[0, 1]$
11. Enclosed by $y = x$ and $y = x^4$
12. Enclosed by $y = x$ and $y = -x^4$
13. Enclosed by $y = x^3$ and $y = x^4$
14. Enclosed by $y = x$ and $y = x^3$
15. Enclosed by $y = x^2$ and $y = x^4$
16. Enclosed by $y = x^4 - x^2$ and $y = x^2 - x^4$
17. Enclosed by $y = e^x$, $y = 2$, and the y-axis
18. Enclosed by $y = e^{-x}$, $y = 3$, and the y-axis
19. Enclosed by $y = \ln x$, $y = 2 - \ln x$, and $x = 4$
20. Enclosed by $y = \ln x$, $y = 1 - \ln x$, and $x = 4$
21. **T** Enclosed by $y = e^x$ and $y = 2x + 1$
22. **T** Enclosed by $y = e^x$ and $y = x + 2$
23. **T** Enclosed by $y = \ln x$ and $y = x/2 - 1/2$
24. **T** Enclosed by $y = \ln x$ and $y = x - 2$

Calculate the consumers' surplus at the indicated unit price $\overline{p}$ for each of the demand equations in Exercises 25–36.

25. $p = 10 - 2q; \overline{p} = 5$
26. $p = 100 - q; \overline{p} = 20$
27. $p = 100 - 3\sqrt{q}; \overline{p} = 76$
28. $p = 10 - 2q^{1/3}; \overline{p} = 6$
29. $p = 500e^{-2q}; \overline{p} = 100$
30. $p = 100 - e^{0.1q}; \overline{p} = 50$
31. $q = 100 - 2p; \overline{p} = 20$
32. $q = 50 - 3p; \overline{p} = 10$
33. $q = 100 - 0.25p^2; \overline{p} = 10$
34. $q = 20 - 0.05p^2; \overline{p} = 5$
35. $q = 500e^{-0.5p} - 50; \overline{p} = 1$
36. $q = 100 - e^{0.1p}; \overline{p} = 20$

Calculate the producers' surplus for each of the supply equations in Exercises 37–48 at the indicated unit price $\overline{p}$.

37. $p = 10 + 2q; \overline{p} = 20$
38. $p = 100 + q; \overline{p} = 200$
39. $p = 10 + 2q^{1/3}; \overline{p} = 12$
40. $p = 100 + 3\sqrt{q}; \overline{p} = 124$

41. $p = 500e^{0.5q}; \bar{p} = 1000$

42. $p = 100 + e^{0.01q}; \bar{p} = 120$

43. $q = 2p - 50; \bar{p} = 40$

44. $q = 4p - 1000, \bar{p} = 1000$

45. $q = 0.25p^2 - 10; \bar{p} = 10$

46. $q = 0.05p^2 - 20; \bar{p} = 50$

47. $q = 500e^{0.05p} - 50; \bar{p} = 10$

48. $q = 10(e^{0.1p} - 1); \bar{p} = 5$

Applications

49. ***College Tuition*** A study of about 1800 U.S. colleges and universities resulted in the demand equation $q = 9859.39 - 2.17p$, where q is the enrollment at a college or university and p is the average annual tuition (plus fees) it charges.[1] Officials at Enormous State University have developed a policy whereby the number of students it will enroll per year at a tuition level of p dollars is given by $q = 100 + 0.5p$. Find the equilibrium tuition price $\bar{p}$ and the consumers' and producers' surpluses at this tuition level. What is the total social gain at the equilibrium price? (Round $\bar{p}$ to the nearest \$1 and all other answers to the nearest \$1000.)

50. ***Fast Food*** A fast-food outlet finds that the demand equation for its new side dish, "Sweetdough Tidbit," is given by

$$p = \frac{128}{(q+1)^2}$$

where p is the price (in cents) per serving and q is the number of servings that can be sold per hour at this price. At the same time, the franchise is prepared to sell $q = 0.5p - 1$ servings per hour at a price of p cents. Find the equilibrium price $\bar{p}$ and the consumers' and producers' surpluses at this price level. What is the total social gain at the equilibrium price?

51. ***Linear Demand*** Given a linear demand equation $q = -mp + b$ $(m > 0)$, find a formula for the consumers' surplus at a price level of $\bar{p}$ per unit.

52. ***Linear Supply*** Given a linear supply equation of the form $q = mp + b$ $(m > 0)$, find a formula for the producers' surplus at a price level of $\bar{p}$ per unit.

53. ***Foreign Trade*** Annual U.S. imports from China in the years 1988–1998 could be approximated by

$$I = 3.0 + 6.2t \qquad (0 \le t \le 10)$$

billion dollars per year, where t represents time in years since 1988. During the same period, annual U.S. exports to China could be approximated by

$$E = 4.1 + 0.94t \qquad (0 \le t \le 10)$$

billion dollars per year.[2]

a. What does the area between the graphs of I and E over the interval [1, 10] represent?

b. Compute the area in part (a), and interpret your answer.

54. ***On-Line Revenue*** America Online's total revenue (including revenue from subscriptions) was flowing in at a rate of approximately

$$R_t = 0.4t^2 + 0.1t + 2 \qquad (0 \le t \le 2.5)$$

billion dollars per year, where t is time in years since January 1997. During the same period, revenue from subscriptions alone flowed in at a rate of approximately

$$R_s = 0.2t^2 + 0.2t + 1 \qquad (0 \le t \le 2.5)$$

billion dollars per year.[3]

a. What does the area between the graphs of R_t and R_s over the interval [0, 1] represent?

b. Compute the area in part (a), rounded to one significant digit, and interpret your answer.

55. ***Cost*** Snapple Beverage Corp.'s annual revenue R and profit P in millions of dollars for the period 1989 through 1993 can be approximated by

$$R = 16.15e^{0.87t} \qquad \text{and} \qquad P = 3.93e^{t}$$

where t is the number of years since 1989.[4]

a. Use these models to estimate Snapple's accumulated costs for this period.

b. How does the answer to part (a) relate to the area between two curves?

c. What does a comparison of the exponents in the formulas for R and P tell you about Snapple?

56. ***Cost*** Microsoft Corp.'s annual revenue R and profit P in billions of dollars for the period 1986 through 1994 can be approximated by

$$R = 26.27e^{0.37t} \qquad \text{and} \qquad P = 0.6256e^{0.37t}$$

where t is the number of years since 1986.[5]

a. Use these models to estimate Microsoft's accumulated costs for this period.

b. How does the answer to part (a) relate to the area between two curves?

c. What does a comparison of the exponents in the formulas for R and P tell you about Microsoft?

---

[1] Based on a study by A. L. Ostrosky, Jr., and J. V. Koch, as cited in their book, *Introduction to Mathematical Economics* (Waveland Press; Prospect Heights, IL 1979), p. 133

[2] The models are based on linear regression of data from the Census Bureau/*New York Times*, April 8, 1999, p. C1.

[3] The models are based on quadratic regression of data from America Online/*New York Times*, July 4, 1999, p. BU6.

[4] Obtained using exponential regression on graphical data published in the *New York Times*, July 10, 1994, Section 13, p. 1. (Raw data were estimated from a graph.) *Source:* Snapple Beverage Corp./Datastream/Nielsen North America.

[5] Obtained using exponential regression on graphical data published in the *New York Times*, July 18, 1994, p. D1. (Raw data were estimated from a graph.) *Source:* Computer Intelligence Infocorp./Microsoft company records.

57. ***Subsidizing Emission Control*** The marginal cost to the utilities industry of reducing sulfur emissions at several levels of reduction is shown in the following table:[6]

Reduction (millions of tons)	8	10	12
Marginal cost (dollars per ton)	270	360	780

Source: Congress of the United States, Congressional Budget Office, *Curbing Acid Rain: Cost, Budget, and Coal Market Effects* (Washington, DC: Government Printing Office, 1986), xx, xxii, 23, 80.

a. Show by substitution that these data can be modeled by

$$C'(q) = 1{,}000{,}000(41.25q^2 - 697.5q + 3210)$$

where q is the reduction in millions of tons of sulfur and $C'(q)$ is the marginal cost of reducing emissions (in dollars per million tons of reduction).[7]

b. If the government subsidizes sulfur emissions at a rate of \$400 per ton, find the total net cost to the utilities industry of removing 12 million tons of sulfur.

c. The answer to part (b) is represented by the area between two graphs. What are the equations of these graphs?

58. ***Variable Emissions Subsidies*** Refer back to the marginal cost equation in Exercise 57. Assume that the government subsidy for sulfur emissions is given by the formula

$$S'(q) = 500{,}000{,}000q$$

where q is the level of reduction in millions of tons and $S'(q)$ is the marginal subsidy in dollars per million tons.

a. Find the total net cost to the utilities industry of removing 12 million tons of sulfur. Interpret your answer.

b. The answer to part (a) is represented by the net area between two graphs. What are the equations of these graphs?

Communication and Reasoning Exercises

59. ***Foreign Trade*** The following graph shows Canada's monthly exports and imports for the year ending February 1, 1994. What does the area between the export and import curves represent?

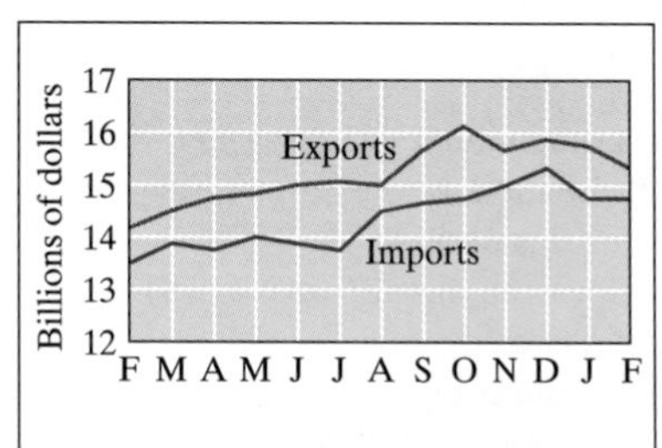

Source: Statistics Canada, *Globe and Mail,* April 20, 1994, p. B3.

60. ***Foreign Trade*** The following graph shows a fictitious country's monthly exports and imports for the year ending February 1, 2003. What does the total area enclosed by the export and import curves represent? What does the definite integral of the difference, Exports − Imports, represent?

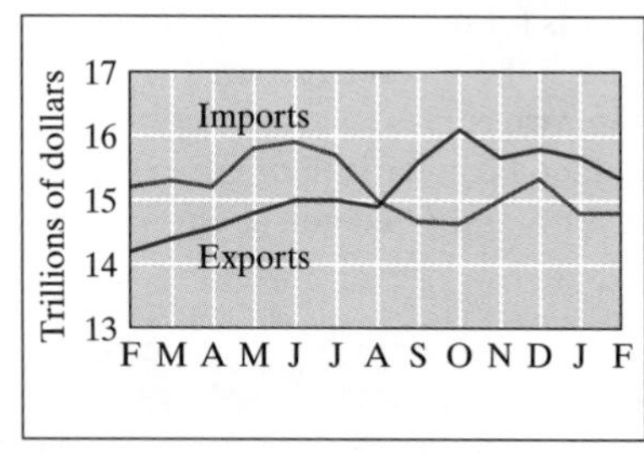

61. What is wrong with the following claim: "I own 100 units of Abbot Laboratories, Inc. stock that originally cost me \$22 per share. My net income from this investment over the period March 5–April 27, 1993, is represented by the area between the stock price curve and the purchase price curve as shown on the following graph."

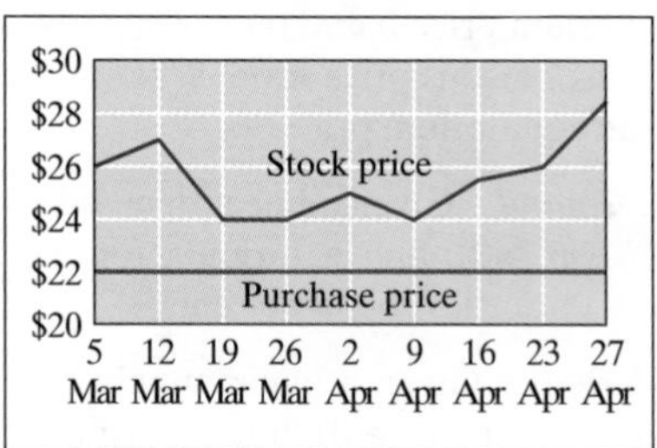

Source: News reports, *Chicago Tribune,* April 28, 1993, Section 3, p. 6.

62. Is it always true that a company's total profit over a 1-year period is represented by the area between the daily revenue and daily cost curves? Illustrate your answer by means of an example.

[6] These figures were produced in a computerized study of reducing sulfur emissions from the 1980 level by the given amounts.

[7] You can obtain this equation directly from the data (as we did) by assuming that $C'(q) = aq^2 + bq + c$, substituting the values of $C'(q)$ for the given values of q, and then solving for a, b, and c.

7.3 Averages and Moving Averages

Averages

We all know how to find the average of a collection of numbers. If we want to find the average of, say, 20 numbers, we simply add them up and divide the sum by 20. More generally, if we want to find the **average,** or **mean,** of the n numbers $y_1, y_2, y_3, \ldots, y_n$, we add them up and divide the sum by n.

Average or Mean of a Set of Values

$$\bar{y} = \frac{y_1 + y_2 + \cdots + y_n}{n}$$

But, we also use the word *average* in other senses. For example, we speak of the average speed of a car during a trip.

example 1

Average Speed

Over the course of 2 hours, my speed varied from 50 mph to 60 mph, following the function $v(t) = 50 + 2.5t^2$, where $0 \le t \le 2$. What was my average speed over those 2 hours?

Solution

Recall that average speed is simply the total distance traveled divided by the time it took. Recall, also, that we can find the distance traveled by integrating the speed:

$$\begin{aligned}\text{Distance traveled} &= \int_0^2 v(t)\,dt \\ &= \int_0^2 (50 + 2.5t^2)\,dt \\ &= \left[50t + \frac{2.5}{3}t^3\right]_0^2 = 100 + \frac{20}{3} \approx 106.67 \text{ miles}\end{aligned}$$

It took 2 hours to travel this distance, so the average speed was

$$\text{Average speed} \approx \frac{106.67}{2} \approx 53.3 \text{ mph}$$

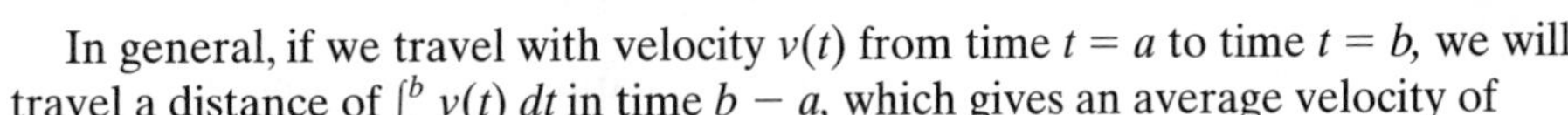

In general, if we travel with velocity $v(t)$ from time $t = a$ to time $t = b$, we will travel a distance of $\int_a^b v(t)\,dt$ in time $b - a$, which gives an average velocity of

$$\text{Average velocity} = \frac{1}{b - a}\int_a^b v(t)\,dt$$

Thinking of this calculation as finding the average value of the velocity function, we generalize and make the following definition.

Average Value of a Function

The **average,** or **mean,** of a function $f(x)$ on an interval $[a, b]$ is

$$\bar{f} = \frac{1}{b - a}\int_a^b f(x)\,dx$$

Quick Example

The average of $f(x) = x$ on $[1, 5]$ is

$$\bar{f} = \frac{1}{b-a}\int_a^b f(x)\,dx$$
$$= \frac{1}{5-1}\int_1^5 x\,dx$$
$$= \frac{1}{4}\left[\frac{x^2}{2}\right]_1^5 = \frac{1}{4}\left(\frac{25}{2} - \frac{1}{2}\right) = 3$$

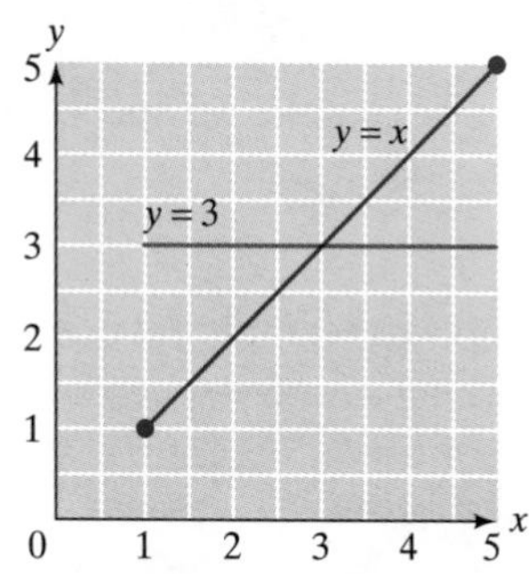

Figure 13

Note: The average of a function has a geometric interpretation. Referring to the Quick Example above, we can compare the graph of $f(x)$ with the graph of $y = 3$, both over the interval $[1, 5]$ (Figure 13). We can find the area under the graph of $f(x)$ by geometry or by calculus; it is 12. The area in the rectangle under $y = 3$ is also 12.

In general, the average $\bar{f}$ of a positive function over the interval $[a, b]$ gives the height of the rectangle over the interval $[a, b]$ that has the same area as the area under the graph of $f(x)$. The equality of these areas follows from the equation

$$(b-a)\bar{f} = \int_a^b f(x)\,dx$$

example 2

Average Balance

A savings account at the People's Credit Union pays 3% interest, compounded continuously, and at the end of the year you get a bonus of 1% of the average balance in the account during the year. If you deposit \$10,000 at the beginning of the year, how much interest and how large a bonus will you get?

Solution

We can use the continuous compound interest formula to calculate the amount of money you have in the account at time t:

$$A(t) = 10{,}000e^{0.03t}$$

where t is measured in years. At the end of 1 year the account will have

$$A(1) = \$10{,}304.55$$

so you will have earned \$304.55 interest. To compute the bonus, we need to find the average amount in the account, which is the average of $A(t)$ over the interval $[0, 1]$. Thus,

$$\bar{A} = \frac{1}{b-a}\int_a^b A(t)\,dt$$
$$= \frac{1}{1-0}\int_0^1 10{,}000e^{0.03t}\,dt$$
$$= \frac{10{,}000}{0.03}\left[e^{0.03t}\right]_0^1 \approx \$10{,}151.51$$

The bonus is 1% of this, or \$101.52.

Before we go on . . . The 1% bonus was one-third of the total interest. Why did this happen? What fraction of the total interest would the bonus be if the interest rate was 4%, 5%, or 10%?

Moving Averages

Suppose you follow the performance of a company's stock by recording the daily closing prices. The graph of these prices may seem jagged or "jittery" due to almost random day-to-day fluctuations. In order to see any trends, you would like a way to "smooth out" these data. The **moving average** is one common way to do that.

example 3

Stock Prices

The following table shows Colossal Conglomerate Corp.'s closing stock prices for 20 consecutive trading days:

Day	1	2	3	4	5	6	7	8	9	10
Price	20	22	21	24	24	23	25	26	20	24
Day	11	12	13	14	15	16	17	18	19	20
Price	26	26	25	27	28	27	29	27	25	24

Plot these prices and also the 5-day moving average.

Solution

The 5-day moving average is the average of each day's price together with the prices of the preceding 4 days. We can compute the 5-day moving averages starting on the fifth day. We get these numbers:

Day	1	2	3	4	5	6	7	8	9	10
Moving average					22.2	22.8	23.4	24.4	23.6	23.6
Day	11	12	13	14	15	16	17	18	19	20
Moving average	24.2	24.4	24.2	25.6	26.4	26.6	27.2	27.6	27.2	26.4

The closing stock prices and moving averages are plotted in Figure 14.

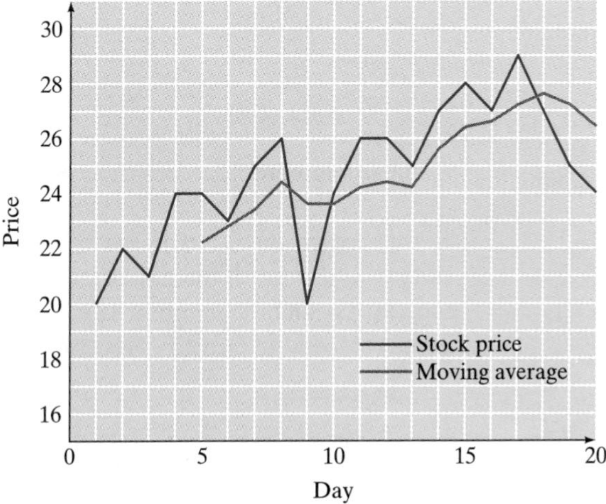

Figure 14

As you can see, the moving average is less volatile than the closing price. Since the moving average averages the stock's performance over 5 days, a single day's fluctuation is smoothed out. Look at day 9 in particular. The moving average also tends to lag behind the actual performance because it takes past history into account. Look at the downturns at days 6 and 18 in particular.

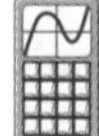

Graphing Calculator

To automate this calculation on a TI-83, first use

seq(X,X,1,20)$\to L_1$ [2nd] [STAT] → OPS → 5

to enter the sequence of numbers 1 through 20 into the list L_1, representing the trading days. Then, using the list editor accessible through the [STAT] menu, enter the daily stock prices in list L_2. You can now calculate the list of 5-day moving averages using the following command:

seq((L_2(X)+L_2(X-1)+L_2(X-2)+L_2(X-3)+L_2(X-4))/5,X,5,20)$\to L_3$

This has the effect of putting the moving averages into elements 1 through 15 of list L_3. If you wish to plot the moving average on the same graph as the daily prices, you will want the averages in L_3 to match up with the prices in L_2. One way to do this is to put four more entries at the beginning of L_3—say, copies of the first four entries of L_2. The following command accomplishes this:

augment(seq(L_2(X),X,1,4),L_3)$\to L_3$ [2nd] [STAT] → OPS → 9

You can now graph the prices and moving averages by creating an xyLine scatter plot through the [STAT PLOT] menu, with L_1 being the Xlist and L_2 being the Ylist for Plot1, and L_1 being the Xlist and L_3 the Ylist for Plot2.

Spreadsheet

In Excel you can compute the moving averages in a column next to the daily prices, as shown here:

	A	B	C
1	1	20	
2	2	22	
3	3	21	
4	4	24	
5	5	24	=AVERAGE(B1:B5)
6	6	23	
20	20	24	

You can then create a facsimile of Figure 14 using a scatter plot.

Before we go on . . . The period of 5 days is arbitrary. Using a longer period of time would smooth the data more but increase the lag. For data used as economic indicators, such as housing prices or retail sales, it is common to compute the 1-year moving average to smooth out seasonal variations.

It is also sometimes useful to compute moving averages of continuous functions. We may want to do this if we use a mathematical model of a large collection of data. Also, some physical systems have the effect of converting an input function (an electrical signal, for example) into its moving average. By an ***n*-unit**

moving average of a function $f(x)$, we mean the function $\bar{f}$ for which $\bar{f}(x)$ is the average of the value of $f(x)$ on $[x - n, x]$. Using the formula for the average of a function, we get the following formula.

***n*-Unit Moving Average of a Function**

The n-unit moving average of a function f is

$$\bar{f}(x) = \frac{1}{n}\int_{x-n}^{x} f(t)\, dt$$

Quick Example

The 2-unit moving average of $f(x) = x^2$ is

$$\bar{f}(x) = \frac{1}{2}\int_{x-2}^{x} t^2\, dt = \frac{1}{6}\left[t^3\right]_{x-2}^{x} = x^2 - 2x + \frac{4}{3}$$

The graphs of $f(x)$ and $\bar{f}(x)$ are shown in the figure.

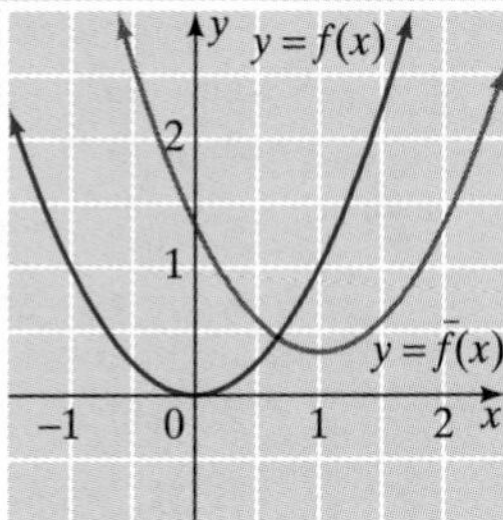

example 4

T Moving Average via Technology

Plot the 3-unit moving average of

$$f(x) = \frac{x}{1 + |x|} \qquad (-5 \le x \le 5)$$

Solution

This function is a little tricky to integrate analytically because $|x|$ is defined differently for positive and negative values of x, so instead we use technology to approximate the integral. Here is how we could do it using a TI-83 graphing calculator. We enter the following:

```
Y1 = X/(1+abs(X))
Y2 = (1/3)fnInt(Y1(T),T,X-3,X)
```

The `Y1` entry is $f(x)$, and the `Y2` entry is a numerical approximation of the 3-unit moving average of $f(x)$:

$$\bar{f}(x) = \frac{1}{3}\int_{x-3}^{x} \frac{t}{1 + |t|}\, dt$$

We set the viewing window ranges to $-5 \le x \le 5$ and $-1 \le y \le 1$ and plot these curves. (Be patient—the calculator has to do a numerical integration to obtain each point on the graph of the moving average.) The result is shown in Figure 15.

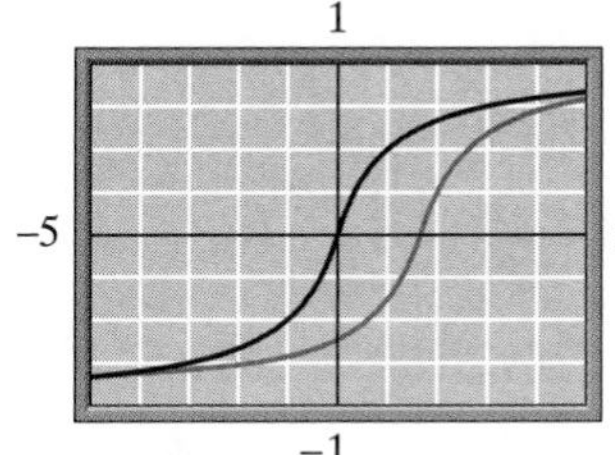

Figure 15

Before we go on . . . Of course, once we've entered `Y1` and `Y2` as above, we can use the calculator to evaluate the moving average at any value of x. For instance, to calculate $\bar{f}(1.2)$, we enter `Y2(1.2)` on the home screen and find

$$\bar{f}(1.2) \approx -0.1196$$

7.3 exercises

Find the averages of the functions in Exercises 1–8 over the given intervals. Plot each function and its average on the same graph (as in the note after the Quick Example in this section).

1. $f(x) = x^3$ over $[0, 2]$ **2.** $f(x) = x^3$ over $[-1, 1]$

3. $f(x) = x^3 - x$ over $[0, 2]$ **4.** $f(x) = x^3 - x$ over $[0, 1]$

5. $f(x) = e^{-x}$ over $[0, 2]$ **6.** $f(x) = e^x$ over $[-1, 1]$

Δ **7.** $f(x) = \sin x$ over $[0, \pi]$

Δ **8.** $f(x) = \cos(2x)$ over $[0, \pi/4]$

In Exercises 9 and 10, complete the given table with the values of the 3-unit moving average of the given function.

9.

x	0	1	2	3	4	5	6	7
$r(x)$	3	5	10	3	2	5	6	7
$\bar{r}(x)$								

10.

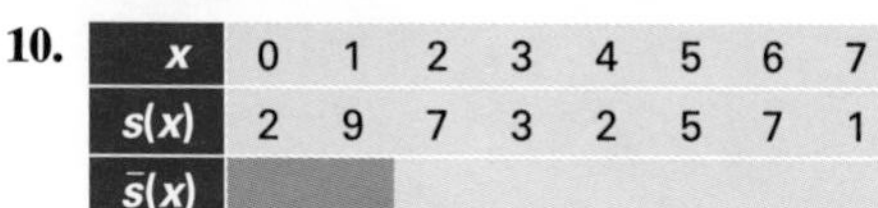

x	0	1	2	3	4	5	6	7
$s(x)$	2	9	7	3	2	5	7	1
$\bar{s}(x)$								

In Exercises 11 and 12, some values of a function and its 3-unit moving average are given. Supply the missing information.

11.

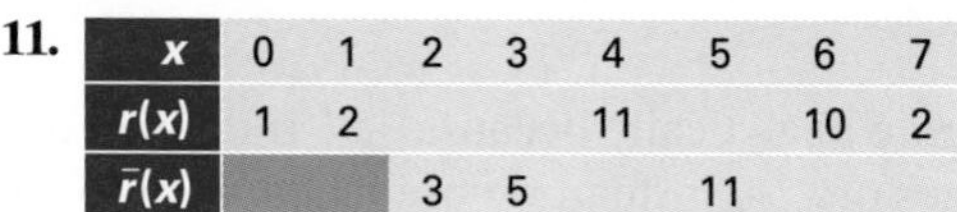

x	0	1	2	3	4	5	6	7
$r(x)$	1	2			11		10	2
$\bar{r}(x)$			3	5		11		

12.

x	0	1	2	3	4	5	6	7
$s(x)$	1	5		1				
$\bar{s}(x)$			5		5	2	3	2

Calculate the 5-unit moving average of each function in Exercises 13–22. Plot each function and its moving average on the same graph, as in Example 4. (You may use graphing technology for these plots, but you should compute the moving averages analytically.)

13. $f(x) = x^3$ **14.** $f(x) = x^3 - x$

15. $f(x) = x^{2/3}$ **16.** $f(x) = x^{2/3} + x$

17. $f(x) = e^{0.5x}$ **18.** $f(x) = e^{-0.02x}$

19. $f(x) = \sqrt{x}$ **20.** $f(x) = x^{1/3}$

Δ **21.** $f(x) = \cos(\pi x/10)$ Δ **22.** $f(x) = \sin(x/100)$

In Exercises 23–26, use graphing technology to plot the given functions together with their 3-unit moving averages.

T **23.** $f(x) = \dfrac{10x}{1 + 5|x|}$ T **24.** $f(x) = \dfrac{1}{1 + e^x}$

T **25.** $f(x) = \ln(1 + x^2)$ T **26.** $f(x) = e^{1-x^2}$

Applications

27. *Spending on Corrections* The following table shows the annual spending by all states in the United States on corrections ($t = 0$ represents the year 1990):

Year, t	0	1	2	3
Spending (billions)	\$16	\$18	\$18	\$20
Year, t	4	5	6	7
Spending (billions)	\$22	\$26	\$28	\$30

Source: National Association of State Budget Officers/*New York Times,* February 28, 1999, p. A1. Data are rounded.

What was the average annual expenditure on corrections for the period January 1992–December 1997?

28. *Profit* After-tax profits of South African Breweries (SAB) from 1991 through 1997 are shown in the following table:

Year, t	1 (1991)	2	3	4
Profit (millions)	\$350	\$360	\$375	\$400
Year, t	5	6	7 (1997)	
Profit (millions)	\$500	\$580	\$600	

Source: Company reports/Bloomberg Financial Markets/*New York Times,* August 27, 1997, p. D4.

What was the average after-tax profit for the period January 1992–December 1996?

29. ***Venture Capital*** Venture capital firms raise money to assist promising entrepreneurs until they can sell shares to the public. During the period 1980–1994, the number of venture capital firms could be approximated by

$$N(t) = 31t + 145 \qquad (0 \le t \le 14)$$

where t is time in years since 1980.[1] What was the average number of venture capital firms during the given period?

30. ***Venture Capital*** During the period 1980–1994, the amount of money that venture capital firms control could be approximated by

$$R(t) = 2.4t + 7.3 \text{ billion dollars} \qquad (0 \le t \le 14)$$

where t is time in years since 1980. What was the average amount of money controlled by venture capital firms during the given period?

31. ***Investments*** If you invest \$10,000 at 8% interest compounded continuously, what is the average amount in your account over 1 year?

32. ***Investments*** If you invest \$10,000 at 12% interest compounded continuously, what is the average amount in your account over 1 year?

33. ***Average Balance*** Suppose you have an account (paying no interest) into which you deposit \$3000 at the beginning of each month. You withdraw money continuously so that the amount in the account decreases linearly to zero by the end of the month. Find the average amount in the account over a period of several months. (Assume that the account starts at \$0 at $t = 0$ months.)

34. ***Average Balance*** Suppose you have an account (paying no interest) into which you deposit \$4000 at the beginning of each month. You withdraw \$3000 during the course of each month in such a way that the amount decreases linearly. Find the average amount in the account in the first 2 months. (Assume that the account starts at \$0 at $t = 0$ months.)

35. ***Spending on Corrections*** Refer back to Exercise 27. Complete the following table by computing the 4-year moving average of spending on corrections:

Year, t	0	1	2	3
Spending (billions)	\$16	\$18	\$18	\$20
Moving average (billions)				
Year, t	4	5	6	7
Spending (billions)	\$22	\$26	\$28	\$30
Moving average (billions)				

Was the moving average increasing linearly over the given period? Explain.

36. ***Profit*** Refer back to Exercise 28. Complete the following table by computing the 4-year moving average of South African Breweries' annual profits:

Year, t	1	2	3	4
Profit (millions)	\$350	\$360	\$375	\$400
Moving average (millions)				
Year, t	5	6	7	
Profit (millions)	\$500	\$580	\$600	
Moving average (millions)				

Is the average of the moving averages the same as the overall average? Explain.

T 37. ***Vacation Spending*** The following table shows approximate annual tourist expenditures, in millions of dollars, in Bermuda for the years 1976–1998:

1976	1977	1978	1979	1980	1981	1982	1983
150	180	180	200	270	300	280	310
1984	1985	1986	1987	1988	1989	1990	1991
340	350	350	430	460	450	470	490
1992	1993	1994	1995	1996	1997	1998	
450	450	520	530	470	470	470	

Source: Bermuda Ministry of Finance/*New York Times,* April 28, 1999, p. C1.

a. Use technology to compute and plot the 5-year moving average of these data.
b. The graph of the moving average should appear almost linear over the range 1981–1991. Use the 1981 and 1991 moving average figures to give an estimate of the rate of change of tourist expenditures in Bermuda for the period 1981–1991. Interpret the result.

T 38. ***Business Spending*** The following table shows approximate annual international business expenditures, in millions of dollars, in Bermuda for the years 1976–1998:

1976	1977	1978	1979	1980	1981	1982	1983
60	90	60	70	100	120	150	160
1984	1985	1986	1987	1988	1989	1990	1991
160	200	240	260	260	310	320	330
1992	1993	1994	1995	1996	1997	1998	
340	360	400	440	500	600	760	

Source: Bermuda Ministry of Finance/*New York Times,* April 28, 1999, p. C1.

a. Use technology to compute and plot the 5-year moving average of these data.
b. The graph of the moving average should appear almost linear over the range 1982–1990. Use the 1982 and 1990 moving average figures to give an estimate of

[1] The (approximate) models in Exercises 29 and 30 are based on a linear regression of data from Venture Economics Information Services, Boston/*New York Times,* May 25, 1995, p. D5.

the rate of change of tourist expenditures in Bermuda for the period 1982–1990. Interpret the result.

39. ***Sport Utility Vehicles*** The average weight of a sport utility vehicle (SUV) can be approximated by

$$W = 3t^2 - 90t + 4200 \qquad (5 \le t \le 27)$$

where t is time in years ($t = 0$ represents 1970) and W is the average weight in pounds of an SUV produced in year t.[2]

a. Compute the average weight of an SUV over the period 1975–1997, to the nearest 10 pounds.
b. Compute the 2-year moving average of W. (You need not simplify the answer.)
c. Without simplifying the answer in part (b), say what kind of function the moving average is.

40. ***Sedans*** The average weight of a sedan can be approximated by

$$W = 6t^2 - 240t + 4800 \qquad (5 \le t \le 27)$$

where t is time in years ($t = 0$ represents 1970) and W is the average weight of a sedan in pounds. Repeat Exercise 39 as applied to sedans.

41. ***Expansion of Fast-Food Outlets*** The following chart shows the approximate number of McDonald's® restaurants in the United States at year-end, in thousands:

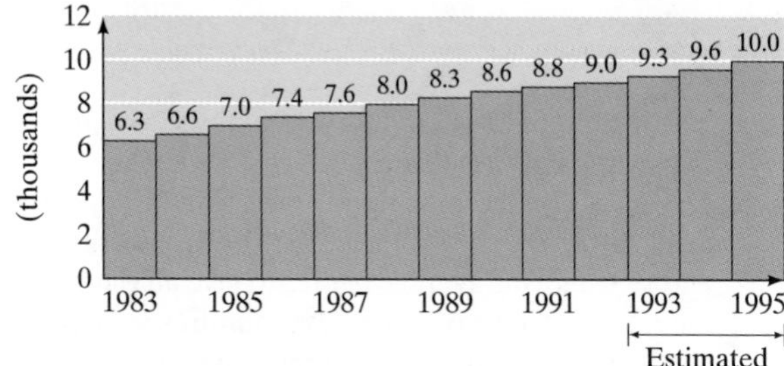

Source: Technomics/U.S. Department of Commerce/*New York Times*, January 9, 1994, p. F5.

a. Use the 1985 and 1995 figures to model these data with a linear function.
b. Find the 4-year moving average of your model.
c. What can you say about the slope of the moving average?
T d. Use graphing technology to plot the model and the 4-year moving average.

42. ***Fast-Food Customers*** The following graph shows the declining number of Americans (year-end figures in thousands) per fast-food outlet in the United States:

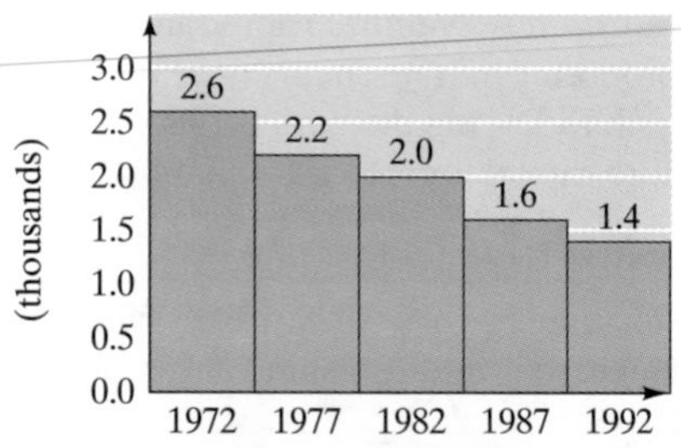

Source: Technomics/U.S. Department of Commerce/*New York Times*, January 9, 1994, p. F5.

a. Use the 1972 and 1992 figures (2.6 Americans per store and 1.4 Americans per store, respectively) to model these data with a linear function.
b. Calculate the 5-year moving average of your model.
c. What can you say about the slope of the moving average?
T d. Use graphing technology to plot the model and the 5-year moving average.

43. ***Moving Average of a Linear Function*** Find a formula for the a-unit moving average of a general linear function $f(x) = mx + b$.

44. ***Moving Average of an Exponential Function*** Find a formula for the a-unit moving average of a general exponential function $f(x) = Ae^{kx}$.

45. ***Market Share*** The following graph shows the market share of McDonald's® restaurants in the fast-food business in the United States:

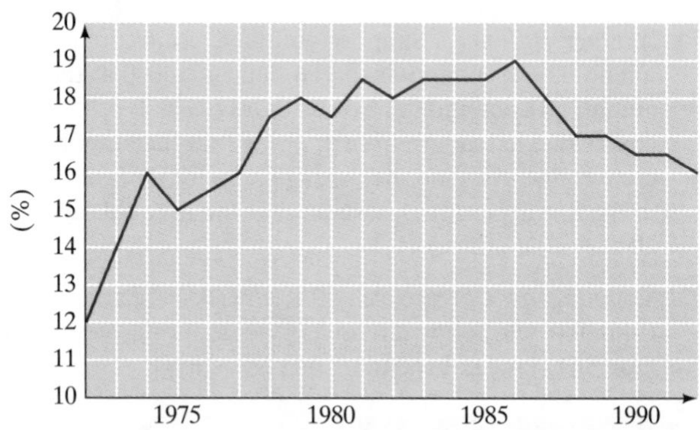

Source: Technomics/U.S. Department of Commerce/*New York Times*, January 9, 1994, p. F5.

a. With $t = 0$ representing 1972, obtain a quadratic model

$$p(t) = at^2 + bt + c$$

that approximates the percentage market share enjoyed by McDonald's. (Use the data for $t = 0$, $t = 10$, and $t = 20$, rounded to the nearest percentage.)
b. Use your model to estimate McDonald's average market share over the period shown in the graph.

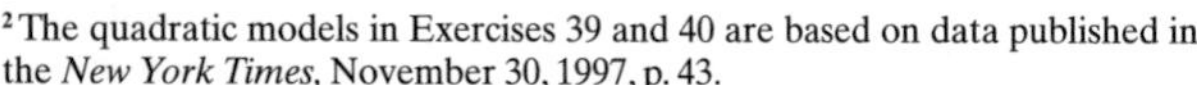

[2] The quadratic models in Exercises 39 and 40 are based on data published in the *New York Times*, November 30, 1997, p. 43.

c. Use your model to estimate McDonald's average market share for the 3 years ending in the year 2000.

46. ***Growth in Prozac Sales*** The following graph shows the annual growth in new prescriptions for Prozac® since its introduction in 1988.

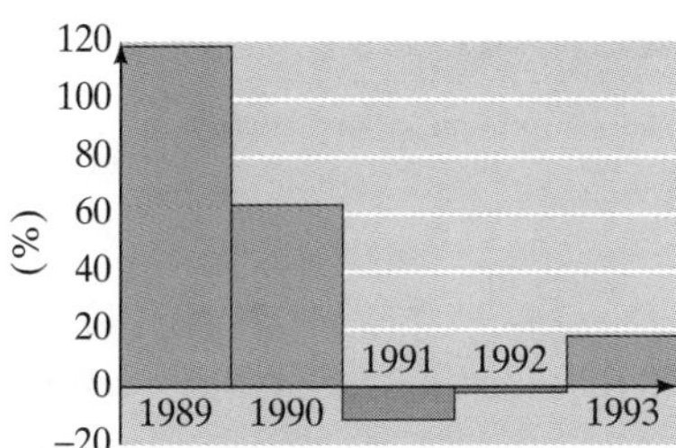

Sources: Mehta and Isaly, Smith Barney Shearson; Lehman Brothers/*New York Times,* January 9, 1994, p. F7.

a. With $t = 0$ representing 1989, obtain a quadratic model

$$p(t) = at^2 + bt + c$$

that approximates the percentage growth in Prozac prescriptions. (Use the data for $t = 0$, $t = 2$, and $t = 4$, rounded to the nearest 10%.)

b. Use your model to estimate the average annual rate of growth in Prozac prescriptions over the period shown in the graph.

c. Use your model to predict the average rate of growth of Prozac prescriptions for the 3 years ending in 1996.

47. ***Fair Weather***[3] The Cancun Royal Hotel's advertising brochure features the following chart, showing the year-round temperature:

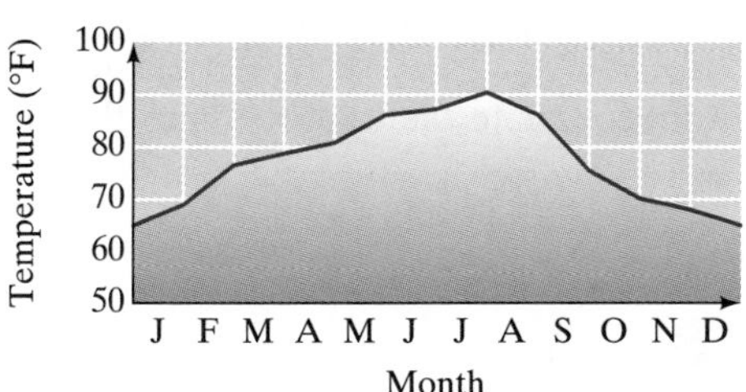

a. Estimate and plot the 2- and 3-month moving averages. (Use graphing technology, if available.)

b. What can you say about the 24-month moving average?

c. Comment on the limitations of a quadratic model for these data.

48. ***Foul Weather*** Repeat Exercise 47 using the following data from the Tough Traveler Lodge in Frigidville:

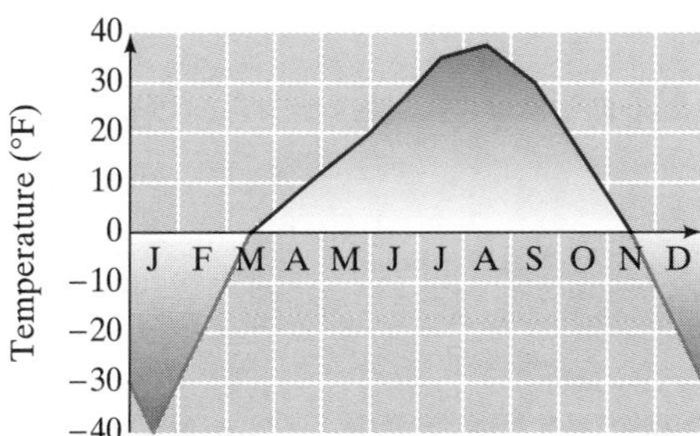

Δ 49. ***Electrical Current*** The typical voltage V supplied by an electrical outlet in the United States is given by

$$V(t) = 165 \cos(120\pi t)$$

where t is time in seconds.

a. Find the average voltage over the interval $\left[0, \frac{1}{6}\right]$. How many times does the voltage reach a maximum in 1 second? (This is referred to as the number of **cycles per second.**)

b. Plot the function $S(t) = (V(t))^2$ over the interval $\left[0, \frac{1}{6}\right]$.

c. The **root mean square** voltage is given by the formula

$$V_{rms} = \sqrt{\bar{S}}$$

where $\bar{S}$ is the average value of $S(t)$ over one cycle. Estimate V_{rms}.

Δ 50. ***Tides*** The depth of water at my favorite surfing spot varies from 5 feet to 15 feet, depending on the tide. Last Sunday high tide occurred at 5:00 A.M. and the next high tide occurred at 6:30 P.M. Use the cosine function to model the depth of water as a function of time t in hours since midnight on Sunday morning. What was the average depth of the water between 10:00 A.M. and 2:00 P.M?

Communication and Reasoning Exercises

51. What property does a (nonconstant) function have if its average value over an interval is zero? Sketch a graph of such a function.

52. Can the average value of a function f on an interval be greater than its value at every point in that interval? Explain.

53. Explain why it is sometimes more useful to consider the moving average of a stock price rather than the stock price itself.

54. Your monthly salary has been increasing steadily for the past year, and your average monthly salary over the past year was x dollars. Would you have earned more money if you were paid x dollars per month? Explain your answer.

55. Criticize the following claim: The average value of a function on an interval is midway between its highest and lowest values.

[3] Inspired by an exercise in *Calculus* by D. Hughes-Hallett et al. New York: John Wiley & Sons, Inc., 1994.

56. Your manager tells you that 12-month moving averages give at least as much information as shorter-term moving averages and very often more. How would you argue that he is wrong?

57. Which of the following most closely approximates the original function: **a.** its 10-unit moving average, **b.** its 1-unit moving average, or **c.** its 0.8-unit moving average. Explain your answer.

58. Is an increasing function larger or smaller than its 1-unit moving average? Explain.

7.4 Continuous Income Streams

A company with a high volume of sales receives money almost continuously. For purposes of calculation, it is convenient to assume that the company literally does receive money continuously. In such a case, we have a function $R(t)$ that represents the rate at which money is being received by the company at time t.

example 1

Continuous Income

An ice cream store's business peaks in late summer. The store's summer revenue is approximated by

$$R(t) = 300 + 4.5t - 0.05t^2 \text{ dollars per day} \qquad (0 \le t \le 92)$$

where t is measured in days after June 1. What is its total revenue for the months of June, July, and August?

Solution

Let us approximate the total revenue by breaking up the interval [0, 92] representing the 3 months into n subintervals $[t_{k-1}, t_k]$, each with length Δt. In the interval $[t_{k-1}, t_k]$ the store receives money at a rate of approximately $R(t_{k-1})$ dollars per day for Δt days, so it receives a total of $R(t_{k-1})\,\Delta t$ dollars. Over the whole summer, then, the store receives approximately

$$R(t_0)\,\Delta t + R(t_1)\,\Delta t + \cdots + R(t_{n-1})\,\Delta t \text{ dollars}$$

As we let n become large to better approximate the total revenue, this Riemann sum approaches the integral

$$\text{Total revenue} = \int_0^{92} R(t)\,dt$$

Substituting the function we are given, we get

$$\text{Total revenue} = \int_0^{92} (300 + 4.5t - 0.05t^2)\,dt$$

$$= \left[300t + 2.25t^2 - \frac{0.05}{3}t^3\right]_0^{92} \approx \$33{,}666$$

Before we go on . . . We could approach this calculation another way: $R(t) = S'(t)$, where $S(t)$ is the total revenue earned up to day t. By the Fundamental Theorem of Calculus,

$$\text{Total revenue} = S(92) - S(0) = \int_0^{92} R(t)\,dt$$

We did the calculation using Riemann sums mainly as practice for the next example.

Generalizing this example, we can give a formula.

Total Value of a Continuous Income Stream

If the rate of receipt of income is $R(t)$ dollars per unit of time, then the total income received from time $t = a$ to $t = b$ is

$$\text{Total value} = TV = \int_a^b R(t)\,dt$$

example 2

Future Value

Suppose that the ice cream store in the preceding example deposits its receipts in an account paying 5% interest per year compounded continuously. How much money will it have in its account at the end of August?

Solution

Now we have to take into account not just the revenue but also the interest it earns in the account. Again, we break the interval [0, 92] into n subintervals. During the interval $[t_{k-1}, t_k]$, approximately $R(t_{k-1})\,\Delta t$ dollars are deposited in the account. That money will earn interest until the end of August, a period of $92 - t_{k-1}$ days, or $(92 - t_{k-1})/365$ days. Remembering that 5% is the *annual* interest rate, we know the formula for continuous compounding tells us that by the end of August those $R(t_{k-1})\,\Delta t$ dollars will have turned into

$$R(t_{k-1})\,\Delta t e^{0.05(92-t_{k-1})/365} = R(t_{k-1})e^{0.05(92-t_{k-1})/365}\,\Delta t \text{ dollars}$$

Adding up the contributions from each subinterval, we see that the total in the account at the end of August will be approximately

$$R(t_0)e^{0.05(92-t_0)/365}\,\Delta t + R(t_1)e^{0.05(92-t_1)/365}\,\Delta t + \cdots + R(t_{n-1})e^{0.05(92-t_{n-1})/365}\,\Delta t$$

This is a Riemann sum; as n gets large, the sum approaches the integral

$$\text{Future value} = FV = \int_0^{92} R(t)e^{0.05(92-t)/365}\,dt$$

Substituting $R(t) = 300 + 4.5t - 0.05t^2$, we obtain

$$FV = \int_0^{92} (300 + 4.5t - 0.05t^2)e^{0.05(92-t)/365}\,dt$$

$$\approx \$33{,}880$$ Using technology or integration by parts.

Before we go on . . . The interest earned in the account was fairly small (compare this answer to the preceding example). Not only was the money in the account for only 3 months, but much of it was put in the account toward the end of that period, and so had very little time to earn interest.

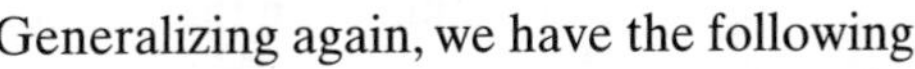

Generalizing again, we have the following.

Future Value of a Continuous Income Stream

If the rate of receipt of income from time $t = a$ to $t = b$ is $R(t)$ dollars per unit of time and the income is deposited as it is received in an account paying interest

r per unit of time, compounded continuously, then the amount of money in the account at time $t = b$ is

$$\text{Future value} = FV = \int_a^b R(t)e^{r(b-t)}\,dt$$

example 3

Present Value

You are thinking of buying the ice cream store discussed in the preceding two examples. What is its income stream worth to you on June 1?

Solution

If we assume that you have access to the same account paying 5% per year compounded continuously, then the value of the income stream on June 1 is the amount of money that, if deposited on June 1, would give you the same future value as the income stream will. If we let PV denote this "present value," its value after 92 days will be

$$PVe^{0.05 \times 92/365}$$

If we equate this with the future value of the income stream, we get

$$PVe^{0.05 \times 92/365} = e^{0.05 \times 92/365} \int_0^{92} R(t)e^{-0.05t/365}\,dt$$

so

$$PV = \int_0^{92} R(t)e^{-0.05t/365}\,dt$$

Substituting the formula for $R(t)$ and integrating using technology or integration by parts, we get

$$PV \approx \$33{,}455$$

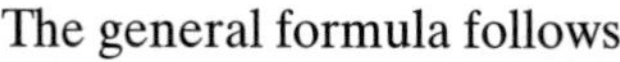

The general formula follows.

Present Value of a Continuous Income Stream

If the rate of receipt of income from time $t = a$ to $t = b$ is $R(t)$ dollars per unit of time and the income is deposited as it is received in an account paying interest r per unit of time, compounded continuously, then the value of the income stream at time $t = a$ is

$$\text{Present value} = PV = \int_a^b R(t)e^{r(a-t)}\,dt$$

We can derive this formula from the relation

$$FV = PVe^{r(b-a)}$$

because the present value is the amount that would have to be deposited at time $t = a$ to give a future value of FV at time $t = b$.

These formulas are more general than we've said. They still work when $R(t) < 0$ if we interpret negative values as money flowing *out* rather than in. That is, we can use these formulas for income we receive or for payments that we make, or for situations where we sometimes receive money and sometimes pay it out. These formulas can also be used for flows of quantities other than money. For example, if we use an exponential model for population growth and we let $R(t)$ represent the rate of immigration $[R(t) > 0]$ or emigration $[R(t) < 0]$, then the future value formula gives the future population.

7.4 exercises

In Exercises 1–8, find the total value of the given income stream and also find its future value (at the end of the given interval) using the given interest rate.

1. $R(t) = 30{,}000, 0 \le t \le 10$, at 7%

2. $R(t) = 40{,}000, 0 \le t \le 5$, at 10%

3. $R(t) = 30{,}000 + 1000t, 0 \le t \le 10$, at 7%

4. $R(t) = 40{,}000 + 2000t, 0 \le t \le 5$, at 10%

5. $R(t) = 30{,}000e^{0.05t}, 0 \le t \le 10$, at 7%

6. $R(t) = 40{,}000e^{0.04t}, 0 \le t \le 5$, at 10%

Δ **7.** $R(t) = 100{,}000 \sin(2\pi t), 0 \le t \le 1$, at 7%

Δ **8.** $R(t) = 100{,}000 \sin(\pi t), 0 \le t \le 2$, at 10%

In Exercises 9–16, find the total value of the given income stream and also find its present value (at the beginning of the given interval) using the given interest rate.

9. $R(t) = 20{,}000, 0 \le t \le 5$, at 8%

10. $R(t) = 50{,}000, 0 \le t \le 10$, at 5%

11. $R(t) = 20{,}000 + 1000t, 0 \le t \le 5$, at 8%

12. $R(t) = 50{,}000 + 2000t, 0 \le t \le 10$, at 5%

13. $R(t) = 20{,}000e^{0.03t}, 0 \le t \le 5$, at 8%

14. $R(t) = 50{,}000e^{0.06t}, 0 \le t \le 10$, at 5%

Δ **15.** $R(t) = 100{,}000 \cos(2\pi t), 0 \le t \le 1$, at 8%

Δ **16.** $R(t) = 100{,}000 \cos(\pi t), 0 \le t \le 2$, at 5%

Applications

17. *Revenue* The rate of receipt of revenue earned by South African Breweries (SAB) from 1991 to 1997 can be modeled by the function[1]

$$R(t) = -0.13t^2 + 0.54t + 5.22 \text{ billion dollars per year} \quad (1 \le t \le 7)$$

where t is time in years since January 1990. (Thus, $t = 1$ represents January 1991.) Estimate SAB's total revenue from January 1991 through January 1997.

18. *Profit* The rate of receipt of after-tax profit earned by South African Breweries (SAB) from 1991 to 1997 can be modeled by the function

$$P(t) = 7.0t^2 - 9.5t + 350 \text{ million dollars per year} \quad (1 \le t \le 7)$$

where t is time in years since January 1990. (Thus, $t = 1$ represents January 1991.) Use this as a model of a continuous income stream to estimate SAB's total after-tax profits from January 1991 through January 1997.

19. *Hawaiian Tourism* The rate of receipt of revenue in Hawaii during the years 1985 to 1993 can be approximated by[2]

$$R(t) = -0.164t^2 + 1.60t + 6.71 \text{ billion dollars per year}$$

where $t = 0$ represents June 30, 1985. Estimate how much revenue Hawaii earned from visitor spending between June 30, 1985, and June 30, 1990. (Give your answer to the nearest billion dollars.)

[1] The models in Exercises 17 and 18 are based on a quadratic regression of graphical data published in the *New York Times,* August 27, 1997, p. D4.

[2] The model is based on a best-fit quadratic for (approximate) tourism data, as measured in constant 1993 dollars, from the Hawaii Visitors Bureau/*New York Times,* September 5, 1995, p. A12.

20. ***Revenue*** The EMC Corporation, which specializes in the manufacture of computer data storage systems, underwent rapid growth during the period 1989–1994, and its annual revenue was approximately

$$R(t) = 0.13e^{0.44t} \text{ billion dollars per year} \qquad (0 \le t \le 5)$$

where t is the number of years since June 1989. Use this model to estimate the total revenue (to the nearest \$0.1 billion) earned by EMC from June 1989 to June 1994.
Source: The authors' model based on data from company reports/IDC/*New York Times,* November 9, 1994, p. D8.

21. ***Revenue*** Refer back to Exercise 17. Suppose that, from January 1991, SAB invested its revenue in an investment yielding 6% compounded continuously. What would the total value of SAB's revenues have been by January 1997?

22. ***Profit*** Refer back to Exercise 18. Suppose that, from January 1991, SAB invested its after-tax profit in an investment yielding 8% compounded continuously. What would the total value of SAB's after-tax profits have been by January 1997?

23. ***Hawaiian Tourism*** Refer back to Exercise 19. Suppose that, from June 30, 1985, the state of Hawaii invested all its tourism revenue at 6% compounded continuously. Estimate the present value on June 30, 1985 of the accumulated revenue through June 30, 1990. (Give your answer to the nearest billion dollars.)

24. ***Revenue*** Refer back to Exercise 20. Suppose that, from June 1989, EMC invested all its revenue at 8% compounded continuously. Estimate the present value in June 1989 of its accumulated revenue through June 1994. (Give your answer to the nearest \$0.1 billion.)

25. ***Saving for Retirement*** You are saving for your retirement by investing \$700 per month in an annuity with a guaranteed interest rate of 6% per year. With a continuous stream of investment and continuous compounding, how much will you have accumulated in the annuity by the time you retire in 45 years?

26. ***Saving for College*** When your first child is born, you begin to save for college by depositing \$400 per month in an account paying 12% interest per year. With a continuous stream of investment and continuous compounding, how much will you have accumulated in the account by the time your child enters college 18 years later?

27. ***Saving for Retirement*** You begin saving for your retirement by investing \$700 per month in an annuity with a guaranteed interest rate of 6% per year. You increase the amount you invest at the rate of 3% per year. With continuous investment and compounding, how much will you have accumulated in the annuity by the time you retire in 45 years?

28. ***Saving for College*** When your first child is born, you begin to save for college by depositing \$400 per month in an account paying 12% interest per year. You increase the amount you save by 2% per year. With continuous investment and compounding, how much will have accumulated in the account by the time your child enters college 18 years later?

29. ***Bonds*** The U.S. Treasury issued a 30-year bond on November 16, 1998, paying 5.25% interest. Thus, if you bought \$100,000 worth of these bonds, you would receive \$5250 per year in interest for 30 years. An investor wishes to buy the rights to receive the interest on \$100,000 worth of these bonds. The amount the investor is willing to pay is the present value of the interest payments, assuming a 6% rate of return. If we assume (incorrectly, but approximately) that the interest payments are made continuously, what will the investor pay?
Source: The Bureau of the Public Debt's web site: http://www.publicdebt.treas.gov/.

30. ***Bonds*** The Megabucks Corporation is issuing a 20-year bond paying 7% interest (see the preceding exercise). An investor wishes to buy the rights to receive the interest on \$50,000 worth of these bonds and seeks a 6% rate of return. If the interest payments are made continuously, what will the investor pay?

31. ***Valuing Future Income*** Inga was injured and can no longer work. As a result of a lawsuit, she is to be awarded the present value of the income she would have received over the next 20 years. Her income at the time she was injured was \$100,000 per year, increasing by \$5000 per year. What will be the amount of her award, assuming continuous income and a 5% interest rate?

32. ***Valuing Future Income*** Max was injured and can no longer work. As a result of a lawsuit, he is to be awarded the present value of the income he would have received over the next 30 years. His income at the time he was injured was \$30,000 per year, increasing by \$1500 per year. What will be the amount of his award, assuming continuous income and a 6% interest rate?

Communication and Reasoning Exercises

33. Your study group friend says that the future value of a continuous stream of income is always greater than the total value. Is she correct? Why?

34. Your other study group friend says that the present value of a continuous stream of income can sometimes be greater than the total value, depending on the interest rate. Is he correct? Explain.

35. Arrange from smallest to largest: total value, future value, present value of a continuous stream of income.

36. a. Arrange the following functions from smallest to largest: $R(t)$, $R(t)e^{r(b-t)}$, and $R(t)e^{r(a-t)}$, where $a \le t \le b$ and r is positive.

b. Use the result from part (a) to justify your answers in Exercises 33–35.

7.5 Improper Integrals and Applications

All the definite integrals we have seen so far have had the form $\int_a^b f(x)\,dx$, with a and b finite and $f(x)$ piecewise continuous on the closed interval $[a, b]$. We can relax these requirements somewhat. When we do so, we obtain what are called **improper integrals.** There are various types of improper integrals.

Integrals in Which a Limit of Integration Is Infinite

These integrals are written as

$$\int_a^{+\infty} f(x)\,dx \qquad \int_{-\infty}^{b} f(x)\,dx \qquad \text{or} \qquad \int_{-\infty}^{+\infty} f(x)\,dx$$

Let us concentrate for a moment on the first form, $\int_a^{+\infty} f(x)\,dx$. What does the $+\infty$ mean here? As it often does, it means that we are to take a limit as something gets large. Specifically, it means the limit as the upper bound of integration gets large.

Improper Integral with an Infinite Limit of Integration

We define

$$\int_a^{+\infty} f(x)\,dx = \lim_{M\to+\infty} \int_a^{M} f(x)\,dx$$

provided the limit exists. If the limit exists, we say that $\int_a^{+\infty} f(x)\,dx$ **converges.** Otherwise, we say that $\int_a^{+\infty} f(x)\,dx$ **diverges.** Similarly, we define

$$\int_{-\infty}^{b} f(x)\,dx = \lim_{M\to-\infty} \int_M^{b} f(x)\,dx$$

provided the limit exists. Finally, we define

$$\int_{-\infty}^{+\infty} f(x)\,dx = \int_{-\infty}^{a} f(x)\,dx + \int_a^{+\infty} f(x)\,dx$$

for some convenient a, provided *both* integrals on the right converge.

Quick Examples

1. $\displaystyle \int_1^{+\infty} \frac{dx}{x^2} = \lim_{M\to+\infty} \int_1^{M} \frac{dx}{x^2} = \lim_{M\to+\infty} \left[-\frac{1}{x}\right]_1^M = \lim_{M\to+\infty} \left(-\frac{1}{M} + 1\right) = 1$ Converges

2. $\displaystyle \int_1^{+\infty} \frac{dx}{x} = \lim_{M\to+\infty} \int_1^{M} \frac{dx}{x} = \lim_{M\to+\infty} \left[\ln|x|\right]_1^M = \lim_{M\to+\infty} (\ln M - \ln 1) = +\infty$ Diverges

3. $\displaystyle \int_{-\infty}^{-1} \frac{dx}{x^2} = \lim_{M\to-\infty} \int_M^{-1} \frac{dx}{x^2} = \lim_{M\to-\infty} \left[-\frac{1}{x}\right]_M^{-1} = \lim_{M\to-\infty} \left(1 + \frac{1}{M}\right) = 1$ Converges

4. $\int_{-\infty}^{+\infty} e^{-x}\,dx = \int_{-\infty}^{0} e^{-x}\,dx + \int_{0}^{+\infty} e^{-x}\,dx$

$$= \lim_{M\to-\infty}\int_{M}^{0} e^{-x}\,dx + \lim_{M\to+\infty}\int_{0}^{M} e^{-x}\,dx$$

$$= \lim_{M\to-\infty} -\left[e^{-x}\right]_M^0 + \lim_{M\to+\infty} -\left[e^{-x}\right]_0^M$$

$$= \lim_{M\to-\infty} (e^{-M}-1) + \lim_{M\to+\infty} (1-e^{-M})$$

$$= +\infty + 1$$

Diverges

5. $\int_{-\infty}^{+\infty} xe^{-x^2}\,dx = \int_{-\infty}^{0} xe^{-x^2}\,dx + \int_{0}^{+\infty} xe^{-x^2}\,dx$

$$= \lim_{M\to-\infty}\int_{M}^{0} xe^{-x^2}\,dx + \lim_{M\to+\infty}\int_{0}^{M} xe^{-x^2}\,dx$$

$$= \lim_{M\to-\infty}\left[-\frac{1}{2}e^{-x^2}\right]_M^0 + \lim_{M\to+\infty}\left[-\frac{1}{2}e^{-x^2}\right]_0^M$$

$$= \lim_{M\to-\infty}\left(-\frac{1}{2}+\frac{1}{2}e^{-M^2}\right) + \lim_{M\to+\infty}\left(-\frac{1}{2}e^{-M^2}+\frac{1}{2}\right)$$

$$= -\frac{1}{2}+\frac{1}{2} = 0$$

Converges

Question We learned that the integral can be interpreted as the area under the curve. Is this still true for improper integrals?

Answer Yes. Figure 16 illustrates how we can represent an improper integral as the area of an infinite region.

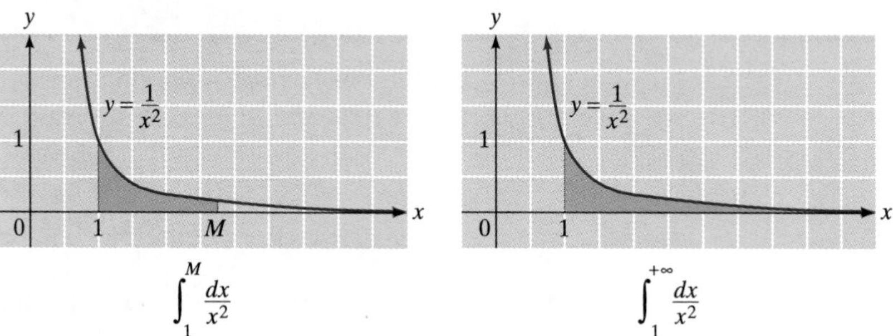

Figure 16

On the left we see the area represented by $\int_1^M dx/x^2$. As M gets larger, the integral approaches $\int_1^{+\infty} dx/x^2$. In the picture, think of M being moved farther and farther along the x-axis in the direction of increasing x, resulting in the region shown on the right.

Question Wait! We calculated $\int_1^{+\infty} dx/x^2 = 1$. Does this mean that the infinitely long area in Figure 16 has an area of only 1 square unit?

Answer That is exactly what it means. If you had enough paint to cover 1 square unit, you would never run out of paint while painting the region in Figure 16. This is one of the places where mathematics seems to contradict common sense. But common sense is notoriously unreliable when dealing with infinities.

example 1

Future Sales of Freon

It is estimated that, beginning in 2000, sales of Freon will decrease at a rate of about 15% per year. Sales of Freon at the start of 2000 were predicted to be 35 million pounds per year.[1] How much Freon, total, will be sold from 2000 on?

Solution

Recall that the total sales between two dates can be computed as the definite integral of annual sales. So, if we wanted the sales between the year 2000 and a year far in the future, we would compute $\int_0^M s(t)\,dt$ with a large M, where $s(t)$ is the annual sales t years after 2000. Since we want to know *all* the Freon sold after 2000, we let $M \to +\infty$; that is, we compute $\int_0^{+\infty} s(t)\,dt$. Since Freon sales are decreasing by 15% per year, we can model $s(t)$ by

$$s(t) = 35(0.85)^t$$

where t is the number of years since 2000. Now

$$\begin{aligned}
\text{Total sales beginning in 2000} &= \int_0^{+\infty} 35(0.85)^t\,dt \\
&= \lim_{M\to+\infty} \int_0^M 35(0.85)^t\,dt \\
&= \frac{35}{\ln 0.85} \lim_{M\to+\infty} \left[0.85^t\right]_0^M \\
&= \frac{35}{\ln 0.85} \lim_{M\to+\infty} (0.85^M - 0.85^0) \\
&= \frac{35}{\ln 0.85}(-1) \approx 215.4 \text{ million pounds}
\end{aligned}$$

Integrals in Which the Integrand Becomes Infinite

We can sometimes compute integrals $\int_a^b f(x)\,dx$ in which $f(x)$ becomes infinite. The first case to consider is when $f(x)$ approaches $\pm\infty$ at either a or b.

example 2

Integrand Infinite at One Endpoint

Calculate $\displaystyle\int_0^1 \frac{1}{\sqrt{x}}\,dx$.

Solution

Notice that the integrand approaches $+\infty$ as x approaches zero from the right and is not defined at zero. This makes the integral an improper integral. Figure 17

[1]These figures are approximations based on published data from the Automobile Consulting Group/*New York Times,* December 26, 1993, p. F23.

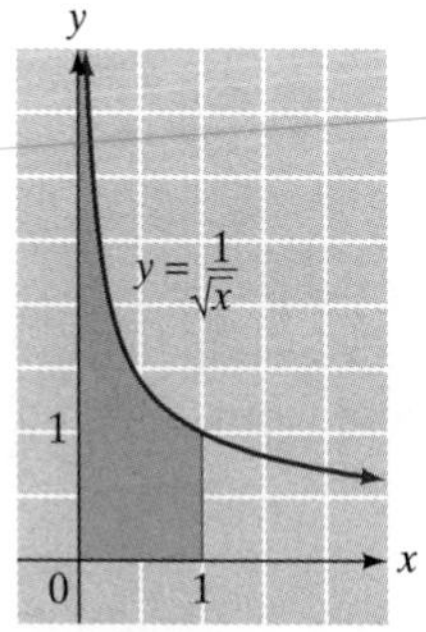

Figure 17

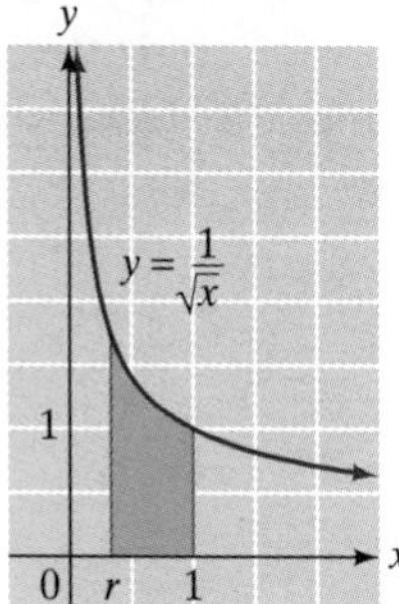

Figure 18

shows the region whose area we are trying to calculate; it extends infinitely vertically rather than horizontally. Now, if $0 < r < 1$, the integral $\int_r^1 (1/\sqrt{x})\, dx$ is a proper integral because we avoid the bad behavior at zero. This integral gives the area shown in Figure 18. If we let r approach zero from the right, the area in Figure 18 will approach the area in Figure 17. So, we calculate

$$\begin{aligned}\int_0^1 \frac{1}{\sqrt{x}}\, dx &= \lim_{r \to 0^+} \int_r^1 \frac{1}{\sqrt{x}}\, dx \\ &= \lim_{r \to 0^+} \left[2\sqrt{x}\right]_r^1 \\ &= \lim_{r \to 0^+} (2 - 2\sqrt{r}) = 2\end{aligned}$$

Thus, just as before, we have an infinitely long region with finite area.

Generalizing, we make the following definition.

Improper Integral in Which the Integrand Becomes Infinite

If $f(x)$ is defined for all x with $a < x \le b$ but approaches $\pm\infty$ as x approaches a, we define

$$\int_a^b f(x)\, dx = \lim_{r \to a^+} \int_r^b f(x)\, dx$$

provided the limit exists. Similarly, if $f(x)$ is defined for all x with $a \le x < b$ but approaches $\pm\infty$ as x approaches b, we define

$$\int_a^b f(x)\, dx = \lim_{r \to b^-} \int_a^r f(x)\, dx$$

provided the limit exists. In either case, if the limit exists, we say that $\int_a^b f(x)\, dx$ **converges.** Otherwise, we say that $\int_a^b f(x)\, dx$ **diverges.**

example 3

Testing for Convergence

Does $\displaystyle\int_{-1}^{3} \frac{x}{x^2 - 9}\, dx$ converge? If so, to what?

Solution

We first check to see where, if anywhere, the integrand approaches $\pm\infty$. That will happen where the denominator becomes zero, so we solve $x^2 - 9 = 0$:

$$\begin{aligned}x^2 - 9 &= 0 \\ x^2 &= 9 \\ x &= \pm 3\end{aligned}$$

The solution $x = -3$ is outside the range of integration, so we ignore it. The solution $x = 3$ is, however, the right endpoint of the range of integration, so the integral is improper. We need to investigate the following limit:

$$\int_{-1}^{3} \frac{x}{x^2-9}\,dx = \lim_{r\to 3^-} \int_{-1}^{r} \frac{x}{x^2-9}\,dx$$

Now, to calculate the integral we use a substitution:

$$u = x^2 - 9$$

$$\frac{du}{dx} = 2x$$

$$dx = \frac{1}{2x}\,du$$

When $x = r$, $u = r^2 - 9$.

When $x = -1$, $u = (-1)^2 - 9 = -8$.

Thus,

$$\int_{-1}^{r} \frac{x}{x^2-9}\,dx = \int_{-8}^{r^2-9} \frac{1}{2u}\,du$$

$$= \frac{1}{2}\big[\ln|u|\big]_{-8}^{r^2-9} = \frac{1}{2}(\ln|r^2-9| - \ln 8)$$

Now we take the limit:

$$\int_{-1}^{3} \frac{x}{x^2-9}\,dx = \lim_{r\to 3^-} \int_{-1}^{r} \frac{x}{x^2-9}\,dx$$

$$= \lim_{r\to 3^-} \frac{1}{2}(\ln|r^2-9| - \ln 8) = -\infty$$

because, as $r \to 3$, $r^2 - 9 \to 0$, and so $\ln|r^2-9| \to -\infty$. Thus, this integral diverges.

---

example 4

An Integrand Infinite Between the Endpoints

Does $\displaystyle\int_{-2}^{3} \frac{1}{x^2}\,dx$ converge? If so, to what?

Solution

Again we check to see whether there are any points at which the integrand approaches $\pm\infty$. There is such a point, at $x = 0$. This is between the endpoints of the range of integration. To deal with this, we break the integral into two integrals:

$$\int_{-2}^{3} \frac{1}{x^2}\,dx = \int_{-2}^{0} \frac{1}{x^2}\,dx + \int_{0}^{3} \frac{1}{x^2}\,dx$$

Each integral on the right is an improper integral with the integrand approaching $\pm\infty$ at an endpoint. If both of the integrals on the right converge, we take the sum as the value of the integral on the left. So now we compute

$$\int_{-2}^{0} \frac{1}{x^2}\,dx = \lim_{r\to 0^-} \int_{-2}^{r} \frac{1}{x^2}\,dx$$

$$= \lim_{r\to 0^-} \left[-\frac{1}{x}\right]_{-2}^{r} = \lim_{r\to 0^-} \left(-\frac{1}{r} - \frac{1}{2}\right)$$

which diverges to $+\infty$. There is no need now to check $\int_0^3 (1/x^2)\,dx$; since one of the two pieces of the integral diverges, we simply say that $\int_{-2}^3 (1/x^2)\,dx$ diverges.

Before we go on . . . What if we had been sloppy and not checked first whether the integrand approached $\pm\infty$ somewhere? Then we probably would have done the following:

$$\int_{-2}^{3} \frac{1}{x^2}\,dx = \left(-\frac{1}{x}\right)_{-2}^{3} = \left(-\frac{1}{3} - \frac{1}{2}\right) = -\frac{5}{6} \qquad \text{✗ WRONG!}$$

Notice that the answer this "calculation" gives is patently ridiculous. Since $1/x^2 > 0$ for all x for which it is defined, any definite integral of $1/x^2$ must give a positive answer. *Moral:* Always check to see whether the integrand blows up anywhere in the range of integration.

We end with an example of what to do if an integral is improper for more than one reason.

example 5

An Integral Improper in Two Ways

Does $\displaystyle\int_0^{+\infty} \frac{1}{\sqrt{x}}\,dx$ converge? If so, to what?

Solution

This integral is improper for two reasons. First, the range of integration is infinite. Second, the integrand blows up at the endpoint zero. In order to separate these two problems, we break up the integral at some convenient point:

$$\int_0^{+\infty} \frac{1}{\sqrt{x}}\,dx = \int_0^{1} \frac{1}{\sqrt{x}}\,dx + \int_1^{+\infty} \frac{1}{\sqrt{x}}\,dx$$

We chose to break the integral at 1. Any positive number would have worked, but 1 is easier to use in calculations.

The first piece, $\int_0^1 (1/\sqrt{x})\,dx$, we discussed earlier; it converges to 2. For the second piece, we have:

$$\int_1^{+\infty} \frac{1}{\sqrt{x}}\,dx = \lim_{M\to+\infty} \int_1^{M} \frac{1}{\sqrt{x}}\,dx$$

$$= \lim_{M\to+\infty} \left[2\sqrt{x}\right]_1^{M} = \lim_{M\to+\infty} (2\sqrt{M} - 2)$$

which diverges to $+\infty$. Since the second piece of the integral diverges, we conclude that $\int_0^{+\infty} (1/\sqrt{x})\,dx$ diverges.

7.5 exercises

Decide whether each integral in Exercises 1–30 converges. If the integral converges, compute its value.

1. $\int_1^{+\infty} x\,dx$
2. $\int_0^{+\infty} e^{-x}\,dx$
3. $\int_{-2}^{+\infty} e^{-0.5x}\,dx$
4. $\int_1^{+\infty} \frac{1}{x^{1.5}}\,dx$
5. $\int_{-\infty}^{2} e^{x}\,dx$
6. $\int_{-\infty}^{-1} \frac{1}{x^{1/3}}\,dx$
7. $\int_{-\infty}^{-2} \frac{1}{x^2}\,dx$
8. $\int_{-\infty}^{0} e^{-x}\,dx$
9. $\int_0^{+\infty} x^2e^{-6x}\,dx$
10. $\int_0^{+\infty} (2x-4)e^{-x}\,dx$
11. $\int_0^{5} \frac{2}{x^{1/3}}\,dx$
12. $\int_0^{2} \frac{1}{x^2}\,dx$
13. $\int_{-1}^{2} \frac{3}{(x+1)^2}\,dx$
14. $\int_{-1}^{2} \frac{3}{(x+1)^{1/2}}\,dx$
15. $\int_{-1}^{2} \frac{3x}{x^2-1}\,dx$
16. $\int_{-1}^{2} \frac{3}{x^{1/3}}\,dx$
17. $\int_{-2}^{2} \frac{1}{(x+1)^{1/5}}\,dx$
18. $\int_{-2}^{2} \frac{2x}{\sqrt{4-x^2}}\,dx$
19. $\int_{-1}^{1} \frac{2x}{x^2-1}\,dx$
20. $\int_{-1}^{2} \frac{2x}{x^2-1}\,dx$
21. $\int_{-\infty}^{+\infty} xe^{-x^2}\,dx$
22. $\int_{-\infty}^{\infty} xe^{1-x^2}\,dx$
23. $\int_0^{+\infty} \frac{1}{x\ln x}\,dx$
24. $\int_0^{+\infty} \ln x\,dx$
25. $\int_0^{+\infty} \frac{2x}{x^2-1}\,dx$
26. $\int_{-\infty}^{0} \frac{2x}{x^2-1}\,dx$
27. ▲ $\int_0^{+\infty} \sin x\,dx$
28. ▲ $\int_0^{+\infty} \cos x\,dx$
29. ▲ $\int_0^{+\infty} e^{-x}\cos x\,dx$
30. ▲ $\int_0^{+\infty} e^{-x}\sin x\,dx$

Applications

31. ***Sales*** My financial adviser has predicted that annual sales of BATMAN® T-shirts will continue to decline by 10% each year. At the moment, I have 3200 of the shirts in stock, and I am selling them at a rate of 200 per year. Will I ever sell them all?

32. ***Revenue*** Alarmed about the sales prospects for my BATMAN T-shirts (see the preceding exercise), I will try to make up lost revenues by increasing the price by \$1 each year. I now charge \$10 per shirt. What is the total amount of revenue I can expect to earn from sales of my T-shirts, assuming the sales levels described in Exercise 31? (Give your answer to the nearest \$1000.)

33. ***Advertising*** Spending on cigarette advertising in the United States declined from close to \$600 million per year at the start of 1991 to half that amount 3 years later. Use a continuous decay model to forecast the total revenue that will be spent on cigarette advertising beginning in January 1991. (Give your answer to the nearest \$100 million.)
Sources: Advertising Age/Competitive Media Reporting/*New York Times*, March 3, 1994, p. D1.

34. ***Sales*** Sales of the text *Calculus and You* have been declining continuously at an annual rate of 5%. Assuming that *Calculus and You* currently sells 5000 copies per year and that sales will continue this pattern of decline, calculate total future sales of the text. [*Hint:* Use the formula for continuously compounded interest with a negative rate.]

35. ***Variable Sales*** The value of your Chateau Petit Mont Blanc 1963 vintage burgundy is increasing continuously at an annual rate of 40%, and you have a supply of 1000 bottles worth \$85 each at today's prices. To ensure a steady income, you have decided to sell your wine at a diminishing rate—starting at 500 bottles per year and then decreasing this figure continuously at a fractional rate of 100% per year. How much income (to the nearest dollar) can you expect to generate by this scheme? [*Hint:* Use the formula for continuously compounded interest.]

36. ***Panic Sales*** Unfortunately, your large supply of Chateau Petit Mont Blanc is continuously turning to vinegar at a

fractional rate of 60% per year! You have thus decided to sell off your Chateau Petit Mont Blanc at $50 per bottle, but the market is a little thin, and you can sell only 400 bottles per year. Since you have no way of knowing which bottles now contain vinegar until they are opened, you will have to give refunds for all the bottles of vinegar. What will your net income be before all the wine turns to vinegar?

37. ***Foreign Investments*** According to data published by the World Bank, the annual flow of private investment to developing countries from more developed countries is approximately

$$q(t) = 10 + 1.56t^2$$

where $q(t)$ is the annual investment in billions of dollars and t is time in years since the start of 1986. Assuming a worldwide inflation rate of 5% per year, find the value of all private aid to developing countries from 1986 on in constant dollars. [The constant dollar value of $q(t)$ dollars t years from now is given by $q(t)e^{-rt}$, where r is the fractional rate of inflation.]

Source: The authors' approximation based on data published by the World Bank/*New York Times,* December 17, 1993, p. D1.

38. ***Foreign Investments*** Repeat Exercise 37 using the following model for government loans and grants to developing countries:

$$q(t) = 38 + 0.6(t - 3)^2$$

Source: See the source in Exercise 37.

Newspaper Circulation *Exercises 39–42 are based on the following chart, which shows annual declines in the circulations of three major metropolitan newspapers for the period June 1994–June 1995:*[2]

	San Francisco Chronicle	Washington Post	New York Times
Rate of decrease (% per year)	4.9%	1.4%	1.3%
June 1995 circulation (newspapers sold per day)	450,000	840,000	1,170,000

Source: Audit Bureau of Circulations, *New York Times,* May 2, 1995, p. D8

39. Assuming that circulations continue to decline at the rate shown, estimate the total sales of the *New York Times* from June 1995 to the indefinite future. (Round your answer to the nearest million newspapers.)

40. Assuming that circulations continue to decline at the rate shown, estimate the total sales of the *Washington Post* from June 1995 to the indefinite future. (Use annual compounding and round your answer to the nearest million newspapers.)

41. If $w(t)$ represents the annual circulation of the *Washington Post* t years since 1995, and $n(t)$ represents the annual circulation of the *New York Times* t years since 1995, calculate and interpret

$$\int_5^{+\infty} \left[n(t) - w(t)\right] dt$$

(Use annual compounding and round your answer to the nearest 1 million.)

42. Repeat Exercise 41, using $\int_{10}^{+\infty} \left[w(t) - s(t)\right] dt$, where $s(t)$ represents the annual circulation of the *San Francisco Chronicle.*

43. ***Pork Exports*** According to figures released by the U.S. Department of Agriculture, the value of U.S. pork exports can be modeled by the equation

$$q(t) = \frac{460e^{(t-4)}}{1 + e^{(t-4)}}$$

where t is time in years since the start of 1985 and $q(t)$ is annual exports in millions of dollars. Investigate the integrals $\int_0^{+\infty} q(t)\, dt$ and $\int_{-\infty}^0 q(t)\, dt$ and interpret your answers.

44. ***Sales*** The weekly demand for your company's Lo-Cal Mousse is modeled by the equation

$$q(t) = \frac{50e^{2t-1}}{1 + e^{2t-1}}$$

where t is time from now in weeks and $q(t)$ is the number of gallons sold per week. Investigate the integrals $\int_0^{+\infty} q(t)\, dt$ and $\int_{-\infty}^0 q(t)\, dt$ and interpret your answers.

45. ***Meteor Impacts*** The frequency of meteor impacts on Earth can be modeled by

$$n(k) = \frac{1}{5.6997k^{1.081}}$$

where $n(k) = N'(k)$ and $N(k)$ is the average number of meteors of energy less than or equal to k megatons that will hit the earth in 1 year. (A small nuclear bomb releases on the order of 1 megaton of energy.)

a. How many meteors of energy at least $k = 0.2$ hit the earth each year?

b. Investigate and interpret the integral $\int_0^1 n(k)\, dk$.

Source: The authors' model based on data published by NASA International Near-Earth-Object Detection Workshop/*New York Times,* January 25, 1994, p. C1.

46. ***Meteor Impacts*** (continuing Exercise 45)

a. Explain why the integral $\int_a^b kn(k)\, dk$ computes the total energy released each year by meteors with energies between a and b megatons.

---

[2] The figures are approximate and are for circulation of weekday editions only.

b. Compute and interpret $\int_0^1 kn(k)\,dk$.

c. Compute and interpret $\int_1^{+\infty} kn(k)\,dk$.

47. ***The Gamma Function*** The gamma function is defined by the formula

$$\Gamma(x) = \int_0^{+\infty} t^{x-1}e^{-t}\,dt$$

a. Find $\Gamma(1)$ and $\Gamma(2)$ (assume that $\lim_{M\to+\infty} Me^{-M} = 0$).

b. Use integration by parts to show that for a positive integer n, $\Gamma(n+1) = n\Gamma(n)$.

c. Deduce that $\Gamma(n) = (n-1)! = (n-1)\times(n-2)\times\cdots\times 2\times 1$ for every positive integer n.

48. ***Laplace Transforms*** The Laplace transform $F(x)$ of a function $f(t)$ is given by the formula

$$F(x) = \int_0^{+\infty} f(t)e^{-xt}\,dt$$

a. Find $F(x)$ for $f(t) = 1$ and for $f(t) = t$.

b. Find a formula for $F(x)$ if $f(t) = t^n$ $(n = 1, 2, 3, \ldots)$.

c. Find a formula for $F(x)$ if $f(t) = e^{at}$ (a constant).

T ***The Normal Curve*** *Exercises 49–52 require the use of a graphing calculator or computer programmed to do numerical integration. The* normal distribution curve, *which models the distributions of data in a wide range of applications, is given by the function*

$$p(x) = \frac{1}{\sqrt{2\pi}\sigma}e^{-(x-\mu)^2/2\sigma^2}$$

where $\pi = 3.14159265\ldots$ and σ and μ are constants called the standard deviation and the mean, respectively. Its graph (for $\sigma = 1$ and $\mu = 2$) is shown in the figure.

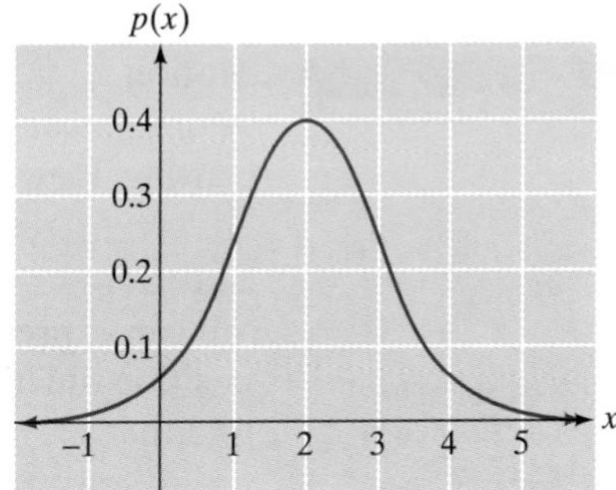

T **49.** With $\sigma = 4$ and $\mu = 1$, approximate $\int_{-\infty}^{+\infty} p(x)\,dx$.

T **50.** With $\sigma = 1$ and $\mu = 0$, approximate $\int_0^{+\infty} p(x)\,dx$.

T **51.** With $\sigma = 1$ and $\mu = 0$, approximate $\int_1^{+\infty} p(x)\,dx$.

T **52.** With $\sigma = 1$ and $\mu = 0$, approximate $\int_{-\infty}^{1} p(x)\,dx$.

Communication and Reasoning Exercises

53. It sometimes happens that the Fundamental Theorem of Calculus (FTC) gives the correct answer for an improper integral. Does the FTC give the correct answer for improper integrals of the form

$$\int_{-a}^{a} \frac{1}{x^{1/r}}\,dx$$

if $r = 3, 5, 7, \ldots$?

54. Does the FTC give the correct answer for improper integrals of the form

$$\int_{-a}^{a} \frac{1}{x^{1/r}}\,dx$$

if $r = 2, 4, 6, \ldots$?

55. Why can't the Fundamental Theorem of Calculus be used to evaluate $\int_{-1}^{1} 1/x\,dx$?

56. Why can't the Fundamental Theorem of Calculus be used to evaluate $\int_1^{+\infty} 1/x^2\,dx$?

T **57.** How could you use technology to approximate improper integrals? (Your discussion should refer to each type of improper integral.)

58. Make up an interesting application with the solution $\int_{10}^{+\infty} 100te^{-0.2t}\,dt = \$1015.01$.

59. Make up an interesting application with the solution $\int_{100}^{+\infty} 1/r^2\,dr = 0.01$.

7.6 Differential Equations and Applications

A **differential equation** is an equation that involves a derivative of an unknown function. A **first-order differential equation** involves only the first derivative of the unknown function. If the differential equation involves the second derivative of the unknown function, we refer to it as a **second-order differential equation.** Higher-order differential equations are defined similarly. In this book, we deal only with first-order differential equations.

To **solve** a differential equation means to find that unknown function. Many of the laws of science and other fields describe how things change. When expressed mathematically, these laws take the form of equations involving derivatives—that is, differential equations. The field of differential equations is a large and very active area of study in mathematics, and we see only a small part of it in this section.

example 1

Motion

A drag racer accelerates from a stop so that its speed is $40t$ ft/s t seconds after starting. How far will the car go in 8 seconds?

Solution

We can express this problem as a differential equation. We wish to find the car's position function $s(t)$. We are told about its speed, which is ds/dt. Precisely, we are told that

$$\frac{ds}{dt} = 40t$$

This is the differential equation we have to solve to find $s(t)$. But we know how to solve this kind of differential equation; we integrate:

$$s(t) = \int 40t\, dt = 20t^2 + C$$

We now have the **general solution** to the differential equation. By letting C take on different values, we get all the possible solutions. We can specify the one **particular solution** that gives the answer to our problem by imposing the **initial condition** that $s(0) = 0$. Substituting into $s(t) = 20t^2 + C$, we get

$$0 = s(0) = 20(0)^2 + C = C$$

so $C = 0$ and $s(t) = 20t^2$. To answer the question, we find the car travels $20(8)^2 = 1280$ feet in 8 seconds.

We did not have to work hard to solve the differential equation in Example 1. In fact, any differential equation of the form $dy/dx = f(x)$ can (in theory) be solved by integrating. (Whether we can actually carry out the integration is another matter!)

Simple Differential Equations

A **simple** differential equation has the form

$$\frac{dy}{dx} = f(x)$$

Its general solution is

$$y = \int f(x)\, dx$$

Quick Example
The differential equation

$$\frac{dy}{dx} = 2x^2 - 4x^3$$

is simple and has the general solution

$$y = \int f(x)\,dx = \frac{2x^3}{3} - x^4 + C$$

Not all differential equations are simple, as the next example shows.

example 2

Consider the differential equation $\dfrac{dy}{dx} = \dfrac{x}{y^2}$.

a. Find the general solution.

b. Find the particular solution that satisfies the initial condition $y(0) = 2$.

Solution

a. This is not a simple differential equation because the right-hand side is a function of both x and y. We cannot solve this equation by just integrating because we would not know what to do with the y if we integrated the right-hand side with respect to x. The solution to this problem is to "separate" the variables.

Step 1. *Separate the variables algebraically.* We rewrite the equation as

$$y^2\,dy = x\,dx$$

Step 2. *Integrate both sides.*

$$\int y^2\,dy = \int x\,dx$$

giving

$$\frac{y^3}{3} = \frac{x^2}{2} + C$$

Step 3. *Solve for the dependent variable.* We solve for y:

$$y^3 = \frac{3}{2}x^2 + 3C = \frac{3}{2}x^2 + D$$

(rewriting $3C$ as D, an equally arbitrary constant), so

$$y = \left(\frac{3}{2}x^2 + D\right)^{1/3}$$

This is the general solution of the differential equation.

b. We need to find the value for D that gives us the solution satisfying the condition $y(0) = 2$. Substituting 0 for x and 2 for y in the general solution, we get

$$2 = \left(\frac{3}{2}(0)^2 + D\right)^{1/3} = D^{1/3}$$

so

$$D = 2^3 = 8$$

Thus, the particular solution we are looking for is

$$y = \left(\frac{3}{2}x^2 + 8\right)^{1/3}$$

Before we go on . . . We can check the general solution by calculating both sides of the differential equation and comparing:

$$\frac{dy}{dx} = \frac{d}{dx}\left(\frac{3}{2}x^2 + D\right)^{1/3} = x\left(\frac{3}{2}x^2 + D\right)^{-2/3}$$

$$\frac{x}{y^2} = \frac{x}{\left(\frac{3}{2}x^2 + 8\right)^{2/3}} = x\left(\frac{3}{2}x^2 + D\right)^{-2/3} \quad ✔$$

Question Above, we wrote $y^2\,dy$ and $x\,dx$. What do they mean?
Answer Although it is possible to give meaning to these symbols, for us they are just a notational convenience. We could have done the following instead:

$$y^2\frac{dy}{dx} = x$$

Now we integrate both sides with respect to x:

$$\int y^2\frac{dy}{dx}\,dx = \int x\,dx$$

On the left we can use substitution to write

$$\int y^2\frac{dy}{dx}\,dx = \int y^2\,dy$$

which brings us back to the equation

$$\int y^2\,dy = \int x\,dx$$

We were able to separate the variables in the preceding example because the right-hand side, x/y^2, was a *product* of a function of x and a function of y—namely,

$$\frac{x}{y^2} = x\left(\frac{1}{y^2}\right)$$

In general, we can say the following.

Separable Differential Equation

A **separable** differential equation has the form

$$\frac{dy}{dx} = f(x)g(y)$$

We solve a separable differential equation by separating the xs and the ys algebraically, writing

$$\frac{1}{g(y)}\,dy = f(x)\,dx$$

and then integrating:

$$\int \frac{1}{g(y)}\,dy = \int f(x)\,dx$$

example 3

Rising Medical Costs

The cost of medical care in the New York City metropolitan area was going up continuously at an instantaneous rate of 7.3% per year in the 1980s and early 1990s. Find a formula for medical cost y as a function of time t.

Source: Federal Bureau of Labor Statistics/*New York Times,* December 25, 1993.

Solution

We discussed this kind of growth when we first discussed exponential functions. By looking at this problem as one that leads to a differential equation, we can see why the exponential function comes in. When we say that the medical cost y was going up continuously at an instantaneous rate of 7.3% per year, we mean that

The instantaneous rate of increase of y was 7.3% of its value

or

$$\frac{dy}{dt} = 0.073y$$

This is a separable differential equation. Separating the variables gives

$$\frac{1}{y}\,dy = 0.073\,dt$$

Integrating both sides, we get

$$\int \frac{1}{y}\,dy = \int 0.073\,dt$$

so

$$\ln y = 0.073t + C$$

(We should write $\ln|y|$, but we know that medical costs are positive.) We now solve for y:

$$y = e^{0.073t+C} = e^{C}e^{0.073t} = Ae^{0.073t}$$

where A is a positive constant. This is the formula we used before for continuous percentage growth. We can determine A if we know, for example, the medical costs at time $t = 0$; in fact, A is equal to $y(0)$.

example 4

Newton's Law of Cooling
Newton's Law of Cooling states that a hot object cools at a rate proportional to the difference between its temperature and the temperature of the surrounding environment. If a hot cup of coffee, at 170°F, is left to sit in a room at 70°F, how will the temperature of the coffee change over time?

Solution

We let $H(t)$ denote the temperature of the coffee at time t. Newton's Law of Cooling tells us that $H(t)$ *decreases* at a rate proportional to the difference between $H(t)$ and 70°F, the temperature of the surrounding environment (also called the **ambient temperature**). In other words,

$$\frac{dH}{dt} = -k(H - 70)$$

where k is some positive constant.[1] Note that $H \geq 70$: The coffee will never cool to less than the ambient temperature. The variables here are H and t, which we can separate as follows:

$$\frac{dH}{H - 70} = -k\,dt$$

Integrating, we get

$$\int \frac{dH}{H - 70} = \int (-k)\,dt$$

so $$\ln(H - 70) = -kt + C$$

(Note that $H - 70$ is positive, so we don't need absolute values.) We now solve for H:

$$H - 70 = e^{-kt+C} = e^{C}e^{-kt} = Ae^{-kt}$$

so $$H(t) = 70 + Ae^{-kt}$$

where A is some positive constant. We can determine the constant A using the initial condition $H(0) = 170$:

$$170 = 70 + Ae^{0} = 70 + A$$

so $$A = 100$$

Therefore,

$$H(t) = 70 + 100e^{-kt}$$

Question But what is k?
Answer The constant k determines the rate of cooling. Its value depends on the thing cooling, in this case the coffee and its container. Figure 19 shows two possible graphs, one with $k = 1$ and the other with $k = 0.1$.

---

[1] When we say that a quantity Q is *proportional* to a quantity R, we mean that $Q = kR$ for some constant k. The constant k is referred to as the **constant of proportionality.**

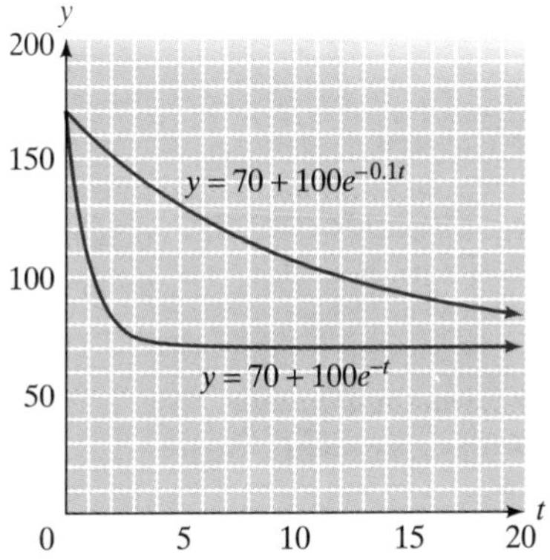

Figure 19

We can see from the graph or the formula for $H(t)$ that the temperature of the coffee will approach the air temperature exponentially.

7.6 exercises

Find the general solution of each differential equation in Exercises 1–10. Where possible, solve for y as a function of x.

1. $\frac{dy}{dx} = x^2 + \sqrt{x}$
2. $\frac{dy}{dx} = \frac{1}{x} + 3$
3. $\frac{dy}{dx} = \frac{x}{y}$
4. $\frac{dy}{dx} = \frac{y}{x}$
5. $\frac{dy}{dx} = xy$
6. $\frac{dy}{dx} = x^2y$
7. $\frac{dy}{dx} = (x+1)y^2$
8. $\frac{dy}{dx} = \frac{1}{(x+1)y^2}$
9. $x\frac{dy}{dx} = \frac{1}{y}\ln x$
10. $\frac{1}{x}\frac{dy}{dx} = \frac{1}{y}\ln x$

For each differential equation in Exercises 11–20, find the particular solution indicated.

11. $\frac{dy}{dx} = x^3 - 2x$; $y = 1$ when $x = 0$
12. $\frac{dy}{dx} = 2 - e^{-x}$; $y = 0$ when $x = 0$
13. $\frac{dy}{dx} = \frac{x^2}{y^2}$; $y = 2$ when $x = 0$
14. $\frac{dy}{dx} = \frac{y^2}{x^2}$; $y = \frac{1}{2}$ when $x = 1$
15. $x\frac{dy}{dx} = y$; $y(1) = 2$
16. $x^2\frac{dy}{dx} = y$; $y(1) = 1$
17. $\frac{dy}{dx} = x(y+1)$; $y(0) = 0$
18. $\frac{dy}{dx} = \frac{y+1}{x}$; $y(0) = 2$
19. $\frac{dy}{dx} = \frac{xy^2}{x^2+1}$; $y(0) = -1$
20. $\frac{dy}{dx} = \frac{xy}{(x^2+1)^2}$; $y(0) = 1$

Applications

21. ***Sales*** Your monthly sales of Tofu Ice Cream are falling at an instantaneous rate of 5% per month. If you currently sell 1000 quarts per month, find the differential equation that describes your change in sales, and then solve it to predict your monthly sales.

22. ***Profit*** Your monthly profit on sales of Avocado Ice Cream are rising at an instantaneous rate of 10% per month. If you currently make a profit of \$15,000 per month, find the differential equation that describes your change in profit, and solve it to predict your monthly profits.

23. ***Cooling*** A bowl of soup at 190°F is placed in a room whose air temperature is 75°F. After 10 minutes the soup has cooled to 150°F. Find the temperature of the soup as a function of time. (Refer to Example 4 for Newton's Law of Cooling.)

24. ***Heating*** Newton's Law of Heating is just the same as his Law of Cooling: The rate of change of temperature is proportional to the difference between the temperature of an object and its surroundings, whether the object is hotter or colder than its surroundings. Suppose that a pie, at 20°F, is put in an oven at 350°F. After 15 minutes

its temperature has risen to 80°F. Find the temperature of the pie as a function of time.

25. ***Market Saturation*** You have just introduced a new computer to the market. You predict that you will eventually sell 100,000 computers and that your monthly rate of sales will be 10% of the difference between the saturation value and the total number you have sold up to that point. Find a differential equation for your total sales (as a function of the month) and solve. (What are your total sales at the moment when you first introduce the computer?)

26. ***Market Saturation*** Repeat Exercise 25, assuming that monthly sales will be 5% of the difference between the saturation value (of 100,000 computers) and the total sales to that point, and assuming that you sell 5,000 computers to corporate customers before placing the computer on the open market.

27. ***Approach to Equilibrium*** The Extrasoft Toy Co. has just released its latest creation, a plush platypus named "Eggbert." The demand function for Eggbert dolls is $D(p) = 50{,}000 - 500p$ dolls per month when the price is p dollars. The supply function is $S(p) = 30{,}000 + 500p$ dolls per month when the price is p dollars. This makes the equilibrium price \$20. The Evans price adjustment model assumes that if the price is set at a value other than the equilibrium price, it will change over time in such a way that its rate of change is proportional to the shortage $D(p) - S(p)$.
   a. Write down the differential equation given by the Evans price adjustment model for the price p as a function of time.
   b. Find the general solution of the differential equation you wrote in part (a). (You will have two unknown constants, one being the constant of proportionality.)
   c. Find the particular solution in which Eggbert dolls are initially priced at \$10 and the price rises to \$12 after 1 month.

28. ***Approach to Equilibrium*** Spacely Sprockets has just released its latest model, the Dominator. The demand function is $D(p) = 10{,}000 - 1000p$ sprockets per year when the price is p dollars. The supply function is $S(p) = 8000 + 1000p$ sprockets per year when the price is p dollars.
   a. Using the Evans price adjustment model described in the preceding exercise, write the differential equation for the price $p(t)$ as a function of time.
   b. Find the general solution of the differential equation you wrote in part (a).
   c. Find the particular solution in which Dominator sprockets are initially priced at \$5 each but fall to \$3 each after 1 year.

29. ***Determining Demand*** Nancy's Chocolates estimates that the elasticity of demand for its dark chocolate truffles is $E = 0.05p - 1.5$, where p is the price per pound. Nancy's sells 20 pounds of truffles per week when the price is \$20 per pound. Find the formula expressing the demand q as a function of p. Recall that the elasticity of demand is given by

$$E = -\frac{dq}{dp} \times \frac{p}{q}$$

30. ***Determining Demand*** Nancy's Chocolates estimates that the elasticity of demand for its chocolate strawberries is $E = 0.02p - 0.5$, where p is the price per pound. Nancy's sells 30 pounds of chocolate strawberries per week when the price is \$30 per pound. Find the formula expressing the demand q as a function of p.

31. ***Logistic Equation*** There are many examples of growth in which the rate of growth is slow at first, becomes faster, and then slows again as a limit is reached. This pattern can be described by the differential equation

$$\frac{dy}{dt} = ay(L - y)$$

where a is a constant and L is the limit of y. Show by substitution that

$$y = \frac{CL}{e^{-aLt} + C}$$

is a solution to this equation, where C is an arbitrary constant.

32. ***Logistic Equation*** Using separation of variables and integration with a table of integrals or a symbolic algebra program, solve the differential equation in Exercise 31 to derive the solution given there.

Exercises 33–36 require the use of technology.

T 33. ***Market Saturation*** You have just introduced a new model of TV. You predict that the market will saturate at 2,000,000 TVs and that your total sales will be governed by the equation

$$\frac{dS}{dt} = \frac{1}{4}S(2 - S)$$

where S is the total sales in millions of TVs and t is measured in months. If you give away 1000 TV sets when you first introduce the TV, what will S be? Sketch the graph of S as a function of t. About how long will it take to saturate the market? (See Exercise 31.)

T 34. ***Epidemics*** A certain epidemic of influenza is predicted to follow the function defined by

$$\frac{dA}{dt} = \frac{1}{10}A(20 - A)$$

where A is the number of people infected (in millions) and t is the number of months after the epidemic starts. If 20,000 cases are reported initially, find $A(t)$ and sketch its graph. When is A growing fastest? How many people will eventually be affected? (See Exercise 31.)

T **35.** ***Growth of Tumors*** The growth of tumors in animals can be modeled by the Gompertz equation:

$$\frac{dy}{dt} = -ay \ln\left(\frac{y}{b}\right)$$

where y is the size of a tumor, t is time, and a and b are constants that depend on the type of tumor and the units of measurement.

a. Solve for y as a function of t.

b. If $a = 1$, $b = 10$, and $y(0) = 5$ cm^3 (with t measured in days), find the specific solution and graph it.

T **36.** ***Growth of Tumors*** Refer back to Exercise 35. Suppose that $a = 1$, $b = 10$, and $y(0) = 15$ cm^3. Find the specific solution and graph it. Comparing its graph to the one obtained in the previous exercise, what can you say about tumor growth in these instances?

Communication and Reasoning Exercises

37. What is the difference between a particular solution and the general solution of a differential equation? How do we get a particular solution from the general solution?

38. Why is there always an arbitrary constant in the general solution of a differential equation? Why are there not two or more arbitrary constants in a first-order differential equation?

39. Show by example that a **second-order** differential equation, one involving the second derivative y'', usually has two arbitrary constants in its general solution.

40. Find a differential equation that is not separable.

41. Find a differential equation with the general solution $y = 4e^{-x} + 3x + C$.

42. Explain how, knowing the elasticity of demand as a function of either price or demand, you can find the demand equation (see Exercise 29).

You're the Expert

Estimating Tax Revenues

You have just been hired by the incoming administration of your country as chief consultant for national tax policy, and you have been getting conflicting advice from the finance experts on your staff. Several of them have come up with plausible suggestions for new tax structures, and your job is to choose the plan that results in more revenue for the government.

Before you can evaluate the plans, you realize that it is essential to know your country's income distribution—that is, how many people earn how much money.[1] One might think that the most useful way of specifying income distribution would be to use a function that gives the exact number $f(x)$ of people who earn a given salary x. This would necessarily be a discrete function—it makes sense only if x happens to be a whole number of cents. There is, after all, no one earning a salary of exactly \$22,000.142567! Furthermore, this function would behave rather erratically because there are, for example, probably many more people who make a salary of exactly \$30,000 than exactly \$30,000.01. Given these problems, it is far more convenient to start with the function defined by

$$N(x) = \text{the total number of people earning between 0 and } x \text{ dollars}$$

Actually, you would want a "smoothed" version of this function. The graph of $N(x)$ might look like the one shown in Figure 20.

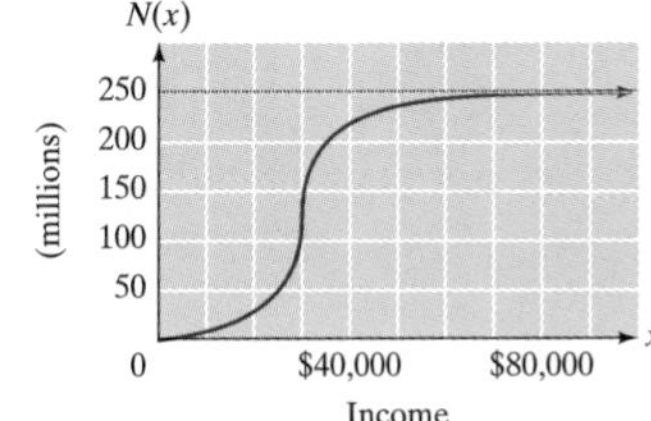

Figure 20

[1] To simplify our discussion, we are assuming that (1) all tax revenues are based on earned income and (2) everyone in the population we consider earns some income.

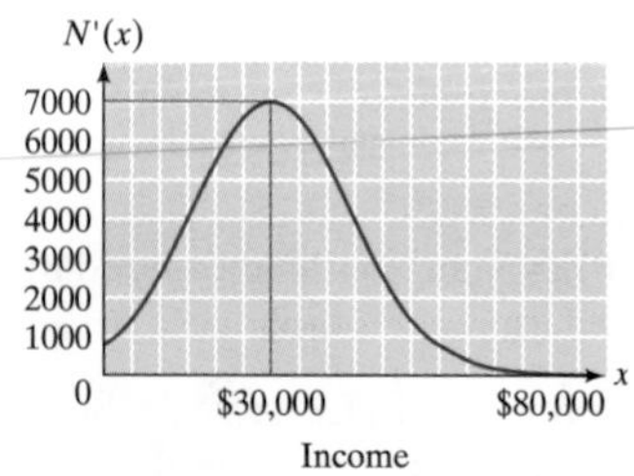

Figure 21

If you take the *derivative* of $N(x)$, you get an **income distribution function.** Its graph might look like the one shown in Figure 21.

Since the derivative measures the rate of change, its value at x is the additional number of taxpayers per $1 increase in salary. Thus, the fact that $N'(20{,}000) \approx 5500$ tells us that approximately 5500 people are earning a salary between $20,000 and $20,001. In other words, N' shows the distribution of incomes among the population—hence, the name "distribution function."[2]

You thus send a memo to your experts requesting the income distribution function for the nation. After much collection of data, they tell you that the income distribution function is

$$N'(x) = 7000\, e^{-(x-30{,}000)^2/400{,}000{,}000}$$

This is in fact the function whose graph is shown in Figure 21 and is an example of a **normal distribution.** Notice that the curve is symmetric around the median income of $30,000 and that about 7000 people are earning between $30,000 and $30,001 annually.[3]

Given this income distribution, your financial experts have come up with two possible tax policies illustrated in Figures 22 and 23.

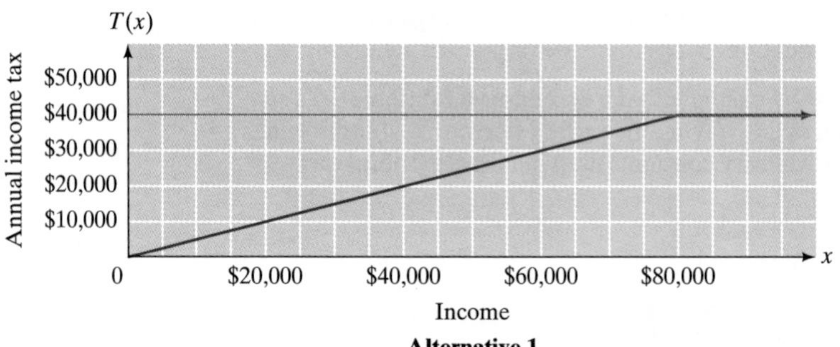

Figure 22

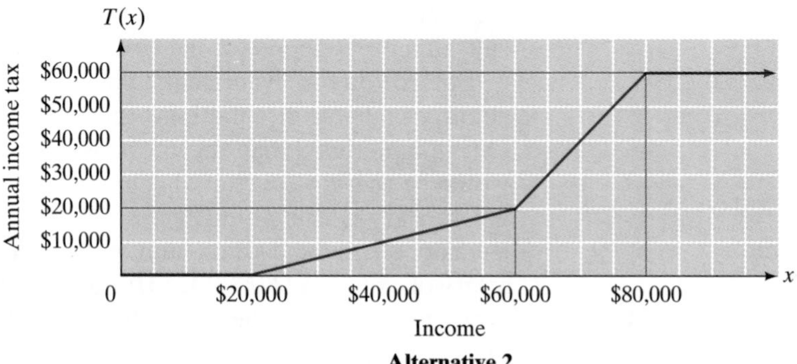

Figure 23

[2] A very similar idea is used in probability. See the optional chapter on Calculus Applied to Probability and Statistics at the web site.

[3] You might find it odd that you weren't given the original function N, but it will turn out that you don't need it. How would you compute it?

In the first alternative, all taxpayers pay half of their income in taxes, except that no one pays more than \$40,000 in taxes. In the second alternative, there are four tax brackets, described by the following table:

Income	Marginal tax rate
\$0–20,000	0%
\$20,000–60,000	50%
\$60,000–80,000	200%
Above \$80,000	0%

Now you must determine which alternative will generate more annual tax revenue.

Each of Figures 22 and 23 is the graph of a function, T. Rather than using the formulas for these particular functions, you begin by working with the general situation. You have an income distribution function N' and a tax function T, both functions of annual income. You need to find a formula for total tax revenues. First you decide to use a cutoff so that you need to work only with incomes in some finite bracket $[0, M]$; you might use, for example, $M = \$10$ million (later you will let M approach $+\infty$). Next you subdivide the interval $[0, M]$ into a large number of intervals of small width, Δx. If $[x_{k-1}, x_k]$ is a typical such interval, you wish to calculate the approximate tax revenue from people whose total incomes lie between x_{k-1} and x_k. You will then sum over k to get the total revenue.

You need to know how many people are making incomes between x_{k-1} and x_k. Because $N(x_k)$ people are making incomes *up to* x_k and $N(x_{k-1})$ people are making incomes up to x_{k-1}, the number of people who have incomes between x_{k-1} and x_k is $N(x_k) - N(x_{k-1})$. Because x_k is very close to x_{k-1}, the incomes of these people are all approximately equal to x_{k-1} dollars, so each of these taxpayers is paying an annual tax of about $T(x_{k-1})$. This gives a tax revenue of

$$\left[N(x_k) - N(x_{k-1})\right]T(x_{k-1})$$

Now you do a clever thing. You write $x_k - x_{k-1} = \Delta x$ and replace $N(x_k) - N(x_{k-1})$ by

$$\frac{N(x_k) - N(x_{k-1})}{\Delta x}\,\Delta x$$

This gives you a tax revenue of about

$$\frac{N(x_k) - N(x_{k-1})}{\Delta x}\,T(x_{k-1})\,\Delta x$$

from wage earners in the bracket $[x_{k-1}, x_k]$. Summing over k gives an approximate total revenue of

$$\sum_{k=1}^{n} \frac{N(x_k) - N(x_{k-1})}{\Delta x}\,T(x_{k-1})\,\Delta x$$

where n is the number of subintervals. The larger n is, the more accurate your estimate will be, so you take the limit of the sum as $n \to \infty$. When you do this, two things happen. First, the quantity

$$\frac{N(x_k) - N(x_{k-1})}{\Delta x}$$

approaches the derivative, $N'(x_k)$. Second, the sum, which you recognize as a Riemann sum, approaches the integral

$$\int_0^M N'(x)T(x)\,dx$$

You now take the limit as $M \to +\infty$ to get

$$\text{Total tax revenue} = \int_0^{+\infty} N'(x)T(x)\,dx$$

This improper integral is fine in theory, but the actual calculation will have to be done numerically, so you stick with the upper limit of $10 million for now. You will have to check that it is reasonable at the end (notice that, by the graph of N', it appears that extremely few, if any, people earn that much). Now you already have a formula for $N'(x)$, but you still need to write formulas for the tax functions $T(x)$ for both alternatives.

Alternative 1 The graph in Figure 22 rises linearly from 0 to 40,000 as x ranges from 0 to 80,000, and then stays constant at 40,000. The slope of the first part is $40{,}000/80{,}000 = 1/2$. The taxation function is therefore

$$T(x) = \begin{cases} \dfrac{x}{2} & \text{if } 0 \le x \le 80{,}000 \\ 40{,}000 & \text{if } x \ge 80{,}000 \end{cases}$$

To perform the integration, you need to break the integral into two pieces, the first from 0 to 80,000 and the second from 80,000 to 10,000,000. In other words,

$$R_1 = \int_0^{80{,}000} (7000e^{-(x-30{,}000)^2/400{,}000{,}000})\frac{x}{2}\,dx + \int_{80{,}000}^{10{,}000{,}000} (7000e^{-(x-30{,}000)^2/400{,}000{,}000})\,40{,}000\,dx$$

You decide not to attempt this by hand![4] You use numerical integration software to obtain a grand total of $R_1 = \$3{,}732{,}760{,}000{,}000$, or $3.73276 trillion (rounded to six significant digits).

Alternative 2 The graph in Figure 23 rises linearly from 0 to 20,000 as x ranges from 20,000 to 60,000, then rises from 20,000 to 60,000 as x ranges from 60,000 to 80,000, and then stays constant at 60,000. The slope of the first incline is $\frac{1}{2}$ and the slope of the second incline is 2 (this is why the *marginal* tax rates are 50% and 200%, respectively). The taxation function is therefore

$$T(x) = \begin{cases} 0 & \text{if } 0 \le x \le 20{,}000 \\ \dfrac{x - 20{,}000}{2} & \text{if } 20{,}000 \le x \le 60{,}000 \\ 20{,}000 + 2(x - 60{,}000) & \text{if } 60{,}000 \le x \le 80{,}000 \\ 60{,}000 & \text{if } x \ge 80{,}000 \end{cases}$$

[4] The first integral requires a substitution and integration by parts. The second cannot be done in elementary terms at all.

Values of x between 0 and 20,000 do not contribute to the integral, so

$$R_2 = \int_{20,000}^{60,000} (7000e^{-(x-30,000)^2/400,000,000})\left(\frac{x-20,000}{2}\right) dx + \int_{60,000}^{80,000} (7000e^{-(x-30,000)^2/400,000,000})\left[20,000 + 2(x-60,000)\right] dx$$
$$+ \int_{80,000}^{10,000,000} (7000e^{-(x-30,000)^2/400,000,000})\, 60,000\, dx$$

Numerical integration software gives R_2 = \$1.52016 trillion—considerably less than Alternative 1. Thus, even though Alternative 2 taxes the wealthy more heavily, it yields less total revenue.

Now what about the cutoff at \$10 million annual income? If you try either integral again with an upper limit of \$100 million, you will see no change in either result to six significant digits. There simply are not enough taxpayers earning an income above \$10,000,000 to make a difference. You conclude that your answers are sufficiently accurate and that the first alternative provides more tax revenue.

exercises

In Exercises 1–6, calculate the total tax revenue for a country with the given income distribution and tax policies (all currency in dollars).

T **1.** $N'(x) = 3000e^{-(x-10,000)^2/10,000}$; 25% tax on all income

T **2.** $N'(x) = 3000e^{-(x-10,000)^2/10,000}$; 45% tax on all income

T **3.** $N'(x) = 5000e^{-(x-30,000)^2/100,000}$; no tax on an income below \$30,000 and \$10,000 tax on any income of \$30,000 or above

T **4.** $N'(x) = 5000e^{-(x-30,000)^2/100,000}$; no tax on an income below \$50,000 and \$20,000 tax on any income of \$50,000 or above

T **5.** $N'(x) = 7000e^{-(x-30,000)^2/400,000,000}$; $T(x)$ with the following graph:

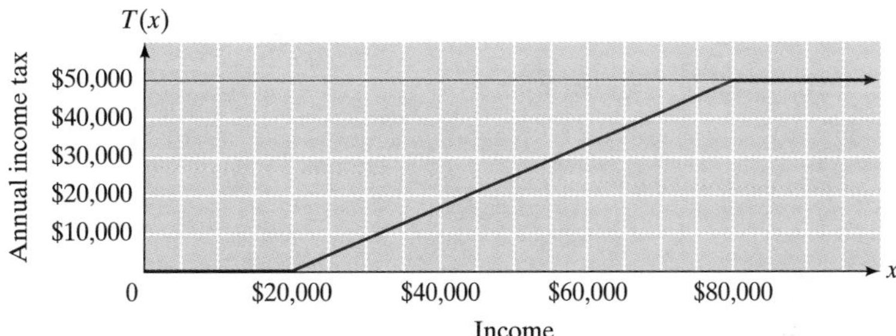

T **6.** $N'(x) = 7000e^{-(x-30,000)^2/400,000,000}$; $T(x)$ with the following graph:

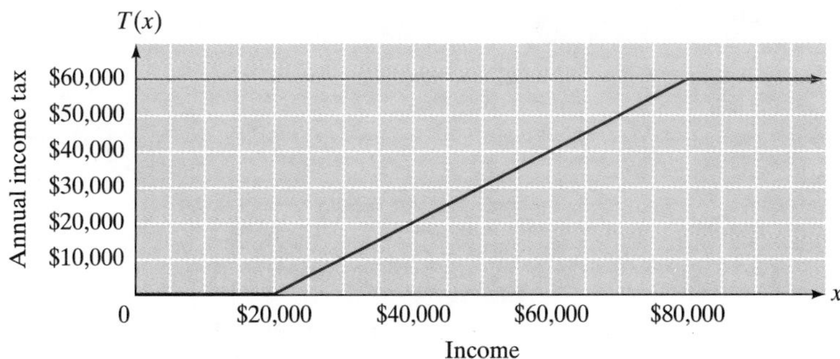

7. Let $P(x)$ be the number of people who earn more than x dollars.

a. What is $N(x) + P(x)$?

b. Show that $P'(x) = -N'(x)$.

c. Use integration by parts to show that, if $T(0) = 0$, then the total tax revenue is $\int_0^{+\infty} P(x)T'(x)\,dx$. [*Note:* You may assume that $T'(x)$ is continuous, but the result is still true if we assume only that $T(x)$ is continuous and piecewise continuously differentiable.]

8. Income tax functions T are most often described, as in the text, by tax brackets and marginal tax rates.

a. If one tax bracket is $a < x \le b$, show that $\int_a^b P(x)\,dx$ is the total income earned in the country that falls into that bracket (P as in the preceding exercise).

b. Use part (a) to explain directly why $\int_0^{+\infty} P(x)T'(x)\,dx$ gives the total tax revenue in the case where T is described by tax brackets and constant marginal tax rates in each bracket.

chapter 7 review test

1. Evaluate the following integrals.

a. $\int (x^2 + 2)e^x\,dx$

b. $\int x^2 \ln 2x\,dx$

c. $\int_{-2}^{2} (x^3 + 1)e^{-x}\,dx$

d. $\int_1^e x^2 \ln x\,dx$

e. $\int x^2 \sin x\,dx$

f. $\int e^x \sin 2x\,dx$

g. $\int_1^{+\infty} \frac{1}{x^5}\,dx$

h. $\int_0^1 \frac{1}{\sqrt{1-x}}\,dx$

2. Find the areas of the following regions.

a. Between $y = x^3$ and $y = 1 - x^3$ for x in $[0, 1]$

b. Between $y = e^x$ and $y = e^{-x}$ for x in $[0, 2]$

c. Enclosed by $y = 1 - x^2$ and $y = x^2$

d. Between $y = x$ and $y = xe^{-x}$ for x in $[0, 2]$

3. Find the average values of the following functions over the indicated intervals.

a. $f(x) = x^2e^x$ over $[0, 1]$

b. $f(x) = (x + 1)\ln x$ over $[1, 2e]$

4. Find the 2-unit moving averages of the following functions.

a. $f(x) = x^{4/3}$

b. $f(x) = \ln x$

5. Solve the following differential equations.

a. $\frac{dy}{dx} = x^2y^2$

b. $\frac{dy}{dx} = xy + 2x$

c. $xy\frac{dy}{dx} = 1; y(1) = 1$

d. $y(x^2 + 1)\frac{dy}{dx} = xy^2; y(0) = 2$

Applications

6. Sales of the bestseller *A River Burns Through It* are dropping at OHaganBooks.com. To try to bolster sales, the company is lowering the price of the book, now \$40, at a rate of \$2 per week. As a result, this week OHaganBooks.com will sell 5000 copies, and it estimates that sales will fall continuously at a rate of 10% per week. How much revenue will it earn on sales of this book over the next 8 weeks?

7. OHaganBooks.com is about to start selling a new coffee table book, *Computer Designs of the Late Twentieth Century.* It estimates the demand curve to be $q = 1000\sqrt{200 - 2p}$, and its willingness to order books from the publisher is given by the supply curve $q = 1000\sqrt{10p - 400}$.

a. Find the equilibrium price and demand.

b. Find the consumers' and producers' surpluses at the equilibrium price.

8. OHaganBooks.com keeps its cash reserves in a bank account paying 6% compounded continuously. It starts a year with \$1 million in reserves and does not withdraw or deposit any money.

a. What is the average amount it will have in the account over the course of 2 years?

b. Find the 1-month moving average of the amount it has in the account.

c. Suppose instead that the account is initially empty but that OHaganBooks.com deposits money continuously into it starting at the rate of \$100,000 per month and increasing continuously by \$10,000 per month. How much money will the company have in the account at the end of 2 years?

d. How much of the amount you found in part (c) was principal deposited and how much was interest earned?

9. The Megabucks Corporation is considering buying OHaganBooks.com. They estimate OHaganBooks.com's revenue stream at \$50 million per year, growing continuously at a 10% rate. Assuming interest rates of 6%, how much is OHaganBooks.com's revenue for the next year worth now?

10. OHaganBooks.com is shopping around for a new bank. A junior executive at one bank offers the following interesting deal: The bank will pay interest continuously at a rate equal to 0.01% of the square of the amount of money they have in the account at any time. By considering what would happen if \$10,000 was deposited in such an account, explain why the junior executive was fired shortly after this offer was made.

Additional On-Line Review

If you follow the path

Web site → Everything for Calculus → Chapter 7

you will find the following additional resources to help you review:

- A comprehensive chapter summary (including examples and interactive features)
- Additional review exercises (including interactive exercises and many with help)
- A true–false chapter quiz
- A numerical integration utility

You're the Expert

Modeling Household Income

The Millennium Real Estate Development Corporation is interested in developing housing projects for medium-sized families that have high household incomes. To decide which income bracket to target, the company has asked you, a paid consultant, for an analysis of the relationship of household size to household income and the effect of increasing household size on household income. How can you analyze the relevant data?

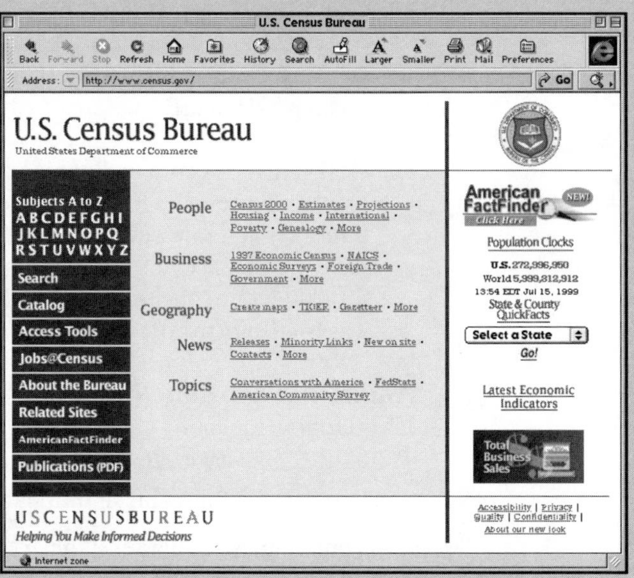

Internet Resources for This Chapter

At the web site, follow the path

Web site → Everything for Calculus → Chapter 8

where you will find links to step-by-step tutorials for the main topics in this chapter, a detailed chapter summary, a true–false quiz, and a collection of sample test questions. You will also find an on-line surface grapher and other useful resources.

Functions of Several Variables

chapter 8

8.1 Functions of Several Variables from the Numerical and Algebraic Viewpoints
8.2 Three-Dimensional Space and the Graph of a Function of Two Variables
8.3 Partial Derivatives
8.4 Maxima and Minima
8.5 Constrained Maxima and Minima and Applications
8.6 Double Integrals

Introduction

We have studied functions of a single variable extensively. But not every useful function is a function of only one variable. In fact, most are not. For example, if you operate an on-line bookstore that competes with Amazon.com, BN.com, and Borders.com, your sales may depend on those of your competitors. Your company's daily revenue might be modeled by a function such as

$$R(x, y, z) = 10{,}000 - 0.01x - 0.02y - 0.01z + 0.00001yz$$

where x, y, and z are the on-line daily revenues of Amazon.com, BN.com, and Borders.com, respectively. Here, R is a function of three variables because it *depends on* x, y, and z. As we shall see, the techniques of calculus extend readily to such functions. Among the applications we look at is optimization: finding, where possible, the maximum or minimum of a function of two or more variables.

8.1 Functions of Several Variables from the Numerical and Algebraic Viewpoints

Recall that a function of one variable is a rule for manufacturing a new number $f(x)$ from a single independent variable x. A function of two or more variables is similar, but the new number now depends on more than one independent variable.

Function of Several Variables

A **real-valued function, f, of $x, y, z, \ldots$** is a rule for manufacturing a new number, written $f(x, y, z, \ldots)$, from the values of a sequence of independent variables $(x, y, z, \ldots)$. The function f is called a **real-valued function of two variables** if there are two independent variables, a **real-valued function of three variables** if there are three independent variables, and so on.

As with functions of one variable, functions of several variables can be represented numerically (using a table of values), algebraically (using a formula), and sometimes graphically[1] (using a graph).

Quick Examples

1. $f(x, y) = x - y$ — Function of two variables
   $f(1, 2) = 1 - 2 = -1$ — Substitute 1 for x and 2 for y.
   $f(2, -1) = 2 - (-1) = 3$ — Substitute 2 for x and -1 for y.
   $f(y, x) = y - x$ — Substitute y for x and x for y.
2. $g(x, y) = x^2 + y^2$ — Function of two variables
   $g(-1, 3) = (-1)^2 + 3^2 = 10$ — Substitute -1 for x and 3 for y.
3. $h(x, y, z) = x + y + xz$ — Function of three variables
   $h(2, 2, -2) = 2 + 2 + 2(-2) = 0$ — Substitute 2 for x, 2 for y, and -2 for z.

Figure 1 illustrates the concept of a function of two variables: In goes a pair of numbers and out comes a single number.

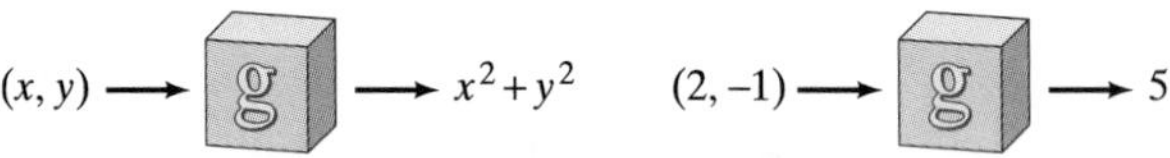

Figure 1

Let's now look at examples of interesting functions of several variables.

example 1

Cost Function

Suppose you own a company that makes two models of speakers: the Ultra Mini and the Big Stack. Your total monthly cost (in dollars) to make x Ultra Minis and y Big Stacks is given by

$$C(x, y) = 10{,}000 + 20x + 40y$$

What is the significance of each term in this formula?

Solution

The terms have meanings similar to those we saw for linear cost functions of a single variable. Let us look at the terms one at a time.

- *Constant term:* Consider the monthly cost of making no speakers at all ($x = y = 0$). We find

  $C(0, 0) = 10{,}000$ — Cost of making no speakers is \$10,000.

  Thus, the constant term 10,000 is the **fixed cost,** the amount you have to pay each month even if you make no speakers.

- *Coefficients of x and y:* Suppose you make a certain number of Ultra Minis and Big Stacks one month and the next month you increase production by one Ultra Mini. The costs are

---

[1] See the next section.

$$C(x, y) = 10{,}000 + 20x + 40y \qquad \text{First month}$$

$$C(x + 1, y) = 10{,}000 + 20(x + 1) + 40y \qquad \text{Second month}$$

$$= 10{,}000 + 20x + 20 + 40y$$

$$= C(x, y) + 20$$

Thus, each Ultra Mini adds \$20 to the total cost. We say that \$20 is the **marginal cost** of each Ultra Mini. Similarly, because of the term $40y$, each Big Stack adds \$40 to the total cost. The marginal cost of each Big Stack is \$40.

Before we go on . . . This is an example of a **linear** function of two variables. The coefficients of x and y play roles similar to that of the slope of a line. In particular, they give the rates of change of the function as each variable increases while the other stays constant (think about it). We say more about linear functions below.

Graphing Calculator

You can have a TI-83 compute $C(x, y)$ numerically as follows: In the "Y=" screen, enter

```
Y1= 10000 + 20*X + 40*Y
```

Then, to evaluate, say, $C(10, 30)$ (the cost to make 10 Ultra Minis and 30 Big Stacks), enter

```
10→X
30→Y
Y1
```

and the calculator will evaluate the function and give the answer, $C(10, 30) = 11{,}400$.

Spreadsheet

Spreadsheets handle functions of several variables easily. The following setup shows how a table of values of C can be created using values of x and y you enter:

	A	B	C
1	10	10	=10000+20*A1+40*B1
2	20	30	
3	15	0	
4	0	30	
5	30	30	
	values of x	values of y	values of $C(x, y)$

A disadvantage of this layout is that it's not easy to enter values of x and y systematically in two columns. Can you find a way to remedy this? (See Example 3 for one method.)

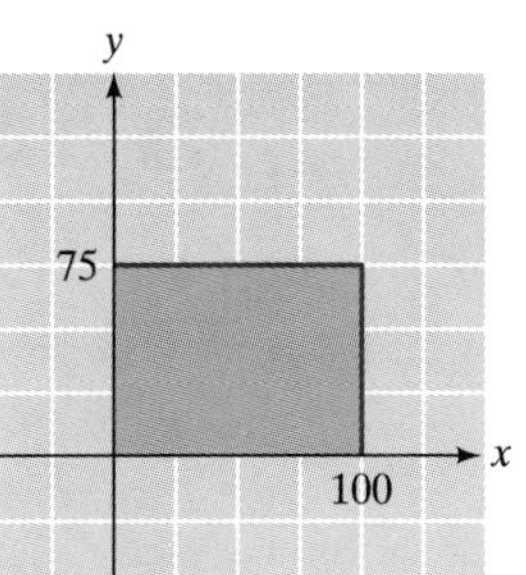

Figure 2

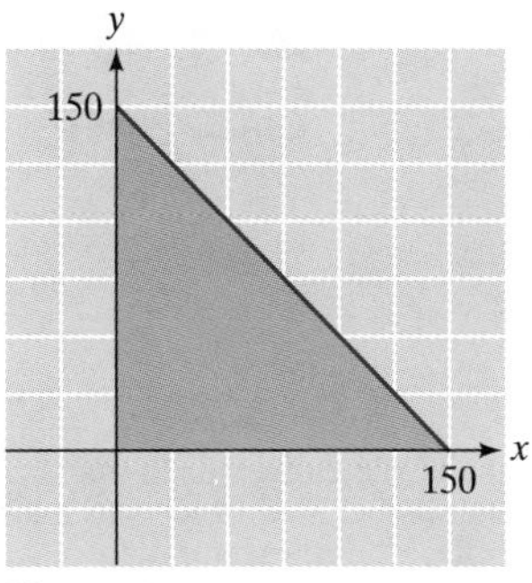

Figure 3

Before we go on . . . Which values of x and y may we substitute into $C(x, y)$? Certainly we must have $x \geq 0$ and $y \geq 0$ because it makes no sense to speak of manufacturing a negative number of speakers. Also, there is certainly some upper bound to the number of speakers that can be made in a month. The bound might take one of several forms. The number of each model may be bounded—say, $x \leq 100$ and $y \leq 75$. The inequalities $0 \leq x \leq 100$ and $0 \leq y \leq 75$ describe the region in the plane shaded in Figure 2. Another possibility is that the *total* number of speakers is bounded—say, $x + y \leq 150$. This, together with $x \geq 0$ and $y \geq 0$, describes the region shaded in Figure 3. In either case, the region shown represents the pairs (x, y) for which $C(x, y)$ is defined. Just as with a function of one variable, we call this region the **domain** of the function. As before, when the domain is not given explicitly, we agree to take the largest possible domain.

example 2

Faculty Salaries

David Katz came up with the following function for the salary of a professor with 10 years of teaching experience in a large university:

$$S(x, y, z) = 13{,}005 + 230x + 18y + 102z$$

where S is the 1969–1970 salary in dollars per year, x is the number of books the professor has published, y is the number of articles published, and z is the number of "excellent" articles published.[2] What salary do you expect a professor with 10 years of experience earned in 1969–1970 if she published 2 books, 20 articles, and 3 "excellent" articles?

Solution

All we need to do is calculate

$$S(2, 20, 3) = 13{,}005 + 230(2) + 18(20) + 102(3) = \$14{,}131$$

Before we go on . . . In Example 1 we gave a linear function of two variables. Here we have an example of a linear function of three variables. Katz came up with his model by surveying a large number of faculty members and then finding the linear function "best" fitting the data. Such models are called **linear regression** models. We see later how to find the coefficients in such a model given the data obtained in the survey.

What does this model say about the value of a single book or a single article? If a book takes 15 times as long to write as an article, how would you recommend a professor spend her writing time?

Here are two simple kinds of functions of several variables.

Linear Function

A **linear function of the variables $x_1, x_2, \ldots, x_n$** is a function of the form

$$f(x_1, x_2, \ldots, x_n) = a_0 + a_1x_1 + \cdots + a_nx_n \qquad (a_0, a_1, a_2, \ldots, a_n \text{ constants})$$

Quick Examples

1. $f(x, y) = 3x - 5y$ — Linear function of x and y
2. $C(x, y) = 10{,}000 + 20x + 40y$ — Example 1
3. $S(x, y, z) = 13{,}005 + 230x + 18y + 102z$ — Example 2

[2] See David A. Katz, "Faculty Salaries, Promotions and Productivity at a Large University," *American Economic Review,* June 1973, pp. 469–477. Prof. Katz's equation actually included other variables, such as the number of dissertations supervised; our equation assumes that all of these are zero.

Interaction Function

If we add to a linear function one or more terms of the form bx_ix_j (b constant), we get a **second-order interaction function.**

Quick Examples

1. $C(x, y) = 10{,}000 + 20x + 40y + 0.1xy$
2. $R(x, y, z) = 10{,}000 - 0.01x - 0.02y - 0.01z - 0.00001yz$

So far, we have specified functions of several variables **algebraically**—by using algebraic formulas. If you have ever studied statistics, you are probably familiar with statistical tables. These tables may also be viewed as representing functions **numerically,** as the next example shows.

example 3

Function Represented Numerically: Body Mass Index

The following table lists some values of the "body mass index," which gives a measure of the massiveness of your body, taking height into account.[3] The variable w represents weight in pounds, and h represents height in inches. A body mass index of 25 or higher is generally considered overweight.

	w	→								
h		130	140	150	160	170	180	190	200	210
↓	60	25.2	27.1	29.1	31.0	32.9	34.9	36.8	38.8	40.7
	61	24.4	26.2	28.1	30.0	31.9	33.7	35.6	37.5	39.4
	62	23.6	25.4	27.2	29.0	30.8	32.7	34.5	36.3	38.1
	63	22.8	24.6	26.4	28.1	29.9	31.6	33.4	35.1	36.9
	64	22.1	23.8	25.5	27.2	28.9	30.7	32.4	34.1	35.8
	65	21.5	23.1	24.8	26.4	28.1	29.7	31.4	33.0	34.7
	66	20.8	22.4	24.0	25.6	27.2	28.8	30.4	32.0	33.6
	67	20.2	21.8	23.3	24.9	26.4	28.0	29.5	31.1	32.6
	68	19.6	21.1	22.6	24.1	25.6	27.2	28.7	30.2	31.7
	69	19.0	20.5	22.0	23.4	24.9	26.4	27.8	29.3	30.8
	70	18.5	19.9	21.4	22.8	24.2	25.6	27.0	28.5	29.9
	71	18.0	19.4	20.8	22.1	23.5	24.9	26.3	27.7	29.1
	72	17.5	18.8	20.2	21.5	22.9	24.2	25.6	26.9	28.3
	73	17.0	18.3	19.6	20.9	22.3	23.6	24.9	26.2	27.5
	74	16.6	17.8	19.1	20.4	21.7	22.9	24.2	25.5	26.7
	75	16.1	17.4	18.6	19.8	21.1	22.3	23.6	24.8	26.0
	76	15.7	16.9	18.1	19.3	20.5	21.7	22.9	24.2	25.4

Sources: Shape Up America/National Institutes of Health/*New York Times,* May 2, 1999, p. WK3.

As the table shows, the value of the body mass index depends on two quantities: w and h. Let us write $M(w, h)$ for the body mass index function. What are $M(140, 62)$ and $M(210, 63)$?

[3] It is interesting that weight-lifting competitions are usually based on weight, rather than on body mass index. As a consequence, taller people are at a significant disadvantage in weight-lifting competitions because they must compete with shorter, stockier people of the same weight. (An extremely thin, very tall person can weigh as much as a muscular short person, although his body mass index would be significantly lower.)

Solution

We can read the answers from the table:

$$M(140, 62) = 25.4 \qquad w = 140 \text{ lb}, h = 62 \text{ in.}$$

$$M(210, 63) = 36.9 \qquad w = 210 \text{ lb}, h = 63 \text{ in.}$$

Spreadsheet

The function $M(w, h)$ is actually given by the formula

$$M(w, h) = \frac{0.45w}{(0.0254h)^2}$$

[The factor 0.45 converts the weight to kilograms, and 0.0254 converts the height to meters. If w is in kilograms and h is in meters, the formula is simpler: $M(w, h) = w/h^2$.] We can use this formula to recreate the table in Excel, as follows:

	A	B	C	D	E
1		130	140	150	160
2	60	=(B$1*0.45)/($A2*0.0254)^2			
3	61				
4	62				
5	63				

Here we have used B$1 instead of B1 for the w-coordinate because we want all references to w to use the same row (1). Similarly, we want all references to h to refer to the same column (A), so we used $A2 instead of A2.

Distance and Related Functions

Newton's Law of Gravity states that the gravitational force exerted by one particle on another depends on their masses and the distance between them. The distance between two particles in the xy-plane can be expressed as a function of their coordinates, as follows:

Distance Formula

The distance between the points $P(x_1, y_1)$ and $Q(x_2, y_2)$ is

$$d = \sqrt{(x_2 - x_1)^2 + (y_2 - y_1)^2} = \sqrt{(\Delta x)^2 + (\Delta y)^2}$$

Derivation

The distance d is shown in the figure.

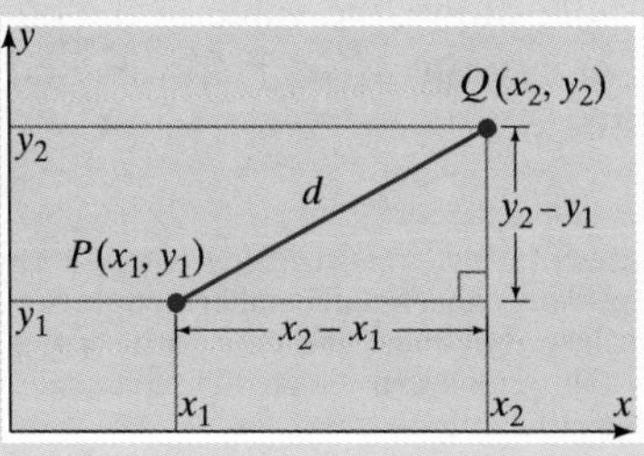

By the Pythagorean theorem applied to the right triangle shown, we get

$$d^2 = (x_2 - x_1)^2 + (y_2 - y_1)^2$$

Taking square roots (d is supposed to be a distance, so we take the positive square root), we get the distance formula. Notice that if we switch x_1 with x_2 or y_1 with y_2, we get the same result.

Quick Examples

1. The distance between the points $(3, -2)$ and $(-1, 1)$ is

$$d = \sqrt{(-1 - 3)^2 + (1 + 2)^2} = \sqrt{25} = 5$$

2. The distance from (x, y) to the origin $(0, 0)$ is

$$d = \sqrt{(x - 0)^2 + (y - 0)^2} = \sqrt{x^2 + y^2} \qquad \text{Distance to the origin}$$

The set of all points (x, y) whose distance from the origin $(0, 0)$ is a fixed quantity r is a circle centered at the origin with radius r. From the second Quick Example, we get the following equation for the circle centered at the origin with radius r:

$$\sqrt{x^2 + y^2} = r \qquad \text{Distance from the origin} = r$$

Squaring both sides gives the following equation, which we use in later sections.

Equation of the Circle of Radius *r* Centered at the Origin

$$x^2 + y^2 = r^2$$

Quick Examples

1. The circle of radius 1 centered at the origin has equation $x^2 + y^2 = 1$.

2. The circle of radius 2 centered at the origin has equation $x^2 + y^2 = 4$.

3. The circle of radius 3 centered at the origin has equation $x^2 + y^2 = 9$.

An interesting application of the distance formula is to Newton's Law of Gravity. According to Newton's law, the gravitational force exerted on a particle with mass m by another particle with mass M is given by the following function of distance:

$$F(r) = G\frac{Mm}{r^2}$$

where r is the distance between the two particles in meters, the masses M and m are given in kilograms, $G \approx 6.67 \times 10^{-11}$, and the resulting force is measured in newtons.[4]

example 4

Newton's Law of Gravity

Find the gravitational force exerted on a particle with mass m situated at the point (x, y) by another particle with mass M situated at the point (a, b). Express the answer as a function of the coordinates of the particle with mass m.

[4] A newton is the force that will cause a 1-kilogram mass to accelerate at 1 m/s^2.

Solution

The formula above for gravitational force is expressed as a function of the distance, r, between the two particles. Since we are given the coordinates of the two particles, we can express r in terms of these coordinates using the formula for distance:

$$r = \sqrt{(x-a)^2 + (y-b)^2}$$

Substituting for r, we get

$$F(x, y) = G\frac{Mm}{(x-a)^2 + (y-b)^2}$$

Before we go on . . . Notice that $F(a, b)$ is not defined because substituting $x = a$ and $y = b$ makes the denominator equal zero. Thus, the largest possible domain of F excludes the point (a, b). Since (a, b) is the only value of (x, y) for which F is not defined, we deduce that *the domain of F consists of all points (x, y) except for (a, b)*. In other words, the domain of F is the whole xy-plane with the single point (a, b) missing.[5]

Question Why have we expressed F as a function of x and y only, and not also as a function of a and b?

Answer It's a matter of interpretation. When we write F as a function of x and y, we are thinking of a and b as *constants*. For example, (a, b) could be the coordinates of the sun—which we often assume to be fixed in space—and (x, y) could be the coordinates of Earth—which is moving around the sun. In that case it is most natural to think of x and y as variable and a and b as constant. In another context we may want to consider F as a function of four variables, x, y, a, and b.

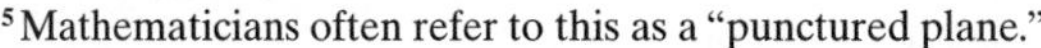

[5] Mathematicians often refer to this as a "punctured plane."

8.1 exercises

For each of the functions in Exercises 1–4, evaluate **a.** $f(0,0)$; **b.** $f(1,0)$; **c.** $f(0,-1)$; **d.** $f(a,2)$; **e.** $f(y,x)$; **f.** $f(x+h, y+k)$.

1. $f(x, y) = x^2 + y^2 - x + 1$

2. $f(x, y) = x^2 - y - xy + 1$

3. $f(x, y) = 0.2x + 0.1y - 0.01xy$

4. $f(x, y) = 0.4x - 0.5y - 0.05xy$

For each of the functions in Exercises 5–8, evaluate **a.** $g(0,0,0)$; **b.** $g(1,0,0)$; **c.** $g(0,1,0)$; **d.** $g(z,x,y)$; **e.** $g(x+h, y+k, z+l)$, *provided such a value exists.*

5. $g(x, y, z) = e^{x+y+z}$

6. $g(x, y, z) = \ln(x + y + z)$

7. $g(x, y, z) = \dfrac{xyz}{x^2 + y^2 + z^2}$

8. $g(x, y, z) = \dfrac{e^{xyz}}{x + y + z}$

9. Let $f(x, y, z) = 1.5 + 2.3x - 1.4y - 2.5z$. Complete the following sentences.

a. f _____ by _____ units for every 1 unit of increase in x.

b. f _____ by _____ units for every 1 unit of increase in y.

c. _____ by 2.5 units for every _____.

10. Let $g(x, y, z) = 0.01x + 0.02y - 0.03z - 0.05$. Complete the following sentences.

a. g _____ by _____ units for every 1 unit of increase in z.

b. g _____ by _____ units for every 1 unit of increase in x.

c. _____ by 0.02 unit for every _____.

In Exercises 11–18, classify each function as linear, interaction, or neither.

11. $L(x, y) = 3x - 2y + 6xy - 4y^2$

12. $L(x, y, z) = 3x - 2y + 6xz$

13. $P(x_1, x_2, x_3) = 0.4 + 2x_1 - x_3$

14. $Q(x_1, x_2) = 4x_2 - 0.5x_1 - x_1^2$

15. $f(x, y, z) = \dfrac{x + y - z}{3}$

16. $g(x, y, z) = \dfrac{xz - 3yz + z^2}{4z} \quad (z \neq 0)$

17. $g(x, y, z) = \dfrac{xz - 3yz + z^2y}{4z} \quad (z \neq 0)$

18. $f(x, y) = x + y + xy + x^2y$

In Exercises 19 and 20, use the given tabular representation of the function f to compute the quantities asked for.

19.

		x →			
		10	20	30	40
y	10	−1	107	162	−3
↓	20	−6	194	294	−14
	30	−11	281	426	−25
	40	−16	368	558	−36

a. $f(20, 10)$ **b.** $f(40, 20)$ **c.** $f(10, 20) - f(20, 10)$

20.

		x →			
		10	20	30	40
y	10	162	107	−5	−7
↓	20	294	194	−22	−30
	30	426	281	−39	−53
	40	558	368	−56	−76

a. $f(10, 30)$ **b.** $f(20, 10)$ **c.** $f(10, 40) + f(10, 20)$

In Exercises 21 and 22, use a spreadsheet or some other method to complete the given tables.

21. $P(x, y) = x - 0.3y + 0.45xy$

		x →			
		10	20	30	40
y	10				
↓	20				
	30				
	40				

22. $Q(x, y) = 0.4x + 0.1y - 0.06xy$

		x →			
		10	20	30	40
y	10				
↓	20				
	30				
	40				

23. The following statistical table lists some values of the "inverse F distribution" ($\alpha = 0.05$):

	n	→				
d		1	2	3	4	5
↓	1	161.4	199.5	215.7	224.6	230.2
	2	18.51	19.00	19.16	19.25	19.30
	3	10.13	9.552	9.277	9.117	9.013
	4	7.709	6.944	6.591	6.388	6.256
	5	6.608	5.786	5.409	5.192	5.050
	6	5.987	5.143	4.757	4.534	4.387
	7	5.591	4.737	4.347	4.120	3.972
	8	5.318	4.459	4.066	3.838	3.688
	9	5.117	4.256	3.863	3.633	3.482
	10	4.965	4.103	3.708	3.478	3.326
	n	→				
d		6	7	8	9	10
↓	1	234.0	236.8	238.9	240.5	241.9
	2	19.33	19.35	19.37	19.39	19.40
	3	8.941	8.887	8.812	8.812	8.785
	4	6.163	6.094	5.999	5.999	5.964
	5	4.950	4.876	4.772	4.772	4.735
	6	4.284	4.207	4.099	4.099	4.060
	7	3.866	3.787	3.677	3.677	3.637
	8	3.581	3.500	3.388	3.388	3.347
	9	3.374	3.293	3.179	3.179	3.137
	10	3.217	3.135	3.020	3.020	2.978

In Excel, you can compute the value of this function at (n, d) by the formula

`=FINV(0.05,n,d)` The 0.05 is the value of alpha (α).

Use Excel to re-create this table.

24. The formula for the body mass index $M(w, h)$, if w is given in kilograms and h is given in meters, is

$$M(w, h) = \frac{w}{h^2}$$ See Example 3.

Use this formula to complete the following table in Excel:

	w	→						
h		100	110	120	130	140	150	160
↓	2							
	2.05							
	2.1							
	2.15							
	2.2							
	2.25							
	2.3							
	2.35							
	2.4							
	2.45							
	2.5							

T *In Exercises 25–28, use either a graphing calculator or a spreadsheet to complete each table. Express all answers as decimals rounded to four decimal places.*

25.

x	y	$f(x, y) = x^2\sqrt{1 + xy}$
3	1	
1	15	
0.3	0.5	
56	4	

26.

x	y	$f(x, y) = x^2e^y$
0	2	
−1	5	
1.4	2.5	
11	9	

27.

x	y	$f(x, y) = x\ln(x^2 + y^2)$
3	1	
1.4	−1	
e	0	
0	e	

28.

x	y	$f(x, y) = \dfrac{x}{x^2 - y^2}$
−1	2	
0	0.2	
0.4	2.5	
10	0	

29. Brand Z's annual sales are affected by the sales of related products X and Y as follows: Each \$1 million increase in sales of brand X causes a \$2.1 million decline in sales of brand Z, whereas each \$1 million increase in sales of brand Y results in an increase of \$0.4 million in sales of brand Z. Currently, brands X, Y, and Z are each selling \$6 million per year. Model the sales of brand Z using a linear function.

30. Let $f(x, y, z) = 43.2 - 2.3x + 11.3y - 4.5z$. Complete the following: An increase of 1 in the value of y causes the value of f to _____ by _____, whereas increasing the value of x by 1 and _____ the value of z by _____ causes a decrease of 11.3 in the value of f.

In Exercises 31–34, find the distance between the given pairs of points.

31. $(1, -1)$ and $(2, -2)$

32. $(1, 0)$ and $(6, 1)$

33. $(a, 0)$ and $(0, b)$

34. (a, a) and (b, b)

35. Find the value of k such that $(1, k)$ is equidistant from $(0, 0)$ and $(2, 1)$.

36. Find the value of k such that (k, k) is equidistant from $(-1, 0)$ and $(0, 2)$.

37. Describe the set of points (x, y) such that $(x - 2)^2 + (y + 1)^2 = 9$.

38. Describe the set of points (x, y) such that $(x + 3)^2 + (y - 1)^2 = 4$.

Applications

39. ***Marginal Cost*** Your weekly cost (in dollars) to manufacture x cars and y trucks is

$$C(x, y) = 240{,}000 + 6000x + 4000y$$

What is the marginal cost of a car? Of a truck?

40. ***Marginal Cost*** Your weekly cost (in dollars) to manufacture x bicycles and y tricycles is

$$C(x, y) = 24{,}000 + 60x + 20y$$

What is the marginal cost of a bicycle? Of a tricycle?

41. ***Marginal Cost*** Your sales of on-line video and audio clips are booming. Your Internet provider, Mindbrook.com, wants to get in on the action and has offered you unlimited technical assistance and consulting if you agree to pay Mindbrook 3¢ for every video clip and 4¢ for every audio clip you sell on the site. Furthermore, Mindbrook agrees to charge you only \$10 per month to host your site. Set up a (monthly) cost function for the scenario and describe each variable.

42. ***Marginal Cost*** Your Cabaret nightspot "Jazz on Jupiter" has become an expensive proposition: You are paying monthly costs of \$50,000 just to keep the place running. On top of that, your regular cabaret artist is charging \$3000 per performance, and your jazz ensemble is charging \$1000 per hour. Set up a (monthly) cost function for the scenario and describe each variable.

43. ***Minivan Sales*** Chrysler's percentage share of the U.S. minivan market in the period 1993–1994 could be approximated by the linear function

$$c(x, y, z) = 72.3 - 0.8x - 0.2y - 0.7z$$

where x is the percentage (0–100) of the market held by foreign manufacturers, y is General Motors' percentage share, and z is Ford's percentage share.

a. Results for the third quarter of 1994 showed Chrysler's share as 38.8%, GM's share as 20.1%, and Ford's share as 32.9%. According to the model, what was the share held by foreign manufacturers?

b. Which of the three competitors would you regard as representing the greatest potential damage to Chrysler's minivan sales? Why?

Source: The authors' model from data by Ford Motor Company/*New York Times,* November 9, 1994, p. D5.

44. ***Minivan Sales*** Refer back to the model in Exercise 43. Use the fact that the variables x, y, z, and c together account for 100% of all minivan sales in the United States to find c as a function of y and z only. What does your model say about Chrysler's domestic competitors?

45. ***On-Line Revenue*** Let us look once again at the example we used to introduce the chapter. Your on-line bookstore is in direct competition with Amazon.com, BN.com, and Borders.com. Your company's daily revenue in dollars is given by

$$R(x, y, z) = 10{,}000 - 0.01x - 0.02y - 0.01z + 0.00001yz$$

where x, y, and z are the on-line daily revenues of Amazon.com, BN.com, and Borders.com, respectively.

a. If, on a certain day, Amazon.com shows a revenue of \$12,000, while BN.com and Borders.com each show \$5000, what does the model predict for your company's revenue that day?

b. If Amazon.com and BN.com each show a daily revenue of \$5000, give an equation showing how your daily revenue depends on that of Borders.com.

46. ***On-Line Revenue*** Repeat Exercise 45 using the revised revenue function

$$R(x, y, z) = 20{,}000 - 0.02x - 0.04y - 0.01z + 0.00001yz$$

Exercises 47–50 involve Cobb-Douglas productivity functions. These functions have the form

$$P(x, y) = Kx^a y^{1-a}$$

where P stands for the number of items produced per year, x is the number of employees, and y is the annual operating budget. (The numbers K and a are constants that depend on the situation we are looking at, with $0 \le a \le 1$.*)*

47. ***Productivity*** How many items will be produced per year by a company that has 100 employees and an annual operating budget of \$500,000 if $K = 1000$ and $a = 0.5$? (Round your answer to one significant digit.)

48. ***Productivity*** How many items will be produced per year by a company that has 50 employees and an annual operating budget of \$1,000,000 if $K = 1000$ and $a = 0.5$? (Round your answer to one significant digit.)

49. ***Modeling Production with Cobb-Douglas*** Two years ago my piano manufacturing plant employed 1000 workers, had an operating budget of \$1 million, and turned out 100 pianos. Last year I slashed the operating budget to \$10,000 and production dropped to 10 pianos.

a. Use the data for each of the 2 years and the Cobb-Douglas formula to obtain two equations in K and a.

b. Take the logs of both sides in each equation and obtain two linear equations in a and $\log K$.

c. Solve these equations to obtain values for a and K.

d. Use these values in the Cobb-Douglas formula to predict production if I increase the operating budget back to \$1 million but lay off half the work force.

50. ***Modeling Production with Cobb-Douglas*** Repeat Exercise 49 using the following data: Two years ago: 1000 employees, \$1 million operating budget, 100 pianos. Last year: 1000 employees, \$100,000 operating budget, 10 pianos.

51. ***Modeling Spending with a Linear Function*** The following table shows total U.S. personal income and consumer spending (in trillions of dollars) for 3 months in 1992:[6]

	April	July	November
Income	5.05	5.05	5.1
Spending	4	4.05	4.15

Source: National Association of Purchasing Management/Department of Commerce/*New York Times,* June 2, 1993, Section 3, p. 1.

Model monthly spending as a function of monthly income and time using a linear function of the form

$$s(i, t) = ai + bt + c \qquad (a, b, c \text{ constants})$$

where s represents monthly consumer spending, i represents monthly income (both in trillions of dollars), and t represents time in months since April 1992. (Round the constants to two decimal places.)

52. ***Modeling International Investments with a Linear Function*** The following table shows the annual flow of private investment and government loans (in billions of dollars) to developing countries in the indicated years:

	1986	1990	1992
Private	20	40	100
Government	40	60	60

Source: World Bank/*New York Times,* December 17, 1993, p. D1.

Model annual private investment as a function of annual government investment and time using a linear function of the form

$$p(g, t) = ag + bt + c \qquad (a, b, c \text{ constants})$$

where p represents annual private investment, g represents annual government investment, and t represents the time in years since 1986.

53. ***Television Ratings*** Based on data for the years 1991–1995, Fox TV's prime-time rating could be approximated by the linear function

$$F(a, c, n) = 3.1a - 0.27c + 0.87n - 36.7$$

---

[6] We have rounded all figures to the nearest \$0.05 trillion.

where a was ABC's prime-time rating, c was CBS's, and n was NBC's.[7]

a. Based on this model, which of Fox's competitors represents the greatest potential threat to its prime-time ratings?

b. Based on this model, which of Fox's competitors represents the least competition for its prime-time ratings?

c. Results for the beginning of the 1994–1995 season showed these ratings: ABC: 12.2; CBS: 11.3; NBC: 11.3. Use the model to estimate Fox's ratings for the same period.

d. Find an equation that expresses ABC's ratings as a function of those of Fox, NBC, and CBS.

Source: A. C. Nielsen Company/*New York Times,* February 13, 1995, p. D1.

54. ***Television Ratings*** Repeat Exercise 53, this time using the following model for prime-time ratings among adults aged 18–49:

$$F(a, c, n) = 3a - 0.2c + n - 23.3$$

and the following data for part (c): ABC: 7.5; CBS: 5.4; NBC: 7.6.

Source: A. C. Nielsen Company/*New York Times,* February 13, 1995, p. D1.

55. ***Demand for Beer*** Economist Richard Stone obtained a demand function of the following form for beer in pre-World War II Great Britain:

$$Q(y, p, r) = Ky^{-0.023}p^{-1.040}r^{0.939}$$

where Q is the value of total annual sales of beer, y is the total real income in Great Britain, p is the average retail price of beer, and r is the average retail price of all other commodities. K is a positive constant that depends on the units of beer and currency.

a. Does the demand for beer increase or decrease with increasing values of r?

b. If $K = 200$, find $Q(2 \times 10^8, 0.5, 500)$ to the nearest whole number, and interpret your answer.

Source: Richard Stone, "The Analysis of Market Demand," *Journal of the Royal Statistical Society,* Vol. 108, 1945, pp. 286–382.

56. ***The Logistic Function*** One form of the logistic equation[8] is

$$f(r, a, t) = \frac{K}{1 + e^{-r(t-a)}} + L$$

where $K > 0$ and $L \geq 0$ are constants and r is restricted to be positive.

a. Does the value of f increase or decrease with increasing t?

b. Does the value of f increase or decrease with increasing a?

c. Assume that $K = 1$ and $L = a = 0$. Use a graphing calculator (or some other method) to determine what the effect of increasing r is on the graph of f versus t.

57. ***Utility Function*** Suppose your newspaper is trying to decide between two competing desktop publishing software packages, Macro Publish and Turbo Publish. You estimate that if you purchase x copies of Macro Publish and y copies of Turbo Publish, your company's daily productivity will be

$$U(x, y) = 6x^{0.8}y^{0.2} + x$$

where $U(x, y)$ is measured in pages per day (U is called a *utility function*). If $x = y = 10$, calculate the effect of increasing x by one unit, and interpret the result.

58. ***Housing Costs***[9] The cost C (in dollars) of building a house is related to the number k of carpenters used and the number e of electricians used by

$$C(k, e) = 15{,}000 + 50k^2 + 60e^2$$

If $k = e = 10$, compare the effects of increasing k by one unit and of increasing e by one unit. Interpret the result.

59. ***Volume*** The volume of an ellipsoid with cross-sectional radii a, b, and c is $V(a, b, c) = \frac{4}{3}\pi abc$.

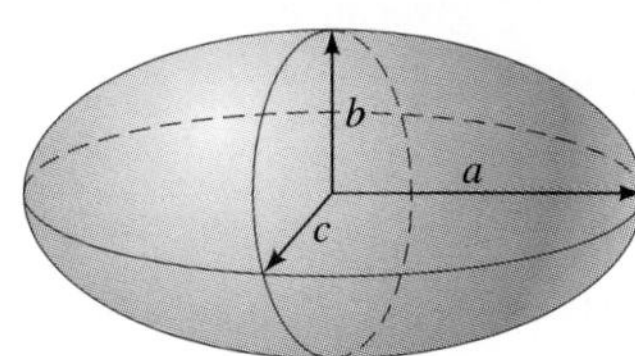

a. Find at least two sets of values for a, b, and c such that $V(a, b, c) = 1$.

b. Find the value of a such that $V(a, a, a) = 1$, and describe the resulting ellipsoid.

60. ***Volume*** The volume of a right elliptical cone with height h and radii a and b of its base is $V(a, b, h) = \frac{1}{3}\pi abh$.

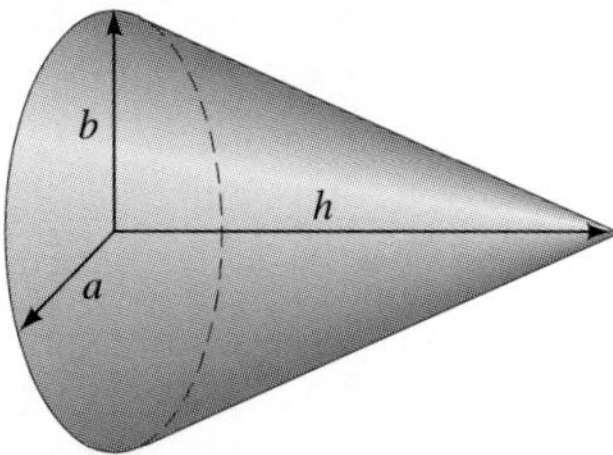

a. Find at least two sets of values for a, b, and h such that $V(a, b, h) = 1$.

b. Find the value of a such that $V(a, a, a) = 1$, and describe the resulting cone.

[7] For Exercises 53 and 54, each rating point represents 1% of all households with television sets, whether or not they are in use. At the start of 1995, this corresponded to 954,000 households. The model is a linear regression.

[8] See *You're the Expert* in Chapter 2 for a discussion of the logistic equation.

[9] Based on an exercise in *Introduction to Mathematical Economics* by A. L. Ostrosky, Jr., and J. V. Koch (Waveland Press, Illinois, 1979).

T 61. ***Level Curves*** The height of each point in a hilly region is given as a function of its coordinates by the formula

$$f(x, y) = y^2 - x^2$$

a. Use graphing technology to plot the curves on which the height is 0, 1, and 2 on the same set of axes. These are called **level curves of *f*.**
b. Without using graphing technology, sketch the curve $f(x, y) = 3$.
c. Without using graphing technology, sketch the curves $f(y, x) = 1$ and $f(y, x) = 2$.

T 62. ***Isotherms*** The temperature (in degrees Fahrenheit) at each point in a region is given as a function of the coordinates by the formula

$$T(x, y) = 60.5(x - y^2)$$

a. Use a graphing calculator to sketch the curves on which the temperature is 0°, 30°, and 90°. These curves are called **isotherms.**
b. Without using a graphing calculator, sketch the isotherms corresponding to 20°, 50°, and 100°.
c. What do the isotherms corresponding to negative temperatures look like?

63. ***Pollution*** The burden of human-made aerosol sulfate in the earth's atmosphere, in grams per square meter, is

$$B(x, n) = \frac{xn}{A}$$

where x is the total weight of aerosol sulfate emitted into the atmosphere per year and n is the number of years it remains in the atmosphere. A is the surface area of the earth, approximately 5.1×10^{14} square meters.
a. Calculate the burden, given the 1995 estimated values of $x = 1.5 \times 10^{14}$ grams per year and $n = 5$ days.
b. Read the problem again. If x and n are as stated, what does the function $W(x, n) = xn$ measure?
Source: Robert J. Charlson and Tom M. L. Wigley, "Sulfate Aerosol and Climatic Change," *Scientific American,* February 1994, pp. 48–57.

64. ***Pollution*** The amount of aerosol sulfate (in grams) was approximately 45×10^{12} grams in 1940 and has been increasing exponentially ever since, with a doubling time of approximately 20 years. Use the model from Exercise 63 to give a formula for the atmospheric burden of aerosol sulfate as a function of the time t in years since 1940 and the number of years n it remains in the atmosphere.
Source: See the source in Exercise 63.

65. ***Alien Intelligence*** Frank Drake, an astronomer at the University of California at Santa Cruz, devised the following equation to estimate the number of planet-based civilizations in our Milky Way galaxy willing and able to communicate with Earth:

$$N(R, f_p, n_e, f_l, f_i, f_c, L) = R f_p n_e f_l f_i f_c L$$

where

R = the number of new stars formed in our galaxy each year

f_p = the fraction of those stars that have planetary systems

n_e = the average number of planets in each such system that can support life

f_l = the fraction of such planets on which life actually evolves

f_i = the fraction of life-sustaining planets on which intelligent life evolves

f_c = the fraction of intelligent-life–bearing planets on which the intelligent beings develop the means and the will to communicate over interstellar distances

L = the average lifetime of such technological civilizations (in years)

a. What would be the effect on N if any one of the variables were doubled?
b. How would you modify the formula if you were interested only in the number of intelligent-life–bearing planets in the galaxy?
c. How could one convert this function into a linear function?
d. (For discussion) Try to come up with an estimate of N.
Source: "First Contact," *New York Times,* October 6, 1992, p. C1

66. ***More Alien Intelligence*** The formula in Exercise 65 restricts attention to planet-based civilizations in our galaxy. Give a formula that includes intelligent planet-based aliens from the galaxy Andromeda. (Assume that all the variables used in the formula for the Milky Way have the same values for Andromeda.)

Communication and Reasoning Exercises

67. Illustrate by means of an example how a real-valued function of the two variables x and y gives different real-valued functions of one variable when we restrict y to be different constants.

68. Give an example of a function of the two variables x and y with the property that interchanging x and y has no effect.

69. Give an example of a function f of the two variables x and y with the property that $f(x, y) = -f(y, x)$.

70. Suppose that $C(x, y)$ represents the cost of x CDs and y cassettes. If $C(x, y + 1) < C(x + 1, y)$ for every $x \geq 0$ and $y \geq 0$, what does this tell about the cost of CDs and cassettes?

8.2 Three-Dimensional Space and the Graph of a Function of Two Variables

Just as functions of a single variable have graphs, so do functions of two or more variables. Recall that the graph of $f(x)$ consisted of all points $(x, f(x))$ in the xy-plane. By analogy, we would like to say that the graph of a function of *two* variables, $f(x, y)$, consists of all points of the form $(x, y, f(x, y))$. Thus, we need three axes: the x-, y-, and z-axes. In other words, our graph lives in **three-dimensional space,** or **3-space.**[1]

Just as we had two mutually perpendicular axes in two-dimensional space (the "xy-plane"), so we have three mutually perpendicular axes in three-dimensional space (Figure 4). In both 2-space and 3-space, the axis labeled with the last letter goes up. Thus, the z direction is the "up" direction in 3-space, rather than the y direction.

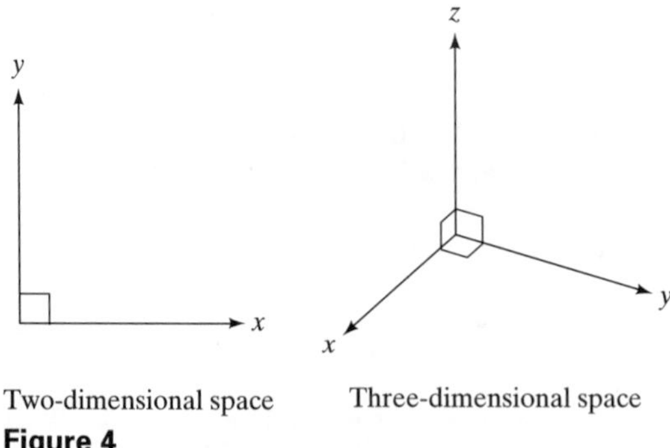

Figure 4

Three important planes are associated with these axes: the xy-plane, the yz-plane, and the xz-plane. These planes are shown in Figure 5.

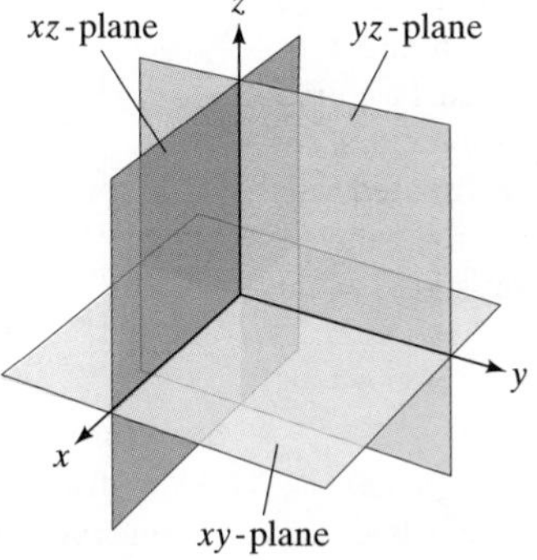

Figure 5

Any two of these planes intersect in one of the axes (for example, the xy- and xz-planes intersect in the x-axis) and all three meet at the origin. Notice that the xy-plane consists of all points with z-coordinate zero, the xz-plane consists of all points with $y = 0$, and the yz-plane consists of all points with $x = 0$.

---

[1] If we were dealing instead with a function of *three* variables, then we would need to go to *four-dimensional* space. Here we run into visualization problems (to say the least!), so we don't discuss the graphs of functions of three or more variables in this text.

In 3-space, each point has *three* coordinates, as you might expect: the x-coordinate, the y-coordinate, and the z-coordinate. To see how this works, look at the following examples.

example 1

Plotting Points in Three Dimensions
Locate the points $P(1, 2, 3)$, $Q(-1, 2, 3)$, $R(1, -1, 0)$, and $S(1, 2, -2)$ in 3-space.

Solution
To locate P, the procedure is similar to the one we used in 2-space: Start at the origin, proceed 1 unit in the x direction, then proceed 2 units in the y direction, and finally proceed 3 units in the z direction. We wind up at the point P shown in Figure 6(a). Here is another, extremely useful way of thinking about the location of P. First, look at the x- and y-coordinates and get the point $(1, 2)$ in the xy-plane. The point we want is then three units vertically above the point $(1, 2)$ because the z-coordinate of a point is just its height. This strategy is shown in Figure 6(b).

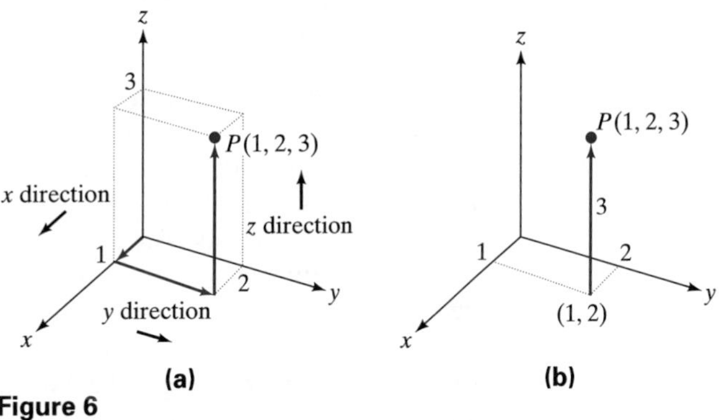

Figure 6

Plotting the points Q, R, and S is similar, using the convention that negative coordinates correspond to moves back, left, or down (see Figure 7).

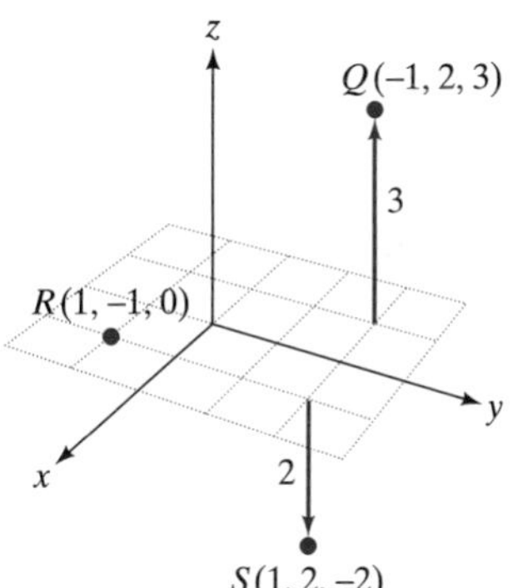

Figure 7

Before we go on . . . Here is yet another approach to locating points in three dimensions. Consider again the point $P(1, 2, 3)$ and imagine a rectangular box situated with one of its corners at the origin, as shown in Figure 8. The corner opposite the origin is the point P.

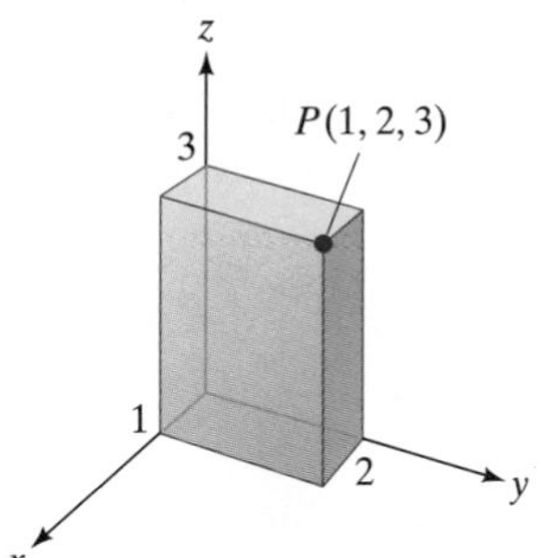

Figure 8

Remember this:

The z-coordinate of a point is its height above the xy-plane.

Our next task is to describe the graph of a function $f(x, y)$ of two variables.

Graph of a Function of Two Variables

The **graph of the function f of two variables** is the set of all points $(x, y, f(x, y))$ in three-dimensional space where we restrict the values of (x, y) to lie in the domain of f. In other words, the graph is the set of all the points (x, y, z) with $z = f(x, y)$.

For *every* point (x, y) in the domain of f, the z-coordinate of the corresponding point on the graph is given by evaluating the function at (x, y). Thus, there is a point on the graph above *every* point in the domain of f, so that the graph is usually a *surface* of some sort.

example 2

Graph of a Function of Two Variables
Describe the graph of $f(x, y) = x^2 + y^2$.

Solution

Your first thought might be to make a table of values. You could choose some values for x and y and then, for each such pair, calculate $z = x^2 + y^2$. For example, you might get the following table:

		$x \to$		
		−1	0	1
y	−1	2	1	2
↓	0	1	0	1
	1	2	1	2

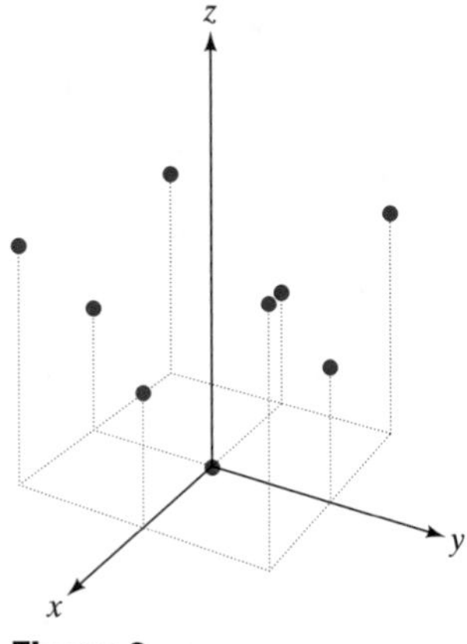

Figure 9

This table gives the following nine points on the graph of f: $(-1, -1, 2)$, $(-1, 0, 1)$, $(-1, 1, 2)$, $(0, -1, 1)$, $(0, 0, 0)$, $(0, 1, 1)$, $(1, -1, 2)$, $(1, 0, 1)$, and $(1, 1, 2)$. These points are plotted in Figure 9. The points on the xy-plane we chose for our table are the grid points in the xy-plane, and the corresponding points on the graph are marked with solid dots. The problem is that this small number of points hardly tells us what the surface looks like, and even if we plotted more points, it is not clear that we would get anything more than a mass of dots on the page.

What can we do? There are several alternatives. One place to start is to use technology to draw the graph.[2] We then obtain something like Figure 10. This particular surface is called a **paraboloid.**

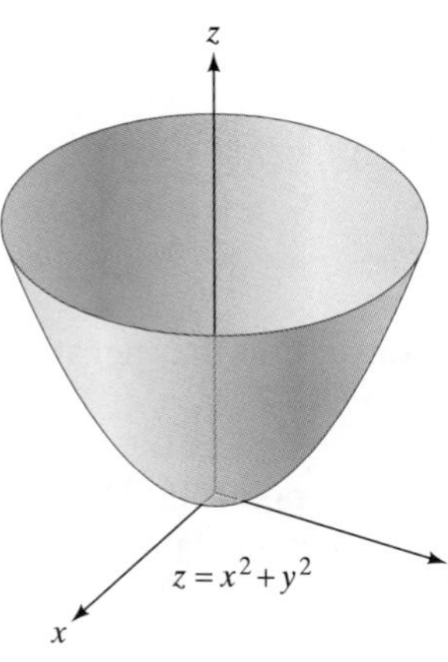

Figure 10

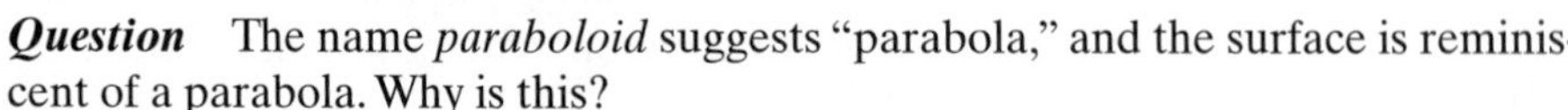

Question The name *paraboloid* suggests "parabola," and the surface is reminiscent of a parabola. Why is this?

Answer In answering this question, we see a useful technique for analyzing a graph. The technique is to slice through the surface with various planes. This takes a little while to explain.

If we slice vertically through this surface along the yz-plane, we get the picture in Figure 11. The front edge, where we cut, looks like a parabola, and it is one. To see why, note that the yz-plane is the set of points where $x = 0$. To get the intersection of $x = 0$ and $z = x^2 + y^2$, we substitute $x = 0$ in the second equation, getting $z = y^2$. This is the equation of a parabola in the yz-plane. Similarly, we can slice through the surface with the xz-plane by setting $y = 0$. This gives the parabola $z = x^2$ in the xz-plane (Figure 12).

We can also look at horizontal slices through the surface—that is, slices by planes parallel to the xy-plane. These are given by setting $z = c$ for various numbers c. For example, if we set $z = 1$, we see only the points with height 1. Substituting in the equation $z = x^2 + y^2$ gives the equation

$$1 = x^2 + y^2$$

which is the equation of a circle of radius 1. If we set $z = 4$, we get the equation of a circle of radius 2:

$$4 = x^2 + y^2$$

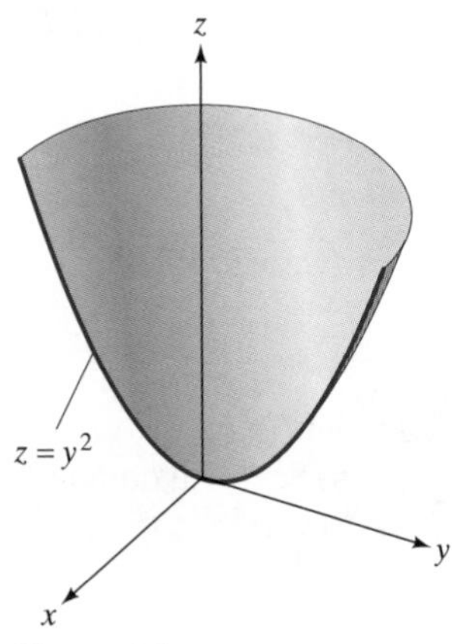

Figure 11

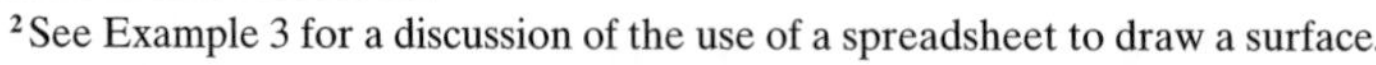

[2] See Example 3 for a discussion of the use of a spreadsheet to draw a surface.

In general, if we slice through the surface at height $z = c$, we get a circle (of radius $\sqrt{c}$). Figure 13 shows several of these circles.

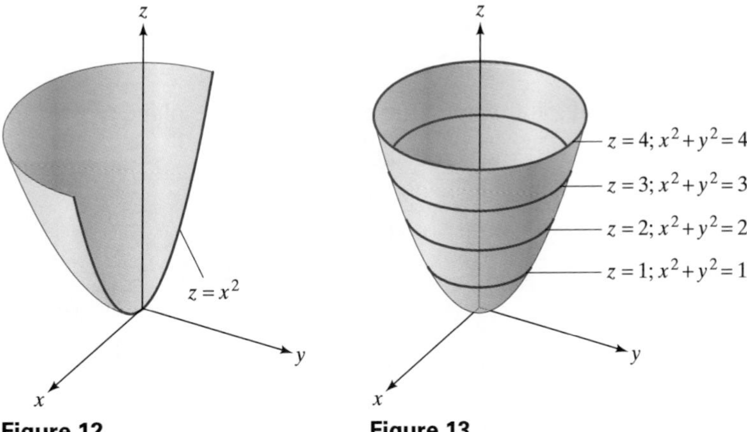

Figure 12 **Figure 13**

Looking at these circular slices, we see that this surface is the one we get by taking the parabola $z = x^2$ and spinning it around the z-axis. This is an example of what is known as a **surface of revolution.**

Before we go on . . . Notice that each horizontal slice through the surface was obtained by putting $z =$ *constant.* This gave us an equation in x and y that describes a curve. These curves are called the **level curves** of the surface $z = f(x, y)$. In this example the equations are of the form $x^2 + y^2 =$ *constant,* and so the level curves are circles. Figure 14 shows the level curves for $c = 0, 1, 2, 3$, and 4.

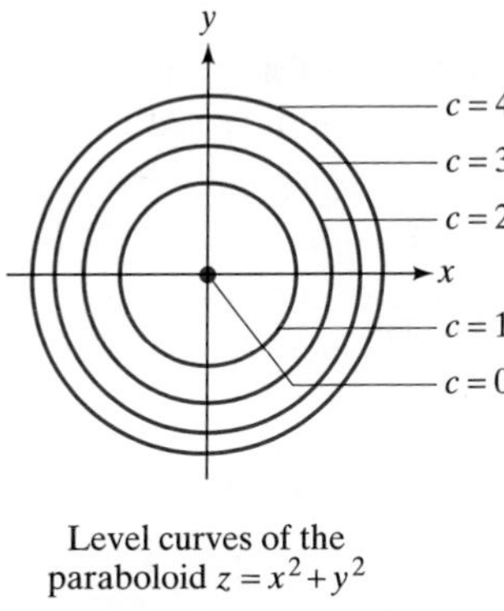

Level curves of the paraboloid $z = x^2 + y^2$

Figure 14

The level curves give a contour map or topographical map of the surface. Each curve shows all of the points on the surface at a particular height c. You can use this contour map to visualize the shape of the surface. In your mind's eye, move the contour at $c = 1$ to a height of 1 unit above the xy-plane, the contour at $c = 2$ to a height of 2 units above the xy-plane, and so on. You end up with something like Figure 13.

The following summary includes the techniques we have just used plus some additional ones.

Analyzing the Graph of a Function of Two Variables

If possible, use technology to render the graph of a given function $z = f(x, y)$. Given the function $z = f(x, y)$, you can analyze its graph as follows.

Step 1
Obtain the x-, y-, and z-intercepts (the places where the plane crosses the coordinate axes).

x-intercept(s): Set $y = 0$ and $z = 0$ and solve for x.
y-intercept(s): Set $x = 0$ and $z = 0$ and solve for y.
z-intercept: Set $x = 0$ and $y = 0$ and solve for z.

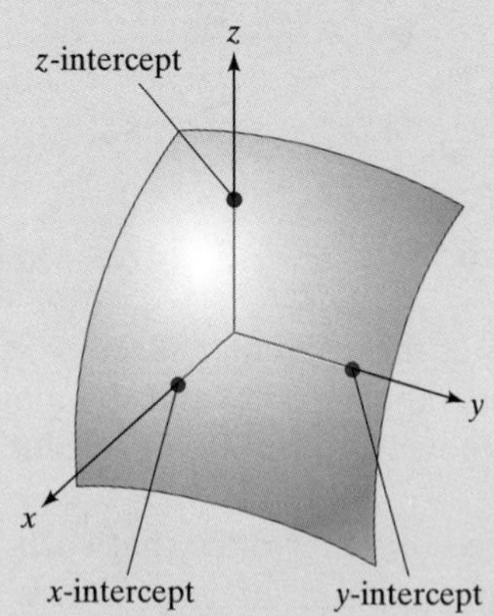

Step 2
Slice the surface along planes parallel to the xy-, yz-, and xz-planes.

Level curves

$z = constant$ Set $z = constant$ and analyze the resulting curves.
These are the curves that result from horizontal slices.

$x = constant$ Set $x = constant$ and analyze the resulting curves.
These are the curves that result from slices parallel to the yz-plane.

$y = constant$ Set $y = constant$ and analyze the resulting curves.
These are the curves that result from slices parallel to the xz-plane.

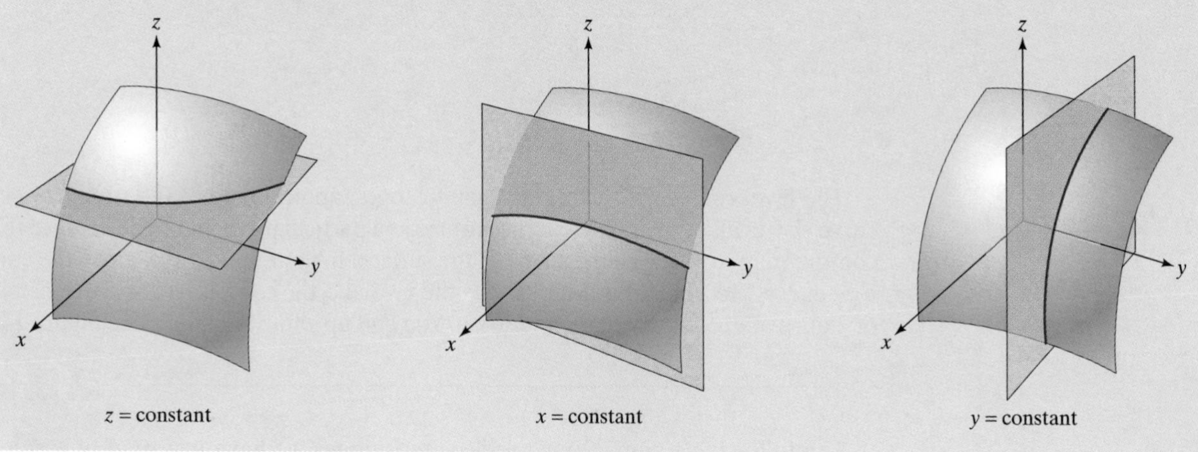

Spreadsheets often have built-in features to render surfaces such as the paraboloid in Example 2. In the next example, we use Excel to graph another surface and then analyze it as above.

example 3

T Analyzing a Surface

Describe the graph of $f(x, y) = x^2 - y^2$.

Solution

Generating the Graph with Excel

First, set up a table showing a range of values of x and y and the corresponding values of the function (see Example 3 in the preceding section):

	A	B	C	D	E	F	G	H
1		-3	-2	-1	0	1	2	3
2	-3	= B1^2-A2^2						
3	-2							
4	-1							
5	0							
6	1							
7	2							
8	3							

(Remember to paste the formula throughout the shaded rectangle.) Next, highlight the shaded rectangle and insert a chart, with the "Surface" option selected and "Series in Columns" selected as the data option. The result is shown in Figure 15, which shows an example of a "saddle point" at the origin (we return to this idea in a later section).

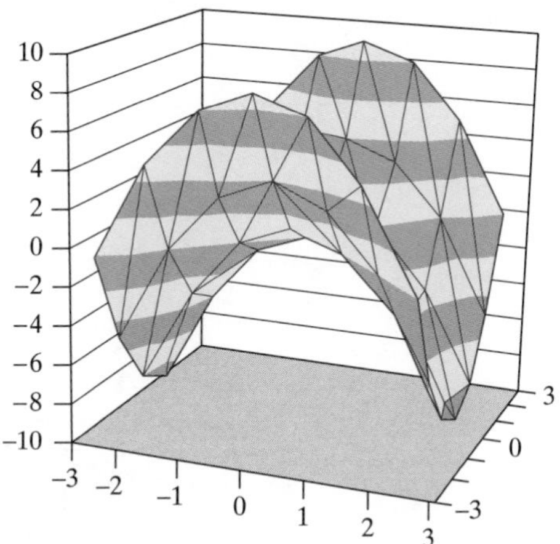

Figure 15

To analyze the graph for the features shown in the box above, replace $f(x, y)$ by z to obtain

$$z = x^2 - y^2$$

Step 1 We find the intercepts. Setting any two of the variables x, y, and z equal to zero results in the third also being zero, so the x-, y-, and z-intercepts are all zero. In other words, the surface touches all three axes in exactly one point, the origin.

Step 2 Slices in various directions show more interesting features.

- *Slice by* $x = c$. This gives $z = c^2 - y^2$, which is the equation of a parabola that opens downward. You can see two of these slices ($c = -3, c = 3$) as the front and back edges of the surface in Figure 15. [More are shown in Figure 16(a).]
- *Slice by* $y = c$. This gives $z = x^2 - c^2$, which is the equation of a parabola once again—this time opening upward. You can see two of these slices ($c = -3, c = 3$) as the left and right edges of the surface in Figure 15. [More are shown in Figure 16(b).]
- *Slice by* $z = c$. This gives $x^2 - y^2 = c$, which is a hyperbola. The level curves for various values of c are visible in Figure 15 as the boundaries between the different shadings in the graph. [See Figure 16(c).] The case $c = 0$ is interesting: The equation $x^2 - y^2 = 0$ can be rewritten as $x = \pm y$ (why?), which represents two lines at right angles to each other.

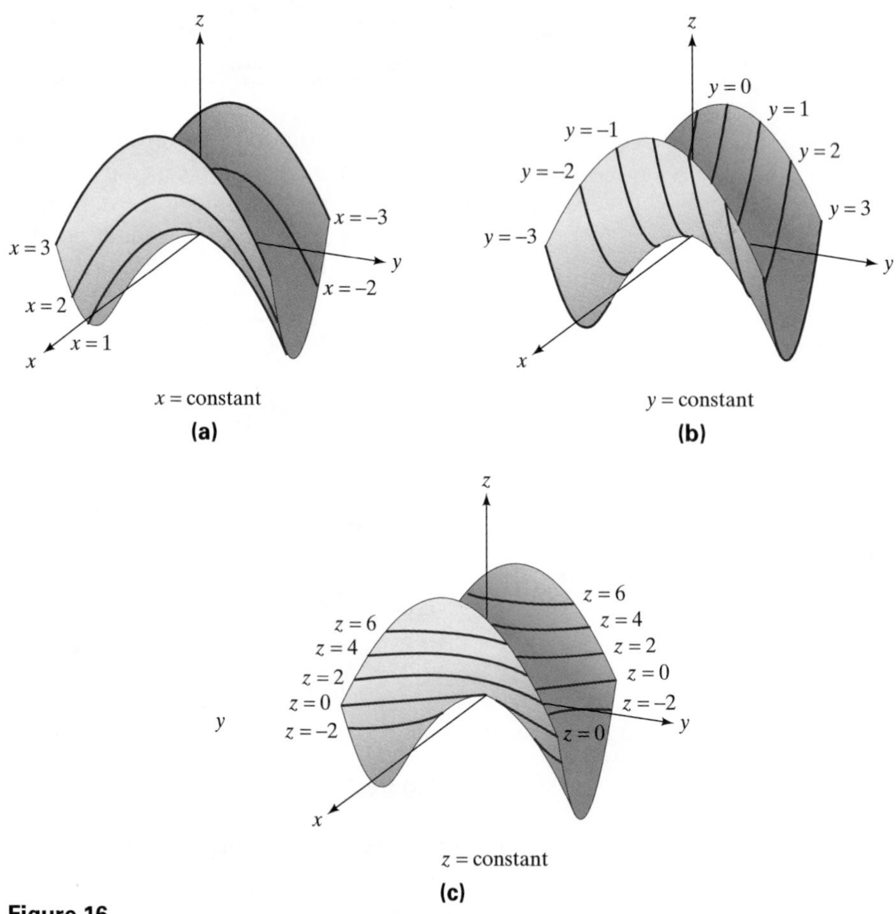

Figure 16

Web Site

For a utility that does on-line graphing of three-dimensional surfaces, follow the path

Web site → Everything for Calculus → Chapter 8 → Surface Graphing Utility

Other Technology

To obtain really beautiful renderings of surfaces, you could use one of the commercial computer algebra software packages, such as Mathematica® or Maple®. These packages can do much more than render surfaces and can be used, for example, to compute derivatives and antiderivatives, to solve equations algebraically, and to perform a variety of algebraic computations.

example 4

Graph of a Linear Function

Describe the graph of $g(x, y) = \frac{1}{2}x + \frac{1}{3}y - 1$.

Solution

Notice first that g is a linear function of x and y. Figure 17 shows a portion of the graph, which is a plane.

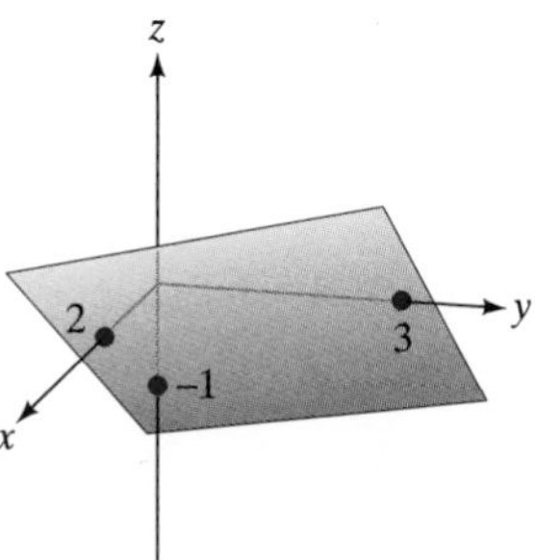

Figure 17

We can get a good idea of what plane this is by looking at the x-, y-, and z-intercepts.

- *x-intercept:* Since points on this axis have both their y- and z-coordinates zero, we set $y = 0$ and $z = 0$, which gives

  $$0 = \frac{1}{2}x - 1$$

  Hence, $x = 2$ is the place where the plane crosses the x-axis.

- *y-intercept:* Setting $x = 0$ and $z = 0$ gives

  $$0 = \frac{1}{3}y - 1$$

  This equation gives $y = 3$ as the place where the plane crosses the y-axis.

- *z-intercept:* We set $x = 0$ and $y = 0$ to get $z = -1$ as the place where the plane crosses the z-axis.

Three points are enough to define a plane, so we can say that the plane is the one passing through the three points $(2, 0, 0)$, $(0, 3, 0)$, and $(0, 0, -1)$.

Before we go on . . . What do the level curves of a linear function of two variables look like?

8.2 exercises

1. Sketch the cube with vertices (0, 0, 0), (1, 0, 0), (0, 1, 0), (0, 0, 1), (1, 1, 0), (1, 0, 1), (0, 1, 1), and (1, 1, 1).

2. Sketch the cube with vertices (−1, −1, −1), (1, −1, −1), (−1, 1, −1), (−1, −1, 1), (1, 1, −1), (1, −1, 1), (−1, 1, 1), and (1, 1, 1).

3. Sketch the pyramid with vertices (1, 1, 0), (1, −1, 0), (−1, 1, 0), (−1, −1, 0), and (0, 0, 2).

4. Sketch the solid with vertices (1, 1, 0), (1, −1, 0), (−1, 1, 0), (−1, −1, 0), (0, 0, −1), and (0, 0, 1).

Sketch the planes in Exercises 5–10.

5. $z = -2$
6. $z = 4$
7. $y = 2$
8. $y = -3$
9. $x = -3$
10. $x = 2$

Match each equation in Exercises 11–18 with one of the graphs below. (If necessary, use technology to render the surfaces.)

11. $f(x, y) = 1 - 3x + 2y$
12. $f(x, y) = 1 - \sqrt{x^2 + y^2}$
13. $f(x, y) = 1 - (x^2 + y^2)$
14. $f(x, y) = y^2 - x^2$
15. $f(x, y) = -\sqrt{1 - (x^2 + y^2)}$
16. $f(x, y) = 1 + (x^2 + y^2)$
17. $f(x, y) = \dfrac{1}{x^2 + y^2}$
18. $f(x, y) = 3x - 2y + 1$

a.

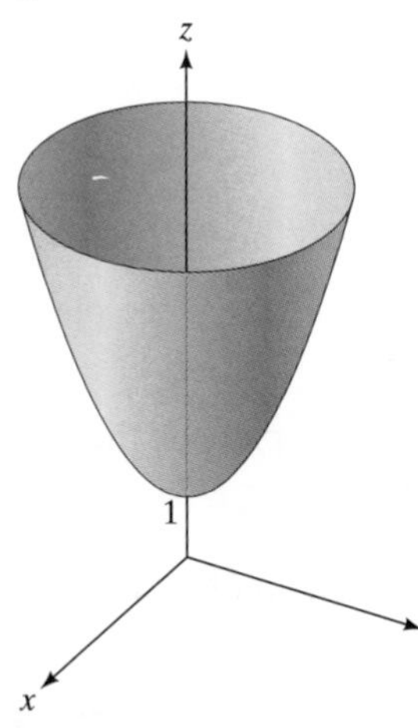

b.

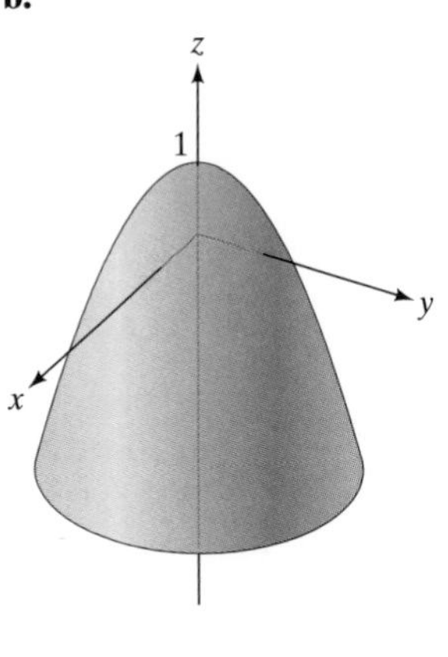

c.

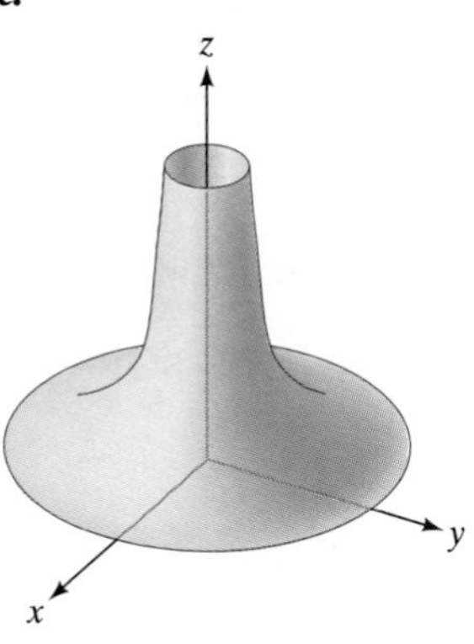

d.

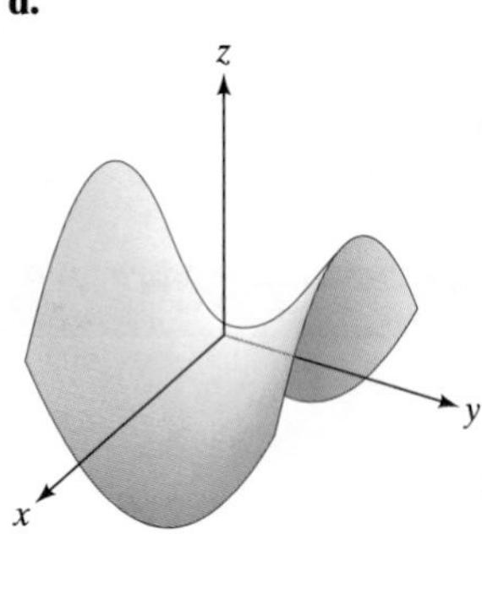

e.

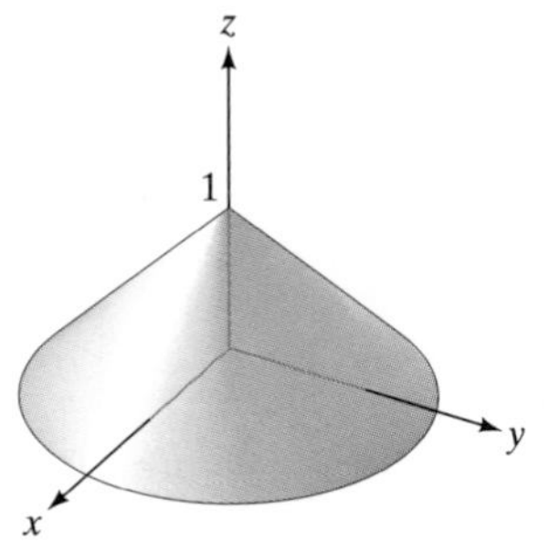

f.

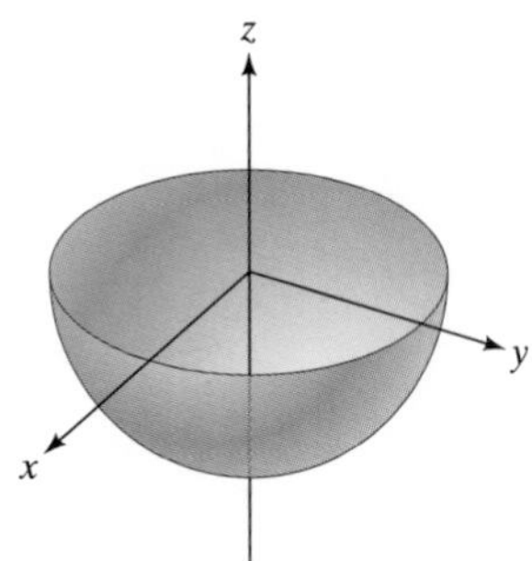

g.

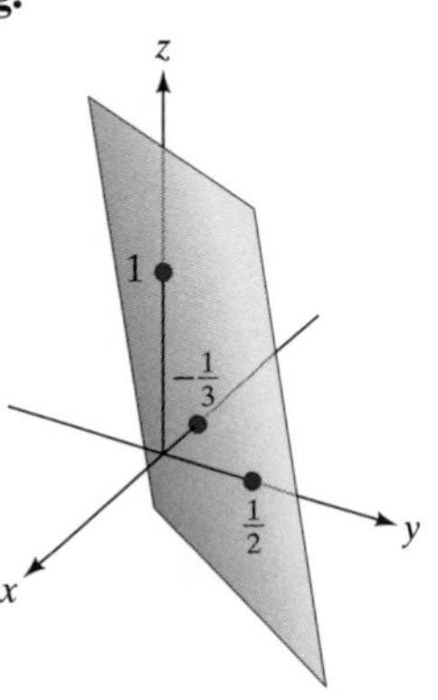

h.

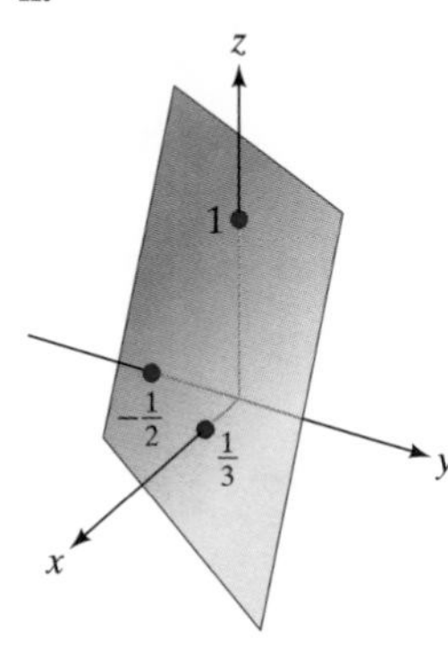

Sketch the graphs of the functions in Exercises 19–40.

19. $f(x, y) = 1 - x - y$
20. $f(x, y) = x + y - 2$
21. $g(x, y) = 2x + y - 2$
22. $g(x, y) = 3 - x + 2y$
23. $h(x, y) = x + 2$
24. $h(x, y) = 3 - y$
25. $r(x, y) = x + y$
26. $r(x, y) = x - y$

T *Use of technology is suggested in Exercises 27–40.*

27. $s(x, y) = 2x^2 + 2y^2$. Show cross sections at $z = 1$ and $z = 2$.

28. $s(x, y) = -(x^2 + y^2)$. Show cross sections at $z = -1$ and $z = -2$.

29. $t(x, y) = x^2 + 2y^2$. Show cross sections at $x = 0$ and $z = 1$.

30. $t(x, y) = \frac{1}{2}x^2 + y^2$. Show cross sections at $x = 0$ and $z = 1$.

31. $f(x, y) = 2 + \sqrt{x^2 + y^2}$. Show cross sections at $z = 3$ and $y = 0$.

32. $f(x, y) = 2 - \sqrt{x^2 + y^2}$. Show cross sections at $z = 0$ and $y = 0$.

33. $f(x, y) = -2\sqrt{x^2 + y^2}$. Show cross sections at $z = -4$ and $y = 1$.

34. $f(x, y) = 2 + 2\sqrt{x^2 + y^2}$. Show cross sections at $z = 4$ and $y = 1$.

35. $f(x, y) = y^2$

36. $g(x, y) = x^2$

37. $h(x, y) = 1/y$

38. $k(x, y) = e^y$

39. $f(x, y) = e^{-(x^2+y^2)}$

40. $g(x, y) = 1/\sqrt{x^2 + y^2}$

Applications

41. ***Marginal Cost (Linear Model)*** Your weekly cost (in dollars) to manufacture x cars and y trucks is

$$C(x, y) = 240{,}000 + 6000x + 4000y$$

a. Describe the graph of the cost function C.
b. Describe the slice by $x = 10$. What cost function does this slice describe?
c. Describe the level curve $z = 480{,}000$. What does this curve tell you about costs?

42. ***Marginal Cost (Linear Model)*** Your weekly cost (in dollars) to manufacture x bicycles and y tricycles is

$$C(x, y) = 24{,}000 + 60x + 20y$$

a. Describe the graph of the cost function C.
b. Describe the slice by $y = 100$. What cost function does this slice describe?
c. Describe the level curve $z = 72{,}000$. What does this curve tell you about costs?

43. ***Market Share (Cars and Light Trucks)*** Based on data from 1980–1998, the relationship among the domestic market shares of three major U.S. manufacturers of cars and light trucks is

$$x_3 = 0.66 - 2.2x_1 - 0.02x_2$$

where x_1, x_2, and x_3 are, respectively, the fractions of the market held by Chrysler, Ford, and General Motors. Thinking of General Motors' market share as a function of the shares of the other two manufacturers, describe the graph of the resulting function. How are the different slices by $x_1 = constant$ related to one another? What does this say about market share?

Source: Ward's AutoInfoBank/*New York Times,* July 29, 1998, p. D6. Model based on linear regression.

44. ***Market Share (Cereals)*** Based on data from 1993–1998, the relationship among the domestic market shares of three major manufacturers of breakfast cereal is

$$x_1 = -0.4 + 1.2x_2 + 2x_3$$

where x_1, x_2, and x_3 are, respectively, the fractions of the market held by Kellogg, General Mills, and General Foods. Thinking of Kellogg's market share as a function of the shares of the other two manufacturers, describe the graph of the resulting function. How are the different slices by $x_2 = constant$ related to one another? What does this say about market share?

Source: Bloomberg Financial Markets/*New York Times,* November 28, 1998, p. C1. Model based on linear regression.

45. ***Marginal Cost (Interaction Model)*** Your weekly cost (in dollars) to manufacture x cars and y trucks is

$$C(x, y) = 240{,}000 + 6000x + 4000y - 20xy$$

(Compare Exercise 41.)

a. Describe the slices $x = constant$ and $y = constant$.
b. Is the graph of the cost function a plane? How does your answer relate to part (a)?
c. What are the slopes of the slices $x = 10$ and $x = 20$? What does this say about cost?

46. ***Marginal Cost (Interaction Model)*** Repeat Exercise 45 using the weekly cost to manufacture x bicycles and y tricycles given by

$$C(x, y) = 24{,}000 + 60x + 20y + 0.3xy$$

(Compare Exercise 42.)

47. ***Housing Costs***[3] The cost C of building a house is related to the number k of carpenters used and the number e of electricians used by

$$C(k, e) = 15{,}000 + 50k^2 + 50e^2$$

[3]Exercises 47 and 48 are based on exercises in *Introduction to Mathematical Economics* by A. L. Ostrosky, Jr., and J. V. Koch (Waveland Press, Illinois, 1979).

Describe the level curves $C = 30{,}000$ and $C = 40{,}000$. What do these level curves represent?

48. ***Housing Costs*** The cost C of building a house (in a different area from that in the previous exercise) is related to the number k of carpenters used and the number e of electricians used by

$$C(k, e) = 15{,}000 + 70k^2 + 40e^2$$

Describe the slices by the planes $k = 2$ and $e = 2$. What do these slices represent?

49. ***Area*** The area of a rectangle of height h and width w is $A(h, w) = hw$. Sketch a few level curves of A. If the perimeter $h + w$ of the rectangle is constant, which h and w give the largest area? (We suggest you draw in the line $h + w = c$ for several values of c.)

50. ***Area*** The area of an ellipse with semimajor axis a and semiminor axis b is $A(a, b) = \pi ab$. Sketch the graph of A. If $a^2 + b^2$ is constant, what a and b give the largest area?

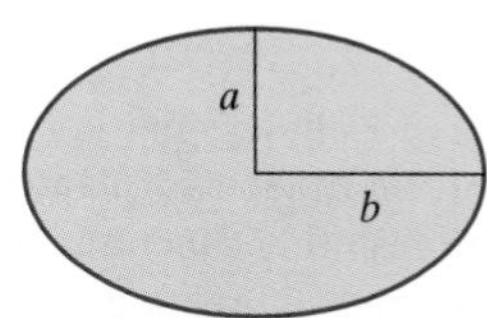

Graphing technology is suggested for Exercises 51–54.

T 51. ***Production (Cobb-Douglas Model)*** Graph the level curves at $z = 0, 1, 2$, and 3 of $P(x, y) = Kx^a y^{1-a}$ if $K = 1$ and $a = 0.5$. Here, x is the number of workers, y is the operating budget, and $P(x, y)$ is the productivity. Interpret the level curve at $z = 3$.

T 52. ***Production (Cobb-Douglas Model)*** Graph the level curves at $z = 0, 1, 2$, and 3 of $P(x, y) = Kx^a y^{1-a}$ if $K = 1$ and $a = 0.25$. Here, x is the number of workers, y is the operating budget, and $P(x, y)$ is the productivity. Interpret the level curve at $z = 0$.

T 53. ***Utility Function*** Suppose your newspaper is trying to decide between two competing desktop publishing software packages, Macro Publish and Turbo Publish. You estimate that if you purchase x copies of Macro Publish and y copies of Turbo Publish, your company's daily productivity will be

$$U(x, y) = 6x^{0.8}y^{0.2} + x$$

where $U(x, y)$ is measured in pages per day (U is called a *utility function*). Graph the level curves at $z = 0, 10, 20$, and 30. What does the level curve at $z = 0$ tell you?

T 54. ***Utility Function*** Suppose your small publishing company is trying to decide between two competing desktop publishing software packages, Macro Publish and Turbo Publish. You estimate that if you purchase x copies of Macro Publish and y copies of Turbo Publish, your company's daily productivity will be given by

$$U(x, y) = 5x^{0.2}y^{0.8} + x$$

where $U(x, y)$ is measured in pages per day. Graph the level curves at $z = 0, 10, 20$, and 30. Give a formula for the level curve at $z = 30$ specifying y as a function of x. What does this curve tell you?

Communication and Reasoning Exercises

55. Your study partner Slim just told you that the surface $z = f(x, y)$ you have been trying to graph must be a plane because you've already found that the slices by $x = constant$ and $y = constant$ are all straight lines. Do you agree or disagree? Explain.

56. Your other study partner Shady claims that, since the surface $z = f(x, y)$ you have been studying is a plane, it follows that all the slices by $x = constant$ and $y = constant$ are straight lines. Do you agree or disagree? Explain.

57. Show that the distance between the points (x, y, z) and (a, b, c) is given by the following **three-dimensional distance formula:**

$$d = \sqrt{(x - a)^2 + (y - b)^2 + (z - c)^2}$$

$$\text{or } d = \sqrt{(\Delta x)^2 + (\Delta y)^2 + (\Delta z)^2}$$

The diagram should be of assistance.

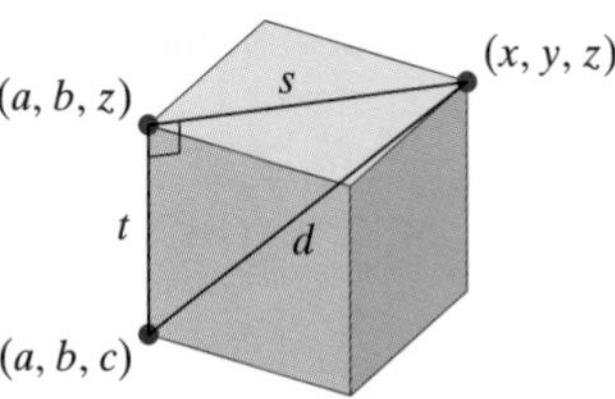

58. Use the result of Exercise 57 to show that the sphere of radius r centered at the origin is given by the equation $x^2 + y^2 + z^2 = r^2$.

59. Why is three-dimensional space used to represent the graph of a function of two variables?

60. Why is it that we can sketch the graphs of functions of two variables on the two-dimensional flat surfaces of these pages?

8.3 Partial Derivatives

Recall that if f is a function of x, then the derivative df/dx measures how fast f changes as x increases. If f is a function of two or more variables, we can ask how fast f changes as each variable increases (and the others remain fixed). These rates of change are called the "partial derivatives of f," and they measure how each variable contributes to the change in f. Here is a more precise definition.

Partial Derivatives

The **partial derivative of f with respect to x** is the derivative of f with respect to x, when all other variables are treated as constant. Similarly, the **partial derivative of f with respect to y** is the derivative of f with respect to y, with all other variables treated as constant, and so on for other variables. The partial derivatives are written as $\partial f/\partial x$, $\partial f/\partial y$, and so on. The symbol ∂ is used (instead of d) to remind us that there is more than one variable and that we are holding the other variables fixed.

Quick Examples

1. Let $f(x, y) = x^2 + y^2$.

$$\frac{\partial f}{\partial x} = 2x + 0 = 2x$$ Because y^2 is treated as constant

$$\frac{\partial f}{\partial y} = 0 + 2y = 2y$$ Because x^2 is treated as constant

2. Let $z = x^2 + xy$.

$$\frac{\partial z}{\partial x} = 2x + y$$ $\frac{\partial}{\partial x}(xy) = \frac{\partial}{\partial x}(x \cdot \text{constant}) = \text{constant} = y$

$$\frac{\partial z}{\partial y} = 0 + x$$ $\frac{\partial}{\partial x}(xy) = \frac{\partial}{\partial x}(\text{constant} \cdot y) = \text{constant} = x$

3. Let $f(x, y) = x^2y + y^2x - xy + y$.

$$\frac{\partial f}{\partial x} = 2xy + y^2 - y$$ y is treated as a constant.

$$\frac{\partial f}{\partial y} = x^2 + 2xy - x + 1$$ x is treated as a constant.

Interpretation

$\partial f/\partial x$ is the rate at which f changes as x changes, for a fixed (constant) y.
$\partial f/\partial y$ is the rate at which f changes as y changes, for a fixed (constant) x.

example 1

Marginal Cost: Linear Model

We return to Example 1 from Section 8.1. Suppose you own a company that makes two models of stereo speakers, the Ultra Mini and the Big Stack. Your total monthly cost (in dollars) to make x Ultra Minis and y Big Stacks is given by

$$C(x, y) = 10{,}000 + 20x + 40y$$

What is the significance of $\partial C/\partial x$ and of $\partial C/\partial y$?

Solution

First we compute these partial derivatives:

$$\frac{\partial C}{\partial x} = 20 \qquad \text{and} \qquad \frac{\partial C}{\partial y} = 40$$

We interpret the results as follows: $\partial C/\partial x = 20$ means that the cost is increasing at a rate of $20 per additional Ultra Mini (if production of Big Stacks is held constant). $\partial C/\partial y = 40$ means that the cost is increasing at a rate of $40 per additional Big Stack (if production of Ultra Minis is held constant). In other words, these are the **marginal costs** of each model of speaker: $\partial C/\partial x$ is the marginal cost of an Ultra Mini and $\partial C/\partial y$ is the marginal cost of a Big Stack.

Before we go on . . . How much does the cost increase if you increase x by Δx and y by Δy? In this example, the change in cost is given by

$$\Delta C = 20\,\Delta x + 40\Delta y = \frac{\partial C}{\partial x}\,\Delta x + \frac{\partial C}{\partial y}\,\Delta y$$

This suggests the **chain rule for several variables.** Part of this rule says that if x and y are both functions of t, then C is a function of t through them, and the rate of change of C with respect to t can be calculated as

$$\frac{dC}{dt} = \frac{\partial C}{\partial x} \cdot \frac{dx}{dt} + \frac{\partial C}{\partial y} \cdot \frac{dy}{dt}$$

We do not have a chance to use this interesting result in this book.

---

example 2

Marginal Cost: Interaction Model

Another possibility for the cost function in the preceding example is the interaction model

$$C(x, y) = 10{,}000 + 20x + 40y + 0.1xy$$

a. Now what are the marginal costs of the two models of speakers?

b. What is the marginal cost of manufacturing Big Stacks at a production level of 100 Ultra Minis and 50 Big Stacks per month?

Solution

a. We compute the partial derivatives:

$$\frac{\partial C}{\partial x} = 20 + 0.1y$$

$$\frac{\partial C}{\partial y} = 40 + 0.1x$$

Thus, the marginal cost of manufacturing Ultra Minis increases by $0.1 or 10¢ for each Big Stack that is manufactured. Similarly, the marginal cost of manufacturing Big Stacks increases by 10¢ for each Ultra Mini that is manufactured.

b. From part (a), the marginal cost of manufacturing Big Stacks is

$$\frac{\partial C}{\partial y} = 40 + 0.1x$$

At a production level of 100 Ultra Minis and 50 Big Stacks per month, we have $x = 100$ and $y = 50$. Thus,

$$\left.\frac{\partial C}{\partial y}\right|_{(100,50)} = 40 + 0.1(100) = \$50 \text{ per Big Stack}$$

Partial derivatives of functions of three variables are obtained in the same way as those for functions of two variables, as the following example shows.

example 3

Function of Three Variables

Calculate $\frac{\partial f}{\partial x}, \frac{\partial f}{\partial y}$, and $\frac{\partial f}{\partial z}$ if $f(x, y, z) = xy^2z^3 - xy$.

Solution

Although we now have three variables, the calculation remains the same: $\partial f/\partial x$ is the derivative of f with respect to x, with *both* other variables, y and z, held constant:

$$\frac{\partial f}{\partial x} = y^2z^3 - y$$

Similarly, $\partial f/\partial y$ is the derivative of f with respect to y, with both x and z held constant:

$$\frac{\partial f}{\partial y} = 2xyz^3 - x$$

Finally, to find $\partial f/\partial z$, we hold both x and y constant and take the derivative with respect to z:

$$\frac{\partial f}{\partial z} = 3xy^2z^2$$

Before we go on . . . The procedure is the same for any number of variables: To get the partial derivative with respect to any one variable, we treat all the others as constants.

Geometric Interpretation of Partial Derivatives

Question We know that if f is a function of one variable x, then the derivative df/dx gives the slopes of the tangent lines to its graph. What about the *partial* derivatives of a function of several variables?

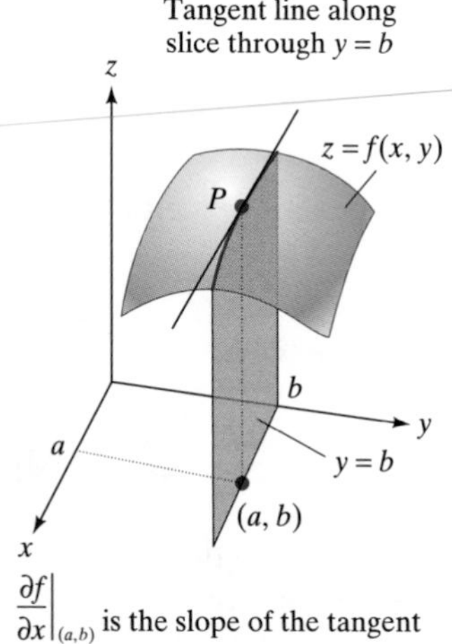

$\left.\dfrac{\partial f}{\partial x}\right|_{(a,b)}$ is the slope of the tangent line at the point $P(a, b, f(a, b))$ along the slice through $y = b$.

Figure 18

Answer Suppose that f is a function of x and y. By definition, $\partial f/\partial x$ is the derivative of the function of x we get by holding y fixed. If we evaluate this derivative at the point (a, b), we are holding y fixed at the value b, taking the ordinary derivative of the resulting function of x, and evaluating this at $x = a$. Now, holding y fixed at b amounts to slicing through the graph of f along the plane $y = b$, resulting in a curve. Thus, the partial derivative is the slope of the tangent line to this curve at the point where $x = a$ and $y = b$ along the plane $y = b$ (Figure 18). This fits with our interpretation of $\partial f/\partial x$ as the rate of increase of f with increasing x when y is held fixed at b. The other partial derivative, $\partial f/\partial y|_{(a,b)}$ is, similarly, the slope of the tangent line at the same point $P(a, b, f(a, b))$ but along the slice by the plane $x = a$. You should draw the corresponding picture for this on your own.

example 4

Geometry of the Interaction Model

We return to the interaction model of cost in Example 2:

$$C(x, y) = 10{,}000 + 20x + 40y + 0.1xy$$

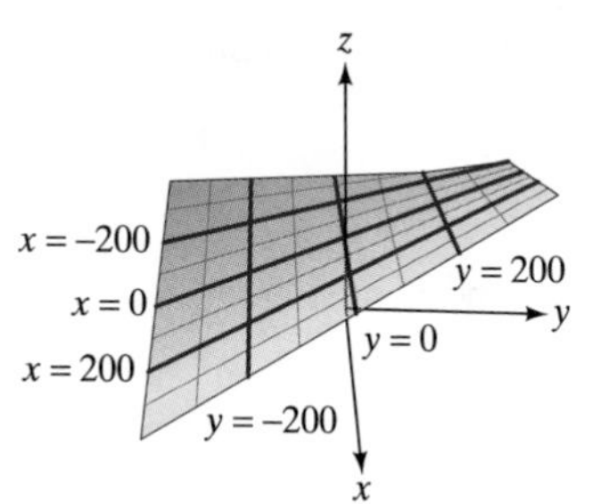

Figure 19

A portion of the surface $z = 10{,}000 + 20x + 40y + 0.1xy$ is shown in Figure 19, together with some of the slices through $x =$ *constant* and $y =$ *constant.* Since these slices give straight lines, they are already tangent to the surface, and their slopes are given by the partial derivatives (which represent the marginal costs when x and y are nonnegative):

$$y = -200: \quad \text{Slope} = \left.\frac{\partial z}{\partial x}\right|_{(x,-200)} = 20 + 0.1(-200) = 0$$

$$y = 0: \quad \text{Slope} = \left.\frac{\partial z}{\partial x}\right|_{(x,0)} = 20 + 0.1(0) = 20$$

$$y = 200: \quad \text{Slope} = \left.\frac{\partial z}{\partial x}\right|_{(x,200)} = 20 + 0.1(200) = 40$$

$$x = -200 \quad \text{Slope} = \left.\frac{\partial z}{\partial y}\right|_{(-200,y)} = 40 + 0.1(-200) = 20$$

$$x = 0 \quad \text{Slope} = \left.\frac{\partial z}{\partial y}\right|_{(0,y)} = 40 + 0.1(0) = 40$$

$$x = 200 \quad \text{Slope} = \left.\frac{\partial z}{\partial y}\right|_{(200,y)} = 40 + 0.1(200) = 60$$

Second-Order Partial Derivatives

Just as for functions of a single variable, we can calculate second derivatives. Suppose we have a function of x and y, say, $f(x, y) = x^2 - x^2y^2$. We know that

$$\frac{\partial f}{\partial x} = 2x - 2xy^2.$$

If we want to take the partial derivative with respect to x once again, there is nothing to stop us:

$$\frac{\partial}{\partial x}\left(\frac{\partial f}{\partial x}\right) = 2 - 2y^2 \qquad \text{Take } \frac{\partial}{\partial x} \text{ of } \frac{\partial f}{\partial x}.$$

(The symbol $\partial/\partial x$ means "the partial derivative with respect to x," just as d/dx stands for "the derivative with respect to x.") This is called the **second partial derivative** and is written $\partial^2 f/\partial x^2$. We get the following derivatives similarly:

$$\frac{\partial f}{\partial y} = -2x^2y$$

$$\frac{\partial^2 f}{\partial y^2} = -2x^2 \qquad \text{Take } \frac{\partial}{\partial y} \text{ of } \frac{\partial f}{\partial y}.$$

Now what if we instead take the partial derivative with respect to y of $\partial f/\partial x$?

$$\frac{\partial^2 f}{\partial y\, \partial x} = \frac{\partial}{\partial y}\left(\frac{\partial f}{\partial x}\right) \qquad \text{Take } \frac{\partial}{\partial y} \text{ of } \frac{\partial f}{\partial x}.$$

$$= \frac{\partial}{\partial y}(2x - 2xy^2) = -4xy$$

Here, $\partial^2 f/(\partial y\, \partial x)$ means "take the partial derivative first with respect to x and then with respect to y" and is called a **mixed partial derivative.** If we differentiate in the opposite order, we get

$$\frac{\partial^2 f}{\partial x\, \partial y} = \frac{\partial}{\partial x}\left(\frac{\partial f}{\partial y}\right) = \frac{\partial}{\partial x}(-2x^2y) = -4xy$$

the same as $\partial^2 f/(\partial y\, \partial x)$. This is no coincidence: The mixed partial derivatives $\partial^2 f/(\partial x\, \partial y)$ and $\partial^2 f/(\partial y\, \partial x)$ are always the same as long as the first partial derivatives are both differentiable functions of x and y and the mixed partial derivatives are continuous. Since all the functions we use are of this type, we can take the derivatives in any order we like when calculating mixed derivatives.

Before we go to the exercises, here is another notation for partial derivatives:

f_x means $\dfrac{\partial f}{\partial x}$

f_y means $\dfrac{\partial f}{\partial y}$

f_{xy} means $(f_x)_y = \dfrac{\partial^2 f}{\partial y\, \partial x}$ (Note the order in which the derivatives are taken.)

f_{yx} means $(f_y)_x = \dfrac{\partial^2 f}{\partial x\, \partial y}$

We sometimes use this more convenient notation, especially for second-order partial derivatives.

8.3 exercises

In Exercises 1–18, calculate $\frac{\partial f}{\partial x}, \frac{\partial f}{\partial y}, \left.\frac{\partial f}{\partial x}\right|_{(1,-1)}$, *and* $\left.\frac{\partial f}{\partial y}\right|_{(1,-1)}$ *when defined.*

1. $f(x, y) = 10{,}000 - 40x + 20y$
2. $f(x, y) = 1000 + 5x - 4y$
3. $f(x, y) = 3x^2 - y^3 + x - 1$
4. $f(x, y) = x^{1/2} - 2y^4 + y + 6$
5. $f(x, y) = 10{,}000 - 40x + 20y + 10xy$
6. $f(x, y) = 1000 + 5x - 4y - 3xy$
7. $f(x, y) = 3x^2y$
8. $f(x, y) = x^4y^2 - x$
9. $f(x, y) = x^2y^3 - x^3y^2 - xy$
10. $f(x, y) = x^{-1}y^2 + xy^2 + xy$
11. $f(x, y) = (2xy + 1)^3$
12. $f(x, y) = 1/(xy + 1)^2$
13. $f(x, y) = e^{x+y}$
14. $f(x, y) = e^{2x+y}$
15. $f(x, y) = 5x^{0.6}y^{0.4}$
16. $f(x, y) = -2x^{0.1}y^{0.9}$
17. $f(x, y) = e^{0.2xy}$
18. $f(x, y) = xe^{xy}$

In Exercises 19–28, find $\frac{\partial^2 f}{\partial x^2}, \frac{\partial^2 f}{\partial y^2}, \frac{\partial^2 f}{\partial x\, \partial y}, \frac{\partial^2 f}{\partial y\, \partial x}$, *and evaluate them all at* $(1, -1)$ *if possible.*

19. $f(x, y) = 10{,}000 - 40x + 20y$
20. $f(x, y) = 1000 + 5x - 4y$
21. $f(x, y) = 10{,}000 - 40x + 20y + 10xy$
22. $f(x, y) = 1000 + 5x - 4y - 3xy$
23. $f(x, y) = 3x^2y$
24. $f(x, y) = x^4y^2 - x$
25. $f(x, y) = e^{x+y}$
26. $f(x, y) = e^{2x+y}$
27. $f(x, y) = 5x^{0.6}y^{0.4}$
28. $f(x, y) = -2x^{0.1}y^{0.9}$

In Exercises 29–40, find $\frac{\partial f}{\partial x}, \frac{\partial f}{\partial y}, \frac{\partial f}{\partial z}$, *and their values at* $(0, -1, 1)$ *if possible.*

29. $f(x, y, z) = xyz$
30. $f(x, y, z) = xy + xz - yz$
31. $f(x, y, z) = -\frac{4}{x + y + z^2}$
32. $f(x, y, z) = \frac{6}{x^2 + y^2 + z^2}$
33. $f(x, y, z) = xe^{yz} + ye^{xz}$
34. $f(x, y, z) = xye^z + xe^{yz} + e^{xyz}$
35. $f(x, y, z) = x^{0.1}y^{0.4}z^{0.5}$
36. $f(x, y, z) = 2x^{0.2}y^{0.8} + z^2$
37. $f(x, y, z) = e^{xyz}$
38. $f(x, y, z) = \ln(x + y + z)$
39. $f(x, y, z) = \frac{2000z}{1 + y^{0.3}}$
40. $f(x, y, z) = \frac{e^{0.2x}}{1 + e^{-0.1y}}$

Applications

41. ***Marginal Cost (Linear Model)*** Your weekly cost (in dollars) to manufacture x cars and y trucks is

$$C(x, y) = 240{,}000 + 6000x + 4000y$$

Calculate and interpret $\partial C/\partial x$ and $\partial C/\partial y$.

42. ***Marginal Cost (Linear Model)*** Your weekly cost (in dollars) to manufacture x bicycles and y tricycles is

$$C(x, y) = 24{,}000 + 60x + 20y$$

Calculate and interpret $\partial C/\partial x$ and $\partial C/\partial y$.

43. ***Market Share (Cars and Light Trucks)*** Based on data from 1980–1998, the relationship among the domestic market shares of three major U.S. manufacturers of cars and light trucks is

$$x_3 = 0.66 - 2.2x_1 - 0.02x_2$$

where x_1, x_2, and x_3 are, respectively, the fractions of the market held by Chrysler, Ford, and General Motors. Calculate $\partial x_3/\partial x_1$ and $\partial x_1/\partial x_3$. What do they signify, and how are they related to each other?

Source: Ward's AutoInfoBank/*New York Times,* July 29, 1998, p. D6. Model based on linear regression.

44. ***Market Share (Cereals)*** Based on data from 1993–1998, the relationship among the domestic market shares of three major manufacturers of breakfast cereal is

$$x_1 = -0.4 + 1.2x_2 + 2x_3$$

where x_1, x_2, and x_3 are, respectively, the fractions of the market held by Kellogg, General Mills, and General Foods. Compute and interpret $\partial x_2/\partial x_3$.

Source: Bloomberg Financial Markets/*New York Times,* November 28, 1998, p. C1. Model based on linear regression.

45. ***Marginal Cost (Interaction Model)*** Your weekly cost (in dollars) to manufacture x cars and y trucks is

$$C(x, y) = 240{,}000 + 6000x + 4000y - 20xy$$

(Compare Exercise 41.) Compute the marginal cost of manufacturing cars at a production level of 10 cars and 20 trucks.

46. ***Marginal Cost (Interaction Model)*** Your weekly cost (in dollars) to manufacture x bicycles and y tricycles is

$$C(x, y) = 24{,}000 + 60x + 20y + 0.3xy$$

(Compare Exercise 42.) Compute the marginal cost of manufacturing tricycles at a production level of 10 bicycles and 20 tricycles.

47. ***Brand Loyalty*** The fraction of Mazda car owners who chose another new Mazda can be modeled by the following function:[1]

$$M(c, f, g, h, t) = 1.1 - 3.8c + 2.2f + 1.9g - 1.7h - 1.3t$$

where c is the fraction of Chrysler car owners who remained loyal to Chrysler, f is the fraction of Ford car owners who remained loyal to Ford, g the corresponding figure for General Motors, h the corresponding figure for Honda, and t for Toyota.

a. Calculate $\partial M/\partial c$ and $\partial M/\partial f$ and interpret the answers.

b. In 1995 it was observed that $c = 0.56, f = 0.56, g = 0.72, h = 0.50$, and $t = 0.43$. According to the model, what percentage of Mazda owners remained loyal to Mazda? (Round your answer to the nearest percentage point.)

Source: Chrysler, Maritz Market Research, Consumer Attitude Research, and Strategic Vision/*New York Times,* November 3, 1995, p. D2.

48. ***Brand Loyalty*** The fraction of Mazda car owners who chose another new Mazda can be modeled by the following function:[2]

$$M(c, f) = 9.4 + 7.8c + 3.6c^2 - 38f - 22cf + 43f^2$$

where c is the fraction of Chrysler car owners who remained loyal to Chrysler and f is the fraction of Ford car owners who remained loyal to Ford.

a. Calculate $\partial M/\partial c$ and $\partial M/\partial f$ evaluated at the point $(0.7, 0.7)$, and interpret the answers.

b. In 1995 it was observed that $c = 0.56$ and $f = 0.56$. According to the model, what percentage of Mazda owners remained loyal to Mazda? (Round your answer to the nearest percentage point.)

Source: See the source in Exercise 47.

49. ***Minivan Sales*** Chrysler's percentage share of the U.S. minivan market in the period 1993–1994 could be approximated by the linear function

$$c(x, y, z) = 72.3 - 0.8x - 0.2y - 0.7z$$

where x is the percentage share of the market held by foreign manufacturers, y is General Motors' percentage share, and z is Ford's percentage share. Calculate and interpret the partial derivatives $\partial c/\partial x, \partial c/\partial y$, and $\partial c/\partial z$.

Source: The authors' model from data by Ford Motor Company/*New York Times,* November 9, 1994, p. D5.

50. ***Minivan Sales*** Refer back to Exercise 49. If we take into account the fact that the variables x, y, z, and c together account for 100% of all minivan sales in the United States, we obtain

$$c = -38.5 + 3.0y + 0.5z$$

where y is General Motors' percentage share of the market and z is Ford's percentage share. Calculate and interpret the partial derivatives $\partial c/\partial y$ and $\partial c/\partial z$.

51. ***Marginal Cost*** Your weekly cost (in dollars) to manufacture x cars and y trucks is

$$C(x, y) = 200{,}000 + 6000x + 4000y - 100{,}000e^{-0.01(x+y)}$$

What is the marginal cost of a car? Of a truck? How do these marginal costs behave as total production increases?

52. ***Marginal Cost*** Your weekly cost (in dollars) to manufacture x bicycles and y tricycles is

$$C(x, y) = 20{,}000 + 60x + 20y + 50\sqrt{xy}$$

What is the marginal cost of a bicycle? Of a tricycle? How do these marginal costs behave as x and y increase?

53. ***Average Cost*** If you average your costs over your total production, you get the **average cost,** written $\bar{C}$:

$$\bar{C}(x, y) = \frac{C(x, y)}{x + y}$$

Find the average cost for the cost function in Exercise 51. Then find the marginal average cost of a car and the marginal average cost of a truck at a production level of 50 cars and 50 trucks. Interpret your answers.

54. ***Average Cost*** Find the average cost for the cost function in Exercise 52 (see the preceding exercise). Then find the marginal average cost of a bicycle and the marginal average cost of a tricycle at a production level of five bicycles and five tricycles. Interpret your answers.

55. ***Marginal Revenue*** As manager of an auto dealership, you offer a car rental company the following deal: You will charge \$15,000 per car and \$10,000 per truck, but you will then give the company a discount of \$5000 times the square root of the total number of vehicles it buys from you. If you consider your marginal revenue, is this a good deal for the rental company?

56. ***Marginal Revenue*** As marketing director for a bicycle manufacturer, you come up with the following scheme: You will offer to sell a dealer x bicycles and y tricycles for

$$R(x, y) = 3500 - 3500e^{-0.02x-0.01y} \text{ dollars}$$

Find your marginal revenue for bicycles and for tricycles. Are you likely to be fired for your suggestion?

57. ***Research Productivity*** Here we apply a variant of the Cobb-Douglas function to the modeling of research

---

[1] The model is an approximation of a linear regression based on data from the period 1988–1995.

[2] The model is an approximation of a second-order regression based on data from the period 1988–1995.

productivity. A mathematical model of research productivity at a particular physics laboratory is

$$P = 0.04x^{0.4}y^{0.2}z^{0.4}$$

where P is the annual number of groundbreaking research papers produced by the staff, x is the number of physicists on the research team, y is the laboratory's annual research budget, and z is the annual National Science Foundation subsidy to the laboratory. Find the rate of increase of research papers per government-subsidy dollar at a subsidy level of \$1,000,000 per year and a staff level of 10 physicists if the annual budget is \$100,000.

58. ***Research Productivity*** A major drug company estimates that the annual number P of patents for new drugs developed by its research team is best modeled by the formula

$$P = 0.3x^{0.3}y^{0.4}z^{0.3}$$

where x is the number of research biochemists on the payroll, y is the annual research budget, and z is the size of the bonus awarded to discoverers of new drugs. Assuming that the company has 12 biochemists on the staff, has an annual research budget of \$500,000, and pays \$40,000 bonuses to developers of new drugs, calculate the rate of growth in the annual number of patents per new research staff member.

59. ***Utility Function*** Your newspaper is trying to decide between two competing desktop publishing software packages, Macro Publish and Turbo Publish. You estimate that if you purchase x copies of Macro Publish and y copies of Turbo Publish, your company's daily productivity will be

$$U(x, y) = 6x^{0.8}y^{0.2} + x$$

where $U(x, y)$ is measured in pages per day.

a. Calculate $\left.\frac{\partial U}{\partial x}\right|_{(10,5)}$ and $\left.\frac{\partial U}{\partial y}\right|_{(10,5)}$ to two decimal places, and interpret the results.

b. What does the ratio $\left.\frac{\partial U}{\partial x}\right|_{(10,5)} \Big/ \left.\frac{\partial U}{\partial y}\right|_{(10,5)}$ tell about the usefulness of these products?

60. ***Grades***[3] A production formula for a student's performance on a difficult English examination is given by

$$g(t, x) = 4tx - 0.2t^2 - x^2$$

where g is the grade the student can expect to get, t is the number of hours of study for the examination, and x is the student's grade point average.

a. Calculate $\left.\frac{\partial g}{\partial t}\right|_{(10,3)}$ and $\left.\frac{\partial g}{\partial x}\right|_{(10,3)}$ and interpret the results.

b. What does the ratio $\left.\frac{\partial g}{\partial t}\right|_{(10,3)} \Big/ \left.\frac{\partial g}{\partial x}\right|_{(10,3)}$ tell about the relative merits of study and grade point average?

61. ***Electrostatic Repulsion*** If positive electric charges of Q and q coulombs are situated at positions (a, b, c) and (x, y, z), respectively, then the force of repulsion they experience is given by

$$F = K\frac{Qq}{(x-a)^2 + (y-b)^2 + (z-c)^2}$$

where $K \approx 9 \times 10^9$, F is given in newtons, and all positions are measured in meters. Assume that a charge of 10 coulombs is situated at the origin and that a second charge of 5 coulombs is situated at (2, 3, 3) and moving in the y direction at 1 meter per second. How fast is the electrostatic force it experiences decreasing?

62. ***Electrostatic Repulsion*** Repeat Exercise 61 assuming that a charge of 10 coulombs is situated at the origin and a second charge of 5 coulombs is situated at (2, 3, 3) and moving in the negative z direction at 1 meter per second.

63. ***Investments*** Recall that the compound interest formula for annual compounding is

$$A(P, r, t) = P(1 + r)^t$$

where A is the future value of an investment of P dollars after t years at an interest rate of r.

a. Calculate $\partial A/\partial P$, $\partial A/\partial r$, and $\partial A/\partial t$, all evaluated at (100, 0.10, 10). (Round answers to two decimal places.) Interpret your answers.

b. What does the function $\left.\frac{\partial A}{\partial P}\right|_{(100,\,0.10,\,t)}$ of t tell about your investment?

64. ***Investments*** Repeat Exercise 63 using the formula for continuous compounding:

$$A(P, r, t) = Pe^{rt}$$

65. ***Modeling with the Cobb-Douglas Formula*** Assume you are given a production formula of the form

$$P(x, y) = Kx^a y^b \qquad (a + b = 1)$$

a. Obtain formulas for $\partial P/\partial x$ and $\partial P/\partial y$, and show that $\partial P/\partial x = \partial P/\partial y$ precisely when $x/y = a/b$.

b. Let x be the number of workers a firm employs and let y be its monthly operating budget in thousands of dollars. Assume that the firm currently employs 100 workers and has a monthly operating budget of \$200,000. If each additional worker contributes as much to productivity as each additional \$1000 per month, find values of a and b that model the firm's productivity.

[3] Based on an exercise in *Introduction to Mathematical Economics* by A. L. Ostrosky, Jr., and J. V. Koch (Waveland Press, Illinois, 1979).

66. ***Housing Costs*** [4] The cost C of building a house is related to the number k of carpenters used and the number e of electricians used by

$$C(k, e) = 15{,}000 + 50k^2 + 60e^2$$

If three electricians are currently employed in building your new house and the marginal cost per additional electrician is the same as the marginal cost per additional carpenter, how many carpenters are being used? (Round your answer to the nearest carpenter.)

67. ***Nutrient Diffusion*** Suppose that 1 cubic centimeter of nutrient is placed at the center of a circular petri dish filled with water. We might wonder how the nutrient is distributed after a time of t seconds. According to the classical theory of diffusion, the concentration of nutrient (in parts of nutrient per part of water) after a time t is given by

$$u(r, t) = \frac{1}{4\pi Dt} e^{-r^2/(4Dt)}$$

where D is the *diffusivity,* which we will take to be 1, and r is the distance from the center in centimeters. How fast is the concentration increasing at a distance of 1 cm from the center 3 seconds after the nutrient is introduced?

68. ***Nutrient Diffusion*** Refer back to Exercise 67. How fast is the concentration of nutrient increasing at a distance of 4 cm from the center 4 seconds after the nutrient is introduced?

[4] Based on an exercise in *Introduction to Mathematical Economics* by A. L. Ostrosky, Jr., and J. V. Koch (Waveland Press, Illinois, 1979.)

Communication and Reasoning Exercises

69. Given that $f(a, b) = r$, $f_x(a, b) = s$, and $f_y(a, b) = t$, complete the following: _____ is increasing at a rate of _____ units per unit of x, _____ is increasing at a rate of _____ units per unit of y, and the value of _____ is _____ when $x =$ _____ and $y =$ _____.

70. A firm's productivity depends on two variables, x and y. Currently, $x = a$, $y = b$, and the firm's productivity is 4000 units. Productivity is increasing at a rate of 400 units per unit *decrease* in x and is decreasing at a rate of 300 units per unit *increase* in y. What does all of this information tell you about the firm's productivity function $g(x, y)$?

71. Give an example of a function $f(x, y)$ with $f(1, 1) = 10$, $f_x(1, 1) = -2$, and $f_y(1, 1) = 3$.

72. Give an example of a function $f(x, y, z)$ that has all of its partial derivatives nonzero constants.

73. The graph of $z = b + mx + ny$ (b, m, and n constants) is a plane.
   a. Explain the geometric significance of the numbers b, m, and n.
   b. Show that the equation of the plane passing through (h, k, l) with slope m in the x direction (in the sense of $\partial/\partial x$) and slope n in the y direction is

$$z = l + m(x - h) + n(y - k)$$

74. The **tangent plane** to the graph of $f(x, y)$ at $P(a, b, f(a, b))$ is the plane that contains the lines tangent to the slice through the graph by $y = b$ (as in Figure 18) and the slice through the graph by $x = a$. Use the result of Exercise 73 to show that the equation of the tangent plane is

$$z = f(a, b) + f_x(a, b)(x - a) + f_y(a, b)(y - b)$$

8.4 Maxima and Minima

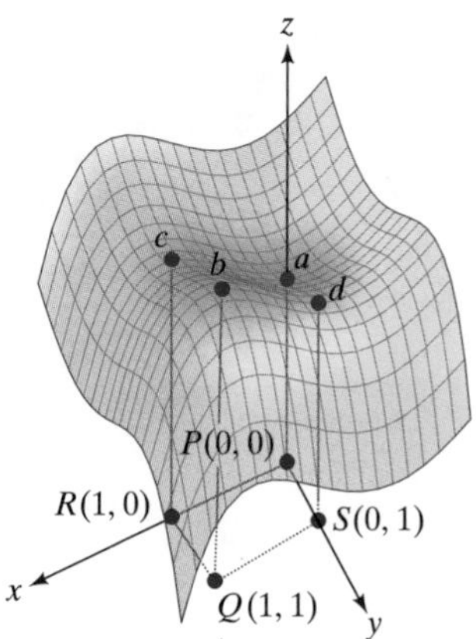

Figure 20

In Chapter 5 on applications of the derivative we saw how to locate the relative extrema of a function of a single variable. In this section we extend our methods to functions of two variables. Similar techniques work for functions of three or more variables.

Figure 20 shows a portion of the graph of the function $f(x, y) = 2(x^2 + y^2) - (x^4 + y^4) + 3$. The graph resembles a "flying carpet," and several interesting points, labeled a, b, c, and d are shown. The point a has coordinates $(0, 0, f(0, 0))$, is directly above the origin $(0, 0)$, and is the lowest point on the portion of the graph shown. Thus, we say that f has a **relative minimum** at $(0, 0)$ because $f(0, 0)$ is smaller than $f(x, y)$ for any (x, y) near $(0, 0)$. Similarly, the point b is higher than any point in its vicinity. Thus, we say that f has a **relative maximum** at $(1, 1)$. The points c and d represent a new phenomenon and are called **saddle points.** They are neither relative maxima nor relative minima but seem to be a little of both.

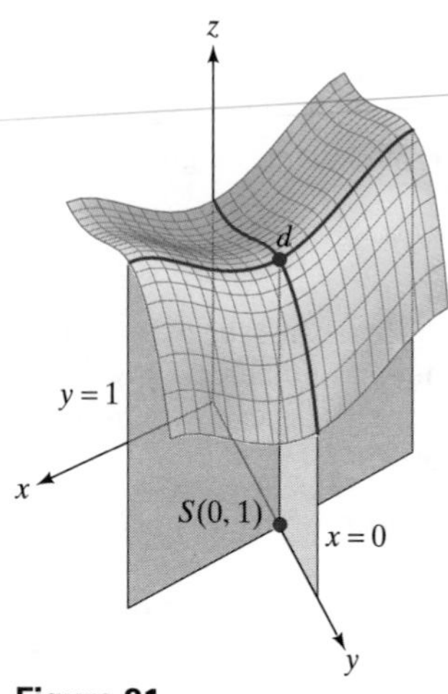

Figure 21

To see more clearly what features a saddle point has, look at Figure 21, which shows a portion of the graph near the point d. It is easy to see where the term *saddle point* comes from. If we slice through the graph along $y = 1$, we get a curve on which d is the *lowest* point. Thus, d looks like a relative minimum along this slice. On the other hand, if we slice through the graph along $x = 0$, we get another curve, on which d is the *highest* point, so d looks like a relative maximum along this slice. This kind of behavior characterizes a saddle point: f has a **saddle point** at (r, s) if f has a relative minimum at (r, s) along some slice through that point and a relative maximum along another slice through that point. If you look at the other saddle point, c, in Figure 20, you see the same characteristics.

Although numerical information can help us locate the approximate positions of relative extrema and saddle points, calculus permits us to locate these points accurately as we did for functions of a single variable. Look once again at Figure 20 and notice the following:

- The points P, Q, R, and S are all in the **interior** of the domain of f; that is, none of them lies on the boundary of the domain. Said another way, we can move some distance in any direction from any of these points without leaving the domain of f.
- The tangent lines along the slices through these points parallel to the x- and y-axes are *horizontal.* Thus, the partial derivatives $\partial f/\partial x$ and $\partial f/\partial y$ are zero when evaluated at any of the points P, Q, R, and S. This gives us a way of locating candidates for relative extrema and saddle points.

Locating Candidates for Relative Extrema and Saddle Points in the Interior of the Domain of f

First set $\partial f/\partial x = 0$ and $\partial f/\partial y = 0$ simultaneously and solve for x and y. Then check that the resulting points (x, y) are in the interior of the domain of f.

Points at which all the partial derivatives of f are zero are called **critical points.** Thus, the critical points are the only candidates for relative extrema and saddle points in the interior of the domain of f.[1]

Quick Examples

1. Let $f(x, y) = x^3 + (y - 1)^2$. Then $\partial f/\partial x = 3x^2$ and $\partial f/\partial y = 2(y - 1)$. Thus, we solve the system

$$3x^2 = 0 \qquad \text{and} \qquad 2(y - 1) = 0$$

The first equation gives $x = 0$, and the second gives $y = 1$. Thus, the only critical point is $(0, 1)$. Since the domain of f is the whole Cartesian plane, the point $(0, 1)$ is interior and hence a candidate for a relative extremum or saddle point.[2]

[1] We look at extrema on the *boundary* of the domain of a function in the next section. What we are calling critical points correspond to the *stationary* points of a function of one variable. We do not consider the analogs of the singular points.

[2] In fact, it is a saddle point. (Can you see why?)

2. Let $f(x, y) = e^{-(x^2+y^2)}$. Taking partial derivatives and setting them equal to zero give

$$-2xe^{-(x^2+y^2)} = 0 \qquad \text{We set } \partial f/\partial x = 0.$$

$$-2ye^{-(x^2+y^2)} = 0 \qquad \text{We set } \partial f/\partial y = 0.$$

The first equation implies that $x = 0$,[3] and the second implies that $y = 0$. Thus, the only critical point is $(0, 0)$. This point is interior and hence a candidate for a relative extremum or saddle point.

example 1

Locating Critical Points

Locate all critical points of $f(x, y) = x^2y - x^2 - 2y^2$. Graph the function to classify the critical points as relative maxima, relative minima, saddle points, or none of these.

Solution

The partial derivatives are

$$f_x = 2xy - 2x = 2x(y - 1)$$

$$f_y = x^2 - 4y$$

Setting these equal to zero gives

$$x = 0 \quad \text{or} \quad y = 1$$

$$x^2 = 4y$$

We get a solution by choosing either $x = 0$ or $y = 1$ and substituting into $x^2 = 4y$.

Case 1: $x = 0$. Substituting into $x^2 = 4y$ gives $0 = 4y^2$ and hence $y = 0$. Thus, the critical point for this case is $(x, y) = (0, 0)$.

Case 2: $y = 1$. Substituting into $x^2 = 4y$ gives $x^2 = 4$ and hence $x = \pm 2$. Thus, we get two critical points for this case: $(2, 1)$ and $(-2, 1)$.

Now we have three critical points altogether: $(0, 0)$, $(2, 1)$, and $(-2, 1)$. We get the corresponding points on the graph by substituting for x and y in the equation for f to get the z-coordinates. The points are $(0, 0, 0)$, $(2, 1, -2)$, and $(-2, 1, -2)$.

To classify the critical points graphically, we look at the graph of f shown in Figure 22. Examining the graph carefully, we see that the point $(0, 0, 0)$ is a relative maximum. As for the other two critical points, are they saddle points or relative maxima? They *seem* to be relative maxima along the y direction, but the slice in the x direction (through $x = 1$) seems to be horizontal. However, a diagonal slice (along $x = \pm y$) shows these two points as minima and so they are saddle points. (If you don't believe this, we will get more evidence below and in a later example.)

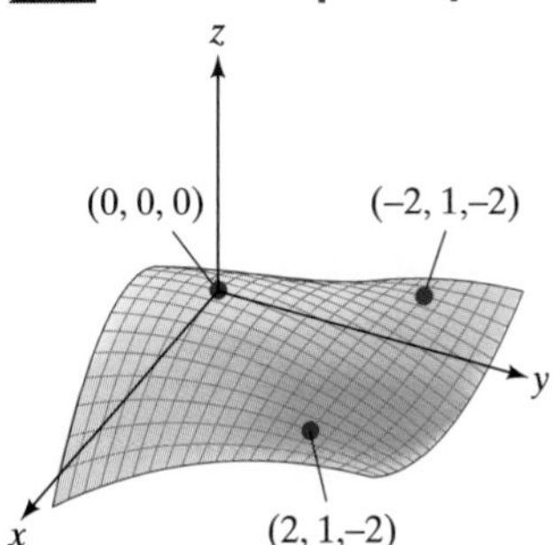

Figure 22

[3] Recall that if a product of two numbers is zero, then one or the other must be zero. In this case the number $e^{-(x^2+y^2)}$ can't be zero (since e^u is never zero), which gives the result claimed.

Classifying the Critical Points Numerically

We can use a spreadsheet to classify the critical points numerically. Following is a tabular representation of the function obtained using Excel. (See Example 3 in Section 8.1 for information on using Excel to generate such a table.)

	x	→						
y		−3	−2	−1	0	1	2	3
↓	−3	−54	−34	−22	−18	−22	−34	−54
	−2	−35	−20	−11	−8	−11	−20	−35
	0	−9	−4	−1	0	−1	−4	−9
	1	−2	−2	−2	−2	−2	−2	−2
	2	1	−4	−7	−8	−7	−4	1
	3	0	−10	−16	−18	−16	−10	0

The shaded and colored cells show rectangular neighborhoods of the three critical points $(0, 0)$, $(2, 1)$, and $(-2, 1)$. (Notice that they overlap.) The values of f at the points are at the centers of these rectangles. Looking at the gray and darker blue neighborhood of $(x, y) = (0, 0)$, we see that $f(0, 0) = 0$ is the largest value of f in the shaded cells, which suggests that f has a maximum at $(0, 0)$. The shaded neighborhood of $(2, 1)$ on the right shows $f(2, 1) = -2$ as the maximum along some slices (e.g., the vertical slice) and a minimum along the diagonal slice from top left to bottom right. This is what results in a saddle point on the graph. The point $(-2, 1)$ is similar, and thus f also has a saddle point at $(-2, 1)$.

Question Is there an algebraic way of deciding whether a given point is a relative maximum, relative minimum, or saddle point?
Answer There is a "second derivative test" for functions of two variables, stated as follows.

Second Derivative Test for Functions of Two Variables

Suppose (a, b) is a critical point in the interior of the domain of the function f of two variables. Let H be the quantity

$$H = f_{xx}(a, b)f_{yy}(a, b) - [f_{xy}(a, b)]^2$$

H is called the *Hessian.*

Then, if H is *positive,*

- f has a relative minimum at (a, b) if $f_{xx}(a, b) > 0$.
- f has a relative maximum at (a, b) if $f_{xx}(a, b) < 0$.

If H is *negative,*

- f has a saddle point at (a, b).

If $H = 0$, the test tells us nothing, so we need to look at the graph or a numerical table to see what is going on.

Quick Examples

1. Let $f(x, y) = x^2 - y^2$. Then

$$f_x = 2x \quad \text{and} \quad f_y = -2y$$

which gives (0, 0) as the only critical point. Also,

$$f_{xx} = 2 \qquad f_{xy} = 0 \qquad \text{and} \qquad f_{yy} = -2$$ Note that these are constant.

which gives $H = (2)(-2) - 0^2 = -2$. Since H is negative, we have a saddle point at (0, 0).

2. Let $f(x, y) = x^2 + 2y^2 + 2xy + 4x$. Then

$$f_x = 2x + 2y + 4 \qquad \text{and} \qquad f_y = 2x + 4y$$

Setting these equal to zero gives a system of two linear equations in two unknowns:

$$x + y = -2$$
$$x + 2y = 0$$

This system has solution $(-4, 2)$, so this is our only critical point. The second partial derivatives are

$$f_{xx} = 2 \qquad f_{xy} = 2 \quad \text{and} \quad f_{yy} = 4$$

so $H = (2)(4) - 2^2 = 4$. Since $H > 0$ and $f_{xx} > 0$, we have a relative minimum at $(-4, 2)$.

Note There is a second derivative test for functions of three or more variables, but it is considerably more complicated. We stick with functions of two variables for the most part in this book. The justification of the second derivative test is beyond the scope of this book.

We can use the second derivative test to analyze the "flying carpet" function we saw at the beginning of this section.

example 2

Using the Second-Derivative Test

Locate and classify all the critical points of

$$f(x, y) = 2(x^2 + y^2) - (x^4 + y^4) + 3$$

Solution

We first calculate the first-order partial derivatives:

$$f_x = 4x - 4x^3 = 4x(1 - x^2)$$
$$f_y = 4y - 4y^3 = 4y(1 - y^2)$$

Setting these equal to zero gives the following simultaneous equations:

$$4x(1 - x^2) = 0$$
$$4y(1 - y^2) = 0$$

The first equation has solutions $x = 0, 1$, or -1, and the second has solutions $y = 0$, 1, or -1. Since these are simultaneous equations, we need values of x and y that satisfy *both* equations. We can choose any one of the values for x that satisfies the first and any one of the values of y that satisfies the second. This gives us a total of *nine* critical points:

$$(0, 0), \quad (0, 1), \quad (0, -1), \quad (1, 0), \quad (1, 1), \quad (1, -1), \quad (-1, 0), \quad (-1, 1), \quad (-1, -1)$$

Four of these are the points P, Q, R, and S on the graph in Figure 20. The other five are also points at which f has extrema or saddle points, but Figure 20 shows only a small portion of the actual graph, and the remaining five points are out of range. (More of the graph is shown in Figure 23.)

We now need to apply the second-derivative test to each point in turn. First, we calculate all the second derivatives we need:

$$f_x = 4x - 4x^3, \quad \text{so} \quad f_{xx} = 4 - 12x^2 \quad \text{and} \quad f_{xy} = 0$$

and $$f_y = 4y - 4y^3, \quad \text{so} \quad f_{yy} = 4 - 12y^2$$

We now look at each critical point in turn.

The point (0, 0): $f_{xx}(0, 0) = 4, f_{yy}(0, 0) = 4$, and $f_{xy}(0, 0) = 0$. Thus, $H = 4(4) - 0^2 = 16$. Since $H > 0$ and $f_{xx}(0, 0) > 0$, we have a relative minimum at $(0, 0, 3)$.

The point (0, 1): $f_{xx}(0, 1) = 4, f_{yy}(0, 1) = -8$, and $f_{xy}(0, 1) = 0$. Thus, $H = 4(-8) - 0^2 = -32$. Since $H < 0$, we have a saddle point at $(0, 1, 4)$.

The point (0, −1): $f_{xx}(0, -1) = 4, f_{yy}(0, -1) = -8$, and $f_{xy}(0, -1) = 0$. These are the same values we got for the last point, so we again have a saddle point at $(0, -1, 4)$.

The point (1, 0): $f_{xx}(1, 0) = -8, f_{yy}(1, 0) = 4$, and $f_{xy}(1, 0) = 0$. Thus, $H = (-8)4 - 0^2 = -32$. Since $H < 0$, we have a saddle point at $(1, 0, 4)$.

The point (1, 1): $f_{xx}(1, 1) = -8, f_{yy}(1, 1) = -8$, and $f_{xy}(1, 1) = 0$. Thus, $H = (-8)^2 - 0^2 = 64$. Since $H > 0$ and $f_{xx}(1, 1) < 0$, we have a relative maximum at $(1, 1, 5)$.

The point (1, −1): These are the same values we got for the last point, so we again have a relative maximum at $(1, -1, 5)$.

The point (−1, 0): These are the same values we got for the point $(1, 0)$, so we have a saddle point at $(-1, 0, 4)$.

The point (−1, 1): These are the same values we got for the point $(1, 1)$, so we have a relative maximum at $(-1, 1, 5)$.

The point (−1, −1): These are again the same values we got for the point $(1, 1)$, so we have a relative maximum at $(-1, -1, 5)$.

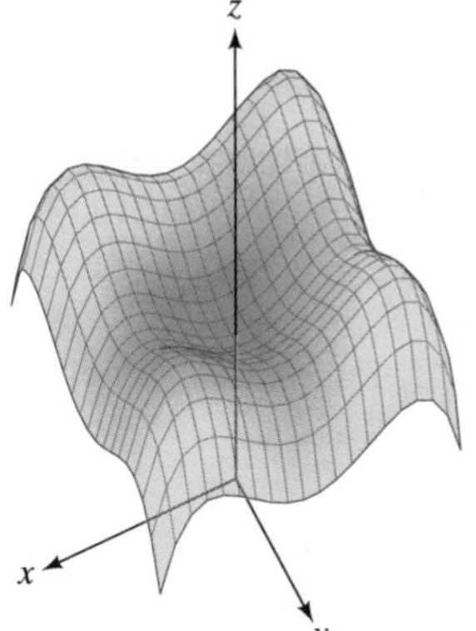

Figure 23

Figure 23 shows a larger portion of the graph of $f(x, y)$ including all nine critical points. We leave it to you to spot their locations on the graph.

---

example 3

The Second Derivative Test Again

Use the second derivative test to analyze the function $f(x, y) = x^2y - x^2 - 2y^2$ discussed in Example 1, and confirm the results we got there.

Solution

We saw in Example 1 that the first-order derivatives are

$$f_x = 2xy - 2x = 2x(y - 1)$$
$$f_y = x^2 - 4y$$

and the critical points are $(0, 0)$, $(2, 1)$, and $(-2, 1)$. We also need the second derivatives:

$$f_{xx} = 2y - 2$$
$$f_{xy} = 2x$$
$$f_{yy} = -4$$

The point (0, 0): $f_{xx}(0, 0) = -2, f_{xy}(0, 0) = 0$, and $f_{yy}(0, 0) = -4$, so $H = 8$. Since $H > 0$ and $f_{xx}(0, 0) < 0$, the second-derivative test tells us that f has a relative maximum at $(0, 0)$.

The point (2, 1): $f_{xx}(2, 1) = 0, f_{xy}(2, 1) = 4$, and $f_{yy}(2, 1) = -4$, so $H = -16$. Since $H < 0$, we know that f has a saddle point at $(2, 1)$.

The point (−2, 1): $f_{xx}(-2, 1) = 0, f_{xy}(-2, 1) = -4$, and $f_{yy}(-2, 1) = -4$, so once again $H = -16$, and f has a saddle point at $(-2, 1)$.

Deriving the Regression Formulas

Back in Section 1.5 we presented the following formulas for the **best-fit** or **regression line** associated with a given set of data points $(x_1, y_1), (x_2, y_2), \ldots, (x_n, y_n)$.

Regression Line

The line that best fits the n data points $(x_1, y_1), (x_2, y_2), \ldots, (x_n, y_n)$ has the form

$$y = mx + b$$

where

$$m = \frac{n(\Sigma xy) - (\Sigma x)(\Sigma y)}{n(\Sigma x^2) - (\Sigma x)^2}$$

$$b = \frac{\Sigma y - m(\Sigma x)}{n}$$

n = number of data points

Question Where do these formulas come from?

Answer We first go back to the definition of the regression line: It is defined to be the line that minimizes the sum of the squares of the vertical distances shown in Figure 24, which shows the regression line $y = mx + b$ associated with $n = 5$ data points.

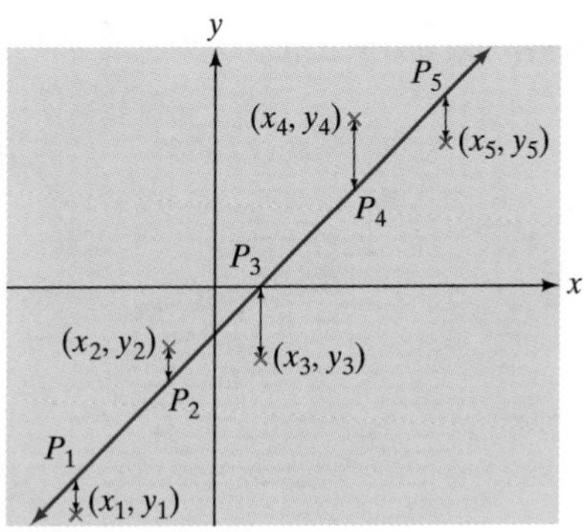

Figure 24

We can see that the points $P_1, \ldots, P_n$ on the regression line have coordinates $(x_1, mx_1 + b), (x_2, mx_2 + b), \ldots, (x_n, mx_n + b)$. Thus, the vertical distances are

$$|mx_1 + b - y_1|, \quad |mx_2 + b - y_2|, \quad \ldots, \quad |mx_n + b - y_n|$$

The sum of the squares of these distances is therefore

$$S(m, b) = (mx_1 + b - y_1)^2 + (mx_2 + b - y_2)^2 + \cdots + (mx_n + b - y_n)^2$$

and this is the quantity we must minimize by choosing m and b. Since we reason that there is a line that minimizes this quantity, there must be a relative minimum at that point. We see in a moment that the function S has at most one critical point, which must therefore be the desired absolute minimum. To obtain the critical points of S, we set the partial derivatives equal to zero and solve:

$$S_m = 0: \quad 2x_1(mx_1 + b - y_1) + \cdots + 2x_n(mx_n + b - y_n) = 0$$

$$S_b = 0: \quad 2(mx_1 + b - y_1) + \cdots + 2(mx_n + b - y_n) = 0$$

Dividing by 2 and gathering terms allows us to rewrite the equations as

$$m(x_1^2 + \cdots + x_n^2) + b(x_1 + \cdots + x_n) = x_1y_1 + \cdots + x_ny_n$$

$$m(x_1 + \cdots + x_n) + b = y_1 + \cdots + y_n$$

We can rewrite these equations more neatly using Σ notation:

$$m(\Sigma x^2) + b(\Sigma x) = \Sigma xy$$

$$m(\Sigma x) + b = \Sigma y$$

This is a system of two linear equations in the two unknowns m and b. It may or may not have a unique solution. When there is a unique solution, we can conclude that the best-fit line is given by solving these two equations for m and b. Alternatively, there is a general formula for the solution of any system of two equations in two unknowns, and if we apply this formula to our two equations, we get the regression formulas above.

8.4 exercises

In Exercises 1–4, classify each labeled point on the graph as one of the following:

a. *a relative maximum*

b. *a relative minimum*

c. *a saddle point*

d. *a critical point but neither a relative extremum nor a saddle point*

e. *none of these*

1.

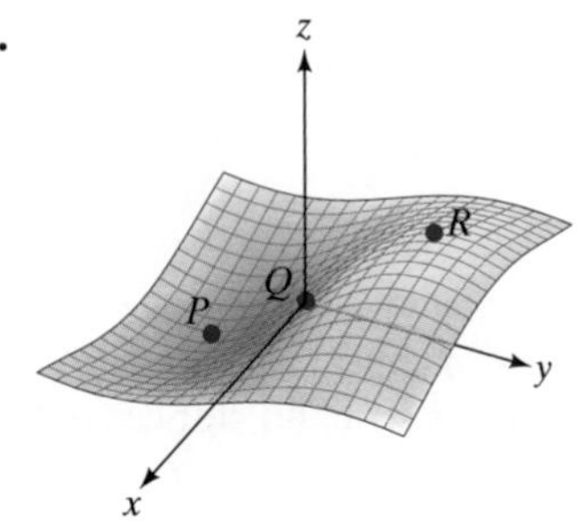

2.

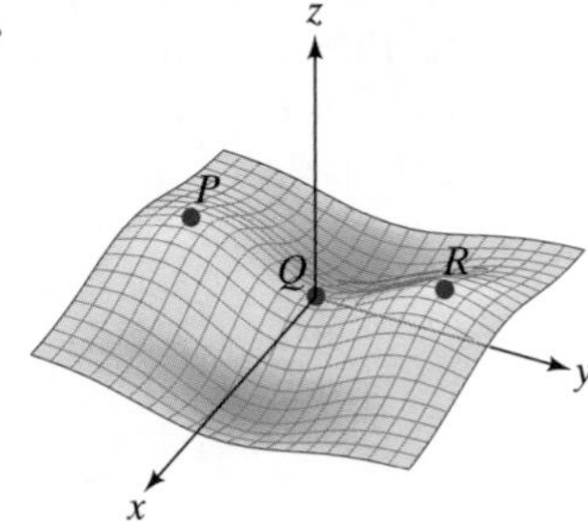

3.

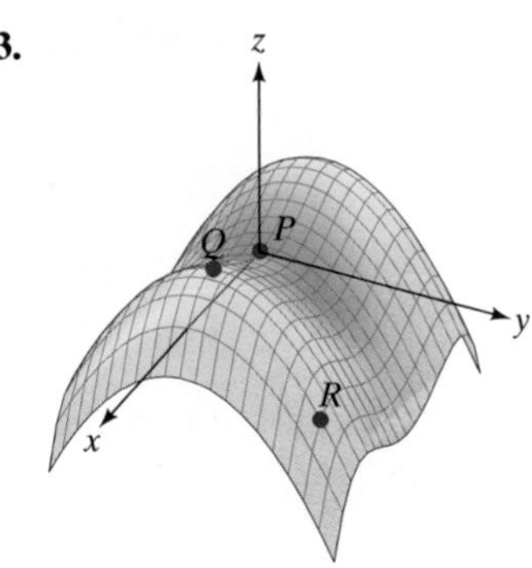

4.

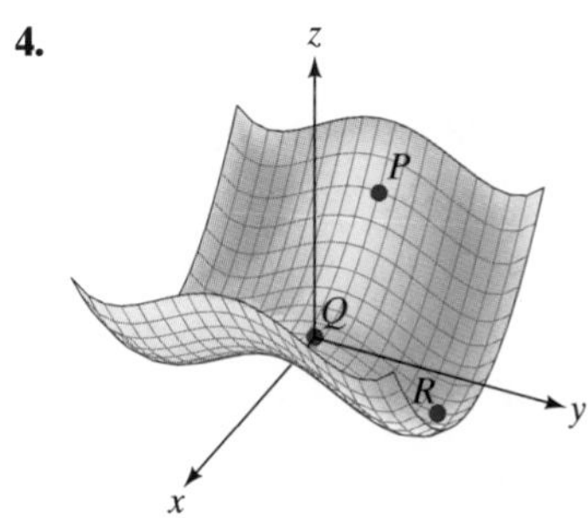

In Exercises 5–10, classify the shaded value in each table as one of the following:

a. a relative maximum
b. a relative minimum
c. a saddle point
d. neither a relative extremum nor a saddle point

5.

	x	→					
y		−3	−2	−1	0	1	2
↓	−3	10	5	2	1	2	5
	−2	9	4	1	0	1	4
	−1	10	5	2	1	2	5
	0	13	8	5	4	5	8
	1	18	13	10	9	10	13
	2	25	20	17	16	17	20
	3	34	29	26	25	26	29

6.

	x	→					
y		−3	−2	−1	0	1	2
↓	−3	5	0	−3	−4	−3	0
	−2	8	3	0	−1	0	3
	−1	9	4	1	0	1	4
	0	8	3	0	−1	0	3
	1	5	0	−3	−4	−3	0
	2	0	−5	−8	−9	−8	−5
	3	−7	−12	−15	−16	−15	−12

7.

	x	→					
y		−3	−2	−1	0	1	2
↓	−3	5	0	−3	−4	−3	0
	−2	8	3	0	−1	0	3
	−1	9	4	1	0	1	4
	0	8	3	0	−1	0	3
	1	5	0	−3	−4	−3	0
	2	0	−5	−8	−9	−8	−5
	3	−7	−12	−15	−16	−15	−12

8.

	x	→					
y		−3	−2	−1	0	1	2
↓	−3	2	3	2	−1	−6	−13
	−2	3	4	3	0	−5	−12
	−1	2	3	2	−1	−6	−13
	0	−1	0	−1	−4	−9	−16
	1	−6	−5	−6	−9	−14	−21
	2	−13	−12	−13	−16	−21	−28
	3	−22	−21	−22	−25	−30	−37

9.

	x	→					
y		−3	−2	−1	0	1	2
↓	−3	4	5	4	1	−4	−11
	−2	3	4	3	0	−5	−12
	−1	4	5	4	1	−4	−11
	0	7	8	7	4	−1	−8
	1	12	13	12	9	4	−3
	2	19	20	19	16	11	4
	3	28	29	28	25	20	13

10.

	x	→					
y		−3	−2	−1	0	1	2
↓	−3	100	101	100	97	92	85
	−2	99	100	99	96	91	84
	−1	98	99	98	95	90	83
	0	91	92	91	88	83	76
	1	72	73	72	69	64	57
	2	35	36	35	32	27	20
	3	−26	−25	−26	−29	−34	−41

11. Sketch the graph of a function that has one relative extremum and no saddle points.

12. Sketch the graph of a function that has one saddle point and one relative extremum.

Locate and classify all the critical points of the functions in Exercises 13–30.

13. $f(x, y) = x^2 + y^2 + 1$

14. $f(x, y) = 4 - (x^2 + y^2)$

15. $g(x, y) = 1 - x^2 - x - y^2 + y$

16. $g(x, y) = x^2 + x + y^2 - y - 1$

17. $h(x, y) = x^2y - 2x^2 - 4y^2$

18. $h(x, y) = x^2 + y^2 - y^2x - 4$

19. $s(x, y) = e^{x^2+y^2}$

20. $s(x, y) = e^{-(x^2+y^2)}$

21. $t(x, y) = x^4 + 8xy^2 + 2y^4$

22. $t(x, y) = x^3 - 3xy + y^3$

23. $f(x, y) = x^2 + y - e^y$

24. $f(x, y) = xe^y$

25. $f(x, y) = e^{-(x^2+y^2+2x)}$

26. $f(x, y) = e^{-(x^2+y^2-2x)}$

27. $f(x, y) = xy + 2/x + 2/y$

28. $f(x, y) = xy + 4/x + 2/y$

29. $g(x, y) = x^2 + y^2 + 2/xy$

30. $g(x, y) = x^3 + y^3 + 3/xy$

Applications

31. ***Brand Loyalty*** Suppose the fraction of Mazda car owners who chose another new Mazda can be modeled by the following function:[4]

$$M(c, f) = 11 + 8c + 4c^2 - 40f - 20cf + 40f^2$$

where c is the fraction of Chrysler car owners who remained loyal to Chrysler and f is the fraction of Ford car owners who remained loyal to Ford. Locate and classify all the critical points and interpret your answer.
Source: Chrysler, Maritz Market Research, Consumer Attitude Research, and Strategic Vision/*New York Times,* November 3, 1995, p. D2.

32. ***Brand Loyalty*** Repeat Exercise 31 using the function

$$M(c, f) = -10 - 8f - 4f^2 + 40c + 20fc - 40c^2$$

33. ***Pollution Control*** The cost of controlling emissions at a firm goes up rapidly as the amount of emissions reduced goes up. Here is a possible model:

$$C(x, y) = 4000 + 100x^2 + 50y^2$$

where x is the reduction in sulfur emissions, y is the reduction in lead emissions (in pounds of pollutant per day), and C is the daily cost to the firm (in dollars) of this reduction. Government clean-air subsidies amount to \$500 per pound of sulfur and \$100 per pound of lead removed. How many pounds of pollutant should the firm remove each day to minimize *net* cost (cost minus subsidy)?

34. ***Pollution Control*** Repeat Exercise 33 using the following information:

$$C(x, y) = 2000 + 200x^2 + 100y^2$$

with government subsidies amounting to \$100 per pound of sulfur and \$500 per pound of lead removed per day.

35. ***Revenue*** Your company manufactures two models of stereo speakers, the Ultra Mini and the Big Stack. Demand for each depends partly on the price of the other. If one is expensive, then more people will buy the other. If p_1 is the price per pair of the Ultra Mini and p_2 is the price of the Big Stack, demand for the Ultra Mini is given by

$$q_1(p_1, p_2) = 100{,}000 - 100p_1 + 10p_2$$

where q_1 represents the number of pairs of Ultra Minis that will be sold in a year. The demand for the Big Stack is given by

$$q_2(p_1, p_2) = 150{,}000 + 10p_1 - 100p_2$$

Find the prices for the Ultra Mini and the Big Stack that will maximize your total revenue.

36. ***Revenue*** Repeat Exercise 35, using the following demand functions:

$$q_1(p_1, p_2) = 100{,}000 - 100p_1 + p_2$$

$$q_2(p_1, p_2) = 150{,}000 + p_1 - 100p_2$$

37. ***Luggage Dimensions*** TransWorld Airlines (TWA) has a rule for checked baggage, "The total dimensions (length + width + height) may not exceed 62 inches for each bag."[5] What are the dimensions of the largest volume bag you can check on a TWA flight? What is its volume?

38. ***Luggage Dimensions*** American Airlines has the same rule as TWA for the first bag you check on a flight, but the second bag you check "may have total outside dimensions (length + width + height) of no more than . . . 55 inches."[6] What are the dimensions of the largest volume second bag you can check on an American flight? What is the volume?

39. ***Package Dimensions*** The U.S. Postal Service (USPS) will accept only packages with length plus girth no more than 108 inches.[7] (See the figure.)

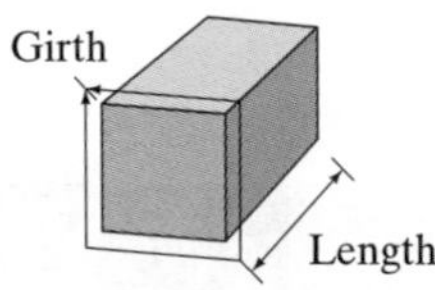

What are the dimensions of the largest volume package the USPS will accept? What is its volume?

40. ***Package Dimensions*** United Parcel Service (UPS) will accept only packages with length no more than 108 inches and length plus girth no more than 130 inches.[8] (See the figure for the preceding exercise.) What are the dimensions of the largest volume package UPS will accept? What is its volume?

[4] This model is not accurate, although it was inspired by an approximation of a second-order regression based on data from the period 1988–1995.

[5] According to information on its web site (http://www.twa.com/) as of March 1999.

[6] According to information on its web site (http://www.aa.com/) as of March 1999.

[7] The requirement at the time this book was written.

[8] The requirement at the time this book was written.

Communication and Reasoning Exercises

41. Let $H = f_{xx}(a, b)f_{yy}(a, b) - f_{xy}(a, b)^2$. What condition on H guarantees that f has a relative extremum at the point (a, b)?

42. Let H be as in Exercise 41. Give an example to show that it is possible to have $H = 0$ and a relative minimum at (a, b).

43. Suppose that when the graph of $f(x, y)$ is sliced by a vertical plane through (a, b) parallel to either the xz-plane or the yz-plane, the resulting curve has a relative maximum at (a, b). Does this mean that f has a relative maximum at (a, b)? Explain your answer.

44. Suppose that f has a relative maximum at (a, b). Does it follow that if the graph of f is sliced by a vertical plane parallel to either the xz-plane or the yz-plane, the resulting curve has a relative maximum at (a, b)? Explain your answer.

45. The tangent plane to a graph was introduced in Exercise 74 in the preceding section. Use the equation of the tangent plane given there to explain why the tangent plane is parallel to the xy-plane at a relative maximum or minimum of $f(x, y)$.

46. Use the equation of the tangent plane given in Exercise 74 in the preceding section to explain why the tangent plane is parallel to the xy-plane at a saddle point of $f(x, y)$.

47. Let $C(x, y)$ be any cost function. Show that when the average cost is minimized, the marginal costs C_x and C_y both equal the average cost. Explain why this is reasonable.

48. Let $P(x, y)$ be any profit function. Show that when the average profit is maximized, the marginal profits P_x and P_y both equal the average profit. Explain why this is reasonable.

8.5 Constrained Maxima and Minima and Applications

So far we have looked only at the relative extrema of f that lie in the interior of the domain of f. There may also be relative extrema on the boundary of the domain (the "endpoints" of the domain may be relative extrema just as for a function of one variable). This situation arises, for example, in optimization problems with constraints, similar to those we saw in Chapter 5 on applications of the derivative. Here is a typical example:

$$\text{Maximize } S = xy + 2xz + 2yz \text{ subject to } xyz = 4 \text{ and } x \geq 0, y \geq 0, z \geq 0.$$

There are two kinds of constraints in this example: equations and inequalities. The inequalities specify a restriction on the domain of S.

Our strategy for solving such problems is essentially the same as the strategy we used earlier. First, we use any equation constraints to eliminate variables. In the examples in this section we are able to reduce to a function of only two variables. The inequality constraints then help define the domain of this function. Next, we locate any critical points in the interior of the domain. Finally, we look at the boundary of the domain. When there is no boundary to worry about, another method, called the *method of Lagrange multipliers,* comes in handy.

We first look at functions of two variables with restricted domains, so we can see how to handle the boundaries.

Restricted Domain

Find the maximum and minimum values of $f(x, y) = xy - x - 2y$ on the triangular region R with vertices $(0, 0)$, $(1, 0)$, and $(0, 2)$.

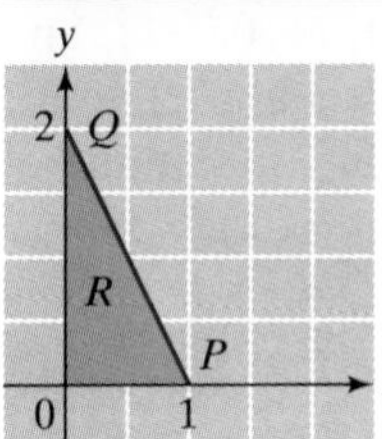

Figure 25

Solution

The domain of f is the region R shaded in Figure 25.

Step 1 *Locate critical points in the interior of the domain.* We take the partial derivatives as usual.

$$f_x = y - 1$$
$$f_y = x - 2$$
$$f_{xx} = 0$$
$$f_{xy} = 1$$
$$f_{yy} = 0$$

The only critical point is thus (2, 1). Since this lies outside the domain (the region R), we ignore it. Thus, there are no critical points in the interior of the domain of f.

Step 2 *Locate relative extrema on the boundary of the domain.* The boundary of the domain consists of three line segments, OP, OQ, and PQ. We deal with these one at a time.

- *Segment OP:* This line segment has equation $y = 0$ with domain $0 \le x \le 1$. The behavior of the function f along this segment is given by substituting $y = 0$ into the expression for x:

  $$f(x, 0) = -x \qquad \text{Substitute } y = 0 \text{ in the expression for } f.$$

  We now find the relative extrema of this function of one variable by the methods we used in the chapter on applications of the derivative. There are no critical points and there are two endpoints, $x = 0$ and $x = 1$. Since $y = 0$, this gives us the following two candidates for relative extrema: $(0, 0, 0)$ and $(1, 0, -1)$.

- *Segment OQ:* This line segment has equation $x = 0$ with $0 \le y \le 2$. Along this segment we see

  $$f(0, y) = -2y \qquad \text{Substitute } x = 0 \text{ in the expression for } f.$$

  We now locate the relative extrema of this function of one variable. Once again, there are only the endpoints, $y = 0$ and $y = 2$. Since $x = 0$, this gives us the two candidates $(0, 0, 0)$ and $(0, 2, -4)$.

- *Segment PQ:* This line segment has equation $y = -2x + 2$ with $0 \le x \le 1$. Along this segment we see

  $$\begin{aligned} f(x, -2x + 2) &= x(-2x + 2) - x - 2(-2x + 2) \\ &= -2x^2 + 5x - 4 \end{aligned}$$

  Substitute $y = -2x + 2$ in the expression for f.

  This function of x (whose graph is an upside down parabola) has a stationary maximum when its derivative, $-4x + 5$, is zero, which occurs when $x = \frac{5}{4}$. Since this is greater than 1, it lies outside the domain $0 \le x \le 1$ and we reject it. There are no other critical points, and the endpoints are $x = 0$ and $x = 1$. When $x = 0$, $y = -2(0) + 2 = 2$, giving the point $(0, 2, -4)$. When $x = 1$, $y = -2(1) + 2 = 0$, giving the point $(1, 0, -1)$.

Thus, the candidates for maxima and minima are $(0, 0, 0)$, $(1, 0, -1)$, and $(0, 2, -4)$ (which happen to lie over the corner points of the domain R). Since their z-coordinates give the value of f, we see that f has an absolute maximum of 0 at the point $(0, 0)$ and an absolute minimum of -4 at the point $(0, 2)$.

Next we look at an example with a constraint equation.

example 2

Minimizing Area

Find the dimensions of an open-top rectangular box that has a volume of 4 cubic feet and the smallest possible surface area.

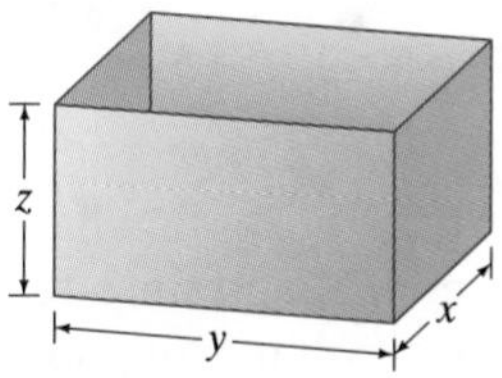

Solution

Our first task is to rephrase this example as a mathematical optimization problem. Figure 26 shows a picture of the box with dimensions x, y, and z. We want to minimize the total surface area, which is given by

$$A = xy + 2xz + 2yz \qquad \text{Base + Sides + Front and back}$$

Figure 26

This is our **objective function.** We can't simply choose x, y, and z to all be zero, however, because the enclosed volume must be 4 cubic feet. So

$$xyz = 4 \qquad \text{Constraint}$$

This is our **constraint** equation. Other unstated constraints are $x \geq 0$, $y \geq 0$, and $z \geq 0$. We now restate the problem as follows:

Minimize $A = xy + 2xz + 2yz$ subject to $xyz = 4$ and $x \geq 0, y \geq 0, z \geq 0$.

As suggested in the discussion before Example 1, we first solve the constraint equation for one of the variables and then substitute into the objective function. Solving the constraint equation for z gives

$$z = \frac{4}{xy}$$

Substituting this into the objective function gives a function of only two variables:

$$A = xy + \frac{8}{y} + \frac{8}{x}$$

Now we minimize the resulting function of two variables. A word about the domain of S: Since we said that x and y must be nonnegative, we already know that (x, y) is restricted to the first quadrant. Moreover, since S is not defined if either x or y is zero, we exclude the x- and y-axes from the domain of S and are left with the interior of the first quadrant as our domain. Since this has no boundary, we look for critical points only in the interior:

$$A_x = y - \frac{8}{x^2} \qquad A_y = x - \frac{8}{y^2}$$

$$A_{xx} = \frac{16}{x^3} \qquad A_{xy} = 1 \qquad A_{yy} = \frac{16}{y^3}$$

We now equate the first partial derivatives to zero:

$$y = \frac{8}{x^2} \qquad \text{and} \qquad x = \frac{8}{y^2}$$

To solve for x and y, we substitute the first of these equations in the second, getting

$$x = \frac{x^4}{8}$$

$$x^4 - 8x = 0$$

$$x(x^3 - 8) = 0$$

The two solutions are $x = 0$, which we reject because x cannot be zero, and $x = 2$. Substituting $x = 2$ in $y = 8/x^2$ gives $y = 2$ also. Thus, the only critical point is (2, 2). To apply the second derivative test, we compute

$$A_{xx}(2, 2) = 2 \qquad A_{xy}(2, 2) = 1 \qquad A_{yy}(2, 2) = 2$$

and find that $H > 0$, so we have a relative minimum at (2, 2).

Question Is this relative minimum an absolute minimum?
Answer Yes. There must be a smallest surface area among all the boxes that hold 4 cubic feet. (Why?) Since this would give a relative minimum of A and since the only possible relative minimum of A occurs at (2, 2), this is the absolute minimum.

To finish the example, we get the value of z by substituting the values of x and y into the constraint equation $z = 4/xy$, which gives $z = 1$. Thus, the required dimensions of the box are

$$x = 2 \text{ ft} \qquad y = 2 \text{ ft} \qquad z = 1 \text{ ft}$$

In the next example we return to the Ultra Mini and Big Stack speaker example and look at an interaction model for profit.

example 3

Maximizing Profit
You own a company that makes two models of stereo speakers, the Ultra Mini and the Big Stack. Each Ultra Mini requires 1 square foot of fabric and 3 feet of wire; each Big Stack requires 5 square feet of fabric and 9 feet of wire. You have each week 100 square feet of fabric and 270 feet of wire to use. Your profit function is estimated to be

$$P(x, y) = 10x + 60y + 0.5xy$$

where x is the number of Ultra Minis, y is the number of Big Stacks, and P is your profit in dollars. Find the number of each model you should make each week to maximize your profit.

Solution
The constraints in this problem come from the limited amount of fabric and wire available. The constraint on fabric is

$$x + 5y \leq 100$$

The constraint on wire is

$$3x + 9y \leq 270$$

There are also the constraints $x \geq 0$ and $y \geq 0$ that are common in applications. Our problem is then the following:

$$\begin{aligned} &\text{Maximize } P = 10x + 60y + 0.5xy \text{ subject to} \\ &\quad x + 5y \leq 100 \\ &\quad 3x + 9y \leq 270 \\ &\quad x \geq 0,\ y \geq 0. \end{aligned}$$

If we graph the domain R, we get Figure 27.

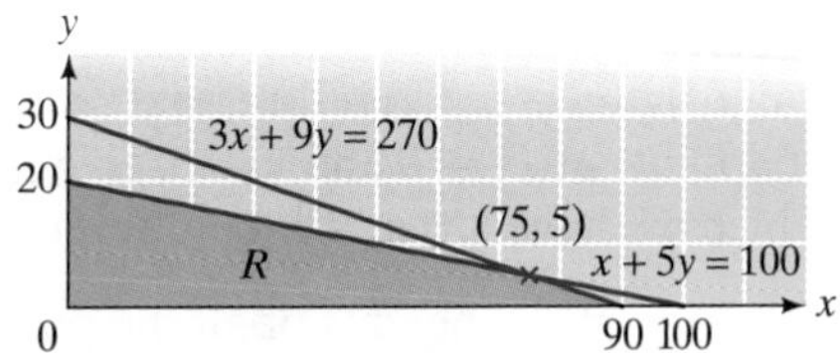

Figure 27

[The point of intersection (75, 5) is found by solving the system of linear equations given by the equations of the two lines.] Since we have no equality constraints, we proceed as before.

Step 1 *Locate critical points in the interior of the domain:*

$$P_x = 10 + 0.5y$$

$$P_y = 60 + 0.5x$$

Setting these equal to zero and solving for x and y gives the only critical point as $(-120, -20)$. However, this critical point is well outside the domain R, so we disregard it.

Step 2 *Locate relative extrema on the boundary of the domain.* The boundary of the domain consists of four line segments, which we consider one at a time.

- *Segment* $y = 0, 0 \leq x \leq 90$: Here

$$P(x, 0) = 10x + 60(0) + 0.5x(0) = 10x$$

$$P' = 10$$

and is therefore never zero. Thus, the only points we need to look at are the endpoints $x = 0$ and $x = 90$, which give $(0, 0, 0)$ and $(90, 0, 900)$.

- *Segment* $x = 0, 0 \leq y \leq 20$: Here

$$P(0, y) = 10(0) + 60y + 0.5(0)y = 60y$$

Again, there are no critical points, only the endpoints $y = 0$ and $y = 20$. This gives one more point, $(0, 20, 1200)$.

- *Segment* $x = 100 - 5y, 5 \leq y \leq 20$: This gives

$$P(100 - 5y, y) = 10(100 - 5y) + 60y + 0.5(100 - 5y)y = 1000 + 60y - 2.5y^2$$

$$P' = 60 - 5y$$

Setting the derivative equal to zero and solving, we get $y = 12$, which is within the domain $5 \leq y \leq 20$. The corresponding point on the graph is $(40, 12, 1360)$. Since $P'' = -5$, this point is a relative maximum. The two endpoints of this segment are $y = 5$ and $y = 20$. The point at $y = 20$ is $(0, 20, 1200)$, which we have already considered, and the other point is $(75, 5, 1237.5)$.

- *Segment* $x = 90 - 3y, 0 \leq y \leq 5$: This gives

$$P(90 - 3y, y) = 10(90 - 3y) + 60y + 0.5(90 - 3y)y = 900 + 75y + 1.5y^2$$

$$P' = 75 - 3y$$

Setting this equal to zero and solving for y give $y = 25$, which is outside the range $0 \le y \le 5$. Therefore, there are no critical points on this line segment. Its endpoints are $y = 0$ and $y = 5$, giving the points (90, 0, 900) and (75, 5, 1237.5), which we have already encountered.

Looking at the heights of all the points we have found, we see that the highest is (40, 12, 1360), so this gives the greatest profit. In other words, you should make 40 Ultra Minis and 12 Big Stacks each week to get the greatest possible profit of $1360 per week.

Before we go on... If you have studied linear programming, this example should remind you of the problems you solved by that technique. However, since the objective function is not linear, the techniques of linear programming fail to solve this problem. In particular, notice that the solution we found is *not* at a corner of the feasible region, as it must be in a linear programming problem, but is in the middle of one edge. The solution to another problem could be in the interior of the region.

The Method of Lagrange Multipliers

Suppose we have an optimization problem in which it is difficult or impossible to solve a constraint equation for one of the variables. Then we can use the method of **Lagrange multipliers** to avoid this difficulty. We restrict our attention to the case of a single constraint equation, although the method generalizes to any number of constraint equations.

Locating Relative Extrema Using the Method of Lagrange Multipliers

To locate the candidates for relative extrema of a function $f(x, y, \ldots)$ subject to the constraint $g(x, y, \ldots) = 0$, we solve the following system of equations for $x, y, \ldots$ and λ:

$$\begin{aligned} f_x &= \lambda g_x \\ f_y &= \lambda g_y \\ &\vdots \\ g &= 0 \end{aligned}$$

The unknown λ is called a **Lagrange multiplier.** The points $(x, y, \ldots)$ that occur in solutions are then the candidates for the relative extrema of f subject to $g = 0$.

example 4

Using Lagrange Multipliers

Use the method of Lagrange multipliers to find the maximum value of $f(x, y) = 2xy$ subject to $x^2 + 4y^2 = 32$.

Solution

We start by rewriting the problem in standard form:

$$\text{Maximize } f(x, y) = 2xy \text{ subject to } x^2 + 4y^2 - 32 = 0.$$

Here, $g(x, y) = x^2 + 4y^2 - 32$, and the system of equations we need to solve is thus

$$f_x = \lambda g_x \quad \text{or} \quad 2y = 2\lambda x$$

$$f_y = \lambda g_y \quad \text{or} \quad 2x = 8\lambda y$$

$$g = 0 \quad \text{or} \quad x^2 + 4y^2 - 32 = 0$$

A convenient way to solve such a system is to solve one of the equations for λ and then substitute in the remaining equations. Thus, we start by solving the first equation to obtain

$$\lambda = \frac{y}{x}$$

(A word of caution: Since we divided by x, we made the implicit assumption that $x \neq 0$, so before continuing, we should check what happens if $x = 0$. If $x = 0$, then the first equation, $2y = 2\lambda x$, tells us that $y = 0$ as well, and this contradicts the third equation: $x^2 + 4y^2 - 32 = 0$. Thus, we can rule out the possibility that $x = 0$.) Substituting in the remaining equations gives

$$x = \frac{4y^2}{x} \quad \text{or} \quad x^2 = 4y^2$$

$$x^2 + 4y^2 - 32 = 0$$

Notice how we have reduced the number of unknowns and also the number of equations by one. We can now substitute $x^2 = 4y^2$ in the last equation, obtaining

$$4y^2 + 4y^2 - 32 = 0$$

$$8y^2 = 32$$

$$y = \pm 2$$

We substitute back to obtain

$$x^2 = 4y^2 = 16$$

$$x = \pm 4$$

We don't need the value of λ, so we won't solve for it. Thus, the candidates for relative extrema are the four points $(-4, -2)$, $(-4, 2)$, $(4, -2)$, and $(4, 2)$. Recall that we are seeking the values of x and y that give the maximum value for $f(x, y) = 2xy$. Since we now have only four points to choose from, we compare the values of f at these four points and conclude that the maximum value of f occurs when $(x, y) = (-4, -2)$ or $(4, 2)$.

Before we go on . . .
Question Something is suspicious here. We didn't check to see whether these candidates were relative extrema to begin with, let alone absolute extrema! How do we justify this omission?
Answer One of the difficulties with using the method of Lagrange multipliers is that it does not provide us with a test analogous to the second derivative test for functions of several variables. However, if we grant that the function in question does have an absolute maximum, then we require no test because one of the candidates must give this maximum.

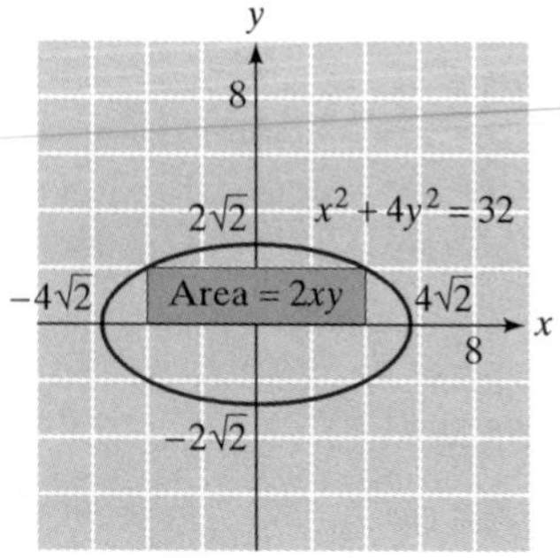

Figure 28

Question But how do we know that the given function has an absolute maximum?
Answer The best way to see this is by giving a geometric interpretation. The constraint $x^2 + 4y^2 = 32$ tells us that the point (x, y) must lie on the ellipse shown in Figure 28. The function $f(x, y) = 2xy$ gives the area of the rectangle shaded in Figure 28. Since there must be a largest such rectangle, the function f must have an absolute maximum for at least one pair of coordinates (x, y).

Question When can I use the method of Lagrange multipliers? When should I use it?
Answer We have discussed the method only when there is a single equality constraint. There is a generalization, which we do not discuss, that works when there are more equality constraints (we need to introduce one multiplier for each constraint). So, if you have a problem with more than one equality constraint, or with any inequality constraints, you must use the method we discussed earlier in this section. On the other hand, if you have one equality constraint and it would be difficult to solve it for one of the variables, then you should use Lagrange multipliers. We note in the exercises where you might use Lagrange multipliers.

Question Why does the method of Lagrange multipliers work?
Answer An adequate answer is beyond the scope of this book.

Software packages like Excel have built-in algorithms that seek absolute extrema whether or not constraints are present. In the next example we use the "Solver" routine in Excel to redo Example 3.[1]

example 5

Using Excel to Solve an Optimization Problem
Use Solver to solve the following problem:

Maximize	$P = 10x + 60y + 0.5xy$	Objective function
subject to	$x + 5y \leq 100$	Constraint 1
	$3x + 9y \leq 270$	Constraint 2
	$x \geq 0$ and $y \geq 0$	Constraint 3 and constraint 4

Solution
First, we set up the problem in spreadsheet form as follows:

	A	B	C	D	
1	0	0	=A1+5*B1	100	constraint 1
2	initial values of x and y		=3*A1+9*B1	270	constraint 2
3			=A1	0	constraint 3
4			=B1	0	constraint 4
5			=10*A1+60*B1+0.5*A1*B1		objective

[1] If "Solver" does not appear in the Tools menu, you should first install it using your Excel installation software. (Solver is one of the "Excel Add-Ins.")

A1 will be the cell that contains the value of x and B1 the cell that contains the value of y. (We have set them both to zero in anticipation that Solver will adjust them to satisfy the constraints and give an optimal solution.) Next we select "Solver" in the Tools menu to bring up the Solver dialog box. We show the dialog box with all the necessary fields completed to solve the problem.

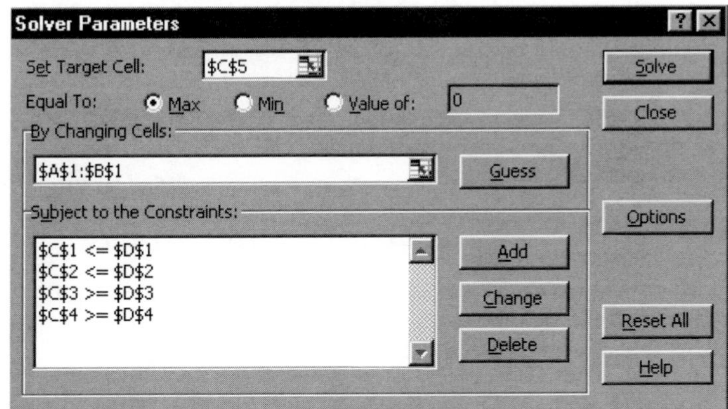

Notes

1. The Target Cell refers to the cell that contains the objective function.
2. "Max" is selected because we are maximizing the objective function.
3. "Changing Cells" are obtained by selecting the cells that contain the current values of x and y.
4. Constraints are added one at a time by pressing the "Add" button and selecting the cells that contain the left- and right-hand sides of each inequality (equality constraints are also permitted).

When we are done, we press "Solve" and an (approximate) optimal solution appears in cells A1 and B1, with the maximum value of P appearing in cell C5. As we saw in Example 3, the optimal solution is $x = 40, y = 12, P = 1360$.

Before we go on . . .
Question Can a software package such as Excel Solver be used interchangeably with the analytic method?
Answer Some optimization problems lead to equations that cannot be solved analytically, and so some form of numerical approach (such as that used in Solver) is essential in those cases. However, with all known numerical approaches there is always a chance of running into one of these problems:

- The solution given is a relative extremum rather than an absolute extremum.
- The solution is not exact.
- Roundoff errors lead to an incorrect solution or prevent finding a solution.
- Only one solution is given even if there is more than one absolute extremum.

So, use Solver with caution.

8.5 exercises

In all the exercises for this section, a software package such as Excel Solver can be used as a check on your analytic work. (Bear in mind, however, the cautions at the end of Example 5.)

Find the maximum and minimum values, and the points at which they occur, for each function in Exercises 1–16.

1. $f(x, y) = x^2 + y^2; 0 \le x \le 2, 0 \le y \le 2$
2. $g(x, y) = \sqrt{x^2 + y^2}; 1 \le x \le 2, 1 \le y \le 2$
3. $h(x, y) = (x - 1)^2 + y^2; x^2 + y^2 \le 4$
4. $k(x, y) = x^2 + (y - 1)^2; x^2 + y^2 \le 9$
5. $f(x, y) = e^{x^2+y^2}; 4x^2 + y^2 \le 4$
6. $g(x, y) = e^{-(x^2+y^2)}; x^2 + 4y^2 \le 4$
7. $h(x, y) = e^{4x^2+y^2}; x^2 + y^2 \le 1$
8. $k(x, y) = e^{-(x^2+4y^2)}; x^2 + y^2 \le 4$
9. $f(x, y) = x + y + 1/(xy); x \ge \frac{1}{2}, y \ge \frac{1}{2}, x + y \le 3$
10. $g(x, y) = x + y + 8/(xy); x \ge 1, y \ge 1, x + y \le 6$
11. $h(x, y) = xy + 8/x + 8/y; x \ge 1, y \ge 1, xy \le 9$
12. $k(x, y) = xy + 1/x + 4/y; x \ge 1, y \ge 1, xy \le 10$
13. $f(x, y) = x^2 + 2x + y^2$; on the region in the figure

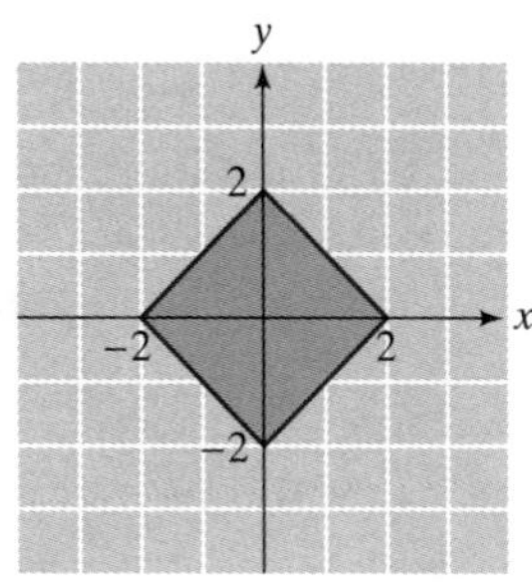

14. $g(x, y) = x^2 + y^2$; on the region in the figure

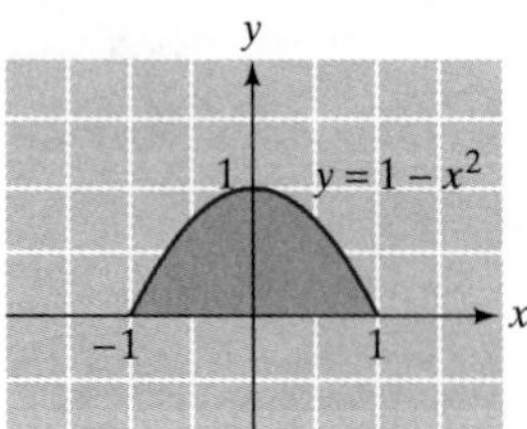

15. $h(x, y) = x^3 + y^3$; on the region in the figure

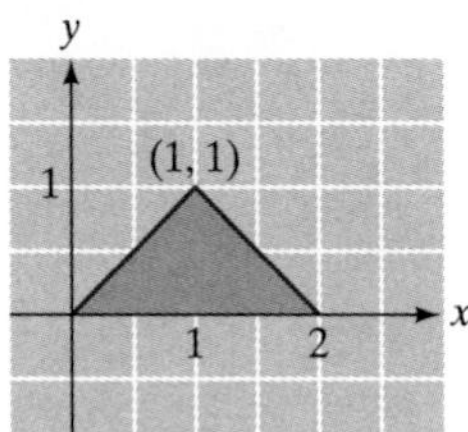

16. $k(x, y) = x^3 + 2y^3$; on the region in the figure

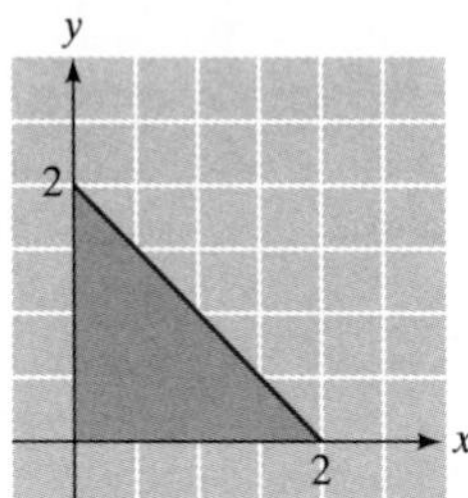

17. At what points on the sphere $x^2 + y^2 + z^2 = 1$ is the product xyz a maximum? (The method of Lagrange multipliers can be used here.)
18. At what point on the surface $z = (x^2 + x + y^2 + 4)^{1/2}$ is the quantity $x^2 + y^2 + z^2$ a minimum? (The method of Lagrange multipliers can be used here.)

Applications

19. ***Cost*** Your bicycle factory makes two models, five-speeds and ten-speeds. Each week your total cost (in dollars) to make x five-speeds and y ten-speeds is

$$C(x, y) = 10{,}000 + 50x + 70y - 0.5xy$$

You want to make between 100 and 150 five-speeds and between 80 and 120 ten-speeds. What combination will cost you the least? What combination will cost you the most?

20. ***Cost*** Your bicycle factory makes two models, five-speeds and ten-speeds. Each week your total cost (in dollars) to make x five-speeds and y ten-speeds is

$$C(x, y) = 10{,}000 + 50x + 70y - 0.46xy$$

You want to make between 100 and 150 five-speeds and between 80 and 120 ten-speeds. What combination will cost you the least? What combination will cost you the most?

21. ***Profit*** Your software company sells two programs, Walls and Doors. Your profit (in dollars) from selling x copies of Walls and y copies of Doors is given by

$$P(x, y) = 20x + 40y - 0.1(x^2 + y^2)$$

If you can sell a maximum of 200 copies of the two programs together, what combination will bring you the greatest profit?

22. ***Profit*** Your software company sells two programs, Walls and Doors. Your profit (in dollars) from selling x copies of Walls and y copies of Doors is given by

$$P(x, y) = 20x + 40y - 0.1(x^2 + y^2)$$

If you can sell a maximum of 400 copies of the two programs together, what combination will bring you the greatest profit?

23. ***Temperature*** The temperature at the point (x, y) on the square with vertices $(0, 0)$, $(0, 1)$, $(1, 0)$, and $(1, 1)$ is given by $T(x, y) = x^2 + 2y^2$. Find the hottest and coldest points on the square.

24. ***Temperature*** The temperature at the point (x, y) on the square with vertices $(0, 0)$, $(0, 1)$, $(1, 0)$, and $(1, 1)$ is given by $T(x, y) = x^2 + 2y^2 - x$. Find the hottest and coldest points on the square.

25. ***Temperature*** The temperature at the point (x, y) on the disc $\{(x, y) \mid x^2 + y^2 \le 1\}$ is given by $T(x, y) = x^2 + 2y^2 - x$. Find the hottest and coldest points on the disc.

26. ***Temperature*** The temperature at the point (x, y) on the disc $\{(x, y) \mid x^2 + y^2 \le 1\}$ is given by $T(x, y) = 2x^2 + y^2$. Find the hottest and coldest points on the disc.

The method of Lagrange multipliers can be used for Exercises 27–46.

27. **Geometry** What point on the surface $z = x^2 + y - 1$ is closest to the origin? [*Hint:* Minimize the square of the distance from (x, y, z) to the origin.]

28. **Geometry** What point on the surface $z = x + y^2 - 3$ is closest to the origin? [*Hint:* Minimize the square of the distance from (x, y, z) to the origin.]

29. **Geometry** Find the point on the plane $-2x + 2y + z - 5 = 0$ closest to $(-1, 1, 3)$. [*Hint:* Minimize the square of the distance from the given point to a general point on the plane.]

30. **Geometry** Find the point on the plane $2x - 2y - z + 1 = 0$ closest to $(1, 1, 0)$.

31. **Construction Cost** A closed rectangular box is made with two kinds of materials. The top and bottom are made with heavy-duty cardboard costing 20¢ per square foot, and the sides are made with lightweight cardboard costing 10¢ per square foot. Given that the box is to have a capacity of 2 cubic feet, what should its dimensions be if the cost is to be minimized?

32. ***Construction Cost*** Repeat Exercise 31 if the heavy-duty cardboard costs 30¢ per square foot, the lightweight cardboard costs 5¢ per square foot, and the box is to have a capacity of 6 cubic feet.

33. ***Construction Cost*** My company wishes to manufacture boxes similar to those described in Exercise 31 as cheaply as possible, but unfortunately the company that manufactures the cardboard is unable to give me price quotes for the heavy-duty and lightweight cardboard. Find formulas for the dimensions of the box in terms of the price per square foot of heavy-duty and lightweight cardboard.

34. ***Construction Cost*** Repeat Exercise 33, assuming that only the bottoms of the boxes are to made with heavy-duty cardboard.

35. ***Package Dimensions*** The U.S. Postal Service (USPS) will accept only packages with length plus girth no more than 108 inches.[2] (See the figure.)

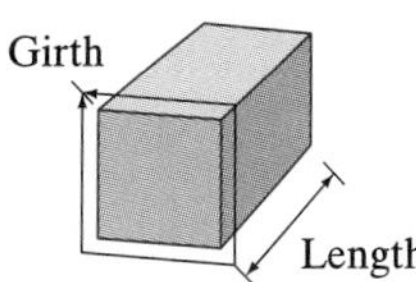

What are the dimensions of the largest volume package the USPS will accept? What is its volume? (This exercise is the same as Exercise 39 in the preceding section. This time, solve it using the methods of this section.)

36. ***Package Dimensions*** United Parcel Service (UPS) will accept only packages with length no more than 108 inches and length plus girth no more than 130 inches.[3] (See the figure for the preceding exercise.) What are the dimensions of the largest volume package UPS will accept? What is its volume? (This exercise is the same as Exercise 40 in the preceding section. This time, solve it using the methods of this section.)

37. ***Geometry*** Find the dimensions of the rectangular box with largest volume that can be inscribed above the xy-plane and under the paraboloid $z = 1 - (x^2 + y^2)$.

38. ***Geometry*** Find the dimensions of the rectangular box with largest volume that can be inscribed above the xy-plane and under the paraboloid $z = 2 - (2x^2 + y^2)$.

39. ***Resource Allocation*** You manage an ice cream factory that makes two flavors: Creamy Vanilla and Continental

[2] The requirement at the time this book was written.

[3] The requirement at the time this book was written.

Mocha. Into each quart of Creamy Vanilla go two eggs and 3 cups of cream. Into each quart of Continental Mocha go one egg and 3 cups of cream. You have in stock 500 eggs and 900 cups of cream. Your profit on x quarts of vanilla and y quarts of mocha is $P(x, y) = 3x + 2y - 0.01(x^2 + y^2)$. How many quarts of each flavor should you produce to make the greatest profit?

40. ***Resource Allocation*** Repeat Exercise 39 using the profit function $P(x, y) = 3x + 2y - 0.005(x^2 + y^2)$.

41. ***Resource Allocation*** Urban Institute of Technology's Math Department offers two courses: Finite Math and Calculus. Each section of Finite Math has 60 students, while each section of Calculus has 50. The department is allowed to offer a total of up to 110 sections. There are no more than 6000 students who would like to take a math course. The university's profit (in dollars) on x sections of Finite Math and y sections of Calculus is

$$P(x, y) = 5{,}000{,}000(1 - e^{-0.02x-0.01y})$$

How many sections of each course should the department offer to make the greatest profit?

42. ***Resource Allocation*** Repeat Exercise 41 using the profit function

$$P(x, y) = 5{,}000{,}000(1 - e^{-0.01x-0.02y})$$

43. ***Nutrition*** Gerber Mixed Cereal for Baby costs 10¢ per serving. Gerber Mango Tropical Fruit Dessert costs 53¢ per serving. For reasons too complicated to explain, you want the product of the number of servings of each to be at least ten per day. How can you do so at the least cost? (Fractions of servings are allowed.)

44. ***Nutrition*** Repeat Exercise 43 if instead (and for even more complicated reasons) you want the product of the number of servings of cereal and the square of the number of servings of dessert to be at least ten.

45. ***Purchasing*** The ESU Business School is buying computers. It has two models to choose from, the Pomegranate and the iZac. Each Pomegranate comes with 400 MB of memory and 80 GB of disk space, while each iZac has 300 MB of memory and 100 GB of disk space. For reasons related to its accreditation, the school would like to be able to say that it has a total of at least 48,000 MB of memory and at least 12,800 GB of disk space. Because of complicated volume pricing, the cost (in dollars) to the school of x Pomegranates and y iZacs is

$$C(x, y) = x^2 + y^2 - 200x - 200y + 320{,}000$$

How many of each kind of computer should the school buy to minimize the average cost per computer?

46. ***Purchasing*** Repeat Exercise 45 assuming that the cost is

$$C(x, y) = x^2 + y^2 - 100x - 100y + 180{,}000$$

Communication and Reasoning Exercises

47. If the partial derivatives of a function of several variables are never zero, is it possible for the function to have relative extrema on some domain? Explain your answer.

48. Suppose we know that $f(x, y)$ has an absolute maximum somewhere in the domain D and that (a, b) is the only point in D such that $f_x(a, b) = f_y(a, b) = 0$. Must it be the case that f has an absolute maximum at (a, b)? Explain.

49. Under what circumstances is it necessary to use the method of Lagrange multipliers?

50. Under what circumstances does the method of Lagrange multipliers not apply?

51. A **linear programming problem in two variables** is a problem of the form: *Maximize (or minimize)* $f(x, y)$ *subject to constraints of the form* $C(x, y) \geq 0$ *or* $C(x, y) \leq 0$. Here, the objective function f and the constraints C are linear functions. There may be several linear constraints in one problem. Explain why the solution cannot occur in the interior of the domain of f.

52. Refer back to Exercise 51. Explain why the solution will actually be at a corner of the domain of f (where two or more of the line segments that make up the boundary meet). This result—or rather a slight generalization of it—is known as the Fundamental Theorem of Linear Programming.

8.6 Double Integrals

When discussing functions of one variable, we computed the area under a graph by integration. The analog for the graph of a function of two variables is the *volume V* under the graph, as in Figure 29. Think of the region R in the xy-plane as the "shadow" under the portion of the surface $z = f(x, y)$ shown.

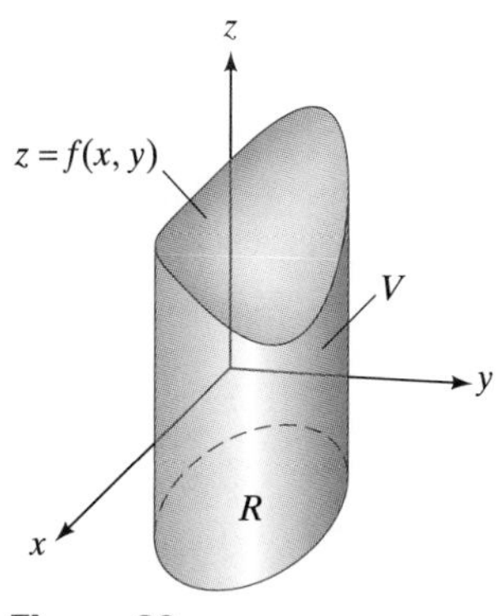

Figure 29

By analogy with the definite integral of a function of one variable, we make the following definition.

Geometric Definition of the Double Integral

The **double integral of $f(x, y)$ over the region R in the xy-plane** is defined as

(Volume *above* the region R and under the graph of f) −
(Volume *below* the region R and above the graph of f)

We denote the double integral of $f(x, y)$ over the region R by $\iint_R f(x, y)\, dx\, dy$.

Quick Example

Take $f(x, y) = 2$ and take R to be the rectangle $0 \le x \le 1, 0 \le y \le 1$. Then the graph of f is a flat horizontal surface, and

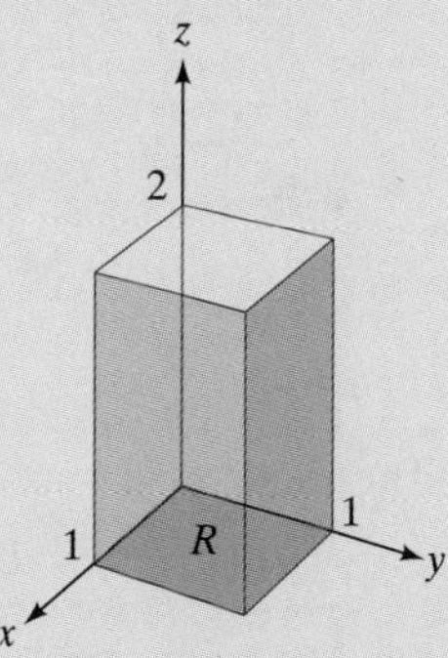

$$\iint_R f(x, y)\, dx\, dy = \text{Volume of box} = \text{Width} \times \text{Length} \times \text{Height} = 1 \times 1 \times 2 = 2$$

As we saw in the case of the definite integral of a function of one variable, we also desire *numerical* and *algebraic* definitions for two reasons: (1) to make the mathematical definition more precise, so as not to rely on the notion of "volume," and (2) for direct computation of the integral using technology or analytical tools.

We start with the simplest case, when the region R is a rectangle $a \le x \le b$ and $c \le y \le d$ (see Figure 30). To compute the volume over R, we mimic what we did to find the area under the graph of a function of one variable. We break up the interval $[a, b]$ into m intervals all of width $\Delta x = (b - a)/m$, and we break up $[c, d]$ into n intervals all of width $\Delta y = (d - c)/n$. Figure 31 shows an example with $m = 4$ and $n = 5$.

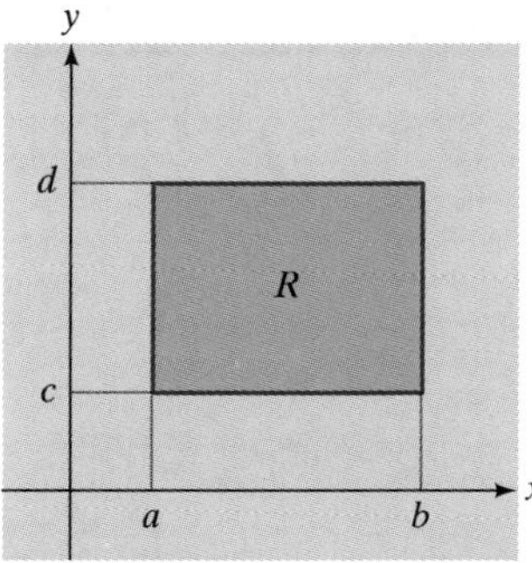

Figure 30

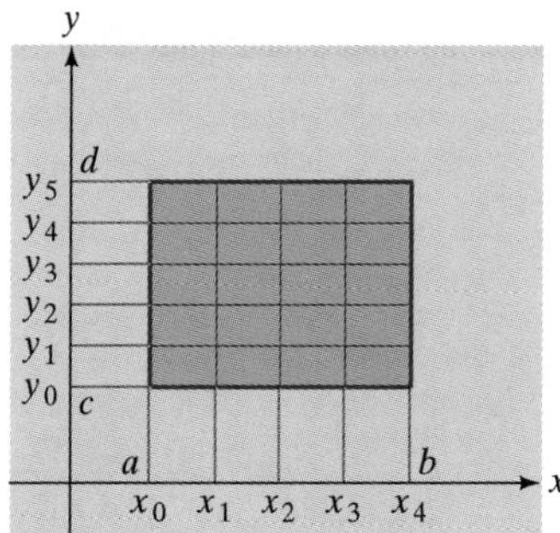

Figure 31

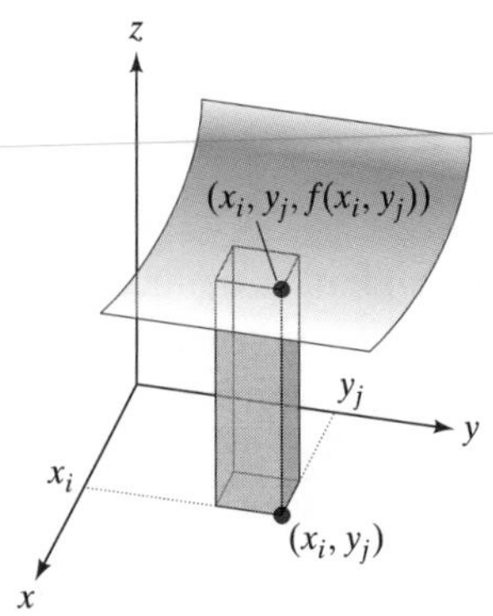

Figure 32

This gives us mn rectangles defined by $x_{i-1} \le x \le x_i$ and $y_{j-1} \le y \le y_j$. Over one of these rectangles, f is approximately equal to its value at one corner—say, $f(x_i, y_j)$. The volume under f over this small rectangle is then approximately the volume of the rectangular brick (size exaggerated) shown in Figure 32. This brick has height $f(x_i, y_j)$, and its base is Δx by Δy. Its volume is therefore $f(x_i, y_j)\,\Delta x\,\Delta y$. Adding together the volumes of all the bricks over the small rectangles in R, we get

$$\iint_R f(x, y)\,dx\,dy \approx \sum_{j=1}^{n} \sum_{i=1}^{m} f(x_i, y_j)\,\Delta x\,\Delta y$$

This double sum is called a **double Riemann sum.** We define the double integral to be the limit of the Riemann sums as m and n go to infinity.

Algebraic Definition of the Double Integral

$$\iint_R f(x, y)\,dx\,dy = \lim_{n\to\infty} \lim_{m\to\infty} \sum_{j=1}^{n} \sum_{i=1}^{m} f(x_i, y_j)\,\Delta x\,\Delta y$$

This definition is adequate (the limit exists) when f is continuous. More elaborate definitions are needed for badly behaved functions.

This definition also gives us a clue about how to compute a double integral. The innermost sum is $\sum_{i=1}^{m} f(x_i, y_j)\,\Delta x$, which is a Riemann sum for $\int_a^b f(x, y_j)\,dx$. The innermost limit is therefore

$$\lim_{m\to\infty} \sum_{i=1}^{m} f(x_i, y_j)\,\Delta x = \int_a^b f(x, y_j)\,dx$$

The outermost limit is then also a Riemann sum, and we get the following way of calculating double integrals.

Computing the Double Integral over a Rectangle

If R is the rectangle $a \le x \le b$ and $c \le y \le d$, then

$$\iint_R f(x, y)\,dx\,dy = \int_c^d \left(\int_a^b f(x, y)\,dx\right) dy = \int_a^b \left(\int_c^d f(x, y)\,dy\right) dx$$

The second formula comes from switching the order of summation in the double sum.

Quick Example

If R is the rectangle $1 \le x \le 2$ and $1 \le y \le 3$, then

$$\iint_R 1\,dx\,dy = \int_1^3 \left(\int_1^2 1\,dx\right) dy$$

$$= \int_1^3 [x]_{x=1}^{2}\,dy \qquad \text{Evaluate the inner integral.}$$

$$= \int_1^3 1\,dy \qquad [x]_{x=1}^{2} = 2 - 1 = 1$$

$$= [y]_{y=1}^{3} = 2$$

The Quick Example used a constant function for the integrand. Here is an example in which the integrand is not constant.

example 1

Double Integral over a Rectangle
Let R be the rectangle $0 \le x \le 1$ and $0 \le y \le 2$. Compute $\iint_R xy\, dx\, dy$.

Solution

$$\iint_R xy\, dx\, dy = \int_0^2 \int_0^1 xy\, dx\, dy$$

(We usually drop the parentheses like this.) As in the Quick Example, we compute this **iterated integral** from the inside out. First we compute

$$\int_0^1 xy\, dx$$

To do this computation, we proceed as when finding partial derivatives: We treat y as a constant. This gives

$$\int_0^1 xy\, dx = \left[\frac{x^2}{2} \cdot y\right]_{x=0}^1 = \frac{1}{2}y - 0 = \frac{y}{2}$$

We can now calculate the outer integral:

$$\int_0^2 \int_0^1 xy\, dx\, dy = \int_0^2 \frac{y}{2}\, dy = \left[\frac{y^2}{4}\right]_0^2 = 1$$

Before we go on . . . We could also reverse the order of integration:

$$\int_0^1 \int_0^2 xy\, dy\, dx = \int_0^1 \left(\left[x \cdot \frac{y^2}{2}\right]_{y=0}^2\right) dx = \int_0^1 2x\, dx = \left[x^2\right]_0^1 = 1$$

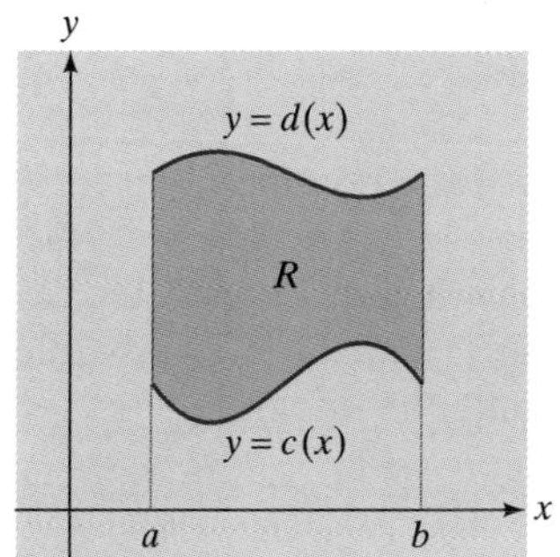

Figure 33

Often we need to integrate over regions R that are not rectangular. There are two cases that come up. The first is a region like the one shown in Figure 33. In this region, the bottom and top sides are defined by the functions $y = c(x)$ and $y = d(x)$, respectively, so that the whole region can be described by the inequalities $a \le x \le b$ and $c(x) \le y \le d(x)$. To evaluate a double integral over such a region, we have the following computation.

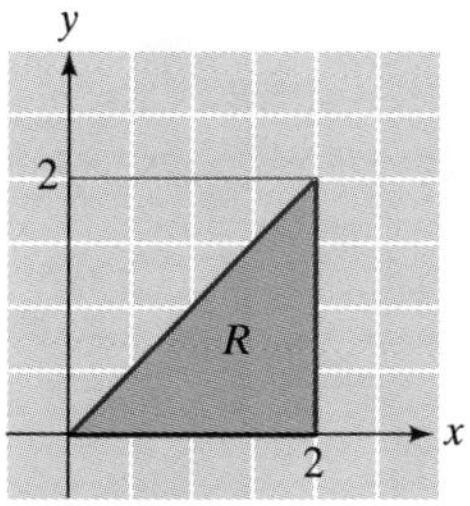

Figure 34

Computing the Double Integral over a Nonrectangular Region

If R is the region $a \le x \le b$ and $c(x) \le y \le d(x)$ (Figure 33), then we integrate over R according to the following equation:

$$\iint_R f(x, y)\, dx\, dy = \int_a^b \int_{c(x)}^{d(x)} f(x, y)\, dy\, dx$$

example 2

Double Integral over a Nonrectangular Region
R is the triangle shown in Figure 34. Compute $\iint_R x\, dx\, dy$.

Solution

R is the region described by $0 \le x \le 2, 0 \le y \le x$. We have

$$\iint_R x\,dx\,dy = \int_0^2 \int_0^x x\,dy\,dx$$
$$= \int_0^2 [xy]_{y=0}^{x}\,dx = \int_0^2 x^2\,dx = \left[\frac{x^3}{3}\right]_0^2 = \frac{8}{3}$$

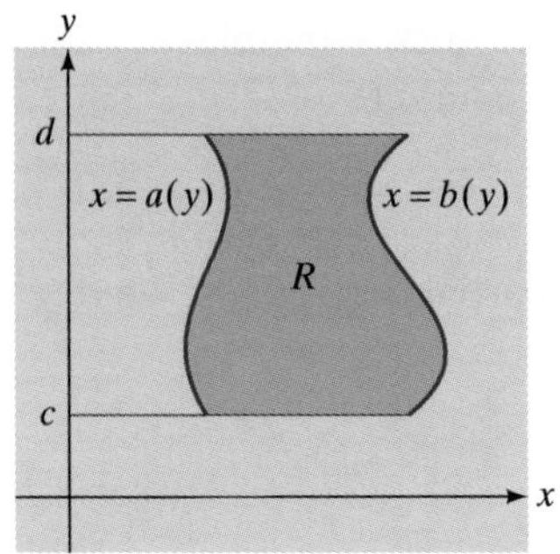

Figure 35

The second type of region is shown in Figure 35. This is the region described by $c \le y \le d$ and $a(y) \le x \le b(y)$. To evaluate a double integral over such a region, we have the following formula.

Computing the Double Integral over a Nonrectangular Region (continued)

If R is the region $c \le y \le d$ and $a(y) \le x \le b(y)$ (Figure 35), then we integrate over R according to the following equation:

$$\iint_R (x, y)\,dx\,dy = \int_c^d \int_{a(y)}^{b(y)} f(x, y)\,dx\,dy$$

example 3

Double Integral over a Nonrectangular Region

Redo Example 2, integrating in the opposite order.

Solution

We can integrate in the opposite order if we can describe the region in Figure 34 in the way shown in Figure 35. In fact, it is the region $0 \le y \le 2$ and $y \le x \le 2$. To see this, we draw a horizontal line through the region, as in Figure 36. The line extends from $x = y$ on the left to $x = 2$ on the right, so $y \le x \le 2$. The possible heights for such a line are $0 \le y \le 2$. We can now compute the integral:

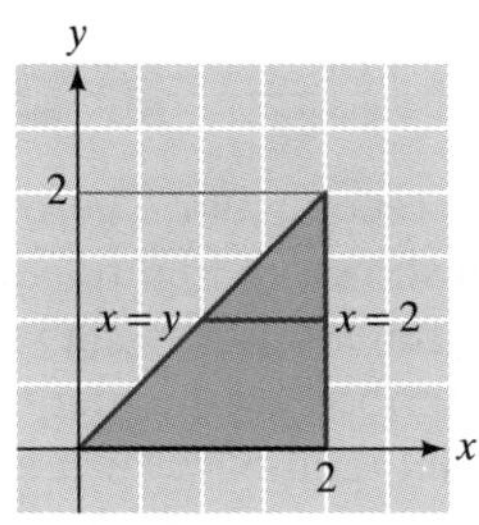

Figure 36

$$\iint_R x\,dx\,dy = \int_0^2 \int_y^2 x\,dx\,dy$$
$$= \int_0^2 \left[\frac{x^2}{2}\right]_{x=y}^{2} dy = \int_0^2 \left(2 - \frac{y^2}{2}\right) dy = \left[2y - \frac{y^3}{6}\right]_0^2 = \frac{8}{3}$$

Before we go on . . . Many regions can be described in two different ways. Sometimes one description is much easier to work with than the other, so it pays to try both.

Applications

There are many applications of double integrals besides finding volumes. We can also use them to find *averages*. Remember that the average of $f(x)$ on $[a, b]$ is given by $\int_a^b f(x)\,dx$ divided by $(b - a)$, the length of the interval.

Average of a Function of Two Variables

The average of $f(x, y)$ on the region R is

$$\bar{f} = \frac{1}{A}\iint_R f(x, y)\, dx\, dy$$

where A is the area of R. We can compute the area A geometrically, or by using the techniques from Chapter 7 on applications of the integral, or by computing

$$A = \iint_R 1\, dx\, dy$$

Quick Example

The average value of $f(x, y) = xy$ on the rectangle given by $0 \le x \le 1$ and $0 \le y \le 2$ is

$$\bar{f} = \frac{1}{2}\iint_R xy\, dx\, dy \qquad \text{The area of the rectangle is 2.}$$

$$= \frac{1}{2}\int_0^2\int_0^1 xy\, dx\, dy$$

$$= \frac{1}{2}\cdot 1 = \frac{1}{2} \qquad \text{We calculated the integral in Example 1.}$$

example 4

Average Revenue

Your company is planning to price its new line of subcompact cars at between \$10,000 and \$15,000. The marketing department reports that if the company prices the cars at p dollars per car, the demand will be between $q = 20{,}000 - p$ and $q = 25{,}000 - p$ cars sold in the first year. What is the average of all the possible revenues your company could expect in the first year?

Solution

Revenue is given by $R = pq$ as usual, and we are told that

$$10{,}000 \le p \le 15{,}000$$

$$20{,}000 - p \le q \le 25{,}000 - p$$

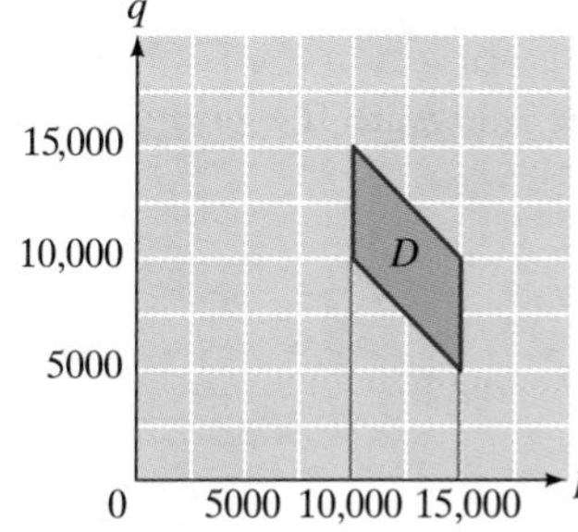

Figure 37

This domain D of prices and demands is shown in Figure 37. To average the revenue R over the domain D we need to compute the area A of D. Using either calculus or geometry, we get $A = 25{,}000{,}000$. We then need to integrate R over D:

$$\iint_D pq\, dp\, dq = \int_{10{,}000}^{15{,}000}\int_{20{,}000-p}^{25{,}000-p} pq\, dq\, dp$$

$$= \int_{10{,}000}^{15{,}000}\left[\frac{pq^2}{2}\right]_{q=20{,}000-p}^{25{,}000-p} dp$$

$$= \frac{1}{2}\int_{10{,}000}^{15{,}000}\left[p(25{,}000 - p)^2 - p(20{,}000 - p)^2\right] dp$$

$$= \frac{1}{2}\int_{10{,}000}^{15{,}000}\left[225{,}000{,}000p - 10{,}000p^2\right] dp \approx 3{,}072{,}916{,}666{,}666{,}667$$

We get the following average.

$$\bar{R} = \frac{3{,}072{,}916{,}666{,}666{,}667}{25{,}000{,}000} \approx \$122{,}900{,}000 \text{ per year}$$

Before we go on . . . To check that this is a reasonable answer, notice that the revenues at the corners of the domain are \$100,000,000 per year, \$150,000,000 per year (at two corners), and \$75,000,000 per year. Some of these are smaller than the average and some larger, as we would expect. You should also check that the maximum possible revenue is \$156,250,000 per year. (What is the minimum possible revenue?)

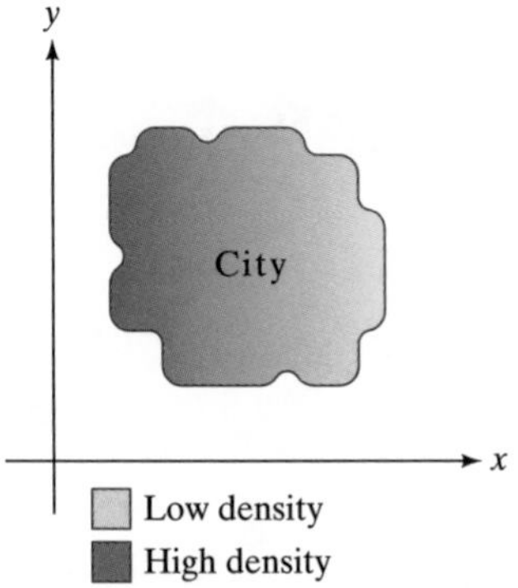

Figure 38

Another useful application involves density. For example, suppose that $P(x, y)$ represents the population density (in people per square mile, say) in the city shown in Figure 38. If we break the city up into small rectangles (for example, city blocks), then the population in the small rectangle $x_{i-1} \leq x \leq x_i$ and $y_{j-1} \leq y \leq y_j$ is approximately $P(x_i, y_j)\,\Delta x\,\Delta y$. Adding up all of these population estimates, we get

$$\text{Total population} \approx \sum_{j=1}^{n} \sum_{i=1}^{m} P(x_i, y_j)\,\Delta x\,\Delta y$$

Since this is a double Riemann sum, when we take the limit as m and n go to infinity, we get the following calculation of the population of the city:

$$\text{Total population} = \iint_{\text{city}} P(x, y)\,dx\,dy$$

example 5

Population

Squaresville is a city in the shape of a square 5 miles on a side. The population density at a distance of x miles east and y miles north of the southwest corner is $P(x, y) = x^2 + y^2$ thousand people per square mile. Find the total population of Squaresville.

Solution

Squaresville is pictured in Figure 39, in which we put the origin in the southwest corner of the city. To compute the total population, we integrate the population density over the city:

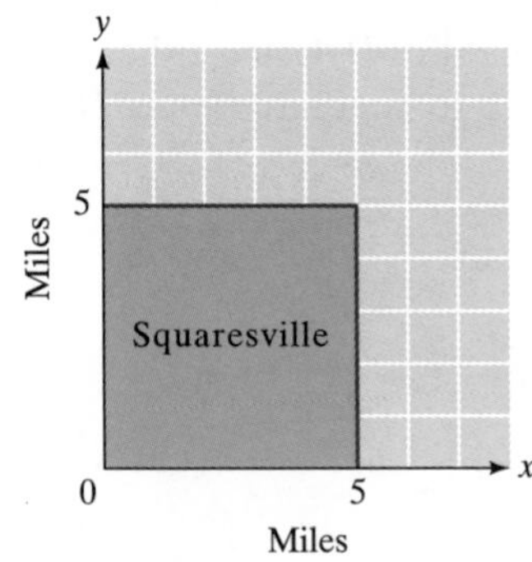

Figure 39

$$\begin{aligned}\text{Population} &= \iint_{\text{Squaresville}} P(x, y)\,dx\,dy \\ &= \int_0^5 \int_0^5 (x^2 + y^2)\,dx\,dy \\ &= \int_0^5 \left[\frac{x^3}{3} + xy^2\right]_{x=0}^{5} dy = \int_0^5 \left[\frac{125}{3} + 5y^2\right] dy = \frac{1250}{3} \approx 417 \text{ thousand people}\end{aligned}$$

Before we go on . . . Note that the average population density is the total population divided by the area of the city, which is about 17,000 people per square mile. Compare this calculation with the calculations of averages in the previous two examples.

8.6 exercises

Compute the integrals in Exercises 1–16.

1. $\int_0^1 \int_0^1 (x - 2y)\, dx\, dy$

2. $\int_{-1}^1 \int_0^2 (2x + 3y)\, dx\, dy$

3. $\int_0^1 \int_0^2 (ye^x - x - y)\, dx\, dy$

4. $\int_1^2 \int_2^3 \left(\frac{1}{x} + \frac{1}{y}\right) dx\, dy$

5. $\int_0^2 \int_0^3 e^{x+y}\, dx\, dy$

6. $\int_0^1 \int_0^1 e^{x-y}\, dx\, dy$

7. $\int_0^1 \int_0^{2-y} x\, dx\, dy$

8. $\int_0^1 \int_0^{2-y} y\, dx\, dy$

9. $\int_{-1}^1 \int_{y-1}^{y+1} e^{x+y}\, dx\, dy$

10. $\int_0^1 \int_y^{y+2} \frac{1}{\sqrt{x+y}}\, dx\, dy$

11. $\int_0^1 \int_{-x^2}^{x^2} x\, dy\, dx$

12. $\int_1^4 \int_{-\sqrt{x}}^{\sqrt{x}} \frac{1}{x}\, dy\, dx$

13. $\int_0^1 \int_0^x e^{x^2}\, dy\, dx$

14. $\int_0^1 \int_0^{x^2} e^{x^3+1}\, dy\, dx$

15. $\int_0^2 \int_{1-x}^{8-x} (x + y)^{1/3}\, dy\, dx$

16. $\int_1^2 \int_{1-2x}^{x^2} \frac{x+1}{(2x+y)^3}\, dy\, dx$

In Exercises 17–24, find $\iint_R f(x, y)\, dx\, dy$, where R is the indicated domain. (Remember that you often have a choice as to the order of integration.)

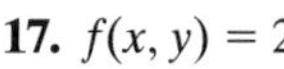

17. $f(x, y) = 2$

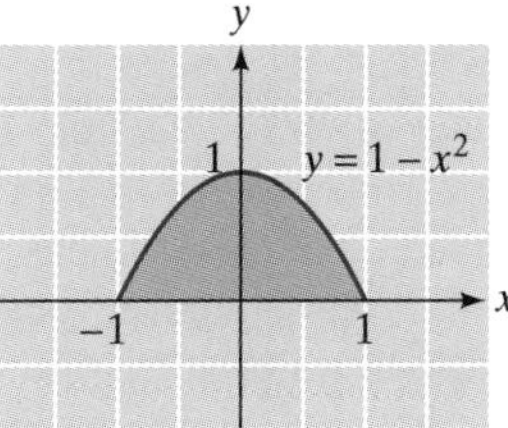

18. $f(x, y) = x$

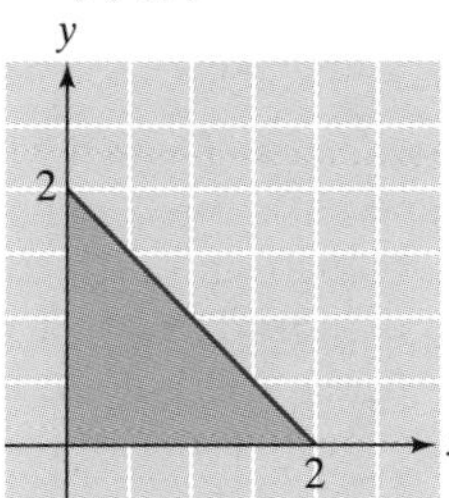

19. $f(x, y) = 1 + y$

20. $f(x, y) = e^{x+y}$

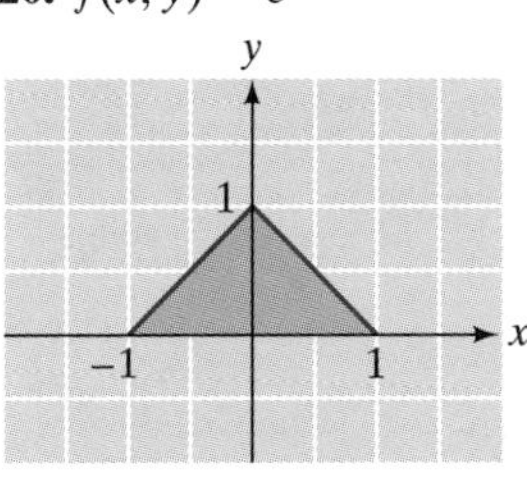

21. $f(x, y) = xy^2$

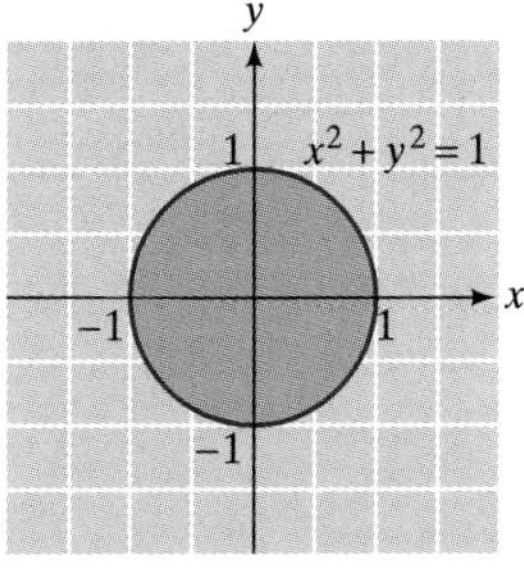

22. $f(x, y) = xy^2$

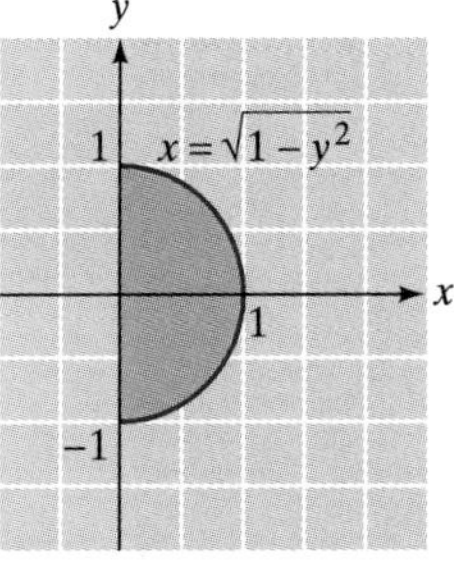

23. $f(x, y) = x^2 + y^2$

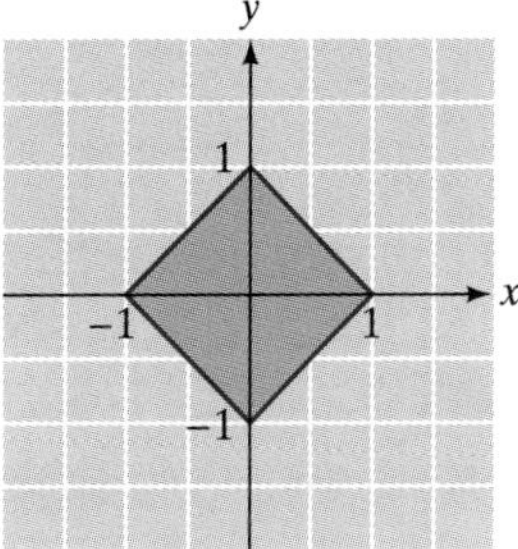

24. $f(x, y) = x^2$

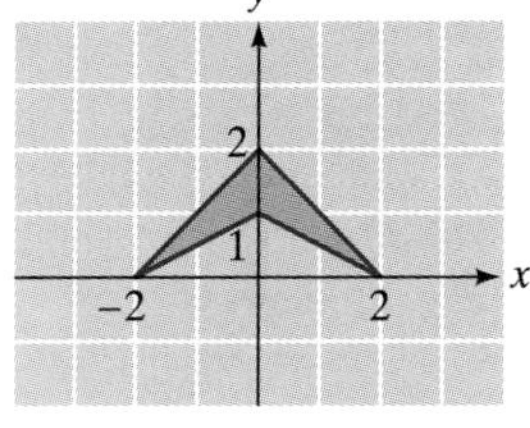

In Exercises 25–30, find the average value of the given function over the indicated domain.

25. $f(x, y) = y$

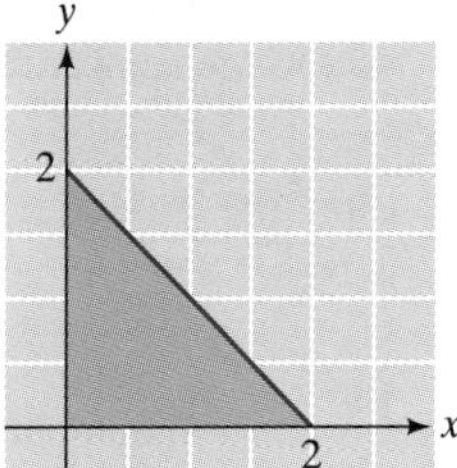

26. $f(x, y) = 2 + x$

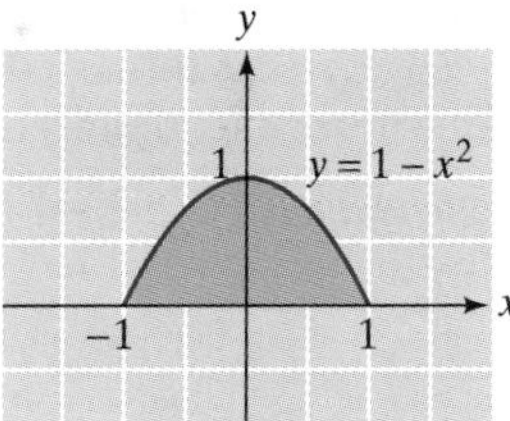

27. $f(x, y) = e^y$

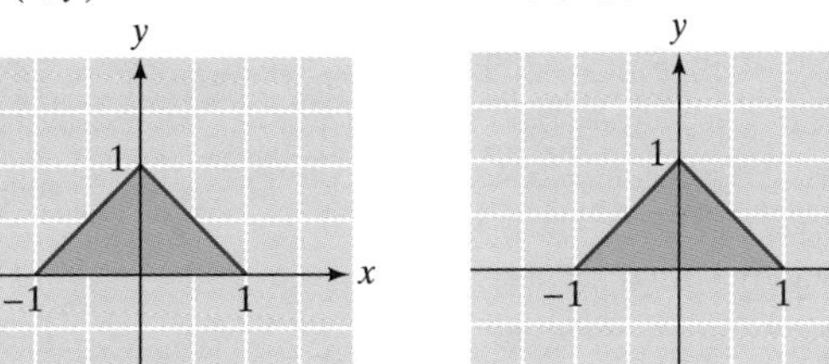

28. $f(x, y) = y$

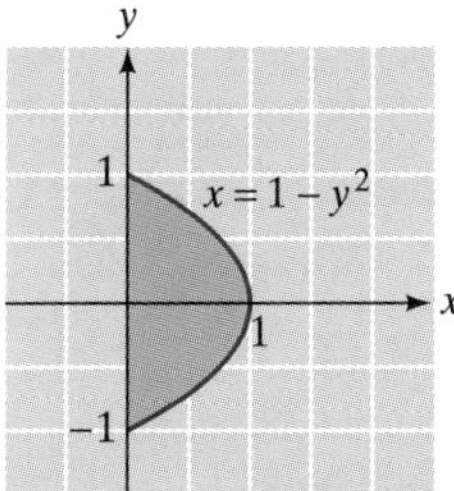

29. $f(x, y) = x^2$

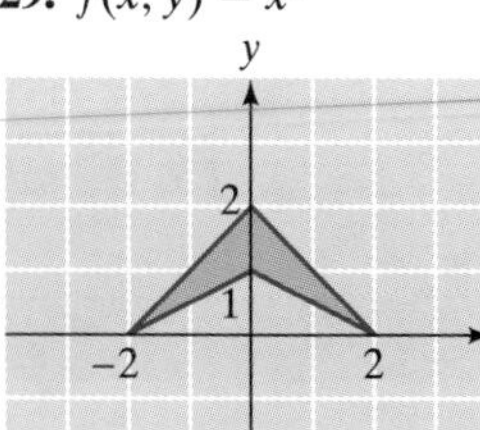

30. $f(x, y) = x^2 + y^2$

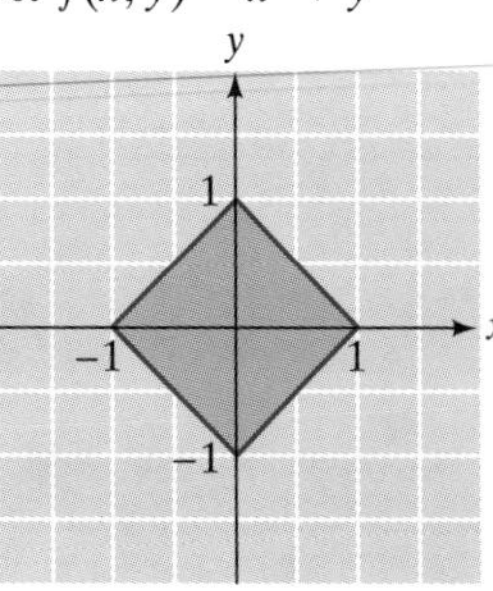

In Exercises 31–36, sketch the region over which you are integrating; then write down the integral with the order of integration reversed (changing the limits of integration as necessary).

31. $\int_0^1 \int_0^{1-y} f(x, y)\, dx\, dy$

32. $\int_{-1}^1 \int_0^{1+y} f(x, y)\, dx\, dy$

33. $\int_{-1}^1 \int_0^{\sqrt{1+y}} f(x, y)\, dx\, dy$

34. $\int_{-1}^1 \int_0^{\sqrt{1-y}} f(x, y)\, dx\, dy$

35. $\int_1^2 \int_1^{4/x^2} f(x, y)\, dy\, dx$

36. $\int_1^{e^2} \int_0^{\ln x} f(x, y)\, dy\, dx$

37. Find the volume under the graph of $z = 1 - x^2$ over the region $0 \le x \le 1$ and $0 \le y \le 2$.

38. Find the volume under the graph of $z = 1 - x^2$ over the triangle $0 \le x \le 1$ and $0 \le y \le 1 - x$.

39. Find the volume of the tetrahedron shown in the figure. Its corners are (0, 0, 0), (1, 0, 0), (0, 1, 0), and (0, 0, 1).

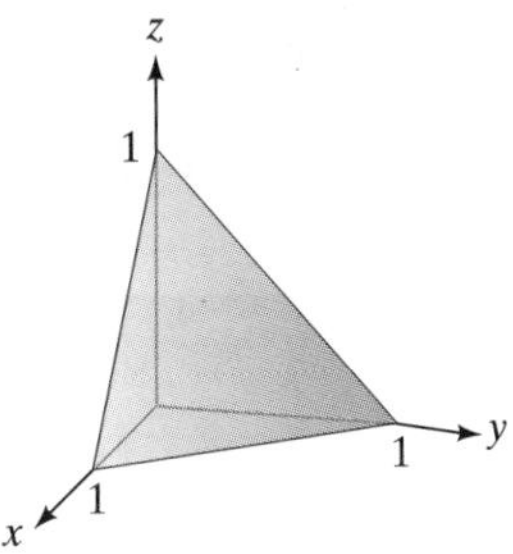

40. Find the volume of the tetrahedron with corners at (0, 0, 0), $(a, 0, 0)$, $(0, b, 0)$, and $(0, 0, c)$.

Applications

41. ***Productivity*** A productivity model at the Handy Gadget Company is

$$P = 10{,}000x^{0.3}y^{0.7}$$

where P is the number of gadgets the company turns out per month, x is the number of employees at the company, and y is the monthly operating budget in thousands of dollars. Because the company hires part-time workers, it uses anywhere between 45 and 55 workers each month, and its operating budget varies from \$8000 to \$12,000 per month. What is the average of the possible numbers of gadgets it can turn out per month? (Round the answer to the nearest 1000 gadgets.)

42. ***Productivity*** Repeat Exercise 41 using the productivity model

$$P = 10{,}000x^{0.7}y^{0.3}$$

43. ***Revenue*** Your latest CD-ROM of clip art is expected to sell between $q = 8000 - p^2$ and $q = 10{,}000 - p^2$ copies if priced at p dollars. You plan to set the price between \$40 and \$50. What are the maximum and minimum possible revenues you can make? What is the average of all the possible revenues you can make?

44. ***Revenue*** Your latest CD-ROM drive is expected to sell between $q = 180{,}000 - p^2$ and $q = 200{,}000 - p^2$ units if priced at p dollars. You plan to set the price between \$300 and \$400. What are the maximum and minimum possible revenues you can make? What is the average of all the possible revenues you can make?

45. ***Revenue*** Your self-published novel has demand curves between $p = 15{,}000/q$ and $p = 20{,}000/q$. You expect to sell between 500 and 1000 copies. What are the maximum and minimum possible revenues you can make? What is the average of all the possible revenues you can make?

46. ***Revenue*** Your self-published book of poetry has demand curves between $p = 80{,}000/q^2$ and $p = 100{,}000/q^2$. You expect to sell between 50 and 100 copies. What are the maximum and minimum possible revenues you can make? What is the average of all the possible revenues you can make?

47. ***Population Density*** The town of West Podunk is shaped like a rectangle 20 miles from west to east and 30 miles from north to south (see the figure). It has a population density of $P(x, y) = e^{-0.1(x+y)}$ hundred people per square mile x miles east and y miles north of the southwest corner of town. What is the total population of the town?

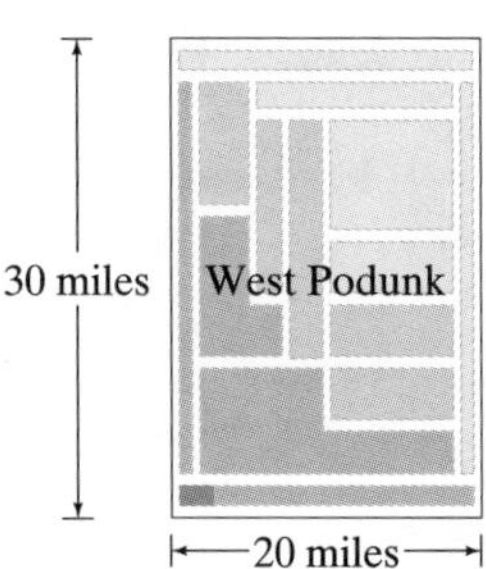

48. ***Population Density*** The town of East Podunk is shaped like a triangle with an east–west base of 20 miles and a north–south height of 30 miles (see the figure). It has a population density of $P(x, y) = e^{-0.1(x+y)}$ hundred people per square mile x miles east and y miles north of the southwest corner of town. What is the total population of the town?

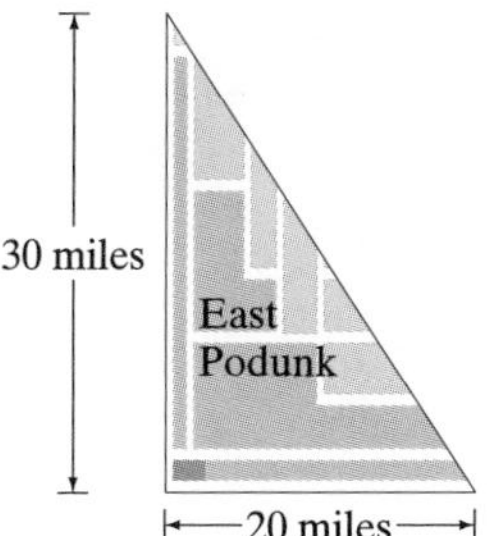

49. ***Temperature*** The temperature at the point (x, y) on the square with vertices $(0, 0)$, $(0, 1)$, $(1, 0)$, and $(1, 1)$ is given by $T(x, y) = x^2 + 2y^2$. Find the average temperature on the square.

50. ***Temperature*** The temperature at the point (x, y) on the square with vertices $(0, 0)$, $(0, 1)$, $(1, 0)$, and $(1, 1)$ is given by $T(x, y) = x^2 + 2y^2 - x$. Find the average temperature on the square.

Communication and Reasoning Exercises

51. Explain how double integrals can be used to compute **a.** the area between two curves in the xy-plane; **b.** the volume of solids in 3-space.

52. Complete the following: The first step in calculating an integral of the form $\int_a^b \int_{r(x)}^{s(x)} f(x, y)\, dy\, dx$ is to evaluate the integral _____, obtained by holding _____ constant and integrating with respect to _____.

53. Show that if a, b, c, and d are constant, then $\int_a^b \int_c^d f(x)g(y)\, dx\, dy = \int_c^d f(x)\, dx \int_a^b g(y)\, dy$. Test this result on the integral $\int_0^1 \int_1^2 ye^x\, dx\, dy$.

54. If the units of $f(x, y)$ are bootlags per square meter and x and y are given in meters, what are the units of $\int_a^b \int_{r(x)}^{s(x)} f(x, y)\, dx\, dy$?

You're the Expert

Modeling Household Income

The Millennium Real Estate Development Corporation is interested in developing housing projects for medium-sized families that have high household incomes. To decide which income bracket to target, the company has asked you, a paid consultant, for an analysis of household income and household size in the United States. In particular, Millennium is interested in these issues:

1. The relationship between household size and household income and the effect of increasing household size on household income
2. The household size that corresponds to the highest household income
3. The change in the relationship between household size and income over time
4. Some near-term projections of household income vs. household size (to, say, 2002)

You decide that a good place to start is with a visit to the Census Bureau's web site at http://www.census.gov. After some time battling with search engines, you discover detailed information on household size vs. household income[1] for the period 1967–1997. The following table summarizes the information on mean household incomes:

[1] Household income is adjusted for inflation and given in 1977 dollars.

Year		Household size →						
		1	2	3	4	5	6	7
↓	1967	16,435	32,622	41,266	45,081	45,957	45,360	42,884
	1968	17,297	34,122	42,641	46,341	47,443	45,936	44,514
	1969	17,655	36,076	43,241	48,248	49,946	49,910	47,132
	1970	18,009	36,192	43,448	48,473	50,808	49,712	46,720
	1971	18,117	36,468	43,320	48,407	50,250	49,770	46,564
	1972	19,155	38,354	46,274	51,993	52,864	51,779	50,088
	1973	19,668	39,482	46,654	52,026	53,771	54,247	52,605
	1974	19,430	38,770	45,487	51,456	53,256	53,215	51,910
	1975	19,100	37,935	44,843	49,961	53,119	51,528	48,761
	1976	19,995	38,785	45,786	51,271	53,797	54,948	52,138
	1977	20,840	39,714	46,743	52,576	54,301	55,362	52,259
	1978	21,766	41,207	48,385	53,647	56,320	56,962	55,336
	1979	21,546	41,693	49,048	54,329	57,739	56,880	56,986
	1980	21,415	40,843	47,559	52,501	54,851	54,371	53,201
	1981	21,925	40,748	47,275	51,714	51,759	53,337	50,790
	1982	22,401	41,216	46,503	51,932	52,008	52,733	49,634
	1983	22,821	41,131	47,277	52,698	52,182	51,038	46,597
	1984	23,676	43,057	49,406	53,898	54,837	51,732	48,091
	1985	23,862	44,041	51,163	55,431	54,437	54,082	50,795
	1986	24,078	45,839	52,671	58,131	57,925	57,209	51,487
	1987	24,814	46,152	53,784	60,470	59,009	55,231	54,587
	1988	25,787	47,267	53,417	60,251	58,154	58,092	52,516
	1989	26,283	48,643	55,299	61,399	59,587	57,785	53,876
	1990	25,351	48,178	53,340	59,218	57,512	55,568	53,927
	1991	24,728	46,811	52,131	57,905	55,453	53,826	51,074
	1992	24,422	45,706	52,263	57,747	55,720	53,862	49,342
	1993	24,677	47,066	53,348	61,663	58,283	56,935	49,317
	1994	25,375	48,252	53,869	61,498	60,010	62,084	53,062
	1995	25,899	48,616	54,854	62,730	62,086	59,224	52,351
	1996	26,940	49,677	56,039	63,388	63,014	57,786	49,689
	1997	27,115	52,221	57,466	66,457	63,043	57,483	55,455

Source: U.S. Bureau of the Census.

You notice that the table is actually a numerical representation of the mean household income I as a function of two variables, the household size n and the year t.

The numbers are a bit overwhelming, so you decide to use Excel to graph the data as a surface (Figure 40). Now you definitely see two trends. First, the household income peaks at five or six people per household, and then drops off at both ends. In fact, the slices through $t = constant$ look parabolic. Second, the household income for all household sizes seems to increase more or less linearly with time (the slices through $n = constant$ are approximately linear).

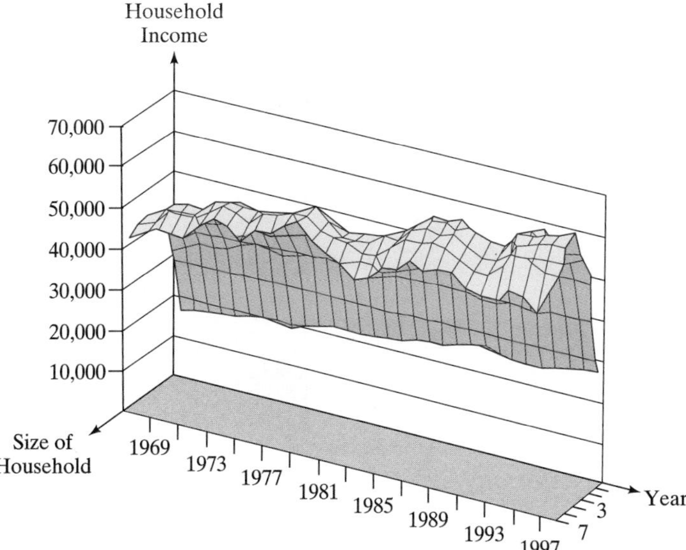

Figure 40

At this point you realize that a mathematical model of these data would be useful; not only would it "smooth out the bumps" but it would also give you a way to complete the project for Millennium. Although many statistics applications will give you a regression model for this type of data, it is up to you to decide on the kind of model. It is in choosing an appropriate model that your analysis of the graph comes in handy. Since I appears to vary quadratically with the household size, you would like a general quadratic of the form

$$I = a + bn + cn^2$$

for each value of time t. Also, since I should vary linearly with time t, you would like

$$I = mt + k$$

for each value of n. Putting these together, you get the following candidate model:

$$I(n, t) = a_1 + a_2n + a_3n^2 + a_4t$$

where a_1, a_2, a_3, and a_4 are constants you need to determine.

You decide to use Excel to generate your model. The specific software tool you need is called the "Analysis Toolpack," which comes with Excel. (It is found in the Tools menu as "Data Analysis." If it is missing, select Add-Ins from the Tools menu and check "Analysis Toolpack.")

Now you are set to do regression analysis. However, the data as shown in the table are not in a form Excel can use for regression; the data need to be organized into columns, as shown on the next page.

	A	B	C	D
1	N	N^2	T	I
2	1	1	0	16,435
3	1	1	1	17,297
4	1	1	2	17,655
32	1	1	30	27,115
33	2	4	0	32,622
34	2	4	1	34,122
216	7	49	28	52,351
217	7	49	29	49,689
218	7	49	30	55,455

The headings of each column show the variables n and t, with the income I in column D. (Instead of using the calendar year for T, we have represented 1967 by $T = 0$.) Notice that the columns of the original table are in column D, one beneath the other. Thus, columns A–C show the independent variables, and column D contains the dependent variable. You now select Data Analysis from the Tools menu. Under "Type of Analysis" you select "Regression," tell it where the dependent and independent variables are (D1-D218 for the Y range and A1-C218 for the X range), check "Labels," and hit "OK."

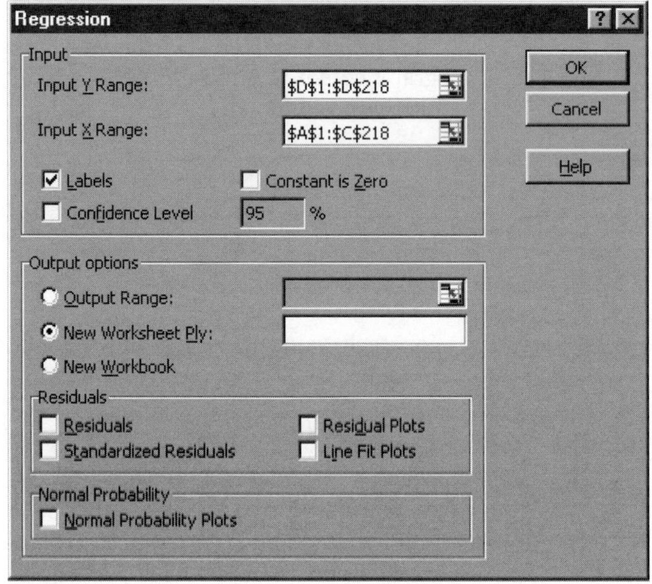

A portion of the output is shown below, with some of the important statistics highlighted.

	A	B	C	D	E	F
220	SUMMARY OUTPUT					
221						
222	*Regression Statistics*					
223	Multiple R	0.9679394				
224	R Square	0.93690669				
225	Adjusted R Sq	0.93601805				
226	Standard Erro	2960.82128				
227	Observations	217				
228						
229	ANOVA					
230		*df*	*SS*	*MS*	*F*	*Significance F*
231	Regression	3	27727902945	9242634315	1054.31742	1.786E-127
232	Residual	213	1867256546	8766462.66		
233	Total	216	29595159490			
234						
235		*Coefficients*	*Standard Error*	*t Stat*	*P-value*	*Lower 95%*
236	Intercept	641.894009	894.6475283	0.71748257	0.4738625	-1121.6048
237	N	20050.0611	474.9294547	42.2169248	1.922E-105	19113.8962
238	N^2	-1994.1371	58.02186759	-34.368716	7.4885E-89	-2108.5078
239	T	401.853456	22.47177269	17.8825882	2.762E-44	357.557861

The desired constants a_1, a_2, a_3, and a_4 appear in the Coefficients column at the bottom left in the correct order: a_1 is the "intercept," a_2 is the coefficient of n, and so on. Thus, if we round to four significant digits, we have

$$a_1 = 641.9 \qquad a_2 = 20{,}050 \qquad a_3 = -1994 \qquad a_4 = 401.9$$

which gives our regression model:

$$I(n, t) = 641.9 + 20{,}050n - 1994n^2 + 401.9t$$

Fine, you say to yourself, now you have the model, but how good a fit is it to the data? That is where the "Multiple R" at the top of the data analysis comes in. R is called the **multiple coefficient of correlation** and generalizes the coefficient of correlation discussed in the section on regression in Chapter 1: The closer R is to 1, the better the fit. We can interpret its square, given in the table as "R Square" with value 0.937, as indicating that approximately 94% of the variation in mean income is explained by the regression model, which is an excellent fit. The "P-values" at the bottom right are also indicators of the appropriateness of the model; a P-value close to zero indicates a high degree of confidence that the corresponding coefficient is really nonzero, whereas a P-value close to 1 indicates low confidence (there is a P-value for each coefficient). Since all the values are extremely tiny, you are confident indeed that the model is an appropriate one. Another statistical indicator is the "F-value" on the right—an indicator of confidence in the model as a whole. The fact that it too is tiny is yet another good sign.[2]

As comforting as these statistics are, nothing can be quite as persuasive as a graph. You turn to your graphing software and notice that the graph of the model appears to be a faithful representation of the data. (See Figure 41.)

[2] We are being deliberately vague about the exact meaning of these statistics, which are discussed fully in many applied statistics texts.

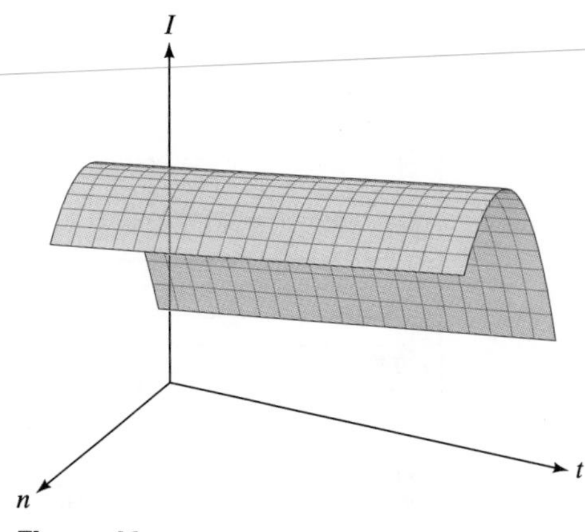

Figure 41

Now you get to work, using the model to address the issues posed by Millennium.

1. *The relationship between household size and household income and the effect of increasing household size on household income.* You already have a quantitative relationship in the regression model. As for the second part of the question, the rate of change of average family income with respect to household size is given by the partial derivative:

$$\frac{\partial I}{\partial n} = 20{,}050 - 3988n \text{ dollars per additional family member}$$

Thus, for example, in a household of five people,

$$\frac{\partial I}{\partial n} = 20{,}050 - 3988(5) = \$110 \text{ per additional family member}$$

On the other hand, when $n = 6$, one has

$$\frac{\partial I}{\partial n} = 20{,}050 - 3988(6) = -\$3878 \text{ per additional family member}$$

Notice that the derivative is independent of time: The rate of change of the average family income with respect to household size is independent of the date (according to your model).

2, 3. *The household size corresponding to the highest household income and the change in the relationship over time.* Although a glance at the graph shows you that there are no relative maxima, holding t constant (that is, on any given year) gives a relative maximum along the corresponding slice when

$$\frac{\partial I}{\partial n} = 0$$

or $20{,}050 - 3988n$, which gives

$$n = \frac{20{,}050}{3988} \approx 5.0$$

In other words, households of five tend to have the highest household incomes. Again, $\partial I/\partial n$ does not depend on t, so this optimal household size seems independent of time t.

4. *Some near-term projections of household income vs. household size.* As we have seen throughout the book, extrapolation can be a risky venture; however, *near-term* extrapolation from a good model can be reasonable. You enter the model in an Excel spreadsheet to obtain the following predicted mean household incomes for the years 1998–2002:

		Household size →						
Year		1	2	3	4	5	6	7
↓	1998	31,160	45,220	55,300	61,400	63,500	61,620	55,740
	1999	31,560	45,630	55,710	61,800	63,900	62,020	56,150
	2000	31,960	46,030	56,110	62,200	64,300	62,420	56,550
	2001	32,360	46,430	56,510	62,600	64,710	62,820	56,950
	2002	32,760	46,830	56,910	63,000	65,110	63,220	57,350

exercises

1. Use Excel to obtain an interaction model of the form

$$I(n, t) = a_1 + a_2n + a_3t + a_4nt$$

Compare the fit of this model with that of the quadratic model above. Comment on the result.

2. How much is to be gained by including a term of the form a_5t^2 in the original model? (Perform the regression and analyze the result by referring to the *P*-value for the resulting coefficient of t^2.)

3. The following table shows some data on U.S. population (in thousands) vs. age and year. This table can be downloaded as an Excel file by following the web path

Web site → Everything for Calculus → Chapter 8 → You're the Expert Excel Data

		Year (0 = 1990) →					
		0	2	4	6	8	8.5
Age	2.5	18,851	19,489	19,694	19,324	19,020	18,974
↓	7.5	18,058	18,285	18,742	19,425	19,912	19,931
	12.5	17,191	18,065	18,666	18,949	19,184	19,291
	17.5	17,763	17,170	17,707	18,644	19,460	19,554
	22.5	19,137	19,085	18,451	17,562	17,685	17,796
	27.5	21,233	20,152	19,142	18,993	18,621	18,513
	32.5	21,909	22,237	22,141	21,328	20,163	19,965
	37.5	19,980	21,092	21,973	22,550	22,600	22,589
	42.5	17,793	18,806	19,714	20,809	21,875	22,014
	47.5	13,823	15,362	16,685	18,438	18,850	19,007
	52.5	11,370	12,059	13,199	13,931	15,727	15,973
	57.5	10,474	10,487	10,937	11,362	12,408	12,631
	62.5	10,619	10,440	10,079	9,997	10,256	10,358
	67.5	10,076	9,973	9,963	9,895	9,575	9,515
	72.5	8,022	8,467	8,733	8,778	8,781	8,780
	77.5	6,146	6,392	6,575	6,873	7,195	7,238
	82.5	3,934	4,135	4,350	4,559	4,712	4,748
	87.5	2,050	2,170	2,287	2,395	2,533	2,560
	92.5	765	860	956	1,024	1,094	1,108
	97.5	206	231	249	287	317	324
	102.5	37	44	50	57	63	62

Use multiple regression to construct **a.** a linear model and **b.** an interaction model for the data. (Round all coefficients to four significant digits.) Does the interaction model give a significantly better fit in terms of the multiple regression coefficient? Referring to the linear model, does the *P*-value for the coefficient of y provide strong evidence that the population profile has been changing with time? (A *P*-value of α indicates that we can be certain with a confidence level of $1 - \alpha$ that the associated coefficient is nonzero.)

4. Graph the data from the preceding exercise and decide whether the linear model gives a faithful representation of the actual data. If not, propose and construct an alternative model. How is the confidence level in the coefficient of y changed?

5. According to your model in the preceding exercise, why does the age group with maximum population not change over time? Propose a model in which it does. Construct such a model and test the additional coefficient.

chapter 8 review test

1.

a. Let $g(x, y, z) = xy(x + y - z) + x^2$. Evaluate $g(0, 0, 0)$, $g(1, 0, 0)$, $g(0, 1, 0)$, $g(x, x, x)$, and $g(x, y + k, z)$.

b. Let $f(x, y, z) = 2.72 - 0.32x - 3.21y + 12.5z$. Complete the following: f ____ by ____ units for every 1 unit of increase in x and ____ by ____ units for every unit of increase in z.

c. Let $h(x, y) = 2x^2 + xy - x$. Complete the following table of values.

		$x \rightarrow$		
		−1	0	1
y	−1			
$\downarrow$	0			
	1			

d. Give a formula for a (single) function f with the property that $f(x, y) = -f(y, x)$ and $f(1, -1) = 3$.

2. For each of the given functions, compute the partial derivatives shown.

a. $f(x, y) = x^2 + xy$; find f_x, f_y, and f_{yy}.

b. $f(x, y) = e^{xy} + e^{3x^2-y^2}$; find $\partial f/\partial x$ and $\partial^2 f/(\partial x\, \partial y)$.

c. $f(x, y, z) = \dfrac{x}{x^2 + y^2 + z^2}$; find $\dfrac{\partial f}{\partial x}, \dfrac{\partial f}{\partial y}, \dfrac{\partial f}{\partial z}$, and $\left.\dfrac{\partial f}{\partial x}\right|_{(0,1,0)}$.

d. $f(x, y, z) = x^2 + y^2 + z^2 + xyz$; find $f_{xx} + f_{yy} + f_{zz}$.

3. In each of the following, locate and classify all critical points.

a. $f(x, y) = (x - 1)^2 + (2y - 3)^2$

b. $g(x, y) = (x - 1)^2 - 3y^2 + 9$

c. $h(x, y) = e^{xy}$

d. $j(x, y) = xy + x^2$

e. $f(x, y) = \ln(x^2 + y^2) - (x^2 + y^2)$

4. Solve the following constrained optimization problems.

a. Find the point on the surface $z = \sqrt{x^2 + 2(y - 3)^2}$ closest to the origin.

b. Find the location and values of the minimum and maximum of $T(x, y) = (x - 1)^2 + y^2$ on the triangle with vertices (0, 1), (3, 0), and (0, 0).

c. Find the minimum and maximum values of $F(x, y) = x^2 + 2y^2 - 2x - 4y + xy$ subject to $-1 \le x \le 1$ and $-2 \le y \le 2$.

5. Compute the given quantities.

a. $\int_0^1 \int_0^2 (2xy)\, dx\, dy$ **b.** $\int_1^2 \int_0^1 xye^{x+y}\, dx\, dy$ **c.** $\int_0^2 \int_0^{2x} \frac{1}{x^2 + 1}\, dy\, dx$

d. The average value of xye^{x+y} over the rectangle $0 \le x \le 1$ and $1 \le y \le 2$

e. $\iint_R (x^2 - y^2)\, dx\, dy$, where R is the region shown in the figure

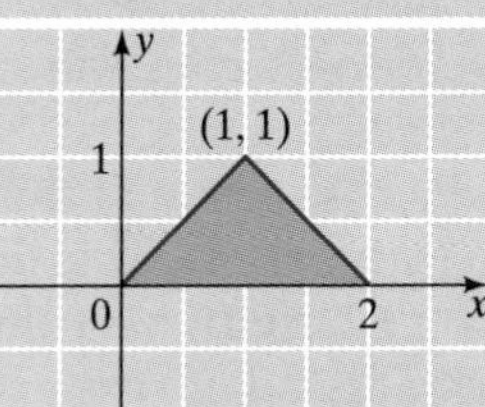

f. Find the volume under the graph of $z = 1 - y$ over the region in the xy-plane between the parabola $y = 1 - x^2$ and the x-axis.

Applications

6. OHaganBooks.com has two principal competitors: JungleBooks.com and FarmerBooks.com. Current web site traffic at OHaganBooks.com is estimated at 5000 hits per day. This number is predicted to decrease by 0.8 for every new customer of JungleBooks.com and by 0.6 for every new customer of FarmerBooks.com.

a. Use this information to model the daily web site traffic at OHaganBooks.com as a linear function of the new customers of its two competitors.

b. According to the model, if Junglebooks.com gets 100 new customers and OHaganBooks.com drops to 4770 hits per day, how many new customers has FarmerBooks.com obtained?

c. The model in part (a) did not take the growth of the total on-line consumer base into account. OHaganBooks.com expects to get approximately one additional hit per day for every 10,000 new Internet shoppers. Modify your model in part (a) so as to include this information.

d. How many new Internet shoppers would it take to offset the effects on traffic at OHaganBooks.com of 100 new customers at each of its competitor sites?

7. To increase business at OHaganBooks.com, you have purchased banner ads at well-known Internet portals and advertised on television. The following interaction model shows the average number h of hits per day as a function of monthly expenditures x on banner ads and y on television advertising (x and y are in dollars).

$$h(x, y) = 1800 + 0.05x + 0.08y + 0.00003xy$$

a. Based on your model, how much traffic can you anticipate if you spend \$2000 per month for banner ads and \$3000 per month on television advertising?

b. Evaluate $\partial h/\partial y$, specify its units of measurement, and indicate whether it increases or decreases with increasing x.

c. How much should the company be spending on banner ads to obtain one hit per day for each \$5 spent per month on television advertising?

d. One or more of the following statements is correct. Identify which one(s).

(A) If nothing is spent on television advertising, one more dollar spent per month on banner ads will buy approximately 0.05 hit per day at OHaganBooks.com.

(B) If nothing is spent on television advertising, one more hit per day at OHaganBooks.com will cost the company about 5¢ per month in banner ads.

(C) If nothing is spent on banner ads, one more hit per day at OHaganBooks.com will cost the company about 5¢ per month in banner ads.

(D) If nothing is spent on banner ads, one more dollar spent per month on banner ads will buy approximately 0.05 hit per day at OHaganBooks.com.

(E) Hits at OHaganBooks.com cost approximately 5¢ per month spent on banner ads, and this cost increases at a rate of 0.003¢ per month per hit.

8. The holiday season is now at its peak and OHaganBooks.com has been understaffed and swamped with orders. The current backlog (orders unshipped for two or more days) has grown to a staggering 50,000, and new orders are coming in at a rate of 5000 per day. Research based on productivity data at OHaganBooks.com results in the following model:

$$P(x, y) = 1000x^{0.9}y^{0.1} \text{ additional orders filled per day}$$

where x is the number of additional personnel hired and y is the daily budget (excluding salaries) allocated to eliminating the backlog.

a. How many additional orders will be filled per day if the company hires ten additional employees and budgets an additional \$1000 per day? (Round the answer to the nearest 100.)

b. In addition to the daily budget, extra staffing costs the company \$150 per day for every new staff member hired. To fill at least 15,000 orders per day at a minimum total daily cost, how many new staff members should the company hire? (Use the method of Lagrange multipliers.)

9. To sell x paperbacks and y hardcover books in a week costs OHaganBooks.com

$$C(x, y) = 0.1x^2 + 0.2y^2 - 200x - 480y + 400{,}000 \text{ dollars}$$

If it sells between 900 and 1100 paperbacks and between 1000 and 1500 hardcover books per week, what combination will cost it the least and what combination will cost it the most?

10. If OHaganBooks.com sells x paperback books and y hardcover books per week, it will make an average weekly profit of

$$P(x, y) = 3x + 10y \text{ dollars}$$

If it sells between 1200 and 1500 paperback books and between 1800 and 2000 hardcover books per week, what is the average of all its possible weekly profits?

Additional On-Line Review

If you follow the path

Web site → Everything for Calculus → Chapter 8

you will find the following additional resources to help you review:

- A comprehensive chapter summary (including examples and interactive features)
- Additional review exercises (including interactive exercises and many with help)
- A true–false chapter quiz
- Several useful utilities, including a three-dimensional grapher and regression utilities

Algebra Review

Appendix

A.1 Real Numbers
A.2 Exponents and Radicals
A.3 Multiplying and Factoring Algebraic Equations
A.4 Rational Expressions
A.5 Solving Polynomial Equations
A.6 Solving Miscellaneous Equations

Introduction

In this appendix we review some topics from algebra that you need to know to get the most out of this book. This appendix can be used either as a refresher course or as a reference.

There is one crucial fact you must always keep in mind: The letters used in algebraic expressions stand for numbers. All the rules of algebra are just facts about the arithmetic of numbers. If you are not sure whether some algebraic manipulation you are about to do is legitimate, try it first with numbers. If it doesn't work with numbers, it doesn't work.

A.1 Real Numbers

The **real numbers** are the numbers that can be written in decimal notation, including those that require an infinite decimal expansion. The set of real numbers includes all integers, positive and negative; all fractions; and the irrational numbers, those with decimal expansions that never repeat. Examples of irrational numbers are

$$\sqrt{2} = 1.414213562373\ldots \qquad \text{and} \qquad \pi = 3.141592653589\ldots$$

It is very useful to picture the real numbers as points on a line. As shown in Figure 1, larger numbers appear to the right, in the sense that if $a < b$, then the point corresponding to b is to the right of the one corresponding to a.

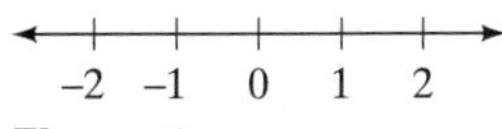

Figure 1

Intervals

Some subsets of the set of real numbers, called **intervals,** show up quite often and so we have a compact notation for them.

Interval Notation

Here is a list of types of intervals along with examples.

Type	Interval	Description	Picture	Example
Closed	$[a, b]$	Set of numbers x with $a \le x \le b$	a ●——● b (includes endpoints)	$[0, 10]$
Open	(a, b)	Set of numbers x with $a < x < b$	a ○——○ b (excludes endpoints)	$(-5, 5)$
Half-open	$(a, b]$	Set of numbers x with $a < x \le b$	a ○——● b	$(-3, 1]$
	$[a, b)$	Set of numbers x with $a \le x < b$	a ●——○ b	$[0, 5)$
Infinite	$[a, +\infty)$	Set of numbers x with $a \le x$	a ●——→	$[10, +\infty)$
	$(a, +\infty)$	Set of numbers x with $a < x$	a ○——→	$(-3, +\infty)$
	$(-\infty, b]$	Set of numbers x with $x \le b$	←——● b	$(-\infty, -3]$
	$(-\infty, b)$	Set of numbers x with $x < b$	←——○ b	$(-\infty, 10)$
	$(-\infty, +\infty)$	Set of all real numbers	←——→	$(-\infty, +\infty)$

Operations

There are five important operations on real numbers: addition, subtraction, multiplication, division, and exponentiation. "Exponentiation" means raising a real number to a power; for instance, $3^2 = 3 \cdot 3 = 9$; $2^3 = 2 \cdot 2 \cdot 2 = 8$.

A note on technology: Most graphing calculators and spreadsheets use an asterisk * for multiplication and a caret ^ for exponentiation. Thus, for instance, 3×5 is entered as `3*5`, $3x$ as `3*x`, and 3^2 as `3^2`.

When we write an expression involving two or more operations, like

$$2 \cdot 3 + 4 \qquad \text{or} \qquad \frac{2 \cdot 3^2 - 5}{4 - (-1)}$$

we need to agree on the order in which to do the operations. Does $2 \cdot 3 + 4$ mean $(2 \cdot 3) + 4 = 10$ or $2 \cdot (3 + 4) = 14$? We all agree to use the following rules for the order in which we do the operations.

Standard Order of Operations

Parentheses and Fraction Bars

First, calculate the values of all expressions inside parentheses or brackets, working from the innermost parentheses out, before using them in other operations. In a fraction, calculate the numerator and denominator separately before doing the division.

Quick Examples

1. $6(2 + [3 - 5] - 4) = 6(2 + (-2) - 4) = 6(-4) = -24$
2. $\dfrac{(4-2)}{3(-2+1)} = \dfrac{2}{3(-1)} = \dfrac{2}{-3} = -\dfrac{2}{3}$
3. $3/(2+4) = \dfrac{3}{2+4} = \dfrac{3}{6} = \dfrac{1}{2}$
4. $(x + 4x)/(y + 3y) = 5x/(4y)$

Exponents

Next, perform exponentiation.

Quick Examples

1. $2 + 4^2 = 2 + 16 = 18$
2. $(2 + 4)^2 = 6^2 = 36$

Note the difference.

3. $2\left(\dfrac{3}{4-5}\right)^2 = 2\left(\dfrac{3}{-1}\right)^2 = 2(-3)^2 = 2 \times 9 = 18$
4. $2(1 + 1/10)^2 = 2(1.1)^2 = 2 \times 1.21 = 2.42$

Multiplication and Division

Next, do all multiplications and divisions, from left to right.

Quick Examples

1. $2(3 - 5)/4 \cdot 2 = 2(-2)/4 \cdot 2$ Parentheses first
$= -4/4 \cdot 2$ Left-most product
$= -1 \cdot 2 = -2$ Multiplications and divisions, left to right
2. $2(1 + 1/10)^2 \times 2/10 = 2(1.1)^2 \times 2/10$ Parentheses first
$= 2 \times 1.21 \times 2/10$ Exponent
$= 4.84/10 = 0.484$ Multiplications and divisions, left to right
3. $4\dfrac{2(4-2)}{3(-2\cdot 5)} = 4\dfrac{2(2)}{3(-10)} = 4\dfrac{4}{-30} = \dfrac{16}{-30} = -\dfrac{8}{15}$

Addition and Subtraction

Last, do all additions and subtractions, from left to right.

Quick Examples

1. $2(3 - 5)^2 + 6 - 1 = 2(-2)^2 + 6 - 1 = 2(4) + 6 - 1 = 8 + 6 - 1 = 13$
2. $\left(\frac{1}{2}\right)^2 - (-1)^2 + 4 = \frac{1}{4} - 1 + 4 = -\frac{3}{4} + 4 = \frac{13}{4}$
3. $3/2 + 4 = 1.5 + 4 = 5.5$
4. $3/(2 + 4) = 3/6 = 1/2 = 0.5$

Note the difference.

5. $4/2^2 + (4/2)^2 = 4/2^2 + 2^2 = 4/4 + 4 = 1 + 4 = 5$

Entering Formulas

Any good calculator or spreadsheet will respect the standard order of operations. However, we must be careful with division and exponentiation and use parentheses as necessary. The following table gives some examples of simple mathematical expressions and their equivalents in the functional format used in most graphing calculators, spreadsheets, and computer programs.

Mathematical expression	Formula	Comments
$\frac{2}{3-x}$	`2/(3-x)`	Note the use of parentheses instead of the fraction bar. If we omit the parentheses, we get the expression shown next.
$\frac{2}{3}-x$	`2/3-x`	The calculator follows the usual order of operations.
$\frac{2}{3\times 5}$	`2/(3*5)`	Putting the denominator in parentheses ensures that the multiplication is carried out first. The asterisk is usually used for multiplication in graphing calculators and computers.
$\frac{2}{x}\times 5$	`(2/x)*5`	Putting the fraction in parentheses ensures that it is calculated first. Some calculators interpret `2/3*5` as $\frac{2}{3\times 5}$ but `2/3(5)` as $\frac{2}{3}\times 5$.
$\frac{2-3}{4+5}$	`(2-3)/(4+5)`	Note once again the use of parentheses in place of the fraction bar.
2^3	`2^3`	The caret ^ is commonly used to denote exponentiation.
2^{3-x}	`2^(3-x)`	Be careful to use parentheses to tell the calculator where the exponent ends. Enclose the *entire exponent* in parentheses.
2^3-x	`2^3-x`	Without parentheses, the calculator will follow the usual order of operations: exponentiation and then subtraction.
3×2^{-4}	`3*2^(-4)`	On some calculators, the negation key is separate from the minus key.
$2^{-4\times 3}\times 5$	`2^(-4*3)*5`	Note once again how parentheses enclose the entire exponent.
$100\left(1+\frac{0.05}{12}\right)^{60}$	`100*(1+0.05/12)^60`	This is a typical calculation for compound interest.
$PV\left(1+\frac{r}{m}\right)^{mt}$	`PV*(1+r/m)^(m*t)`	This is the compound interest formula. *PV* is understood to be a single number (present value) and not the product of *P* and *V* (or else we would have used `P*V`).
$\frac{2^{3-2}\times 5}{y-x}$	`2^(3-2)*5/(y-x)` or `(2^(3-2)*5)/(y-x)`	Notice again the use of parentheses to hold the denominator together. We could also have enclosed the numerator in parentheses, although this is optional. (Why?)
$\frac{2^y+1}{2-4^{3x}}$	`(2^y+1)/(2-4^(3*x))`	Here, it is necessary to enclose both the numerator and the denominator in parentheses.
$2^y+\frac{1}{2}-4^{3x}$	`2^y+1/2-4^(3*x)`	This is the effect of leaving out the parentheses around the numerator and denominator in the preceding expression.

Accuracy and Rounding

When we use a calculator or computer, the results of our calculations are often given to far more decimal places than are useful. For example, suppose we are told that a square has an area of 2.0 square feet and we are asked how long its sides are. Each side is the square root of the area, which the calculator tells us is

$$\sqrt{2} \approx 1.414213562$$

However, the measurement of 2.0 square feet is likely accurate to only two digits, so our estimate of the lengths of the sides can be no more accurate than that. Therefore, we round the answer to two digits:

$$\text{Length of one side} \approx 1.4 \text{ feet}$$

The digits that follow 1.4 are meaningless. The following guide makes these ideas more precise.

Significant Digits, Decimal Places, and Rounding

The number of **significant digits** in the decimal representation of a number is the number of digits that are not leading zeros after the decimal point (as in .0005) or trailing zeros before the decimal point (as in 5,400,000). We say that a value is **accurate to *n* significant digits** if only the first n significant digits are meaningful.

When to Round

After doing a computation in which all the quantities are accurate to no more than n significant digits, we round the final result to n significant digits.

Quick Examples

1. 0.00067 has two significant digits. — The 000 before 67 are leading zeros.
2. 0.000670 has three significant digits. — The 0 after 67 is significant.
3. 5,400,000 has two or more significant digits. — We can't say how many of the zeros are trailing.[1]
4. 5,400,001 has seven significant digits. — The string of zeros is not trailing.
5. Rounding 63,918 to three significant digits gives 63,900.
6. Rounding 63,958 to three significant digits gives 64,000.
7. $\pi = 3.141592653\ldots$ and $\frac{22}{7} = 3.142857142\ldots$. Therefore, $\frac{22}{7}$ is an approximation of π that is accurate to only three significant digits (3.14).
8. $4.02(1 + 0.02)^{1.4} \approx 4.13$ — We rounded to three significant digits.

[1]If we obtained 5,400,000 by rounding 5,401,011, then it has three significant digits because the zero after the 4 is significant. On the other hand, if we obtained it by rounding 5,411,234, then it has only two significant digits. The use of scientific notation avoids this ambiguity: 5.40×10^6 (or 5.40 E6 on a calculator or computer) is accurate to three digits and 5.4×10^6 is accurate to two.

One more point, though: If, in a long calculation, you round the intermediate results, your final answer may be even less accurate than you think. As a general rule,

> *When calculating, don't round intermediate results. Rather, use the most accurate results obtainable or have your calculator or computer store them for you.*

When you are done with the calculation, *then* round your answer to the appropriate number of digits of accuracy.

A.1 exercises

Calculate each expression in Exercises 1–24, giving the answer as a whole number or a fraction in lowest terms.

1. $2(4+(-1))(2\cdot -4)$

2. $3+([4-2]\cdot 9)$

3. `20/(3*4)-1`

4. `2-(3*4)/10`

5. $\dfrac{3+([3+(-5)])}{3-2\times 2}$

6. $\dfrac{12-(1-4)}{2(5-1)\cdot 2-1}$

7. `(2-5*(-1))/1-2*(-1)`

8. `2-5*(-1)/(1-2*(-1))`

9. $2\cdot(-1)^2/2$

10. $2+4\cdot 3^2$

11. $2\cdot 4^2+1$

12. $1-3\cdot(-2)^2\times 2$

13. `3^2+2^2+1`

14. `2^(2^2-2)`

15. $\dfrac{3-2(-3)^2}{-6(4-1)^2}$

16. $\dfrac{1-2(1-4)^2}{2(5-1)^2\cdot 2}$

17. `10*(1+1/10)^3`

18. `121/(1+1/10)^2`

19. $3\left(\dfrac{-2\cdot 3^2}{-(4-1)^2}\right)$

20. $-\left(\dfrac{8(1-4)^2}{-9(5-1)^2}\right)$

21. $3\left(1-\left(-\frac{1}{2}\right)^2\right)^2+1$

22. $3\left(\frac{1}{9}-\left(\frac{2}{3}\right)^2\right)^2+1$

23. `(1/2)^2-1/2^2`

24. `2/(1^2)-(2/1)^2`

Convert each expression in Exercises 25–50 into its technology formula equivalent as in the table in the text.

25. $3\times(2-5)$

26. $4+\frac{5}{9}$

27. $\dfrac{3}{2-5}$

28. $\dfrac{4-1}{3}$

29. $\dfrac{3-1}{8+6}$

30. $3+\dfrac{3}{2-9}$

31. $3-\dfrac{4+7}{8}$

32. $\dfrac{4\times 2}{(2/3)}$

33. $\dfrac{2}{3+x}-xy^2$

34. $3+\dfrac{3+x}{xy}$

35. $3.1x^3-4x^{-2}-\dfrac{60}{x^2-1}$

36. $2.1x^{-3}-x^{-1}+\dfrac{x^2-3}{2}$

37. $\dfrac{\left(\frac{2}{3}\right)}{5}$

38. $\dfrac{2}{\left(\frac{3}{5}\right)}$

39. $3^{4-5}\times 6$

40. $\dfrac{2}{3+5^{7-9}}$

41. $3\left(1+\dfrac{4}{100}\right)^{-3}$

42. $3\left(\dfrac{1+4}{100}\right)^{-3}$

43. $3^{2x-1}+4^x-1$

44. $2^{x^2}-(2^{2x})^2$

45. 2^{2x^2-x+1}

46. $2^{2x^2-x}+1$

47. $\dfrac{4e^{-2x}}{2-3e^{-2x}}$

48. $\dfrac{e^{2x}+e^{-2x}}{e^{2x}-e^{-2x}}$

49. $3\left(1-\left(-\frac{1}{2}\right)^2\right)^2+1$

50. $3\left(\frac{1}{9}-\left(\frac{2}{3}\right)^2\right)^2+1$

A.2 Exponents and Radicals

In the last section we discussed exponentiation, or "raising to a power"; for example, $2^3 = 2 \cdot 2 \cdot 2$. In this section we discuss the algebra of exponentials more fully. First, we look at *integer* exponents: cases where the powers are positive or negative whole numbers.

Integer Exponents

Positive Integer Exponents

If a is any real number and n is any positive integer, then by a^n we mean the quantity $a \cdot a \cdot \cdots \cdot a$ (n times); thus, $a^1 = a$, $a^2 = a \cdot a$, $a^5 = a \cdot a \cdot a \cdot a \cdot a$. In the expression a^n, the number n is called the **exponent** and the number a is called the **base.**

Quick Examples

$3^2 = 9$ $\quad$ $2^3 = 8$

$0^{34} = 0$ $\quad$ $(-1)^5 = -1$

$10^3 = 1000$ $\quad$ $10^5 = 100{,}000$

Negative Integer Exponents

If a is any real number *other than zero* and n is any positive integer, then we define

$$a^{-n} = \frac{1}{a^n} = \frac{1}{a \cdot a \cdot \cdots \cdot a} \quad (n \text{ times})$$

Quick Examples

$2^{-3} = \frac{1}{2^3} = \frac{1}{8}$ $\quad$ $1^{-27} = \frac{1}{1^{27}} = 1$

$x^{-1} = \frac{1}{x^1} = \frac{1}{x}$ $\quad$ $(-3)^{-2} = \frac{1}{(-3)^2} = \frac{1}{9}$

$y^7y^{-2} = y^7 \cdot \frac{1}{y^2} = y^5$ $\quad$ 0^{-2} is not defined

Zero Exponent

If a is any real number other than zero, then we define

$a^0 = 1$

Quick Examples

$3^0 = 1$ $\quad$ $1{,}000{,}000^0 = 1$

0^0 is not defined

When combining exponential expressions, we use the following identities.

Exponent identity	Quick examples
1. $a^m a^n = a^{m+n}$	$2^3 2^2 = 2^{3+2} = 2^5 = 32$ $x^3 x^{-4} = x^{3-4} = x^{-1} = \frac{1}{x}$ $\frac{x^3}{x^{-2}} = x^3 \frac{1}{x^{-2}} = x^3 x^2 = x^5$
2. $\frac{a^m}{a^n} = a^{m-n}$ if $a \neq 0$	$\frac{4^3}{4^2} = 4^{3-2} = 4^1 = 4$ $\frac{x^3}{x^{-2}} = x^{3-(-2)} = x^5$ $\frac{3^2}{3^4} = 3^{2-4} = 3^{-2} = \frac{1}{9}$
3. $(a^n)^m = a^{nm}$	$(3^2)^2 = 3^4 = 81$ $(2^x)^2 = 2^{2x}$
4. $(ab)^n = a^n b^n$	$(4 \cdot 2)^2 = 4^2 2^2 = 64$ $(-2y)^4 = (-2)^4 y^4 = 16y^4$
5. $\left(\frac{a}{b}\right)^n = \frac{a^n}{b^n}$ if $b \neq 0$	$\left(\frac{4}{3}\right)^2 = \frac{4^2}{3^2} = \frac{16}{9}$ $\left(\frac{x}{-y}\right)^3 = \frac{x^3}{(-y)^3} = -\frac{x^3}{y^3}$

Caution

- In the first two identities, the bases of the expressions must be the same. For example, the first gives $3^2 3^4 = 3^6$ but does *not* apply to $3^2 4^2$.
- People sometimes invent their own identities, such as $a^m + a^n = a^{m+n}$, which is wrong! (Try it with $a = m = n = 1$.) If you wind up with something like $2^3 + 2^4$, you are stuck with it; there are no identities to simplify it further. (You can factor out 2^3, but whether that is a simplification depends on what you are going to do with the expression next.)

example 1

Combining the Identities

$$\frac{(x^2)^3}{x^3} = \frac{x^6}{x^3} \quad \text{By 3}$$
$$= x^{6-3} \quad \text{By 2}$$
$$= x^3$$

$$\frac{(x^4y)^3}{y} = \frac{(x^4)^3 y^3}{y} \quad \text{By 4}$$
$$= \frac{x^{12}y^3}{y} \quad \text{By 3}$$
$$= x^{12}y^{3-1} \quad \text{By 2}$$
$$= x^{12}y^2$$

example 2

Eliminating Negative Exponents

Simplify the following and express the answer using no negative exponents.

a. $\dfrac{x^4y^{-3}}{x^5y^2}$ **b.** $\left(\dfrac{x^{-1}}{x^2y}\right)^5$

Solution

a. $\dfrac{x^4y^{-3}}{x^5y^2} = x^{4-5}y^{-3-2} = x^{-1}y^{-5} = \dfrac{1}{xy^5}$

b. $\left(\dfrac{x^{-1}}{x^2y}\right)^5 = \dfrac{(x^{-1})^5}{(x^2y)^5} = \dfrac{x^{-5}}{x^{10}y^5} = \dfrac{1}{x^{15}y^5}$

Radicals

If a is any nonnegative real number, then its **square root** is the nonnegative number whose square is a. For example, the square root of 16 is 4 because $4^2 = 16$. We write the square root of n as $\sqrt{n}$. (Roots are also referred to as **radicals.**) It is important to remember that $\sqrt{n}$ is never negative. Thus, for instance, $\sqrt{9}$ is 3 and not -3, even though $(-3)^2 = 9$. If we want to speak of the "negative square root" of 9, we write it as $-\sqrt{9} = -3$. If we want to write both square roots at once, we write $\pm\sqrt{9} = \pm 3$.

The **cube root** of a real number a is the number whose cube is a. The cube root of a is written as $\sqrt[3]{a}$ so that, for example, $\sqrt[3]{8} = 2$ (since $2^3 = 8$). Note that we can take the cube root of any number, positive, negative, or zero. For instance, the cube root of -8 is $\sqrt[3]{-8} = -2$ because $(-2)^3 = -8$. Unlike square roots, the cube root of a number may be negative. In fact, the cube root of a always has the same sign as a.

Higher roots are defined similarly. The **fourth root** of the *nonnegative* number a is defined as the nonnegative number whose fourth power is a and written $\sqrt[4]{a}$. The **fifth root** of any number a is the number whose fifth power is a, and so on.

Note We cannot take an even-numbered root of a negative number, but we can take an odd-numbered root of any number. Even roots are always positive, whereas odd roots have the same sign as the number we start with.

example 3

***n*th Roots**

$\sqrt{4} = 2$	Since $2^2 = 4$
$\sqrt{16} = 4$	Since $4^2 = 16$
$\sqrt{1} = 1$	Since $1^2 = 1$
If $x \geq 0$, then $\sqrt{x^2} = x$.	Since $x^2 = x^2$
$\sqrt{2} \approx 1.414213562$	$\sqrt{2}$ is not a whole number.
$\sqrt{1+1} = \sqrt{2} \approx 1.414213562$	First add and then take the square root.[1]
$\sqrt{9+16} = \sqrt{25} = 5$	Contrast with $\sqrt{9} + \sqrt{16} = 3 + 4 = 7$.
$\sqrt[3]{27} = 3$	Since $3^3 = 27$

[2]In general, $\sqrt{a+b}$ means the square root of the *quantity* $(a + b)$. The radical sign acts like a pair of parentheses or a fraction bar, telling us to evaluate what is inside before taking the root. (See the *Caution* below.)

$\sqrt[3]{-64} = -4$ Since $(-4)^3 = -64$

$\sqrt[4]{16} = 2$ Since $2^4 = 16$

$\sqrt[4]{-16}$ is not defined Even-numbered root of a negative number

$\sqrt[5]{-1} = -1$ since $(-1)^5 = -1$ Odd-numbered root of a negative number

$\sqrt[n]{-1} = -1$ if n is any odd number

Question In the example we saw that $\sqrt{x^2} = x$ if x is nonnegative. What happens if x is negative?

Answer If x is negative, then x^2 is positive, and so $\sqrt{x^2}$ is still defined as the nonnegative number whose square is x^2. This number must be $|x|$, the **absolute value of x,** which is the nonnegative number with the same size as x. For instance, $|-3| = 3$ while $|3| = 3$ and $|0| = 0$. It follows that

$$\sqrt{x^2} = |x|$$

for every real number x, positive or negative. For instance,

$$\sqrt{(-3)^2} = \sqrt{9} = 3 = |-3|$$

$$\sqrt{3^2} = \sqrt{9} = 3 = |3|$$

In general, we find that

$$\sqrt[n]{x^n} = x \text{ if } n \text{ is odd and } \sqrt[n]{x^n} = |x| \text{ if } n \text{ is even.}$$

We use the following identities to evaluate radicals of products and quotients.

Radicals of Products and Quotients

If a and b are any real numbers (nonnegative in the case of even-numbered roots), then

$\sqrt[n]{ab} = \sqrt[n]{a}\sqrt[n]{b}$ Radical of a product = Product of radicals

$\sqrt[n]{\dfrac{a}{b}} = \dfrac{\sqrt[n]{a}}{\sqrt[n]{b}}$ if $b \neq 0$ Radical of a quotient = Quotient of radicals

Notes

- The first rule is similar to the rule $(a \cdot b)^2 = a^2b^2$ for the square of a product, and the second rule is similar to the rule $\left(\dfrac{a}{b}\right)^2 = \dfrac{a^2}{b^2}$ for the square of a quotient.
- ***Caution:*** There is no corresponding identity for addition:

  $\sqrt{a+b}$ is *not* equal to $\sqrt{a} + \sqrt{b}$

  (consider $a = b = 1$, for example). Equating these expressions is a common error, so be careful!

Quick Examples

1. $\sqrt{9 \cdot 4} = \sqrt{9}\sqrt{4} = 3 \times 2 = 6$ Alternatively, $\sqrt{9 \cdot 4} = \sqrt{36} = 6$

2. $\sqrt{\frac{9}{4}} = \frac{\sqrt{9}}{\sqrt{4}} = \frac{3}{2}$

3. $\sqrt{4(3+13)} = \sqrt{4(16)} = \sqrt{4}\sqrt{16} = 2 \times 4 = 8$

4. $\sqrt[3]{-216} = \sqrt[3]{(-27)8} = \sqrt[3]{-27}\sqrt[3]{8} = (-3)2 = -6$

5. $\sqrt{x^3} = \sqrt{x^2 \cdot x} = \sqrt{x^2}\sqrt{x} = x\sqrt{x}$ if $x \geq 0$

6. $\sqrt{\frac{x^2+y^2}{z^2}} = \frac{\sqrt{x^2+y^2}}{\sqrt{z^2}} = \frac{\sqrt{x^2+y^2}}{|z|}$ We can't simplify the numerator any further.

Rational Exponents

We already know what we mean by expressions such as x^4 and a^{-6}. The next step is to make sense of *rational* exponents: exponents of the form p/q with p and q integers as in $a^{1/2}$ and $3^{-2/3}$.

Question What should we mean by $a^{1/2}$?
Answer The overriding concern here is that all the exponent identities should remain true. In this case the identity to look at is the one that says $(a^m)^n = a^{mn}$. This identity tells us that

$$(a^{1/2})^2 = a^1 = a$$

That is, $a^{1/2}$, when squared, gives us a. But that must mean that $a^{1/2}$ is the *square root* of a, or

$$a^{1/2} = \sqrt{a}$$

A similar argument tells us that if q is any positive whole number, then

$$a^{1/q} = \sqrt[q]{a}, \quad \text{the } q\text{th root of } a$$

Notice that if a is negative, this makes sense only for q odd. To avoid this problem we usually stick to positive a.

Question If p and q are integers (q positive), what should we mean by $a^{p/q}$?
Answer By the exponent identities, $a^{p/q}$ should equal both $(a^p)^{1/q}$ and $(a^{1/q})^p$. The first is the qth root of a^p, and the second is the pth power of $a^{1/q}$, which gives us the following generalization.

Conversion Between Rational Exponents and Radicals

If a is any nonnegative number, then

$$a^{p/q} = \sqrt[q]{a^p} = \left(\sqrt[q]{a}\right)^p$$

↑ Exponential form ↖ ↖ Radical form

In particular,

$$a^{1/q} = \sqrt[q]{a}, \quad \text{the } q\text{th root of } a.$$

Notes

- If a is negative, all of this makes sense only if q is odd.
- All the exponent identities continue to work when we allow rational exponents p/q. In other words, we are free to use all the exponent identities even though the exponents are not integers.

Quick Examples

1. $4^{3/2} = \left(\sqrt{4}\right)^3 = 2^3 = 8$

2. $8^{2/3} = \left(\sqrt[3]{8}\right)^2 = 2^2 = 4$

3. $9^{-3/2} = \frac{1}{9^{3/2}} = \frac{1}{\left(\sqrt{9}\right)^3} = \frac{1}{3^3} = \frac{1}{27}$

4. $\frac{\sqrt{3}}{\sqrt[3]{3}} = \frac{3^{1/2}}{3^{1/3}} = 3^{1/2-1/3} = 3^{1/6} = \sqrt[6]{3}$

5. $2^2 2^{7/2} = 2^2 2^{3+1/2} = 2^2 2^3 2^{1/2} = 2^5 2^{1/2} = 2^5\sqrt{2}$

example 4

Simplifying Algebraic Expressions

Simplify the following.

a. $\frac{(x^3)^{5/3}}{x^3}$ **b.** $\sqrt[4]{a^6}$ **c.** $\frac{(xy)^{-3}y^{-3/2}}{x^{-2}\sqrt{y}}$

Solution

a. $\frac{(x^3)^{5/3}}{x^3} = \frac{x^5}{x^3} = x^2$

b. $\sqrt[4]{a^6} = a^{6/4} = a^{3/2} = a \cdot a^{1/2} = a\sqrt{a}$

c. $\frac{(xy)^{-3}y^{-3/2}}{x^{-2}\sqrt{y}} = \frac{x^{-3}y^{-3}y^{-3/2}}{x^{-2}y^{1/2}} = \frac{1}{x^{-2+3}y^{1/2+3+3/2}} = \frac{1}{xy^5}$

Solving Equations with Exponents

example 5

Solving Equations

Solve the following equations.

a. $x^3 + 8 = 0$ **b.** $x^2 - \frac{1}{2} = 0$ **c.** $x^{3/2} - 64 = 0$

Solution

a. Subtracting 8 from both sides gives $x^3 = -8$. Taking the cube root of both sides gives $x = -2$.

b. Adding $\frac{1}{2}$ to both sides gives $x^2 = \frac{1}{2}$. Thus, $x = \pm\sqrt{\frac{1}{2}} = \pm\frac{1}{\sqrt{2}}$.

c. Adding 64 to both sides gives $x^{3/2} = 64$. Taking the reciprocal (2/3) power of both sides gives

$$(x^{3/2})^{2/3} = 64^{2/3}$$

$$x^1 = \left(\sqrt[3]{64}\right)^2 = 4^2 = 16$$

so $x = 16$.

A.2 exercises

Evaluate the expressions in Exercises 1–16.

1. 3^3 **2.** $(-2)^3$ **3.** $-(2 \cdot 3)^2$ **4.** $(4 \cdot 2)^2$

5. $\left(\frac{-2}{3}\right)^2$ **6.** $\left(\frac{3}{2}\right)^3$ **7.** $(-2)^{-3}$ **8.** -2^{-3}

9. $\left(\frac{1}{4}\right)^{-2}$ **10.** $\left(\frac{-2}{3}\right)^{-2}$ **11.** $2 \cdot 3^0$ **12.** $3 \cdot (-2)^0$

13. $2^3 2^2$ **14.** $3^2 3$ **15.** $2^2 2^{-1} 2^4 2^{-4}$ **16.** $5^2 5^{-3} 5^2 5^{-2}$

Simplify each expression in Exercises 17–30, giving your answer with no negative exponents.

17. x^3x^2 **18.** x^4x^{-1} **19.** $-x^2x^{-3}y$ **20.** $-xy^{-1}x^{-1}$

21. $\frac{x^3}{x^4}$ **22.** $\frac{y^5}{y^3}$ **23.** $\frac{x^2y^2}{x^{-1}y}$ **24.** $\frac{x^{-1}y}{x^2y^2}$

25. $\frac{(xy^{-1}z^3)^2}{x^2yz^2}$ **26.** $\frac{x^2yz^2}{(xyz^{-1})^{-1}}$ **27.** $\left(\frac{xy^{-2}z}{x^{-1}z}\right)^3$

28. $\left(\frac{x^2y^{-1}z^0}{xyz}\right)^2$ **29.** $\left(\frac{x^{-1}y^{-2}z^2}{xy}\right)^{-2}$ **30.** $\left(\frac{xy^{-2}}{x^2y^{-1}z}\right)^{-3}$

Evaluate the expressions in Exercises 31–50, rounding your answer to four significant digits where necessary.

31. $\sqrt{4}$ **32.** $\sqrt{5}$ **33.** $\sqrt{\frac{1}{4}}$ **34.** $\sqrt{\frac{1}{9}}$

35. $\sqrt{\frac{16}{9}}$ **36.** $\sqrt{\frac{9}{4}}$ **37.** $\frac{\sqrt{4}}{5}$ **38.** $\frac{6}{\sqrt{25}}$

39. $\sqrt{9} + \sqrt{16}$ **40.** $\sqrt{25} - \sqrt{16}$ **41.** $\sqrt{9 + 16}$

42. $\sqrt{25 - 16}$ **43.** $\sqrt[3]{8 - 27}$ **44.** $\sqrt[4]{81 - 16}$

45. $\sqrt[3]{27/8}$ **46.** $\sqrt[3]{8 \times 64}$ **47.** $\sqrt{(-2)^2}$

48. $\sqrt{(-1)^2}$ **49.** $\sqrt{\frac{1}{4}(1 + 15)}$ **50.** $\sqrt{\frac{1}{9}(3 + 33)}$

Simplify the expressions in Exercises 51–58 given that x, y, z, a, b, and c are positive real numbers.

51. $\sqrt{a^2b^2}$ **52.** $\sqrt{\frac{a^2}{b^2}}$ **53.** $\sqrt{(x + 9)^2}$

54. $\left(\sqrt{x + 9}\right)^2$ **55.** $\sqrt[3]{x^3(a^3 + b^3)}$ **56.** $\sqrt[4]{\frac{x^4}{a^4b^4}}$

57. $\sqrt{\frac{4xy^3}{x^2y}}$ **58.** $\sqrt{\frac{4(x^2 + y^2)}{c^2}}$

Rewrite the expressions in Exercises 59–66 in exponential form.

59. $\sqrt{3}$ **60.** $\sqrt{8}$ **61.** $\sqrt{x^3}$ **62.** $\sqrt[3]{x^2}$

63. $\sqrt[3]{xy^2}$ **64.** $\sqrt{x^2y}$ **65.** $\frac{x^2}{\sqrt{x}}$ **66.** $\frac{x}{\sqrt{x}}$

Rewrite the expressions in Exercises 67–72 in radical form.

67. $2^{2/3}$ **68.** $3^{4/5}$ **69.** $x^{4/3}$ **70.** $y^{7/4}$

71. $(x^{1/2}y^{1/3})^{1/5}$ **72.** $x^{-1/3}y^{3/2}$

Simplify the expressions in Exercises 73–82.

73. $4^{-1/2}4^{7/2}$ **74.** $2^{1/a}/2^{2/a}$ **75.** $3^{2/3}3^{-1/6}$ **76.** $2^{1/3}2^{-1}2^{2/3}2^{-1/3}$

77. $\frac{x^{3/2}}{x^{5/2}}$ **78.** $\frac{y^{5/4}}{y^{3/4}}$ **79.** $\frac{x^{1/2}y^2}{x^{-1/2}y}$ **80.** $\frac{x^{-1/2}y}{x^2y^{3/2}}$

81. $\left(\frac{x}{y}\right)^{1/3}\left(\frac{y}{x}\right)^{2/3}$ **82.** $\left(\frac{x}{y}\right)^{-1/3}\left(\frac{y}{x}\right)^{1/3}$

Solve each equation in Exercises 83–96 for x, rounding your answer to four significant digits where necessary.

83. $x^2 - 16 = 0$ **84.** $x^2 - 1 = 0$

85. $x^2 - \frac{4}{9} = 0$ **86.** $x^2 - \frac{1}{10} = 0$

87. $x^2 - (1 + 2x)^2 = 0$ **88.** $x^2 - (2 - 3x)^2 = 0$

89. $x^5 + 32 = 0$ **90.** $x^4 - 81 = 0$

91. $x^{1/2} - 4 = 0$ **92.** $x^{1/3} - 2 = 0$

93. $1 - \frac{1}{x^2} = 0$ **94.** $\frac{2}{x^3} - \frac{6}{x^4} = 0$

95. $(x - 4)^{-1/3} = 2$ **96.** $(x - 4)^{2/3} + 1 = 5$

A.3 Multiplying and Factoring Algebraic Expressions

Multiplying Algebraic Expressions

Distributive Law

The **distributive law** for real numbers states that

$$a(b \pm c) = ab \pm ac$$

$$(a \pm b)c = ac \pm bc$$

for any real numbers a, b, and c.

Quick Examples

1. $2(x - 3)$ is *not* equal to $2x - 3$ but is equal to $2x - 2(3) = 2x - 6$.
2. $x(x + 1) = x^2 + x$
3. $2x(3x - 4) = 6x^2 - 8x$
4. $(x - 4)x^2 = x^3 - 4x^2$
5. $(x + 2)(x + 3) = (x + 2)x + (x + 2)3 = (x^2 + 2x) + (3x + 6) = x^2 + 5x + 6$
6. $(x + 2)(x - 3) = (x + 2)x - (x + 2)3 = (x^2 + 2x) - (3x + 6) = x^2 - x - 6$

There is a quicker way of expanding expressions like the last two, called the "FOIL" method (First, Outer, Inner, Last). Consider, for instance, the expression $(x + 1)(x - 2)$. The FOIL method says: Take the product of the first terms: $x \cdot x = x^2$, the product of the outer terms: $x \cdot (-2) = -2x$, the product of the inner terms: $1 \cdot x = x$, and the product of the last terms: $1 \cdot (-2) = -2$, and then add them all up to get $x^2 - 2x + x - 2 = x^2 - x - 2$.

example 1

FOIL

a. $(x - 2)(2x + 5) = 2x^2 + 5x - 4x - 10 = 2x^2 + x - 10$

$2x^2$ ↑ First, $5x$ ↑ Outer, $-4x$ ↑ Inner, -10 ↑ Last

b. $(x^2 + 1)(x - 4) = x^3 - 4x^2 + x - 4$

c. $(a - b)(a + b) = a^2 + ab - ab - b^2 = a^2 - b^2$

d. $(a + b)^2 = (a + b)(a + b) = a^2 + ab + ab + b^2 = a^2 + 2ab + b^2$

e. $(a - b)^2 = (a - b)(a - b) = a^2 - ab - ab + b^2 = a^2 - 2ab + b^2$

The last three identities are particularly important and are worth memorizing.

Special Formulas

$(a - b)(a + b) = a^2 - b^2$	Difference of two squares
$(a + b)^2 = a^2 + 2ab + b^2$	Square of a sum
$(a - b)^2 = a^2 - 2ab + b^2$	Square of a difference

Quick Examples

1. $(2 - x)(2 + x) = 4 - x^2$
2. $(1 + a)(1 - a) = 1 - a^2$
3. $(x + 3)^2 = x^2 + 6x + 9$
4. $(4 - x)^2 = 16 - 8x + x^2$

Here are some longer examples that require the distributive law.

example 2

Multiplying Algebraic Expressions

a. $(x+1)(x^2+3x-4) = (x+1)x^2 + (x+1)3x - (x+1)4$

$= (x^3+x^2) + (3x^2+3x) - (4x+4)$

$= x^3 + 4x^2 - x - 4$

b. $\left(x^2 - \frac{1}{x} + 1\right)(2x+5) = \left(x^2 - \frac{1}{x} + 1\right)2x + \left(x^2 - \frac{1}{x} + 1\right)5$

$= (2x^3 - 2 + 2x) + \left(5x^2 - \frac{5}{x} + 5\right)$

$= 2x^3 + 5x^2 + 2x + 3 - \frac{5}{x}$

c. $(x-y)(x-y)(x-y) = (x^2 - 2xy + y^2)(x-y)$

$= (x^2 - 2xy + y^2)x - (x^2 - 2xy + y^2)y$

$= (x^3 - 2x^2y + xy^2) - (x^2y - 2xy^2 + y^3)$

$= x^3 - 3x^2y + 3xy^2 - y^3$

Factoring Algebraic Expressions

We can think of factoring as applying the distributive law in reverse—for example,

$$2x^2 + x = x(2x+1)$$

which can be checked by using the distributive law. Factoring is an art that you will learn with experience and the help of a few useful techniques.

Factoring Using a Common Factor

To use this technique, locate a **common factor**—a term that occurs as a factor in each of the expressions being added or subtracted (for example, x is a common factor in $2x^2 + x$ because it is a factor of both $2x^2$ and x). Once you have located a common factor, "factor it out" by applying the distributive law.

Quick Examples

1. $2x^3 - x^2 + x$ has x as a common factor, so

$$2x^3 - x^2 + x = x(2x^2 - x + 1)$$

2. $2x^2 + 4x$ has $2x$ as a common factor, so

$$2x^2 + 4x = 2x(x+2)$$

3. $2x^2y + xy^2 - x^2y^2$ has xy as a common factor, so

$$2x^2y + xy^2 - x^2y^2 = xy(2x + y - xy)$$

4. $(x^2+1)(x+2) - (x^2+1)(x+3)$ has x^2+1 as a common factor, so

$$(x^2+1)(x+2) - (x^2+1)(x+3) = (x^2+1)[(x+2)-(x+3)]$$

$$= (x^2+1)(x+2-x-3)$$

$$= (x^2+1)(-1) = -(x^2+1)$$

5. $12x(x^2 - 1)^5(x^3 + 1)^6 + 18x^2(x^2 - 1)^6(x^3 + 1)^5$ has $6x(x^2 - 1)^5(x^3 + 1)^5$ as a common factor, so

$$\begin{aligned} 12x(x^2 - 1)^5(x^3 + 1)^6 + 18x^2(x^2 - 1)^6(x^3 + 1)^5 \\ &= 6x(x^2 - 1)^5(x^3 + 1)^5[2(x^3 + 1) + 3x(x^2 - 1)] \\ &= 6x(x^2 - 1)^5(x^3 + 1)^5(2x^3 + 2 + 3x^3 - 3x) \\ &= 6x(x^2 - 1)^5(x^3 + 1)^5(5x^3 - 3x + 2) \end{aligned}$$

We would also like to be able to reverse calculations such as $(x + 2)(2x - 5) = 2x^2 - x - 10$. That is, starting with the expression $2x^2 - x - 10$, we would like to **factor** it to get the expression $(x + 2)(2x - 5)$. An expression of the form $ax^2 + bx + c$, where a, b, and c are real numbers, is called a **quadratic** expression in x. Thus, given a quadratic expression $ax^2 + bx + c$, we would like to write it in the form $(dx + e)(fx + g)$ for some real numbers d, e, f, and g. There are some quadratics, such as $x^2 + x + 1$, that cannot be factored in this form at all. Here, we consider only quadratics that do factor and in such a way that the numbers d, e, f, and g are integers (whole numbers; other cases are discussed in Section A.5). The usual technique of factoring such quadratics is a "trial and error" approach.

Factoring Quadratics by Trial and Error

To factor the quadratic $ax^2 + bx + c$, factor ax^2 as $(a_1x)(a_2x)$ (with a_1 positive) and c as c_1c_2, and then check whether $ax^2 + bx + c = (a_1x \pm c_1)(a_2x \pm c_2)$. If not, try other factorizations of ax^2 and c.

Quick Examples

1. To factor $x^2 - 6x + 5$, first factor x^2 as $(x)(x)$ and 5 as $(5)(1)$:

$(x + 5)(x + 1) = x^2 + 6x + 5$ No good

$(x - 5)(x - 1) = x^2 - 6x + 5$ Desired factorization

2. To factor $x^2 - 4x - 12$, first factor x^2 as $(x)(x)$ and -12 as $(1)(-12)$, $(2)(-6)$, or $(3)(-4)$. Trying them one by one gives

$(x + 1)(x - 12) = x^2 - 11x - 12$ No good

$(x - 1)(x + 12) = x^2 + 11x - 12$ No good

$(x + 2)(x - 6) = x^2 - 4x - 12$ Desired factorization

3. To factor $4x^2 - 25$, we can follow the above procedure or recognize $4x^2 - 25$ as the difference of two squares:

$$4x^2 - 25 = (2x)^2 - 5^2 = (2x - 5)(2x + 5)$$

Note: Not all quadratic expressions factor. In Section A.5 we look at a test that tells us whether or not a given quadratic factors.

The next examples require either a little more work or a little more thought.

example 3

Factoring Quadratics

Factor each expression.

a. $4x^2 - 5x - 6$ **b.** $x^4 - 5x^2 + 6$

Solution

a. Possible factorizations of $4x^2$ are $(2x)(2x)$ and $(x)(4x)$. Possible factorizations of -6 are $(1)(-6)$ and $(2)(-3)$. We now systematically try out all the possibilities until we come up with the correct one.

$(2x)(2x)$ and $(1)(-6)$:	$(2x + 1)(2x - 6) = 4x^2 - 10x - 6$	No good
$(2x)(2x)$ and $(2)(-3)$:	$(2x + 2)(2x - 3) = 4x^2 - 2x - 6$	No good
$(x)(4x)$ and $(1)(-6)$:	$(x + 1)(4x - 6) = 4x^2 - 2x - 6$	No good
$(x)(4x)$ and $(2)(-3)$:	$(x + 2)(4x - 3) = 4x^2 + 5x - 6$	Almost!
Change signs:	$(x - 2)(4x + 3) = 4x^2 - 5x - 6$	Correct

b. The expression $x^4 - 5x^2 + 6$ is not a quadratic, you say? Correct, it's a quartic (a fourth-degree expression). However, it looks rather like a quadratic. In fact, it is quadratic *in* x^2, meaning that

$$(x^2)^2 - 5(x^2) + 6 = y^2 - 5y + 6$$

where $y = x^2$. The quadratic $y^2 - 5y + 6$ factors as

$$y^2 - 5y + 6 = (y - 3)(y - 2)$$

so $$x^4 - 5x^2 + 6 = (x^2 - 3)(x^2 - 2)$$

This is sometimes a useful technique.

Our last example is here to remind you why we want to factor polynomials in the first place. We return to this in Section A.5.

example 4

Solving a Quadratic Equation by Factoring

Solve the equation $3x^2 + 4x - 4 = 0$.

Solution

We first factor the left-hand side to get

$$(3x - 2)(x + 2) = 0$$

Thus, the product of the two quantities $(3x - 2)$ and $(x + 2)$ is zero. Now, if a product of two numbers is zero, one of the two must be zero. In other words, either $3x - 2 = 0$, giving $x = \frac{2}{3}$, or $x + 2 = 0$, giving $x = -2$. Thus, there are two solutions: $x = \frac{2}{3}$ and $x = -2$.

A.3 exercises

Expand each expression in Exercises 1–22.

1. $x(4x + 6)$

2. $(4y - 2)y$

3. $(2x - y)y$

4. $x(3x + y)$

5. $(x + 1)(x - 3)$

6. $(y + 3)(y + 4)$

7. $(2y + 3)(y + 5)$

8. $(2x - 2)(3x - 4)$

9. $(2x - 3)^2$

10. $(3x + 1)^2$

11. $\left(x + \frac{1}{x}\right)^2$

12. $\left(y - \frac{1}{y}\right)^2$

13. $(2x - 3)(2x + 3)$

14. $(4 + 2x)(4 - 2x)$

15. $\left(y - \frac{1}{y}\right)\left(y + \frac{1}{y}\right)$

16. $(x - x^2)(x + x^2)$

17. $(x^2 + x - 1)(2x + 4)$

18. $(3x + 1)(2x^2 - x + 1)$

19. $(x^2 - 2x + 1)^2$

20. $(x + y - xy)^2$

21. $(y^3 + 2y^2 + y)(y^2 + 2y - 1)$

22. $(x^3 - 2x^2 + 4)(3x^2 - x + 2)$

In Exercises 23–30, factor each expression and simplify as much as possible.

23. $(x + 1)(x + 2) + (x + 1)(x + 3)$

24. $(x + 1)(x + 2)^2 + (x + 1)^2(x + 2)$

25. $(x^2 + 1)^5(x + 3)^4 + (x^2 + 1)^6(x + 3)^3$

26. $10x(x^2 + 1)^4(x^3 + 1)^5 + 15x^2(x^2 + 1)^5(x^3 + 1)^4$

27. $(x^3 + 1)\sqrt{x + 1} - (x^3 + 1)^2\sqrt{x + 1}$

28. $(x^2 + 1)\sqrt{x + 1} - \sqrt{(x + 1)^3}$

29. $\sqrt{(x + 1)^3} + \sqrt{(x + 1)^5}$

30. $(x^2 + 1)\sqrt[3]{(x + 1)^4} - \sqrt[3]{(x + 1)^7}$

*In Exercises 31–48, **a.** factor the given expression; **b.** set the expression equal to zero and solve for the unknown (x in the odd-numbered exercises and y in the even-numbered exercises).*

31. $2x + 3x^2$

32. $y^2 - 4y$

33. $6x^3 - 2x^2$

34. $3y^3 - 9y^2$

35. $x^2 - 8x + 7$

36. $y^2 + 6y + 8$

37. $x^2 + x - 12$

38. $y^2 + y - 6$

39. $2x^2 - 3x - 2$

40. $3y^2 - 8y - 3$

41. $6x^2 + 13x + 6$

42. $6y^2 + 17y + 12$

43. $12x^2 + x - 6$

44. $20y^2 + 7y - 3$

45. $x^2 + 4xy + 4y^2$

46. $4y^2 - 4xy + x^2$

47. $x^4 - 5x^2 + 4$

48. $y^4 + 2y^2 - 3$

A.4 Rational Expressions

Rational Expression

A **rational expression** is an algebraic expression of the form P/Q, where P and Q are simpler expressions (usually polynomials) and the denominator Q is not zero.

Quick Examples

1. $\dfrac{x^2 - 3x}{x}$ $\quad P = x^2 - 3x,\ Q = x$

2. $\dfrac{x + \frac{1}{x} + 1}{2x^2y + 1}$ $\quad P = x + \dfrac{1}{x} + 1,\ Q = 2x^2y + 1$

3. $3xy - x^2$ $\quad P = 3xy - x^2,\ Q = 1$

Algebra of Rational Expressions

We manipulate rational expressions in the same way that we manipulate fractions, using the following rules.

	Algebraic rule	Quick example
Product:	$\dfrac{P}{Q} \cdot \dfrac{R}{S} = \dfrac{PR}{QS}$	$\dfrac{x+1}{x} \cdot \dfrac{x-1}{2x+1} = \dfrac{(x+1)(x-1)}{x(2x+1)} = \dfrac{x^2-1}{2x^2+x}$
Sum:	$\dfrac{P}{Q} + \dfrac{R}{S} = \dfrac{PS + RQ}{QS}$	$\dfrac{2x-1}{3x+2} + \dfrac{1}{x} = \dfrac{(2x-1)x + 1(3x+2)}{x(3x+2)} = \dfrac{2x^2+2x+2}{3x^2+2x}$
Difference:	$\dfrac{P}{Q} - \dfrac{R}{S} = \dfrac{PS - RQ}{QS}$	$\dfrac{x}{3x+2} - \dfrac{x-4}{x} = \dfrac{x^2 - (x-4)(3x+2)}{x(3x+2)}$ $= \dfrac{-2x^2+10x+8}{3x^2+2x}$
Reciprocal:	$\dfrac{1}{\left(\frac{P}{Q}\right)} = \dfrac{Q}{P}$	$\dfrac{1}{\left(\frac{2xy}{3x-1}\right)} = \dfrac{3x-1}{2xy}$
Quotient:	$\dfrac{\left(\frac{P}{Q}\right)}{\left(\frac{R}{S}\right)} = \dfrac{P}{Q} \cdot \dfrac{S}{R} = \dfrac{PS}{QR}$	$\dfrac{\left(\frac{x}{x-1}\right)}{\left(\frac{y-1}{y}\right)} = \dfrac{xy}{(x-1)(y-1)} = \dfrac{xy}{xy-x-y+1}$
Cancellation:	$\dfrac{P\cancel{R}}{Q\cancel{R}} = \dfrac{P}{Q}$	$\dfrac{(x-1)(xy+4)}{(x^2y-8)(x-1)} = \dfrac{xy+4}{x^2y-8}$

Caution: Cancellation of summands is *invalid*. For instance,

$$\frac{\cancel{x} + (2xy^2 - y)}{\cancel{x} + 4y} = \frac{(2xy^2 - y)}{4y}$$ ✘ WRONG! Do *not* cancel a summand.

$$\frac{\cancel{x}(2xy^2 - y)}{4\cancel{x}y} = \frac{(2xy^2 - y)}{4y}$$ ✓ CORRECT Do cancel a factor.

Here are some examples that require several algebraic operations.

example 1

Simplifying Rational Expressions

a. $$\frac{\left(\frac{1}{x+y}-\frac{1}{x}\right)}{y}=\frac{\left(\frac{x-(x+y)}{x(x+y)}\right)}{y}=\frac{\left(\frac{-y}{x(x+y)}\right)}{y}=\frac{-y}{xy(x+y)}=\frac{-1}{x(x+y)}$$

b. $$\frac{(x+1)(x+2)^2-(x+1)^2(x+2)}{(x+2)^4}=\frac{(x+1)(x+2)\left[(x+2)-(x+1)\right]}{(x+2)^4}$$
$$=\frac{(x+1)(x+2)(x+2-x-1)}{(x+2)^4}$$
$$=\frac{(x+1)(x+2)}{(x+2)^4}=\frac{x+1}{(x+2)^3}$$

c. $$\frac{2x\sqrt{x+1}-\frac{x^2}{\sqrt{x+1}}}{x+1}=\frac{\left(\frac{2x\left(\sqrt{x+1}\right)^2-x^2}{\sqrt{x+1}}\right)}{x+1}=\frac{2x(x+1)-x^2}{(x+1)\sqrt{x+1}}$$
$$=\frac{2x^2+2x-x^2}{(x+1)\sqrt{x+1}}$$
$$=\frac{x^2+2x}{\sqrt{(x+1)^3}}=\frac{x(x+2)}{\sqrt{(x+1)^3}}$$

A.4 exercises

Rewrite each expression in Exercises 1–16 as a single rational expression, simplified as much as possible.

1. $\frac{x-4}{x+1}\cdot\frac{2x+1}{x-1}$

2. $\frac{2x-3}{x-2}\cdot\frac{x+3}{x+1}$

3. $\frac{x-4}{x+1}+\frac{2x+1}{x-1}$

4. $\frac{2x-3}{x-2}+\frac{x+3}{x+1}$

5. $\frac{x^2}{x+1}-\frac{x-1}{x+1}$

6. $\frac{x^2-1}{x-2}-\frac{1}{x-1}$

7. $\frac{1}{\left(\frac{x}{x-1}\right)}+x-1$

8. $\frac{2}{\left(\frac{x-2}{x^2}\right)}-\frac{1}{x-2}$

9. $\frac{1}{x}\left[\frac{x-3}{xy}+\frac{1}{y}\right]$

10. $\frac{y^2}{x}\left[\frac{2x-3}{y}+\frac{x}{y}\right]$

11. $\frac{(x+1)^2(x+2)^3-(x+1)^3(x+2)^2}{(x+2)^6}$

12. $\frac{6x(x^2+1)^2(x^3+2)^3-9x^2(x^2+1)^3(x^3+2)^2}{(x^3+2)^6}$

13. $\frac{(x^2-1)\sqrt{x^2+1}-\frac{x^4}{\sqrt{x^2+1}}}{x^2+1}$

14. $\frac{x\sqrt{x^3-1}-\frac{3x^4}{\sqrt{x^3-1}}}{x^3-1}$

15. $\frac{\frac{1}{(x+y)^2}-\frac{1}{x^2}}{y}$

16. $\frac{\frac{1}{(x+y)^3}-\frac{1}{x^3}}{y}$

A.5 Solving Polynomial Equations

Polynomial Equation

A **polynomial equation** in one unknown is an equation that can be written in the form

$$ax^n + bx^{n-1} + \cdots + rx + s = 0$$

where $a, b, \ldots, r$, and s are constants.

We call the largest exponent of x that appears in a nonzero term of a polynomial the **degree** of that polynomial.

Quick Examples

1. $3x + 1 = 0$ has degree 1 because the largest power of x that occurs is $x = x^1$. Degree 1 equations are called **linear** equations.
2. $x^2 - x - 1 = 0$ has degree 2 because the largest power of x that occurs is x^2. Degree 2 equations are also called **quadratic equations,** or just **quadratics.**
3. $x^3 = 2x^2 + 1$ is a degree-3 polynomial (or **cubic**) in disguise. It can be rewritten as $x^3 - 2x^2 - 1 = 0$, which is in the standard form for a degree-3 equation.
4. $x^4 - x = 0$ has degree 4. It is called a **quartic.**

Now comes the question: How do we solve these equations for x? This question was asked by mathematicians as early as 1600 B.C. Let's look at these equations one degree at a time.

Solution of Linear Equations

By definition, a linear equation can be written in the form

$$ax + b = 0 \qquad a \text{ and } b \text{ are fixed numbers with } a \neq 0$$

Solving this equation is a nice mental exercise: Subtract b from both sides and then divide by a, getting $x = -b/a$. Don't bother memorizing this formula, just go ahead and solve linear equations as they arise. If you feel you need practice, see the exercises at the end of the section.

Solution of Quadratic Equations

By definition, a quadratic equation has the form

$$ax^2 + bx + c = 0 \qquad a, b, \text{ and } c \text{ are fixed numbers and } a \neq 0^1$$

The solutions of this equation are also called the **roots** of $ax^2 + bx + c$. We assume you saw quadratic equations somewhere in high school but may be a little hazy about the details of their solution. There are two ways of solving these equations—one works sometimes, and the other works every time.

---

[1]What happens if $a = 0$?

Solving Quadratic Equations by Factoring (works sometimes)

If we can factor[2] a quadratic equation $ax^2 + bx + c = 0$, we can solve the equation by setting each factor equal to zero.

Quick Examples

1. $x^2 + 7x + 10 = 0$
   $(x + 5)(x + 2) = 0$ — Factor the left-hand side.
   $x + 5 = 0$ or $x + 2 = 0$ — If a product is zero, one or both factors is zero.
   Solutions:
   $x = -5$ and $x = -2$
2. $2x^2 - 5x - 12 = 0$
   $(2x + 3)(x - 4) = 0$ — Factor the left-hand side.
   $2x + 3 = 0$ or $x - 4 = 0$
   Solutions:
   $x = -\frac{3}{2}$ and $x = 4$

Test for Factoring

The quadratic $ax^2 + bx + c$, with a, b, and c integers (whole numbers), factors into an expression of the form $(rx + s)(tx + u)$, with r, s, t, and u integers precisely when the quantity $b^2 - 4ac$ is a perfect square (that is, it is the square of an integer). If this happens, we say that the quadratic **factors over the integers.**

Quick Examples

1. $x^2 + x + 1$ has $a = 1, b = 1$, and $c = 1$, so $b^2 - 4ac = -3$, which is not a perfect square. Therefore, this quadratic does not factor over the integers.
2. $2x^2 - 5x - 12$ has $a = 2, b = -5$, and $c = -12$, so $b^2 - 4ac = 121$. Since $121 = 11^2$, this quadratic does factor over the integers (we factored it above).

Solving Quadratic Equations with the Quadratic Formula (works every time)

The solutions of the general quadratic $ax^2 + bx + c = 0$ $(a \neq 0)$ are given by

$$x = \frac{-b \pm \sqrt{b^2 - 4ac}}{2a}$$

We call the quantity $\Delta = b^2 - 4ac$ the **discriminant** of the quadratic (Δ is the Greek letter delta), and we have the following general rules:

- If Δ is positive, there are two distinct real solutions.
- If Δ is zero, there is only one real solution: $x = -b/2a$. (Why?)
- If Δ is negative, there are no real solutions.

[2]See Section A.3 on factoring for a review of how to factor quadratics.

Quick Examples

1. $2x^2 - 5x - 12 = 0$ has $a = 2, b = -5$, and $c = -12$.

$$x = \frac{-b \pm \sqrt{b^2 - 4ac}}{2a} = \frac{5 \pm \sqrt{25 + 96}}{4} = \frac{5 \pm \sqrt{121}}{4} = \frac{5 \pm 11}{4}$$

$$= \frac{16}{4} \text{ or } -\frac{6}{4} = 4 \text{ or } -3/2$$ Δ is positive in this example.

2. $4x^2 = 12x - 9$ can be rewritten as $4x^2 - 12x + 9 = 0$, which has $a = 4$, $b = -12$, and $c = 9$.

$$x = \frac{-b \pm \sqrt{b^2 - 4ac}}{2a} = \frac{12 \pm \sqrt{144 - 144}}{8} = \frac{12 \pm 0}{8} = \frac{12}{8} = \frac{3}{2}$$

Δ is zero in this example.

3. $x^2 + 2x - 1 = 0$ has $a = 1, b = 2$, and $c = -1$.

$$x = \frac{-b \pm \sqrt{b^2 - 4ac}}{2a} = \frac{-2 \pm \sqrt{8}}{2} = \frac{-2 \pm 2\sqrt{2}}{2} = -1 \pm \sqrt{2}$$

The two solutions are $x = -1 + \sqrt{2} = 0.414\ldots$ and $x = -1 - \sqrt{2} = -2.414\ldots$.

Δ is positive in this example.

4. $x^2 + x + 1 = 0$ has $a = 1, b = 1$, and $c = 1$. Since $\Delta = -3$ is negative, there are no real solutions.

Question This is all very useful, but where on earth does the quadratic formula come from?

Answer To see where it comes from, we will solve a general quadratic equation using "brute force." Start with the general quadratic equation

$$ax^2 + bx + c = 0$$

First, divide out the nonzero number a to get

$$x^2 + \frac{bx}{a} + \frac{c}{a} = 0$$

Now we **complete the square:** Add and subtract the quantity $\frac{b^2}{4a^2}$ to get

$$x^2 + \frac{bx}{a} + \frac{b^2}{4a^2} - \frac{b^2}{4a^2} + \frac{c}{a} = 0$$

We do this to get the first three terms to factor as a perfect square:

$$\left(x + \frac{b}{2a}\right)^2 - \frac{b^2}{4a^2} + \frac{c}{a} = 0$$

(Check this by multiplying out.) Adding $\frac{b^2}{4a^2} - \frac{c}{a}$ to both sides gives

$$\left(x + \frac{b}{2a}\right)^2 = \frac{b^2}{4a^2} - \frac{c}{a} = \frac{b^2 - 4ac}{4a^2}$$

Taking square roots gives

$$x + \frac{b}{2a} = \frac{\pm\sqrt{b^2 - 4ac}}{2a}$$

Finally, adding $-\frac{b}{2a}$ to both sides yields the result;

$$x = -\frac{b}{2a} + \frac{\pm\sqrt{b^2 - 4ac}}{2a}$$

or $$x = \frac{-b \pm\sqrt{b^2 - 4ac}}{2a}$$

Solution of Cubic Equations

By definition, a cubic equation can be written in the form

$$ax^3 + bx^2 + cx + d = 0$$ *a, b, c,* and *d* are fixed numbers and $a \neq 0$.

Now we get into something of a bind. Although there is a perfectly respectable formula for the solutions, it is very complicated and involves the use of complex numbers rather heavily.[3] So we discuss instead a much simpler method that *sometimes* works nicely. Here is the method in a nutshell.

Solving Cubics by Finding One Factor

Start with a given cubic equation $ax^3 + bx^2 + cx + d = 0$.

Step 1: By trial and error, find one solution $x = s$. If a, b, c, and d are integers, the only possible *rational* solutions[4] are those of the form $s = \pm$(factor of d)/(factor of a).

Step 2: It is now possible to factor the cubic as

$$ax^3 + bx^2 + cx + d = (x - s)(ax^2 + ex + f) = 0.$$

To find $ax^2 + ex + f$, divide the cubic by $x - s$ using long division.[5]

Step 3: The factored equation says that either $x - s = 0$ or $ax^2 + ex + f = 0$. We already know that s is a solution, and now we see that the other solutions are the roots of the quadratic. Note that this quadratic may or may not have any real solutions, as usual.

Quick Example

To solve the cubic $x^3 - x^2 + x - 1 = 0$, we first find a single solution. Here, $a = 1$ and $d = -1$. Since the only factors of ±1 are ±1, the only possible rational

[3]It was when this formula was discovered in the 16th century that complex numbers were first taken seriously. Although we would like to show you the formula, it is too large to fit in this footnote.

[4]There may be *irrational* solutions, however; for example, $x^3 - 2 = 0$ has the single solution $x = \sqrt[3]{2}$.

[5]Alternatively, use "synthetic division," a shortcut that would take us too far afield to describe.

solutions are $x = \pm 1$. By substitution, we see that $x = 1$ is a solution. Thus, $(x - 1)$ is a factor. Dividing by $(x - 1)$ yields the quotient $(x^2 + 1)$. Thus,

$$x^3 - x^2 + x - 1 = (x - 1)(x^2 + 1) = 0$$

so that either $x - 1 = 0$ or $x^2 + 1 = 0$. Since the discriminant of the quadratic $x^2 + 1$ is negative, we don't get any real solutions from $x^2 + 1 = 0$, so the only real solution is $x = 1$.

Possible Outcomes When Solving a Cubic Equation

If you consider all the cases, there are three possible outcomes when solving a cubic equation:

1. One real solution (as in the Quick Example above)
2. Two real solutions (try, for example, $x^3 + x^2 - x - 1 = 0$)
3. Three real solutions (see the next example)

example 1

Solving a Cubic

Solve the cubic $2x^3 - 3x^2 - 17x + 30 = 0$.

Solution

First we look for a single solution. Here, $a = 2$ and $d = 30$. The factors of a are ± 1 and ± 2, and the factors of d are $\pm 1, \pm 2, \pm 3, \pm 5, \pm 6, \pm 15$, and ± 30. This gives us a large number of possible ratios: $\pm 1, \pm 2, \pm 3, \pm 5, \pm 6, \pm 15, \pm 30, \pm\frac{1}{2}, \pm\frac{3}{2}, \pm\frac{5}{2}$, and $\pm\frac{15}{2}$. Undaunted, we first try $x = 1$ and $x = -1$, getting nowhere. So we move on to $x = 2$, and we hit the jackpot because substituting $x = 2$ gives $16 - 12 - 34 + 30 = 0$. Thus, $(x - 2)$ is a factor. Dividing yields the quotient $2x^2 + x - 15$. Here is the calculation.

$$\begin{array}{r}
2x^2 + x - 15 \\
x - 2 \overline{\smash{)}\,2x^3 - 3x^2 - 17x + 30} \\
\underline{2x^3 - 4x^2} \\
x^2 - 17x \\
\underline{x^2 - 2x} \\
-15x + 30 \\
\underline{-15x + 30} \\
0
\end{array}$$

Thus,

$$2x^3 - 3x^2 - 17x + 30 = (x - 2)(2x^2 + x - 15) = 0$$

Setting the factors equal to zero gives either $x - 2 = 0$ or $2x^2 + x - 15 = 0$. We could solve the quadratic using the quadratic formula, but luckily, we notice that it factors as

$$2x^2 + x - 15 = (x + 3)(2x - 5)$$

Thus, the solutions are $x = 2$, $x = -3$, and $x = \frac{5}{2}$.

Solution of Higher-Order Polynomial Equations

Logically speaking, our next step should be to discuss quartics, then quintics (fifth-degree equations), and so on forever. Well, we've got to stop somewhere, and cubics may be as good a place as any. On the other hand, since we've gotten so far, we ought to at least tell you what is known about higher-order polynomials.

Quartics Just as in the case of cubics, there is a formula to find the solutions of quartics.[6]

Quintics and Beyond All good things must come to an end, we're afraid. It turns out that there is no "quintic formula." In other words, there is no single algebraic formula or collection of algebraic formulas that gives the solutions to all quintics. This question was settled by the Norwegian mathematician Niels Henrik Abel in 1824 after almost 300 years of controversy. (In fact, several notable mathematicians previously claimed to have devised formulas for solving the quintic, but these were all shot down by other mathematicians—this being one of the favorite pastimes of practitioners of our art.) The same negative answer applies to polynomial equations of degree 6 and higher. It's not that these equations don't have solutions, just that they can't be found using algebraic formulas.[7] However, there are certain special classes of polynomial equations that can be solved with algebraic methods. The way of identifying such equations was discovered around 1829 by the French mathematician Évariste Galois.[8]

A.5 exercises

Solve the equations in Exercises 1–12 for x (mentally, if possible).

1. $x + 1 = 0$ **2.** $x - 3 = 1$ **3.** $-x + 5 = 0$

4. $2x + 4 = 1$ **5.** $4x - 5 = 8$ **6.** $\frac{3}{4}x + 1 = 0$

7. $7x + 55 = 98$ **8.** $3x + 1 = x$ **9.** $x + 1 = 2x + 2$

10. $x + 1 = 3x + 1$ **11.** $ax + b = c \quad (a \neq 0)$

12. $x - 1 = cx + d \quad (c \neq 1)$

By any method, determine all possible real solutions of each equation in Exercises 13–30. Check your answers by substitution.

13. $2x^2 + 7x - 4 = 0$ **14.** $x^2 + x + 1 = 0$

15. $x^2 - x + 1 = 0$ **16.** $2x^2 - 4x + 3 = 0$

17. $2x^2 - 5 = 0$ **18.** $3x^2 - 1 = 0$

19. $-x^2 - 2x - 1 = 0$ **20.** $2x^2 - x - 3 = 0$

21. $\frac{1}{2}x^2 - x - \frac{3}{2} = 0$ **22.** $-\frac{1}{2}x^2 - \frac{1}{2}x + 1 = 0$

23. $x^2 - x = 1$ **24.** $16x^2 = -24x - 9$

25. $x = 2 - \dfrac{1}{x}$ **26.** $x + 4 = \dfrac{1}{x - 2}$

27. $x^4 - 10x^2 + 9 = 0$ **28.** $x^4 - 2x^2 + 1 = 0$

29. $x^4 + x^2 - 1 = 0$ **30.** $x^3 + 2x^2 + x = 0$

Find all possible real solutions of each of the equations in Exercises 31–44.

31. $x^3 + 6x^2 + 11x + 6 = 0$ **32.** $x^3 - 6x^2 + 12x - 8 = 0$

33. $x^3 + 4x^2 + 4x + 3 = 0$ **34.** $y^3 + 64 = 0$

35. $x^3 - 1 = 0$ **36.** $x^3 - 27 = 0$

37. $y^3 + 3y^2 + 3y + 2 = 0$ **38.** $y^3 - 2y^2 - 2y - 3 = 0$

39. $x^3 - x^2 - 5x + 5 = 0$ **40.** $x^3 - x^2 - 3x + 3 = 0$

41. $2x^6 - x^4 - 2x^2 + 1 = 0$ **42.** $3x^6 - x^4 - 12x^2 + 4 = 0$

43. $(x^2 + 3x + 2)(x^2 - 5x + 6) = 0$

44. $(x^2 - 4x + 4)^2(x^2 + 6x + 5)^3 = 0$

---

[6] See, for example, *First Course in the Theory of Equations* by L. E. Dickson (New York: Wiley, 1922), or *Modern Algebra* by B. L. van der Waerden (New York: Frederick Ungar, 1953).

[7] What we mean by an "algebraic formula" is a formula in the coefficients using the operations of addition, multiplication, division, and the taking of radicals. Mathematicians call the use of such formulas in solving polynomial equations "solution by radicals." If you were a math major, you would eventually go on to study this under the heading of Galois theory.

[8] Both Abel and Galois died young. Abel died of tuberculosis at the age of 26, and Galois was killed in a duel at the age of 21.

A.6 Solving Miscellaneous Equations

Equations often arise in calculus that are not polynomial equations of low degree. Many of these complicated-looking equations can be solved easily if you remember the following, which we used in the previous section.

Solving an Equation of the Form $P \cdot Q = 0$

If a product is equal to zero, then at least one of the factors must be zero. That is, if $P \cdot Q = 0$, then either $P = 0$ or $Q = 0$.

Quick Examples

1. $x^5 - 4x^3 = 0$

$x^3(x^2 - 4) = 0$ Factor the left-hand side.

Either $x^3 = 0$ or $x^2 - 4 = 0$ Either $P = 0$ or $Q = 0$.

$x = 0, 2, \text{or } -2$ Solve the individual equations.

2. $(x^2 - 1)(x + 2) + (x^2 - 1)(x + 4) = 0$

$(x^2 - 1)[(x + 2) + (x + 4)] = 0$ Factor the left-hand side.

$(x^2 - 1)(2x + 6) = 0$

Either $x^2 - 1 = 0$ or $2x + 6 = 0$ Either $P = 0$ or $Q = 0$.

$x = -3, -1, \text{or } 1$ Solve the individual equations.

example 1

Solving by Factoring

Solve $12x(x^2 - 4)^5(x^2 + 2)^6 + 12x(x^2 - 4)^6(x^2 + 2)^5 = 0$.

Solution

Again, we start by factoring the left-hand side:

$$\begin{aligned} 12x(x^2 - 4)^5(x^2 + 2)^6 + 12x(x^2 - 4)^6(x^2 + 2)^5 &= 12x(x^2 - 4)^5(x^2 + 2)^5[(x^2 + 2) + (x^2 - 4)] \\ &= 12x(x^2 - 4)^5(x^2 + 2)^5(2x^2 - 2) \\ &= 24x(x^2 - 4)^5(x^2 + 2)^5(x^2 - 1) \end{aligned}$$

Setting this equal to zero, we get

$$24x(x^2 - 4)^5(x^2 + 2)^5(x^2 - 1) = 0$$

which means that at least one of the factors of this product must be zero. Now it certainly cannot be the 24, but it could be the x: $x = 0$ is one solution. It could also be that

$$(x^2 - 4)^5 = 0$$

or

$$x^2 - 4 = 0$$

which has solutions $x = \pm 2$. Could it be that $(x^2 + 2)^5 = 0$? If so, then $x^2 + 2 = 0$, but this is impossible because $x^2 + 2 \geq 2$ no matter what x is. Finally, it could be that $x^2 - 1 = 0$, which has solutions $x = \pm 1$. This gives us five solutions to the original equation:

$$x = -2, -1, 0, 1, \text{or } 2$$

example 2

Solving by Factoring

Solve $(x^2 - 1)(x^2 - 4) = 10$.

Solution

Watch out! You may be tempted to say that $x^2 - 1 = 10$ or $x^2 - 4 = 10$, but this does not follow. If two numbers multiply to give 10, what must they be? There are lots of possibilities: 2 and 5, 1 and 10, −500,000 and −0.000002 are just a few. The fact that the left-hand side is factored is nearly useless to us if we want to solve this equation. What we have to do is multiply out, bring the 10 over to the left, and hope that we can factor what we get. Here goes:

$$x^4 - 5x^2 + 4 = 10$$

$$x^4 - 5x^2 - 6 = 0$$

$$(x^2 - 6)(x^2 + 1) = 0$$

(Here we used a sometimes useful trick that we mentioned in Section A.3: We treated x^2 like x and x^4 like x^2, so factoring $x^4 - 5x^2 - 6$ is essentially the same as factoring $x^2 - 5x - 6$.) *Now* we are allowed to say that one of the factors must be zero: $x^2 - 6 = 0$ has solutions $x = \pm\sqrt{6} = \pm 2.449\ldots$ and $x^2 + 1 = 0$ has no real solutions. Therefore, we get exactly two solutions: $x = \pm\sqrt{6} = \pm 2.449\ldots$.

To solve equations that involve rational expressions, the following rule is very useful.

Solving an Equation of the Form $P/Q = 0$

If $P/Q = 0$, then $P = 0$.

How else could a fraction equal zero? If that is not convincing, we multiply both sides by Q (which cannot be zero if the quotient is defined).

Quick Example

$$\frac{(x+1)(x+2)^2 - (x+1)^2(x+2)}{(x+2)^4} = 0$$

$$(x+1)(x+2)^2 - (x+1)^2(x+2) = 0 \qquad \text{If } P/Q = 0, \text{ then } P = 0.$$

$$(x+1)(x+2)[(x+2) - (x+1)] = 0 \qquad \text{Factor.}$$

$$(x+1)(x+2)(1) = 0$$

Now either $x + 1 = 0$ or $x + 2 = 0$, so $x = -1$ or $x = -2$. But $x = -2$ does not make sense in the original equation. It makes the denominator zero. So it is not a solution and $x = -1$ is the only solution.

example 3

Solving a Rational Equation

Solve $1 - \dfrac{1}{x^2} = 0$.

Solution

We can write 1 as $\frac{1}{1}$, so that we now have a difference of two rational expressions:

$$\frac{1}{1} - \frac{1}{x^2} = 0$$

To combine these we can put both over a common denominator of x^2, which gives

$$\frac{x^2 - 1}{x^2} = 0$$

Now we can set the numerator, $x^2 - 1$, equal to zero. Thus,

$$x^2 - 1 = 0$$

so $$(x - 1)(x + 1) = 0$$

giving $x = \pm 1$.

Before we go on . . . This equation could also have been solved by writing

$$1 = \frac{1}{x^2}$$

and then multiplying both sides by x^2.

---

example 4

Solve $\dfrac{2x - 1}{x} + \dfrac{3}{x - 2} = 0$.

Solution

We *could* first perform the subtraction on the left and then set the top equal to zero, but here is another approach. Subtracting the second expression from both sides gives

$$\frac{2x - 1}{x} = \frac{-3}{x - 2}$$

Cross-multiplying [multiplying both sides by both denominators—that is, by $x(x - 2)$], now gives

$$(2x - 1)(x - 2) = -3x$$

so $$2x^2 - 5x + 2 = -3x$$

Adding $3x$ to both sides gives the quadratic equation

$$2x^2 - 2x + 1 = 0$$

The discriminant is $(-2)^2 - 4 \cdot 2 \cdot 1 = -4 < 0$, so we conclude that there is no real solution.

Before we go on . . . Notice that when we said that $(2x - 1)(x - 2) = -3x$, we were *not* allowed to conclude that $2x - 1 = -3x$ or $x - 2 = -3x$.

example 5

Solve $\dfrac{\left(2x\sqrt{x+1} - \dfrac{x^2}{\sqrt{x+1}}\right)}{x+1} = 0$.

Solution

Setting the numerator equal to zero gives

$$2x\sqrt{x + 1} - \frac{x^2}{\sqrt{x + 1}} = 0$$

This still involves fractions. To get rid of the fractions we could put everything over a common denominator ($\sqrt{x+1}$) and then set the top equal to zero, or we could multiply the whole equation by that common denominator in the first place to clear fractions. If we do the second, we get

$$2x(x+1) - x^2 = 0$$
$$2x^2 + 2x - x^2 = 0$$
$$x^2 + 2x = 0$$

Factoring gives

$$x(x+2) = 0$$

so either $x = 0$ or $x + 2 = 0$, giving us $x = 0$ or $x = -2$. Again, one of these is not really a solution. The problem is that $x = -2$ cannot be substituted into $\sqrt{x+1}$ because we would then have to take the square root of -1, and we are not allowing ourselves to do that. Therefore, $x = 0$ is the only solution.

A.6 exercises

Solve the following equations.

1. $x^4 - 3x^3 = 0$
2. $x^6 - 9x^4 = 0$
3. $x^4 - 4x^2 = -4$
4. $x^4 - x^2 = 6$
5. $(x+1)(x+2) + (x+1)(x+3) = 0$
6. $(x+1)(x+2)^2 + (x+1)^2(x+2) = 0$
7. $(x^2+1)^5(x+3)^4 + (x^2+1)^6(x+3)^3 = 0$
8. $10x(x^2+1)^4(x^3+1)^5 - 10x^2(x^2+1)^5(x^3+1)^4 = 0$
9. $(x^3+1)\sqrt{x+1} - (x^3+1)^2\sqrt{x+1} = 0$
10. $(x^2+1)\sqrt{x+1} - \sqrt{(x+1)^3} = 0$
11. $\sqrt{(x+1)^3} + \sqrt{(x+1)^5} = 0$
12. $(x^2+1)\sqrt[3]{(x+1)^4} - \sqrt[3]{(x+1)^7} = 0$
13. $(x+1)^2(2x+3) - (x+1)(2x+3)^2 = 0$
14. $(x^2-1)^2(x+2)^3 - (x^2-1)^3(x+2)^2 = 0$
15. $\dfrac{(x+1)^2(x+2)^3 - (x+1)^3(x+2)^2}{(x+2)^6} = 0$
16. $\dfrac{6x(x^2+1)^2(x^2+2)^4 - 8x(x^2+1)^3(x^2+2)^3}{(x^2+2)^8} = 0$
17. $\dfrac{2(x^2-1)\sqrt{x^2+1} - \dfrac{x^4}{\sqrt{x^2+1}}}{x^2+1} = 0$
18. $\dfrac{4x\sqrt{x^3-1} - \dfrac{3x^4}{\sqrt{x^3-1}}}{x^3-1} = 0$
19. $x - \dfrac{1}{x} = 0$
20. $1 - \dfrac{4}{x^2} = 0$
21. $\dfrac{1}{x} - \dfrac{9}{x^3} = 0$
22. $\dfrac{1}{x} - \dfrac{1}{x+1} = 0$
23. $\dfrac{x-4}{x+1} - \dfrac{x}{x-1} = 0$
24. $\dfrac{2x-3}{x-1} - \dfrac{2x+3}{x+1} = 0$
25. $\dfrac{x+4}{x+1} + \dfrac{x+4}{3x} = 0$
26. $\dfrac{2x-3}{x} - \dfrac{2x-3}{x+1} = 0$

Answers to Selected Exercises

Chapter 1

1.1 exercises

1. a. 2 **b.** 0.5 **3. a.** −1.5 **b.** 8 **c.** −8 **5. a.** −7 **b.** −3 **c.** 1 **d.** $4y - 3$ **e.** $4(a + b) - 3$ **7. a.** 3 **b.** 6 **c.** 2 **d.** 6 **e.** $a^2 + 2a + 3$ **f.** $(x + h)^2 + 2(x + h) + 3$ **9. a.** 2 **b.** 0 **c.** 65/4 **d.** $x^2 + 1/x$ **e.** $(s + h)^2 + 1/(s + h)$ **f.** $(s + h)^2 + 1/(s + h) - (s^2 + 1/s)$ **11. a.** 1 **b.** 1 **c.** 0 **d.** 27 **13. a.** Yes; $f(4) = 63/16$ **b.** Not defined **c.** Not defined **15. a.** Not defined **b.** Not defined **c.** Yes; $f(-10) = 0$

17.

x	0	1	2	3	4	5	6	7	8	9	10
f(x)	5	1.1	−2.6	−6.1	−9.4	−12.5	−15.4	−18.1	−20.6	−22.9	−25

19.

x	0.5	1.5	2.5	3.5	4.5	5.5
h(x)	−0.6000	0.3846	0.7241	0.8491	0.9059	0.9360
x	6.5	7.5	8.5	9.5	10.5	
h(x)	0.9538	0.9651	0.9727	0.9781	0.9820	

21. a. $P(0) = 23$, $P(3) = 55$, $P(1.5) \approx 56$. Thus, Lotus's profit was $23 million in the year beginning January 1990, $55 million in the year beginning January 1993, and $56 million in the year beginning July 1991. **b.** $0 \le t \le 4$ **23. a.** $C(0) = 800$, $C(4) = 2800$, $C(5) = 4100$. Thus, there were 800 coffee shops in 1990, 2800 in 1994, and 4100 in 1995. **b.** 1996

c.

t	0	1	2	3	4	5	6	7	8	9	10
C(t)	800	1300	1800	2300	2800	4100	5400	6700	8000	9300	10,600

25. $T(26{,}000) = 3697.50 + 0.28(26{,}000 - 24{,}650) = \$4075.50$; $T(65{,}000) = 13{,}525.50 + 0.31(65{,}000 - 59{,}750) = \$15{,}153$ **27. a.** 358,600 **b.** 361,200 **c.** $6.00 **29. a.** $0 \le t \le 8$ **b.** $t \ge 0$ is not an appropriate domain because it would predict investments in South Africa into the indefinite future with no basis. (It would also lead to preposterous results for large values of t.) **31. a.** (2) **b.** $36.8 billion **33. a.** $12,000 **b.** $N(q) = 2000 + 100q^2 - 500q$; $N(20) = \$32{,}000$

35. a.

t	9	10	11	12	13	14	15	16	17	18	19	20
p(t)	35.0	59.4	73.5	82.1	87.5	91.0	93.4	95.1	96.2	97.1	97.7	98.2

b. 82.1% **c.** 14 months **37.** If weekly profit P is specified as a function of selling price s, then the independent variable is s and the dependent variable is P. **39.** As the text reminds us: To evaluate f of a quantity (such as $x + h$) replace x everywhere by the *whole quantity* $x + h$, getting $f(x + h) = (x + h)^2 - 1$. **41.** False. Functions with infinitely many points in their domain, such as $f(x) = x^2$, cannot be specified numerically.

1.2 exercises

1. a. 20 **b.** 30 **c.** 30 **d.** 20 **e.** 0 **3. a.** −1 **b.** 1.25 **c.** 0 **d.** 1 **e.** 0 **5. a.** (I) **b.** (IV) **c.** (V) **d.** (VI) **e.** (III) **f.** (II)

7.

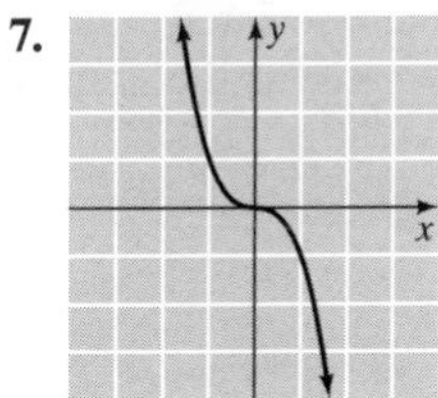

9.

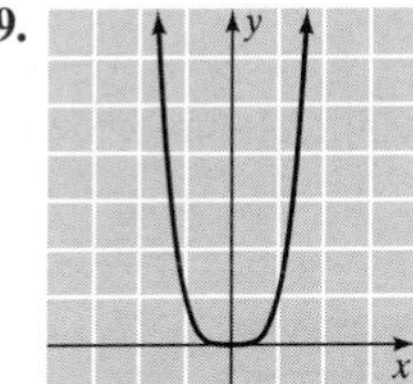

11.

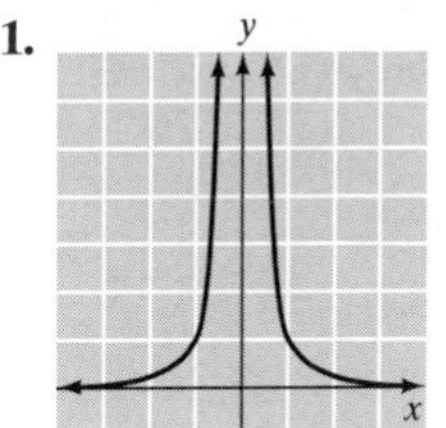

13. a. −1 **b.** 2 **c.** 2

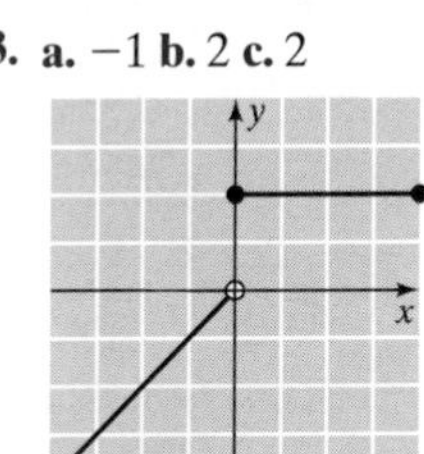

15. a. 0 **b.** 2 **c.** 3 **d.** 3

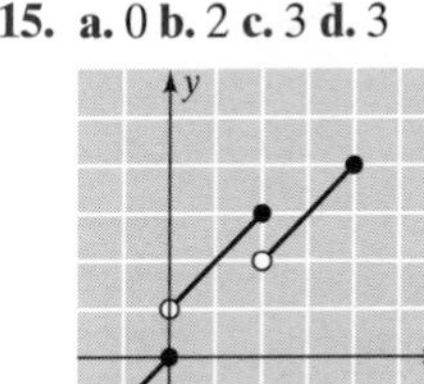

17. a. 1 **b.** 0 **c.** 1

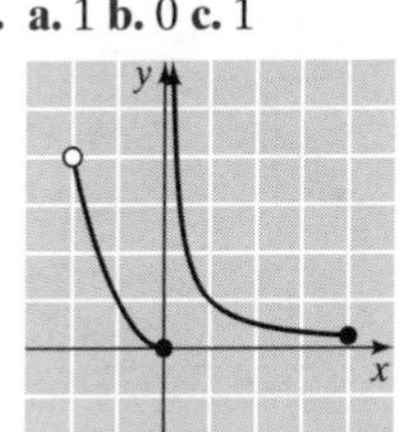

19. $f(2) \approx 200$, $f(6) \approx 800$, $f(11.5) \approx 1000$. In 1982, 200,000 sport utility vehicles were sold; in 1986, 800,000 were sold; and in the year beginning July 1991, 1 million were sold.

21. The largest value is $f(8) \approx 950$. From 1986 through 1990, the year in which the most sport utility vehicles were sold was 1988, when 950,000 were sold. **23.** 1987 and 1991 **25. a.**

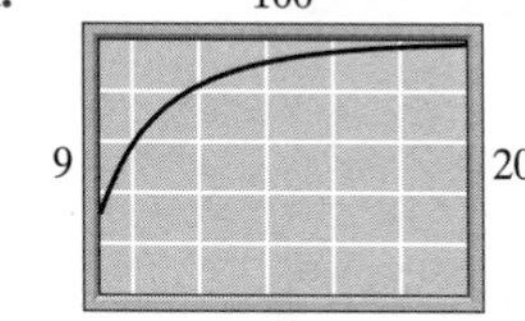

b. 82% **c.** 14 months

27. (A) is the best choice; the other models predict either perpetually increasing tourism or perpetually decreasing tourism. In fact, there is a parabola that passes exactly through the three data points.
29. 1999

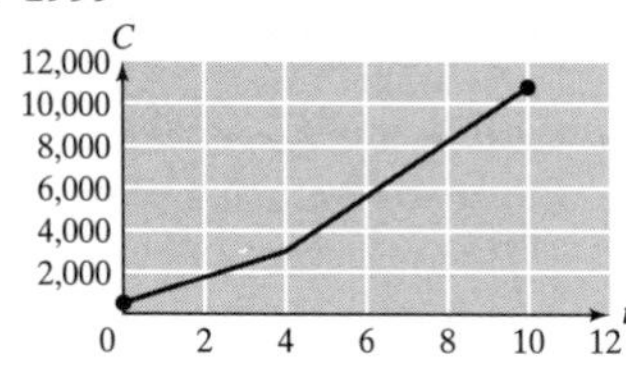

31. True. We can construct a table of values from any graph by reading off a set of values. **33.** False. In a numerically specified function, only certain values of the function are specified, giving only certain points on the graph. **35.** They are different portions of the graph of the associated equation $y = f(x)$. **37.** The graph of $g(x)$ is the same as the graph of $f(x)$ but shifted 5 units to the right.

23.

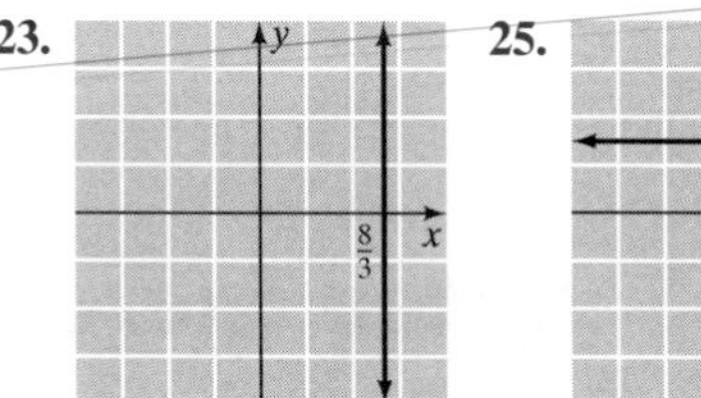

25.

27.

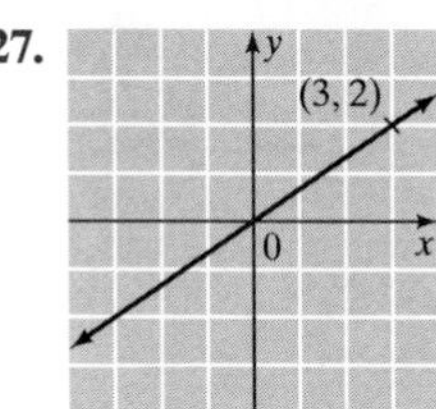

29. 2 **31.** 2 **33.** -2 **35.** Undefined **37.** 1.5 **39.** -0.09 **41.** $\frac{1}{2}$ **43.** $(d - b)/(c - a)$ **45. a.** 1 **b.** $\frac{1}{2}$ **c.** 0 **d.** 3 **e.** $-\frac{1}{3}$ **f.** -1 **g.** Undefined **h.** $-\frac{1}{4}$ **i.** -2 **47.** $y = 3x$ **49.** $y = \frac{1}{4}x - 1$ **51.** $y = 10x - 203.5$ **53.** $y = -5x + 6$ **55.** $y = -3x + 2.25$ **57.** $y = -x + 12$ **59.** $y = 2x + 4$ **61.** $f(x) = -(a/b)x + (c/b)$. If $b = 0$, then a/b is undefined, and y cannot be specified as a function of x. (The graph of the resulting equation would be a vertical line.) **63.** If, in a straight line, y is increasing three times as fast as x, then its slope is 3. **65.** If m is positive, then y will increase as x increases; if m is negative, then y will decrease as x increases; if m is zero, then y will not change as x changes.

1.3 exercises

1.

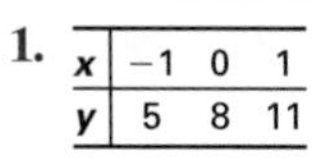

x	−1	0	1
y	5	8	11

$m = 3$

3.

x	2	3	5
f(x)	−1	−2	−4

$m = -1$

5.

x	−2	0	2
f(x)	4	7	10

$m = \frac{3}{2}$

7. $f(x) = -x/2 - 2$ **9.** $f(0) = -5$, $f(x) = -x - 5$ **11.** f is linear, $f(x) = 4x + 6$ **13.** g is linear, $g(x) = 2x - 1$

15.

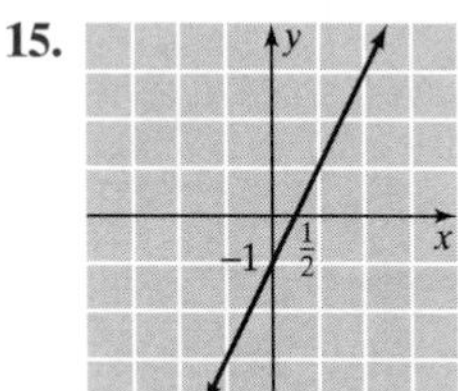

17.

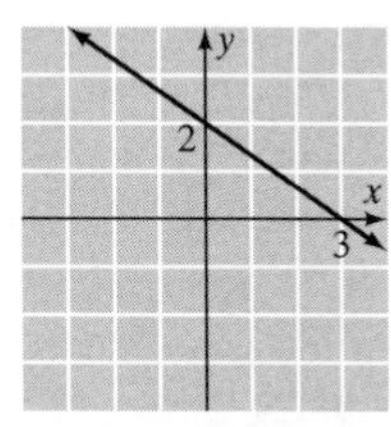

19.

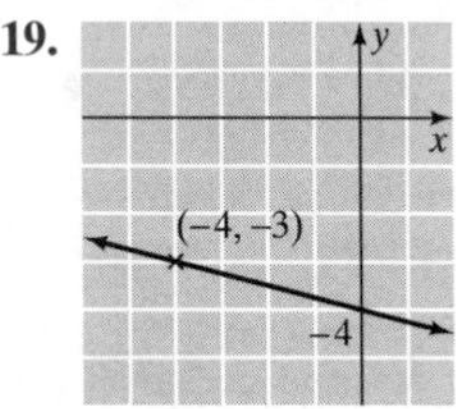

21.

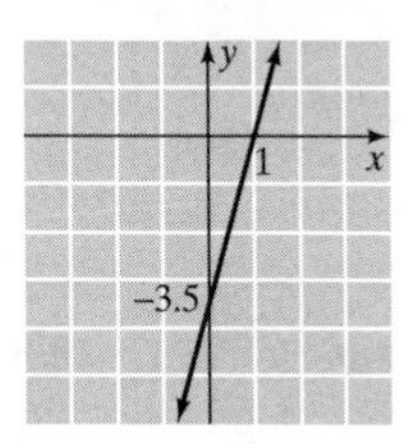

1.4 exercises

1. $C(x) = 1500x + 1200$ per day **a.** \$5700 **b.** \$1500 **c.** \$1500 **3.** Fixed cost = \$8000, marginal cost = \$25 per bicycle **5. a.** $C(x) = 0.4x + 70$, $R(x) = 0.5x$, $P(x) = 0.1x - 70$ **b.** $P(500) = -20$; a loss of \$20 **c.** 700 copies **7.** $q = -40p + 2000$ **9.** $q = -0.15p + 56.25$. The demand equation predicts that demand increases as the price goes down. However, the first-quarter 1998 price was lower than the third quarter 1997 price, and yet the demand was also lower. **11. a.** Demand: $q = -60p + 150$; supply: $q = 80p - 60$ **b.** \$1.50 each **13. a.** The two points corresponding to $x = 9$ and $x = 10$ **b.** Sales revenues at Creative Labs showed the largest quarterly increase (approximately \$150 million) during the second quarter in 1995. **15.** $n = 1600t/3 + 400$. The slope is the rate of change of the number of people who pass the exam, and its units are people per year. **17. a.** $s = 5t + 50$. The rate of increase of Medicare spending was \$5 billion per year. **b.** \$235 billion **19. a.** 2.5 feet per second **b.** 20 feet along the track **c.** After 6 seconds **21. a.** 130 miles per hour **b.** $s = 130t - 1300$ **23.** $F = 1.8C + 32$; 86°F, 72°F, 14°F, 7°F **25.** $I(N) = 0.05N + 50{,}000$; $N = \$1{,}000{,}000$; marginal income is $m = 5¢$ per dollar of net profit. **27. a.** $c = 0.05n + 1.7$ **b.** The slope represents the increase in daily circulation per additional newspaper that a company publishes. **c.** The model predicts that circulation would increase to 6.35 million. **29. a.** $P = 0.2R - 25$ (in millions of dollars) **b.** $C = 0.8R + 25$ (in millions of dollars) **c.** \$25 million **31.** $T(r) = \frac{1}{4}r + 45$; $T(100) = 70°\text{F}$ **33.** $P(x) = 100x - 5132$, with domain [0, 405]. For profit, $x \geq 52$. **35.** 5000 units **37.** $FC/(SP - VC)$ **39.** $P(x) = 579.7x - 20{,}000$, with domain $x \geq 0$; $x = 34.50$ grams per day for break-even **41. a.** The slope 31 in the equation for $N(t)$ is the number of new venture capital firms per year. The slope 2.4 in the equation for $R(t)$ is the additional amount of money controlled by venture capital firms each year. **b.** The intercept 145 in the equation for $N(t)$ represents the number of new venture capital firms at time $t = 0$ (1980). The intercept 7.3 in the equation for $R(t)$ represents the amount of money controlled by venture capital firms at time $t = 0$ (1980). **c.** The reason for the apparent discrepancy is that different scales are used on the two graphs: 100 venture capital firms correspond to \$10 billion. **43.** Decreasing by 3.3 percentage points per year

45. $C(t) = \begin{cases} -1400t + 30{,}000 & \text{if } 0 \leq t \leq 5 \\ 7400t - 14{,}000 & \text{if } 5 < t \leq 10 \end{cases}$

$C(3) = 25{,}800$ students

47. $d(r) = \begin{cases} -40r + 74 & \text{if } r \leq 1.3 \\ \dfrac{130r}{3} - \dfrac{103}{3} & \text{if } r > 1.3 \end{cases}$ $\quad d(1) = 34\%$

49. It must increase by ten units each day, including the third. **51.** Increasing the number of items from the break-even results in a profit. Since the slope of the revenue graph is greater than the slope of the cost graph, it is higher than the cost graph to the right of the point of intersection and hence corresponds to a profit.

1.5 exercises

1. $y = 1.5x - 0.6667$

3. $y = 0.4118x + 0.9706$

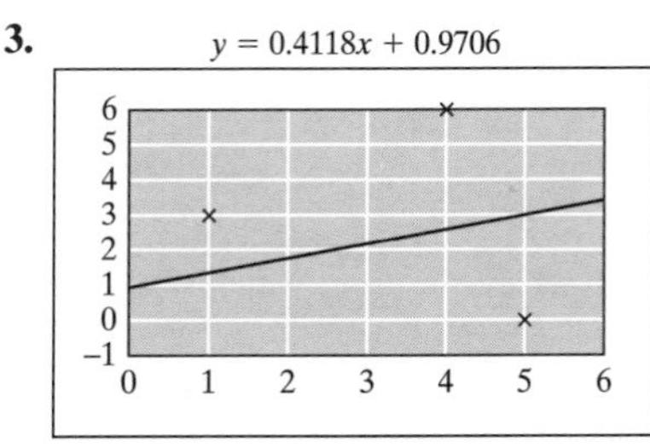

5. a. $r = 0.9959$ (best, not perfect) **b.** $r = 0.9538$ **c.** $r = 0.3273$ (worst) **7.** $y = -0.435x + 7.98$, $y(10) \approx 3.6\%$ **9.** $y = 1.46x + 6.97$, $y(10) \approx 22$ **11.** 9900 phone lines **13.** Increasing at the rate of 0.29 years per year **15.** $P = 101.08t - 330.33$, $P(13) \approx \$984$ million **19.** The regression line is the line that passes through the given points. **21.** No. The regression line through $(-1, 1)$, $(0, 0)$, and $(1, 1)$ passes through none of these points.

Chapter 1 Review Test

1. a. **b.** **c.** **d.**

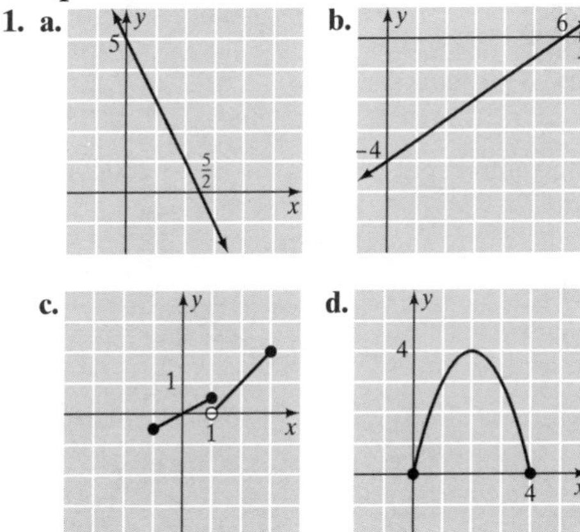

2. a. $y = -3x + 11$ **b.** $y = -x + 1$ **c.** $y = \frac{1}{2}x + \frac{3}{2}$ **d.** $y = \frac{1}{2}x - 1$ **3. a.** 500 hits: 10 books per day; 1000 hits: 20 books per day; 1500 hits: 32.5 books per day **b.** Book sales are increasing at a rate of 0.025 books per hit when the number of hits is between 1000 and 2000 per day **c.** 1400 hits per day **4. a.** $h = 0.05c + 1800$ **b.** 2100 hits per day **c.** \$14,000 per month **5. a.** 2080 hits per day **b.** Probably not. This model predicts that web site traffic will start to decrease as advertising expenditures increase beyond \$8500 per month and then drop toward zero. **6. a.** Cost: $C = 900 + 4x$, revenue: $R = 5.5x$, profit: $P = 1.5x - 900$ **b.** 600 books per month **c.** 900 books per month **7. a.** $q = -60p + 950$ **b.** 50 novels per month **c.** \$10, for a profit of \$1200 **8. a.** $q = -53.3945p + 909.8165$ **b.** 483 novels per month

Chapter 2

2.1 exercises

1. Vertex: $\left(-\frac{3}{2}, -\frac{1}{4}\right)$;
y-intercept: 2;
x-intercepts: $-2, -1$

3. Vertex: $(2, 0)$;
y-intercept: -4;
x-intercept: 2

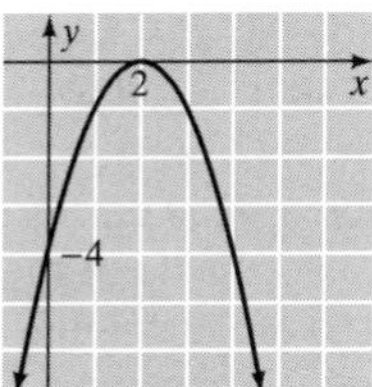

5. Vertex: $(-20, 900)$; y-intercept: 500; x-intercepts: $-50, 10$

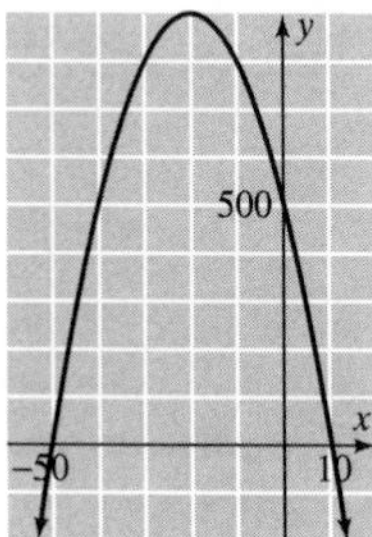

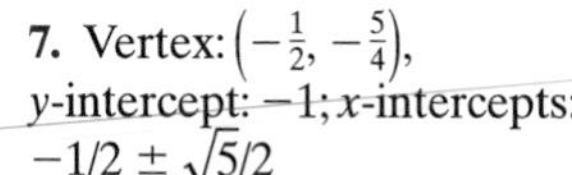

7. Vertex: $\left(-\frac{1}{2}, -\frac{5}{4}\right)$, y-intercept: -1; x-intercepts: $-1/2 \pm \sqrt{5}/2$

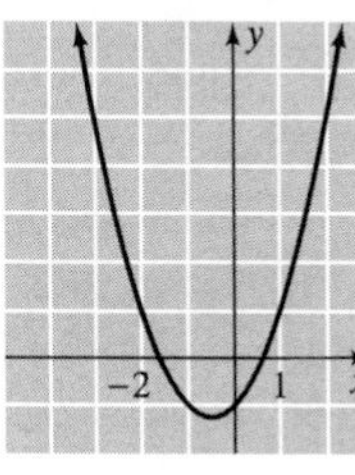

9. Vertex: $(0, 1)$; y-intercept: 1; no x-intercepts

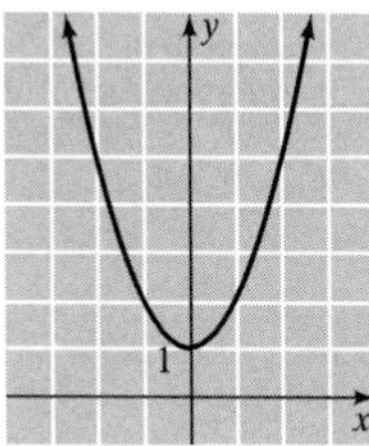

11. $R = -4p^2 + 100p$; maximum revenue when $p =$ \$12.50

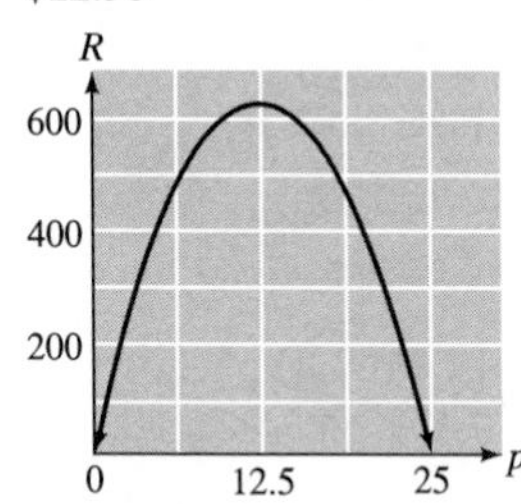

13. $R = -2p^2 + 400p$; maximum revenue when $p =$ \$100

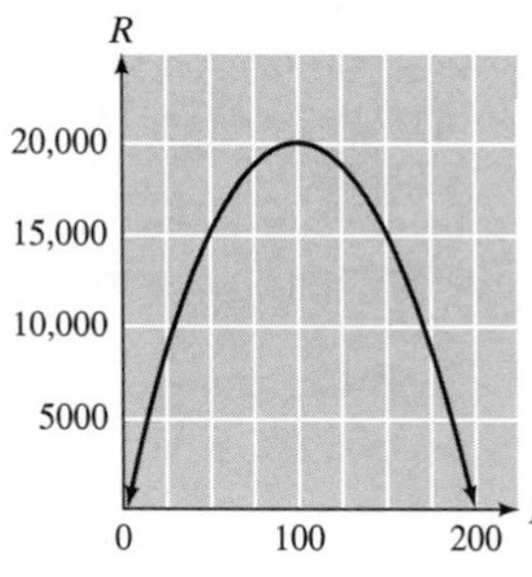

15. 1985 ($t = 15$); 3525 pounds

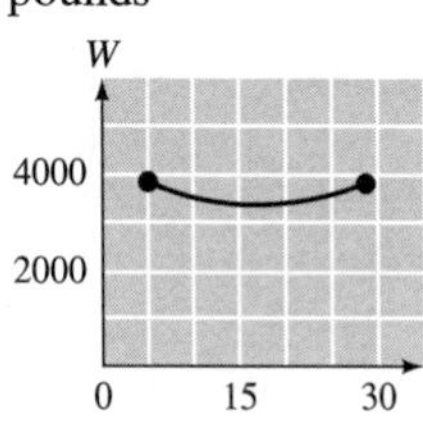

17. 5000 pounds. The model is not trustworthy for vehicles that weigh more than 5000 pounds because it predicts increasing fuel economy with increasing weight, and 5000 is close to the upper limit of the domain of the function. **19.** Maximum revenue when $p =$ \$140; $R =$ \$9800 **21.** Maximum revenue with 70 houses; $R =$ \$9,800,000 **23. a.** $q = -560x + 1400$; $R = -560x^2 + 1400x$ **b.** $P = -560x^2 + 1400x - 30$; $x =$ \$1.25; $P =$ \$845 per month **25.** $C = -200x + 620$; $P = -400x^2 + 1400x - 620$; $x =$ \$1.75 per hit; $P =$ \$605 per month **27. a.** $q = -10p + 400$ **b.** $R = -10p^2 + 400p$ **c.** $C = -30p + 4200$ **d.** $P = -10p^2 + 430p - 4200$; $p =$ \$21.50 **29.** The x-coordinate of the vertex represents the unit price that leads to the maximum revenue, the y-coordinate of the vertex gives the maximum possible revenue, the x-intercepts give the unit prices that result in zero revenue, and the y-intercept gives the revenue resulting from zero unit price (which is obviously zero). **31.** If $q = mp + b$ (with $m < 0$), then the revenue is given by $R = pq = mp^2 + bp$. This is the equation of a parabola with $a = m < 0$ and so is concave down. Thus, the vertex is the highest point on the parabola, showing that there is a single highest value for R—namely, the y-coordinate of the vertex. **33.** Since $R = pq$, the demand must be given by $q = \dfrac{R}{p} = \dfrac{-50p^2 + 60p}{p} = -50p + 60$.

2.2 exercises

1.

x	−3	−2	−1	0	1	2	3
$f(x)$	$\frac{1}{64}$	$\frac{1}{16}$	$\frac{1}{4}$	1	4	16	64

3.

x	−3	−2	−1	0	1	2	3
$f(x)$	27	9	3	1	$\frac{1}{3}$	$\frac{1}{9}$	$\frac{1}{27}$

5.

x	−3	−2	−1	0	1	2	3
$g(x)$	$\frac{1}{4}$	$\frac{1}{2}$	1	2	4	8	16

7.

x	−3	−2	−1	0	1	2	3
$h(x)$	−24	−12	−6	−3	$-\frac{3}{2}$	$-\frac{3}{4}$	$-\frac{3}{8}$

9.

x	−3	−2	−1	0	1	2	3
$r(x)$	$-\frac{7}{8}$	$-\frac{3}{4}$	$-\frac{1}{2}$	0	1	3	7

11.

x	−3	−2	−1	0	1	2	3
$s(x)$	$\frac{1}{16}$	$\frac{1}{8}$	$\frac{1}{4}$	$\frac{1}{2}$	1	2	4

13.

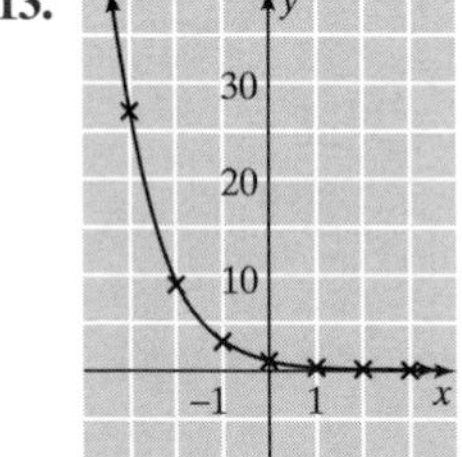

15.

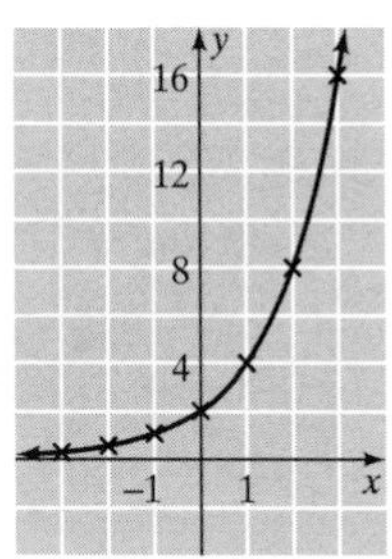

17.

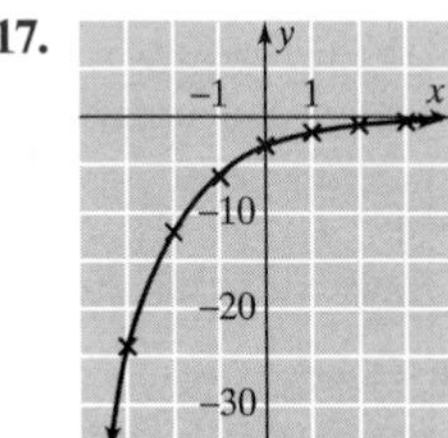

19.

x	−3	−2	−1	0	1	2	3
$f(x)$	403.4	54.60	7.389	1	0.1353	0.01832	0.002479

21.

x	−3	−2	−1	0	1	2	3
$h(x)$	4662	280.0	16.82	1.01	0.06066	0.003643	0.0002188

23.

x	−3	−2	−1	0	1	2	3
$r(x)$	9.781	16.85	29.02	50	86.13	148.4	255.6

25. $f(x) = 500(0.5)^x$ **27.** $f(x) = 10(3)^x$ **29.** $f(x) = 500(0.45)^x$ **31.** $f(x) = -100(1.1)^x$ **33.** `2^(x-1)` **35.** `2/(1-2^(-4*x))` **37.** `(3+x)^(3*x)/(x+1)` or `((3+x)^(3*x))/(x+1)` **39.** `2*e^((1+x)/x))`

41.

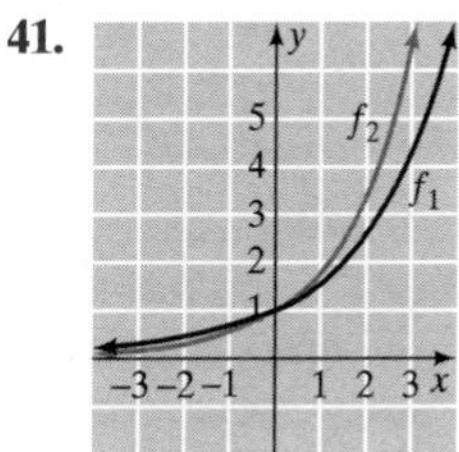

43.

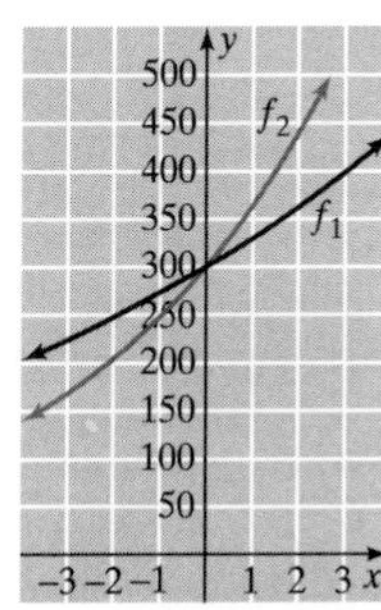

45.

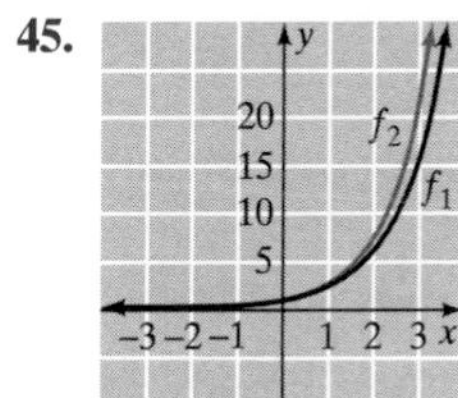

47.

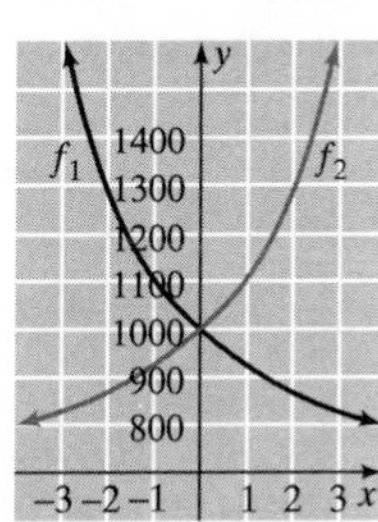

49. $y = 3(2^{(x-1)/2})$ **51.** $y = 5(4^{(x+1)/5})$ **53.** $y = 3\left(\frac{2}{3}\right)^{(x-2)/4}$ **55.** \$7081 **57.** 56.8, 31.0, 16.9, 9.25, 5.05, 2.76, 1.51 grams **59.** 20,000 years **61.** $y = 1000(2^{t/3})$; 65,536,000 bacteria after 2 days **63.** 53 mg **65. a.** $P = (125/3)t + 350$ **b.** $P = 350(12/7)^{t/6}$. The exponential model is a better fit. **67. a.** $y = 50{,}000(1.5^{t/2})$, t = time in years since 2 years ago **b.** 91,856 tags **69.** 311 million **71. a.** \$21.99 **b.** \$22.14, which is 15¢ better than Quarterly S&L can do. **73.** \$838.14 **75. a.** 472, 540, 617, 705 ppm **b.** 2100 ($t = 350$) **77.** 2005 **79.** (b) **81.** Linear functions better: cost functions where there is a fixed cost and a variable cost; simple interest, where interest is paid on the original amount invested. Exponential functions better: compound interest, population growth. (In both of these, the rate of growth depends on the present number of items rather than on some fixed quantity.) **83.** This reasoning is suspect. The bank need not use its computer resources to update all the accounts every minute; it can instead use the continuous compounding formula to calculate the balance in any account at any time.

2.3 exercises

1.

Logarithmic form	$\log_{10} 10{,}000 = 4$	$\log_4 16 = 2$	$\log_3 27 = 3$
Logarithmic form	$\log_5 5 = 1$	$\log_7 1 = 0$	$\log_4 \frac{1}{16} = -2$

3.

Exponential form	$(0.5)^2 = 0.25$	$5^0 = 1$	$10^{-1} = 0.1$
Exponential form	$4^3 = 64$	$2^8 = 256$	$2^{-2} = \frac{1}{4}$

5. 3.36 years **7.** 11 years **9.** 2.03 years **11.** 62,814 years old **13.** 8 years **15.** 132 months **17.** 5 years **19.** 3 years **21.** 13.43 years **23.** 55,300 years old **25.** 3.8 months **27.** 2360 million years **29.** 3.89 days **31.** 3.2 hours **33.** $P = 23{,}200e^{-0.2261439t}$; 2400 people in 1997 **35.** The least-squares model gives a higher employment figure for 1997. **37.** 2002 **39. a.** About 5.012×10^{16} joules of energy **b.** About 2.24% **d.** 31.62 **41. a.** 75 dB, 69 dB, 61 dB **b.** $D = 95 - 20 \log r$ **c.** 57,000 feet **43.** The logarithm of a negative number, if it were defined, would be the power to which a base must be raised to give that negative number. But raising a base to a power never results in a negative number, so there can be no such number as the logarithm of a negative number. **45.** $\log_4 y$ **47.** 8 **49.** x

2.4 exercises

1.

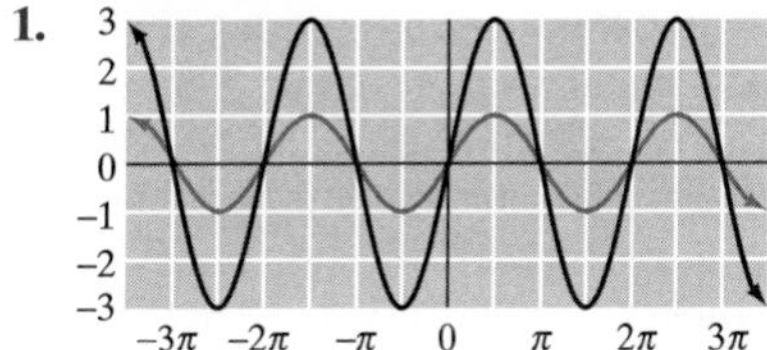

3.

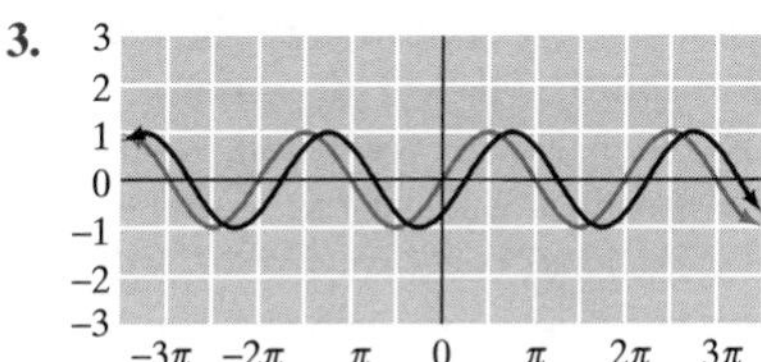

5.

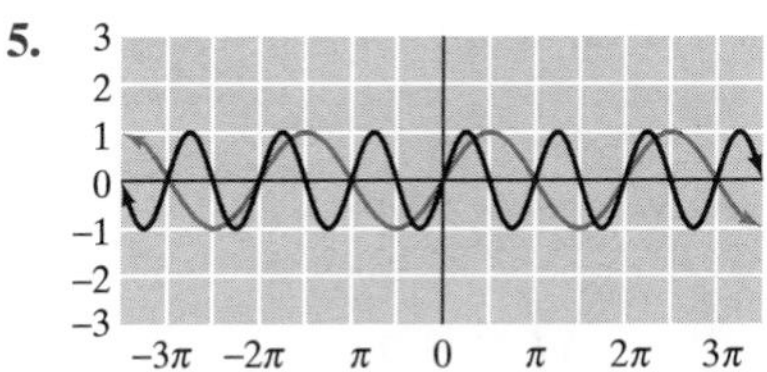

7. 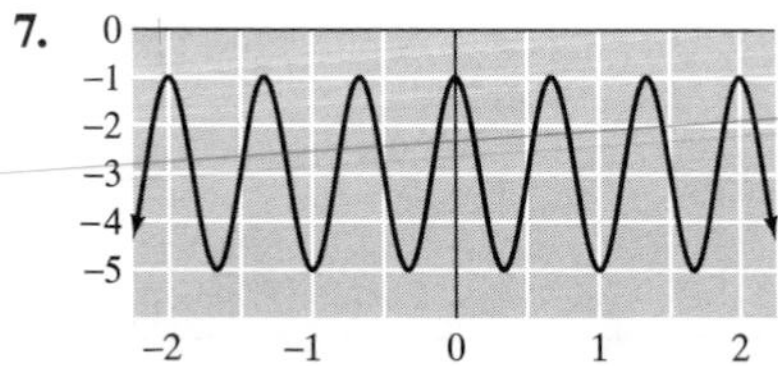

9. 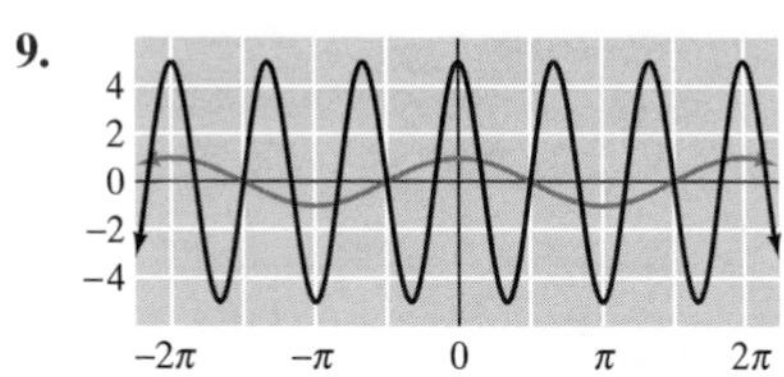

11. 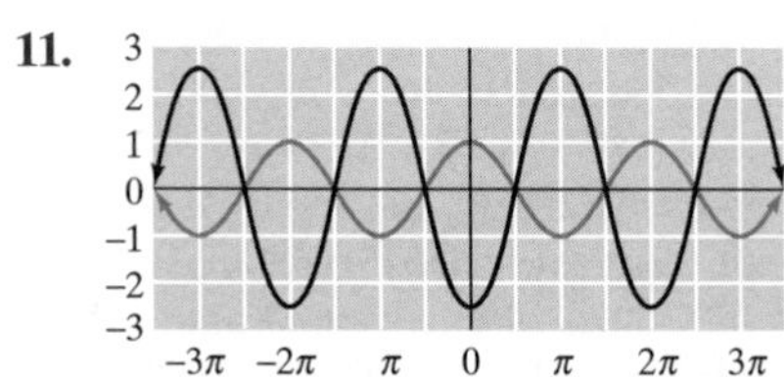

13. $f(x) = 1 + \sin(2\pi x)$ **15.** $f(x) = 1.5 \sin[4\pi(x - 0.25)]$ **17.** $f(x) = 50 \sin[\pi(x - 5)/10] - 50$ **19.** $f(x) = \cos(2\pi x)$ **21.** $f(x) = 1.5 \cos[4\pi(x - 0.375)]$ **23.** $f(x) = 40 \cos[\pi(x - 10)/10] + 40$ **25.** $f(t) = 4.2 \sin(\pi/2 - 2\pi t) + 3$ **27.** $g(x) = 4 - 1.3 \sin[\pi/2 - 2.3(x - 4)]$ **31.** $\frac{\sqrt{3}}{2}$ **37.** $\tan(x + \pi) = \tan(x)$

39. a. Maximum sales occurred when $t \approx 4.5$ (during the first quarter of 1996). Minimum sales occurred when $t \approx 2.2$ (during the third quarter of 1995) and $t \approx 6.8$ (during the third quarter of 1996). **b.** Maximum quarterly revenues were \$0.561 billion; minimum quarterly revenues were \$0.349 billion. **c.** Maximum: $0.455 + 0.106 = 0.561$; minimum: $0.455 - 0.106 = 0.349$ **41.** Amplitude = 0.106, vertical offset = 0.455, phase shift = 1.16, angular frequency = 1.39, period = 4.52. In 1995 and 1996, quarterly revenue from the sale of computers at Computer City fluctuated in cycles of 4.52 quarters about a baseline of \$0.455 billion. Every cycle, quarterly revenue peaked at \$0.561 billion (\$0.106 above the baseline) and dipped to a low of \$0.349 billion. During the second quarter of 1995 ($t = 1.16$) quarterly revenues were at the baseline level and on an upward cycle. **43.** $s(t) = 7.5 \sin[2\pi(t - 9)/12] + 87.5$ **45.** $s(t) = 7.5 \cos(\pi t/6) + 87.5$ **47.** $d(t) = 5 \sin[2\pi(t - 1.625)/13.5] + 10$ **49. a.** $u(t) = 2.5 \sin[2\pi(t - 0.75)] + 7.5$ **b.** $c(t) = 1.04^t[2.5 \sin(2\pi(t - 0.75)) + 7.5]$

51. a. 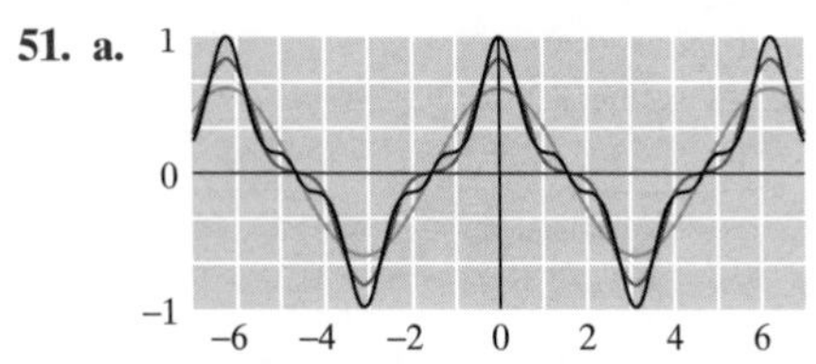

b. $$y_{11} = \frac{2}{\pi}\cos x + \frac{2}{3\pi}\cos 3x + \frac{2}{5\pi}\cos 5x + \frac{2}{7\pi}\cos 7x + \frac{2}{9\pi}\cos 9x + \frac{2}{11\pi}\cos 11x$$

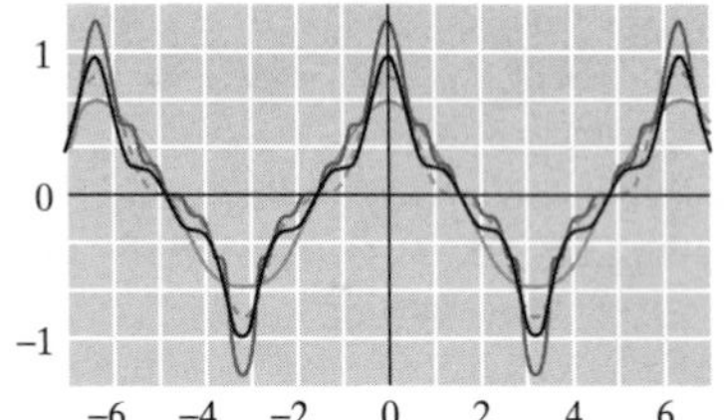

c. $$y_{11} = \frac{6}{\pi}\cos\frac{x}{2} + \frac{6}{3\pi}\cos\frac{3x}{2} + \frac{6}{5\pi}\cos\frac{5x}{2} + \frac{6}{7\pi}\cos\frac{7x}{2} + \frac{6}{9\pi}\cos\frac{9x}{2} + \frac{6}{11\pi}\cos\frac{11x}{2}$$

53. The period is approximately 12.6 units. **55.** Lows: $B - A$; highs: $B + A$ **57.** He is correct. The other trig functions can be obtained from the sine function by first using the formula $\cos x = \sin(x + \pi/2)$ to obtain cosine, and then using the formulas

$$\tan x = \frac{\sin x}{\cos x}, \quad \text{cotan } x = \frac{\cos x}{\sin x},$$

$$\sec x = \frac{1}{\cos x}, \quad \text{cosec } x = \frac{1}{\sin x}$$

to obtain the rest. **59.** The largest B can be is A. Otherwise, if B is larger than A, then the low figure for sales would have the negative value of $A - B$.

Chapter 2 Review Test

1. a. 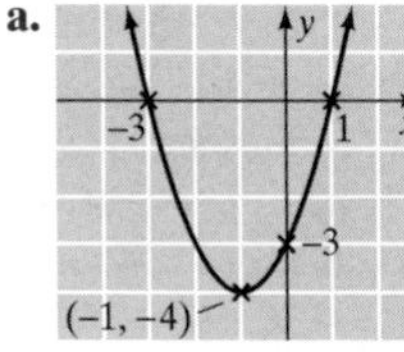

b. 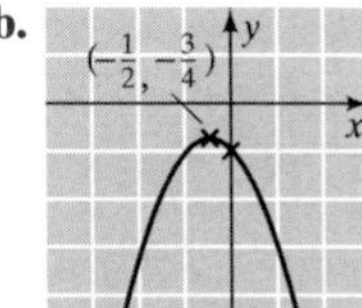

2. a. $f(x) = \frac{2}{3}3^x$ **b.** $f(x) = 10\sqrt{2}(1/\sqrt{2})^x$ **3. a.** $f(x) = 1 + 2\sin x$ **b.** $f(x) = 10 \sin(2x + \pi/2)$ **4. a.** \$8500 per month; an average of approximately 2100 hits per day **b.** \$29,049 per month **c.** The fact that -0.000005, the coefficient of c^2, is negative **5. a.** $R = -60p^2 + 950p$; $p = \$7.92$ per novel; monthly revenue = \$3760.42 **b.** $P = -60p^2 + 1190p - 4700$; $p = \$9.92$ per novel; monthly profit = \$1200.42 **6. a.** \$63,496.04 **b.** 7.0 hours **c.** 32.8 hours **7. a.** 10,527 **b.** At 8.5 weeks **8.** $s(t) = 10{,}500 + 1500 \sin(0.0806t - 1.57)$

Chapter 3

3.1 exercises

1. −3 **3.** 0.3 **5.** −$25,000 per month **7.** −200 items per dollar **9.** $-\frac{1}{3}$% per year **11.** 0.75% increase in unemployment per 1% increase in the deficit **13.** 4 **15.** 2 **17.** $\frac{7}{3}$
19. Difference quotient $(dq) = 2h$

h	dq
1	2
0.1	0.2
0.01	0.02
0.001	0.002
0.0001	0.0002

21. $dq = -\dfrac{1}{2(2+h)}$

h	dq
1	−0.1667
0.1	−0.2381
0.01	−0.2488
0.001	−0.2499
0.0001	−0.24999

23. $dq = 8 + h$

h	dq
1	9
0.1	8.1
0.01	8.01
0.001	8.001
0.0001	8.0001

25. a. 1990–1992. The annual profit was increasing most rapidly, at an average rate of $28.5 million per year, during the period 1990–1992. **b.** 1992–1994. The annual profit was decreasing most rapidly, at an average rate of $50 million per year, during the period 1992–1994. **27.** 32.5 firms per year **29.** 1988–1989; 100 firms per year **31.** 1989–1993 **33.** $R(t)/N(t)$ measures the average amount of money controlled by each venture capital firm. **35.** The index was increasing at an average rate of 300 points per day. **37.** −$0.24 billion per year. SAB's annual revenue was decreasing at an average rate of $0.24 billion per year over the period January 1991 through January 1995. **39.** The average rate of change of f over an interval $[a, b]$ can be determined numerically, using a table of values; graphically, measuring the slope of the corresponding line segment through two points on the graph; or algebraically, using an algebraic formula for the function. Of these, the least precise is the graphical method because it relies on reading coordinates of points on a graph. **41.** 6 units of quantity A per unit of quantity C

3.2 exercises

1.

h	1	0.1	0.01
Average rate	39	39.9	39.99

Instantaneous rate = 40 rupees per day

3.

h	1	0.1	0.01
Average rate	140	66.2	60.602

Instantaneous rate = 60 rupees per day

5.

h	10	1
C_{ave}	4.799	4.7999

$C'(1000) = \$4.8$ per item

7.

h	10	1
C_{ave}	99.91	99.90

$C'(100) = \$99.9$ per item

9. −2 **11.** −1.5 **13.** 3 **15.** −5 **17.** 16 **19.** 0 **21.** −1/400 **23.** 1.000 **25.** 1.000 **27.** 1.000 **29.** 2.000 **31. a.** −$0.24 billion per year. SAB's annual revenue was decreasing at an average rate of $0.24 billion per year over the period January 1991 through January 1995. **b.** $0.28 billion per year. SAB's annual revenue was increasing at a rate of $0.28 billion per year in January 1991. **c.** Annual revenues slowed, and then decreased after January 1991. **33.** $q(100) = 50{,}000$, $\left.\dfrac{dq}{dp}\right|_{p=100} = -500$. A total of 50,000 pairs of sneakers can be sold at a price of $100, but the demand is decreasing at a rate of 500 pairs per $1 increase in the price. **35. a.** −96 ft/s **b.** −128 ft/s **37.** $L(0.95) = 31.2$ meters and $L'(0.95) = -304.2$ meters/warp. Thus, at a speed of warp 0.95, the spaceship has an observed length of 31.2 meters and its length is decreasing at a rate of 304.2 meters per unit warp, or 3.042 meters per increase in speed of 0.01 warp. **39. a.** 60% of children can speak at the age of 10 months. At the age of 10 months, this percentage is increasing by 18.2 percentage points per month. **b.** As t increases, p approaches 100 percentage points (all children eventually learn to speak) and dp/dt approaches zero because the percentage stops increasing. **41.** $A(0) = 4.5$ million; $A'(0) = 60{,}000$ **43.** $S(5) \approx 109$, $\left.\dfrac{dS}{dt}\right|_{t=5} \approx 9.1$. After 5 weeks, sales are 109 pairs of sneakers per week, and sales are increasing at a rate of 9.1 pairs per week each week. **45.** $q(2) \approx 8166$, $\left.\dfrac{dq}{dt}\right|_{t=2} \approx 599.2$. Thus, 2 months after the introduction, 8166 video game units have been sold, and total sales are growing at a rate of 599.2 units per month. **47.** If $f(x) = mx + b$, then its average rate of change over any interval $[x, x+h]$ is $[m(x+h) + b - (mx+b)]/h = m$. Since this does not depend on h, the instantaneous rate is also equal to m. **49.** The difference quotient is not defined when $h = 0$ because there is no such number as 0/0.

3.3 exercises

1. a. R **b.** P **3. a.** P **b.** R **5. a.** Q **b.** P **7. a.** Q **b.** R **c.** P **9. a.** R **b.** Q **c.** P **11. a.** (1,0) **b.** None **c.** (−2, 1) **13. a.** (−2, 0.3), (0, 0), (2, −0.3) **b.** None **c.** None **15.** passing through $(a, f(a))$ with slope $f'(a) = \lim_{h\to 0} \dfrac{f(a+h) - f(a)}{h}$ **17. a.** 7 **b.** 4
19. a. 0 **b.** 2 **21. a.** −2 **b.** −2
23. a. 3 **b.** $y = 3x + 2$

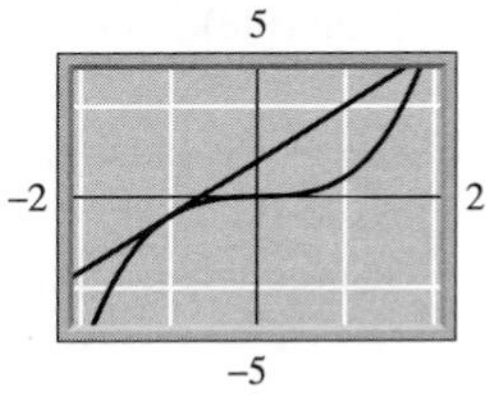

25. a. $\frac{3}{4}$ **b.** $y = \frac{3}{4}x + 1$

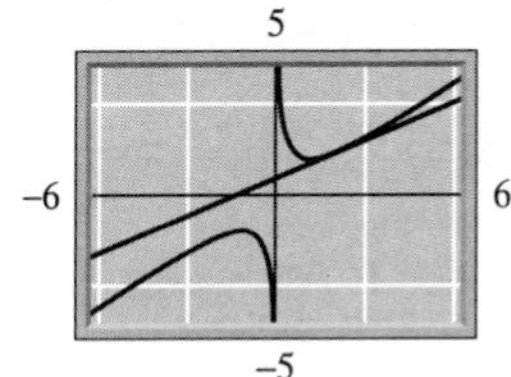

27. a. $\frac{1}{4}$ **b.** $y = \frac{1}{4}x + 1$

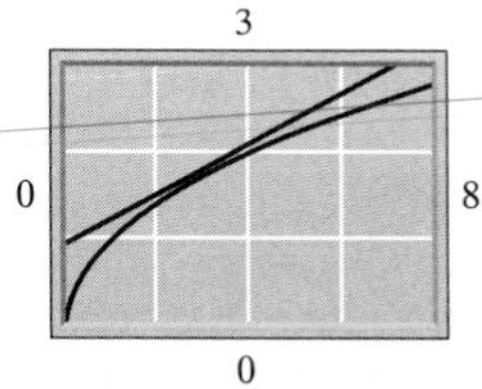

29. -0.12 **31.** -0.58 **33.** $0.623, 0$ **35.** $1.423, 2.577$ **37. a.** 6° per month. The average high temperature in Central Park rises at an average rate of 6° per month during the period April 1–September 1. **b.** 10° per month. The average high temperature in Central Park is rising at a rate of 10° per month on April 1. **c.** $-8°$ per month. The average high temperature in Central Park is falling at a rate of 8° per month on September 1. **d.** Somewhere in the first week of February and again about July 1 **e.** August 1; the temperature is not changing on that date. **f.** About the middle of April **39.** $q(100) = 50{,}000$; $\left.\frac{dq}{dp}\right|_{p=100} = -500$. A total of 50,000 pairs of sneakers can be sold at a price of \$100, but the demand is decreasing at a rate of 500 pairs per \$1 increase in the price. **41.** $L(0.95) = 31.2$ meters and $L'(0.95) = -304.2$ meters/warp. Thus, at a speed of warp 0.95, the spaceship has an observed length of 31.2 meters and its length is decreasing at a rate of 304.2 meters per unit warp, or 3.042 meters per increase in speed of 0.01 warp. **43.** The tangent to the graph is horizontal at that point, and so the graph is almost horizontal near that point.

45.

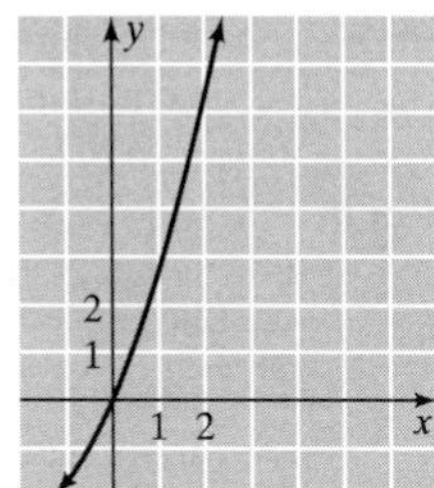

47. The slope of the tangent is always positive (so that the graph moves up as it goes from left to right).

3.4 exercises

1. a. $f'(x) = 2x$ **b.** 4 **3. a.** $f'(x) = 6x + 1$ **b.** 7 **5. a.** $f'(x) = 2 - 2x$ **b.** 4 **7. a.** $f'(x) = 3x^2 + 2$ **b.** 14 **9. a.** $f'(x) = -1/x^2$ **b.** -1 **11.** $5x^4$ **13.** $-4x^{-3}$ **15.** $-0.25x^{-0.75}$ **17.** $8x^3 + 9x^2$ **19.** $-1 - 1/x^2$ **21.** $\frac{dy}{dx} = 10(0) = 0$ (constant multiple and power rule) **23.** $\frac{dy}{dx} = \frac{d}{dx}(x^2) + \frac{d}{dx}(x)$ (sum rule) $= 2x + 1$ (power rule) **25.** $\frac{dy}{dx} = \frac{d}{dx}(4x^3) + \frac{d}{dx}(2x) - \frac{d}{dx}(1)$ (sum and difference) $= 4\frac{d}{dx}(x^3) + 2\frac{d}{dx}(x) - \frac{d}{dx}(1)$ (constant multiples) $= 12x^2 + 2$ (power rule) **27.** $\frac{d}{dx}(x^{104} - 99x^2 + x) = \frac{d}{dx}(x^{104}) - \frac{d}{dx}(99x^2) + \frac{d}{dx}(x)$ (sums and differences) $= 104x^{103} - 99\frac{d}{dx}(x^2) + 1$ (constant multiples and power rule) $= 104x^{103} - 198x + 1$ (power rule) **29.** $f'(x) = 2x - 3$ **31.** $f'(x) = 1 + 0.5x^{-0.5}$ **33.** $g'(x) = -2x^{-3} + 3x^{-2}$ **35.** $g'(x) = -\frac{1}{x^2} + \frac{2}{x^3}$ **37.** $h'(x) = -\frac{0.8}{x^{1.4}}$ **39.** $h'(x) = -\frac{2}{x^3} - \frac{6}{x^4}$ **41.** $r'(x) = -\frac{2}{3x^2} + \frac{0.1}{2x^{1.1}} + \frac{4.4x^{0.1}}{3}$ **43.** $r'(x) = \frac{2}{3} - \frac{0.1}{2x^{0.9}} - \frac{4.4}{3x^{2.1}}$ **45.** $s'(x) = \frac{1}{2\sqrt{x}} - \frac{1}{2x\sqrt{x}}$ **47.** $s'(x) = 3x^2$ **49.** $t'(x) = 1 - 4x$ **51.** $1 - 2/x^3$ **53.** $2.6x^{0.3} + 1.2x^{-2.2}$ **55.** $3at^2 - 4a$ **57.** $5.15x^{9.3} - 99x^{-2}$ **59.** $-\frac{2.31}{t^{2.1}} - \frac{0.3}{t^{0.4}}$ **61.** $4\pi r^2$ **63.** 3 **65.** -2 **67.** $\frac{1}{32}$ **69.** -5 **71.** $y = 3x + 2$

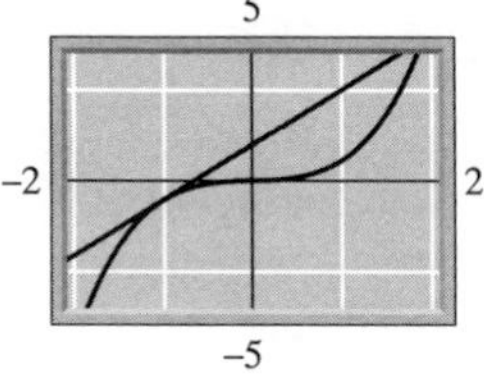

73. $y = \frac{3}{4}x + 1$

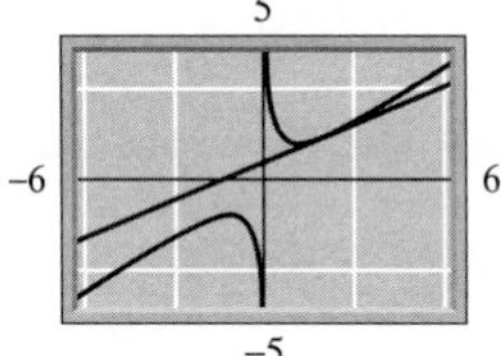

75. $y = \frac{1}{4}x + 1$

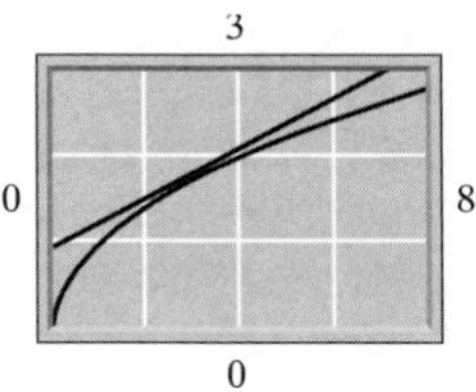

77. $x = -\frac{3}{4}$ **79.** No such values **81.** $x = 1, -1$ **85.** The rate of change of C_2 is twice the rate of change of C_1. **87.** The rates of change of cost and revenue must be equal. **89.** $c'(15) \approx 1.389$; $c'(30) \approx 0.011$. Thus, the hourly oxygen consumption is increasing at a slower rate when the chick hatches. (Notice that the oxygen consumption is still increasing, however.) **91. a.** 0, −32, −64, −96, −128 ft/s **b.** 2.5 seconds; downward at 80 ft/s **93.** After graphing the curve $y = 3x^2$, draw the line that passes through $(-1, 3)$ with slope -6. **95.** The slope of g is twice the slope of f. **97.** The algebraic method because it gives the exact value of the derivative. The other two approaches give only approximate values (except in some special cases). **99.** The left-hand side is not equal to the right-hand side. The *derivative* of the left-hand side is equal to the right-hand side, so your friend should have written $\frac{d}{dx}(3x^4 + 11x^5) = 12x^3 + 55x^4$.

3.5 exercises

1. $C'(1000) = \$4.80$ per item **3.** $C'(100) = \$99.90$ per item **5.** $C'(x) = 4$; $R'(x) = 8 - x/500$; $P'(x) = 4 - x/500$; $P'(x) = 0$ when $x = 2000$. Thus, at a production level of 2000, the profit is stationary (neither increasing nor decreasing) with respect to the production level. This may indicate a maximum profit at a production level of 2000. **7. a.** $C'(x) = 1200 - 0.004x$. The cost is going up at a rate of \$1,199,984 per television commercial. The exact cost of airing the fifth television commercial is $C(5) - C(4) = \$1,199,982$. **b.** $\bar{C}(x) = 150/x + 1200 - 0.002x$; $\bar{C}(4) = \$1,237,492$ per television commercial. The average cost of airing the first four television commercials is \$1,237,492. **9.** The profit on the sale of 1000 videocassettes is \$3000 and is decreasing at a rate of \$3 per additional videocassette sold. **11.** $P \approx \$257.07$ and $dP/dx \approx 5.07$. Your current profit is \$257.07 per month, and this will increase at a rate of \$5.07 per additional magazine in sales. **13.** $P'(50) = \$350$. This means that, at an employment level of 50 workers, the firm's daily profit will increase at a rate of \$350 per additional worker it hires. **15. a.** \$2.50 per pound **b.** $R(q) = 20{,}000/q^{0.5}$ **c.** $R(400) = \$1000$. This is the monthly revenue that will result from setting the price at \$2.50 per pound. $R'(400) = -\$1.25$ per pound of tuna. Thus, at a demand level of 400 pounds per month, the revenue is decreasing at a rate of \$1.25 per pound. **d.** The fishery should raise the price (to reduce the demand). **17. a.** $R(q) = 152.33q - 2.22q^2$; $R'(q) = 152.33 - 4.44q$. **b.** Annual per capita revenue = 2066.5¢; $R'(50) = -69.67$¢ per pound of poultry. Thus, the revenue is decreasing at a rate of 69.67¢ per additional pound of poultry demanded. **c.** $P(q) = 142.33q - 2.22q^2$; $P'(q) = 142.33 - 4.44q$; $P'(50) = -79.67$¢ per pound. Thus, if a farmer lowers the price to increase the per capita demand by 1 pound, the annual profit will decrease by approximately 79.67¢. **19. a.** $C(x) = 500{,}000 + 685{,}000x - 10{,}000\sqrt{x}$; $C'(x) = 685{,}000 - (5000/\sqrt{x})$; $\bar{C}(x) = (500{,}000/x) + 685{,}000 - (10{,}000/\sqrt{x})$ **b.** $C'(3) = \$682{,}000$ per spot; $\bar{C}(3) = \$846{,}000$ per spot. Since the marginal cost is less than the average cost, the cost of the fourth ad is lower than the average cost of the first three, so the average cost will decrease as x increases. **21. a.** $C'(q) = 200q$; $C'(10) = \$2000$ per 1-pound reduction in emissions. **b.** $S'(q) = 500$. Thus, $S'(q) = C'(q)$ when $500 = 200q$, or $q = 2.5$ pounds per day reduction. **c.** $N(q) = C(q) - S(q) = 100q^2 - 500q + 4000$. This is a parabola with its lowest point (vertex) given by $q = 2.5$. The net cost at this production level is $N(2.5) = \$3375$ per day. The value of q is the same as that for part (b). The net cost to the firm is minimized at the reduction level for which the cost of controlling emissions begins to increase faster than the subsidy. This is why we get the answer by setting these two rates of increase equal to each other. **23.** $M'(10) \approx 0.0002557$ mpg/mph. This means that, at a speed of 10 mph, the fuel economy is increasing at a rate of 0.0002557 miles per gallon per 1-mph increase in speed. $M'(60) = 0$ mpg/mph. This means that, at a speed of 60 mph, the fuel economy is neither increasing nor decreasing with increasing speed. $M'(70) \approx -0.00001799$. This means that, at 70 mph, the fuel economy is decreasing at a rate of 0.00001799 miles per gallon per 1-mph increase in speed. Thus, 60 mph is the most fuel-efficient speed for the car. **25. a.** 3.98¢ **b.** Decreasing at a rate of 0.00096¢ per extra ton **27.** **(C)** **29.** **(D)** **31.** Cost is often measured as a function of the number of items x. Thus, $C(x)$ is the cost of producing (or purchasing, as the case may be) x items. **a.** The average cost function $\bar{C}(x)$ is given by $\bar{C}(x) = C(x)/x$. The marginal cost function is the derivative, $C'(x)$, of the cost function. **b.** The average cost $\bar{C}(r)$ is the slope of the line through (0, 0) and the point where $x = r$. The marginal cost of the rth unit is the slope of the tangent to the graph of the cost function at the point where $x = r$. **c.** The average cost function $\bar{C}(x)$ gives the average cost of producing the first x items. The marginal cost function $C'(x)$ is the rate at which cost is changing with respect to the number of items x, or the incremental cost per item, and approximates the cost of producing the $(x + 1)$st item. **33.** The marginal cost **35.** The circumstances described suggest that the average cost function is at a relatively low point at the current production level, and so it would be appropriate to advise the company to maintain current production levels; raising or lowering the production level will result in increasing average costs.

3.6 exercises

1. 0 **3.** 4 **5.** Does not exist **7.** $\frac{3}{2}$ **9.** $\frac{1}{2}$ **11.** Diverges to $+\infty$ **13.** 0 **15.** 1 **17.** 0 **19.** 1 **21. a.** -2 **b.** -1 **23. a.** 1 **b.** 1 **c.** 0 **d.** $+\infty$ **25. a.** 0 **b.** 2 **c.** -1 **d.** Does not exist **e.** 2 **f.** $+\infty$ **27. a.** 1 **b.** 1 **c.** 2 **d.** Does not exist **e.** 1 **f.** 2 **29. a.** 1 **b.** $+\infty$ **c.** $+\infty$ **d.** $+\infty$ **e.** Not defined **f.** -1 **31. a.** -1 **b.** $+\infty$ **c.** $-\infty$ **d.** Does not exist **e.** 2 **f.** 1 **33.** Continuous on its domain **35.** Continuous on its domain **37.** Discontinuous at $x = 0$ **39.** Discontinuous at $x = -1$ **41.** Continuous on its domain **43.** Discontinuous at $x = -1$ and 0 **45.** $\lim_{t\to+\infty} r(t) = -\infty$; $\lim_{t\to+\infty} [r(t)/n(t)] \approx 2.45$. In the long term, visitor spending in Hawaii will fall without bound, but spending per visitor will level off at \$2.45. In the real world, visitor spending can never fall without bound (and thus become negative). Thus, the given models should not be extrapolated far into the future. **47. a.** $\lim_{t\to+\infty} n(t) \approx 2.5$. Book sales can be expected to level off at 2.5 billion per year in the long term. **b.** $\lim_{t\to+\infty} n'(t) \approx 0$. The number of books sold annually can be expected to stop changing in the long term. **49.** 470. This suggests that students whose parents earn an exceptionally large income score an average of 470 on the SAT verbal test. **51.** *Numerical evaluation of limits:* To approximate $\lim_{x\to a} f(x)$ numerically, choose values of x closer and closer to and on either side of $x = a$, and evaluate $f(x)$ for each of them. The limit (if it exists) is then the number that these values of $f(x)$ approach. A disadvantage of this method is that it may never give the exact value of the limit but only an approximation. (However, we can make this as accurate as we like.) *Graphical evaluation* *of limits:* To approximate $\lim_{x \to a} f(x)$ graphically, draw the graph of f either by hand or with technology. Then place a pencil point (or the trace cursor) on the graph to the left of the point where $x = a$, and then move it along the curve toward the point where $x = a$, reading off the y-coordinates as you go. The left limit (if it exists) is the value that the y-coordinates are approaching. Similarly, starting on the right of the point where $x = a$ and moving left along the graph gives the right limit. If both limits exist and are the same, then this is the limit. This method has the same disadvantage as the numerical method: We obtain only an approximate value. **53.** An example is $f(x) = (x - 1)(x - 2)$. **55.** It could, by expanding more and more slowly as it approaches the limit of 130,000 billion light-years.

3.7 exercises

1. 1 **3.** 2 **5.** 0 **7.** 6 **9.** 4 **11.** 2 **13.** 0 **15.** 0 **17.** 12 **19.** Diverges to $+\infty$ **21.** Does not exist; left and right limits differ. **23.** $\frac{3}{2}$ **25.** $\frac{1}{2}$ **27.** Diverges to $+\infty$ **29.** 0 **31.** $\frac{3}{2}$ **33.** $\frac{1}{2}$ **35.** Diverges to $-\infty$ **37.** 0 **39.** 0 **41.** 2 **43.** -4 **45.** $2x - 2$ **47.** $-10x + 2$ **49.** $3t^2 + 1$ **51.** $4t^3 - 1$ **53.** $-6/t^2$ **55.** $1 - 1/x^2$ **57.** $-1/(x - 2)^2$ **59.** Discontinuity at $x = 0$ **61.** Continuous everywhere **63.** Discontinuity at $x = 0$ **65.** Discontinuity at $x = 0$ **67.** $\lim_{t \to +\infty} n(t) = 0$. If the trend were to continue indefinitely, the annual number of DWI arrests in New Jersey would decrease to zero in the long term. **69.** $\lim_{t \to +\infty} p(t) = 100$; $\lim_{t \to +\infty} p'(t) = 0$. The percentage of children who learn to speak approaches 100% as their age increases, with the number of additional children learning to speak approaching zero. **71.** *Numerical and graphical evaluation of limits:* See the answer to Exercise 51 in Section 3.6. *Algebraic evaluation of limits:* To approximate $\lim_{x \to a} f(x)$ algebraically, first check whether $f(x)$ is a closed-form function. Then check whether $x = a$ is in its domain. If so, the limit is just $f(a)$; that is, it is obtained by substituting $x = a$. If not, then try to first simplify $f(x)$ in such a way as to transform it into a new function such that $x = a$ is in its domain, and then substitute. A disadvantage of this method is that it is sometimes extremely difficult to evaluate limits algebraically, and rather sophisticated methods are often needed. **73.** The statement may not be true; for instance, if $f(x) = \begin{cases} x + 2 & \text{if } x < 0 \\ 2x - 1 & \text{if } x \geq 0 \end{cases}$, then $f(0)$ is defined and equals -1 and yet $\lim_{x \to 0} f(x)$ does not exist. The statement can be corrected by requiring that f be a closed-form function: "If f is a closed-form function and $f(a)$ is defined, then $\lim_{x \to a} f(x)$ exists and equals $f(a)$."

75. $f(x) = \begin{cases} 0 & \text{if } x \text{ is any number other than 1 or 2} \\ 1 & \text{if } x = 1 \text{ or } 2 \end{cases}$

Chapter 3 Review Test

1. a.

h	1	0.01	0.001
Average rate of change	-0.5	-0.9901	-0.9990

Slope ≈ -1

b.

h	1	0.01	0.001
Average rate of change	23	6.8404	6.7793

Slope ≈ 6.8

c.

h	1	0.01	0.001
Average rate of change	6.3891	2.0201	2.0020

Slope ≈ 2

d.

h	1	0.01	0.001
Average rate of change	0.6931	0.9950	0.9995

Slope ≈ 1

2. a. i. P **ii.** Q **iii.** R **iv.** S **b. i.** None **ii.** R **iii.** Q **iv.** P **c. i.** Q **ii.** None **iii.** None **iv.** None **d. i.** None **ii.** R **iii.** P and S **iv.** Q **3. a.** (B) **b.** (C) **c.** (B) **d.** (A) **e.** (C) **4. a.** $50x^4 + 2x^3 - 1$ **b.** $-50/x^6 - 2/x^5 + 1/x^2$ **c.** $9x^2 + x^{-2/3}$ **d.** $-4.2/x^{3.1} - 0.1/(2x^{0.9})$ **5. a.** $2x + 1$ **b.** $-1/x^2$ **6. a.** 500 books per week **b.** [3, 4], [4, 5] **c.** [3, 5]; 650 books per week **7. a.** 274 books per week **b.** 636 books per week **c.** No; the function w begins to decrease after $t = 14$. Graph:

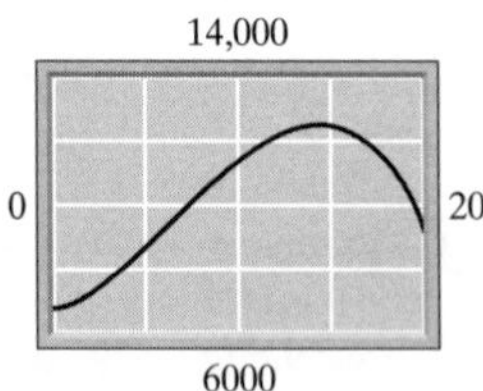

8. a. 553 books per week **b.** 15 books per week **c.** Sales level off at around 10,527 books per week, with a zero rate of change. Graph:

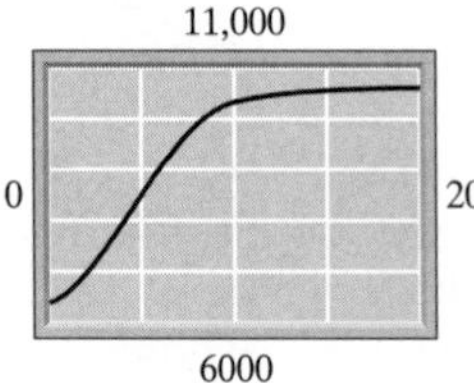

9. a. \$2.88 per book **b.** \$3.715 per book **c.** Approximately $-$\$0.000104 per book **d.** At a sales level of 8000 books per week, the cost is increasing at a rate of \$2.88 per book (so that the 8001st book costs approximately \$2.88 to sell), and it costs an average of \$3.715 per book to sell the first 8000 books. Moreover, the average cost is decreasing at a rate of \$0.000104 per book per additional book sold.

Chapter 4

4.1 exercises

1. 3 **3.** $3x^2$ **5.** $2x+3$ **7.** $210x^{1.1}$ **9.** $-2/x^2$ **11.** $2x/3$
13. $(x^2-1)+2x(x+1)=(x+1)(3x-1)$
15. $(x^{-0.5}+4)(x-x^{-1})+(2x^{0.5}+4x-5)(1+x^{-2})$
17. $8(2x^2-4x+1)(x-1)$
19. $(1/3.2-3.2/x^2)(x^2+1)+2x(x/3.2+3.2/x)$
21. $2x(2x+3)(7x+2)+2x^2(7x+2)+7x^2(2x+3)$
23. $5.3(1-x^{2.1})(x^{-2.3}-3.4)-2.1x^{1.1}(5.3x-1)(x^{-2.3}-3.4)-2.3x^{-3.3}(5.3x-1)(1-x^{2.1})$

25. $\frac{1}{2\sqrt{x}}\left(\sqrt{x}+\frac{1}{x^2}\right)+(\sqrt{x}+1)\left(\frac{1}{2\sqrt{x}}-\frac{2}{x^3}\right)$

27. $\frac{2(3x-1)-3(2x+4)}{(3x-1)^2}=\frac{-14}{(3x-1)^2}$

29. $\frac{(4x+4)(3x-1)-3(2x^2+4x+1)}{(3x-1)^2}=\frac{6x^2-4x-7}{(3x-1)^2}$

31. $\frac{(2x-4)(x^2+x+1)-(x^2-4x+1)(2x+1)}{(x^2+x+1)^2}=\frac{5x^2-5}{(x^2+x+1)^2}$

33. $\frac{(0.23x^{-0.77}-5.7)(1-x^{-2.9})-2.9x^{-3.9}(x^{0.23}-5.7x)}{(1-x^{-2.9})^2}$

35. $\frac{\frac{1}{2}x^{-1/2}(x^{1/2}-1)-\frac{1}{2}x^{-1/2}(x^{1/2}+1)}{(x^{1/2}-1)^2}=\frac{-1}{\sqrt{x}(\sqrt{x}-1)^2}$

37. $-3/x^4$

39. $\frac{[(x+1)+(x+3)](3x-1)-3(x+3)(x+1)}{(3x-1)^2}=\frac{3x^2-2x-13}{(3x-1)^2}$

41. $\{[(x+1)(x+2)+(x+3)(x+2)+(x+3)(x+1)](3x-1)-3(x+3)(x+1)(x+2)\}/(3x-1)^2$
43. $4x^3-2x$ **45.** 64 **47.** 3 **49.** $y=12x-8$ **51.** $y=x/4+1/2$ **53.** $y=-2$ **55.** $S'(5)=10$ (sales are increasing at a rate of 1000 units per month); $p'(5)=-10$ (the price is dropping at a rate of \$10 per stereo system per month); $R'(5)=900{,}000$ (revenue is increasing at a rate of \$900,000 per month) **57.** Decreasing at a rate of \$1 per day **59.** Decreasing at a rate of \$0.10 per month **61.** $M'(x)=(3600x^{-2}-1)/(x+3600x^{-1})^2$; $M'(10)\approx 0.0002557$ mpg/mph—this means that, at a speed of 10 mph, the fuel economy is increasing at a rate of 0.0002557 miles per gallon per 1-mph increase in speed; $M'(60)=0$ mpg/mph—this means that, at a speed of 60 mph, the fuel economy is neither increasing nor decreasing with increasing speed; $M'(70)\approx -0.00001799$—this means that, at 70 mph, the fuel economy is decreasing at a rate of 0.00001799 miles per gallon per 1 mph increase in speed. 60 mph is the most fuel-efficient speed for the car (in the next chapter we discuss how to locate largest values in general). **63.** \$111,870,000 per year **65.** $R'(p)=-5.625/(1+0.125p)^2$; $R'(4)=-2.5$ thousand organisms per hour, per 1000 organisms. This means that the reproduction rate of organisms in a culture containing 4000 organisms is declining at a rate of 2500 organisms per hour, per 1000 organisms. **67.** Oxygen consumption is decreasing at a rate of 1634 milliliters per day. This is due to the fact that the number of eggs is decreasing because $C'(25)$ is positive. **69.** The analysis is suspect; it seems to be asserting that the annual increase in revenue, which we can think of as dR/dt, is the product of the annual increases, dp/dt in price and dq/dt in sales. However, since $R=pq$, the product rule implies that dR/dt is not the product of dp/dt and dq/dt but is instead $dR/dt=(dp/dt)\cdot q+p\cdot(dq/dt)$. **71.** $q=-p+1000$ is one example. **73.** The correct answer is (a).

4.2 exercises

1. $4(2x+1)$ **3.** $-(x-1)^{-2}$ **5.** $2(2-x)^{-3}$ **7.** $(2x+1)^{-0.5}$
9. $-4(4x-1)^{-2}$ **11.** $-3/(3x-1)^2$ **13.** $4(x^2+2x)^3(2x+2)$
15. $-4x(2x^2-2)^{-2}$ **17.** $-5(2x-3)(x^2-3x-1)^{-6}$ **19.** $-6x/(x^2+1)^4$ **21.** $1.5(0.2x-4.2)(0.1x^2-4.2x+9.5)^{0.5}$ **23.** $4(2s-0.5s^{-0.5})(s^2-s^{0.5})^3$ **25.** $-x/\sqrt{1-x^2}$ **27.** $-[(x+1)(x^2-1)]^{-3/2}(3x-1)(x+1)$ **29.** $6.2(3.1x-2)+6.2/(3.1x-2)^3$
31. $2[(6.4x-1)^2+(5.4x-2)^3][12.8(6.4x-1)+16.2(5.4x-2)^2]$
33. $-2(x^2-3x)^{-3}(2x-3)(1-x^2)^{0.5}-x(x^2-3x)^{-2}(1-x^2)^{-0.5}$
35. $-56(x+2)/(3x-1)^3$ **37.** $3z^2(1-z^2)/(1+z^2)^4$
39. $3[(1+2x)^4-(1-x)^2]^2[8(1+2x)^3+2(1-x)]$

41. $-0.43(x+1)^{-1.1}[2+(x+1)^{-0.1}]^{3.3}$ **43.** $-\frac{\left(\frac{1}{\sqrt{2x+1}}-2x\right)}{(\sqrt{2x+1}-x^2)^2}$

45. $54(1+2x)^2\left[1+(1+2x)^3\right]^2\left\{1+\left[1+(1+2x)^3\right]^3\right\}^2$

47. $(100x^{99}-99x^{-2})\,dx/dt$ **49.** $(-3r^{-4}+0.5r^{-0.5})\,dr/dt$
51. $4\pi r^2\,dr/dt$ **53.** $-47/4$ **55.** $c=100(0.42+0.02t)^2-160(0.42+0.02t)+110$; $-\$0.88$ per trade per month
57. a. $R(p)=-4p^2/3+80p$; $\left.\frac{dR}{dp}\right|_{q=60}=\$40$ per \$1 increase in price **b.** $-\$0.75$ per ruby **c.** $-\$30$ per ruby. Thus, at a demand level of 60 rubies per week, the weekly revenue is decreasing at a rate of \$30 per additional ruby demanded.
59. $\left.\frac{dP}{dn}\right|_{n=10}=146{,}454.9$. At an employment level of ten engineers, Paramount will increase its profit at a rate of \$146,454.90 per additional engineer hired. **61.** $dM/dB\approx 0.000116$. This means that approximately 1.16 more manatees are killed each year for each additional 10,000 registered boats.

63. 12π mi²/h **65.** \$200,000π/week = \$628,318.53/week **67. a.** $q'(4) \approx 333$ units per month **b.** $dR/dq = 800$ **c.** $dR/dt \approx$ \$266,000 per month **69.** 3% per year **71.** 8% per year **73.** Following the calculation thought experiment, pretend that you were evaluating the function at a specific value of x. If the last operation you would perform is addition or subtraction, look at each summand separately. If the last operation is multiplication, use the product rule first; if it is division, use the quotient rule first; if it is any other operation (such as raising a quantity to a power or taking a radical of a quantity), use the chain rule first. **75.** An example is

$$f(x) = \sqrt{x + \sqrt{x + \sqrt{x + \sqrt{x + \sqrt{x + 1}}}}}.$$

4.3 exercises

1. $1/(x-1)$ **3.** $1/(x \ln 2)$ **5.** $2x/(x^2+3)$ **7.** e^{x+3} **9.** $-e^{-x}$ **11.** $4^x \ln 4$ **13.** $2^{x^2-1}2x \ln 2$ **15.** $1 + \ln x$ **17.** $2x \ln x + (x^2+1)/x$ **19.** $10x(x^2+1)^4 \ln x + (x^2+1)^5/x$ **21.** $3/(3x-1)$ **23.** $4x/(2x^2+1)$ **25.** $(2x - 0.63x^{-0.7})/(x^2 - 2.1x^{0.3})$ **27.** $-2/(-2x+1) + 1/(x+1)$ **29.** $3/(3x+1) - 4/(4x-2)$ **31.** $1/(x+1) + 1/(x-3) - 2/(2x+9)$ **33.** $5.2/(4x-2)$ **35.** $2/(x+1) - 9/(3x-4) - 1/(x-9)$ **37.** $\dfrac{1}{(x+1)\ln 2}$ **39.** $\dfrac{1 - 1/t^2}{(t + 1/t)\ln 3}$ **41.** $\dfrac{2\ln|x|}{x}$ **43.** $\dfrac{2}{x} - \dfrac{2\ln(x-1)}{x-1}$ **45.** $e^x(1+x)$ **47.** $1/(x+1) + 3e^x(x^3+3x^2)$ **49.** $e^x(\ln|x| + 1/x)$ **51.** $2e^{2x+1}$ **53.** $(2x-1)e^{x^2-x+1}$ **55.** $2xe^{2x-1}(1+x)$ **57.** $4(e^{2x-1})^2$ **59.** $-4/(e^x - e^{-x})^2$ **61.** $5e^{5x-3}$ **63.** $-\dfrac{\ln x + 1}{(x \ln x)^2}$ **65.** $2(x-1)$ **67.** $\dfrac{1}{x \ln x}$ **69.** $\dfrac{1}{2x \ln x}$ **71.** $y = (e/\ln 2)(x-1) \approx 3.92(x-1)$ **73.** $y = x$ **75.** $y = -[1/(2e)](x-1) + e$ **77.** \$451.00 per year **79.** 277,000 people per year **81.** 0.000283 gram per year **83.** \$446.02 per year **85.** $P'(22) \approx -1.4$. This indicates that, in ancient Rome, the percentage of people surviving was decreasing at a rate of 1.4% per year at age 22; that is, 1.4% of the original birth cohort died that year. **87. a.** **(A)** **b.** The verbal SAT increases by approximately 1 point. **c.** $S'(x)$ decreases with increasing x, so that as parental income increases, the effect of income on SAT scores decreases. **89.** 3,110,000 cases/month; 11,200,000 cases/month; 722,000 cases/month **91. a.** $n'(t) = (2{,}243{,}310e^{-0.141t})/(2.90 + 74.0e^{-0.141t})^2$; $n'(13) \approx 1652.4 \approx 1650$ to three significant digits. The constants in the model are specified to three significant digits, so we cannot expect the answer to be accurate to more than that. In other words, all digits from the fourth on are probably meaningless. The answer tells one that, in June 2000 ($t = 13$), the number of phones tapped was increasing at a rate of 1650 wiretaps per year.

b.

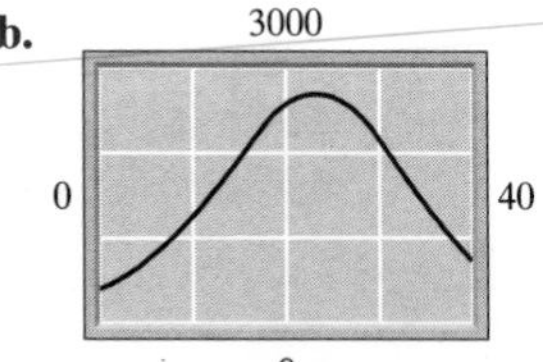

The number of phone lines monitored will be increasing most rapidly in approximately June 2010 ($t \approx 23$).

93.
a.

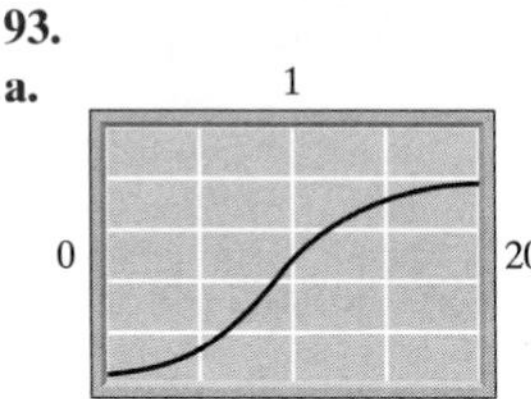

$p'(10) \approx 0.09$, so the percentage of firms using numeric control is increasing at a rate of 9% per year after 10 years.

b. 0.80. Thus, in the long run, 80% of all firms will be using numeric control. **c.** $p'(t) = (0.3816e^{4.46-0.477t})/(1 + e^{4.46-0.477t})^2$; $p'(10) = 0.0931$

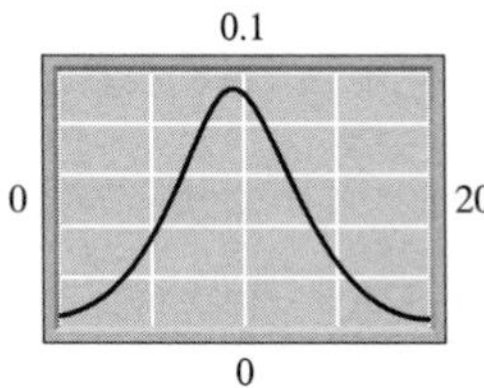

d. 0. Thus, in the long run, the percentage of firms using numeric control will stop increasing.
95. $n'(t) = (0.08e^{0.002t} + 502e^{3-0.498t})/(1 + 25e^{3-0.5t})^2$

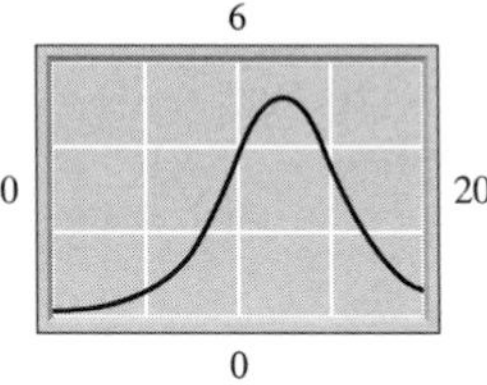

$n'(t)$ is a maximum when $t \approx 12.5$. ($n'(12.5) \approx 5.167$) This means that the rate of growth of enrollment in HMOs reached a maximum of 5.167 million new enrollments per year in about June 1987. **97.** $P(10)$ is the magnitude of the quantity at time $t = 10$, whereas $P'(10)$ is the rate at which the quantity is changing at time $t = 10$. The units of $P'(10)$ are kilograms per day. **99.** **(B), (A)** **101.** If $f(x) = e^{kx}$, then the fractional rate of change is $f'(x)/f(x) = ke^{kx}/e^{kx} = k$, the fractional growth rate. **103.** If $A(t)$ is growing exponentially, then $A(t) = A_0e^{kt}$ for constants A_0 and k. Its percentage rate of change is then $A'(t)/A(t) = kA_0e^{kt}/A_0e^{kt} = k$, a constant.

4.4 exercises

1. $\cos x + \sin x$ **3.** $(\cos x)(\tan x) + (\sin x)(\sec^2 x)$ **5.** $-2 \operatorname{cosec} x \operatorname{cotan} x - \sec x \tan x + 3$ **7.** $\cos x - x \sin x + 2x$

9. $(2x-1)\tan x + (x^2-x+1)\sec^2 x$ **11.** $-[\text{cosec}^2 x(1+\sec x) + \text{cotan}\, x \sec x \tan x]/(1+\sec x)^2$ **13.** $-2\cos x \sin x$ **15.** $2\sec^2 x \tan x$ **17.** $\pi \cos\left[\frac{\pi}{5}(x-4)\right]$ **19.** $-(2x-1)\sin(x^2-x)$ **21.** $(2.2x^{1.2}+1.2)\sec(x^{2.2}+1.2x-1)\tan(x^{2.2}+1.2x-1)$ **23.** $\sec x \tan x \tan(x^2-1) + 2x \sec x \sec^2(x^2-1)$ **25.** $e^x[-\sin(e^x) + \cos x - \sin x]$ **27.** $\sec x$ **33.** $e^{-2x}[-2\sin(3\pi x) + 3\pi\cos(3\pi x)]$ **35.** $1.5[\sin(3x)]^{-0.5}\cos(3x)$ **37.** $\frac{x^4-3x^2}{(x^2-1)^2}\sec\left(\frac{x^3}{x^2-1}\right)\tan\left(\frac{x^3}{x^2-1}\right)$ **39.** $\frac{\text{cotan}(2x-1)}{x} - 2\ln|x|\,\text{cosec}^2(2x-1)$ **41.** $c'(t) = 7\pi\cos[2\pi(t-0.75)]$; $c'(0.75) \approx \$21.99$ per year $\approx \$0.42$ per week **43. a.** $d(t) = 10 + 5\cos(2\pi t/13.5)$ **b.** $d'(t) = -(10\pi/13.5)\sin(2\pi t/13.5)$; $d'(7) \approx 0.270$. At noon, the tide was rising at a rate of 0.270 feet per hour. **45.** $c'(t) = 1.035^t[\ln(1.035)(0.8\sin(2\pi t) + 10.2) + 1.6\pi\cos(2\pi t)]$; $c'(1) = 1.035[10.2\ln|1.035| + 1.6\pi] \approx \$5.57$ per year, or $0.11 per week. **47. a.** (C) **b.** Increasing at a rate of 0.157 degree per thousand years **49.** $f(x) = \sin x$; $f(x) = \cos x$ **51.** $f(x) = e^{-x}$; $f(x) = -2e^{-x}$ **53.** The graph of $\cos x$ slopes down over the interval $(0, \pi)$, so that its derivative is negative over that interval. The function $-\sin x$, and not $\sin x$, has this property.

4.5 exercises

1. $-2/3$ **3.** x **5.** $(y-2)/(3-x)$ **7.** $-y$ **9.** $-y/[x(1+\ln x)]$ **11.** $-x/y$ **13.** $-2xy/(x^2-2y)$ **15.** $-(6+9x^2y)/(9x^3-x^2)$ **17.** $3y/x$ **19.** $(p+10p^2q)/(2p-q-10pq^2)$ **21.** $(ye^x-e^y)/(xe^y-e^x)$ **23.** $se^{st}/(2s-te^{st})$ **25.** $ye^x/(2e^x+y^3e^y)$ **27.** $(y-y^2)/(-1+3y-y^2)$ **29.** $-y/(x+2y-xye^y-y^2e^y)$ **31.** $1/\sec^2 y$ **33.** $-[1+y\cos(xy)]/[1+x\cos(xy)]$ **35.** $(x^3+x)\sqrt{x^3+2}\left(\frac{3x^2+1}{x^3+x} + \frac{1}{2}\frac{3x^2}{x^3+2}\right)$ **37.** $x^x(1+\ln x)$ **39.** 1 **41.** -2 **43.** -0.03314 **45.** 31.73 **47.** -0.1898 **49.** 0 **51. a.** 500 T-shirts **b.** $\left.\frac{dq}{dp}\right|_{p=5} = -125$ T-shirts per dollar. Thus, when the price is set at \$5, the demand is dropping by 125 T-shirts per \$1 increase in price. **53.** $\left.\frac{dk}{de}\right|_{e=15} = -0.307$ carpenters per electrician. This means that, for a \$200,000 house whose construction employs 15 electricians, adding one more electrician would cost as much as approximately 0.307 additional carpenters. In other words, one electrician is worth approximately 0.307 carpenters **55. a.** 22.93 hours (The other root is rejected because it is greater than 30.) **b.** $\left.\frac{dt}{dx}\right|_{x=3.0} \approx -11.2$ hours per grade point. This means that, for a 3.0 student who scores 80 on the examination, 1 grade point is worth approximately 11.2 hours. **57.** $\left.\frac{dy}{dx}\right|_{x=100} = -\frac{3}{2}\frac{y}{x} \approx -\$848{,}528$ per worker. The annual expenditure to maintain production at 20,000 units per day is decreasing at a rate of \$848,528 per additional worker employed. In other words, each additional worker is worth approximately \$848,528 to the company in annual expense. **59.** $\frac{dr}{dy} = 2\frac{r}{y}$, so $\frac{dr}{dt} = 2\frac{r}{y}\frac{dy}{dt}$ by the chain rule **61.** Let $y = f(x)g(x)$. Then $\ln y = \ln f(x) + \ln g(x)$ and $\frac{1}{y}\frac{dy}{dx} = \frac{f'(x)}{f(x)} + \frac{g'(x)}{g(x)}$, so $\frac{dy}{dx} = y\left(\frac{f'(x)}{f(x)} + \frac{g'(x)}{g(x)}\right) = f(x)g(x)\left(\frac{f'(x)}{f(x)} + \frac{g'(x)}{g(x)}\right) = f'(x)g(x) + f(x)g'(x)$. **63.** Writing $y = f(x)$ specifies y as an explicit function of x. This can be regarded as an equation giving y as an *implicit* function of x. The procedure of finding dy/dx by implicit differentiation is then the same as finding the derivative of y as an explicit function of x: We take d/dx of both sides. **65.** True. The answer follows by differentiating both sides of the equation $y = f(x)$ with respect to y, since we get $1 = f'(x)\cdot\frac{dx}{dy}$, giving $\frac{dx}{dy} = \frac{1}{f'(x)} = \frac{1}{dy/dx}$.

Chapter 4 Review Test

1. a. $e^x(x^2+2x-1)$ **b.** $-4x/(x^2-1)^2$ **c.** $20x(x^2-1)^9$ **d.** $-20x/(x^2-1)^{11}$ **e.** $e^x(x^2+1)^9(x^2+20x+1)$ **f.** $6(x-1)^2/(3x+1)^4$ **g.** $2xe^{x^2-1}$ **h.** $2x(x^2+2)e^{x^2-1}$ **i.** $2x/(x^2-1)$ **j.** $\frac{2x-2x\ln(x^2-1)}{(x^2-1)^2}$ **k.** $-2x\sin(x^2-1)$ **l.** $2x[\cos(x^2+1)\cos(x^2-1) - \sin(x^2+1)\sin(x^2-1)]$ **2. a.** $\frac{2x-1}{2y}$ **b.** $-\frac{2y}{2(x+y)-1}$ **c.** $-y/x$ **d.** $\frac{y}{x(1-y)}$ **e.** $\frac{2x}{\cos y - 2y}$ **f.** $\frac{\cos y}{x\sin y}$ **3. a.** $x = (1-\ln 2)/2$ **b.** $x = 0$ **c.** None **d.** $x = 1/3$ **4. a.** $R'(t) = 200(20-t) - (1000+200t)$; rising at a rate of \$3000 per week **b.** 300 books per week **c.** $R = pq$ gives $R' = p'q + pq'$. Thus, $R'/R = R'/(pq) = (p'q + pq')/pq = p'/p + q'/q$. **5. a.** Rising at a rate of 40 units per year **b.** \$110 per year **c.** $Q = P/E$ gives $Q' = (P'E - PE')/E^2$. Thus, $Q'/Q = Q'/(P/E) = (P'E - PE')/PE = P'/P - E'/E$. **6. a.** $s'(t) = \frac{2460.7e^{-0.55(t-4.8)}}{(1+e^{-0.55(t-4.8)})^2}$ **b.** 553 books per week **c.** 15 books per week **7.** 616.8 hits per day per week **8. a.** -16.67 copies per \$1. The demand for the gift edition of *Lord of the Rings* is dropping at a rate of about 16.67 copies per \$1 increase in the price. **b.** $dR/dp = q + p(dq/dp) \approx 1000 + 40(-16.67) = 333.33$ is positive, so the price should be raised.

Chapter 5

5.1 exercises

1. Absolute min: $(-3, -1)$; relative max: $(-1, 1)$; relative min: $(1, 0)$; absolute max: $(3, 2)$ **3.** Absolute min: $(3, -1)$ and $(-3, -1)$; absolute max: $(1, 2)$ **5.** Absolute min: $(-3, 0)$ and $(1, 0)$; absolute max: $(-1, 2)$ and $(3, 2)$ **7.** Relative min:

$(-1, 1)$ **9.** Absolute min: $(-3, -1)$; relative max: $(-2, 2)$; relative min: $(1, 0)$; absolute max: $(3, 3)$ **11.** Relative max: $(-3, 0)$; absolute min: $(-2, -1)$; stationary nonextreme point: $(1, 1)$ **13.** Stationary minimum at $x = -1$ **15.** Stationary minima at $x = -2$ and $x = 2$; stationary maximum at $x = 0$ **17.** Singular minimum at $x = 0$; stationary nonextreme point at $x = 1$ **19.** Stationary minimum at $x = -2$; singular nonextreme point at $x = -1$; singular nonextreme point at $x = 1$; stationary maximum at $x = 2$ **21.** Absolute max: $(0, 1)$; absolute min: $(2, -3)$; relative max: $(3, -2)$ **23.** Absolute min: $(-4, -16)$; absolute max: $(-2, 16)$; absolute min: $(2, -16)$; absolute max: $(4, 16)$ **25.** Absolute min: $(-2, -10)$, absolute max: $(2, 10)$ **27.** Absolute min: $(-2, -4)$; relative max: $(-1, 1)$; relative min: $(0, 0)$ **29.** Relative max: $(-1, 5)$; absolute min: $(3, -27)$ **31.** Absolute min: $(0, 0)$ **33.** Relative min: $\left(-2, \frac{5}{3}\right)$; relative max: $(0, -1)$; relative min $\left(2, \frac{5}{3}\right)$ **35.** Relative max: $(0, 0)$; absolute min: $(1/3, -2\sqrt{3}/9)$ **37.** Relative max: $(0, 0)$; absolute min: $(1, -3)$ **39.** No relative extrema **41.** Absolute min: $(1, 1)$ **43.** Relative max: $(-1, 1 + 1/e)$; absolute min: $(0, 1)$; absolute max: $(1, e - 1)$ **45.** Relative max: $(-6, -24)$; relative min: $(-2, -8)$ **47.** Absolute max: $(1/\sqrt{2}, \sqrt{e/2})$; absolute min: $(-1/\sqrt{2}, -\sqrt{e/2})$ **49.** Relative min of 0 at $x = 0$, absolute max of 1 at $x = \pi/2, 5\pi/2, 9\pi/2$; absolute min of -1 at $x = 3\pi/2, 7\pi/2, 11\pi/2$; relative max of 0 at $x = 6\pi$ **51.** Relative min: $(0.15, -0.52)$ and $(2.45, 8.22)$; relative max: $(1.40, 0.29)$ **53.** Absolute max: $(-5, 700)$; relative max: $(3.10, 28.19)$ and $(6, 40)$; absolute min: $(-2.10, -392.69)$; relative min: $(5, 0)$

55.

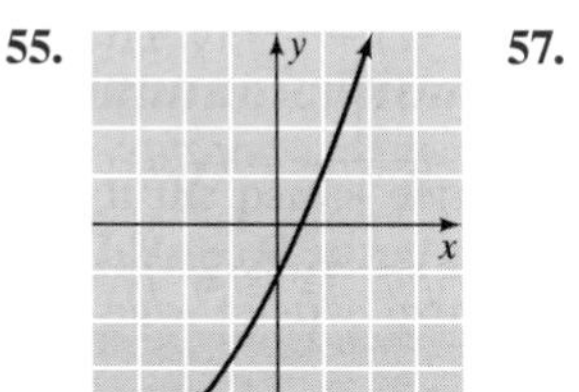

57.

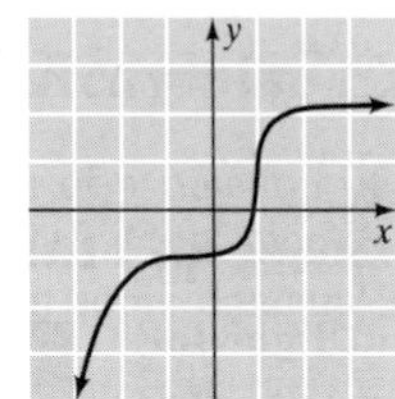

59. Not necessarily; it could be neither a relative maximum nor a relative minimum, as in the graph of $y = x^3$ at the origin.

5.2 exercises

1. $x = y = 5$; $P = 25$ **3.** $x = y = 3$; $S = 6$ **5.** $x = 2, y = 4$; $F = 20$ **7.** $x = 20, y = 10, z = 20$; $P = 4000$ **9.** 5×5 **11.** $5 \times 10 = 50$ square feet **13.** $p = \$10$ **15.** $p = \$30$ **17. a.** \$1.41 per pound **b.** 5000 pounds **c.** \$7071.07 per month **19.** 34.5¢ per pound, for an annual (per capita) revenue of \$5.95 **21.** \$42.50 per ruby, for a weekly profit of \$408.33 **23. a.** 656 headsets, for a profit of \$28,120 **b.** \$143 per headset **25.** $1600/27 \approx 59$ cubic inches **27.** 30 years from now **29.** 55 days **31.** 25 additional trees **33.** 38,730 CD players, giving an average cost of \$28 per CD player **35.** 2.5 **37.** $l = w = h \approx 20.67$ in.; volume ≈ 8827 in.^{3} **39.** $l = 30$ in., $w = 15$ in., $h = 30$ in. **41.** $l = 36$ in., $w = h = 18$ in., $V = 11{,}664$ in.^{3} **43.** 1600 copies. At this value of x, average profit equals marginal profit; beyond this the marginal profit is less than the average profit. **45.** Decreasing most rapidly in 1963; increasing most rapidly in 1989 **47.** Maximum when $t = 17$ days. This means that the embryo's oxygen consumption is increasing most rapidly 17 days after the egg is laid. **49.** $h = r \approx 11.7$ cm **51.** 1992 ($t \approx 2.20$) **53.** Absolute minimum of 5137 at $n = 13.19$; absolute maximum of 34,040 at $n = 15$. Thus, the salary value per extra year of school is increasing most slowly (\$5137 per year) at a level of 13.19 years of schooling and most rapidly (\$34,040 per year) at a level of 15 years of schooling. **55.** You should sell them in 17 years' time, when they will be worth approximately \$3960. **57.** 71 employees **59.** (d) **61.** The company can accomplish the objective by cutting away the entire sheet of cardboard, resulting in a box with surface area zero. **63.** The minimum of dq/dp is the fastest that the demand is dropping in response to increasing price.

5.3 exercises

1. 6 **3.** $4/x^3$ **5.** $-0.96x^{-1.6}$ **7.** $e^{-(x-1)}$ **9.** $2/x^3 + 1/x^2$ **11. a.** $a = -32$ ft/s^2 **b.** $a = -32$ ft/s^2 **13. a.** $a = 2/t^3 + 6/t^4$ ft/s^2 **b.** $a = 8$ ft/s^2 **15. a.** $a = -1/(4t^{3/2}) + 2$ ft/s^2 **b.** $a = 63/32$ ft/s^2 **17.** $(1, 0)$ **19.** $(1, 0)$ **21.** None **23.** $(-1, 0), (1, 1)$ **25.** Points of inflection at $x = -1$ and $x = 1$ **27.** One point of inflection, at $x = -2$ **29.** Points of inflection at $x = -2, x = 0$, and $x = 2$ **31.** One point of inflection, at $x = 0$ **33.** Points of inflection at $x = -2$ and $x = 2$

35. Absolute min at $(-1, 0)$; no points of inflection

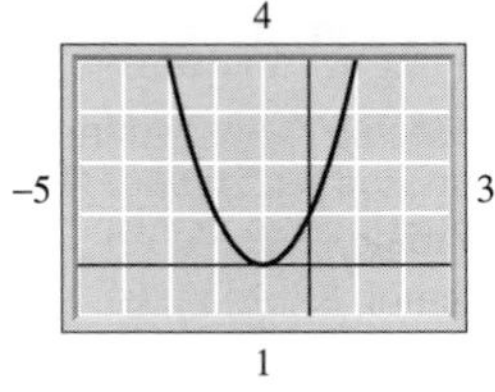

37. Relative max at $(-2, 21)$; relative min at $(1, -6)$; point of inflection at $\left(-\frac{1}{2}, \frac{15}{2}\right)$

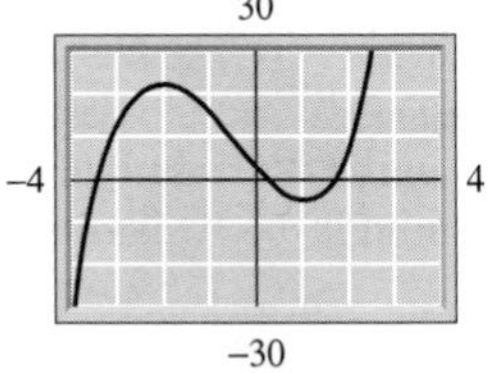

39. Absolute min at $(-4, -16)$ and $(2, -16)$; absolute max at $(-2, 16)$ and $(4, 16)$; point of inflection at $(0, 0)$

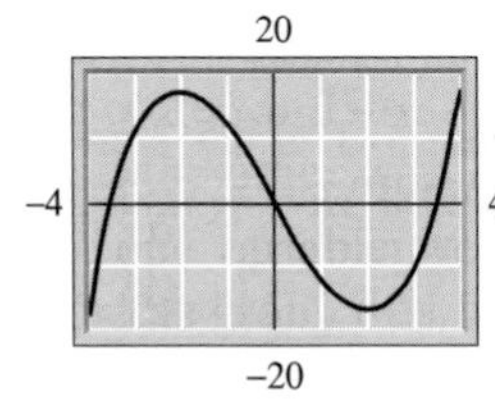

41. Absolute min at $(0, 0)$; points of inflection at $\left(\frac{1}{3}, \frac{11}{324}\right)$ and $\left(1, \frac{1}{12}\right)$

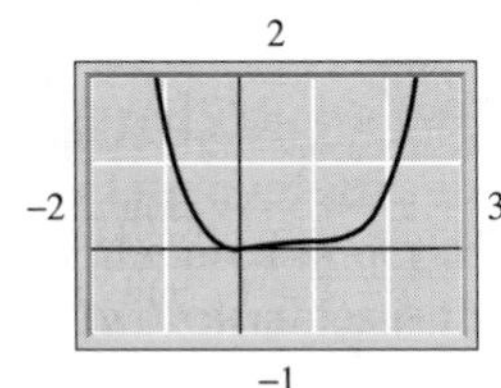

43. Relative min at $\left(-2, \frac{5}{3}\right)$ and $\left(2, \frac{5}{3}\right)$; relative max at $(0, -1)$; vertical asymptotes: $x = \pm 1$

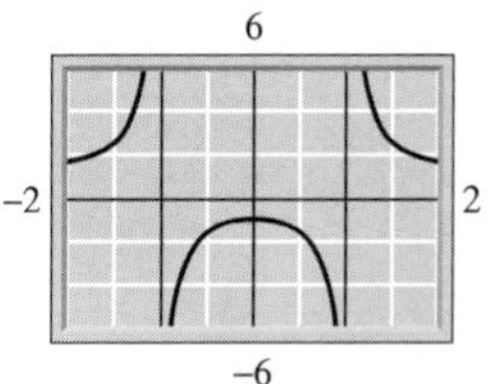

45. No extrema; point of inflection at $(0, 0)$

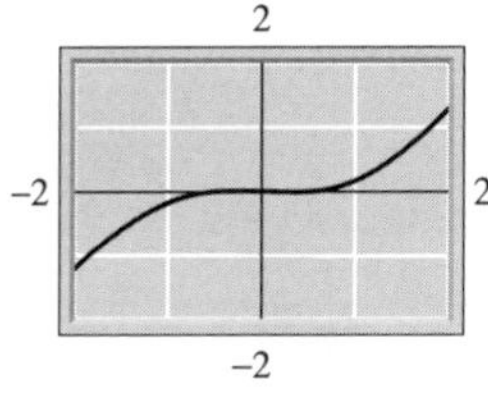

47. Relative min at $(1, 2)$; relative max at $(-1, -2)$; vertical asymptote: $y = 0$

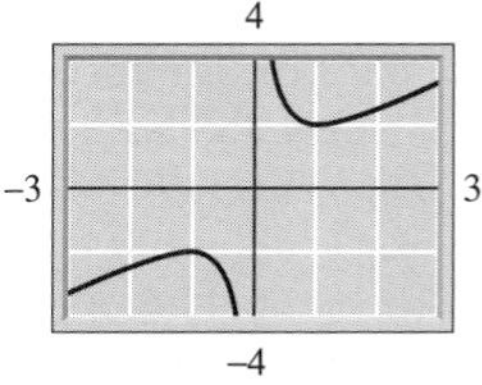

49. Relative max at $(0, 0)$; absolute min at $(1, -2)$

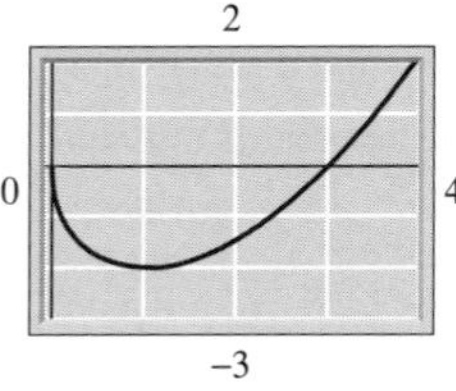

51. Absolute min at $(1, 1)$; vertical asymptote: $x = 0$

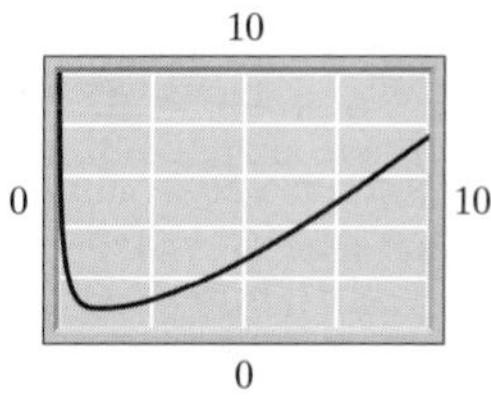

53. No relative extrema; points of inflection at $(1, 1)$ and $(-1, 1)$; vertical asymptote: $x = 0$

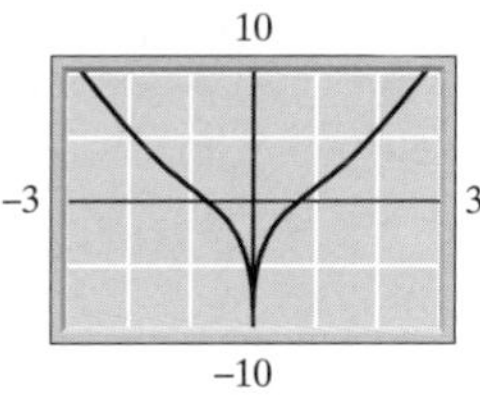

55. Absolute min at $(0, 1)$; absolute max at $(1, e - 1)$; relative max at $(-1, 1 + e^{-1})$

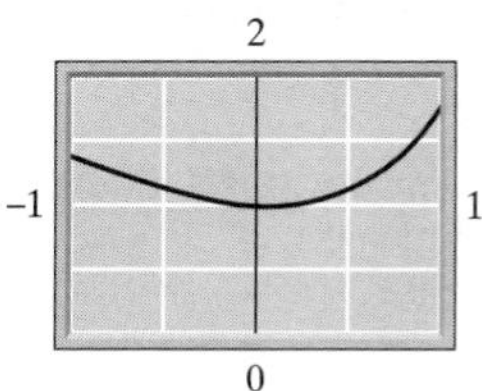

57. Relative max at $(-\pi, 0)$; absolute min at $(-\pi/2, -1)$; absolute max at $(\pi/2, 1)$; relative min at $(\pi, 0)$; point of inflection at $(0, 0)$

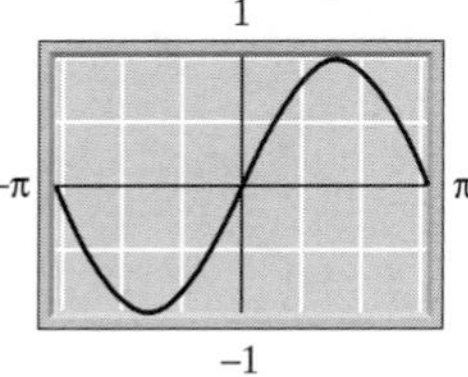

59. Absolute min at $(1.40, -1.49)$; points of inflection at $(0.21, 0.61)$ and $(0.79, -0.55)$

61. Relative min at $(-0.46, 0.73)$; relative max at $(0.91, 1.73)$; absolute min at $(3.73, -10.22)$; points of inflection at $(0.20, 1.22)$ and $(2.83, -5.74)$

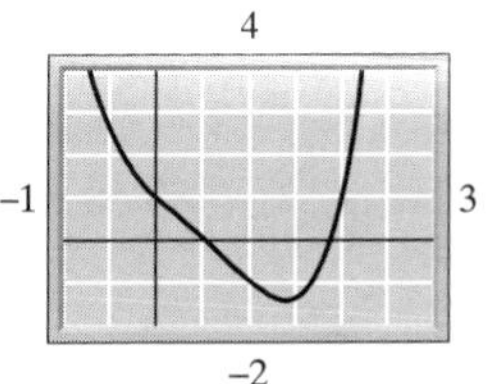

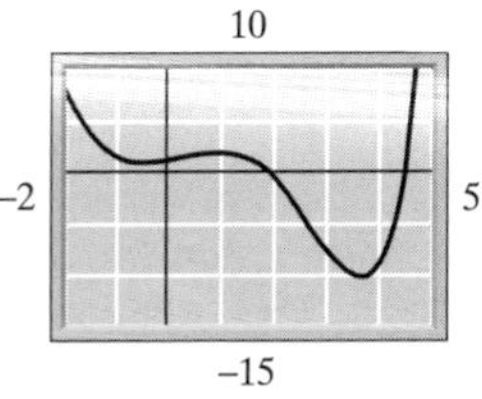

63. Absolute max at $(-6, 4.56)$; absolute min at $(\pm 2.60, -0.18)$; relative max at $(0, 1)$; absolute max at $(6, 4.56)$; points of inflection at $(-4.91, 2.61)$, $(-1.37, 0.39)$, $(1.37, 0.39)$, and $(4.91, 2.61)$

6
−6
6
−2

65. a. 2 years into the epidemic **b.** 2 years into the epidemic **67. a.** 1998 **b.** 2000 **c.** 1996 **69.** Point of inflection at (12.89, 193.8). The prison population was declining most rapidly at $t = 12.89$ (that is, in 1963), at which time the prison population was approximately 193,800. **71. a.** No **b.** Since the graph is concave up, the derivative of S is increasing, and so the rate of *decrease* of SAT scores with increasing numbers of prisoners is diminishing. In other words, the apparent effect of more prisoners is diminishing. **73. a.** $\left.\frac{d^2n}{ds^2}\right|_{s=3} = -21.494$. Thus, for a firm with annual sales of \$3 million, the rate at which new patents are produced decreases with increasing firm size. This means that the returns (as measured in the number of new patents per increase of \$1 million in sales) are diminishing as the firm size increases. **b.** $\left.\frac{d^2n}{ds^2}\right|_{s=7} = 13.474$. Thus, for a firm with annual sales of \$7 million, the rate at which new patents are produced increases with increasing firm size by 13.474 new patents per \$1 million increase in annual sales. **c.** There is a point of inflection when $s \approx 5.4587$, so that in a firm with sales of \$5,458,700 per year, the number of new patents produced per additional \$1 million in sales is a minimum. **75.** About \$570 per year, after about 12 years **77.** Increasing most rapidly in 17.64 years; decreasing most rapidly now (at $t = 0$)

79. a. Approximately 0.077 billion dollars per firm

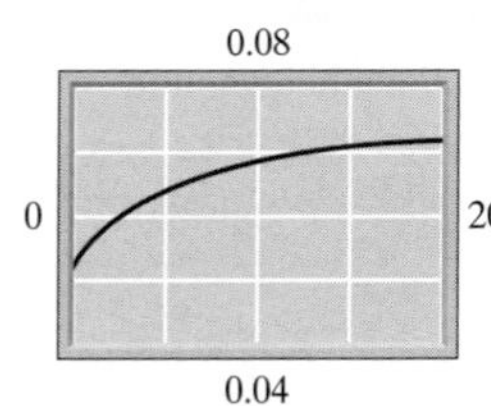

b. $R(t)/N(t)$ represents the average amount of money (in billions of dollars) controlled by a venture capital firm. $\lim_{t\to+\infty} R(t)/N(t)$ is a long-term prediction of the value of the average amount of money controlled by a venture capital firm in the long term. **81.** nonnegative **83.** At a point of inflection, the graph of a function changes from concave up to concave down, or vice versa. If it changes from concave up to concave down, then the derivative changes from increasing to decreasing and hence has a relative maximum. Similarly, if it changes from concave down to concave up, the derivative has a relative minimum.

5.4 exercises

1. $P = 10{,}000$; $dP/dt = 1000$ **3.** Let R be the annual revenue of my company, and let q be annual sales. $R = 7000$ and $dR/dt = 700$. Find dq/dt. **5.** Let p be the price of a pair of shoes, and let q be the demand for shoes. $dp/dt = 5$. Find dq/dt. **7.** Let T be the average global temperature, and let q be the number of Bermuda shorts sold per year. $T = 60$ and $dT/dt = 0.1$. Find dq/dt. **9. a.** $6/(100\pi) \approx 0.019$ km/s **b.** $6/(8\sqrt{\pi}) \approx 0.4231$ km/s **11.** 7.5 ft/s **13.** The price is decreasing at a rate of approximately 31¢ per pound per month. **15.** Monthly sales will drop at a rate of 40 T-shirts per month. **17.** Raise the price by 3¢ per week. **19.** The daily operating budget is dropping at a rate of \$2.40 per year. **21.** $3/(4\pi) \approx 0.24$ ft/min **23.** The y-coordinate is decreasing at a rate of 16 units per second. **25.** $2300/\sqrt{4100} \approx$ 36 mph **27.** \$1814 per year **29.** Their prior experience must increase at a rate of approximately 0.97 years every year **31.** $\frac{2500}{9\pi}\left(\frac{3}{5000}\right)^{2/3} \approx 0.63$ m/s **33.** $\frac{\sqrt{1+128\pi}}{4\pi} \approx 1.6$ cm/s **35.** The average SAT score was 904.71 and decreasing at a rate of 0.11 per year. **37.** Decreasing by 2 percentage points per year **39.** You need to know an equation that relates the two changing quantities, as well as the values of the quantities that appear in the derived equation (obtained by taking the derivative with respect to time t), to solve the derived equation for the desired rate of change. **41.** The section's goal is to compute the rate of change of a quantity based on a knowledge of the rate of change of a related quantity. **43.** Let $x =$ my grades and $y =$ your grades. If $dx/dt = 2\,dy/dt$, then $dy/dt = \left(\frac{1}{2}\right) dx/dt$.

5.5 exercises

1. $E = 1.5$; the demand is going down 1.5% per 1% increase in price at that price level; revenue is maximized when $p =$ \$25; weekly revenue at that price is \$12,500. **3. a.** $E = \frac{6}{7}$; the demand is going down 6% per 7% increase in price at that price level; thus a price increase is in order. **b.** Revenue is maximized when $p = 100/3 \approx \$33.33$. **c.** Demand would be $(100 - 100/3)^2 = (200/3)^2 \approx 4444$ cases per week. **5. a.** $E = 1.71$. Thus, the demand is elastic at the given tuition level, showing that a decrease in fees will result in an increase in revenue. **b.** It should charge an average of \$2271.75 per student, and this will result in an enrollment of about 4930 students, giving a revenue of about \$11,199,000. **7. a.** $E = -\frac{mp}{mp+b}$ **b.** $p = -\frac{b}{2m}$ **9. a.** $E = r$ **b.** E is independent of p. **c.** If $r = 1$, then the revenue is not affected by the price. If $r > 1$, then the revenue is always elastic. If $r < 1$, then the revenue is always inelastic. This is an unrealistic model because there should always be a price at which the revenue is a maximum. **11. a.** $E = 51$. The demand is going down 51% per 1% increase in price at that price level. Thus, a large price decrease is advised. **b.** Revenue is maximized when $p = 0.50$ yen. **c.** Demand would be $100e^{-3/4+1/2} \approx 78$ paint-by-number sets per month. **13. a.** $q = -1500p + 6000$ **b.** \$2 per hamburger, giving a total weekly revenue of \$6000 **15. a.** $f(p) = 1000e^{-0.3p}$ **b.** At $p = \$3$, $E = 0.9$; at $p = \$4$, $E = 1.2$; at $p = \$5$, $E = 1.5$ **c.** $p = \$3.33$ **d.** $p = \$5.36$. Selling at a lower price would increase demand, but you cannot sell more than 200 pounds anyway. You should charge as much as you can and still be able to sell all 200 pounds **17.** $(Y/Q) \cdot (dQ/dY) = \beta$. An increase in income of x% will result in an increase in demand of βx%. (Note that we should *not* take the negative here because we expect an increase in income to produce an *increase* in demand.) **19.** the price is lowered **21.** The distinction is best illustrated by an example. Suppose that q is measured in weekly sales and p is the unit price in dollars. Then the quantity $-dq/dp$ measures the drop in weekly sales per \$1 increase in price. The elasticity of demand E, on the other hand, measures the *percentage* drop in sales per 1% increase in price. Thus, $-dq/dp$ measures absolute change, while E measures fractional, or percentage, change.

Chapter 5 Review Test

1. a. Relative max: $(-1, 5)$; absolute min: $(-2, -3)$ and $(1, -3)$ **b.** Relative min: $(2, 3)$; absolute min: $\left(-2, -\frac{1}{9}\right)$ **c.** Absolute min: $(1, 0)$ **d.** No extrema **e.** Absolute min: $\left(-2, -\frac{1}{4}\right)$ **f.** Absolute min: $(0, 2)$ **2. a.** Relative max at $x = 1$; point of inflection at $x = -1/2$ **b.** Relative min at $x = 2$; point of inflection at $x = -1$ **c.** Relative max at $x = -2$; relative min at $x = 1$; point of inflection at $x = -1$ **d.** Relative min at $x = -2.5$ and $x = 1.5$; relative max at $x = 3$; point of inflection at $x = 2$

3. a.

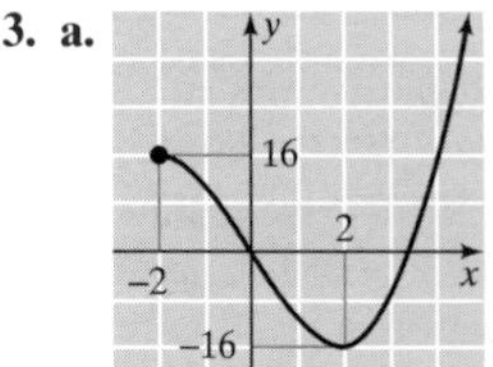

Relative max: $(-2, 16)$; absolute min: $(2, -16)$; point of inflection: $(0, 0)$; no horizontal or vertical asymptotes

b.

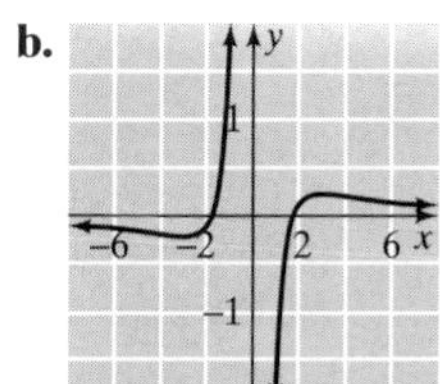

Relative min: $(-3, -2/9)$; relative max: $(3, 2/9)$; points of inflection: $(-3\sqrt{2}, -5\sqrt{2}/36)$ and $(3\sqrt{2}, 5\sqrt{2}/36)$; vertical asymptote: $x = 0$; horizontal asymptote: $y = 0$

c.

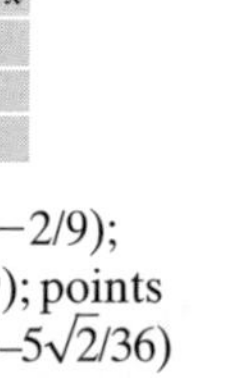

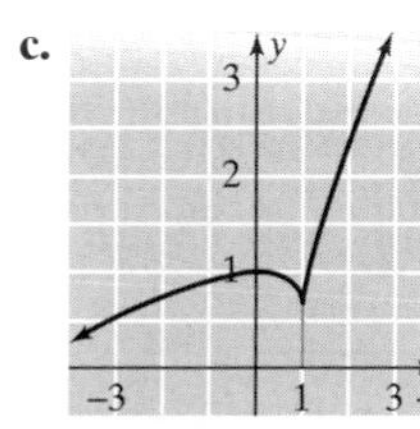

Relative max: $(0, 1)$; relative min: $\left(1, \frac{2}{3}\right)$; no inflection points; no horizontal or vertical asymptotes

4. a. $E = (2p^2 - 33p)/(-p^2 + 33p + 9)$ **b.** 0.52, 2.03. When the price is \$20, demand is dropping at a rate of 0.52% per 1% increase in the price; when the price is \$25, demand is dropping at a rate of 2.03% per 1% increase in the price. **c.** \$22.14 per book **5. a.** Profit $= -p^3 + 42p^2 - 288p - 181$ **b.** \$24 per copy **c.** \$3275 **d.** For maximum revenue, the company should charge \$22.14 per copy. At this price, the cost is decreasing linearly with increasing price, while the revenue is not decreasing (its derivative is zero). Thus, the profit is increasing with increasing price, suggesting that the maximum profit will occur at a higher price. **6. a.** $s'(t) = (2460.7e^{-0.55(t-4.8)})/(1 + e^{-0.55(t-4.8)})^2$. Sales were growing fastest in week 5.

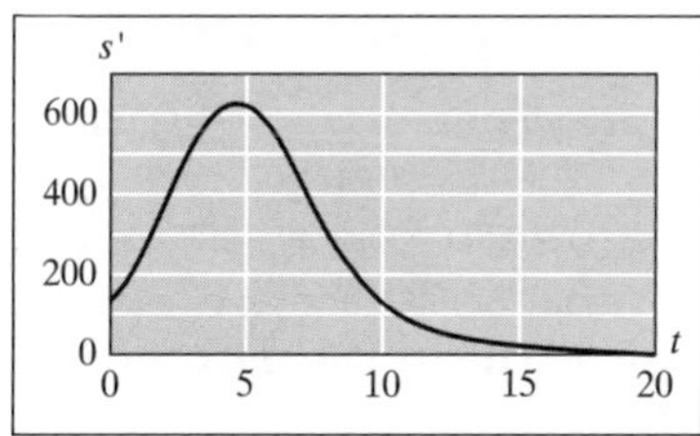

b. Point of inflection **c.** 10,527. If weekly sales continue as predicted by the model, they will level off at around 10,500 books per week in the long term. **d.** 0. If weekly sales continue as predicted by the model, the rate of change of sales approaches zero in the long term. **7.** $-1.5/(1 + 3\pi/2) \approx -0.263$ inches per second

Chapter 6

6.1 exercises

1. $x^6/6 + C$ **3.** $6x + C$ **5.** $x^2/2 + C$ **7.** $x^3/3 - x^2/2 + C$ **9.** $x + x^2/2 + C$ **11.** $-x^{-4}/4 + C$ **13.** $x^{3.3}/3.3 - x^{-0.3}/0.3 + C$ **15.** $u^3/3 - \ln|u| + C$ **17.** $2x^{3/2}/3 + C$ **19.** $3x^5/5 + 2x^{-1} - x^{-4}/4 + 4x + C$ **21.** $\ln|x| - 2/x + 1/(2x^2) + C$ **23.** $3x^{1.1}/1.1 - x^{5.3}/5.3 - 4.1x + C$ **25.** $x^{0.9}/0.3 + 40/x^{0.1} + C$ **27.** $2e^x + 5\ln|x| + x/4 + C$ **29.** $12.2x^{0.5} + x^{1.5}/9 - e^x + C$ **31.** $-\cos x + \sin x + C$ **33.** $2\sin x + 4.3\cos x - 9.33x + C$ **35.** $3.4\tan x + (\sin x)/1.3 - 3.2e^x + C$ **37.** $-1/x - 1/x^2 + C$ **39.** $f(x) = x^2/2 + 1$ **41.** $f(x) = e^x - x - 1$ **43.** $C(x) = 5x - x^2/20{,}000 + 20{,}000$ **45.** $C(x) = 5x + x^2 + \ln x + 994$ **47.** $I(t) = -0.7t + 7.4$; $I(8) = 1.8$ **49.** $N(t) = 25{,}000t^3/3 - 137{,}000t^2/2 + 68{,}000t + 3{,}100{,}000$ **51.** $C'(t) = 5t + 50$; $C(t) = 2.5t^2 + 50t$ **53. a.** $s = t^3/3 + t + C$ **b.** $C = 1$; $s = t^3/3 + t + 1$ **55.** 320 ft/s downward **59.** $(1280)^{1/2} \approx 35.78$ ft/s **61. a.** 80 ft/s **b.** 60 ft/s **c.** 1.25 seconds **63.** $\sqrt{2} \approx 1.414$ times as fast **65.** $\int (f(x) + g(x))\,dx$ is, by definition, an antiderivative of $f(x) + g(x)$. Let $F(x)$ be an antiderivative of $f(x)$, and let $G(x)$ be an antiderivative of $g(x)$. Then, since the derivative of $F(x) + G(x)$ is $f(x) + g(x)$ (by the rule for sums of derivatives), this means that $F(x) + G(x)$ is an antiderivative of $f(x) + g(x)$. In symbols, $\int (f(x) + g(x))\,dx = F(x) + G(x) + C = \int f(x)\,dx + \int g(x)\,dx$, the sum of the indefinite integrals. **67.** $\int x \cdot 1\,dx = \int x\,dx = x^2/2 + C$, whereas $\int x\,dx \cdot \int 1\,dx = (x^2/2 + D) \cdot (x + E)$, which is not the same as $x^2/2 + C$, no matter what values we choose for the constants C, D, and E. **69.** $\int f(x)\,dx$ represents the total cost of manufacturing x items. The units of $\int f(x)\,dx$ are the product of the units of $f(x)$ and the units of x. **71.** If you take the derivative of the indefinite integral of $f(x)$, you obtain $f(x)$ back. On the other hand, if you take the indefinite integral of the derivative of $f(x)$, you obtain $f(x) + C$.

6.2 exercises

1. $(3x + 1)^6/18 + C$ **3.** $(-2x + 2)^{-1}/2 + C$ **5.** $(3x^2 + 3)^4/24 + C$ **7.** $(x^2 + 1)^{2.3}/4.6 + C$ **9.** $x + 3e^{3.1x-2} + C$ **11.** $(7.6/3)\sin(3x - 4) + C$ **13.** $-(1/6)\cos(3x^2 - 4) + C$ **15.** $-2\cos(x^2 + x) + C$ **17.** $(1/6)\tan(3x^2 + 2x^3) + C$ **19.** $-(1/6)\ln|\cos(2x^3)| + C$ **21.** $3\ln|\sec(2x - 4) + \tan(2x - 4)| + C$ **23.** $2(3x^2 - 1)^{3/2}/9 + C$ **25.** $-(1/2)e^{-x^2+1} + C$ **27.** $-(1/2)e^{-(x^2+2x)} + C$ **29.** $(1/2)\sin(e^{2x} + 1) + C$ **31.** $(x^2 + x + 1)^{-2}/2 + C$ **33.** $(2x^3 + x^6 - 5)^{1/2}/3 + C$ **35.** $(x - 2)^7/7 + (x - 2)^6/3 + C$ **37.** $4[(x + 1)^{5/2}/5 - (x + 1)^{3/2}/3] + C$ **39.** $3e^{-1/x} + C$ **41.** $20\ln|1 - e^{-0.05x}| + C$ **43.** $(e^{2x^2-2x} + e^{x^2})/2 + C$ **51.** $-e^{-x} + C$ **53.** $(1/2)e^{2x-1} + C$ **55.** $(2x + 4)^3/6 + C$ **57.** $(1/5)\ln|5x - 1| + C$ **59.** $(1.5x)^4/6 + C$ **61.** $[\cos(-1.1x - 1)]/1.1 + C$ **63.** $f(x) = (x^2 + 1)^4/8 - 1/8$ **65.** $f(x) = (1/2)e^{x^2-1}$ **67.** $C(x) = 5x - 1/(x + 1) + 995.5$ **69.** $N(t) = 25{,}000(t - 1988)^3/3 - 137{,}000(t - 1988)^2/2 + 68{,}000t - 132{,}084{,}000$ **71. a.** $s = (t^2 + 1)^5/10 + t^2/2 + C$ **b.** $C = 9/10$; $s = (t^2 + 1)^5/10 + t^2/2 + 9/10$ **73.** None; the substitution $u = x$ simply replaces the letter x throughout by the letter u and thus does not change the integral at all. For instance, the integral $\int x(3x^2 + 1)\,dx$ becomes $\int u(3u^2 + 1)\,du$ if we substitute $u = x$. **75.** The integral $\int (2x + 1)(x^2 + x)\,dx$ can be calculated by expanding the integrand or by using the substitution $u = x^2 + x$, as follows:

Expanding the integrand: $\int (2x+1)(x^2+x)\,dx = \int (2x^3 + 3x^2 + x)\,dx = \frac{x^4}{2} + x^3 + \frac{x^2}{2} + C$
Substituting $u = x^2 + x$: $\int (2x+1)(x^2+x)\,dx = \int u\,du = \frac{u^2}{2} + C = \frac{(x^2+x)^2}{2} + C$
Since expanding the second answer gives the first, both methods result in the same solution. **77.** The purpose of substitution is to introduce a new variable that is defined in terms of the variable of integration. One cannot say $u = u^2 + 1$ because u is not a new variable. Instead, we define $w = u^2 + 1$ (or any other letter different from u).

6.3 exercises

1. 4 **3.** 6 **5.** 0.7456 **7.** 2.3129 **9.** 2.5048 **11.** 1.8429 **13.** 3.3045, 3.1604, 3.1436 **15.** 0.0275, 0.0258, 0.0256 **17.** 1.9835, 1.9998, 2.0000 **19.** \$99.95 **21.** −150,000 people; 150,000 jobs were lost **23.** \$66.4 billion. This represents the total sales of IBM mainframes and peripherals from 1990 through 1994. **25.** 91.2 ft **27.** 9.0526; Medicare spent \$9.0526 billion for hospice care in the United States from 1989 to 1997. **29 a.** 99.4% **b.** 0 (to at least 15 decimal places) **31.** If increasing n does not change the value of the answer when rounded to three decimal places, then the answer is likely accurate to three decimal places.

6.4 exercises

1. 1 **3.** $\frac{1}{2}$ **5.** $\frac{1}{4}$ **7.** 2 **9.** 0 **11.** 0 **13.** 0.45, 0.495 **15.** 1.9835, 1.9998 **17.** 14.24, 17.0144 **19.** 1.3813, 1.4541 **21.** \$66.4 billion. This represents the total sales of IBM mainframes and peripherals from 1990 through 1994. **23.** The total investment in developing countries in the years 1986 through 1993 **25. a.** Area represents the total amount spent by Medicare for hospice care in the United States from 1990 to 1996.

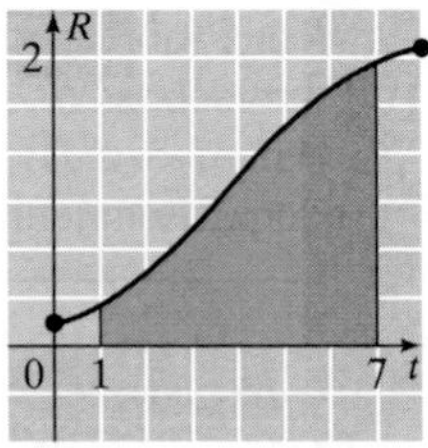

b. 6.84. Medicare spent \$6.84 billion for hospice care in the United States from 1990 to 1996. **27.** Yes. The Riemann sum gives an estimated area of 420 square feet.

6.5 exercises

1. x **3.** $x^2/2 - 2$ **5.** $4(x-a)$ **7.** $x^2/2$ **9.** $x^3/3$ **11.** $e^x - 1$ **13.** $\ln x$ **15.** $x^2/2$ if $0 \le x \le 1$; $1/2 + 2(x-1)$ if $x > 1$ **17.** 14/3 **19.** 0 **21.** 40/3 **23.** −0.9045 **25.** −2 **27.** ln(2) **29.** $2(e-1)$ **31.** 2/3 **33.** $(e^1 - e^{-3})/2$ **35.** $50(e^{-1} - e^{-2})$ **37.** $e^{2.1} - e^{-0.1}$ **39.** 0 **41.** $(5/2)(e^3 - e^2)$ **43.** $3^{5/2}/10 - 3^{3/2}/6 + 1/15$ **45.** 0 **47.** 1 **49.** (1/3)[ln 26 − ln 7] **51.** 1/2 **53.** 16/3 **55.** 56/3 **57.** 1/2 **59.** \$783 **61.** 49,000 people **63.** \$2.4 billion **65.** 296 miles **67. a.** $c(t) = 9.2 - 0.7636t$ **b.** \$55 billion **69.** 61 milliliters **73. a.** **(C)** **b.** \$620 billion **75.** \$5052 million **77. a.** $s(t) = 330e^{0.0872t}$ **b.** \$9044 billion (compared to \$8225 billion) **c.** \$7011 billion **79.** 907 T-shirts **81.** 9 gallons **85.** \$940 million **87.** 3 years **93.** Two things: (1) calculate definite integrals by using antiderivatives and (2) obtain formulas for an antiderivative of a function in terms of a definite integral, or area **95.** $F(x) = \begin{cases} 0 & \text{if } x < 0 \\ x & \text{if } x \ge 0 \end{cases}$; $F'(x) = f(x)$ for every $x \ne 0$ [since $F(x)$ is not differentiable at $x = 0$]. **97.** They are related by the Fundamental Theorem of Calculus, which (briefly) states that the definite integral of a suitable function can be calculated by evaluating the indefinite integral at the two endpoints and subtracting. **99.** $v(t) = t - 5$ **101.** $f(x) = e^{-x}$

Chapter 6 Review Test

1. a. $\frac{x^3}{3} - 5x^2 + 2x + C$ **b.** $e^x + \frac{2}{3}x^{3/2} + C$ **c.** $\frac{1}{22}(x^2+4)^{11} + C$ **d.** $-\frac{1}{3(x^3+3x+2)} + C$ **e.** $-\frac{5}{2}e^{-2x} + C$ **f.** $-e^{-x^2/2} + C$ **2. a.** 1/4 **b.** ln 10 ≈ 2.303 **c.** 52/9 **d.** −2 **3. a.** 32/3 **b.** 97/3 **c.** $(1 - e^{-25})/2$ **d.** 4 **4. a.** 0.77781682, 0.7499786, 0.74714013 **b.** 0.78690744, 0.69719004, 0.68849294 **5. a.** $100{,}000 - 10p^2$ **b.** \$100 **6. a.** About 86,000 books **b.** About 35,800 books **7. a.** 5 seconds **b.** 100 ft/s **c.** 156.25 ft above ground level

Chapter 7

7.1 exercises

1. $2e^x(x-1) + C$ **3.** $-e^{-x}(2+3x) + C$ **5.** $e^{2x}(2x^2 - 2x - 1)/4 + C$ **7.** $-e^{-2x+4}(2x^2 + 2x + 3)/4 + C$ **9.** $-e^{-x}(x^2 + x + 1) + C$ **11.** $\frac{1}{7}x(x+2)^7 - \frac{1}{56}(x+2)^8 + C$ **13.** $-\frac{x}{2(x-2)^2} - \frac{1}{2(x-2)} + C$ **15.** $(x^4 \ln x)/4 - x^4/16 + C$ **17.** $(t^3/3 + t)\ln(2t) - t^3/9 - t + C$ **19.** $\frac{3}{4}t^{4/3}\left(\ln t - \frac{3}{4}\right) + C$ **21.** $-x\cos x + \sin x + C$ **23.** $-\frac{1}{2}e^{-x}\cos x - \frac{1}{2}e^{-x}\sin x + C$ **25.** e **27.** 38,229/286 **29.** $\frac{7}{2}\ln 2 - \frac{3}{4}$ **31.** $\frac{1}{4}$ **33.** $\pi^2 - 4$ **35.** $1 - 11e^{-10}$ **37.** $4\ln 2 - \frac{7}{4}$ **39.** $28{,}800{,}000(1 - 2e^{-1})$ ft **41.** $5001 + 10x - 1/(x+1) - [\ln(x+1)]/(x+1)$ **43.** \$33,598 **45.** \$1478 million

7.2 exercises

1. $\frac{8}{3}$ **3.** 4 **5.** 1 **7.** $e - \frac{3}{2}$ **9.** $\frac{2}{3}$ **11.** $\frac{3}{10}$ **13.** $\frac{1}{20}$ **15.** $\frac{4}{15}$ **17.** $2 \ln 2 - 1$ **19.** $8 \ln 4 + 2e - 16$ **21.** 0.3222 **23.** 0.3222 **25.** \$6.25 **27.** \$512 **29.** \$119.53 **31.** \$900 **33.** \$416.67 **35.** \$326.27 **37.** \$25 **39.** \$0.50 **41.** \$386.29 **43.** \$225 **45.** \$25.50 **47.** \$12,684.63 **49.** $\bar{p} = \$3655$; $CS \approx \$856{,}000$; $PS \approx \$3{,}716{,}000$. The total social gain is approximately \$4,572,000. **51.** $CS = [1/(2m)](b - m\bar{p})^2$ **53. a.** The area represents the accumulated U.S. trade deficit with China (total value of imports over exports) for the 9-year period 1989–1998. **b.** 250.47; the United States accumulated a \$250.47 billion trade deficit with China over the period 1989–1998. **55. a.** \$373.35 million **b.** This is the area of the region between the graphs of $R(t)$ and $P(t)$ for $0 \le t \le 4$. **c.** Since the exponent for P is larger, this tells us that the ratio of profit to revenue was increasing; that is, costs accounted for a decreasing proportion of revenues. **57. b.** \$7260 million **c.** The area between $y = 1{,}000{,}000(41.25q^2 - 697.5q + 3210)$ and $y = 400{,}000{,}000$ for $0 \le q \le 12$ **59.** The area between the monthly export and import curves represents Canada's accumulated trade surplus for the given period; that is, the total value of exports over imports. **61.** The area under a curve can represent income only if the curve is a graph of income *per unit time.* The value of a stock price is not income per unit time; the income can be realized only when the stock is sold, and it amounts to the current market price. The total net income from the given investment would be the stock price on the date of sale minus the purchase price of \$22.

7.3 exercises

1. Average = 2

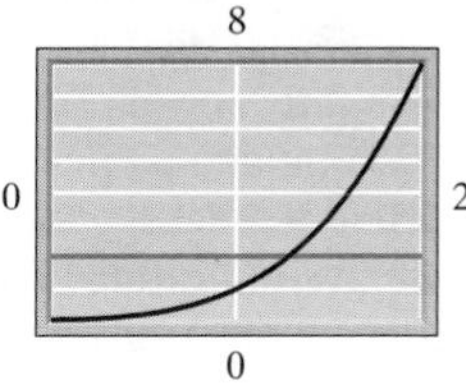

3. Average = 1

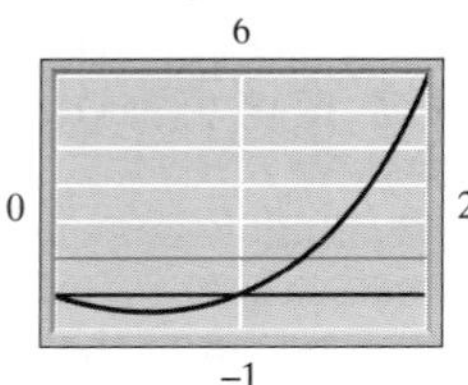

5. Average = $(1 - e^{-2})/2$

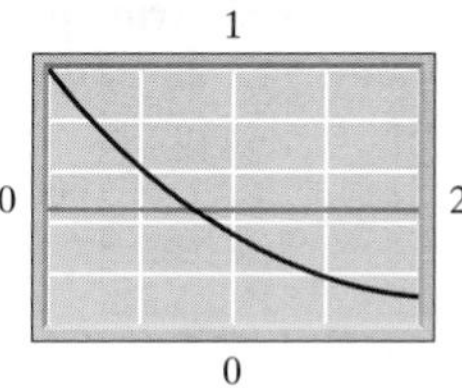

7. Average = $2/\pi$

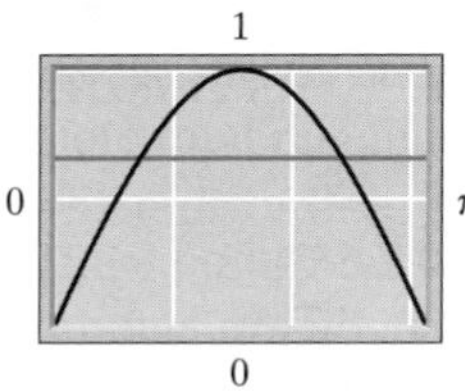

9.

x	0	1	2	3	4	5	6	7
$r(x)$	3	5	10	3	2	5	6	7
$\bar{r}(x)$			6	6	5	10/3	13/3	6

11.

x	0	1	2	3	4	5	6	7
$r(x)$	1	2	6	7	11	15	10	2
$\bar{r}(x)$			3	5	8	11	12	9

13. $\bar{f}(x) = x^3 - \frac{15}{2}x^2 + 25x - \frac{125}{4}$

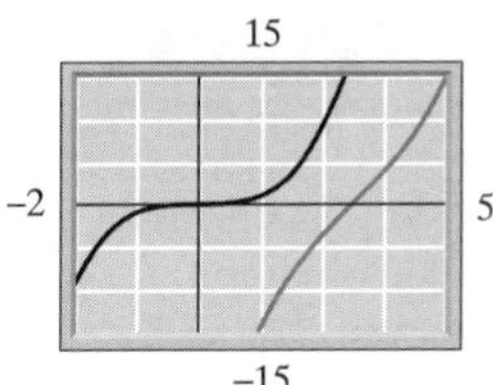

15. $\bar{f}(x) = \frac{3}{25}[x^{5/3} - (x - 5)^{5/3}]$

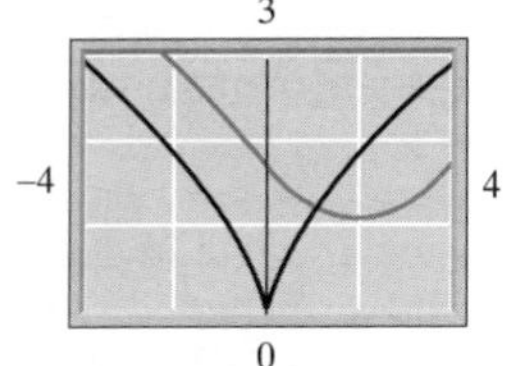

17. $\bar{f}(x) = \frac{2}{5}(e^{0.5x} - e^{0.5(x-5)})$

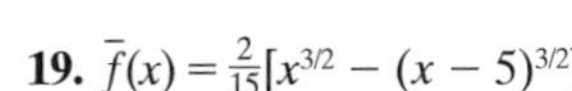

19. $\bar{f}(x) = \frac{2}{15}[x^{3/2} - (x - 5)^{3/2}]$

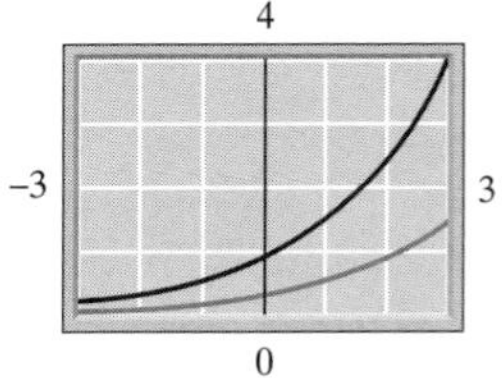

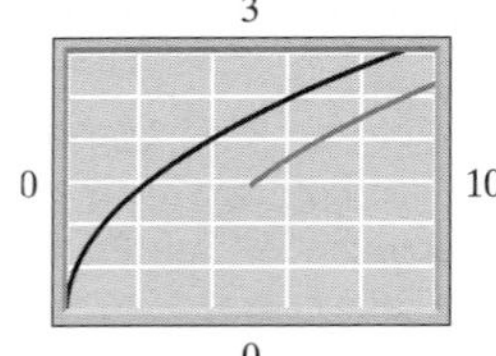

21. $\bar{f}(x) = \frac{2}{\pi}\left[\sin\left(\frac{\pi x}{10}\right) - \sin\left(\frac{\pi x}{10} - \frac{\pi}{2}\right)\right] = \frac{2}{\pi}\left[\sin\left(\frac{\pi x}{10}\right) + \cos\left(\frac{\pi x}{10}\right)\right]$

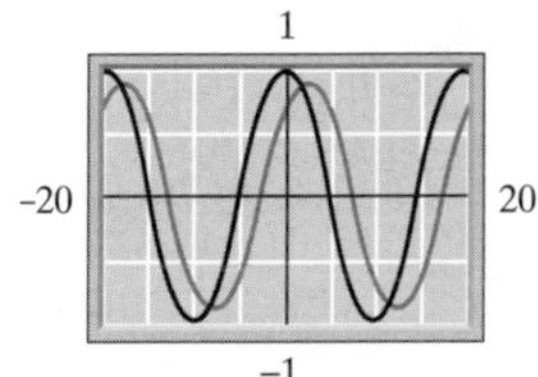

23.

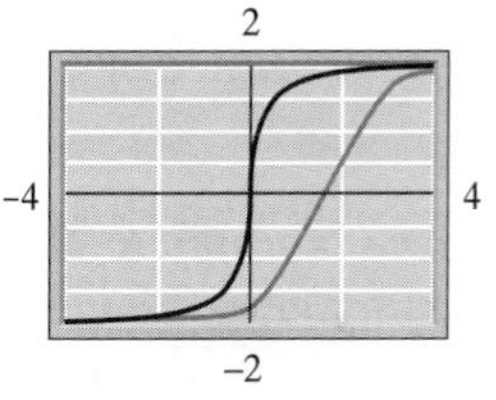

25.

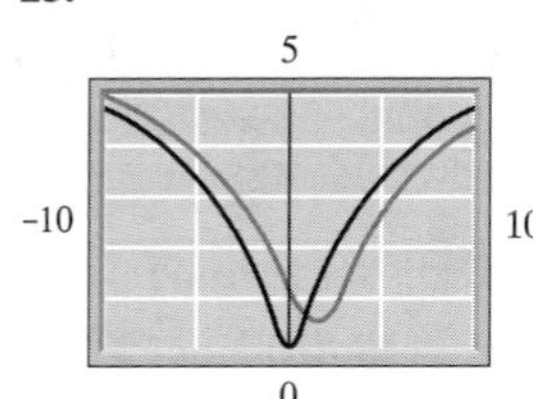

27. \$24 billion per year **29.** 362 firms **31.** \$10,410.88 **33.** \$1500

35.

Year, t	0	1	2	3	4	5	6	7
Spending (billions)	\$16	18	18	20	22	26	28	30
Moving average (billions)				\$18	19.5	21.5	24	26.5

No; the rate of change of the moving average was gradually increasing, showing an acceleration in spending.

37. a.

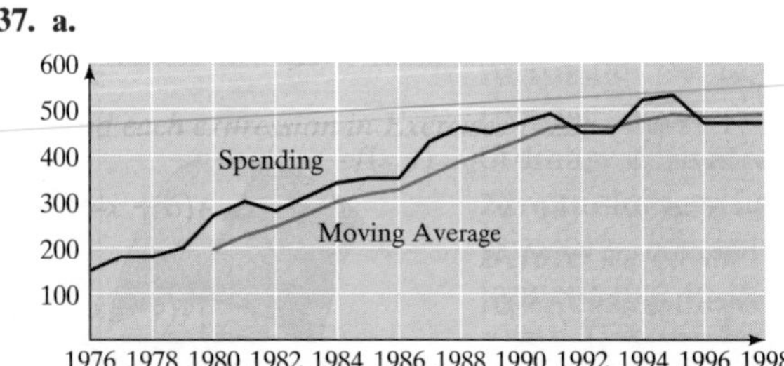

b. \$23.4 million per year. Tourist spending in Bermuda was increasing at a rate of approximately \$23.4 million per year during the given period. **39. a.** 3650 pounds **b.** $\frac{1}{2}[x^3 - (x-2)^3 - 45[x^2 - (x-2)^2] + 8400]$ **c.** Quadratic **41. a.** $n(t) = 6.4 + 0.3t$, where t is the number of years since 1983 **b.** $\overline{n}(t) = 5.8 + 0.3t$ **c.** The slope of the moving average is the same as the slope of the original function.

d.

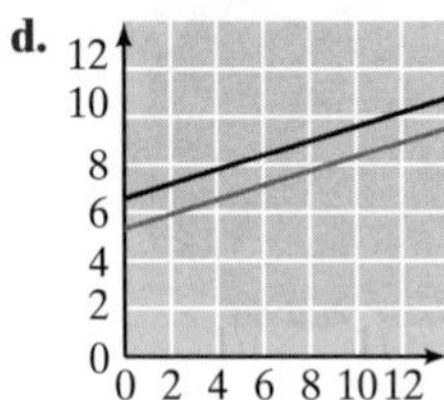

43. $\bar{f}(x) = mx + b - ma/2$ **45. a.** $p(t) = -0.04t^2 + t + 12$ **b.** 16.67% **c.** 10.38%

47. a.

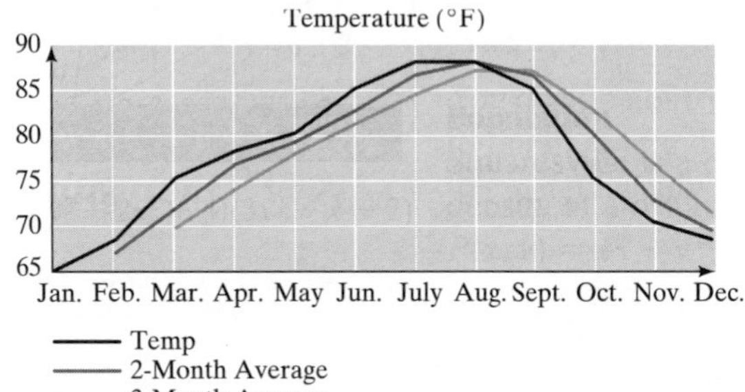

b. The 24-month moving average is constant and equal to the year-long average of approximately 77°. **c.** A quadratic model could not be used to predict temperatures beyond the given 12-month period because temperature patterns are periodic, whereas parabolas are not. **49. a.** Average voltage over $\left[0, \frac{1}{6}\right]$ is zero; 60 cycles per second

b.

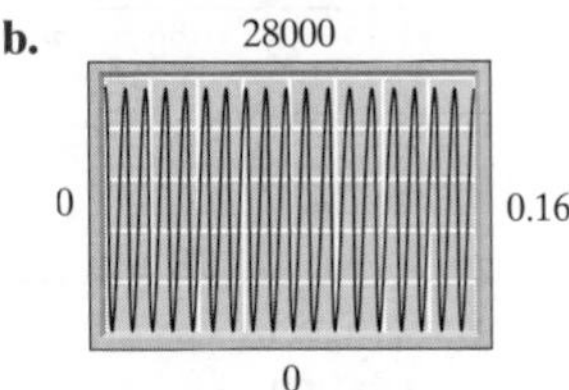

c. 116.673 volts **51.** The area above the x-axis equals the area below the x-axis. Example: $y = x$ on $[-1, 1]$ **53.** The moving average "blurs" the effects of short-term oscillations in the price and shows the longer-term trend of the stock price. **55.** This need not be the case; for instance, the function $f(x) = x^2$ on $[0, 1]$ has average value $\frac{1}{3}$, whereas the value midway between the maximum and minimum is $\frac{1}{2}$. **57. (C)** A shorter-term moving average most closely approximates the original function because it averages the function over a shorter period, and continuous functions change by only a small amount over a small period.

7.4 exercises

1. $TV = \$300{,}000$; $FV = \$434{,}465.45$ **3.** $TV = \$350{,}000$; $FV = \$498{,}496.61$ **5.** $TV = \$389{,}232.76$; $FV = \$547{,}547.16$ **7.** $TV = 0$; $FV = \$1153.86$ **9.** $TV = \$100{,}000$; $PV = \$82{,}419.99$ **11.** $TV = \$112{,}500$; $PV = \$92{,}037.48$ **13.** $TV = \$107{,}889.50$; $PV = \$88{,}479.69$ **15.** $TV = 0$; $PV = \$15.58$ **17.** \$29.46 billion **19.** \$47 billion **21.** \$36.10 billion **23.** \$40 billion **25.** \$1,943,162.44 **27.** \$3,086,245.73 **29.** \$73,036.35 **31.** \$1,792,723.35 **33.** She is correct, provided there is a positive rate of return, in which case the future value (which includes interest) is greater than the total value (which does not). **35.** $PV \le TV \le FV$

7.5 exercises

1. Diverges **3.** Converges to $2e$ **5.** Converges to e^2 **7.** Converges to $\frac{1}{2}$ **9.** Converges to 1/108 **11.** Converges to $3 \times 5^{2/3}$ **13.** Diverges **15.** Diverges **17.** Converges to $\frac{5}{4}(3^{4/5} - 1)$ **19.** Diverges **21.** Converges to 0 **23.** Diverges **25.** Diverges **27.** Diverges **29.** Converges to $\frac{1}{2}$ **31.** No; you will not sell more than 2000 of them. **33.** \$2600 million **35.** \$70,833 **37.** \$25,160 billion **39.** 32,636,000,000 newspapers **41.** 10,303,000,000 newspapers. Starting in June 2000, the *New York Times* will sell 10,303 million more newspapers than the *Washington Post,* assuming the decay models continue to hold. **43.** $\int_0^{+\infty} q(t)\,dt$ diverges, indicating that there is no bound to the expected future exports of pork. $\int_{-\infty}^0 q(t)\,dt$ converges to approximately 8.3490, indicating that total exports of pork prior to 1985 amounted to approximately \$8.3490 million. **45. a.** 2.468 meteors on average **b.** The integral diverges. We can interpret this as saying that the number of impacts by meteors smaller than 1 megaton is very large. (This makes sense because, for example, this number includes meteors no larger than a grain of dust.) **47 a.** $\Gamma(1) = 1$; $\Gamma(2) = 1$ **49.** 1 **51.** 0.1587 **53.** Yes; the integrals converge to zero, and the FTC also gives zero. **55.** The integral does not converge, so the number given by the FTC is meaningless. **57.** In all cases, you need to rewrite the improper integral as a limit and use technology to evaluate the integral of which you are taking the limit. Evaluate the integral for several values of the endpoint approaching the limit. In the case of an integral in which one of the limits of integration is infinite, you may have to instruct the calculator or computer to use more subdivisions as you approach $+\infty$.

7.6 exercises

1. $y = \frac{x^3}{3} + \frac{2x^{3/2}}{3} + C$ **3.** $\frac{y^2}{2} = \frac{x^2}{2} + C$ **5.** $y = Ae^{x^2/2}$ **7.** $y = -\frac{2}{(x+1)^2 + C}$ **9.** $y = \sqrt{(\ln|x|)^2 + C}$ **11.** $y = \frac{x^4}{4} - x^2 + 1$ **13.** $y = (x^3 + 8)^{1/3}$ **15.** $y = 2x$ **17.** $y = e^{x^2/2} - 1$ **19.** $y = -\frac{2}{\ln(x^2+1) + 2}$ **21.** With $s(t)$ = monthly sales after t months, $ds/dt = -0.05s$; $s = 1000$ when $t = 0$. Solution: $s = 1000e^{-0.05t}$. **23.** $H(t) = 75 + 115e^{-0.04274t}$, where t is measured in minutes **25.** With $S(t)$ = total sales after t months, $dS/dt = 0.1(100{,}000 - S)$; $S(0) = 0$. Solution: $S = 100{,}000(1 - e^{-0.1t})$. **27. a.** $dp/dt = k(D(p) - S(p)) = k(20{,}000 - 1000p)$ **b.** $p = 20 - Ae^{-kt}$ **c.** $p = 20 - 10e^{-0.2231t}$ **29.** $q = 0.6078e^{-0.05p}p^{1.5}$ **33.** $S = \frac{2/1999}{e^{-0.5t} + 1/1999}$ Graph:

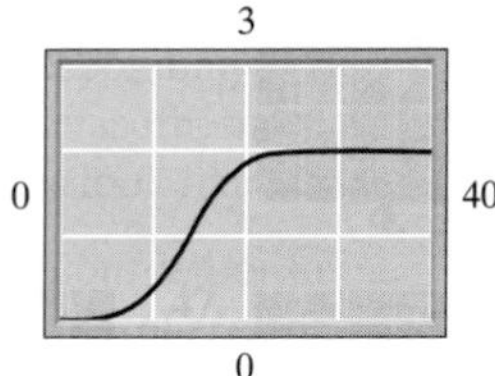

It will take about 27 months to saturate the market. **35. a.** $y = be^{Ae^{-at}}$, A = constant **b.** $y = 10e^{-0.69315e^{-t}}$ Graph:

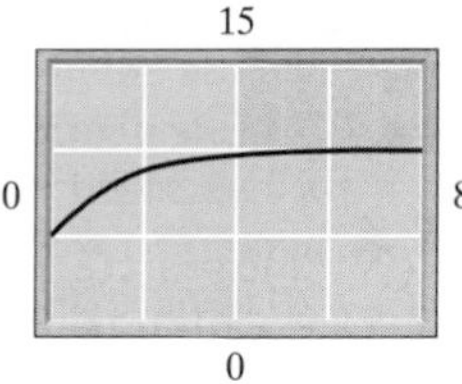

37. A general solution gives all possible solutions to the equation, using at least one arbitrary constant. A particular solution is one specific function that satisfies the equation. We obtain a particular solution by substituting specific values for any arbitrary constants in the general solution. **39.** Example: $y'' = x$ has general solution $y = \frac{1}{2}x^2 + Cx + D$ (integrate twice). **41.** $y' = -4e^{-x} + 3$

Chapter 7 Review Test

1. a. $(x^2 - 2x + 4)e^x + C$ **b.** $\frac{1}{3}x^3 \ln 2x - x^3/9 + C$ **c.** $-e^2 - 39/e^2$ **d.** $(2e^3 + 1)/9$ **e.** $-x^2 \cos x + 2x \sin x + 2 \cos x + C$ **f.** $\frac{1}{5}e^x \sin 2x - \frac{2}{5}e^x \cos 2x + C$ **g.** $\frac{1}{4}$ **h.** 2 **2. a.** $\frac{1}{2}$ **b.** $e^2 + \frac{1}{e^2} - 2$ **c.** $\frac{2\sqrt{2}}{3}$ **d.** $1 + 3/e^2$ **3. a.** $e - 2$ **b.** $\left[2(e^2 + e)\ln 2 + e^2 + \frac{5}{4}\right]/(2e - 1) \approx 5.10548$ **4. a.** $\frac{3}{14}[(x^{7/3} - (x - 2)^{7/3}]$ **b.** $\frac{1}{2}[x \ln x - (x - 2)\ln(x - 2) - 2]$ **5. a.** $y = -\frac{3}{x^3 + C}$ **b.** $y = Ae^{x^2/2} - 2$ **c.** $y = \sqrt{2\ln|x| + 1}$ **d.** $y = 2\sqrt{x^2 + 1}$ **6.** Approximately $\$910{,}000$ **7. a.** $\bar{p} = \$50$; $\bar{q} = 10{,}000$ **b.** $CS \approx \$333{,}000$; $PS \approx \$66{,}700$ **8. a.** $\$1{,}062{,}500$ **b.** $997{,}500e^{0.06t}$ **c.** $\$5{,}549{,}000$ **d.** Principal: $\$5{,}280{,}000$; interest: $\$269{,}000$ **9.** $\$51$ million **10.** The amount in the account would be given by $y = 10{,}000/(1 - t)$, where t is time in years, so it would approach infinity 1 year after the deposit.

Chapter 8

8.1 exercises

1. a. 1 **b.** 1 **c.** 2 **d.** $a^2 - a + 5$ **e.** $y^2 + x^2 - y + 1$ **f.** $(x + h)^2 + (y + k)^2 - (x + h) + 1$ **3. a.** 0 **b.** 0.2 **c.** -0.1 **d.** $0.18a + 0.2$ **e.** $0.1x + 0.2y - 0.01xy$ **f.** $0.2(x + h) + 0.1(y + k) - 0.01(x + h)(y + k)$ **5. a.** 1 **b.** e **c.** e **d.** e^{x+y+z} **e.** $e^{x+h+y+k+z+l}$ **7. a.** Does not exist **b.** 0 **c.** 0 **d.** $xyz/(x^2 + y^2 + z^2)$ **e.** $(x + h)(y + k)(z + l)/[(x + h)^2 + (y + k)^2 + (z + l)^2]$ **9. a.** f increases by 2.3 units **b.** f decreases by 1.4 units **c.** f decreases by 2.5 units for every 1-unit increase in z. **11.** Neither **13.** Linear **15.** Linear **17.** Interaction **19. a.** 107 **b.** -14 **c.** -113 **21.** Reading left to right, starting at the top: 52, 107, 162, 217, 94, 194, 294, 394, 136, 281, 426, 571, 178, 368, 558, 748 **25.** Reading top to bottom: 18, 4, 0.0965, 47,040 **27.** Reading top to bottom: 6.9078, 1.5193, 5.4366, 0 **29.** Let z = annual sales of Z (in millions of dollars), x = annual sales of X, and y = annual sales of Y. The model is $z = -2.1x + 0.4y + 16.2$. **31.** $\sqrt{2}$ **33.** $\sqrt{a^2 + b^2}$ **35.** $\frac{1}{2}$ **37.** Circle with center $(2, -1)$ and radius 3 **39.** The marginal cost of a car is $\$6000$ per car. The marginal cost of a truck is $\$4000$ per truck. **41.** $C(x, y) = 10 + 0.03x + 0.04y$, where C is the cost in dollars; x = number of video clips sold per month, y = number of audio clips sold per month **43. a.** The model predicts 8.06%. (The actual figure was 8.2%, supporting the accuracy of the model.) **b.** Foreign manufacturers because each 1% gain of the market by foreign manufacturers decreases Chrysler's share by 0.8%—the largest of the three. **45. a.** $\$9980$ **b.** $R(z) = 9850 + 0.04z$ **47.** 7,000,000 **49. a.** $100 = K(1000)^a(1{,}000{,}000)^{1-a}$; $10 = K(1000)^a(10{,}000)^{1-a}$ **b.** $\log K - 3a = -4$; $\log K - a = -3$ **c.** $a = 0.5$, $K \approx 0.003162$ **d.** $P = 71$ pianos (to the nearest piano) **51.** $s(i, t) = 0.67i + 0.02t + 0.63$ **53. a.** CBS **b.** ABC **c.** 7.9 **d.** $A(f, n, c) = (f + 0.27c - 0.87n + 36.7)/3.1$, where f is Fox's prime-time rating **55. a.** The demand function increases with increasing values of r. **b.** $Q(2 \times 10^8, 0.5, 500) = 90{,}680$. This means that if the total real income in Great Britain is 2×10^8 units of currency, if the average retail price of beer is 0.5 unit of currency per unit of beer, and if the average retail price of all other commodities is 500 units of currency, then 90,680 units of beer will be sold per year. **57.** $U(11, 10) - U(10, 10) \approx 5.75$. This means that if your company now has ten copies of Macro Publish and ten copies of Turbo Publish, then the purchase of one additional copy of Macro Publish will result in a productivity increase of approximately 5.75 pages per day. **59. a.** $(a, b, c) = (3, 1/4, 1/\pi)$; $(a, b, c) = (1/\pi, 3, 1/4)$ **b.** $a = [3/(4\pi)]^{1/3}$. The resulting ellipsoid is a sphere with radius a.

61. a.

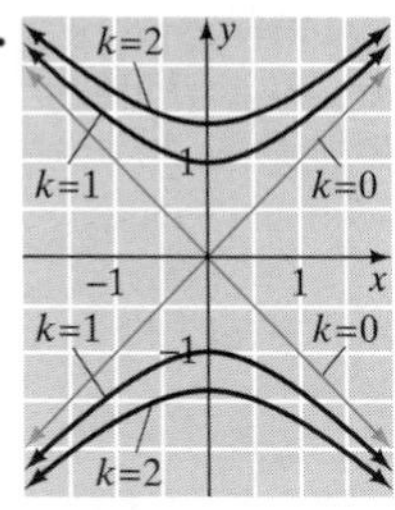

b.

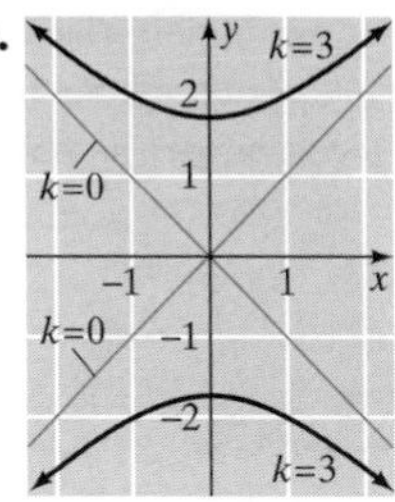

c.

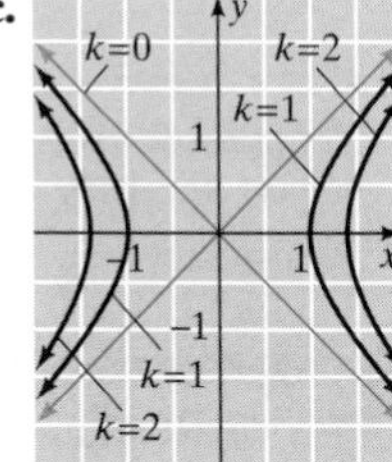

63. a. 4×10^{-3} gram per square meter **b.** The total weight of sulfates in the earth's atmosphere **65. a.** The value of N would be doubled. **b.** $N(R, f_p, n_e, f_l, f_i, L) = Rf_p n_e f_l f_i L$, where here L is the average lifetime of an intelligent civilization **c.** Take the logarithm of both sides, since this would yield the linear function $\ln(N) = \ln(R) + \ln(f_p) + \ln(n_e) + \ln(f_l) + \ln(f_i) + \ln(f_c) + \ln(L)$. **67.** For example, take $f(x, y) = x + y$. Then setting $y = 3$ gives $f(x, 3) = x + 3$. This can be viewed as a function of the single variable x. Choosing other values for y gives other functions of x. **69.** $f(x, y) = x - y$

8.2 exercises

1.

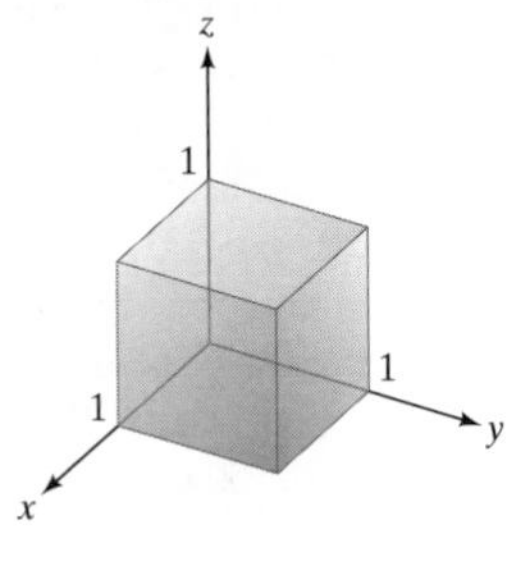

3.

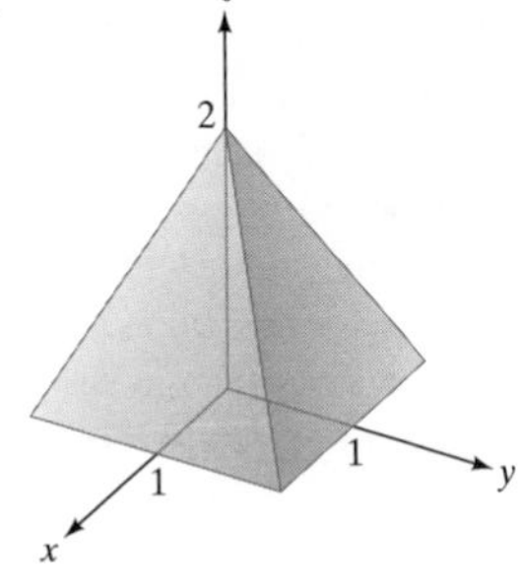

5. **7.**

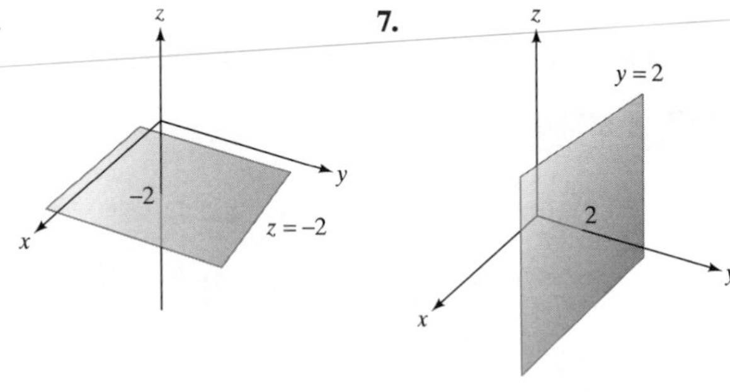

9.

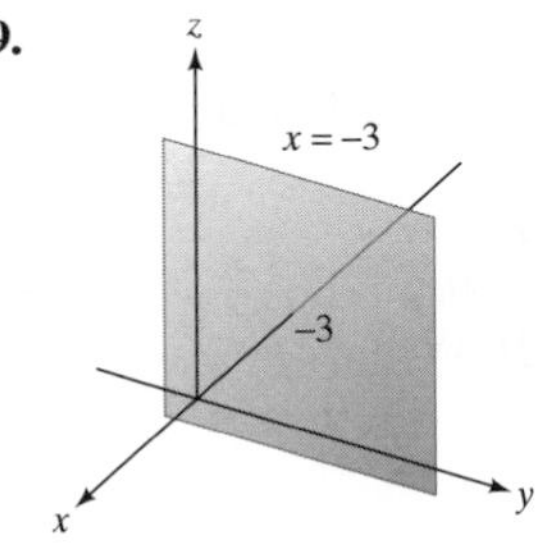

11. (H) **13.** (B) **15.** (F) **17.** (C)

19.

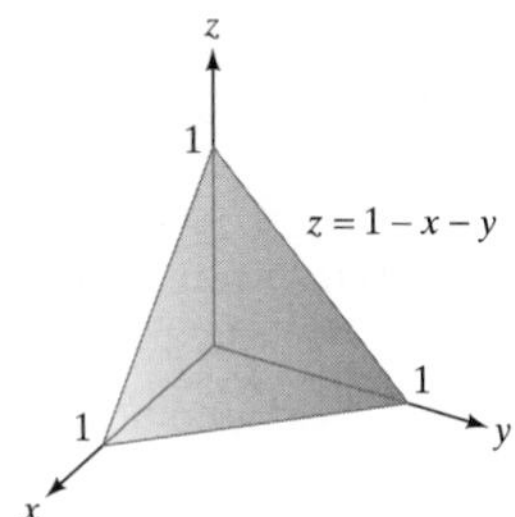

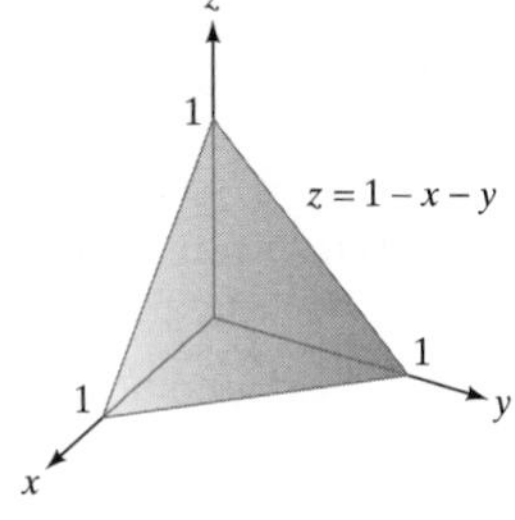

21.

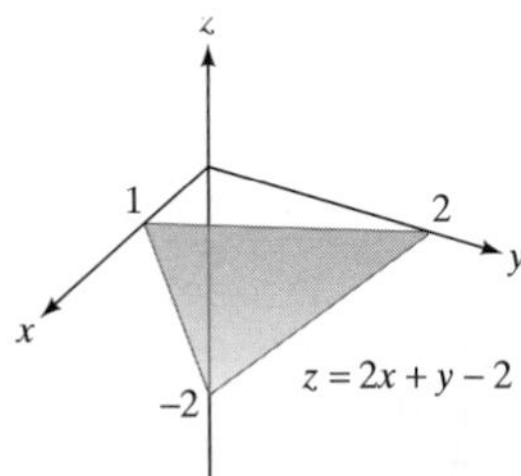

23.

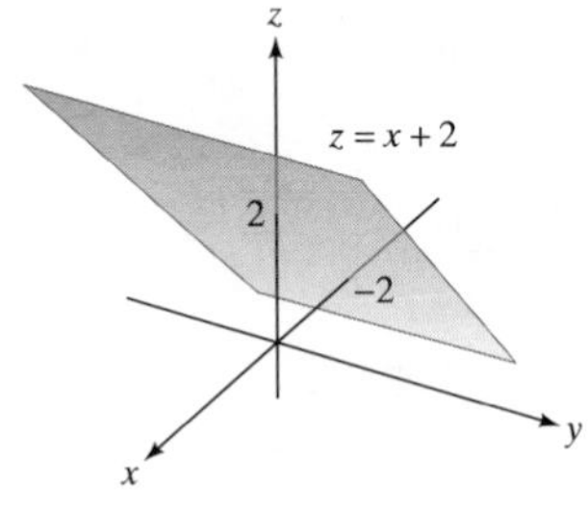

25.

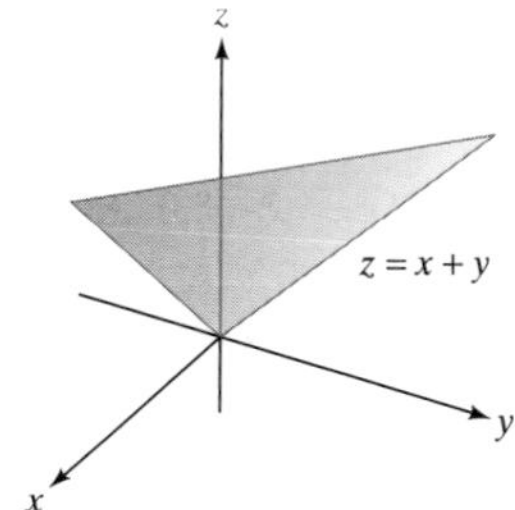

27.

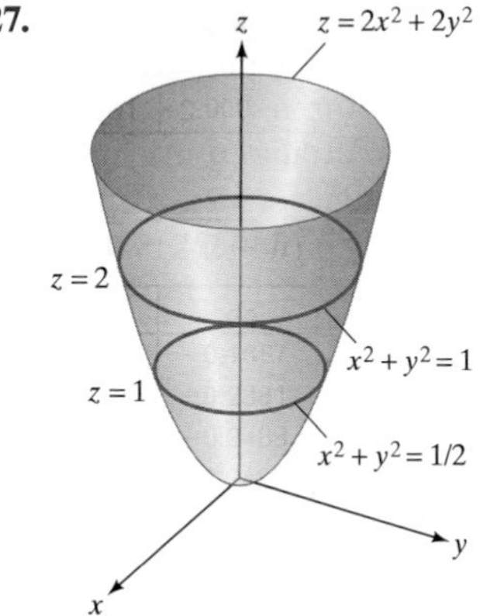

29.

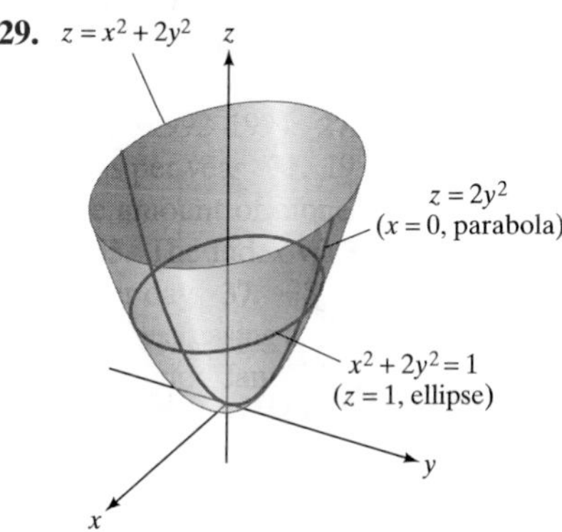

31.

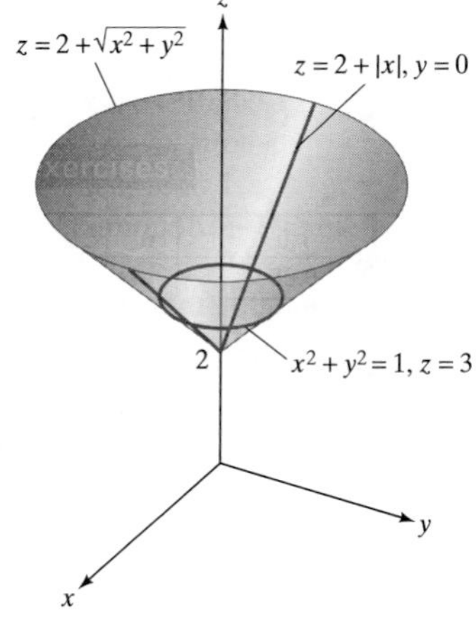

33.

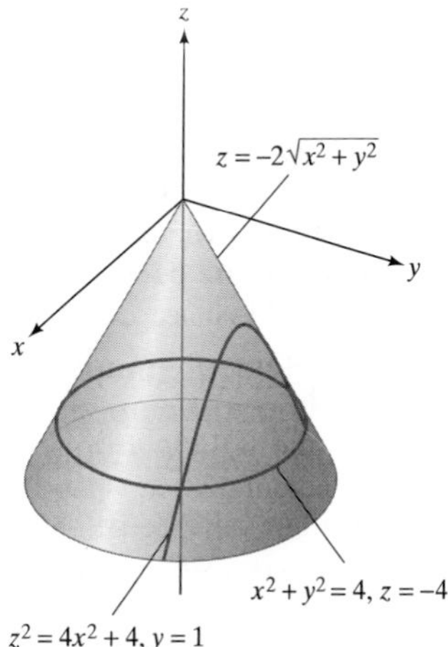

35.

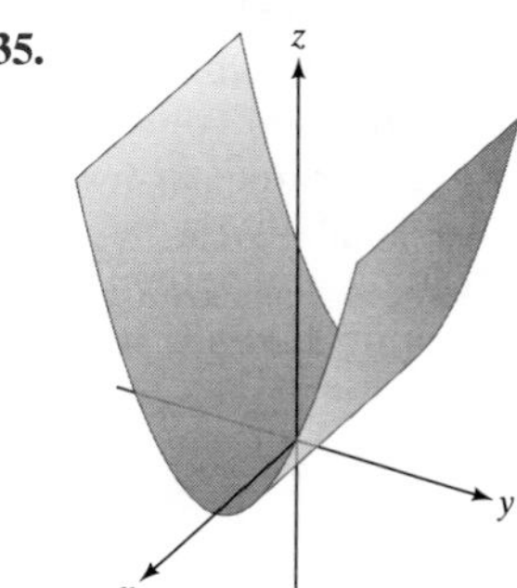

37.

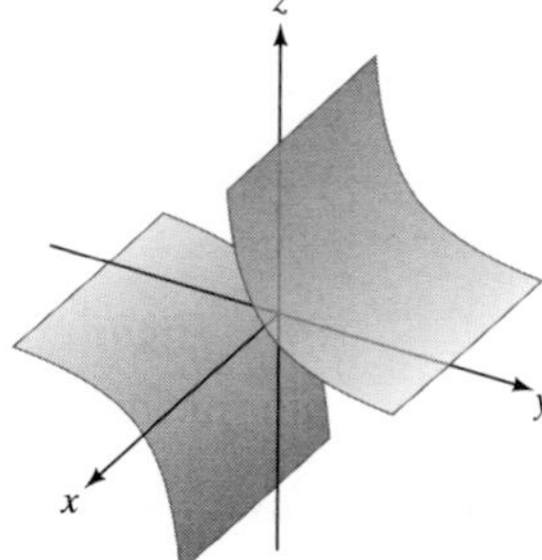

39.

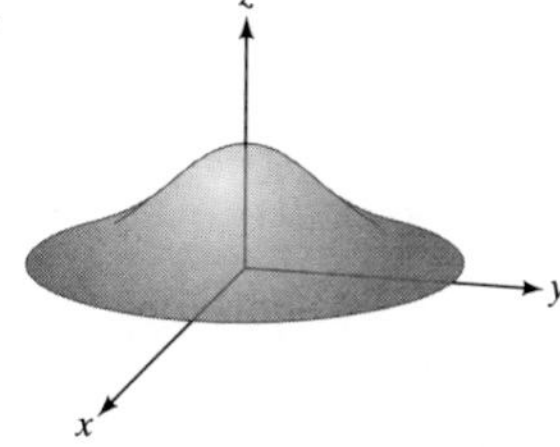

41. a. The graph is a plane with x-intercept -40, y-intercept -60, and z-intercept 240,000. **b.** The slice by $x = 10$ is the straight line with equation $z = 300{,}000 + 4000y$. It describes the cost function for the manufacture of trucks if car production is held fixed at 10 cars per week. **c.** The level curve

$z = 480{,}000$ is the straight line $6000x + 4000y = 240{,}000$. It describes the number of cars and trucks you can manufacture to maintain weekly costs at \$480,000. **43.** The graph is a plane with x_1-intercept 0.3, x_2-intercept 33, and x_3-intercept 0.66. The slices by $x_1 = constant$ are straight lines that are parallel to one another. Thus, the rate of change of General Motors' share as a function of Ford's share does not depend on Chrysler's share. Specifically, GM's share decreases by 0.02 percentage point per 1-percentage-point increase in Ford's market share, regardless of Chrysler's share. **45. a.** The slices $x = constant$ and $y = constant$ are straight lines. **b.** No. Even though the slices $x = constant$ and $y = constant$ are straight lines, the level curves are not, and so the surface is not a plane. **c.** The slice by $x = 10$ has a slope of 3800. The slope through $x = 20$ has a slope of 3600. Manufacturing more cars lowers the marginal cost of manufacturing trucks. **47.** Both slices are quarter-circles. (We see only the portion in the first quadrant because $e \geq 0$ and $k \geq 0$.) The level curve $C = 30{,}000$ represents the relationship between the number of electricians and the number of carpenters used in building a home that costs \$30,000. Similarly for the level curve $C = 40{,}000$. **49.** The following figure shows several level curves together with several lines of the form $h + w = c$.

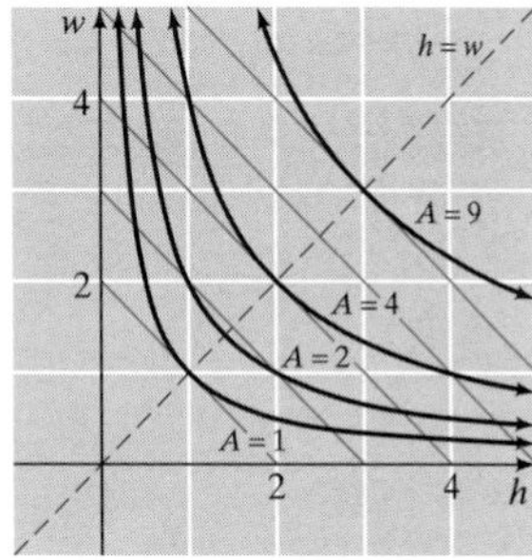

From the figure, thinking of the curves as contours on a map, we see that the largest value of A anywhere along any of the lines $h + w = c$ occurs midway along the line, when $h = w$. Thus, the largest area rectangle with a fixed perimeter occurs when $h = w$ (that is, when the rectangle is a square).

51.

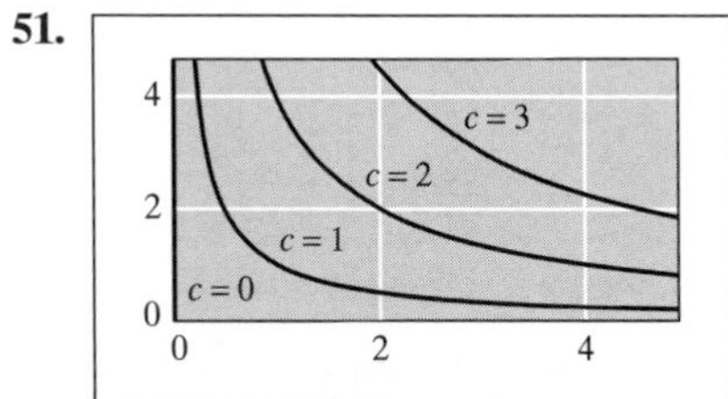

The level curve at $z = 3$ has the form $3 = x^{0.5}y^{0.5}$, or $y = 9/x$, and shows the relationship between the number of workers and the operating budget at a production level of 3 units.

53.

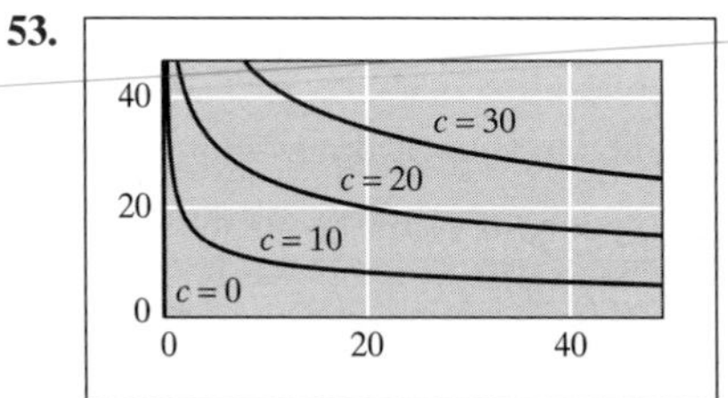

The level curve at $z = 0$ consists of the nonnegative y-axis ($x = 0$) and tells us that zero utility corresponds to zero copies of Macro Publish, regardless of the number of copies of Turbo Publish. (Zero copies of Turbo Publish does not necessarily result in zero utility, according to the formula.) **55.** Disagree, for example, the function $f(x, y) = xy$ has such slices, but its graph is not a plane. **59.** We need one dimension for each of the variables plus one dimension for the value of the function.

8.3 exercises

1. $f_x(x, y) = -40$; $f_y(x, y) = 20$; $f_x(1, -1) = -40$; $f_y(1, -1) = 20$ **3.** $f_x(x, y) = 6x + 1$; $f_y(x, y) = -3y^2$; $f_x(1, -1) = 7$; $f_y(1, -1) = -3$ **5.** $f_x(x, y) = -40 + 10y$; $f_y(x, y) = 20 + 10x$; $f_x(1, -1) = -50$; $f_y(1, -1) = 30$ **7.** $f_x(x, y) = 6xy$; $f_y(x, y) = 3x^2$; $f_x(1, -1) = -6$; $f_y(1, -1) = 3$ **9.** $f_x(x, y) = 2xy^3 - 3x^2y^2 - y$; $f_y(x, y) = 3x^2y^2 - 2x^3y - x$; $f_x(1, -1) = -4$; $f_y(1, -1) = 4$ **11.** $f_x(x, y) = 6y(2xy + 1)^2$; $f_y(x, y) = 6x(2xy + 1)^2$; $f_x(1, -1) = -6$; $f_y(1, -1) = 6$ **13.** $f_x(x, y) = e^{x+y}$; $f_y(x, y) = e^{x+y}$; $f_x(1,-1) = 1$; $f_y(1,-1) = 1$ **15.** $f_x(x, y) = 3x^{-0.4}y^{0.4}$; $f_y(x, y) = 2x^{0.6}y^{-0.6}$; $f_x(1,-1)$ undefined; $f_y(1,-1)$ undefined **17.** $f_x(x, y) = 0.2ye^{0.2xy}$; $f_y(x, y) = 0.2xe^{0.2xy}$; $f_x(1, -1) = -0.2e^{-0.2}$; $f_y(1, -1) = 0.2e^{-0.2}$ **19.** $f_{xx}(x, y) = 0$; $f_{yy}(x, y) = 0$; $f_{xy}(x, y) = f_{yx}(x, y) = 0$; $f_{xx}(1, -1) = 0$; $f_{yy}(1, -1) = 0$; $f_{xy}(1, -1) = f_{yx}(1, -1) = 0$ **21.** $f_{xx}(x, y) = 0$; $f_{yy}(x, y) = 0$; $f_{xy}(x, y) = f_{yx}(x, y) = 10$; $f_{xx}(1, -1) = 0$; $f_{yy}(1, -1) = 0$; $f_{xy}(1, -1) = f_{yx}(1, -1) = 10$ **23.** $f_{xx}(x, y) = 6y$; $f_{yy}(x, y) = 0$; $f_{xy}(x, y) = f_{yx}(x, y) = 6x$; $f_{xx}(1, -1) = -6$; $f_{yy}(1, -1) = 0$; $f_{xy}(1, -1) = f_{yx}(1, -1) = 6$ **25.** $f_{xx}(x, y) = e^{x+y}$; $f_{yy}(x, y) = e^{x+y}$; $f_{xy}(x, y) = f_{yx}(x, y) = e^{x+y}$; $f_{xx}(1, -1) = 1$; $f_{yy}(1,-1) = 1$; $f_{xy}(1, -1) = f_{yx}(1, -1) = 1$ **27.** $f_{xx}(x, y) = -1.2x^{-1.4}y^{0.4}$; $f_{yy}(x, y) = -1.2x^{0.6}y^{-1.6}$; $f_{xy}(x,y) = f_{yx}(x,y) = 1.2x^{-0.4}y^{-0.6}$; $f_{xx}(1, -1)$ undefined; $f_{yy}(1, -1)$ undefined; $f_{xy}(1, -1)$ and $f_{yx}(1, -1)$ undefined **29.** $f_x(x, y, z) = yz$; $f_y(x, y, z) = xz$; $f_z(x, y, z) = xy$; $f_x(0, -1, 1) = -1$; $f_y(0, -1, 1) = 0$; $f_z(0, -1, 1) = 0$ **31.** $f_x(x, y, z) = 4/(x + y + z^2)^2$; $f_y(x, y, z) = 4/(x + y + z^2)^2$; $f_z(x, y, z) = 8z/(x + y + z^2)^2$; $f_x(0, -1, 1)$ undefined; $f_y(0, -1, 1)$ undefined; $f_z(0, -1, 1)$ undefined **33.** $f_x(x, y, z) = e^{yz} + yze^{xz}$; $f_y(x, y, z) = xze^{yz} + e^{xz}$; $f_z(x, y, z) = xy(e^{yz} + e^{xz})$; $f_x(0, -1, 1) = e^{-1} - 1$; $f_y(0, -1, 1) = 1$; $f_z(0,-1,1) = 0$

35. $f_x(x, y, z) = 0.1x^{-0.9}y^{0.4}z^{0.5}$; $f_y(x, y, z) = 0.4x^{0.1}y^{-0.6}z^{0.5}$; $f_z(x, y, z) = 0.5x^{0.1}y^{0.4}z^{-0.5}$; $f_x(0, -1, 1)$ undefined; $f_y(0, -1, 1)$ undefined, $f_z(0, -1, 1)$ undefined **37.** $f_x(x, y, z) = yze^{xyz}$; $f_y(x, y, z) = xze^{xyz}$; $f_z(x, y, z) = xye^{xyz}$; $f_x(0, -1, 1) = -1$; $f_y(0, -1, 1) = f_z(0, -1, 1) = 0$ **39.** $f_x(x, y, z) = 0$; $f_y(x, y, z) = -\frac{600z}{y^{0.7}(1 + y^{0.3})^2}$; $f_z(x, y, z) = \frac{2000}{1 + y^{0.3}}$; $f_x(0, -1, 1)$ undefined; $f_y(0, -1, 1)$ undefined; $f_z(0, -1, 1)$ undefined **41.** $\partial C/\partial x = 6000$; the marginal cost to manufacture each car is \$6000. $\partial C/\partial y = 4000$; the marginal cost to manufacture each truck is \$4000. **43.** $\partial x_3/\partial x_1 = -2.2$; General Motors' market share decreases by 2.2 percentage points per 1-percentage-point increase in Chrysler's market share if Ford's share is unchanged. $\partial x_1/\partial x_3 = -1/2.2$; Chrysler's market share decreases by 1 percentage point per 2.2-percentage-point increase in General Motors' market share if Ford's share is unchanged. The two partial derivatives are reciprocals of each other. **45.** \$5600 per car **47. a.** $\partial M/\partial c = -3.8$, $\partial M/\partial f = 2.2$. For every 1-point increase in the percentage of Chrysler owners who remain loyal, the percentage of Mazda owners who remain loyal decreases by 3.8 points. For every 1-point increase in the percentage of Ford owners who remain loyal, the percentage of Mazda owners who remain loyal increases by 2.2 points. **b.** 16% **49.** $\partial c/\partial x = -0.8$; Chrysler's percentage of the market decreases by 0.8 percentage points for every 1-percentage-point rise in foreign manufacturers' share. $\partial c/\partial y = -0.2$; Chrysler's percentage of the market decreases by 0.2 percentage point for every 1-percentage-point rise in Ford's share. $\partial c/\partial z = -0.7$; Chrysler's percentage of the market decreases by 0.7 percentage point for every 1-percentage-point rise in GM's share. **51.** The marginal cost of cars is $\$6000 + 1000e^{-0.01(x+y)}$ per car. The marginal cost of trucks is $\$4000 + 1000e^{-0.01(x+y)}$ per truck. Both marginal costs decrease as production rises. **53.** $\bar{C}(x, y) = (200{,}000 + 6000x + 4000y - 100{,}000e^{-0.01(x+y)})/(x + y)$; $\bar{C}_x(50, 50) = -\$2.64$ per car. This means that at a production level of 50 cars and 50 trucks per week, the average cost per vehicle is decreasing by \$2.64 for each additional car manufactured. $\bar{C}_y(50, 50) = -\$22.64$ per truck. This means that at a production level of 50 cars and 50 trucks per week, the average cost per vehicle is decreasing by \$22.64 for each additional truck manufactured. **55.** No; your marginal revenue from the sale of cars is $\$15{,}000 - 2500/\sqrt{x + y}$ per car and $\$10{,}000 - 2500/\sqrt{x + y}$ per truck from the sale of trucks. These increase with increasing x and y. In other words, you will earn more revenue per vehicle with increasing sales, and so the rental company will pay more for each additional vehicle it buys. **57.** $P_z(10, 100{,}000, 1{,}000{,}000) \approx 0.0001010$ paper per dollar **59. a.** $U_x(10, 5) = 5.18$, $U_y(10, 5) = 2.09$. This means that if 10 copies of Macro Publish and 5 copies of Turbo Publish are purchased, the company's daily productivity is increasing at a rate of 5.18 pages per day for each additional copy of Macro purchased and by 2.09 pages per day for each additional copy of Turbo purchased. **b.** $\frac{U_x(10, 5)}{U_y(10, 5)} \approx 2.48$ is the ratio of the usefulness of one additional copy of Macro to one of Turbo. Thus, with 10 copies of Macro and 5 copies of Turbo, the company can expect approximately 2.48 times the productivity per additional copy of Macro compared to Turbo. **61.** 6×10^9 N/s **63. a.** $A_P(100, 0.1, 10) = 2.59$; $A_r(100, 0.1, 10) = 2357.95$; $A_t(100, 0.1, 10) = 24.72$. Thus, for a \$100 investment at 10% interest, after 10 years the accumulated amount is increasing at a rate of \$2.59 per \$1 of principal, at a rate of \$2357.95 per increase of 1 in r (note that this would correspond to an increase in the interest rate of 100%), and at a rate of \$24.72 per year. **b.** $A_P(100, 0.1, t)$ tells you the rate at which the accumulated amount in an account bearing 10% interest with a principal of \$100 is growing per \$1 increase in the principal, t years after the investment. **65. a.** $P_x = Ka(y/x)^b$ and $P_y = Kb(x/y)^a$. They are equal precisely when $a/b = (x/y)^b(x/y)^a$. Substituting $b = 1 - a$ now gives $a/b = x/y$. **b.** The given information implies that $P_x(100, 200) = P_y(100, 200)$. By part (a), this occurs precisely when $a/b = x/y = 100/200 = 1/2$. But $b = 1 - a$, so $a/(1 - a) = 1/2$, giving $a = 1/3$ and $b = 2/3$. **67.** Decreasing at 0.0075 cc/s **69.** f is increasing at a rate of s units per unit of x, f is increasing at a rate of t units per unit of y, and the value of f is r when $x = a$ and $y = b$. **71.** One example is $f(x, y) = -2x + 3y + 9$. Another is $f(x, y) = xy - 3x + 2y + 10$. **73. a.** b is the z-intercept of the plane; m is the slope of the intersection of the plane with the xz-plane; n is the slope of the intersection of the plane with the yz-plane. **b.** Write $z = b + rx + sy$. We are told that $\partial z/\partial x = m$, so $r = m$. Similarly, $s = n$. Thus, $z = b + mx + ny$. We are also told that the plane passes through (h, k, l). Substituting gives $l = b + mh + nk$. This gives b as $l - mh - nk$. Substituting in the equation for z therefore gives $z = l - mh - nk + mx + ny = l + m(x - h) + n(y - k)$, as required.

8.4 exercises

1. P: relative minimum; Q: none of these; R: relative maximum **3.** P: saddle point; Q: relative maximum; R: none of these **5.** Relative minimum **7.** Neither **9.** Saddle point

11.

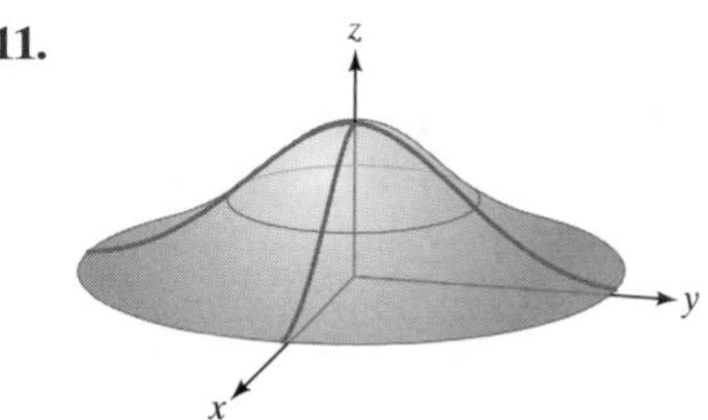

13. Minimum at $(0, 0, 1)$ **15.** Maximum at $\left(-\frac{1}{2}, \frac{1}{2}, \frac{3}{2}\right)$ **17.** Maximum at $(0, 0, 0)$; saddle points at $(\pm 4, 2, -16)$ **19.** Minimum at $(0, 0, 1)$ **21.** Minimum at $(-2, \pm 2, -16)$; $(0, 0)$ a critical point that is not a relative extremum **23.** Saddle point at $(0, 0, -1)$ **25.** Maximum at $(-1, 0, e)$ **27.** Minimum at $(2^{1/3}, 2^{1/3}, 3(2^{2/3}))$ **29.** Minima at $(1, 1, 4)$ and $(-1, -1, 4)$ **31.** Minimum of $\frac{1}{3}$ at $(c, f) = \left(\frac{2}{3}, \frac{2}{3}\right)$. Thus, at least $\frac{1}{3}$ of all Mazda owners would choose another new Mazda, and this lowest loyalty occurs when $\frac{2}{3}$ of Chrysler and Ford owners remain loyal to their brands. **33.** It should remove 2.5 pounds of sulfur and 1 pound of lead per day. **35.** They should charge \$580.81 for the Ultra Mini and \$808.08 for the Big Stack. **37.** $l = w = h = 20.67$ in.; volume $= 8826.96$ cubic inches **39.** 18 in. × 18 in. × 36 in.; volume $= 11{,}664$ cubic inches **41.** H must be positive. **43.** No. In order for there to be a relative maximum at (a, b), *all* vertical planes through (a, b) should yield a curve with a relative maximum at (a, b). It could happen that a slice by another vertical plane through (a, b) (such as $x - a = y - b$) does not yield a curve with a relative maximum at (a, b). [An example is $f(x, y) = x^2 + y^2 - \sqrt{xy}$ at the point $(0, 0)$. Look at the slices through $x = 0$, $y = 0$, and $y = x$.] **45.** The equation of the tangent plane at the point (a, b) is $z = f(a, b) + f_x(a, b)(x - a) + f_y(a, b)(y - b)$. If f has a relative extremum at (a, b), then $f_x(a, b) = 0 = f_y(a, b)$. Substituting these into the equation of the tangent plane gives $z = f(a, b)$, a constant. But the graph of $z =$ *constant* is a plane parallel to the xy-plane. **47.** $\bar{C}_x = \frac{\partial}{\partial x}\left(\frac{C}{x + y}\right) = \frac{(x + y)C_x - C}{(x + y)^2}$. If this is zero, then $(x + y)C_x = C$, or $C_x = C/(x + y) = \bar{C}$. Similarly, if $\bar{C}_y = 0$, then $C_y = \bar{C}$. This is reasonable because if the average cost is decreasing with increasing x, then the average cost is greater than the marginal cost C_x. Similarly, if the average cost is increasing with increasing x, then the average cost is less than the marginal cost C_x. Thus, if the average cost is stationary with increasing x, then the average cost equals the marginal cost C_x. (The situation is similar for the case of increasing y.)

8.5 exercises

1. Maximum value of 8 at $(2, 2)$; minimum value of 0 at $(0, 0)$ **3.** Maximum value of 9 at $(-2, 0)$; minimum value of 0 at $(1, 0)$ **5.** Maximum value of e^4 at $(0, \pm 2)$; minimum value of 1 at $(0, 0)$ **7.** Maximum value of e^4 at $(\pm 1, 0)$; minimum value of 1 at $(0, 0)$ **9.** Maximum value of 5 at $\left(\frac{1}{2}, \frac{1}{2}\right)$; minimum value of 3 at $(1, 1)$ **11.** Maximum value of 161/9 at $(1, 9)$ and $(9, 1)$; minimum value of 12 at $(2, 2)$ **13.** Maximum value of 8 at $(2, 0)$; minimum value of -1 at $(-1, 0)$ **15.** Maximum value of 8 at $(2, 0)$; minimum value of 0 at $(0, 0)$ **17.** $(1/\sqrt{3}, 1/\sqrt{3}, 1/\sqrt{3})$, $(-1/\sqrt{3}, -1/\sqrt{3}, 1/\sqrt{3})$, $(1/\sqrt{3}, -1/\sqrt{3}, -1/\sqrt{3})$, $(-1/\sqrt{3}, 1/\sqrt{3}, -1/\sqrt{3})$ **19.** For minimum cost of \$16,600, make 100 five-speeds and 80 ten-speeds. For maximum cost of \$17,400, make 100 five-speeds and 120 ten-speeds. **21.** For a maximum profit of \$4500, sell 50 copies of Walls and 150 copies of Doors. **23.** Hottest point: $(1, 1)$; coldest point: $(0, 0)$ **25.** Hottest points: $(-1/2, \pm\sqrt{3}/2)$; coldest point: $(1/2, 0)$ **27.** $\left(0, \frac{1}{2}, -\frac{1}{2}\right)$ **29.** $\left(-\frac{5}{9}, \frac{5}{9}, \frac{25}{9}\right)$ **31.** $l \times w \times h = 1 \times 1 \times 2$ **33.** $(2l/h)^{1/3} \times (2l/h)^{1/3} \times 2^{1/3}(h/l)^{2/3}$, where $l =$ cost of lightweight cardboard and $h =$ cost of heavy-duty cardboard per square foot **35.** 18 in. × 18 in. × 36 in.; volume $= 11{,}664$ cubic inches **37.** $1 \times 1 \times \frac{1}{2}$ **39.** 150 quarts of vanilla and 100 quarts of mocha, for a profit of \$325 **41.** Offer 100 sections of Finite Math and no sections of Calculus. **43.** Use 7.280 servings of Mixed Cereal and 1.374 servings of Tropical Fruit Dessert. **45.** Buy 100 of each. **47.** Yes. There may be relative extrema at points on the boundary of the domain of the function. The partial derivatives of the function need not be zero at such points. **49.** If the only constraint is an equality constraint, and if it is impossible to eliminate one of the variables in the objective function by substitution (solving the constraint equation for a variable or some other method). **51.** In a linear programming problem, the objective function is linear, and so the partial derivatives can never be zero. (We are ignoring the simple case in which the objective function is constant.) It follows that the extrema cannot occur in the interior of the domain (since the partial derivatives must be zero at such points).

8.6 exercises

1. $-\frac{1}{2}$ **3.** $e^2/2 - \frac{7}{2}$ **5.** $(e^3 - 1)(e^2 - 1)$ **7.** $\frac{7}{6}$ **9.** $(e^3 - e - e^{-1} + e^{-3})/2$ **11.** $\frac{1}{2}$ **13.** $(e - 1)/2$ **15.** $\frac{45}{2}$ **17.** $\frac{8}{3}$ **19.** $\frac{4}{3}$ **21.** 0 **23.** $\frac{2}{3}$ **25.** $\frac{2}{3}$ **27.** $2(e - 2)$ **29.** $\frac{2}{3}$

31. $\int_0^1 \int_0^{1-x} f(x, y)\, dy\, dx$

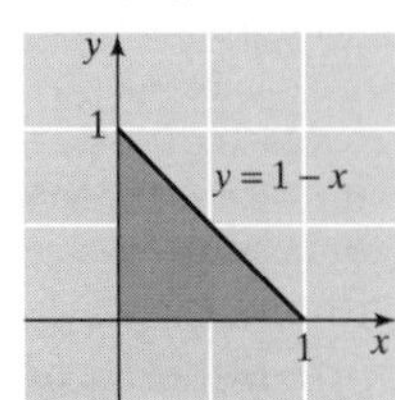

33. $\int_0^1 \int_{x^2-1}^{1} f(x, y)\, dy\, dx$

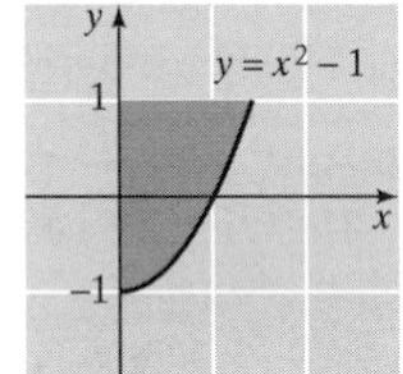

35. $\int_1^4 \int_1^{2/\sqrt{y}} f(x, y)\, dx\, dy$

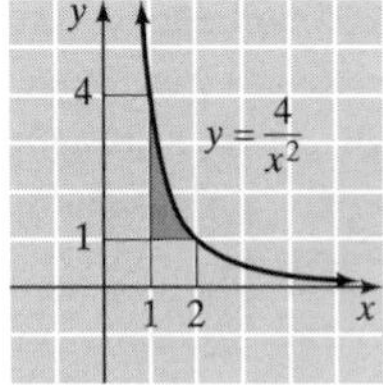

37. $\frac{4}{3}$ **39.** $\frac{1}{6}$ **41.** 162,000 gadgets **43.** Maximum revenue is \$375,500; minimum revenue is \$256,000; average revenue is \$312,750. **45.** Maximum revenue is \$20,000; minimum revenue is \$15,000; average revenue is \$17,500. **47.** 8216 **49.** 1 degree **51. a.** The area between the curves $y = r(x)$ and $y = s(x)$ and the vertical lines $x = a$ and $x = b$ is given by $\int_a^b \int_{r(x)}^{s(x)} dy\, dx$. **b.** The area of the solid above the region in part (a) and under the surface $z = f(x, y)$ is given by

$\int_a^b \int_{r(x)}^{s(x)} f(x, y)\, dy\, dx$. Hence, double integrals can be used for both types of calulation. **53.** Left-hand side is

$$\int_a^b \int_c^d f(x)g(y)\, dx\, dy = \int_a^b \left(g(y) \int_c^d f(x)\, dx\right) dy$$

[since $g(y)$ is treated as a constant in the inner integral]

$$= \left(\int_c^d f(x)\, dx\right) \int_a^b g(y)\, dy$$

[since $\int_c^d f(x)\, dx$ is a constant and can therefore be taken outside the integral]

$\int_0^1 \int_1^2 ye^x\, dx\, dy = \frac{1}{2}(e^2 - e)$ no matter how we compute it.

Chapter 8 Review Test

1. a. 0; 1; 0; $x^3 + x^2$; $x(y + k)(x + y + k - z) + x^2$ **b.** Decreases by 0.32 unit; increases by 12.5 units **c.** Reading left to right, starting at the top: 4, 0, 0, 3, 0, 1, 2, 0, 2 **d.** Answers may vary; two examples are $f(x, y) = 3(x - y)$ and $f(x, y) = 3(x - y)^3$ **2. a.** $f_x = 2x + y$; $f_y = x$; $f_{yy} = 0$ **b.** $\partial f/\partial x = ye^{xy} + 6xe^{3x^2-y^2}$; $\partial^2 f/(\partial x\, \partial y) = (xy + 1)e^{xy} - 12xye^{3x^2-y^2}$ **c.** $\dfrac{\partial f}{\partial x} = \dfrac{-x^2 + y^2 + z^2}{(x^2 + y^2 + z^2)^2}$; $\dfrac{\partial f}{\partial y} = -\dfrac{2xy}{(x^2 + y^2 + z^2)^2}$; $\dfrac{\partial f}{\partial z} = -\dfrac{2xz}{(x^2 + y^2 + z^2)^2}$; $\left.\dfrac{\partial f}{\partial x}\right|_{(0,1,0)} = 1$ **d.** 6 **3. a.** Absolute minimum at $\left(1, \frac{3}{2}\right)$ **b.** Saddle point at $(1, 0)$ **c.** Saddle point at $(0, 0)$ **d.** Saddle point at $(0, 0)$ **e.** Absolute maximum at each point on the circle $x^2 + y^2 = 1$ **4. a.** $(0, 2, \sqrt{2})$ **b.** Minimum of 0 at $(1, 0)$; maximum of 4 at $(3, 0)$ **c.** Minimum of $-\frac{16}{7}$ at $\left(\frac{4}{7}, \frac{6}{7}\right)$; maximum of 21 at $(-1, -2)$ **5. a.** 2 **b.** e^2 **c.** ln 5 **d.** e^2 **e.** 1 **f.** $\frac{4}{5}$ **6. a.** $h(x, y) = 5000 - 0.8x - 0.6y$ hits per day (x = number of new customers at JungleBooks.com, y = number of new customers at FarmerBooks.com) **b.** 250 **c.** $h(x, y, z) = 5000 - 0.8x - 0.6y + 0.0001z$ (z = number of new Internet shoppers) **d.** 1.4 million **7. a.** 2320 hits per day **b.** $0.08 + 0.00003x$ hits (daily) per dollar spent on television advertising per month; increases with increasing x **c.** \$4000 per month **d.** (A) **8. a.** About 15,800 orders per day **b.** 11 **9.** Selling 1000 paperbacks and 1200 hardcover books will cost the least. Selling 900 or 1100 paperbacks and 1500 hardcover books will cost the most. **10.** \$23,050

Appendix: Algebra Review

A.1 exercises

1. −48 **3.** $\frac{2}{3}$ **5.** −1 **7.** 9 **9.** 1 **11.** 33 **13.** 14 **15.** $\frac{5}{18}$ **17.** 13.31 **19.** 6 **21.** $\frac{43}{16}$ **23.** 0 **25.** `3*(2-5)` **27.** `3/(2-5)` **29.** `(3-1)/(8+6)` **31.** `3-(4+7)/8` **33.** `2/(3+x)-x*y^2` **35.** `3.1x^3-4x^(-2)-60/(x^2-1)` **37.** `(2/3)/5` **39.** `3^(4-5)*6` **41.** `3*(1+4/100)^(-3)` **43.** `3^(2*x-1)+4^x-1` **45.** `2^(2x^2-x+1)` **47.** `4*e^(-2*x)/(2-3e^(-2*x))` or `4*(e^(-2*x))/(2-3e^(-2*x))` **49.** `3(1-(-1/2)^2)^2+1`

A.2 exercises

1. 27 **3.** −36 **5.** $\frac{4}{9}$ **7.** $-\frac{1}{8}$ **9.** 16 **11.** 2 **13.** 32 **15.** 2 **17.** x^5 **19.** $-\dfrac{y}{x}$ **21.** $\dfrac{1}{x}$ **23.** x^3y **25.** $\dfrac{z^4}{y^3}$ **27.** $\dfrac{x^6}{y^6}$ **29.** $\dfrac{x^4y^6}{z^4}$ **31.** 2 **33.** $\frac{1}{2}$ **35.** $\frac{4}{3}$ **37.** $\frac{2}{5}$ **39.** 7 **41.** 5 **43.** −2.668 **45.** $\frac{3}{2}$ **47.** 2 **49.** 2 **51.** ab **53.** $x + 9$ **55.** $x\sqrt[3]{a^3 + b^3}$ **57.** $2y/\sqrt{x}$ **59.** $3^{1/2}$ **61.** $x^{3/2}$ **63.** $(xy^2)^{1/3}$ **65.** $x^2/x^{1/2}$ **67.** $\sqrt[3]{2^2}$ **69.** $\sqrt[3]{x^4}$ **71.** $\sqrt[5]{\sqrt{x}\sqrt[3]{y}}$ **73.** 64 **75.** $\sqrt{3}$ **77.** $1/x$ **79.** xy **81.** $(y/x)^{1/3}$ **83.** ±4 **85.** $\pm\frac{2}{3}$ **87.** $-1, -\frac{1}{3}$ **89.** −2 **91.** 16 **93.** ±1 **95.** 33/8

A.3 exercises

1. $4x^2 + 6x$ **3.** $2xy - y^2$ **5.** $x^2 - 2x - 3$ **7.** $2y^2 + 13y + 15$ **9.** $4x^2 - 12x + 9$ **11.** $x^2 + 2 + 1/x^2$ **13.** $4x^2 - 9$ **15.** $y^2 - 1/y^2$ **17.** $2x^3 + 6x^2 + 2x - 4$ **19.** $x^4 - 4x^3 + 6x^2 - 4x + 1$ **21.** $y^5 + 4y^4 + 4y^3 - y$ **23.** $(x + 1)(2x + 5)$ **25.** $(x^2 + 1)^5(x + 3)^3(x^2 + x + 4)$ **27.** $-x^3(x^3 + 1)\sqrt{x + 1}$ **29.** $(x + 2)\sqrt{(x + 1)^3}$ **31. a.** $x(2 + 3x)$ **b.** $x = 0, -\frac{2}{3}$ **33. a.** $2x^2(3x - 1)$ **b.** $x = 0, \frac{1}{3}$ **35. a.** $(x - 1)(x - 7)$ **b.** $x = 1, 7$ **37. a.** $(x - 3)(x + 4)$ **b.** $x = 3, -4$ **39. a.** $(2x + 1)(x - 2)$ **b.** $x = -\frac{1}{2}, 2$ **41. a.** $(2x + 3)(3x + 2)$ **b.** $x = -\frac{3}{2}, -\frac{2}{3}$ **43. a.** $(3x - 2)(4x + 3)$ **b.** $x = \frac{2}{3}, -\frac{3}{4}$ **45. a.** $(x + 2y)^2$ **b.** $x = -2y$ **47. a.** $(x^2 - 1)(x^2 - 4)$ **b.** $x = \pm 1, \pm 2$

A.4 exercises

1. $\dfrac{2x^2 - 7x - 4}{x^2 - 1}$ **3.** $\dfrac{3x^2 - 2x + 5}{x^2 - 1}$ **5.** $\dfrac{x^2 - x + 1}{x + 1}$ **7.** $\dfrac{x^2 - 1}{x}$ **9.** $\dfrac{2x - 3}{x^2y}$ **11.** $\dfrac{(x + 1)^2}{(x + 2)^4}$ **13.** $\dfrac{-1}{\sqrt{(x^2 + 1)^3}}$ **15.** $\dfrac{-(2x + y)}{x^2(x + y)^2}$

A.5 exercises

1. −1 **3.** 5 **5.** 13/4 **7.** 43/7 **9.** −1 **11.** $(c - b)/a$ **13.** $-4, \frac{1}{2}$ **15.** no solutions **17.** $\pm\sqrt{\frac{5}{2}}$ **19.** −1 **21.** −1, 3 **23.** $\dfrac{1 \pm \sqrt{5}}{2}$ **25.** 1 **27.** ±1, ±3 **29.** $\pm\sqrt{\dfrac{-1 + \sqrt{5}}{2}}$ **31.** −1, −2, −3 **33.** −3 **35.** 1 **37.** −2 **39.** $1, \pm\sqrt{5}$ **41.** $\pm 1, \pm\dfrac{1}{\sqrt{2}}$ **43.** −2, −1, 2, 3

A.6 exercises

1. 0, 3 **3.** $\pm\sqrt{2}$ **5.** $-1, -\frac{5}{2}$ **7.** −3 **9.** 0, −1 **11.** $x = -1$ ($x = -2$ is not a solution.) **13.** $-2, -\frac{3}{2}, -1$ **15.** −1 **17.** $\pm\sqrt[4]{2}$ **19.** ±1 **21.** ±3 **23.** $\frac{2}{3}$ **25.** $-4, -\frac{1}{4}$

Index

Δ (change in), 27, 29, 30, 124, 125, 138. *See also* Δ notation
Δ notation, 138
Σ notation, 56

A

ABC, 442
Abel, Niels Henrik (1802–1829), A26
absolute value, 23, A10
acceleration, 283, 322
  due to gravity, 283
  of sales, 284
accuracy and rounding, A5
acidity, 103
acquisition of language, 14, 26, 191
advertising, 2, 26, 63, 67, 79, 170, 413
agriculture, 280
AIDS, 85, 226
Air Transportation Association of America, 52
alcohol, 93, 102
algebraic model, 7, 10. *See also* model
algebraically specified function, 4, 5, 6
  common types, 11
alien intelligence, 443
Amazon.com, 441
Amerada Hess Corp., 231, 281
America Online (AOL), 391
American Airlines, 280
amplitude, 106, 107
angular frequency, 107
antiderivative, 315. *See also* integral
area
  between two graphs, 380, 382, 383, 384
  ellipse, 454
  function, 354
  rectangle, 454
  under a curve, 346, 359
aspirin, 92, 102
asset appreciation, 282, 293
asymptote, 188, 289, 290
automation, 298
average
  balance, 394, 399
  cost, 164, 168, 169, 280, 297, 461
  moving, 395, 397, 400
  of a set of values, 393
  profit, 281
  revenue, 489
  speed, 393
average rate of change, 124, 126
  algebraic point of view, 128
  graphical point of view, 127
  numerical point of view, 124
  with technology, 129
aviation, 52, 53

B

bacteria, 92
balloons, 300
best-fit. *See* regression
Beyond Interactive, 2
bicycles, 49
Big Brother, 230
biology, 52, 208. *See also* embryo development
BN.com, 441
body mass index, 435, 439
bodybuilding, 306
Boeing, 52
bonds, 101, 406
book sales, 67, 182, 281, 294

Borders.com, 441
Boston Chicken, 124
box design, 279
brand loyalty, 461, 472
break-even point, 40, 49, 52, 53, 76, 77
bus travel, 208
business spending, 399

C

calculation thought experiment, 204, 205, 209, 211
calculus, 3, 152
  discovery of, 123
  fundamental theorem of, 354, 356, 360
carbon dating, 88, 92, 99, 101
carbon. *See* carbon dating
career choices, 53
Carnival Corp., 363
cash flow, 111
CBS, 442
Celsius, 51
Centers for Disease Control (CDC), 291
centigrade, 51
Certified Financial Planners, 50, 324
Cerulli Associates, 50
chain rule, 209, 210
  derivation of, 211
  in differential notation, 216
  verbal form of, 210
change
  linear, 48
  rate of, 3, 47, 48, 49, 294
chocolates, 422
Chrysler Corp., 440, 453, 460, 461, 472
clearing a market, 46
closed-form function, 184, 185, 186
Cobb-Douglas model, 243, 244, 276, 282, 300, 440, 454, 462, 492
Coca-Cola®, 118
coefficient of correlation, 59, 63
coffee shops, 12, 26
College Board, 54, 62
college tuition, 308, 391
communication among bees, 229
compound interest, 81, 87
compounding continuously. *See* continuous compounding
computer sales, 114
concavity
  and second derivative, 286
  concave up/down, 286
  point(s) of inflection, 286
cone, 301
constraint(s), 270
consumer satisfaction, 345
consumer's
  expenditure, 385
  surplus, 385, 386
  willingness to spend, 385
continuity, 183, 187. *See also* continuous function
continuous
  compounding, 89, 90
  function(s), 177, 178
  growth, 90
  income streams, 402, 403
  on domain, 178
correlation coefficient, 59, 63
cos. *See* cosine function
cosec, csc. *See* cosecant function
cosecant function, 112
  derivative of, 235
cosine function, 104, 109, 110
  amplitude, 111
  derivative of, 232, 233, 235
  general, 110
  graph, 109, 110
cost, cost function, 49, 162, 163, 170, 171, 238, 334, 344, 362, 379, 391, 432, 482, 483
  acceleration of, 172
  average, 164, 168, 169, 280, 461
  fixed, 39, 40, 432
  linear, 38, 39, 40, 433, 455
  marginal, 39, 40, 163, 169, 171, 321, 324, 363, 440, 453, 455, 456, 460, 461
  minimum, 483
  total, 335, 350, 358, 391, 362, 363
  variable, 39, 40
cot, cotan. *See* cotangent function
cotangent function, 112
  derivative of, 235
CPA exam, 52
Creative Labs, Inc., 50
crime, 292, 301
  disorganized, 53
  organized, 52
critical point, 260, 464, 465, 466
csc. *See* cosecant function
cubic function, 11, 22
currency, 128, 133
current, electrical, 107, 401
cyclical, cyclical behavior. *See* periodic
cylinder, 301

D

Datastream, 12, 51, 52
daylight, 150
decibel, 103

definite integral, 335, 338, 341, 356. *See also* integral, Riemann sum
as a sum, 335, 338
as a total, 341
as area, 346, 348, 359
computing with the fundamental theorem of calculus, 356
deflation, 114, 238
degrees of freedom, 65
delta notation, 138
demand, demand function, 13, 42, 43, 44, 49, 50, 55, 68, 74, 75, 141, 142, 151, 170, 179, 231, 235, 300, 306, 307, 308, 391, 422, 442
elastic/inelastic, 302, 303
elasticity of, 302, 303
equilibrium, 45
income elasticity of, 308
dental plans, 14
dependent variable, 4, 18
derivative(s)
algebraic point of view, 152, 154, 186
approximation/estimation of, 137, 147, 148
chain rule, 209, 210
definition of, 134, 186
first derivative, 283
function, 152, 153
graph of, 154, 267
graphical point of view, 143, 147, 148, 154
numerical point of view, 133, 154
of a composite. *See* chain rule
of a constant, 158
of a product. *See* product rule
of a quotient. *See* quotient rule
of exponential functions, 224, 225
of implicit functions, 239, 241, 242
of logarithmic functions, 220, 221, 223
of sines and cosines, 232, 233, 235
of sums, differences, and constant multiples, 156, 157
of trigonometric functions, 232, 233, 235, 236
partial. *See* partial derivatives
power rule, 154, 155
power rule, generalized, 211
product rule, 199, 201, 205
quotient rule, 199, 201, 206
second derivative, 283, 286
shortcuts, 154
derived equation, 295
difference quotient, 126. *See also* average rate of change
balanced, 137, 143, 149
differentiable, 135, 159
differential equation(s), 415
first-order, 415
general solution of, 416
particular solution of, 416
second-order, 415, 423
separable, 418
simple, 416
solving, 416
differential notation, 155, 156, 157, 216
differentiation, differentiate, 134
implicit, 239, 241, 242
logarithmic, 243
diffusion, 463
of new technology, 231
diminishing returns, point of, 286, 287
discontinuity, 178
discontinuous function, 177, 178
discord, 115
discriminant, 73
displacement. *See* position, motion
distance formula, 436, 454
divorce rates, 54, 302
dolls, 305
domain, 4, 5
natural, 4, 5, 7
double integral, 485, 487
algebraic definition of, 486
geometric definition of, 485
over a nonrectangular region, 487, 488
over a rectangle, 487
Dow Jones & Co., 51
dummy variable, 338

E

e (base of natural logarithms), 88, 89
$E = mc^2$, 366
Earth, the, 238
eccentricity of earth's orbit, 238
ecology, 218
education, 281, 292, 301
Einstein, Albert (1879–1955). *See* relativity
elasticity of demand, 302, 303
electrical current, 107, 401
Electronic Privacy Information Center, 62
electrostatic repulsion, 462
Eli Lilly Corp., 365
embryo development, 162, 208, 209, 281, 363
EMC Corp., 362, 406
emission control, 170, 191, 280, 392, 472
employment, 94, 102, 108, 182, 231, 281, 301, 334, 344
energy, 366

entering formulas, A4
EPA. *See* U.S. Environmental Protection Agency
epidemics, 70, 85, 116, 226, 230, 291, 422
equation. *See also* function
- notation, 7, 18
- of a circle, 437

equilibrium, 46, 388
- demand, 45, 46
- price, 44, 46, 50, 389
- supply, 45, 46

erf. *See* error function
error function, 366
errors (in regression), 55
expansion, 208
exponential form (exponent form), 95, 158
- exponential function, 11, 23, 71, 79, 80
- derivative of, 224, 225
- fitting to data, 84
- graph, 82
- identities, 80
- trendline, 85
- viewed numerically, 81

exponential model, 84. *See also* exponential function
exponents, A7
- integer, A7
- identities, 80, A8
- rational, A11
- solving equations with, A12

exports, 364
extrapolation, 6, 57
extrema, 258. *See also* maxima and minima

F

factoring algebraic expressions, A15
factoring quadratics, A17
faculty salaries, 434
Fahrenheit, 51
fast
- cars, 51
- food, 391, 400

FBI, 62, 230
fences, 278
fixed cost, 39, 40
flies, 93
FOIL (First, Outer, Inner, Last), A14
Ford Motor Company, 26, 440, 453, 460, 461, 472
foreign trade, 391, 392, 414
Fox TV, 441
Frankenstein
- bride of, 278
- son of, 278

freon, demand for, 253
frogs, 93
fuel economy, 78, 171, 208, 365
fuel efficiency. *See* fuel economy
function, 3, 4
- antiderivative of. *See* antiderivative
- closed-form, 184, 186
- common types 11, 22
- composite, 210
- continuous. *See* continuous function
- cosecant, 112
- cosine. *See* cosine function
- cost. *See* cost function
- cotangent, 112
- cubic, 11, 22
- demand. *See* demand function
- derivative of. *See* derivative
- discontinuous. *See* discontinuous function
- domain of, 4, 5, 434
- error, 366
- evaluating using a graphic calculator, 8, 9
- evaluating using a spreadsheet, 8, 9, 437
- evaluating using technology, 7
- evaluating using the web site, 8, 9
- exponential, 11, 23, 71, 79. *See also* exponential function
- graph of, 16, 17, 445, 446
- graphing, 17, 20
- graphing using a graphing calculator, 18, 22
- graphing using a spreadsheet, 18, 22, 83
- graphing using the web site, 19
- graphs of common types, 22
- implicit. *See* implicit function
- integral of. *See* integral
- interaction. *See* interaction function
- linear, 3, 11, 22, 27, 434, 441. *See also* linear function
- logarithmic, 71, 94. *See also* logarithmic function
- logistic, 71, 116, 118, 253, 365, 422, 442
- natural domain, 4, 5, 7
- nonlinear, 11
- notation, 6, 7, 18
- of a real variable, 4
- of several variables, 431, 446, 457
- piecewise-defined, 9, 20
- polynomial, 11
- profit. *See* profit function
- quadratic, 11, 17, 22, 71
- rational, 11, 23, 188
- real-valued, 4
- revenue. *See* revenue function
- secant, 112

sine. *See* sine function
specified algebraically, 4, 5, 6
specified graphically, 4, 15, 445
specified numerically, 4, 5, 435
specified verbally, 4
supply. *See* supply function
tangent, 112
trigonometric, 103, 232. *See also* sine function, cosine function, etc.
fundamental theorem of calculus, 354, 356, 360
future value, 87, 403

G

Galilei, Galileo (1564–1642), 284
Galois, Évariste (1811–1832), A26
gamma function, 415
Gannett Company, 51
gas heating demand, 235
Genentech, 365
General Foods, 453, 460
General Mills, Inc., 453, 460
General Motors, 440, 453, 460, 461
general solution, 416
Genzyme, 365
Global Positioning System (GPS), 47
global warming, 78, 93
Government bonds. *See* bonds
grades, 301, 462
graph, 16, 17
analyzing, 288
area between, 380, 382, 383, 384
area under, 346, 348
common types, 22
features of, 288
of a function of two variables, 445, 446, 448
of a linear function, 29, 30
of a linear function of two variables, 451
of the derivative, 149
of the integral, 354
slope of tangent to, 143, 144, 147
graphically specified function, 4, 15
graphing
an exponential function, 82
a function, 17, 18, 20
a linear equation, 32
a quadratic function, 73
graphing calculator
computing Riemann sums, 340
distortion in graph, 31
estimating the derivative, 148, 154
evaluating a function, 8, 9
ExpReg, 85
finding average rate of change, 129
finding correlation coefficient, 60
finding regression exponential curve, 85
finding regression line, 58
graphing a function, 18, 22
graphing the derivative, 154
graphing the integral, 354
LEFTSUM program, 343
LinReg, 58, 60
LIST functions, 340
locating extrema, 267
nDeriv, 137
moving average, 397
plotting points, 18
GRE economics exam, 171, 172, 209, 219, 282
growth of tumors, 423
Grumman Corp. *See* Northrop Grumman Corp.

H

half-life, 99, 101, 102
harmony, 115
harvesting forests, 282, 293
Hawaii Visitors Bureau, 26
Health Care Financing Administration, 50
health care spending, 364
health clubs, 42
heating, 421
Hermite, Charles (1822–1901), 380
HIV. *See* AIDS
HMOs, 231
home sales, 54, 57
horizontal line, 33, 36
housing costs, 170, 442, 453, 454, 463

I

identities
exponential, 80
logarithmic, 98
trigonometric, 110, 114, 236
implicit differentiation, 239, 241, 242
implicit function, 239, 242
improper integrals, 407, 410, 412
income taxes, 12, 13, 171, 423
income, 13, 51, 230, 301, 406
household, 493
independent variable, 4, 18
industrial output, 292
infant mortality, 62
inflation, 115, 238, 367
inflection, point(s) of, 286, *See also* concavity

Information Highway, 142, 219
initial condition, 416
initial quantity, 48
instantaneous
 rate of change, 133, 134, 135. *See also* derivative
 velocity, 138, 139
integral(s), 315. *See also* integration
 algebraic point of view, 354
 as area, 345, 348, 349
 as a total, 341, 350
 change of variables, 326
 definite, 335, 338, 341, 345, 348, 356
 double, 484, 485. *See also* double integral
 fundamental theorem of calculus, 354, 356
 using, 356
 geometric point of view, 345, 348, 349
 graph of, 354
 improper, 407, 410, 412
 indefinite, 315, 316
 limits of, 338
 numerical point of view, 335
 of a polynomial times a function, 376
 of exponential and trigonometric functions, 318, 332
 of constant multiples, 319
 of sums and differences, 318, 319
 power rule for, 317, 318
 Riemann sum, 337
 shortcuts, 333
 substitution, 326, 358
integrand, 316, 338
integration, 316. *See also* integral
 by parts, 373, 374, 378
 change of variables, 326
 constant of, 316
 numerical, 343
 variable of, 316, 320, 338
interaction function, interaction model, 435, 453, 456, 460, 499
 graph of, 458
intercept, 28, 29, 30, 32, 44, 49
interest rates, 324
intermediate results, don't round, 58
International Data Corp., 49
Internet advertising, 2, 63, 67
interpolation, 5
intervals, A1, A2
investment firms, 50
investments, 13, 87, 92, 93, 98, 100, 101, 229, 293, 352, 363, 398, 414, 462
Investor Responsibility Research Center, 13
isotherms, 443

K

Kellogg, 453, 460
Kepler, Johannes (1571–1630), 94
Kiwi Airlines, 53
Knight Ridder, Inc., 51

L

labor, 276
ladders, sliding/falling, 295, 299, 300
Lagrange Multipliers, 478
Laplace Transform, 415
learning to read, 142
learning to speak, 142
least-squares line. *See* regression line
Leibniz, Gottfried (1646–1716), 123, 138
Leibniz notation, 156
level curves, 443, 447
life expectancy, 62, 229
life span. *See* life expectancy
limit(s)
 algebraic point of view, 183
 at infinity, 175, 187, 189
 definition of, 174
 divergent to $+\infty$, 177
 estimating graphically, 176, 177
 estimating numerically, 173
 existence, nonexistence, 174, 175
 graphical point of view, 176, 177
 infinite, 178
 numerical point of view, 172, 173
 of a closed form function, 184
 of a rational function, 188
 of integration, 338
 one-sided, 175
 two-sided, 175
line. *See* lines, linear functions
linear change, 48
linear function, 3, 11, 22, 27, 48, 49
 cost, 38, 39, 40, 433, 455
 demand, 43, 44, 50, 55, 68
 fitting to data, 34, 54
 graph, 29, 30, 32
 intercept, 28, 29, 30, 32, 44, 49
 numerical point of view, 27, 28
 of several variables, 434, 441
 profit, 40
 revenue, 40
 slope. *See* slope of a linear function
 supply, 46, 50
linear model, 34, 38, 49, 54, 57, 63. *See also* linear function
 reliability, 43
linear programming, 484

linear regression, 3, 54, 56, 63, 64, 68, 434
lines. *See also* linear function
  horizontal, 33, 36
  parallel, 36
  secant, 145, 147
  tangent, 143, 144, 147
  vertical, 33, 36
local extrema, maxima, and minima. *See* relative extrema
logarithm(s), 94, 95. *See also* logarithmic function
  base of, 95, 96
  calculating, 95
  common, 95
  identities, 98
  natural, 95
  of an absolute value, 222, 223
logarithmic differentiation, 243
logarithmic function, 71, 94. *See also* logarithm
  derivative of, 220, 221, 223, 366
  graph, 97, 98
logistic function (logistic curve, logistic model), 71, 116, 118, 142, 365
loss, 40, 42
lot size, 308, 309
Lotus Development Corp., 12, 132
luggage dimensions, 280, 472
Luxotica Group S.p.A., 51, 52

M

magazine advertising, 26
magazine circulation, 15
marginal
  analysis, 163
  cost, 39, 40, 163, 169, 171, 209, 453, 455, 456, 460, 461
  loss, 169
  product, 166, 170, 172, 209, 214, 218
  profit, 40, 165, 169
  revenue, 40, 165, 216, 218, 461
market
  clearing, 46
  growth, 247
  saturation, 422
  share, 400, 453, 460
marketing strategy, 279, 280
marriage, 282
mathematical model, 3, 10. *See also* model
maxima and minima, 257. *See also* maximum, minimum
  absolute, 259, 265
  applications of, 269
  constrained, 473
  endpoint, 260, 261
  locating, 261, 465
  locating with technology, 267
  local, 259
  global, 259
  of functions of several variables, 463
  on a closed interval, 265
  optimization problems, 273
  relative, 258, 259, 463, 464
  singular, 260, 261
  stationary, 260, 261
maximum. *See also* minimum, maxima and minima
  area, 271
  profit, 278, 279, 476, 483
  revenue, 74, 273, 278, 304, 305
  volume, 483
Mazda, 461, 472
McDonald's, 400
McDonnell Douglas, 52
medical costs, 419
Medicare, 50, 324, 330, 344, 352
meteor impacts, 414
microprocessors, 49
Microsoft Corp., 391
minima. *See* maxima and minima
minimum. *See also* maximum, maxima and minima
  area for a given volume, 274, 281, 475
  average cost, 270
  cost, 483
minivan sales, 440, 442, 461
model. *See also* function
  algebraic, 7, 10
  Cobb-Douglas. *See* Cobb-Douglas model
  exponential, 71, 79
  linear, 34, 38, 49, 54, 57, 63
  logistic, 71, 116, 118, 253, 365, 422, 442
  mathematical, 3, 10
  nonlinear, 71
  numerical, 5, 10
  quadratic, 7
  trigonometric, 103. *See also* sine function, cosine function
modeling household income, 493
model trains, 50
mold, 218
money stock, 219
motion
  along a graph, 300
  around a circle, 300
  in a straight line, 322, 324, 325, 334, 344, 363
  vertical, 283, 325, 342, 379

movement. *See* motion
moving average, 395, 397, 400
multiplying algebraic expressions, A14
muscle recovery time, 52
music, 115

N

Napier, John (1550–1617), 94
National Association of State Budget Officers, 13
natural domain, 4, 5, 7
natural gas reserves, 62
NBC, 442
New York Times, 414
Newhouse Newspapers, Inc., 51
Newspaper Association of America, 51
newspapers, 49, 50, 51, 414
Newton, Isaac (1642–1727), 123, 420
Newton's law of cooling, 420, 421
Newton's law of gravity, 437
Nielsen Co., 63
nightclub management, 79
nonlinear function, model, 11, 71
normal curve/distribution, 345, 415, 424
Northrop Grumman Corp., 94, 102
numerical integration, 343. *See also* Riemann sum
numerical model, 5, 10
numerically specified function, 4, 5, 81
nutrient diffusion, 463
nutrition, 484

O

objective function, 269
observed values, 55
Office of Technology Assessment, 61
oil
  exploration costs, 363
  recovery, 61
on-line
  revenue, 391, 441
  trading, 217
optimization problem, 269, 273
order of operations, A2

P

panic sales, 413
parabola, 17, 18, 22, 72
  features of, 72
  vertex, 72
  x-intercepts, 72
paraboloid, 447
parallel lines, 36
partial derivative(s), 455
  interpretation of, 455
  mixed, 459
  second-order, 459
  viewed geometrically, 457
particular solution, 416
partition, 338
pasta imports, 50
patents, 292
pens, 49
percentage rate of change, 219, 232
perfume, 52, 53
period, 107
periodic behavior, 104
  employment patterns, 108
  seasonal fluctuations, 121, 235
  temperature fluctuations, 104
pH level, 103
phase shift, 107
pianos, 49
piecewise-defined function, 9
  graphing, 20
plantation management, 279, 293
plutonium, 102
point-slope formula, 34, 35
pollution, 61, 122, 191, 218, 280, 353, 443, 472. *See also* emission control
polynomial equations, A21
polynomial function, 11
population
  density, 490, 492, 493
  growth, 229
  Lower Anchovia, 229
  prison, 281, 292, 301
  Squaresville, 490
  U.S., 93
  Upper Anchovia, 229
  world, 93
position, 48, 322, 379
power rule, 154, 155
  generalized, 211
predicted values, 55
present value, 87, 404
price
  elasticity. *See* elasticity of demand
  equilibrium, 44, 50, 422
prison population, 281, 292
producer's surplus, 387, 388
product rule, 199, 201
  derivation of, 205
  verbal form of, 201
production function (Cobb-Douglas), 244. *See also* Cobb-Douglas model
production lot size management, 308

productivity, 441, 461, 462, 492. *See also* Cobb-Douglas model
profit, profit function, 12, 39, 40, 42, 50, 51, 52, 62, 68, 76, 77, 78, 93, 133, 162, 281, 292, 421
  average, 398, 399
  marginal, 40, 165, 169
  maximum, 278, 279, 476, 483
  total, 352, 344, 405, 406
Prozac, 342, 401
Publishers Information Bureau, 16, 26
publishing, 68
purchasing, 484
p-value, 65

Q

quadratic function, 11, 17, 22, 71
  fitting to data, 64, 68
quadratic model, 7, 64, 68, 71
quotient rule, 199, 201
  derivation of, 206
  verbal form of, 201

R

radicals, A9
  of products and quotients, A10
radioactive decay, 88, 101, 102, 229
rate of change, 3, 47, 48, 49
  average. *See* average rate of change
  instantaneous. *See* instantaneous rate of change
  percentage, 219, 232
  related rates, 294, 295
rational expression, A19
rational function, 11, 23, 188
real estate, 324
real numbers, A1
refrigerators, 40
regression
  correlation coefficient, 59, 63
  deriving formulas for, 469
  exponential, 85
  formula, 56
  linear, 3, 54, 56, 63, 64, 68, 469
  quadratic, 64, 68
  using a graphing calculator, 58, 60
  using a spreadsheet, 57, 58, 60, 247
  using the web site, 59
reinvested interest, 81. *See also* compound interest
related rates, 294, 295
relative extrema, maxima, and minima, 258, 259, 463, 464
relativity, theory of
  energy formula, 366
  Lorenz contraction, 141, 151
  time dilation, 142, 151
resource allocation, 483, 484
retirement, 54
revenue, revenue function, 12, 13, 17, 39, 40, 50, 51, 52, 73, 78, 79, 93, 141, 207, 208, 219, 300, 379, 380, 391, 423, 472
  marginal, 40, 165, 170, 171, 216, 218, 461
  maximum, 74, 273, 278, 282, 304, 305, 492
  total, 362, 363, 405, 406, 413
Richter scale, 102
Riemann sum, 337
  computing, 338
  left, 337
  midpoint, 353
  right, 353
  using technology, 339
Riemann, Georg Friedrich Bernhard, 337
rise, 29, 30
romance novels, 51
roots, A9
rounding, A5
run, 29, 30

S

saddle point(s), 463, 464
sales, 12, 25, 47, 49, 67, 114, 132, 142, 182, 226, 281, 284, 291, 352, 364, 365, 409, 413, 414, 421
San Francisco Chronicle, 414
SAT scores, 62, 182, 183, 230, 281, 292, 301
saving
  for college, 406
  for retirement, 406
sawtooth wave, 115
seasonal fluctuations. *See also* cyclical 121
sec. *See* secant function
secant function, 112
  derivative of, 235
secant line, 145, 147
second derivative test, 466, 467, 468
Securities and Exchange Commission (SEC), 50
sedans, 77, 400
semiconductor manufacture, 9, 10, 12, 20, 26, 53
ships, 300
shower curtains, 49
significant digits, A5
simple differential equation, 416
sin. *See* sine function

sine function, 104, 105
  amplitude, 106, 107
  derivative of, 232, 233, 235
  general, 107
  graph, 105
slope of a linear function, 3, 28, 29, 30, 32
  formula, 33
  interpreting, 39, 40, 44, 48, 49
  undefined, 33
  units, 48
  zero, 33
Snapple Beverage Corp., 391
snowtree crickets, 52
soap bubbles, 219
social gain, total, 391
social ills, 190
Social Security Administration, 54
soft drinks, 49
solving miscellaneous equations, A27
solving polynomial equations, A21
Sony, 13
sound intensity, 103
South African Breweries (SAB), 93, 133, 141, 398, 399, 405, 406
special theory of relativity. *See* relativity
Specialty Coffee Association of America, 12, 26
specifying a function, 4
speed. *See* velocity
spending on corrections, 13, 398, 399
sport utility vehicles (SUVs), 25, 77, 127, 344, 352, 400
spreadsheet
  Analysis Toolpack, 495
  computing Riemann sums, 339
  curve fitting, 250
  Data Analysis, 495
  distortion in graph, 31
  estimating derivatives, 137
  estimating limits, 174
  evaluating a function, 8, 9
  evaluating a function of several variables, 436
  finding a regression exponential curve, 85
  finding average rate of change, 130
  finding correlation coefficient, 60, 497
  finding maximum value of a function, 480
  finding p-value, 65
  finding regression line, 57, 58
  FINV, 439
  Goal Seek, 42
  graphing a function, 18, 22, 83
  graphing a function of two variables, 449
  graphing the integral, 355
  LINEST, 58, 60
  plotting points, 18
  recognizing linear data, 29
  regression, 57, 58, 85, 495
  Solver, 250, 481
  trendline, 58, 60, 85
square root, 23
square wave, 115
steepness of a graph, 144
stocks, stock funds, 62, 111, 395
strength, 325
study time, 171
sum of the squares of the errors, sum of squares error (SSE), 56, 249, 250
summation notation, 56
sun spots, 299
supply, supply function, 44, 46, 50, 300, 391
  equilibrium, 45
surface of revolution, 447
surveying, 353
SUV. *See* sport utility vehicles

T

tan. *See* tangent function
tangent function, 112
  derivative of, 235
tangent line, 143, 144, 147
  definition of, 144, 147
  slope of, 145
tangent plane, 463
taxicab, 38
tax revenues, 423
television advertising, 79, 170
television ratings, 441, 442
temperature, 51, 483, 492
test scores, 345
thought experiment, calculation. *See calculation thought experiment*
three-dimensional space, 444, 445
tides, 114, 115, 238, 401
tilt of earth's axis, 238
total
  change in a quantity, 341
  cost, 335, 350, 358, 362, 363, 391
  definite integral as, 341
  profit, 352, 344, 405, 406
  revenue, 362, 363, 405, 406, 413
  sales, 352, 363, 364, 365, 409
total social gain, 390
tourism, 5, 6, 7, 10, 26, 182, 324, 334, 344, 362, 405, 406
toxic waste treatment, 14

Toyota, 461
transportation costs, 171
trigonometric function, 103. *See also* sine function, cosine function, etc.
  cosine (cos), 104, 109, 232, 233, 233, 318, 332
  cosecant (cosec, csc), 112, 235, 332
  cotangent (cotan, cot), 112, 235, 332
  derivatives of, 232, 233, 235, 236
  identities, 110, 114, 236
  integrals of, 318, 332
  secant (sec), 112, 235, 332
  sine (sin), 104, 105, 232, 233, 235, 318, 332
  tangent (tan), 112, 235, 332
trigonometric identities. *See* identities
T-shirts, 40, 44, 78, 79
tuxedos, 49
TWA, 280, 472

U

U.S. Department of Commerce, 50
U.S. Department of Justice, 62
U.S. Department of Transportation, 5
U.S. Energy Information Administration, 62
U.S. Environmental Protection Agency (EPA), 191
U.S. Postal Service, 280, 472, 483
U.S. Treasury, 406
United States Global Positioning System Industry Council, 47
UPS, 280, 472
uranium, 101
utility function, 442, 454, 462

V

vacation spending, 399
variable
  dependent, 4, 18
  independent, 4, 18
variable cost, 39, 40
velocity, 48, 50, 51, 141, 162, 322
  average, 139
  instantaneous, 138, 139
venture capital, 53, 132, 293, 399
Venture Economics Information Services, 53
verbally specified function, 4
vertex, 17
vertical line test, 20
vertical line, 33, 36
vertical offset, 107
VLSI Research, 9, 12, 20, 26, 53
volume
  cone, 442
  ellipsoid, 442

W

wage inflation, 367
Walt Disney Co., 62
warp, 141, 142
Washington Post, 414
wave
  sawtooth, 115
  square, 115
wavelength, 107
weakness, 325. *See also* strength
weather, 150, 401
web site
  profit, 78
  traffic, 119
Web Site (Student Web Site http://www.AppliedCalc.com)
  function evaluator and grapher, 8, 9, 19, 80, 83, 130
  graphing functions, 19
  numerical integration, 343
  numerical integration utility, 341
  simple regression, 59, 85
weight of a child, 3, 15
World Bank, 414

X

x-intercept, 32

Y

y-intercept, 28, 29, 30, 32, 44, 49